Proceedings of the Second Conference on the Use of Computers in the Coal Industry

SOCIETY OF MINING ENGINEERS

AIME

PROCEEDINGS OF THE SECOND CONFERENCE ON THE
Use of Computers in the Coal Industry

April 15-17
The University of Alabama
Tuscaloosa, Alabama

Thomas Novak
Richard L. Sanford
Y.J. Wang

Editors

Sponsored by

Department of Mineral Engineering
College of Engineering
The University of Alabama

and

Department of Mining Engineering
College of Mineral and Energy Resources
West Virginia University

Published by
Society of Mining Engineers
of
American Institute of Mining, Metallurgical, and Petroleum Engineers, Inc.
New York, NY • 1985

Library of Congress Catalog Card Number 85-70438
ISBN 0-89520-437-1

ORGANIZING COMMITTEE

Thomas Novak
The University of Alabama
Chairman

Richard L. Sanford
The University of Alabama
Co-Chairman

Y.J. Wang
West Virginia University
Co-Chairman

Ray Hollub
The University of Alabama

Syd Peng
West Virginia University

James E. Barker
North River Energy Company

Bradley A. Brasfield
Drummond Coal Company, Inc.

Charles A. Dixon
Jim Walter Resources, Inc.

William J. Douglas
Roy F. Weston, Inc.

Michael D. Loy
Skelly and Loy

H. Kenneth Sack
U.S. Bureau of Mines

R. Bruce Tippin
The University of Alabama

FOREWORD

When one considers the evolution from picks and shovels to the modern machines that remove minerals from the earth and the relatively short period in which this occurred, some appreciation of the impact of technological changes on the Coal Industry—any industry, in fact—can be realized. We often describe such rapid changes as being *revolutionary* to emphasize how new technology uproots long-standing traditional practices and sounds a death-knell for those members of an industry that do not adopt new technology.

Computers are currently among the most visible products of rapid technological change. The evolution of computers—strangers to the work place only a few years ago—as well as their adoption by industry and their adaptation to new and innovative labor-saving and time-saving (cost-saving) applications is currently so rapid as to be revolutionary. Like technological advances in the past, this revolutionary product is going to permit new levels of productivity and profits for some companies, while decline is almost certain for those who resist incorporating it into their operations. The highly competitive Coal Industry is certainly no exception, and for this reason the College of Engineering and the Department of Mineral Engineering appreciate the opportunity to present this Second Conference on the Use of Computers in the Coal Industry.

The College of Engineering has assumed a special responsibility for supporting the industries of our state and nation because, through them, we enjoy the fruits of the highest standard of living in the world. Playing a role in aiding in the incorporation of new technology and practices into our industries in a timely manner is a part of this responsibility, and we are proud to present such an important conference.

Robert F. Barfield
Dean, College of Engineering
The University of Alabama

PREFACE

The First Conference on Use of Computers in the Coal Industry, held at West Virginia University, was a success from several standpoints. First, it brought nearly 400 individuals from industry, education, and government together in an organized meeting for the exchange of ideas. Participants made new acquaintances and became familiar with the current applications of computers in the coal industry. Second, it resulted in a published volume of proceedings containing 78 papers that not only provides a reference to computer users, but set the stage for a continuing collection of information to be exchanged in future conferences. Finally, it demonstrated that a conference on practical applications is feasible—it drew papers and participants from all segments of the coal industry.

The Second Conference, being held 18 months after the first on The University of Alabama campus, has basically kept the same purpose and format. This volume contains 58 papers, plus the Keynote Address, with approximately the same types of subjects being addressed as in the first. Continuing interest demonstrates continuing need, and the Third Conference will be held at West Virginia University in 1986.

The editors wish to thank the staff of the Society of Mining Engineers of AIME, especially Marianne Snedeker, for their technical help and for publishing the proceedings. The assistance of the Organizing Committee is also gratefully acknowledged.

Thomas Novak
Richard L. Sanford
Y.J. Wang
Editors

TABLE OF CONTENTS

KEYNOTE ADDRESS

Chapter 1

MAXIMIZING THE MICRO OPPORTUNITY

L. Michael Kaas

U.S. Bureau of Mines
Chief, Division of Minerals Information Systems
Washington, D.C.

In the last few years, a computing revolution has taken place. Millions of microcomputers have invaded our homes, our schools, and our offices. While there is some indication that the sale of home computers may have leveled off for the time being, micros are certainly here to stay in the business world. They offer a number of opportunities to us as professionals and to the mining industry as well.

The low cost of microcomputer hardware has made computing more easily accessible and affordable. There is no reason why even a small mine, plant, or office cannot have microcomputing capability. Traditional desktop information processing tools such as adding machines, calculators, Kardex files, and even less powerful computer terminals are candidates for replacement by micros. Many functions involving production, cost, personnel, equipment, and engineering data may now utilize a microcomputer to provide better management information that can help increase mining productivity.

The increasing availability of micros has been instrumental in greatly expanding the number of computer users and potential users, many of whom had never dreamed of using a computer. As these new computer users gain experience with micros, we can expect that many new and productive applications will emerge. Some of the successful mainframe and minicomputer applications will also migrate into the realm of the micro.

Micros have acted as a catalyst to expand the so-called "computer literacy" throughout our population. This has resulted in a far greater awareness of what computers can do and a greater willingness to accept them as tools for getting jobs done more productively. This awareness has helped reduce the fear of the computer felt by many. In spite of George Orwell's predictions, we did survive 1984 after all! For many, that survival may have come from the simple realization that if children can master computers, surely adults can also learn to use them effectively.

Advancements in electronics and manufacturing will make increasing amounts of computing power available on our desktops in the years ahead. Introduction of any new technology usually has some pitfalls. Microcomputers are no exception. Several common pitfalls have been encountered by microcomputer users, managers and data processing staffs.

Picking the Right Tool for the Job

The first pitfall is failure to pick the right tool for the job. Picking the right computer for the job is a function of the hardware and software. Let's look at hardware first. The very name microcomputer implies a very small computer with lower processing capacity than is available on most mini and mainframe computers. For many users, processing capacity is not a serious problem. The easier accessability of a micro may far outweigh disadvantages of reduced capacity. However, for some applications in geophysics, geostatistics and simulation, lower machine throughput can be a problem.

Peripheral equipment may be a more serious limitation on a micro. The single keyboard may mean that the end user becomes his own data entry technician. If large files of data must be entered, data entry becomes a poor use of the professional's time. Disk storage capacity may pose another limitation for large sets of data. Even though high capacity hard disks are available, the responsibility for data protection, backup, and recovery falls squarely on the end user, not on some operator in a distant computer center.

Specialized peripherals such as digitizers and plotters capable of handling large engineering drawings usually cost many times the price of a microcomputer. Consequently, it may not be economically feasibile to use them with a single micro. These devices are usually most productive in a location where they may be shared by many users and operated by lower cost technicians.

The popular synonym for microcomputer is "personal" computer. It is easy to see that most microcomputers are indeed personal workstations. They are not designed to be shared by more than one individual and they are not really designed for jobs requiring large volumes of data input or processing, or the use of unusual peripheral equipment. Picking the right tool for the job means that these limitations of the micro must be weighed against the capabilities of minicomputers or mainframes that may also be available for use.

Do microcomputers foreshadow the end of minis and mainframes? Probably not. There is no single "best" hardware solution to all computer applications. It is a matter of balancing the trade-offs among the implementation options available. For the foreseeable future, the optimum hardware configuration for many users will be a microcomputer workstation connected to a more powerful mini or mainframe host which can supplement the micro with larger data storage and processing capability.

Software Availability

The limited availability of mining-related software for microcomputers may pose a more serious problem. In many instances, having a reliable software package will determine if a micro is installed or not. Much excellent general purpose software is available at low cost for applications such as word processing, data file management, electronic spreadsheets, and project management. However, software packages for unique mining applications are relatively few in number and have a wide range of price and quality.

The high development cost and limited market for mining software has always inhibited its development. The perception by many that micro packages should also carry a micro price tag is another inhibiting factor. Even when a package has been developed, finding out about its availability is a major problem. Advertising and promotional budgets for new mining programs are frequently minimal.

Computer Applications Take People Time

Another frequently encountered pitfall is underestimation of the time and effort required to develop applications on a micro. In spite of the impression given by TV advertising, effective computer use requires a substantial investment in personnel time, particularly for a first time user. Learning to do word processing or operate a spreadsheet package is relatively easy; however, other more complicated applications usually take a great deal more time. Unfortunately, becoming productive with a microcomputer may mean being somewhat unproductive as an engineer or geologist while this new technology is being mastered. Both the end user and management should be committed to make the proper investment of people time to insure a successful installation.

That investment should ultimately be expected to bear fruit by producing tangible gains in productivity. However, when hardware or software limitations turn a professional into a data entry technician, computer operator, or even a programmer, the result may be counterproductive unless the resulting operational gains are substantial. When expectations are too high, the novelty of the microcomputer wears off rapidly leaving hardware and software to stand idle. This kind of experience may discourage management or end users from attempting useful applications in the future.

Conflicts with the Data Processing Department

Perhaps the most unfortunate pitfall of all occurs when microcomputers become the cause of conflict between computer users and the established data processing (DP) department. Some DP groups view the micro as a threat to their very existence and are quickly labeled "anti-micro" by potential users. This is an unfortunate situation. As indicated earlier, there is no indication that the availability of microcomputers is causing the demise of the mainframe computer and there are excellent reasons for choosing a mini or mainframe computer over a micro for many applications. The DP departments who are fostering micro usage believe that in the long term, the micros will increase the demand for larger scale mainframe applications.

The best strategy for a DP department is to follow the end users' example and invest some time and effort in learning what is available in the microcomputer field. In that way it can more effectively advise and support the full gamut of available data processing technology.

Another conflict occurs when micros are used, either intentionally or unintentionally, to circumvent corporate data processing plans and hardware procurement policies. In an effort to avoid proliferation of incompatible hardware and software, many companies have established procurement guidelines or controls policed by the DP department. Unfortunately, most companies with established mainframe DP operations also have a substantial backlog of planned application development projects. For an end user with projects trapped in the backlog, a micro may appear to be a tempting solution to the problem. The hardware cost may even be low enough to escape the DP department's scrutiny on new equipment purchases. However, if the user's efforts subvert the corporate DP plan or result in the use of incompatible hardware and software, the organization is not being well served.

Data Management Considerations

Even when the DP department encourages the use of micros, serious conflicts may still arise in the management of data resources. After nearly 30 years of large scale computer usage, most data processing professionals and many non-DP managers have come to appreciate the "asset value" of data. Like equipment and personnel, information is a vital asset of the business and it must be cared for in an appropriate manner. Most companies with mainframe or minicomputers have implemented many data base applications. These applications utilize elaborate data base management systems (DBMS) which facilitate data sharing among applications and users, provide security to protect the data from unauthorized access and maintain data integrity by preventing its alteration or loss.

DBMS software packages are also available for microcomputers. Their descriptions use much of the same jargon as those of the larger computer packages. Unfortunately, the difference between most micro and mainframe DBMS's is like night and day. Most microcomputer DBMS's are designed to aid a single user in the implementation of simple data files for his own use. This is a far less complicated hardware and software environment than that found in a system where many people concurrently access the same large data base without getting in each others' way. Yet many a micro user has, somewhat naively, attempted to build large scale data base applications. These efforts usually fail!

The micro user must assume all responsibility for data entry and file updating, data security, programming and report generation plus backup protection against accidental or intentional data loss or alteration. Once the magnitude of these responsibilities is fully understood, many micro users will opt for the DP department to relieve them of the burden by putting the application on the mainframe.

A final aspect of data management is the question of data sharing. Every organization has countless sets of data that require routine access by more than one individual. When a new application with new data files is implemented on a micro, one question that should be asked is, "Who else will need or want access to this data?" If other potential users can be identified, the requirement to share data may dictate that the application be developed for a larger machine where DBMS software facilitates data sharing.

In the enthusiasm for implementing micro applications, data management considerations are frequently overlooked. These oversights breed more conflict between users and the DP department and greatly impede effective use of micros. The DP staff can help by informing potential microcomputer users of the data bases maintained or planned to be maintained centrally. Information can also be made available on software packages for retrieval and transfer of data between a mini or mainframe and a micro.

Avoiding the Pitfalls

Maximizing the micro opportunity comes down to providing an environment in which the end users of microcomputers are able to learn, experiment and ultimately develop ways to improve operational productivity. They should be able to do this with a minimum number of constraints, and with the support of the DP department. There needs to be just enough management control to insure that the pitfalls discussed earlier are avoided.

While it can be expected that some potential new computer users may be sold on a microcomputer as the only way to solve a problem, the data processing organization must have a broader understanding of available technology - mainframe, mini and micro - on which to base advice to users and management. If the DP staff is not knowledgeable on micros, training should be a high priority objective.

Having a knowledgeable DP staff is the first step toward insuring that micros have been given full consideration in information system (I/S) planning. If current I/S strategies and plans were formulated before the widespread acceptance of microcomputers, they should be reexamined. An organization operating with an I/S plan that arbitrarily excludes the use of micros is missing an opportunity and has placed the DP staff and end users on a collision course.

What if the organization does not have an information system plan? An I/S plan is a means of establishing and communicating the direction the organization intends to take to provide the information required for achievement of organizational goals and objectives. The rapid advancements in information technology and the growing dependence of mining companies on information systems emphasize the need for planning. Furthermore, management studies clearly correlate the successful use of computer technology with a formal information system planning process that is linked closely to organizational and business planning. If an organization is operating without an I/S plan, there is strong justification for creating one and for involving end users in the planning process. The literature contains numerous references on how to develop plans[1]. Seminars and consultants are also available to aid the process.

An information system plan should clearly describe the organization's objectives and priorities for system development. From the plan, potential microcomputer users can see the direction being taken to install computer applications and the associated data bases, hardware, and software. For example, if the development of a central, shared data base of drill hole and reserve data is already planned, it should be clear that a micro application requiring such data should not attempt to develop and maintain a separate data base. Instead, provisions should be made to transmit working copies of data from the central data base to the micro whenever necessary.

A set of guidelines for microcomputer usage can help communicate the hardware, software and support policies of the organization. Many organizations have established a list of "approved" microcomputer hardware. This is a sensible approach to insure a level of compatibility not only among machines but also among the users. With the continuous availability of improved hardware, it is important that such a list be kept current. Similarly, lists of "approved" software packages may also be included in the guidelines. The software marketplace is even more dynamic than that for hardware, therefore, some software lists may be limited to general purpose packages for applications such as spreadsheets, word processing, and micro-to-mainframe communications. Alternatively, a working group or committee of users and DP personnel could be given responsibility to review new software packages, particularly those of a more technical nature, before they are included on the software list. Finally, since most commercial software packages require a formal licensing agreement, the guidelines should include instructions for handling these licenses and for fulfilling their terms and conditions including copyright protection.

Both hardware and software require support. Responsibility for providing different types of support should also be stated in the guidelines. The geographic dispersion of mining operations as well as the personal nature of microcomputers may dictate that only do-it-yourself support is to be available. Alternatively, the DP department might be charged with responsiblity for various forms of support from training in the "approved" hardware and software, to central procurement of hardware and software, to on-site installation, to post installation support via a telephone "hot line." Clearly stating support responsibilities in the guidelines removes another cause for conflict between users and the DP staff.

The I/S plan and a set of microcomputer guidelines will help an organization avoid some of the pitfalls. However, one pitfall that may require industry-wide action is the difficulty of finding out about what software is available. One approach that seems to be quite effective is the establishment of user groups and special interest groups.

Microcomputer user groups have been established to provide support and foster the exchange of information and software for most brands of hardware. The membership of some of these groups has grown very large. They hold periodic meetings, publish newsletters or magazines, operate electronic bulletin boards, and provide "hot lines" to experts in particular aspects of hardware and software. Some sell low cost public domain software from their program libraries.

Industry specific user groups are also being organized. There are two such groups of potential interest to mining professionals. COGS, the Computer Oriented Geological Society is based in Denver, Colorado, and has several local sections starting up around the country.

COGS holds monthly meetings, publishes a monthly newsletter and a software catalog, and sells several disks of geologic software. A second and similar user group has been organized by members of the Society of Petroleum Engineers (SPE) in Dallas, Texas. The success of these groups raises the obvious question, "Why not a mining users group?"

Summary

The applications for microcomputers in the mining industry are many. Whether the user is an engineer designing a mine section layout, a maintenance supervisor developing a PM schedule, a warehouseman keeping track of a repair parts inventory, or a mine manager developing a budget spreadsheet, these applications provide an opportunity to improve mining productivity. Mainframes and minicomputers are not dead, but micros will make computing available to an even larger group of users. With a reasonable amount of planning, the common pitfalls that many have encountered with microcomputers can be avoided, and the users will be well on their way to maximizing the micro opportunity.

Reference

(1) Kaas, L. Michael, "Planning for Mining Information Systems," Computer Methods for the 80's in the Mineral Industry, Society of Mining Engineers of the American Institute of Mining, Metallurgical and Petroleum Engineers, Inc., New York, New York, 1979, pp. 3-14.

COMPUTER SYSTEMS AND SELECTION

Chapter 2

Mining Software Trends - Avoiding the Risks
of Technological Obsolescence

William L. Hodgson[1] and Gordon H. Jardine[2]

ABSTRACT

The rapid technological advances in computer hardware has major implications for the design of mining software and, in turn, for those involved in the evaluation and selection of mining software. Unless software engineering and design techniques evolve as rapidly as hardware capabilities, the users of mining software face the risks of either limiting their technology to ageing software or frequently replacing software to avoid technological obsolescence.

This paper examines current trends in mining software development, and outlines the emerging features of software engineering and design, aimed at avoiding such risks. An outline of the software criteria which users can apply, in practice, in evaluating and selecting mining software is also presented.

1.   President - North American Operations, Mincom Pty. Ltd., Atlanta, Ga.
2.   Director of Mining Technology, Mincom Pty. Ltd., Australia.

INTRODUCTION

Recent developments in computer hardware technology dramatically improving price/performance have set the scene for equally dramatic changes in computer usage by the coal industry. The relatively low cost of micro and mini computer hardware, the promises of unrivalled productivity from their use, and the problems derived from weak world markets for coal have all contributed to a compulsion among most companies to "computerise".

Thus increasingly the geologist and mining engineer, once interested only in proving and developing reserves, are now faced with the need to appreciate and evaluate the computer hardware and software they will use to "assist them".

Computer science, a phenomenon of the 60's and 70's, has propagated new technocrats with the ability and confidence to reshape the way we work, think and play.

The physicists who brought us the early computers have been replaced by a new genre of computer scientists. The results of their specialised efforts have brought about the hardware and software revolutions that characterise modern computing. However, the area of technical computing for the coal industry is still largely dominated by software developed by geologists and mining engineers embodying a deep understanding for the problems and (sometimes) a means to their solution, but rarely the flexibility and efficiency that is the hallmark of the computer scientist. For the coal industry, most of the revolution is yet to come.

HARDWARE TRENDS - WHAT THEY MEAN

The now almost ubiquitous PC exemplifies the rate of technological advancement in computer hardware. With new brands and models announced weekly, it also represents the coming of age of the computing industry which can now "boast" marketing expertise worthy of the pharmaceutical industry.

However, for reasons discussed hereafter, the PC is less than the complete answer to the profitable application of computers to coal mining industry problems. Of greater significance in this regard is the inclusion of new microprocessor chips in communication devices, process control units and cheap, high resolution graphics devices.

Nonetheless, the PC dramatically exemplifies advances in hardware technology. Cheaper, vastly more powerful CPU's, ever more compact memory, both on chips and non-volatile storage devices, all provide the means for constantly improving cost effectiveness and productivity to whatever purpose they are applied.

Technical areas within the coal mining industry to which computers offer major productivity benefits are:

- Data capture (logging information in the field)
- Database (efficient retrieval of ALL information relevant to a particular property/task)
- Deposit Modelling
- Mine Planning
- Process Simulation (Mining/Ore Processing Plant Simulation)
- Mine Scheduling (Production Simulation)

A common theme to all areas is a strong requirement for graphics, and here the microcomputer revolution has generated significant impact. The heavy computer processing requirements for manipulating and displaying graphical data, which once implied a high cost (both in central computer resources and the expense of medium-high resolution display or plotting equipment), now can be devolved onto powerful, but relatively inexpensive, graphics workstations.

Communications is a further area where cost and reliability breakthroughs have provided options in the way in which distribution of computing resources to end-users of a system can be performed. The provision of a complete range of processing services at remote sites from a well supported city-based computing facility is a cost justifiable option being increasingly used within Australia. The particular attraction of this option over the trend in the "1970's" to distribute processing power by physically locating computing equipment at minesites, lies in the cost of maintaining staff and computer equipment at remote locations, and the poor availability resulting from both higher component failure rates (a consequence of harsher environments), and slower maintenance response time.

Sharing of data, but not CPU, is a promise offered by networking. However, while the ideal computing environment may comprise a central database machine feeding a swarm of fast (graphics oriented) satellite processors, further advances will be required in cost/speed/reliability of local and wide area communications to cost effectively justify it. This is the problem faced by purchasers of PC based "solutions", where geologists, mining engineers, draftspeople etc. need to share common data. Centralised data management which heeds all aspects of currency and security is key to any workstation oriented approach.

SOFTWARE TRENDS

"Software" - The very name implies a flexibility that is frequently difficult for users to identify. Much of the software used today in the coal mining industry was written by mining engineers, plant engineers and geologists. These people perceived that computer solutions, generally mimicing manual methods, could be profitably employed to answer problems in deposit modelling, mine scheduling, and maintenance planning. Not being computer scientists, they produced practical results which were neither easy to use nor general in their applicability. Consequently, most software is now being written for the industry by computer scientists (under the guidance of geologists and engineers) with a view to the generality and maintainability of their products and an awareness of the rapidly growing sophistication/expectations among target users.

Increasingly sophisticated computing environments, designed to provide the end-user with simpler, more secure data management, complicate the task of providing single applications to perform specific tasks in the same manner across a range of different computers. The results from one application are frequently input to another in a trend towards complete information (the so-called decision support) systems. The task of developing and maintaining applications in such environments requires specialisation beyond that of the geologist - turned - programmer. Software development is emerging from a cottage industry stage and becoming specialist manufacturing.

The tools of the specialist applications developer include code generators, database managers, data dictionaries, report writers and specialised libraries. Software "factories" provide for structured design and development of products built from standard parts. Software engineering is for real, and provides the methodology to ensure the reliability of the products now flowing to market.

The mining industry, because of its small size and specialist requirements, has largely missed the early benefits of the factory approach to software development. However, in investigating new software, some indication of its genesis and compatibility with other systems (hardware and software) should be sought. Hand tooled, individually crafted software, like an individually designed and built automobile, is likely to be expensive to buy, tune and maintain. A factory approach brings together specialist designers, builders, quality control staff, documenters and sales staff to produce products rather than works of creative genius, well packaged and supported functionality rather than undocumented dreams. It also provides a design base from which advances can be made and measured.

As well as the small market size (cf. retailing and banking), the coal mining industry confronts the software developer with a number of non-trivial challenges. While the commercial aspects of coal mining are reasonably standard, the technical measurement aspects of deposit modelling and production simulation involve representation of geological surfaces in, effectively, three dimensions, the association of ore "quality" information within these, and the interaction of equipment models with the complete model to simulate exploitation over time.

If this all sounds a little complex, it is! Numerous user requirements are superimposed on this picture. A modelling system, to be useful, must allow the user to define the geological environment (structure and stratigraphy) and thereafter work in terms of that environment. Mining simulation must further conform to the specific equipment on site (or planned), and the operating techniques intended for it.

What we have is the requirement for a level of user tailoring, which, to date, has largely been supplied by the production of client (site) specific applications (handmade autos). The attendant costs and difficulties both of initial modification and of ongoing maintenance detract from this approach. By contrast, the use of easy-to-master languages, embodied within applications, provides both user and supplier with a more desirable alternative. A single, basic application framework is developed, documented and maintained, while users needing extended, specialist functions can develop these within the application framework.

It is instructive to put software into an evolutionary context; early systems were card oriented applications designed to run remotely as batch jobs on batch oriented hardware. Results generally required intermediate interpretation by either the author or a DP staff member before they could be successfully utilised by decision-makers at large. As the personal creations of an expert, the systems required a high level of expertise to operate and interpret.

Most current technical applications are interactive, utilise online graphics, and endeavour to communicate using the vocabulary of the end-users profession. The use of interactive graphics was a significant improvement, constituting the simplest way for a person to define point locations and surface geometrics to a computer. Where some internal language capability exists, the applications are partially user configurable. However, in the main, present applications have limited configurability, their graphical interactivity largely restricted to display and pointing input.

Newly released products demonstrate the trend to increased use of graphics (drastically reducing conventional VDU terminal input), and user configurability. Such software meets today's typical coal mining industry hardware setting of mid-range mini-computers, but will readily migrate to the graphics workstation within a network as this technology develops. The ability to migrate across differing hardware and into different software environments will flow from highly modular design, incorporating

- Database
- Graphics Sub-system
- Independent Report Compilers
- Independent Screen Compilers
- Data Dictionaries

which can usually themselves migrate with minimal changes at the Operating System level. The interfaces between the application and the subsystems are simple to maintain at the higher level, or to replace with calls to functional equivalents that may have been written specifically for the particular hardware.

EVALUATION CRITERIA

Computer hardware can be more or less benchmarked in CPU and I/O performance under a simulated load. Peripheral equipment, graphics terminals, VDU's, printers and digitisers are really the units that users interact with, and thus care in selecting value, based on user requiremnts for these, will be rewarded.

The purpose in acquiring the computer equipment is to achieve improved profitability via more efficient operation. The software purchased will define how well that main objective is satisfied. Inefficient software will create resource problems with hardware.

However, the two overriding criteria in choosing software are:

1.   Will it solve my perceived problems:

    (a)   today; and
    (b)   thereafter?

and

2.   How well will it be supported?

Given the rate of technological change as evidenced by hardware obsolescence, and the unlikeliness of a dramatic drop in the price of specialised technical application software in such a small market as coal mining, it is certainly prudent to hedge all bets by ensuring that all applications can migrate to other hardware environments, and that it will support the full range of peripheral equipment that may be envisaged. This is particularly relevant in the coal mining industry given the high reliance that will be placed on graphics devices.

Since all software controls the function of hardware to many users, the application software becomes the measure of the entire system. How well users then perceive the "system" to be meeting their needs will change as those needs change. Consequently, systems need to be tailorable to permit user-extension within the constraints of an overall objective. Fortunately geological modelling and mine scheduling systems mandate a degree of user tailorability (don't they?), to allow their fit to specific mine structure, stratigraphy, mining equipment and methods. Once configured to suit the mine, most software will perform in accordance with its specifications. How well these specifications accord to the users expectations as they become experienced in its use is the ultimate test of good buying. Productive software is well used software (given that the reasons for its purchase were sound).

AVOIDING SOFTWARE OBSOLESCENCE

There is little reason for carefully purchased software to "wear out". If the checklist:

- modular and well maintained
- machine independent
- wide range of supported peripherals
- supplier's long term viability

has been followed, applications may well outlive generations of machines upon which they run.

Ultimately, the commitment of the supplier to maintain and enhance the modules of an application to a state that supports the pantheon of advancing hardware will decide how fast it ages. If well engineered, with the same supporting modules across a wide range (and age) of products, applications will keep pace with the state-of-the-art of hardware and with the ever-increasing experience and sophistication of users.

Chapter 3

TURNING DATA INTO TIMELY INFORMATION

R. Wayne Horscroft

Emery Mining Corporation
31 North Main
Huntington, UT 84528

Abstract

In the past 30 years the growth of electronics has greatly affected the data gathering and information handling activities of our business communities. We have seen the acronym for this influence change from EDP (electronic data processing) to DP (data processing) to MIS (management information services) to ?. This rapidly changing environment has resulted in lost perspectives, backlogs, cynicism, and general disenchantment for those who have placed their data into the electronic data processing resources of a company only to find rearranged data being reported back to them rather than useable information.

The MIS pyramid presented in this paper stands for a philosophy and approach which successfully blends the major data producing, data processing, and information receiving functions of a company into an environment capable of "turning data into timely information."

Company Environment

Emery Mining Corporation is a coal mining management company located in Emery County, Utah. The company was formed in 1980 to manage 5 coal mines, and currently employs 850 people, producing 4.5 million tons per year of coal for 5 steam powered electrical generating units.

Aside from the coal mines, there is a centralized warehouse facility and an administrative office, with each site being 10-15 road miles separated from any of the other sites.

Management Information Services (MIS) Environment

An IBM 4361 MOD 5 mainframe is located at the administrative office using dedicated phone lines for data transmissions to and from each remote site. Each location has several installed terminals to provide mine staff with on-line access to the MIS applications. A printer has also been installed at each site to provide a local printing facility for reports required on a rapid turn around basis.

A Personal Computer/Workstation has been installed at one remote site for use as a multi-purpose work station to support wordprocessing, spreadsheet calculations, data list management in a 'stand-alone' desktop mode; at the same time this workstation can act as a terminal into the mainframe applications. The network is maintained in an "up" status 20 hours per day, (Monday through Friday) and 24 hours on Saturday and Sunday. Each evening during the week the 4 hour slot from 8:00 p.m. to midnight is used for daily control reports, housekeeping (backups, etc) and general status reporting to close out a business day.

MIS Objective and Philosophy

The Management Information Services charter encompasses:

*   Development of timely information systems
*   Operational support of computer facilities and equipment
*   Operational support of communications facilities and services
*   Consulting and developing of personal office systems equipment and facilities
*   Coordination of office services support, facilities, and equipment

Bearing in mind that data in its most elemental form is a fact which requires assimilation and processing with other facts to represent information, the MIS charter elements have been pursued in an integrated fashion to produce timely information.

The MIS long range plan was developed around the MIS charter to provide the ways and means to gather data, to process that data, and to produce meaningful information for all users of the MIS facilities.

A diagrammatic view (the MIS Pyramid) of the data processing functions, business objectives and information levels was developed and became the cornerstone for all activities within MIS. The pyramid and base used in this illustration grew from a study of Nolan's information levels triange along with the additional dimensions of Business Systems Planning, MIS Systems Development Methodology, and Technology/Equipment Analysis.

The foundation of the MIS pyramid addresses the 4 major data producing areas in a business environment:

* Products
* People
* Materials
* Money

After dividing the foundation into 4 segments, one for each of the significant areas, we conducted a broad analysis which resulted in each data producing system within our business being assigned to its most logical segment.

MIS Pyramid Base

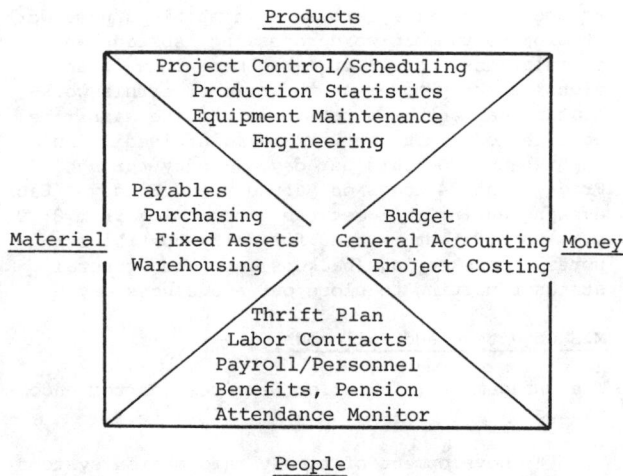

Products

```
           Project Control/Scheduling
             Production Statistics
            Equipment Maintenance
                 Engineering
         Payables
         Purchasing          Budget
Material Fixed Assets  General Accounting  Money
         Warehousing       Project Costing
                Thrift Plan
              Labor Contracts
             Payroll/Personnel
             Benefits, Pension
             Attendance Monitor
```

People

Upon the base of the MIS pyramid with its 4 defined business support segments, major support functions can be placed to illustrate filters for all data to flow upward and through:

1. Business planning support
2. Information support
3. Development methodologies support
4. Technology and equipment support

MIS Pyramid Body

The four sides of the MIS Pyramid are defined in these next views.

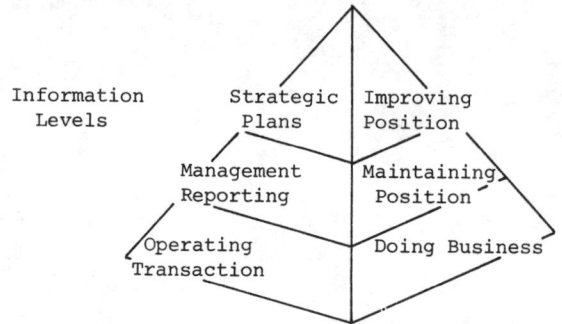

Front VIEW

Business Plan Levels

Side - <u>Business Plan Levels</u> -
This function begins with
    "Doing business in the marketplace"
progresses to
    "Maintaining our business position"
and then becomes
    "Improving our business position".

By conducting analysis and reviews in each of these levels a set of business plan tasks and objectives can be developed, which naturally build one upon the other.

Side - <u>Information Levels</u> -
Facts and reports which define
    "Operating transactions for doing
    business"
continues into
    "Management reporting to support
    decisions and maintain our
    business position"
concluding with
    "Strategic decision models to
    substantiate improving our
    business position approaches"

Technology/   Strategy   Retrieve
Equipment   Presentation   Postulate
Levels   (Workstations)   Correlate

User   Manipulate
Presentation   Update
(Network)   Report

(Terminal)   Edit, Validate,
Physical Capture   Capture

Development
Methodology Levels

Rear VIEW

Side -   Development Methodologies Levels -
provide the
"scientific methods" approach for
developing systems and procedures
to Edit, Validate, and Capture
data items in the "Operating
transactions area"
progressing into systems to
Manipulate, Update, and Report data
information in the "Management
reporting area"
which provide a strong clean base of data
to Retrieve, Postulate, and correlate information into models for
"Strategic decision models"

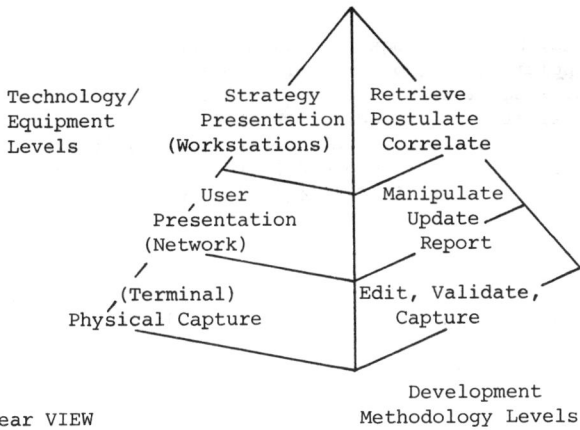

Side -   Technology and Equipment Levels -
is the assemblage of physical tools and
facilities which allow
- Data capture at their source and
verification of that capture
(terminals)
progressing to a level of
- Information production and presentation suitable for the re-
questor's needs (network)
and finally encouraging the
- Strategy extrapolation and presentation needed for decision
support. (workstations)

MIS Plan Implementation

Systems development priorities were aligned with
the 4 major segments shown in the base of the
pyramid and pursued in a logical sequence; the
base, then level 1, level 2, and finally level 3
for each subsystem. As in the construction of a
building a natural sequence was established to
provide a strong foundation for the succeeding
levels. It took very little time to realize all
segments for one level do not have to be in place
before starting the next level.

In the same way that a bricklayer begins with the
corners and builds layers toward the center while
going upwards, so the information subsystems can
be constructed within each level and across
levels.

Tools and techniques used to construct the indepth
dimensions of each segment of the pyramid are
illustrated by this relationships chart:

SIDE	LEVEL	TOOLS/TECHNIQUES
Business Plan Levels	Doing Business Maintaining Business Improving Business	Business Systems Planning process as used by IBM
Information Levels	Operating Transactions Management Reporting Strategic Decision Support	Structured Analysis and Systems Specifications by Tom DeMarco
Development Methodologies Levels	Edit, Validate, Capture Manipulate, Update Report Retrieve, Postulate, and Correlate	Modeling, Simulation, Decision Support System, Report Writers, Prototyping, Screen Painter, Structured Analysis, Struc- tured Programming, Database Management
Technology and Equipment Levels	Data Capture and Verification Information Production and presentation Strategy Extrapolation	Evaluative Checklists and criteria weighting

MIS PYRAMID LEVEL 1

Examples of the data elements and documents found in each segment of the MIS pyramid base of systems at the Operating Transactions level: (note - these lists are not all encompassing, but they will illustrate the grouping.)

People

Daily Crew Sheets, Attendance exceptions, Employment Application, Insurance Claims, Thrift Plan deductions, Pension Plan deductions, Tax deductions, Personal data, Accident report, Citations

Materials

Issue Requisitions, Work Orders, Warehouse Transfers, Material Receipts, Vendor Invoices, Purchase Orders, Purchase Requisitions, Quotation Request, Vendor data Returns, Inventory Counts, Non-Receiver Invoices, Supplements

Money

Payroll Checks, Payable Checks, Vouchers, Bank Statements, Cancelled Checks, Material Costs, Equipment Costs, Credit Vouchers,

Products

Faceboss report, Citations, Daily tonnage slips, Production data, Daily Time lost statistics, Ash Content Analysis,

Analysis of these items using Structured Analysis techniques, resulted in the start of a logical data dictionary of terms for the corporate culture within each of the major areas. Data element accountability, accessibility, and reporting has been a natural outgrowth for our data security administration function.

MIS PYRAMID LEVEL 2

The refinement processes applied as data becomes information to support the 'maintaining position' level is illustrated in these Management Reports (note - these lists are not all encompassing, but they do illustrate the typical grouping)

People

Payroll register, Personnel Actions, Pension Statement, Thrift Account Statements, Attendance Statistics, Accidents Analysis, Citations Analysis, Mailing List, Workload Statistics, Labor Utilization,

Materials

Daily Activity logs, Weekly Analysis and Reviews, Supplements, Vendor Performance Review, Expediting and Encumbrances, Stocking levels and Movement, Receiver Reports, Selective transaction reports, Transfer Analysis, Cycle Count Reports, Active Requisitions, Recommended Reorders, Material/Vendor Reports, Materials Catalog, Parts Cross-referencing, Physical Inventory/Tags,

Money

Authorized invoices by account, Dollar distributions, Uninvoiced dollars Analysis, JV Entries report, Balance Sheet, Budget Reports, Daily Invoiced Expenses, Paid Invoices Reports, Aging In-Progress Invoices, Yearly Ledger, Allocations, Major Account Summary, Balances Outstanding Report, Consolidations, Actual Repair Costs, Held Invoices, Tax Authority, Income Statement, Cash Requirements, Analysis, Chart of Accounts, Trial Balance,

Product

Tons Mined Report, Consolidated Operations Reports, Gas Analysis, Vent Simulation, Pump Simulation, Management Statistics, Budget/Production Statistics, Shift time Analysis,

MIS PYRAMID LEVEL 3

With the strong foundation put in place through the two preceding levels our attention can be focussed upon Strategic Decisions Planning in these respective areas.

People

Turnover statistics, Labor Contract Analysis, Pension Growth

Materials

Material Turnover Rates, Vendor Performance Analysis, Stocking Levels

Money

Budget/Threshold performance

Production

Mine mapping, Project Analysis/Planning, Preventive Maintenance Reporting

In closing, consider this quote:

"The problems of the world cannot possibly be solved by skeptics or cynics whose horizons are limited by the obvious realities. We need men who can dream of things that never were and ask, Why Not?"

                                    Spencer W. Kimball

Current Status of MIS Environment

Equipment

Main Office

1	4361 mod 5 CPU, 10 3370 Disk drives,
2	3430 tape drives, 1 3203 Hispeed printer
2	3268 printers, 35 3279 Color terminals
2	3279 Color graphics terminals
6	3270-PC Workstations with printers

Network
(supporting 4 remote sites)

4	9600 baud leased lines - one joins into microwave network
3	3268 printers, 20 3279 Color terminals
1	3270-PC Workstation with printer

Programs

EMC Developed

Mine Maintenance	- heavy on-line, medium batch loads
Mine Production	- light on-line, heavy batch loads
Time Management	- heavy on-line, light batch loads
Employee Benefits	- light on-line, medium batch loads
Gas Analysis	- heavy on-line
Safety Analysis	- self-contained on a PC Workstation

Purchased or Leased

Payroll/Personnel	- light on-line, heavy batch loads
Accounts Payable	- heavy on-line, medium batch loads
Materials Management	- heavy on-line, light batch loads
Purchasing	- heavy on-line, light batch load
General Ledger	- medium on-line, medium batch load
Engineering Simulation	- heavy batch

Notes

(a) Time Management, Payroll/Personnel are integrated to share data wherever possible and provide automated transfer of data to the General Ledger.

(b) Accounts Payable, Materials Management, Purchasing, and Mine Maintenance are integrated to share data wherever possible and provide automated transfer of data to the General Ledger.

Recommended Readings which have influenced my directions and attitudes in MIS management and style:

Blanchard and Johnson: 1982, The One Minute Manager, William Morrow
Drucker, Peter: 1980, Managing in Turbulent Times, Harper and Row
Durst, Michael: 1982, Management by Responsibility
DeMarco, Tom: 1978, Structured Analysis and Systems Specifications, Yourdon
Eckols, Steve: 1983, How to Design and Develop Business Systems, Mike Murach
Martin, James: 1984, An Information Systems Manifesto, Prentice-Hall
McGowan and Kelly: 1975, Top Down Structured Programming Techniques, Petrocelli/Charter
Naisbitt, John: 1983, Megatrends, Warner Communications
Orr, Kenneth: 1977, Structured Systems Development, Yourdon
Pascarella, Perry: 1984, Humanagement in the Future Corporation, Van Nostrand Reinhold
Rockart, J.F: 1979, A New Approach to Defining the Chief Executive's Information Needs, CISR #37 Harvard Business School

Chapter 4

INTEGRATING MICROCOMPUTER COMMERCIAL SOFTWARE
WITH MINING APPLICATIONS

W. Charles Patrick, Research Associate, Mining & Energy Resources, University of Pittsburgh, Pgh. PA

Ronald Wyatt, Mining Engineering Consultant, Pittsburgh, PA

Dr. Robert H. Trent, Director, Mining & Energy Resources, University of Pittsburgh, Pittsburgh, PA

ABSTRACT

The purpose of this paper is to outline a project to integrate microcomputer technology into mining operations by utilizing commercially available fourth generation software. Fourth generation is a descriptive term for application software which requires some knowledge of programming structure but doesn't require detailed knowledge of computer language.

Mining operations are becoming more and more complex and the information management required to run a modern operation efficiently has prompted better data processing methods. Just as in other industries, a vast amount of data is available, but it must be transformed into useful information. The mining industry has traditionally depended on complicated high-level programs to handle its data-processing needs. This is costly and requires computer specialist for operation of the system.

Commercially available software packages can take information management out of the hands of the specialists and give it back to the user, namely the mining professional. This project utilizes software packages to manage areas such as minng production, planning and scheduling, equipment selection and maintenance, and associated cash flow analyses.

INTRODUCTION

The integration of computer software to mining applications requires an indepth look into the procedures used to accomplish complex data management in the mining environment. The process through which data is manipulated to yield useful information must be analyzed and matched to the capabilities of the available software. In recent years with the development of relatively inexpensive yet powerful microcomputer hardware and software, the pairing of data management processes and data processing equipment has been made much easier.

The authors utilize mainly two types of fourth generation software to integrate software and mining applications:
   A. Database Management (R:Base 4000 and d-Base II and III),
   B. Electronic Spreadsheet Management (LOTUS-1,2,3).

The authors have developed several applications, as follows, with each of the above two types of software and also, are integrating the LOTUS spreadsheet manager with the R:Base 4000 database manager to permit a detailed database, report generator, and spreadsheet which transforms information intermittently.

Database Management System

A database manager is a system that stores information in two-dimensional tables of columns and rows. The system can retrieve, rearrange, sort, and format information across several tables as requested, both quickly and effectively. A fully relational design gives the power to compare, combine, and manipulate information stored in the database, creating new tables containing just the information needed. Through menu-driven prompts, the database manager can perform all functions interactively using a concise command language similar to plain English.

Equipment Maintenance Application

STATEMENT OF THE PROBLEM: The harsh environment and stringent governmental regulations in underground mining necessitate an extremely complex technology. Maintaining this technology has become a difficult task because of the vast amount of data collection and interpretation required for preventive and predictive maintenance. A maintenance decision support system was designed and developed with the objective of aiding management in making decisions regarding the maintenance of underground mining machinery.

The foundation for this decision support system is a microcomputer database manager entitled "R:Base 4000" by Microrim, Inc. The system was designed to assemble collected data, perform analyses on these data, recognize and report problems, and suggest alternatives to aid in improving the availability of mining machinery.

Of the modern mining machinery, the "continuous miner" has the poorest on-line availability. This is greatly due to the inherent interdependency of the machine components. The continuous miner availability problem can be addressed and to a certain extent controlled. Decomposing this complex problem into subproblems; then, combining the solutions of these sub-problems can lead to a solution for the original complex problem.

The major thrust of this project was to assess the continuous miner situation and provide the user (i.e., the mining maintenance management) with information that can aid in making maintenance decisions.

ARRANGEMENT OF THE DATABASE: A database is a collection of organized data in tables or files called relations. The headings or columns in the relations are called fields or attributes. This project utilizes a database called MDS (Maintenance Data System) for storing and retrieving information in a systematic order. The structure of decision support system is depicted in Fig. 1. Two types of relations were used in this database: data storage and report/analysis relations.

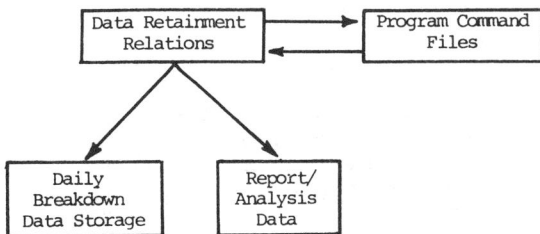

Figure 1. Decision Support System Structure

Data storage relations are used to store daily data on individual component breakdowns for all machines involved in the system. Two types of storage relations exist: SYSTEM1 and SYS10. SYSTEM1 contains the following breakdown data:

1. Date of the breakdown.
2. Time of the breakdown.
3. Part and ID number.
4. Cost of the part.
5. Machine and ID number.
6. Description of work done.
7. Delay time
   A. Time to gather parts and tools.
   B. Time to perform the work.
8. Forced or scheduled downtime.
9. Size of maintenance crew required.

The relation called SYS10 contains similar breakdown data for the current week only. A weekly analysis is performed using this relation. In this analysis, data within the SYS10 relation is appended or added to the SYSTEM1 relation.

The program functions through the report/analysis relations. These relations are used to retrieve data from the storage relations and perform calculations. Analyses are performed on a weekly, monthly, and annual basis.

DATA ANALYSES: Within the scope of this paper, it is not possible to explain all of the individual analyses. Thus, an explanation of the weekly analysis is provided, since the monthly and annual analyses are similar in form to the weekly, except for different time frames.

Within the weekly analysis two (2) command files are addressed, each with a separate function. The first file collects the data entered for the week, addresses each individual machine and its corresponding components, and calculates several criteria. The criteria for each component within each machine includes the total delay time (SDLY), the number of breakdowns (NOCC), and the number of replacements (NREP) for the week.

An important feature of this system is the procedure used to advance the program through each machine and its corresponding components. The program advances from one machine to another collecting data on each component within the individual machine. The machine ID number, entered with each breakdown, refers the program to a particular file which contains information on components for each machine. The importance of this is that, by adding a machine's components to the master component relation, the program will have the ability to analyze that machine at any time.

The second command file within the weekly analysis uses conditional statements to check collected data against decision parameters set by the authors. The values of the parameters were arrived at after the authors' consultation with industry maintenance professionals. However, these values can be easily changed to fit individual applications. Fig. 2 shows the sequential analysis of the data. The data is analyzed by looking at individual components on each machine and advanced only after each component is fulled checked. Once the components for each machine with breakdowns that week are analyzed, the total number of breakdowns and total delay time for each machine is calculated and checked against another set of decision parameters. This is illustrated in the lower section of Fig. 2 under "Total Machine".

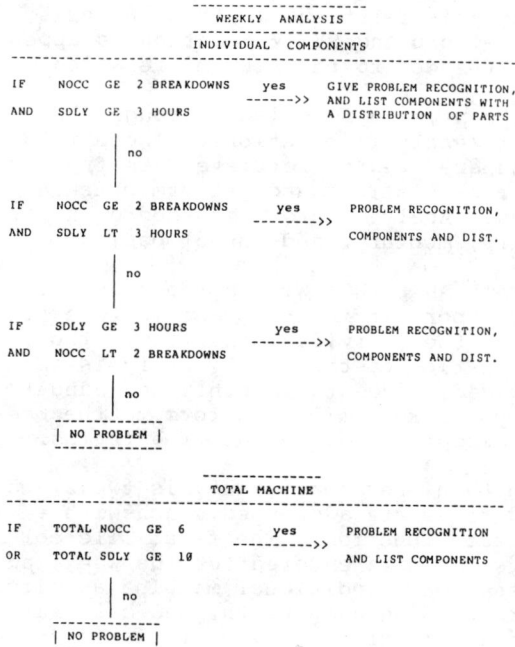

```
-----------------------
       WEEKLY ANALYSIS
-----------------------
      INDIVIDUAL COMPONENTS
---------------------------------------------------------

IF   NOCC GE 2 BREAKDOWNS      yes      GIVE PROBLEM RECOGNITION,
                            ------>>    AND LIST COMPONENTS WITH
AND  SDLY GE 3 HOURS                   A DISTRIBUTION OF PARTS


            | no


IF   NOCC GE 2 BREAKDOWNS      yes     PROBLEM RECOGNITION,
                            -------->>
AND  SDLY LT 3 HOURS                   COMPONENTS AND DIST.


            | no


IF   SDLY GE 3 HOURS          yes     PROBLEM RECOGNITION,
                            -------->>
AND  NOCC LT 2 BREAKDOWNS              COMPONENTS AND DIST.


            | no
      ---------------
      | NO PROBLEM |
      ---------------

             ----------------
---------------- TOTAL MACHINE -------------------------------

IF  TOTAL NOCC GE 6            yes    PROBLEM RECOGNITION
                            ----------->>
OR  TOTAL SDLY GE 10                  AND LIST COMPONENTS


            | no
      ---------------
      | NO PROBLEM |
      ---------------
```

Figure 2. Weekly Analysis Decision Parameter Flowchart

Upon completion of this analysis, the program moves to another command file. This file shifts the weekly data, input by the user and stored in SYS10, to the main file (SYSTEM1) for later analysis. This clears the weekly file for data to be input for the following week.

IMPLEMENTATION: Introducing a new system and gaining acceptance in an organization is a challenging problem, especially if the system deals with the management decision making aspects.

For the system to be effective, the user must build an information base from which machine behavior can be found. A one (1) year period is the minimum data collection time to create a sufficient base. The user must make a commitment to update the database and execute the analyses periodically. Timely, correct information is the key element for making this system effective.

For a company to utilize this system, the following equipment must be available:

1. IBM Microcomputer (or comparable model) with the following capabilities:
   A. 256K Ram Memory
   B. 2 Disk Drives

2. Printer w/132 columns

3. "R:Base, Series 4000 - Relational Database Management System", by Microrim, Inc.

The current price of the above equipment is approximately $4000.00.

A ten (10) megabyte hard disk is recommended for this system. This device has mass storage to enable the user to retain information on up to 100 machines comparable to the continuous miner for a five (5) year period. It would greatly enhance the speed of the program by taking advantage of the high speed data access capabilities of the hard disk.

The use of this system requires a commitment on the part of the mining company to input accurate data and execute the program in a timely manner. Once an information base is created, the system becomes cost effective by efficiently pin-pointing machinery problems. With the number and complexity of mining machines increasing year after year, maintenance management must be computerized to make the mining industry competitive.

LOTUS APPLICATIONS IN MINING

A series of three application models is presented to show the flexibility of Lotus and its use in mining applications. The three application models are:
(1.) Production Planning and Scheduling
(2.) Coal Quality
(3.) Cash Flow Analysis

The interaction of the three application models can be used to determine to a first approximation the viability of a surface coal mine prospect The three models linked together permits a rapid determination not possible by hand calculator techniques.

In this paper a complete determination of the economic viability to a first degree is, of course, not presented. Only part of this is given   here.

In addition, changes that an evaluator would like to add to a preliminary feasibility study would normally require all new calculations. This is not the case with Lotus as the calculations are interconnected. Thus, one change causes a

cascade effect changing all other calculated values. This permits a "what if
"type of evaluation not really possible
with the hand calculator.

After a preliminary evaluation is
made of a prospect the application can be
readily refined by a stepwise process to
permit more detailed second and later
approximations of the prospect's
viability. Only partial segments of this
evaluation is included in this paper.

Production Planning and Scheduling

The application model starts with a
determination of in-place coal reserves.
The reserves are calculated based on coal
acres, coal thickness, coal density,
overburden limits and likely coal
deductions. In-place coal reserves are
then calculated with the spreadsheet
segment of Lotus.

Coal washabilities of the reserves
which are discussed in part two of the
Lotus applications are used to predict the
quantity of coal rejected in considering
the reserves with the proposed (user)
contract specifications.

The user sets the contract
specifications. The specifications can
contain alternative sets of conditions.
These alternative are handled with the
data table command procedures available in
Lotus. The spreadsheet segment of Lotus
permits an almost instanteous determin-
tion of the marketable coal from the
reserve base. With the marketable coal
quantity determined and a proposed monthly
coal requirement the life of the reserve
base is then determined.

The Lotus spreadsheet illustrating
this is given below.

Reserve determination

The quantity of in-place coal
reserves along with additional parameters
for monthly overburden production
requirements are given in fig. 1. Note the
letters across the the top boundary
(columns) and the numbers down the left
boundary (rows). The intersection of the
row and column are used to identify the
numeric value or label located there.
These intersections are referred to as
"cells".For example, in cell C6 is the
value of 2023500, and in cell D6 is the
label sq m.

Overburden requirements to meet
the monthly coal requirements are then
calculated by Lotus' spreadsheet.

```
A       B         C       D       E           F        G
 2                       Reserves/Overburden
 3 -----------------------------------------------------------------
 4 Property description
 5 -----------------------------------------------------------------
 6 size              2023500 sq m  coal density  12894 n/cu m
 7 coal thickness    1.83 m
 8 overburden max.   27.43 m       monthly coal  18144000 kg
 9 overburden min.   12.19 m
10
11                                 Required
12        Reserves                 Monthly overburden
13 ---------------                 ------------
14  4860700000 kg                   149631 cu m
15                                 ------------
16 -----------------------------------------------------------------
```

FIGURE 3. PARAMETERS IN DETERMINING
IN-PLACE RESERVES AND REQUIRED MONTHLY
OVERBURDEN

In the above spreadsheet the
"Reserves"in cell B14 is calculated by the
following Lotus formula statement:

Reserves in place

$$43560*C7*F6*C6/2000*907$$

where: C7 = coal thickness
F6 = coal density
C6 = coal area

The above Lotus formula is entered
directly in cell B14. Any changes to the
cell references in the formula causes the
value of "Reserves" to be instantly
recalculated.

"Required monthly overburden" is
calculated by use of the following Lotus
formula statement:

Monthly overburden

$$ROUND(F8/(B14/C6)*((C8+C9)/2)*43560/27,0)*0.765$$

where: F8 = monthly coal
B14 = reserves
C6 = coal area
C8 = o.b. max.
C9 = o.b. min.

and: @ROUND = Lotus function to round to 0
decimals places.

Overburden Equipment Size

The user selects for overburden
removal either a specific sized dragline or
shovel. The "Monthly Overburden removed"
with the suggested equipment is
calculated. This value is compared with
the "Required Overburden" for the"Monthly
Coal" requirement. If the "Required
Overburden" exceeds that possible with the
specific piece of equipment - the user is
prompted to try another dragline or
shovel.

After selecting another dragline or shovel, the user views almost instantly whether the equipment is sized large enough. If not, the user selects the next larger size and continues the procedure until the appropiate size is selected.

Fig. 4 below summarizes the above calculations for the "Required Overburden","Produced Overburden" from the dragline or shovel and the "Job - Management Factors".

```
J       K       L       M       N       O       P
2       Overburden Required/Overburden Production/Job Factors
3  ----------------------------------------------------------
4  Parameters        Values        Parameters        Values
5  ----------------------------------------------------------
6  bucket/shovel cap.  7.65 cu m   overburden max.   27.43 m
7  fill factor         0.75        overburden min.   12.19 m
8  wt loose/wt solid   0.75
9  time avail. month   720 hr      Calculations:
10 swings/hr           55
11 job-factor          1           Required o.b.     5543   m
12 management-factor   1           -------------------------
13                                 -------------------------
14                                 Produced o.b.     5224   m
15                                 -------------------------
16                     Deficiency                    319   n.
17                                 -------------------------
18                     Excess                        0 CU FT
19                                 -------------------------
20                          JOB TABLE
21                         (job factors)
22                     -------------------------------------
23                     | excellent  good        fair   poor
24                     -------------------------------------
25                          1      2         3      4
26                         0.91   0.88      0.83   0.76
27                            management factors)
28                     -------------------------------------
29                         0.91   0.85      0.78   0.68
30
31 ---------------------------------------------------------
```

FIGURE 4. SUMMARY OF PARAMETERS NECESSARY TO CALCULATE MONTHLY OVERBURDEN REMOVAL WITH DRAGLINE OR SHOVEL

In the above spreadsheet two additional calculations are included besides the Lotus formulas for "Reserves" and "Required Overburden"; they are: "Produced Overburden" and "Defficiency" or "Excess". The Lotus formula for Produced overburden is:

Produced overburden

@ROUND((L6*L7*L8*L9*L10*@HLOOKUP(L11,L25. .029,1)* @HLOOKUP(L12,L25. .029,4)),0)

where: L6 = bucket capacity
       L7 = fill factor
       L8 = weight ratio
       L9 = time avail. month
       L10 = swings per hour

and: @HLOOKUP selects the job and management factors for the tables of that label.

The advantage of Lotus spreadsheet capability in the above calculations is the speed with which the spreadsheet recalculates for new values. In addition if the user later decides to change some or all of the parameters of the reserve base, the calculations for the selection of the dragline or shovel is automatically adjusted.

Truck Sizing and Selection

Additional equipment is selected based on the user defined parameters. For example, truck selection is determined by the monthly coal requirement, haul length, loading and spotting times, and other factors. If these should change, the size and number of trucks may also change.

The user of the application model is prompted then for haulage parameters. When these have been entered into the spreadsheet the trucks' specified sizes are almost instantly displayed. Truck delays and loader delays are also determined and the results are available to the user so that he may choose other sized trucks or loaders.

```
AT    AU      AV    AW AX   AY      AZ      BA   BB
2                  Truck & Loader Sizing and Scheduling
3  ----------------------------------------------------------
4  Truck parameters          Loader parameters
5  ----------------------------------------------------------
6  capacity      18144 kg    capacity        3.8 cu m
7  truck cycl    25 min.     fill fact.      90 %
8  hourly pro   169646 kg    time/pass       30 sec.
9  efficiency    92 %
10 shift hour    7.5 hrs.    LOADER PASSES/TRUCK
11 swell fact    0.8         --------------------------
12 equipment     80 %                        4.0
13 --------------------------------------------------
14 TRUCKS REQUIRED           TRUCK LOAD TIME  (minutes)
15                           --------------------------
16       7                                    2.0
17 --------------------------------------------------
18 TRUCK SPACING             LOADER WAIT      (minutes)
19                           --------------------------
20       4.2 min.                             2.2
21 --------------------------------------------------
22                           TRUCK WAIT       (minutes)
23                           --------------------------
24              |                             0.0
25 --------------------------------------------------
```

FIGURE 5. TRUCK AND LOADER PARAMETERS WITH RESULTING REQUIRED CAPACITIES AND NUMBERS

The spreadsheet shown above is tied into the previous two spreadsheets by "hourly production" required in cell AV8. The other calculations on the spreadsheet are all related to the value calculated in cell AV8. This is illustrated in the Lotus formulas as given below.

Trucks required

@Round((AV8/((AV9/100*60)/AV&*AV6*AV11)* 1/(AV12/100))+0.5,0)

where: AV8 = hourly production
AV9 = efficiency
AV7 = truck cycle time
AV6 = truck capacity
AV11 = swell factor
AV12 = equipment avail.

and: @ROUND(+0.5,0) round up

Truck load time

+BA12*AZ8/60

where: BA12 = loader passes per truck
AZ8 = time per pass

Truck spacing

@ROUND(AV7/(AV16-1),1)

where: AV16 -1 = one less than number of required trucks

and: @ROUND(,1) round to the nearest 0.1 minutes

Loader wait

@ROUND(@IF(AV20-BA16>0,AV20-BA16,0),1)

where: AV20 = truck spacing
BA16 = truck loading time
AV20-BA20 = loader waiting time

and: @IF(AV20-BA16>0 = if the truck spacing is greater than the truck loading time the loader

Loader passes per truck

@ROUND((AV6*AV11*1/((AZ6*(AZ7/100))/
(((2000/(COAL_DENSITY*AV11))*1 /27)))),1)

where: COAL_DENSITY is a range name for a cell on a previous spreadsheet. (note: all of the above spreadsheet are segments of a larger spreadsheet and as such are interrelated with the use of connecting formulas.

Drilling

Drilling requirements are determined per shift based on the previous entered parameters such as overburden maximum, minimums, coal requirements, and other factors.

The user is then assisted by the
model in selecting the appropiate size and number of drills.

```
BO    BP       BQ      BR  BS BT    BU      BV  BW  BX
2                    Drilling
3 Parameters              Resulting values
4 -------------------------|-------------------|----------
5  drill capacity    183 m |HOLES PER SHIFT    |    9
6  hole spacing            |-------------------|----------
7      7.6  m  by   7.6 m |SHIFT DRILLING      |  180 m
8  avg drill depth   19.8 m|-------------------|----------
9  drill diameter    0.25 m|DRILLING HOURS     |    8
10 hours per shift    7.5 hr|-------------------|----------
11                         |NUMBER DRILLS      |    1
12                         |-------------------|----------
13                         |NUMBER SHIFTS      |    1
14 -------------------------|-------------------|----------
```

FIGURE 6. DRILLING PARAMETERS SELECTED BY USER AND RESULTING DERIVED VALUES

Certain decisions concerning scheduling of drills, and the maximum number of hours per drill before additional drills are required, are summarized in the spreadsheet below.

This spreadsheet is used with the table lookup functions available on Lotus to schedule the number of drills, drilling hours, and number of shifts given in the above spreadsheet.

```
CA        CB         CC CD   CE   CF   CG   CH   CI
2                    Drill Schedule Table
3 -------------------------------------------------
4 drilling hours        0    8   16   24   32   40
5 shifts                1    1    2    2    2    2
6 drills                1    1    1    1    2    2
7 max. hours per drill  8    8   12   12    8   10
8
```

FIGURE 7. DRILLING SCHEDULE USED TO DETERMINE NUMBER OF SHIFTS, AND DRILLS

Coal Quality

Washability data for a coal is entered into a database. The data base is then used with standard calculations for individual fractions and cumulative float and sink. Parabolic interpolation is used to determine values between actual washabilitiy data. Based on stipulated contract specifications the resulting payments for btu, ash, and sulfur qualities are used to determine an adjusted price for the coal washed at any specific gravity within the wash range.

The value per ton of run of mine coal is also determined based on the resulting yield at the user selected specific gravity. This resulting yield per cent is then back substituted into the first spreadsheet to give the washability reject per cent.

The model automatically tries new specific gravities for the entire range of the washability specific gravities. The resulting values are summarized in a table with the most economic choice listed below. The user then has the option of disregarding that option and selecting the next most economic. This progression can continue until the user terminates the process. A summary of the results of the automated section of the model is listed below in the table of washability economics. Graphs of the specific gravities versus the value per ton of run of mine coal is then printed to illustrate the changing economics as the coal is washed at the varying specific gravities. Other graphics are provided for the user if requested. One of these graphs is given below as an illustration of the available graphing techniques in Lotus.

A segments of the database spreadsheet for the coal quality calculations is given below. This spreadsheet gives the

cumulative float percentages for the
following coal qualities:
(1.) ash
(2.) btu
(3.) sulfur

Also given below is the tabular
values of the coal qualities for specific
gravities from 1.30 to 1.81.

Coal Washabilities

	Individual Fractions															Cumulative Sink						
(A)	(B)	(C)	(D)	(E)	(F)	(G)	(H)	(I)	(J)	(K)	(L)	(M)	(N)	(O)	(P)	(Q)	(R)	(S)	(T)	(U)	(V)	(W)
Sp. gav.	wt.%	ash%	sul%	btu	wt.%	ash prod.	lf prodbtu	prod ash	prod.sul	prod btu	btu prod	ash%	sulf%	btu	wt%	ash prod	sulf prod	btu prod	ash%	sulf%	btu	
1.27 float 1.2	34.50	2.80	1.80	13000	34.50	96.60	62.10	448500	96.60	62.10	448500	2.80	1.80	13000	100.00	1292.97	214.69	1207320	12.93	2.15	12073	17.25
1.30 1.27 x 1.	28.40	3.90	2.00	12000	62.90	110.76	56.80	363520	207.36	118.90	812020	3.30	1.89	12910	65.50	1196.37	152.59	758820	18.27	2.33	11585	40.70
1.30 1.30 x 1.	16.90	8.00	2.40	12900	79.80	148.72	40.56	218010	356.08	159.46	1030030	4.46	2.00	12908	37.10	1085.61	95.79	395300	29.26	2.58	10655	71.35
1.50 1.30 x 1.	5.40	16.90	2.60	11500	85.20	91.26	14.04	62100	447.34	173.50	1092130	5.25	2.04	12818	20.20	936.89	55.23	177290	46.30	2.73	8777	82.50
1.70 1.50 x 1.	3.30	30.60	2.30	9800	88.50	100.98	7.59	32340	540.32	181.09	1124470	6.20	2.05	12706	14.80	845.63	41.19	115190	57.14	2.78	7783	86.85
1.90 1.70 x 1.	3.00	46.20	2.70	9200	91.50	130.60	4.10	27600	686.92	189.19	1152070	7.51	2.07	12591	11.50	744.65	33.60	82850	64.75	2.92	7204	90.00
2.00 sink 1.90	8.50	71.30	3.00	6500	100.00	606.05	25.50	55250	1292.97	214.69	1207320	12.93	2.15	12073	8.50	686.05	25.50	55250	71.30	3.00	6500	95.75

FIGURE 8. COAL WASHABILITY DATABASE

WASHABILITY ECONOMICS
----

s.g.	yield	sulfur	ash	btu	sul pay	ash pay	btu pay	adj pay	value/ton(rom)
1.30	62.90	1.89	3.30	12910	-0.04	-0.20	30.98	30.74	19.34
1.31	65.59	1.91	3.47	12922	-0.05	-0.24	30.99	30.70	20.14
1.32	68.12	1.92	3.64	12926	-0.05	-0.28	30.99	30.66	20.89
1.33	70.48	1.94	3.79	12927	-0.05	-0.32	30.99	30.62	21.59
1.34	72.68	1.95	3.94	12925	-0.05	-0.36	30.99	30.59	22.23
1.35	74.71	1.97	4.09	12922	-0.05	-0.40	30.99	30.55	22.82
1.36	76.57	1.98	4.22	12918	-0.05	-0.43	30.99	30.51	23.36
1.37	78.27	1.99	4.34	12913	-0.05	-0.46	30.98	30.47	23.85
1.38	79.80	2.00	4.46	12908	-0.05	-0.49	30.98	30.44	24.29
1.39	80.35	2.00	4.53	12895	-0.05	-0.51	30.96	30.40	24.43
1.40	80.88	2.01	4.61	12886	-0.05	-0.53	30.94	30.36	24.56
1.41	81.39	2.01	4.67	12877	-0.05	-0.54	30.92	30.33	24.68
1.42	81.89	2.01	4.74	12870	-0.05	-0.56	30.90	30.29	24.80
1.43	82.36	2.02	4.81	12863	-0.05	-0.58	30.88	30.26	24.92
1.44	82.82	2.02	4.88	12856	-0.05	-0.59	30.87	30.22	25.03
1.45	83.26	2.02	4.94	12849	-0.05	-0.61	30.85	30.19	25.13
1.46	83.69	2.03	5.01	12843	-0.05	-0.63	30.83	30.15	25.23
1.47	84.09	2.03	5.07	12837	-0.05	-0.64	30.81	30.12	25.33
1.48	84.48	2.03	5.13	12831	-0.05	-0.66	30.80	30.09	25.42
1.49	84.85	2.03	5.19	12824	-0.05	-0.67	30.78	30.06	25.50
1.50	85.20	2.04	5.25	12818	-0.05	-0.69	30.76	30.02	25.58
1.51	85.37	2.04	5.29	12813	-0.05	-0.70	30.75	30.00	25.61
1.52	85.54	2.04	5.33	12808	-0.05	-0.71	30.74	29.98	25.64
1.53	85.71	2.04	5.37	12803	-0.05	-0.72	30.72	29.95	25.68
1.54	85.88	2.04	5.41	12797	-0.05	-0.73	30.71	29.93	25.71
1.55	86.05	2.04	5.45	12791	-0.05	-0.74	30.70	29.91	25.74
1.56	86.22	2.04	5.50	12786	-0.05	-0.75	30.68	29.88	25.77
1.57	86.39	2.04	5.54	12780	-0.05	-0.76	30.67	29.86	25.79
1.58	86.56	2.04	5.58	12774	-0.05	-0.77	30.66	29.83	25.82
1.59	86.72	2.04	5.63	12769	-0.05	-0.78	30.64	29.81	25.85
1.60	86.89	2.04	5.68	12763	-0.05	-0.79	30.63	29.78	25.88
1.61	87.05	2.04	5.73	12757	-0.05	-0.81	30.62	29.76	25.91
1.62	87.22	2.04	5.77	12752	-0.05	-0.82	30.60	29.73	25.93
1.63	87.38	2.04	5.82	12746	-0.05	-0.83	30.59	29.71	25.96
1.64	87.54	2.04	5.87	12740	-0.05	-0.84	30.58	29.68	25.98
1.65	87.70	2.04	5.93	12735	-0.05	-0.86	30.56	29.65	26.01
1.66	87.86	2.04	5.98	12729	-0.05	-0.87	30.55	29.63	26.03
1.67	88.02	2.04	6.03	12723	-0.05	-0.88	30.53	29.60	26.06
1.68	88.18	2.04	6.08	12717	-0.05	-0.90	30.52	29.57	26.08
1.69	88.34	2.05	6.14	12712	-0.05	-0.91	30.51	29.55	26.10
1.70	88.50	2.05	6.20	12706	-0.05	-0.92	30.49	29.52	26.12
1.71	88.21	2.04	5.96	12919	-0.05	-0.86	30.55	29.63	26.14
1.72	87.96	2.04	5.75	13026	-0.05	-0.81	30.60	29.73	26.15
1.73	87.76	2.04	5.58	13080	-0.05	-0.77	30.64	29.82	26.17
1.74	87.61	2.04	5.44	13104	-0.05	-0.74	30.67	29.89	26.18
1.75	87.50	2.03	5.33	13109	-0.05	-0.71	30.70	29.94	26.20
1.76	87.44	2.03	5.25	13101	-0.05	-0.69	30.72	29.98	26.21
1.77	87.43	2.03	5.21	13084	-0.05	-0.68	30.73	30.00	26.23
1.78	87.46	2.03	5.20	13062	-0.05	-0.67	30.74	30.01	26.25
1.79	87.54	2.03	5.21	13034	-0.05	-0.68	30.73	30.00	26.27
1.80	87.67	2.03	5.26	13002	-0.05	-0.69	30.72	29.98	26.28
1.81	87.84	2.04	5.34	12968	-0.05	-0.71	30.71	29.94	26.30

FIGURE 9. SUMMARY OF PARABOLIC
INTERPOLATION OF COAL WASABILITY DATABASE

Specific Gravity Curve
Individual fractions

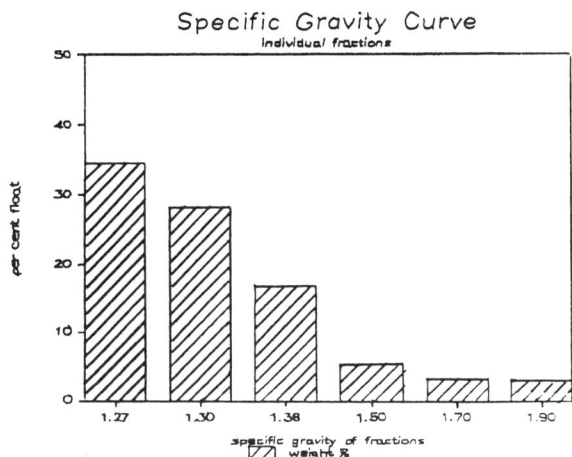

FIGURE 10. WEIGHT PERCENT OF INDIVIDUAL FLOAT FRACTIONS VERSUS SPECIFIC GRAVITY

Economic Analysis and Cash Flow

The spreadsheet segment for this section is not included. A description of the mechanics is as follows. The equipment thus far selected is copied to the an economic spreadsheet.

In addition, other necessary equipment that the user enters is also copied to the economic spreadsheet. At the same time, the selected equipment is used as the criterion to search an equipment database that has previously been prepared.

The equipment database permits the user to extract immediately details about the equipment so as to make an economic determination of the viability of the suggested equipment mix. Copying of the equipment from the equipment spreadsheet to the economic spreadsheet is accomplished automatically by the use of macros. The macros in Lotus permit personalized automated control of the application model. The macros provide the option of menu driven commands or manual commands to complete the economic spreadsheet.

The cash flow analysis is determined by the data generated from production, planning and scheduling along with data from the coal qualities application model. Graphics are used to highlight the production scheduling, coal qualities and cash flow analysis, and can be viewed immediately as changes to the data are made. In addition all graphs are printed so that they may be included with the print out of the spreadsheets and incorporated into a report summary. The report summary is prepared by capturing sections of the spreadsheets into a word processor and including explanatory text with spreadsheets and graphics.

Lotus/R:Base Interface: Maintenance History Program

Lotus is interfaced with R:Base 4000 to produce a complete maintenance history report with accompanying graphics. Lotus is interfaced with R:Base provide a major advantage over R:Base alone in the ease of data entry and accompanying changes. In addition, Lotus permits full screen editing with split screen viewing not available with R:Base. In addition to the superior data entry, changing, and viewing capabilities, Lotus permits the display of data utilizing graphics. This capability is not available with R:Base.

Data is entered by the user either by menu request or in free format of individual fields, individual records or some combination of either. This data entry is a function of the Lotus software. The data entered is saved to a file automatically without user intervention and is transparent to the user. R:Base then accesses the data, also automatically. R:Base then manipulates the data, producing an analysis as requested by the user. R:Base then produces a report to the user specifications. The data that is generated from the analysis by the maintenance history program is captured from a saved file. Lotus performs the capture and displays the analysis as a series of graphs. The graphs are of the following types:

1. Line graph
2. Bar graph
3. Stacked bar graph
4. Pie graph
5. X-Y graph

The interfaced Lotus and R:base software is programmed to interact without user intervention after the data entry and report form specifications. The automated aspect is controlled in Lotus by macros and in R:Base by command files. Additional interfacing also exists between the DOS (disk operating system) and both softwares. This interaction will be controlled by command files in the DOS.

Conclusion

The authors' intent is to aid the mining community in the realization that microcomputer applications in mining are not for computer specialists only. With the advent of fourth generation software, the mining professional now has enormous capabilities not available only a few years ago.

   This paper gives only a few of the
possible data processing arrangements.
This technology is applicable in any
engineering application with repetitious
calcuations or "what-if" situations,
limited only by the user's knowledge of
the software and judgement as to whether
the amount of processing is worthy of
computerization. The mining professional
must now make the choice between present
procedures or utilizing an extremely
powerful tool that will save both time
and money.

References

Cassidy, Samuel, 1973, Elements of Prac-
    tical Coal Mining, John Wiley Pub.,
    pp. 155.

Cummins, Authur, 1973. SME Mining Engi-
    neering Handbook, Volume 1, John Wiley
    Pub., pp. 12-43.

"Logistic Performance of Continuous Min-
    ers", 1976, Bendix Corp. in a contract
    for the U.S. Bureau of Mines, p. 2-1.

McCall, J.J., 1965, "Management Policies
    For Stochastically Failing Equipment:
    A Survey," Management Science. pp.
    493-624.

Posner, John et.al., 1983, Lotus 1-2-3
    User's Manual, Lotus Development Corp.
    pp. 31-274.

Pfleider,E.P., ed., 1972, Surface Mining,
    AIME Pub., pp.139-224, 425-637.

"R:Base 4000 - Relational Database Man-
    agement System, 1984, Microrim Corp.,
    p. 1-1.

Chapter 5

COMPUTER FACILITIES AND THE EDUCATION OF MINING ENGINEERS

James A. Procarione and Robert E. Cameron

Department of Mining Engineering
University of Utah
Salt Lake City, Utah

ABSTRACT

Due to the influx of computers into the every-day life, it is essential that an engineer's education include exposure to these machines. This is especially true of mining engineers due to the growing complexity of the industry and to the emerging sophistication of mining software. One objective of an educational plan should be the introduction of computers to all majors. It is the experience of using a computer which is of the most value to engineering students not the sophistication of the hardware or the software. This experience will result in a substantial growth in confidence and understanding on the part of the mining engineer whose field is just beginning to truly computerize.

The presense of specialized computer hardware, such as a drafting plotter, graphics terminal, and digitizer, and the associated software can greatly increase the value of the educational training a person receives. Their use in the classroom will force the student to become acquainted with the power of computers and, at the same time, provide an excellent opportunity for a professor to present examples of practical applications. In addition, exposure of students to various computer languages will allow them to make future decisions in the proper selection, evaluation, and production of software. The knowledge of the merits of BASIC, FORTRAN, PASCAL, and others will be invaluable when the inevitable explosion of software hits the mining industry.

There does exist a minimum computer configuration which is basic for the majority of mining applications. The chosen computer should be equipped with a multiuser/multitasking operating system. The advantages of a graphics terminal, drafting plotter, and digitizer are evident. Addition of a magnetic tape subsystem greatly enhances a computer facility by allowing the acquisition and transfer of software and satisfying long term storage requirements for programs and data. Printers, both dot matrix and letter quality, are a necessity. All of these items and other often overlooked elements are justified in a number of ways; most of which are obvious by

simply considering their benefits.

Engineers who use computers on a day-to-day basis quickly learn that developing input data for a program requires the same judgement and knowledge as if the solution were to be "hand" generated. The advantage of computer assistance lies in the machine's ability to do the tedious, time-consuming calculations quickly and accurately. The important theme which educators should convey to their students is that engineers must still ask the proper questions and must evaluate the results according to their education and experience. This is one thing that no computer or program can yet do.

INTRODUCTION

It is probably an understatement to say that computers have vastly changed our lives. Industry employs millions of people whose jobs are entirely dependent upon the control and computational power only these machines can provide. Large American corporations, such as IBM and AT&T, spend millions of dollars to place their computer into our economy and into our homes. The day would be considered highly unusual if there were no contact, either actively or passively, with these machines. The computer industry is a vast and complex entity.

One important niche in which computers have found a permanent home is in the areas of science and engineering. Physics, chemistry, and other physical sciences have long realized the value of quick, accurate mathematical calculations. These disciplines have helped push the maturing computer industry to point it is today. Engineering has benefited from utilizing and developing sophisticated software as well as contributing to the growth of this technology. There are programs for doing stress analysis in materials, electrical load simulations, and water flow distributions. Some problems in mechanical, electrical, and civil engineering would be extremely difficult, if not impossible, to analyze if it were not for the power of computers. Many mechanical systems are being controlled by computers; from automated data

collection systems to the actual operation of devices (robotics). The development and manufacture
of new products has been simplified by utilizing
computer aided design and computer aided manufacturing (CAD/CAM) systems.

Mining stands upon the threshold of a technological explosion as the specific needs of this
profession attract the attention of the computer
industry. In the past, computer programs used for
analyzing mining problems were developed by
universities and companys for internal consumption. Some software, such as the Pennsylvania
State University Mine Ventilation Simulator
(PSUMVS), have found wide acceptance outside its
development sphere. But its use is due mainly to
diffusion through faculty and former students
rather than by a concerted marketing effort. Both
the computer and mineral industries are beginning
to realize that mining can draw upon a vast quantity of software developed for other engineering
professions. This opens up a new market for some
very sophisticated software. For example, a mine
electrical distribution network only differs in
scale from that constructed by a utility, and so
the analytical programs developed by the electric
company can be applied by the mining company. The
minerals industry has also started to explore the
possibility of computer based data acquisition
systems for mine environment monitoring and the
possible use of robotics to increase productivity.
This has resulted in an increased demand for computer oriented training of mining professionals.

To meet the growing challenge of computerization, educators must integrate computer hardware
and software into the classroom. In order to pru-
dently select suitable computer facilities, educators must first ask themselves, "What facilities
and training is vital to the education of mining
engineers?" Next, it must be decided how to best
integrate this technology into current curriculum
as well as introduce new courses which simultaneously stress engineering fundamentals and computer
applications. Finally, mining departments are
faced with the problem of increasingly limited
financial resources for computer acquisitions due
to strained state, federal, and university budgets. This makes it extremely important that new
purchases fulfill the computational requirements
of several specialties within the department.

Once careful consideration of some important
concerns are made, a mining department can select
an "ideal" computer system. It is important to
remember that the specific equipment will vary
with each academic program. Hopefully, this paper
will help educators evaluate and purchase facilities by addressing some of the problems, challenges, and considerations which must be addressed
before selecting new or expanding present computer
facilities. To do so, four areas will be examined: the academic environment; its educational
goals; the needs of the mining department; and the
philosophy involved in computerization. Finally,
in light of some additional considerations, a
workable computer facility will be described which
is capable of fulfilling the goal of educating
mining engineers.

UNIVERSITY ENVIRONMENT

All computer facilities must be chosen to
operate under a specific set of conditions; these
are sometimes termed "the environment". The
environment will dictate, to a large extent, the
type of facilities which will be required and
effectively utilized. It is important that there
exist a good understanding of the environment surrounding a specific university before attempting
to integrate current, or before trying to select
new, computer facilities.

Needless to say, all universities are expansive
and dynamic organisms which makes it difficult to
make anything but general statements about the
environment. Each institution has its own unique
and diverse personality, with individual educational and administrative philosophies. Various
policies establish options available to the educator in selecting and utilizing computer facilities. For example, one university may have a cost
recovery system which classifies the purchase of
time on an existing computer as a non-budgetable
item. But at the same time, maintenance contracts
and operating costs are acceptable line budget
items. In contrast, accounting procedures may be
established which heavily favor existing computer
facilities and virtually prohibit establishment of
additional sites. These different practices will
affect the composition of the selected system.

At most universities, it is the responsibility
of faculty members to determine the content of
course work and academic programs. It is here
that the movement to integrate computer facilities
into the curriculum must reside. Regardless of
the external pressures exerted by the mining community or university administration to incorporate
computer training, the curriculum will only change
as faculty members adapt course content and thrust
computer technology into the classroom. This
means many courses must be reviewed and rewritten;
requiring a substantial effort on the part of the
professor.

Faculty interests must be precariously balanced
between the classroom and research. It is notori-
ously understood that the application of the "publish or perish" principle is utmost in the concerns of university departments. To publish,
faculty members must be almost constantly engaged
in research. This limits the amount of time that
can be spent working towards the integration of
computer technology into the classroom. If it is
realized that a large block of time and effort
must be expended for a usable computer program to
be developed or to be evaluated and purchased from
a second party, it is clear why there is some
reluctance on the part of professors to use computers in the classroom. This is due mainly to
constraints on their time and the vast effort
involved.

On the student's side of the situation, the
view is quite different as they are forced to
utilize whatever computer facilities are offered.
It is up to each individual student to become familiar and comfortable with computers and their

capabilities. This is best accomplished by sitting at a terminal and using the machine. While a student can be pushed towards utilization by professors assigning computer oriented problems, those taking the initiative will ultimately feel more at ease and be more successful in exploiting the power of this technology. Professors and students must share the burden in making sure that the advantages of computers are exploited at the university to the fullest benefit of each group.

Students also have time constraints. The average university student is working about 20 hours a week along with trying to maintain a full schedule of classes. He must somehow budget his time between various course and work commitments, and somehow fit in a social life. This combination of faculty and student limitations often means that the amount of time available to spend with computers may be very limited.

Ultimately, this means that any attempt to integrate computer facilities into university curriculums will require a very detailed consideration of these time constraints. It also indicates that this introduction may require more effort than many people wish to budget for the task. However, if research programs are designed with the idea of integrating the work into the classroom and professors use computers to replace some of the current course work, a reasonable and eloquent solution may be developed which is acceptable to all parties.

Graduate students are heavily involved with research and it is in the laboratory where many receive the majority of their university instruction. Computers can be very instrumental to their education and to the quality of research they can perform. However, careful consideration must be made as to the type of equipment purchased for the specific research. A project determining the effects of skips moving in a ventilation shaft requires high speed data acquisition facilities while a project on numerical modeling of rock behavior might need array processing capabilities. Current and future research projects should bear significantly on the development of the computer facilities.

As stated earlier, faculty members have a huge responsibility for research and computers are becoming an intergal part of such efforts. Strip charts and paper tapes have been virtually eliminated as recording media for data collection by many equipment manufacturers. Instead, microprocessors are used to collect and store discrete pieces of information which can be examined and manipulated with computers and other digital devices. Increasing utilization combined with limited financial resources for computer equipment acquisition has resulted in a trend towards employing the same computer in the classroom as was purchased for the laboratory. Therefore, computer facilities should be simultaneously oriented towards class work and research.

These are just a few of the conditions which exist in a university environment that have

bearing upon the evaluation of computer purchases and their introduction into the college curriculum. It will require consideration of multiple use of this equipment in both the laboratory and classroom. Any plan must take into account the time constraints of the faculty who will have to either write or select software (and possibly evaluate future hardware additions) and the time constraints of students who will be expected to use the computer facility. Additionally, the process requires careful review of administrative policies.

EDUCATIONAL GOALS

Educational goals is another area which must be given careful consideration before attempting to select a computer facility. While many non-educators feel that the simple introduction of computer technologies into the classroom will benefit graduating seniors, this could be a wrong conclusion if the use does not conform to the primary principles employed in training mining engineers. If not approached correctly, it could actually impede their education by consuming valuable class time in the pursuit of technologies which do not properly prepare students in the basic fundamentals important to good engineering practices.

Educational goals, as translated by many universities, usually contain a statement which may be roughly stated as, "releasing the creative potential of each individual student". While the statement is not the one usually sited by most engineering departments, it does convey their goals in a somewhat abstract form. The primary idea behind the university education of mining engineers is to produce people capable of developing and evaluating good solutions to problems. This involves the process of teaching fundamentals in mathematics, physics, chemistry, material behaviors, management, and a host of other topics, and encouraging the creative synthesis of solutions. Engineering generally combines design and economic criteria to arrive at some "best" solution which is both technically feasible and cost effective. (The term best is used rather loosely to mean the better of several alternatives which were defined by an organization's particular perspective.)

Companies and educators alike have discovered that good computer facilities can greatly enhance the ability of engineers to arrive at good solutions through independent and innovative thinking. Often the large number of calculations involved in solving even some of the most simple problems have acted as a barrier in the path of this process. However, the computer can be employed to do the mundane task of routine mathematical analysis quickly and accurately. This allows a student or engineer to spend more time analyzing the problem or it allows him to explore more feasible solutions. Although this problem was partially overcome by the appearance of inexpensive, hand-held calculators in the early 1970's, the potential now exists for this constraint to be of minimal importance as access to computers increases.

Care must be taken that computer analysis of a problem does not replace teaching design fundamentals in the classroom. Although such a course of action may make a student very capable at using certain design programs, it can lead to deficiencies in the primary concepts vital to the engineering of a mining system. For example, if we teach the design of mine water systems strictly from the standpoint of running a pipe network program, the student can lose perspective as to what constitutes good design practises (such as when parallel pumps in networks may be useful and how different layouts effect power requirements). This will lead to handicaping the design phase of the problem solution since there is insufficient comprehension of design principles to result in adequate input for the computer program or to properly interpret the output. The result is a situation where the software is actually used to help choose the best of several, possibly poor designs.

The primary objective of education is to produce engineers capable of using their knowledge of the physical behavior of systems to solve problems and to implement new designs. The introduction of computer technologies into an engineering curriculum must insure that this goal is being met.

PERCEIVED NEEDS

Another area which must be considered before selecting computer equipment to integrate into the academic experience is the primary use of those facilities. As stated earlier, students need computer facilities to ease their computational limitations while simultaneously improving the quality of education. If computer facilities are provided which would reduce their workload, students will very quickly take the initiative. Therefore, the acquisition of equipment could encourage independent action in applying such hardware and software outside of the classroom. Two such obvious areas of wide appeal are word processing facilities and statistical packages.

The access to word processing software can be a valuable teaching aid. An area often neglected in an engineer's education is that of report writing. A word processor can simplify the task of revising a paper and result in much higher quality work. In addition, such a program package should include a spelling checker to catch typographical errors as well as misspellings. Students are usually very willing to use a word processor to produce their class reports if they have easy access to suitable facilities. It saves them time, energy, and money in trying to commercially obtain typed reports. Students are also less reluctant to edit and change drafts if they are using a word processing system. Its value for graduate students, research appointments, and professors cannot be stressed enough. Word processing software can improve the quality of publications and reports released by a department when text processing capabilities are placed into the hands of these

individuals and not just their secretaries. The inclusion of this capacity should be considered in any educational computer facility.

Statistical packages are useful under many different conditions at a university, but it is in the area of research where they become virtually indispensable. With the advent of automated collection systems, data has become easier to accumulate. Vast amounts are generated by many research projects which makes hand reduction impossible. Often the search for correlations and trends can only be done effectively by using a computer. Statistical software transforms this immense task into the act of choosing what data is to be included and the desired statistic. This leaves more time for analysis and creative interpretation.

Careful considerations must be made as to the performance required to run the desired applications. As previously stated, mining can draw upon programs developed for other fields to solve many of its problems. Those simulation, structural analysis, database management, and financial packages which professors wish to use in course work have size and speed requirements. The proposed facility should be able to adequately service these needs.

In addition, there are definitely unique features in the mining industry which call for specialized software. Take for example mine ventilation. Airflow in ducts is only a rough approximation of the complexity of an underground mine ventilation network. Few, if any, building ventilation systems have leakage between separate ducts, booster fans, large dimensions, distant delivery points, gas injections, heat and moisture changes, and other such complications. This is one reason for the appearance of a number of ventilation network simulators. Other programs of a specialized nature are also available and should be given due consideration.

Another large group of software specific to mining falls under the general heading of economics. Programs abound which will compute the expected worth of a mineral deposit in a number of ways. They provide many different and useful capabilities such as stochastic modeling, variable costing based upon mining method, and automatic forecasting of product revenues. The programs which are desirable for classroom use must be able to be implemented on the selected computer facilities.

Mapping and the generation of grade-tonnage reports is an area of potential classroom utilization of computers. The two tasks are intimately related since surveying data is utilized in each. Therefore, there exists a definite benefit in linking the two projects together. This requires a system with enough mass storage to accept an application of this type. Computer generated mine maps will mean a large savings in time and manpower in an academic as well as an industrial setting. But implementing this feature will require specialized hardware such as a plotter and a graphics terminal.

Simulation is a field of computer usage which has tremendous potential benefits to the mining industry. Powerful languages now exist which allow continuous or discrete modeling of most physical systems. Presently, simulation is being used at open-pit operations in a variety of areas including production, dispatching, maintenance, and scheduling. This technology requires the computer facilities to have some form of a high level simulation language so that students and faculty members can become better acquainted with modeling techniques.

Coal mines are also taking advantage of this new technology in the areas of mine planning, scheduling, and face simulation. Programs are available which allow an engineer to examine a working section in detail. Basic input data is obtained employing time study techniques and is used by the program to model the face under present conditions. Questions concerning procedures, new equipment performance, and different mining conditions can be addressed by changing certain parameters and exercising the model. Mine planning and scheduling is simplified due to the application of production models. The results can detail tonnage, machine and man requirements, and supply consumption, which may be compared to desired or expected amounts. This same information can also be used in the financial planning aspect of the project. It is relatively easy to reschedule a mine or to examine a new mining plan using a computer than to actually implement the idea. The result is a better solution then can be generated with traditional mine planning techniques.

The final area of needs which must be considered are those of mining companies. It is important for educators to understand the educational expectations of the mining industry before selecting any sort of computer facilities and introducing them into the classroom. Three very important needs of the industry as viewed in our region are: engineers who can use computer aided design systems; engineers who have some idea as to when computer solutions are appropriate; and, finally, engineers who can assist in the selection of facilities for the engineering environment. These areas, and others specific to a region, should be addressed in the academic training of the mining engineer.

The special needs of mining are varied and often specific to a company or property. This leads to a mixture of externally and internally acquired programs. Some software vendors recognize this and offer services in programming as well as simply selling software. The number of mining programs is growing rapidly due to a vacuum which presently exists. It is expected that, in the future, the number of programs on the market will cause the level of evaluation that is now observed in the selection of a spread-sheet or word processor program.

OPERATIONAL PHILOSOPHY

A computer facility is more than just a collection of hardware and software. It is also very much a clear and workable philosophy. The educator's perception of computers will determine how they are implemented into a mining curriculum. While it would be naive to think that any one philosophy is more appropriate than another, this section will deal primarily with the views held by the authors.

At the core of any operational philosophy should be the idea that students, researchers, and faculty must be provided the opportunity to work with computers. Access must be as free and unstructured as possible. One problem seen in many students and faculty (although less so now than in the past since the proliferation of the personal computers) is a feeling of unease when using computers. Utilization of computer facilities allows one to become acquainted with their benefits and capabilities. People overcome their anxiety by becoming acquainted with a computer system, and they realize that it can be used effectly to increase their productivity.

A key disagreement among computer experts is whether it is best to use a central university computer facility, to use an in-house facility on a department or college level, or to use PCs (personal computers) in combination with a university wide system. If an institution has a central computer facility, it must cater to a diverse and changing clientele. Science people want fast, accurate computations while business people prefer data management and manipulation abilities. Though there is an area of overlap, it is easy to recognize the difficulty in providing for the special needs of any one discipline, group, or individual. The response by the university to this problem usually manifests itself in the acquisition of computer programmers. These people then work with the department or individual to develop the disired application. The programmers, though highly skill, must be trained in often specialized fields before any software can even be considered; a tremendous cost in the time to all those involved. Therefore, there is a tendency by colleges or departments to purchase mini and microcomputers and place the task of program development into the hands of the individual with the necessary professional expertise; the researcher.

It is believed, and often found true, that this sort of in-house computer facilities can better serve specialized needs. This is due, as stated above, to the expert in the discipline and the programmer being the same person. Communication of the problems, needs, and expectations of the profession to another person is eliminated. The result is a workable, if not elegant, computer program which does exactly what is wanted. Unless specifically written for outside consumption, the developed software is usually limited in sophistication and applicability; a task at which a professional programmer can be of assistance. But this method is still successful in many situations and is often employed.

In recent years the personal computer has exploded into the market place. Many times these computers are a very cost effective alternative to providing unstructured access for students. Quite often 10 - 15 PCs can be purchased for less than half of the cost of a minicomputer system and handle the same work load. These machines will also effectively handle data collection and reduction in the research environment. These small computers work extremely well in certain environments where multiple use and sharing of resources is not critical and they are found to be very valuable within the university environment. Our experience, however, has shown that they do not work extremely well for classroom assignments and general student access. Several PCs mean that when a professor changes or introduces new programs for a course, he must update many copies of the software (one for each student). There may be the additional complication of copyright violations as students make duplicates of purchased software for their own use. It also means that each of the systems must have their own dedicated printer. If word processing is provided, each machine must have separate access to a letter quality printer or some sort of sharing arrangement worked out, which can create volatile situations.

Exposure to different types of software is important. If a student runs "good", "bad", interactive, and batch programs, it will help him to better evaluate acquisitions in the future. Most people are quick to learn what they like and dislike about a computer program when they are exposed to a variety of approaches. If they have never run anything but one application, they will tend to believe that its approach is sufficient. A variety of experiences will help formulate in an engineer's mind those features which are appropriate for a given situation; enabling him to better select a software package.

The degree of sophistication of the software probably contributes little to the overall education of an engineer. Undergraduates do not need to be exposed to the latest in finite element analysis, computer aided design, or computer aided engineering. The important concept is for educators to teach student how to use these systems to better solve problems. A computer algorithm which is inefficient still exposes the student to its use. This means that it is not necessary to have the latest or the best computer facilities for educational purposes, it only requires equipment be in-line with the educational goals.

Sophisticated software and hardware, however, may be a requirement for graduate research. Here an inefficient algorithm may hamper, prolong, or invalidate research objectives. At our institution we use smaller in-house facilities for educational and noncritical research applications since their use is relatively inexpensive. When computational speed and sophistication is needed, we utilize the much larger university facility.

Many computer systems only offer the choice of one or two computer languages. The ability to run several types is recommended since certain applications are best suited towards a particular language. For example, a rather advanced truck haulage simulator can be easily constructed with one of the general purpose simulation languages such as SLAM. Different computer languages are important to the overall efficiency of a computer installation. A student's exposure to various codes will broaden his knowledge base and improve his ability to select good software packages for the solution of mining problems.

The final comment on philosophy is in the area of introducing computer usage into the curriculum. There are essentially two methods of approaching this task; provide access with little or no software support, or provide access with software support. We believe that an educational plan for computer utilization should incorporate both techniques and should, roughly, take the following shape. First, general usage should be introduced to freshmen and sophomores. This can primarily take the shape of running "user friendly" programs which are employed to check hand generated homework calculations. Second, juniors should be steered into running simple computer aided design software with some simple programming problems. Here the approach should primarily emphasize computer replacement of of the hand calculations. Third, seniors must be encouraged to independently utilize available software, write their own programs for homework solutions, and move into the application of software to solve or analyze real mining problems. This introduces the graduating engineer to the complexity of the mining industry as opposed to the highly controlled exposure he had in previous, lower class experiences.

Computer usage on the graduate level however should take on a slightly different form. The major objective behind graduate education differs from that of undergraduate education. Since each institution's objective and each individual's graduate programs differ, computer education must be custom tailored to the research topic. However, we feel some common areas should be included in the interest of developing research skills in graduate students. These include advanced programming and real world computer interfacing. Programming should encompass writing applications suitable to solve actual mining problems while computer interfacing should involve either the design and utilization of data acquisition systems for research or the computer control of equipment. It is very important to insure that any deficiencies in primary computer skills are corrected before entering more advanced concepts.

PURCHASING CONSIDERATIONS

There are three major areas which must be carefully considered before a computer facility is developed. These are the selection of the hardware, the choice of the software, and the installation of the system. Each equally affects the computerization plan and, therefore, should be viewed simultaneously.

There are a great many pieces of equipment under the heading of hardware. They include the processing unit (sometimes referred to as the computer), disk drives, printers, terminals, tape drives, plotters, digitizers, and modems. The speed of the computer is only one factor to examine when looking at the specification sheet. Total memory, empty bus slots, bus definition (type of bus), available operating systems, and dimensions are other factors to consider. Each of these items has an impact on the decision to purchase a certain brand. The memory capacity will indicate not only how large a program can be implemented (this is additionally influenced by the operating system), but also how effectively it will handle a multiuser or multitasking environment. If there is insufficient memory, time is lost in loading and unloading between memory and disk storage (this is termed "swapping").

Expansion is usually accomplished by adding new boards and, therefore, both the number of empty slots and the bus definition specifications are extremely important. The selected computer system will very seldom come equipped with everything you want now or will want in the future and so, these two specifications must be examined. The bus definition influences the availability of boards and their cost. This is a great consideration when trying to get the most performance for the available money. On the other hand, a bus may not presently be well supported but does have a good future potential. The arguements concerning present support or future potential will depend upon the situation.

The choice of an operating system is dependent upon the intended use of the computer, but it is nice to have as wide a variety as possible. It is not uncommon for objectives to change and necessitate the replacement of this piece of software. The more types of operating systems available the better as it indicates the computer's acceptability by the software industry, and results in a much wider selection of programs.

The computer system will often come with disk drives which will primarily be used to load the operating system (this is commonly referred to as "booting the system"). The drives are usually small, 10 to 20 megabytes total. Normally, this is insufficient mass storage for many applications including multitasking. A disk addition requires the purchase of two pieces of equipment; the controller board and the drive. Besides compatibility with the computer and disk, the number of drives this board can handle is of interest to the purchaser. Disk drives come in two types; fixed and removable. Fixed drives usually have a greater capacity but suffer from the inability to be transported. The actual size (capacity) will depend upon the use of the computer facility; with one megabyte generally being sufficient for the ordinary user.

Terminals are items in a computer facility whose choice is often neglected. Usually price is the sole consideration. A terminal is the connection between the user and the computer, and to fill this role effectively it must be rugged and reliable. Special features, such as "arrow" and programmable keys, may be required for the operation of some programs. The terminal itself must conform to the standards of the terminal control board (usually current-loop or RS232C). Even the choice of screen color (black, green, or amber) should be considered. A well selected computer terminal, based upon intended use and user acceptability, will be well worth the additional cost over the purchase of a "cheaper model".

Another common piece of hardware is a printer. There are usually two type deemed necessary for an educational system; a general use (high speed) and a letter quality printer. Band and dot matrix printers are examples of general use machines. They are fast, rugged, reliable, and cost effective. Some dot matrix printers have a dot addressable feature which allows limited graphics capabilities. Their main function is in the areas of program listing generation, printing of program results, and the production of rough drafts of papers.

Letter quality printers, on the other hand, are slow and expensive in comparison. But they are invaluable in the final production of papers and reports. In addition, students may employ them in the preparation of resumes. The cost and availability of different letter types and sizes (usually in the form of interchangeable wheels) are major considerations in the printer purchase, but the ability of the software to use provided features should not be forgotten.

The final category in the area of hardware contains equipment such as tape drives, digitizers, and plotters. They are placed in a separate group since this equipment is not normally required for day-to-day operations, or are viewed as having specialized tasks not required by the average user. In a way this is correct but it does not diminish their importance in a computer facility being used in education. The major advantages of a tape drive are in long term storage and program protection. It is not always required that data and programs remain active at all times. Research results may not be used again for a number of years or a teaching program will be used only once a year. In such situations, a magnetic tape is an excellent storage media.

This is not the only reason for utilizing a tape, protection of valuable programs and data is the other benefit. For example, an operating system is often modified by literally thousands of man-hours of effort and so becomes a major investment. If something happens to this sole copy, such as a disk head crash, it would be difficult to bring the computer back to the previous level of operation in a reasonable amount of time. A second copy on tape would be very valuable. Tape drives are also an excellent means of exchanging programs. If the advantages and uses of this piece of equipment are examined, a tape drive becomes a necessity rather than a luxury.

Digitizers, plotters, modems, and other such devices have special uses and will have to be chosen on a case by case basis. The research program can probably justify many of these types of additions but the purchase should also account for possible educational uses. Digitizers and plotters are especially useful in areas where mapping plays a major role, such as in surveying and mine layout. Modems allow remote connection to the computer facility which is often useful to researchers in the field. A decision must be made as to the desirability of providing this type of access to the general user.

Software considerations are varied and dependent upon the intended application but all have user sophistication and machine implementation in common. Some programs are necessarily complex in order to perform their function efficiently. Unfortunately, this makes them rather difficult to use without special training and has lead to the current phenomenon of "user friendly" software. It has also contributed to the inclusion of tutorials in many application packages. Word processing software usually includes a "help" facility to aid in the use of its features. The applications chosen for an educational computer system must be appropriate for the sophistication of the intended user. The more training a student receives in the use of a program, the more complex it may be, but this will require more effort on the part of the professor. Finite element software is extremely complex but is not intended for the uninformed user. But it is entirely appropriate for an advanced rock mechanics class where the required expertise can be taught. Both the input and output should be examined for suitability. There is little benefit from the difficulty in either data development or result interpretation.

The second area of concern is the operation of the application on a particular machine; commonly termed implementation. Implementation is a concept which has been used many times in the previous sections and has many meanings. First it must be realized that almost any program can be forced to operate if enough of its features are deleted or modified. This can ruin a good program if it is not meant to operate under the present computer configuration. Examples of this abound in the software market, especially at the personal computer level.

One common modification for most programs is the "down dimensioning" of the arrays. To be useful, the software must be of sufficient size to handle fairly complex, real world problems. If not, hand solutions might as well be employed. The advantage to having the computer solve the problems is to free the student from the calculations and let him explore the various interactions of the real system. Computers incapable of including complexities because of insufficient size do not fulfill this educational requirement and will have very little impact upon the training in mining engineering.

Another concern in implementation is turnaround time. This is usually defined as the time between the start of a program and its completion. In many people's minds, the shorter the turnaround the better and so there is a push for bigger and faster machines for fulfill this desire. However, not every application requires the same amount of time between start and finish. This is due to the iterative nature of computer generated solutions in the engineering field. The amount of effort that it takes to evaluate the output has a tremendous bearing on how much time the computer can take in determining the next answer. For example, a financial model may require less than an hour to formulate another question and so many runs can be generated in one day. On the other hand, how many finite element solutions can be adequately interpreted in one hour's time? This indicates that the largest, fastest computer on the market is not necessarily the one really needed in a facility. A balance must be created in turn-around time to insure an appropriate response for the particular application.

A final area of consideration concerns itself with the installation of the computer facility and involves often overlooked or underestimated factors. There are a variety of requirements necessary for the physical implementation of a computer. Space is usually the first of these problems to be faced. There is a vast difference between a microcomputer sitting on a table and a mainframe which requires an entire room. Space must also be allocated for the terminals and other peripherals. There are three areas to examine when housing a facility: operation, service, and controlled access. A cabinet containing a piece of hardware can very seldom be placed against a wall due to the need of having the back available for service and the front for operation. At the very least it should be mobile enough to allow such access.

Controlling the availability of certain equipment is also desirable in assigning space. The general user does not require physical contact with those devices that require special instruction on their operation. This equipment includes the computer, disk drives, tape drives, plotters, and digitizers. Usually some type of control should be placed on their use due to their complexity. In no way should controlled access be translated to restricted access; students should still be allowed to use the devices but only after being properly trained.

Besides space requirements, special consideration may have to be given to power and air conditioning needs of the facility. Except for microcomputers, more than a standard three-prong outlet will be necessary. The installation of the special plug may require stringing a separate power line with its own circuit breaker (a good idea anyway due to voltage drops caused by other equipment on the line). Power filtering and transient suppression may also be required depending upon computer sensitivity and the acceptability of the electrical distribution system. Sometimes a battery backup or an uninterruptable power supply is used to keep the system in operation during short power drops or to allow the computer to be "brought down gracefully" upon complete electrical failure. This prevents the total loss of all work presently residing in the computer.

In addition to power sensitivity, computer components are also affected by temperature, humidity, and dust. These may have to be controlled by cooling, dehumidifying, and filtering of the air entering the room. It is often better not to rely upon the building's air conditioning facility unless the space was specifically designed to contain a computer system. One practice which will help keep dust at a minimum is to place the room at a higher pressure than that of the surrounding building. This will prevent dust and other particulate matter from migrating into the room. Remember, cleanliness is important in the successful operation of the entire system, therefore, eating, smoking, and other such activities should be prohibited around the equipment.

The very complexity of a computer system calls for some level of expertise to be secured. This could be a faculty member, graduate student, or outside employee. There is a tremendous amount of work involved in keeping a computer facility in operation. New application implementation, additional equipment selection, user addition, operating system updating, and many other tasks require a certain amount of knowledge concerning the hardware and software. Usually this job fall to the purchaser of the system or a student using it for his research. The first person has limited time and graduate students are transient in nature. The result is periods of relatively high productivity followed by the forming of a backlog of tasks which will be accomplished "real soon now". The required commitment will depend upon the size of the facility and its goals. A programmer, system or general, may be required for a facility with an varied role to play in a number of classes and research projects. The need for personnel or expertise cannot be overlooked when considering a facility.

The requirement of extensive cabling is often overlooked and greatly underestimated when connecting peripherals, especially terminals, to a computer. Most of the equipment additions, such as disk and tape drives, are supplied with the appropriate wire connections. It is the terminals, printers, modems, and other such devices which are most often forgotten when it comes to cabling. Just a glance at a computer supply catalog shows how expensive two connectors and a length of wire really are. The method usually employed is to purchase these components separately and supply the labor in-house to construct them in the required lengths. A suggestion developed from experience is to lay the wire before attaching the connector; there may be some tight places along the cable's route. Distances are sometimes limited by the transmission power of the terminal board and so specifications should be checked against needs.

These "hidden" considerations can be major influences when planning a computer facility. They have additional costs in time and money which cannot be avoided, only minimized. One final member of this group bears mentioning and that is the time it takes to obtain a fully operational computer facility. This not only includes bidding and shipping time (which is often significant) but also accounts for set-up time after arrival of the equipment. The period can be anywhere from a few weeks if purchased as a unit, to many months if separate pieces must be integrated. In any case, there is more involved than just uncrating the machine and turning on the power. All of the peripherals must communicate with the central processing unit (CPU) and each other. The hardware must be connected and the software configured for the various devices. It is unreasonable to expect a computer system to work without problems, even if purchased as a package. This should be made clear to every potential user. Usually, the larger, more complex installations have the least problems; it is the medium sized facilities (minicomputers) which seem to have the most. This is due to the tendency to treat mainframes and super minicomputers with more care when they are installed.

MINIMUM SYSTEM CONFIGURATION

Considering mining's special needs and the requirements for young mining engineers to be trained in the use of computers, there is minimum hardware configuration which meets many of the previously discussed criteria. In general the computer must be fast, have the ability to accept moderately sized programs, have graphics capabilities, be equipped with high speed and letter quality printers, be multiuser and multitasking, have available a variety of languages, and be expandable. The list below presents such a system and accessories which is felt to provide the performance desired in an educational setting. Though brand names are given, they should not be viewed as an endorsement as other machines may have the same or superior performance. Instead the manufacturer's designation should be viewed as a type of shorthand notation for the equipment specifications.

- Digital Equipment Company VAX 750 with 2 MB of memory, a floating point unit (FPU), and the VMS operating system
- Digital Equipment Company RA81, 400 MB disk
- Digital Equipment Company TU77 tape drive
- 2 Datasouth DS180 dot matrix printers
- Diablo 630 letter quality printer
- 20 Hazeltine Esprit III terminals
- Megatek Corporation 1645 graphics terminal
- Hayes Smartmodem 2400
- Numonics Corporation digitizer
- Hewlett Packard 7585B plotter

Such a computer facility is capable of efficiently handling fifteen to twenty simultaneous users and will cost somewhere around $200,000. It has the disk capacity to adequately support 100 active users (this includes individual and class accounts) and 20 large research projects. This system can meet the computational requirements of a mining department whose total enrollment (graduate plus undergraduate students) is between 100 and 150 students.

The central processing unit (CPU) has a 32-bit architecture which can operate in a multitasking and multiuser environment and has virtual memory capabilities. (Virtual memory allows program sizes larger than the actual physical memory space to be implemented). The physical memory size of 2 megabytes will adequately handle the expected number of simultaneous users without degrading response time due to swapping. If it is anticipated that a number of large applications will be simultaneously employed on the system, more memory could be easily added.

Due to the large number of floating point calculations in engineering applications, a floating point unit (sometimes referred to as an accelerator) is included at part of the computer package. Some applications, such as finite element analysis, could be greatly enhanced if an array processor were also purchased. This piece of equipment must not be confused with a floating point unit as its use is more specialized. The floating point unit is used in such mundane tasks as converting characters to their ASCII representations and in character string comparisons as well as in mathematical computations. Graphics applications all but demand a floating point accelerator for acceptable response times. The increase in performance due to this addition is well worth the cost.

The 400 megabyte (MB) hard disk subsystem is considered a bare minimum for the intended use of the facility. Our experience has shown that most users consume about one megabyte of storage. On our much smaller, 120 megabyte disk, there are about 60 active computer accounts which occupy approximately one-half of the total disk space. The rest is reserved for the operating system and selected large scale users where the one megabyte limit is entirely inadequate. For example, a database presently under development may occupy as much as 15 megabytes of storage. It is important to take stock of what actual disk needs may be encountered, not forgetting that the operating system has certain space requirements where temporary files are kept (notice also that this is a dynamic process and the requirements change with computer load). If large virtual memory programs are run, then the minimum requirement will be at least 2.5 to 3 times the largest program size. Remember that although a CPU may be capable of handling 4 gigabyte programs with only 1 megabyte of physical memory, you must still be able to store the complied version, the source, and the swapping image on disk.

The general purpose printer chosen is capable of 180 characters per second (this specification is not hard to meet). It is anticipated that undergraduates, graduate students, research personnel, and professors will be using the facilities in different locations, so two printers were chosen to accept the expected load. It is not hard to justify this number in light of their inexpensive nature. The letter quality printer will be available for general use, however, our experience has shown it is used more effectively when its access is somewhat controlled.

Computer languages will include as many as can be utilized by the mining department. FORTRAN is considered the first and foremost required language on a system to be used in an engineering setting. The vast majority of the existing software and most of the programs currently in production use FORTRAN in one of its many versions (mainly the 1966 or the 1977 standard). Regardless of how antiquated people may accuse this language of being, it is still the standard language for engineering applications. If this compiler is not obtained, a tremendous amount of time will be dedicated to rewriting programs in another code.

A structured high level language is also considered as part of the software package. It would be either C, ALGOL, PASCAL, or MODULA-2 as there seems to be a great deal of excellent computer work being done utilizing these languages. There is the additional benefit of learning a second programming language to compare and contrast with FORTRAN. This experience is helpful in judging the relative benefits of writting a certain application in a certain computer code.

BASIC is a language which is considered too simple by many to include as part of an engineering computer facility. This is not totally true and its very simplicity can be a great encouragement. The main drawback in writting applications in this code is that a standard does not exist. Each computer usually has its own particular dialect and so students who learn this language on one machine may find it difficult to adjust to its implementation on another. It is far better to learn a language with a standard (such as C, FORTRAN, or PASCAL) which is virtually independent of the computer brand. For this reason, the use of BASIC is discouraged in our computational environment.

For even lower level support, an assembler is included. This allows the production of device drivers and fast data collection routines. In addition, a database management system would be helpful, but is not required initially. This along with text editors, spelling checkers, graphics packages, simulation languages, statistics packages, and other useful software are readily available for the type of computer facility selected.

CONCLUSIONS

Education is in the unique and enviable position to fill the primary goal of training young
mining engineers to think creatively. Computers
are contributing to this task by freeing students
from tedious calculations. They are then better
able to study the underlying principles and use
them to explore alternate solutions to real world
problems. To take advantage of the power and
freedom a computer can give, a machine must be
chosen to meet the needs of the institution,
faculty members, and students. Hardware and
software purchases must be considered carefully.
Time and expertise requirements must be effectively met while concurrently providing sufficient
computational power and flexibility to meet future
needs. The use of computers in the classroom can
only be successful if both faculty and students
are sufficiently motivated.

The development of the computer facility must
fit into the environment which will surround it
and meet some very specific requirements. Only by
the careful consideration of a variety of areas
can academia successfully respond to the need for
mining engineers who can utilize computers and yet
maintain a high level of creativity. The growing
use of these machines will result in a higher
level of expertise in the mining profession.

BIBLIOGRAPHY

Bideaux, Richard A., 1981, "Minicomputer Applica-
    tions for Today's Mining Problems," Mining
    Engineering, Vol. 33, No. 11, Nov., pp. 1584-
    1587.

Chorover, Stephen L., 1984, "Cautions on Computers
    in Education," Byte, Vol. 9, No. 6, June, pp.
    223-226.

    Just, Louis C. and Mathieson, Graham A., 1981,
    "High-Powered Minicomputers Gain Mineral Indus-
    try Acceptance," Mining Engineering, Vol. 33,
    No. 11, Nov., pp. 1579-1583.

Khosrow, Badiozamni, 1984, "Mining By Computer -
    Is it Really the Way of the Future?," Mining
    Engineering, Vol. 36, No. 9, Sept., pp. 1302-
    1304.

Meyer, William L., 1984, "Computers: A Tool to
    Aide Mine Productivity,", Mining Engineering,
    Vol. 35, No. 11, Nov., pp. 1526-1528.

Osgood, Donna, 1984, "A Computer on Every Desk,",
    Byte, Vol. 9, No. 6, June, pp. 162-184.

Van Cleave, David and Rome, David L., 1981,
    "Selecting a Minicomputer From a Manufacturer,"
    Mining Engineering, Vol. 33, No. 11, pp. 1598-
    1601.

Chapter 6

THE PERSONAL COMPUTER AS AN ENGINEERING WORKSTATION

THE REMOTE DATA PROCESSING CONCEPT

D. David Marston

Richard R. Marston

Marston & Marston, Inc.

St. Louis, Missouri

ABSTRACT

The micro-computer and attendant technology have had unusually rapid and widespread acceptance by the mining industry. Personal computers are presently utilized in a variety of functions including accounting, inventory, maintenance and engineering. The low initial capital required and the resultant availability of local processing capabilities are the main reasons behind the surge in applications of such units.

Confusion over the actual processing capabilities of the eight (8) to thirty-two (32) bit micro-processing units is a phenomenon as widespread as the units themselves, as computer marketing personnel are liable to promise anything in the pursuit of sales. Unfortunately, the likely consequences of discovery of processing or storage limitations of a personal computer are either the shelving of the unit or the compromising of work quality. Neither need be the case, as technology is currently available to allow engineering functions of any size, complexity, or variability to be performed accurately, rapidly, and completely.

The following comprises a description of a cost-effect solution to the problem of micro-processor limitations through the use of a personal computer (PC) workstation and remote data processing.

INTRODUCTION

The rapid advancement in micro-computer technology in the past few years has resulted in a deluge of new engineering applications for the PC. In addition to the familiar utilization as a word processing tool, many have found that combining faster, more accurate processors with comprehensive software packages allows the PC to become an innovative local workstation.

Computer-aided tasks for which micros are nearly ideal include data compilation and editing, rudimentary sort functions, rapid organization of information into presentable form, and some surprisingly complex data manipulations (i.e. spread-sheeting). As the tasks for mine evaluation grow more complex, however, the limitations of the processing capacity of the PC grow more evident.

Proper configuration of micro-processor hardware and software can result in the continued utilization of the unit as a local workstation. In tandem with time-sharing access to a well-designed engineering computer center, in-depth analytical and design functions may be completed on-site, with a minimum of incurred main-frame processing charges. This presentation will deal with the options available through usage of a properly arranged personal computer working in tandem with remote data processing on a main-frame computer facility.

THE PERSONAL COMPUTER: A GRAPHICAL DESIGN CENTER

The configuration of the personal computing unit to be used as an engineering workstation determines the capabilities available to the station user. The following comprises a brief summary of some of the suggested hardware and software requirements for a comprehensive PC engineering workstation.

Workstation Configuration

An ideal mine design or analysis workstation should incorporate the following capabilities:

* Local database facilities for ongoing compilation and revision of geological, financial, and planning data.

* Modem and software allowing access and "dumb" terminal emulation for communication with mainframe computer facilities.

* A graphics monitor with at least 640 by 480 pixel resolution.

* At least two (2) 360K floppy disk drives or one (1) 360K floppy disk drive and one (1) hard disk drive (10-32MB).

* At least 320K of RAM memory.

Simply stated, proper configuration of a PC design station requires several additions to a basic configuration. Assume, for instance, that an office PC to be used as a workstation was previously in use as a word processing unit with the following hardware:

1) 64K system board
2) Color graphics card
3) Keyboard
4) Color monitor
5) One (1) 320K floppy disk drive
6) Parallel printer port
7) Parallel printer

The hardware additions to the unit would be:

1) One (1) memory board (256K)
2) One (1) modem, capable of at least 1200 baud (bits per second) communications
3) One (1) color monitor with at least 640 by 480 pixel resolution
4) One electronic digitizer compatible with the micro-processor

These hardware additions enable the PC, with certain software applications, to perform as an on-line storage tube, "dumb" terminal, and data transmission unit as well as a local processing facility.

Software Requirements

Communications software technology has progressed with rapid strides in recent years. As a result, several inexpensive programs are available to run the various "smart" modems. The main criteria in selecting proper modem software may be summarized in two words, Speed and Emulation.

The system must be capable of multiple baud rate data exchange. Such a capability requirement is primarily due to the numerous types of modems encountered in industry usage. As well, the software should be flexible enough to handle any "smart" modem attached to the computer.

Terminal emulation facilities are extremely important in choosing communications software. Different mainframe software packages occasionally require certain types of terminals to obtain proper screen formatting. Ideally, the communications software should be capable of translating a large variety of screen positioning control codes into "normal" PC screen presentation. At the very least, a facility should be available for custom installation of communications parameters.

Database software for micro-computers is numerous and selection of any one type is more a matter of personal preference than system design. Proper interfacing of micro and main-frame databases requires that the output from one be easily readable to the other. Utilization of intelligently patterned ASCII data files solves most of the problems encountered in this area.

However, it is essential that the micro database package be powerful enough to hold at least the minimum amount of data which is required by the main-frame. This raises the requirement of programs which access multiple data files in the course of file preparation, and which sort data based on various criteria.

Recently, software has been introduced which allows storage tube emulation on the IBM PC. This feature, coupled with a graphics display tube of as great a resolution as available, enables the workstation to act as a powerful graphics workstation. Therefore, visual editing facilities are available on a local terminal without the high cost of storage tube technology.

Data presentation is the task for which personal computers were originally designed and implemented. Word processing and spread sheeting programs are readily available from a myriad of manufacturers. Many are quite similar and selection is again a matter of personal preference and budget limitations.

ELEMENTS OF COMPUTER-AIDED MINE DESIGN

Proper definition of the tasks involved in the design, sequencing, and costing analysis of a mining project is necessary to emphasize the complexity of tasks involved in the processing, accounting, and presentation of data. The following comprises a brief summary of the activities involved in project analysis and the respective uses of a PC workstation working in conjunction with a remote data processing facility. A graphical representation of the proposed interface is shown in Figure 1.

Geological Modelling

The initialization of any mine planning process requires the building of a comprehensive, extremely accurate model of the geological characteristics of the area under scrutiny. Accuracy is the key in this process, as every subsequent stage of the planning is dependent upon the data obtained through manipulation of the geological model.

A system of discrete, keyed databases for geological, quality, and interpretive input forms a flexible and interactive means for data storage. The Marston & Marston, Inc., database networks are based upon the QUEO-IV database manager and manipulative language. The advantages to combined database management and programming facilities are ease and variety of data input, simplification of data extraction and "flagging", and flexibility in the adaptation of other software or database systems.

Topographical, ownership, and operations limiting surface or subterranean data are stored either on a discretionary database similar to those previously mentioned or within the files of a computer aided design (CAD) system adapted by Marston & Marston, Inc., especially for mining

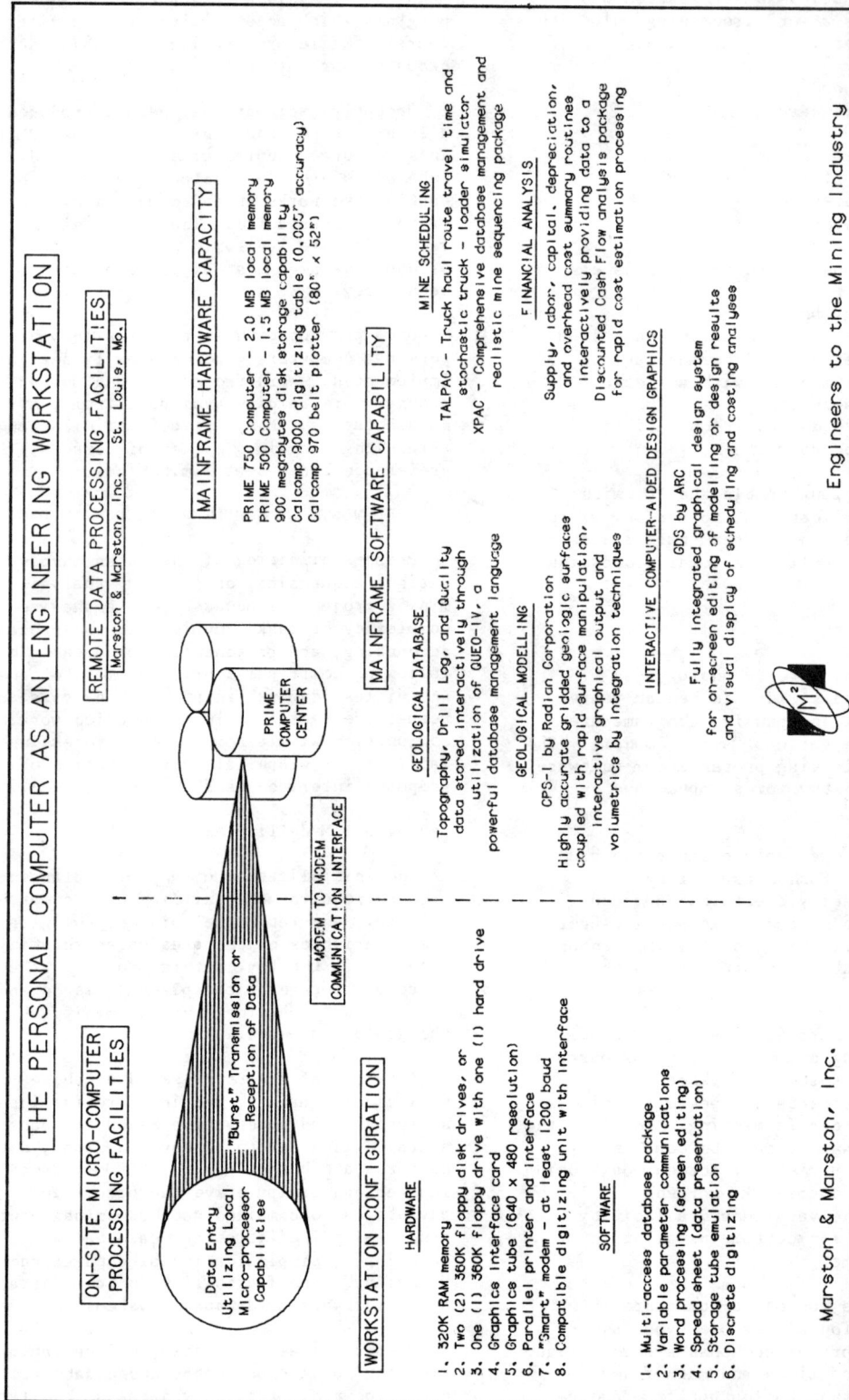

THE PERSONAL COMPUTER AS AN ENGINEERING WORKSTATION

ON-SITE MICRO-COMPUTER PROCESSING FACILITIES

Data Entry Utilizing Local Micro-processor Capabilities

"Burst" Transmission or Reception of Data

WORKSTATION CONFIGURATION

HARDWARE

1. 320K RAM memory
2. Two (2) 360K floppy disk drives, or
3. One (1) 360K floppy drive with one (1) hard drive
4. Graphics interface card
5. Graphics tube (640 x 480 resolution)
6. Parallel printer and interface
7. "Smart" modem - at least 1200 baud
8. Compatible digitizing unit with interface

SOFTWARE

1. Multi-access database package
2. Variable parameter communications
3. Word processing (screen editing)
4. Spread sheet (data presentation)
5. Storage tube emulation
6. Discrete digitizing

Marston & Marston, Inc.

REMOTE DATA PROCESSING FACILITIES
Marston & Marston, Inc.   St. Louis, Mo.

MAINFRAME HARDWARE CAPACITY

PRIME 750 Computer - 2.0 MB local memory
PRIME 500 Computer - 1.5 MB local memory
900 megabytes disk storage capability
Calcomp 9000 digitizing table (0.005" accuracy)
Calcomp 970 belt plotter (80" x 52")

MODEM TO MODEM COMMUNICATION INTERFACE

PRIME COMPUTER CENTER

MAINFRAME SOFTWARE CAPABILITY

GEOLOGICAL DATABASE

Topography, Drill Logs, and Quality data stored interactively through utilization of QUED-IV, a powerful database management language

GEOLOGICAL MODELLING

CPS-I by Radian Corporation
Highly accurate gridded geologic surfaces coupled with rapid surface manipulation, interactive graphical output and volumetrics by integration techniques

MINE SCHEDULING

TALPAC - Truck haul route travel time and stochastic truck - loader simulator
XPAC - Comprehensive database management and realistic mine sequencing package

FINANCIAL ANALYSIS

Supply, labor, capital, depreciation, and overhead cost summary routines interactively providing data to a Discounted Cash Flow analysis package for rapid cost estimation processing

INTERACTIVE COMPUTER-AIDED DESIGN GRAPHICS

GDS by ARC
Fully integrated graphical design system for on-screen editing of modelling or design results and visual display of scheduling and costing analyses

M²

Engineers to the Mining Industry

FIGURE 1

Proposed Mining Engineering personal computer workstation linked to Remote Data Processing center.

applications. The design system, called GDS (by ARC), provides visual editing through on-line storage tube display during digitization as well as accuracy to five thousandths of an inch, depending upon the digitizer used. Data extraction is provided by a version of the BASIC programming language, which outputs standard ASCII data files for volumetric, geological interpretation or model fencing purposes.

In addition to the advantages in accuracy and visual presentation, the maintenance of a graphical database provides final engineering drawings as a direct result of ongoing analysis tasks. GDS incorporates an enormous library of character fonts (including Leroy) and linestyles with facilities for creation of additional fonts or styles. An engineering workstation may thus become a highly productive "drafting table", simply through completion of necessary layout and/or design functions.

Geological data compilation on the PC may be completed in two ways. Programs designed for the database package chosen for the unit allow on-site data entry, incurring no connect or communications charges to the main-frame computer center. Such programs may involve manual keypunching or digitization of interpretive or limitational characteristics. Gathering of data locally allows most of the editing functions to be performed on-site, depending on the flexibility of the micro database and/or word processing software. The second method of data entry would be via usage of the GDS software for on-line digitization of pertinent data. With this option, the resultant data files would be on file at the Marston & Marston, Inc., computer center immediately, ready for subsequent processing.

Transfer of information gleaned on the local workstation to be remotely processed on the main-frame facilities involves two (2) steps:

1) Creation of sequential ASCII files with pre-determined file parameters

2) "Burst" transmission of the files to the PRIME computer center along with associated instructions or explanations

The data would be re-assembled and processed on the main-frame facilities utilizing CPS-1 geological modelling software, by Radian Corporation. The CPS software enables creation of three-dimensional mathematical arrays representing geological features such as (to name but a few):

* Structure isopleths
* Strata thickness isopachs
* Quality isopleths
* Weighted block value models (kriging)
* Ratio isopleths
* Faults, washouts, or other subterranean features

The arrays of values, or grids, representing the assorted surfaces are calculated according to pre-selected gridding algorithms. Included among these algorithms are linear least squares, quadratic least squares, linear and quadratic projections and kriging.

The geological modelling system incorporates fully interactive graphical displays for all surficial output. The kriging function includes associated variogram plotting for analysis of the data. As surfaces become finalized, bench plans on any scale or contour increment desired may be produced. These plots may then be added to or edited either on-line through utilizing the design capabilities of GDS or through manipulation of the geological database or digitization of desired trend lines. Again, the latter may be performed at the workstation and subsequently "burst fed" to the main-frame for processing. Errors in data input or interpretive geological decisions are readily detected in the resultant plots. Statistical analyses of each surface are output for rapid identification of maximums, minimums, and standard deviation.

Subsequent to finalization of the "stand-alone" representation of the model, the surfaces may be manipulated to provide further information on the project area. These manipulations may be as simple as the subtraction of two structure grids in the obtaining of interburden waste thicknesses, or as complex as the repeated factoring of several grids in the calculation of waste or quality ratio grids and representations of mining costs as a function of spatial location. Isometric views of sample topographical and complex, faulted top of coal structure are illustrated in Figures 2 and 3. Figure 4 shows a portion of the contour plot of the coal structure surface.

Mine Design - Cut and Infrastructure Layout

The capability of the CAD and geologic modelling system to interact allow the design of an optimum mine plan to be an interactive process, with the implementation of a variety of ore release options prior to final plan selection. The layout phase of the engineering process is greatly enhanced by the flexibility of the design system in the location and relocation of pits, roads, buildings, ponds, powerlines or other required surface features, and panels, stopes, drifts, ramps, inclines, shafts, etc., in underground designs.

Mining pit, cut, block, or panel digitization is performed subsequent to selection of a target sequencing philosophy, or more simply, the economic limitational parameters have been identified and incorporated into the mine model. INitial mining blocks are digitized and named according to the scheduling model (to be discussed later). Following blocks may be digitized manually or generated through tracing functions available in the CAD system. These trace routines generate identical blocks along either boundary of the existing cuts or benches. In the case of underground layouts, entire blocks

COMPUTER GENERATED TOPOGRAPHY

FIGURE 2

Isometric view of gridded topographic model
created with CPS-1 software.

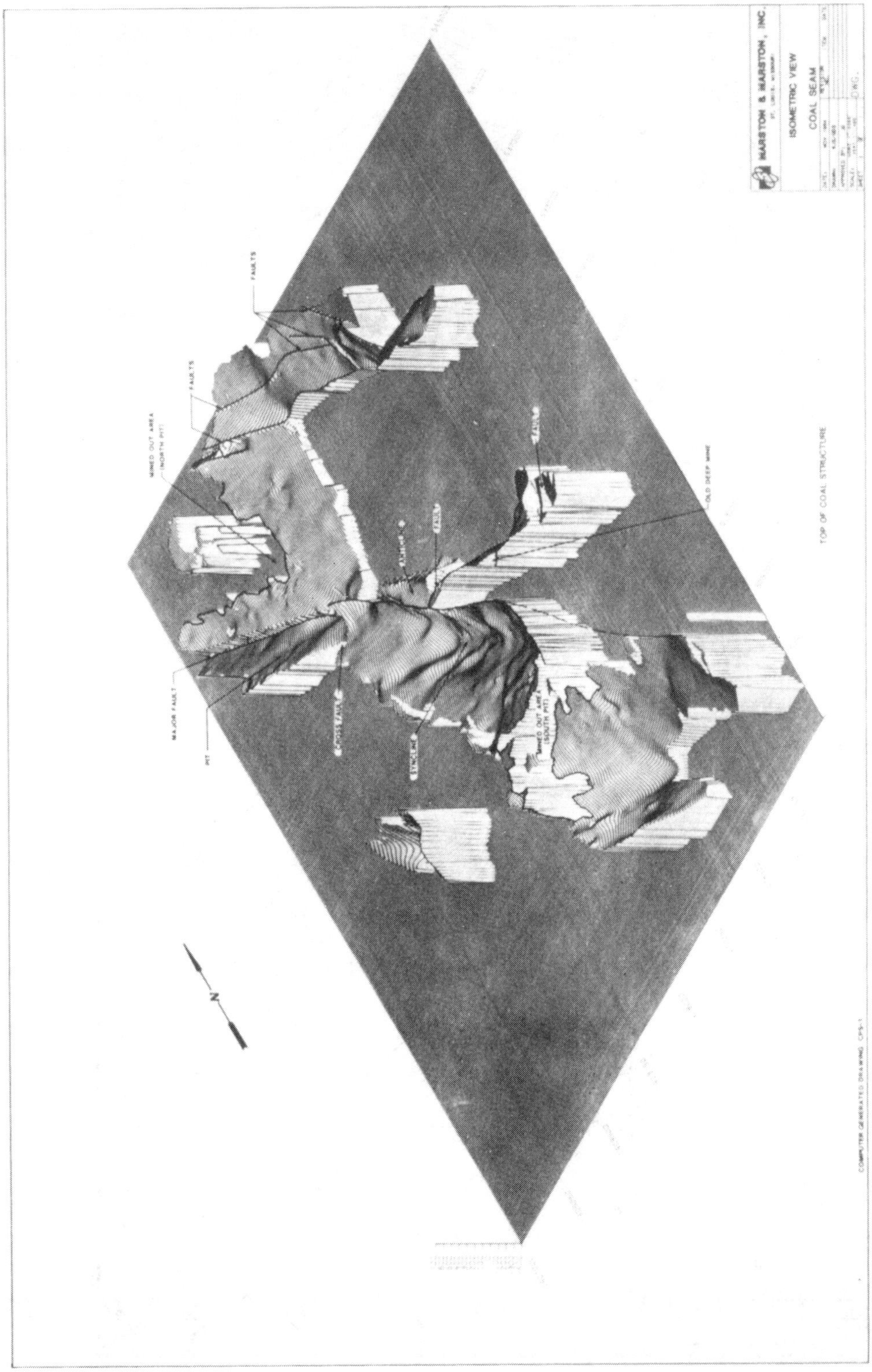

Marston & Marston, Inc.   Engineers to the Mining Industry

FIGURE 3

Isometric view of complex, faulted top of coal
structure.

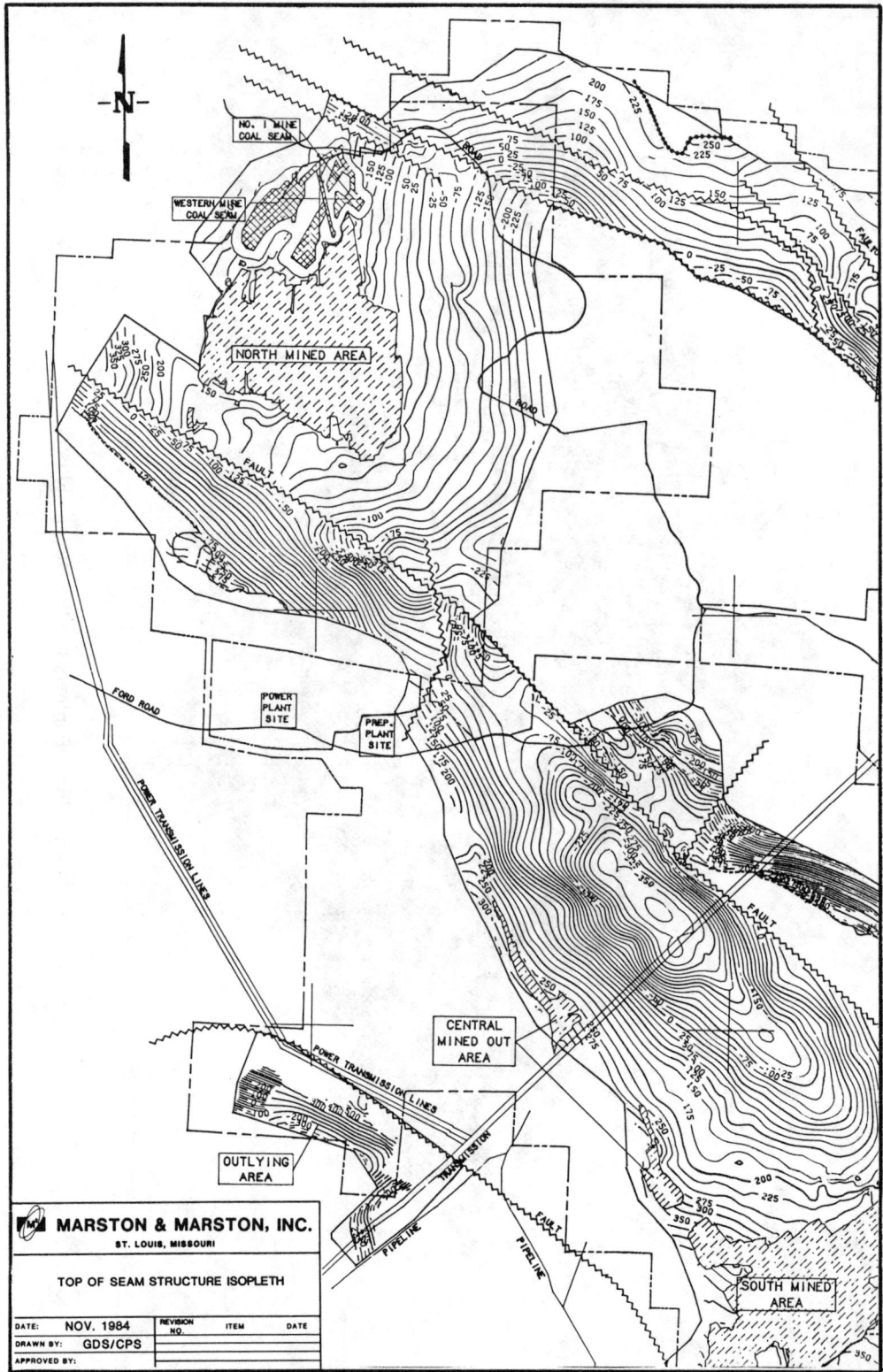

Figure 4 - Structure by CPS-1, mining limits and infrastructure by GDS.

may be repeated onscreen, which in the case of room and pillar applications proves extremely rapid. In any case, once the full range of cuts are created in a drawing file, the updating of all or any portion of the plan is simplified in that comprehensive graphical editing facilities allow creation, modification, replacement, or deletion of any facet of the layout while providing visual evidence of the effect of the change on the overall plan.

Utilization of the CAD system would require the workstation facilities to be on-line at the Marston & Marston, Inc., computer center. The rapidity and flexibility of design functions attained with the software would more than offset the incurred costs. However, CPS software allows input of any sequential values forming polygons as data for volumetric or blanking operations. Digitization of mining or limitational blocks could therefore be performed off-line using micro software capable of identifying the data points received (i.e., appending pertinent information to said points). Data transmission would then be as mentioned previously, with subsequent processing on the PRIME facilities.

Mine Scheduling

The cut, panel, or pit layout is output as ASCII data files which form blocks for volumetric calculations on the mine model. Accurate, rapid volumetrics are essential to provide raw data to the mine scheduling model, which accepts the data according to specific allocations to pre-determined areas. These data are usually in the form of total volumes of waste, ore, and/or interburden with associated data on quality or toxic characteristics.

Marston & Marston, Inc., utilizes a pair of software packages known as XPAC and TALPAC, by Mining Software Associates, for mine sequencing and truck-loader productivity analyses, respectively. The XPAC system incorporates a comprehensive database management package with realistic scheduling functions in the preparation of mining schedules. The TALPAC software provides haul route travel time simulation for any required haul unit as well as stochastic simulation of single loader and multiple haulage unit configurations.

The XPAC database consists of various levels of data constructed to report data to the limit of accuracy of the available information. For example, a typical breakdown of a multiple level surface operation could be as follows:

LEVEL 1:   TOTAL MINE
LEVEL 2:   BENCH
LEVEL 3:   STRIP
LEVEL 4:   BLOCK

The database is designed to be cumulative, i.e., the values reported in the STRIP level are the sums of values in the BLOCK level, and so forth. However, certain items such as ratio or quality are "weighted" against accumulating data items,

thus resulting in accurate summations in higher levels. A single XPAC database may have up to eight (8) levels, with levels from two (2) down containing as many records as required. Each record may carry up to eighty (80) data rows, depending upon scheduling needs.

Raw CPS volumes are entered into the lowest level of the database. Prior to scheduling activity, the volumes must be manipulated into a form which reflects a realistic material allocation. For instance, a single pit may initially be assigned a block waste volume and pit area. From these data an average thickness of waste material may be derived. Assume that the thickness totals 130 feet, but previous analysis of the primary waste removal unit indicated an operating limit of 100 feet in bench thickness. Further, a volume corresponding to three (3) feet in waste thickness was to be moved in bench preparation operations. THrough proper outlining of the material allocations required, the raw data may be manipulated into a format by which the required volumes are assigned to the correct materials handling machinery. These allocations are by no means universal, as material allocations may be tailored to suite expected operating conditions. An example of a mining bench layout is shown in Figure 5.

Data manipulations are not restricted to the allocations of volumes alone. Given equipment operating characteristics, attributes such as operating hours for equipment may be assigned to specified blocks. An example of such allocation is operating drill hours, which may be varied through the usage of another database sub-dividing descriptor. Appropriately enough, such sub-dividers are called Descriptive Attributes. To illustrate the example of drill operating hours allocation, one Descriptive Attributes may pertain to blocks of material within a previously defined hard overburden area. Another may encompass those blocks in a relatively softer, easier drilling area. Based upon these database flags, different drill production rates may be used in the assignation of drill operating hours. This is, of course, a simplified example of the use of these Attributes.

Once the data are allocated to the satisfaction of the mine planners, the scheduling files are initialized. The first step is the construction of an equipment database. This database may consist of individual pieces of equipment or operating spreads, each assigned a productivity rate based on either time or volume. It should be noted that these rates may be varied at any stage of the scheduling process, to reflect any anticipated fluctuations in the actual rates encountered.

The next step is the construction of scheduling categories. These categories must be arranged in such fashion that the sequence of material removal is logical. For example, a logical series of scheduling categories may be of the following format:

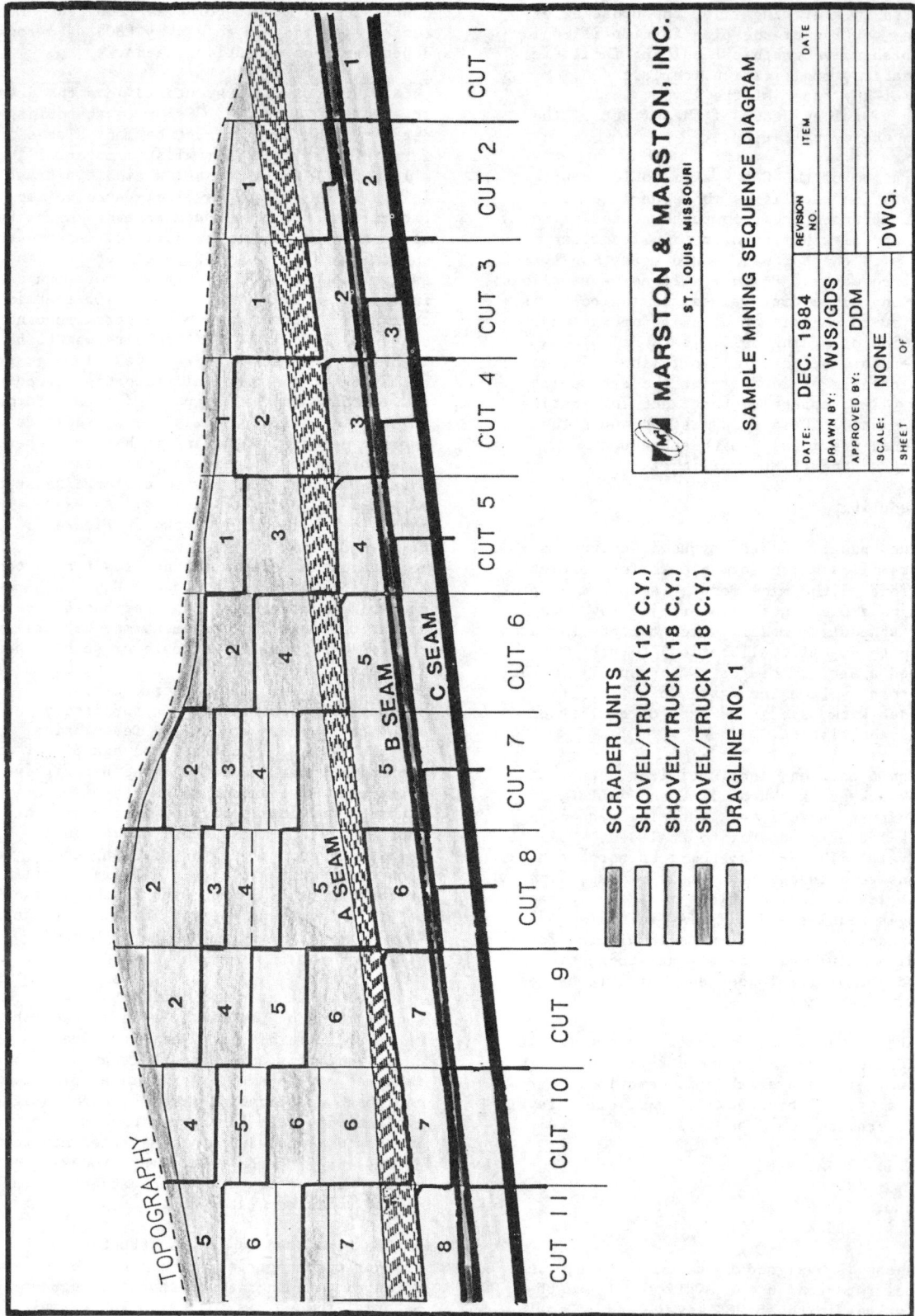

Figure 5 - Cross-sectional view of sample XPAC cut layout.

Subsequent to compilation of the operational costs of mining for the project, a discounted cash flow (DCF) analysis is performed. Three (3) reports are generated in the course of this analysis, the Income Derivation, the Summary Sheet, and the Discounted Cash Flow Sheet. The reports may reflect values tabulated using any chosen depreciation method or combination thereof.

The DCF analysis is an interactive process, using incrementally increased values from an input value to achieve the desired result. For each time period in the course of the analysis, the net present value of the net income is discounted at rates varying from five (5) to one hundred (100) percent. In this fashion, the required revenue per ore unit to give a pre-determined rate of return on investment may be calculated. Conversely, the rate of return given the revenue may be quickly obtained. The output from these analyses may be reassembled as input to graphical display charts to more easily display the study results.

The Income Derivation presents totals on a time period basis of the factors utilized in the calculation of after tax income. These include total revenue, total costs, and a variety of pertinent allowances. The Summary Sheet tabulates cash outlays and unit cost or revenue values throughout the period studied.

The Discounted Cash Flow printout summarizes the capital outlay, total income, and net income for each time period in the study. Subsequent columns indicate the net present value of that income at various discount rates. The software than extrapolates to the cost at which the net present value is zero (0), providing the rate of return on investment at the indicated revenue level.

Advantages to the PC Workstation Concept

Computer-aided mine design techniques have been proven to increase productivity on initial project analyses to as much as two hundred (200) percent of manual techniques. However, the greatest advantage to the creation and maintenance of geological, design, and financial models is encountered in subsequent revisions to any or all of the models. The availability of instantaneous access to a truly staggering amount of data is a substantial asset to any planning operation.

The advantages to time-sharing on proven mining oriented computer facilities with a PC workstation may be summarized in two (2) categories, Time and Money.

The time advantage may be further classified as savings in design and in implementation. A truly interactive mine design system is under constant modification and innovation. On the front end, an investment of five (5) to ten (10) years is not uncommon. A considerable portion of the time is spent on verification of computer

results in anticipation of the classic computer adage, "garbage in = garbage out". All of this developmental period is pure overhead, with negligible returns until a working system is obtained. Purchase of a complete system, the capital costs of which will be discussed subsequently, also involves a substantial training period before realistic numbers may be obtained. Time-sharing with a firm comprised of mining professionals allows training to be performed while ongoing analyses are performed. The results generated in learning the system, due to the fact that mining personnel are the training staff, are valid analytical data.

Cost savings occur due to limited capital expenditures and the capability of rapid data transmission via "burst" feed techniques between computers. Capital expenditures for super mini-computer processors with sufficient memory to adequately compile and manipulate the data required for project analysis may be in the range of $200,000 to $500,000 including peripheral devices such as digitizers, plotters, and storage tube terminals. Additionally, software for geologic modelling, computer-aided design, mine scheduling and project costing may incur costs of up to $400,000, depending on the complexity of the systems chosen. These costs do not include overhead charges for the computer personnel required to maintain the hardware and develop or install the software. Including peripheral devices and software, a properly organized PC workstation should cost in the range of $7,500 to $15,000. Any interested and adequately prepared engineer or geologist should be able to operate this system.

Communication charges are based on current phone charges in the locale of the workstation. These costs may be minimized in a variety of ways, including:

1) Usage of one of the variety of long-distance service organizations.

2) Transmission of data through "burst" feed techniques, eliminating the need for constant on-line time and providing error-checking on emitted data.

3) Formation of command files to run when the data are sent, providing rapid execution of desired tasks.

4) Transmission of data during non-prime phone rate periods.

Intelligent usage of the various techniques of data transmission can greatly reduce the total communication charges incurred. On-line user sessions could be kept at a minimum, while maximizing processing results.

Computer access on a time-sharing basis involves charges based on connect time, central processor unit usage, and I/O device usage. As well, licensing charges on software are computed on time increments of actual utilization. Given

a comprehensive training period, an accomplished
computer-aided mine planner can achieve enormous
productivity at a reasonable cost of services.
The utilization of a PC workstation linked to a
remote data processing facility should be
considered by any company interested in detailed
mine planning or evaluation. A tremendous
savings in capital costs and development time
with no loss in engineering quality makes this a
viable alternative to either a larger computing
facility or merely a PC alone.

a comprehensive training period, an accomplished
computer-aided mine planner can achieve enormous
productivity at a reasonable cost of services.

Chapter 7

IMPOSING HIERARCHICAL DATA STRUCTURES ON COMMERCIAL SOFTWARE FOR MINING APPLICATIONS

by

John C. Lascola
Physical Scientist

David M. Hyman
Geologist
Bureau of Mines
U.S. Department of the Interior
Pittsburgh, Pennsylvania

ABSTRACT

Table-oriented decision-support software is widely available for most computer systems. Typically, these software packages include various graphic utilities as well as capabilities for processing data on a row or column basis and work best with some form of list-type logical data structure. However, large amounts of data gathered during several projects requiring considerable amounts of repetitive computation and file organization are best represented logically by a tree or hierarchical structure. In the Bureau of Mines' case, existing table-oriented software provided the required computational, maintenance, and graphic capabilities, but lacked an inherently hierarchical data structure. The solution was to impose a dynamic hierarchical data structure onto a set of tables and graphs generated by a commercial decision-support software package. A general purpose procedure was written to produce a multileveled tree structure of subdirectories, where the names of the subdirectories describe the source of the data. The heart of this system is the use of table entries as nodes of a tree that point to the next level of branching until the actual data are finally reached. This menu-driven procedure is presently being used to input, reduce, organize, and output data from diverse projects, ranging from oil shale and coal core gas desorption analysis to ground and aerial surveys of mining-induced subsidence.

INTRODUCTION

The sustenance of research is data. Data are facts collected in the form of numbers or words describing such things as shape, size, conditions, or events. In a business situation, the data collections might be customer accounts, inventories, or mining progress and projections. Small collections of data are easy to manage but usually provide little useful insight. Large collections of data potentially contain the answers to important questions, but their size and interrelations make them difficult to manage. The computer is an obvious answer to the problems associated with managing large data collections, but it is the software that makes the computer useful.

SOFTWARE

The collector of data needs decision-support software. In its simplest form, decision-support software is a computer program that provides a means of entering and storing data, and then rearranges or converts them into a form that aids the user in making decisions. Two fairly well known examples are database management and spread sheet software.

Database management software typically works best where the main need is to store, manipulate, and retrieve information. Large amounts of data can be efficiently organized through the use of search, sort, and retrieve routines. The final product is usually a report containing a selected subset of the database. Only a few basic calculations such as addition, subtraction, multiplication, and division can normally be performed, and graphics may not be available.

Spread sheet software typically works best where the main need is to make decisions based upon simple numerical models. Numerous repetitive calculations can be performed on tabular data and the results analyzed to answer "what if" questions. Mathematical, statistical, logical, and graphic capabilities are usually available. Words, however, are normally treated as labels for numbers.

A full-function decision-support software package integrates the concepts of database and spread sheet, along with extensive graphic, file management, and programming capabilities. This kind of software is normally found on mini- and mainframe computers used in time share operations, but is becoming increasingly available for desk top microcomputers (personal computers).

PROBLEM

A simple model for a data collection would be a notebook filled with observations. In real life, however, data sets are often more complicated. They typically are made up of sets of information of the same type describing similar events or objects. An example would be the maintenance records for various kinds of mining machinery, located at three or more different company-owned mines.

Unfortunately, most computer database management systems are used much like a notebook. A large number of observations are written into a single file and a large group of similar files accumulate without logical relationships or a connecting logical structure. The problem is to effectively store and manipulate large amounts of data gathered during several projects requiring considerable amounts of common repetitive computations, graphics, and file organization. A solution to this problem is to impose a dynamic hierarchical (tree-like) data structure onto a set of tables and graphs generated by a commercially available, full-function decision-support software package.

SIMPLE EXAMPLE

Consider mine equipment maintenance data as an example. A paper file system could use a single form to record the maintenance items performed on a single date for a given machine. All of the records for that machine could be stored in a single file folder. All of the file folders for similar machines could be stored within a single file drawer divider section. All of the file sections for a given mine could be stored in a single file cabinet drawer. All of the drawers, each representing different mines, could be located in a single file cabinet. This is a tree-like or hierarchical filing system. The levels are maintenance items (data items), all maintenance completed on a given date (form), individual machine (folder), type of machine (drawer sections), mine (drawer), and company (cabinet).

Typical microcomputer software could handle the same maintenance data by setting up tables where the rows represent records and the columns hold attributes (the maintenance items). Each table could be a single computer file representing one individual machine. The files for one type of machine could be exclusively saved on a single floppy disk. The floppy disks for each mine could be stored in a separate file box, and the individual file boxes could be kept on the same table near the computer. This is also a hierarchical filing system, but much of the filing is done by the operator and not the computer. In addition, it is difficult to perform simultaneous operations or comparisons on data contained in separate files.

VAX-RS/1 ENVIRONMENT

The Bureau of Mines' Pittsburgh Research Center (PRC) has several centrally located minicomputers and a wide variety of microcomputers, and uses a large amount of commercial and inhouse written software. The particular full-function decision-support software package discussed here is RS/1 (TM),[1] a

_____
[1] Reference to specific equipment, trade names, or manufacturers does not imply endorsement by the Bureau of Mines.

computer-aided analysis system for scientists and engineers supplied by BBN Software Products. RS/1 was used because it is installed on PRC's Digital Equipment Corporation (DEC (TM)) VAX (TM) minicomputer, the only computer that is readily available through time share to all of the Center's research personnel, and because it adequately meets their needs. While it is true that the larger dynamic memory and storage capacity of a minicomputer provides more versatility than is available on a typical micro-computer, other hardware-software combinations could have worked equally well.

The VAX/VMS (TM) operating system supports hierarchical file directories, and RS/1 takes advantage of that capability. A set of subdirectories can be created within the main directory and each subsequent subdirectory. The full name of an RS/1 table, graph, or directory is its full path name representing the tree traversal to obtain it. Specifically, it is the concatenated name of each subdirectory name separated by the delimiting character "@" and followed by the final object name. An example is @SUB1@SUB2@TABLE, where the access path is from the first subdirectory (SUB1) to the second subdirectory (SUB2) and finally to the end object (TABLE). When viewing a directory, only the names of the items stored in that directory are displayed. Many, or even all, of these items can be pointers to subdirectories. Using this feature, each level of names in the filing system hierarchy can be a separate subdirectory containing the names of the next level of subdirectories. The individual data items fill the cells of the tables located at the bottom of the tree. In this manner, the entire filing system takes place within the computer and can be stored on a single disk.

The main disadvantages of using RS/1 subdirectories occur in the RS/1 command level interactive mode. Tables are buried by layers of subdirectories. The operator must traverse down through these layers of subdirectories until, at the bottom, a subdirectory is reached that contains the actual list of tables needed. Each path through the layers leads to similar collections or levels of tables. An alternative approach is to use complete path names to access tables or graphs. However, path names are usually long and not convenient to remember, while typographical errors produce unexpected results.

Using RS/1's built-in procedural language, a general purpose menu-driven procedure was written to manage the filing system and direct the operator to all of the functions needed to manipulate a particular set of data. The operator calls up the procedure for the desired database. Only those objects in the top level directory that are subdivisions of the database will be displayed. A subdirectory is selected from the menu and the contents of the new subdirectory are accessed and displayed. An input error routine automatically questions any name not already listed. If the entry was a typographical error, a correction can be made. If the entry really was a new name, it is entered into the directory, and a new subdirectory, with any other needed structures, is created. This process is repeated until the lowest level is reached, where the actual data manipulation takes place.

EXAMPLES FROM MINING RESEARCH

The GASSITY procedure is an example of a hierarchical data structure imposed on commercial software. It is used by PRC to store, manipulate, and analyze coal chip and oil shale core gas desorption data for mining applications. A number of core holes, vertical or horizontal, are drilled at various mine properties, and samples from each core are analyzed for gas content. The hierarchical levels are database (material type), mine, core, and sample. See figure 1. The data collected include geographical location, lithology, sample position within the core, sample size, time when cutting of sample begins and ends, time when sample is sealed in container, desorption time and volumes, and gas and chemical analysis. From these data, desorption volumes of seven constituent gases must be calculated, their graphs constructed, and reports generated. These are used as aids in mine ventilation design.

The actual operation is simple and direct. The command "CALL GASSITY" runs the GASSITY procedure within the RS/1 environment. The names of two available databases, COAL_CHIPS and OIL_SHALE, are displayed. Both are subdirectories within the main directory. The operator is prompted to enter the new or existing name of the database needed to work with. The database subdirectory containing the names of all available mines in that level is then displayed, and the operator is again prompted to make a choice. The mine subdirectories contain the lists of all available cores from that mine. If a new core name is selected, a table to hold all of the drilling data for all samples of that core is also created within the new core subdirectory. The DRILL table will eventually contain the name, lithology, position within the core, size, drilling and container closure time, and volumes of all samples named in the core subdirectory. The DRILL table, with its additional information, is then displayed instead of the core subdirectory. See figures 2 and 3. If a new sample name is selected, a new sample subdirectory is created and listed in the core subdirectory, and two tables, ASSAY and DESORB, are created in the new sample subdirectory. The ASSAY table will eventually contain the lab number, position within the sample, description, and analysis of any portion of the sample analyzed. The DESORB table will eventually contain all of the desorption data such as the time, temperature, and pressure data needed to calculate the various gas volumes as well as the cumulative volumes. All of this is simply done by entering the database, mine, core, and sample names.

Once in the sample subdirectory, data are manipulated through the use of menus. The main menu options are

1. Exit from the GASSITY procedure;

2. Display tables and directories;

3. Enter new data;

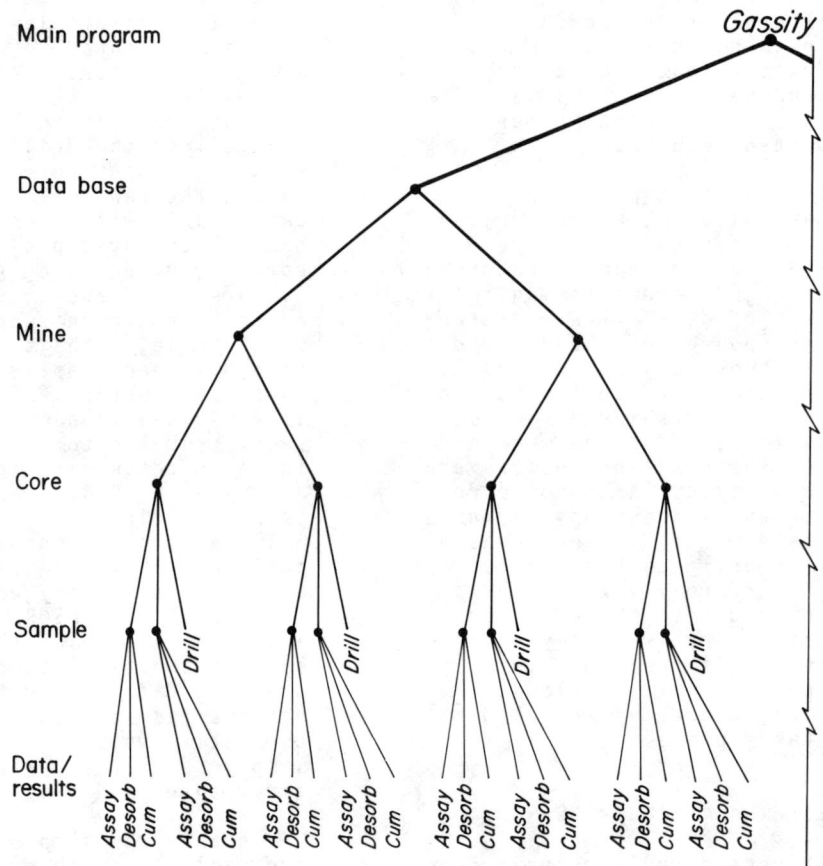

FIGURE 1. - Diagram of the GASSITY procedure with two subdirectory branches from each node for brevity. Note that each core subdirectory contains a DRILL table and each sample subdirectory contains an ASSAY table, a DESORB table, and a CUM graph.

```
Directory Listing: @COAL_CHIPS@ALPHA_1@V1

0 Name    1 Type
------------------
1 DRILL   Table
2 _505    Directory
3 _523    Directory
```

FIGURE 2. - An example of a typical core subdirectory. The name @COAL_CHIPS@ALPHA_1@V1
refers to the V1 core from the ALPHA_1 mine in the COAL_CHIPS database.
Only two samples are shown for brevity.

```
DRILL   2R x 13C

        DRILL DATA OF CORE @COAL_CHIPS@ALPHA_1@V1
```

0 SAMPLE NAME	1 LITHOLOGY (text)	2 BEGINS (ft)	3 ENDS (ft)	4 LENGTH (ft)	5 DIAM (in)
1 _505	COAL CHIPS	1300	1306	6	0.811
2 _523	COAL CHIPS	1967	1975	8	0.811

0 SAMPLE NAME	6 BEGINS (d-m-y h:m:s)	7 ENDS (d-m-y h:m:s)	8 CANNED (d-m-y h:m:s)
1 _505	18-APR-84 14:20:00	18-APR-84 14:35:00	18-APR-84 14:40:00
2 _523	20-APR-84 15:20:00	20-APR-84 15:35:00	20-APR-84 15:45:00

0 SAMPLE NAME	9 LOST TIME (h:m:s)	10 CONTAINER (text)	11 Vc (cm3)	12 Vsample (cm3)	13 Vg (cm3)
1 _505	00:12:30	AL 505 L=12	2471	609.49	1861.51
2 _523	00:17:30	AL 523 L=12	2471	812.65	1658.35

FIGURE 3. - An example of a typical DRILL table. Notice that this is the actual table
named in the core subdirectory.

4. Edit or change values;

5. Perform all calculations;

6. Produce all graphics;

7. Print tables on line printer;

8. Choose another sample to work with.

Each of these options produces its own menu. Each of these menus contains an exit and a return to main menu routine. Help is available at every prompt by pressing the "?" key. Each option in the data entry routine contains simple escape and error correction routines. The graphic option draws the graph CUM of the seven DESORB table cumulative gas volumes. Three available options are

1. Time in hours vs linear volume;

2. Time in square root of hours vs linear volume;

3. Time in square root of hours vs logarithmic volume.

Three supporting tables, DATA OF CUM, CURVES OF CUM, and AXES OF CUM, are created and updated as needed. To save core memory space, no more than one form of the graph exists at any time, but all forms are always available through program-controlled transformation of the supporting tables. An almost identical procedure is available for regular coal core desorption data that stores the results of proximate and ultimate analyses, instead of the Fischer assay.

The SUBSIDE procedure, a modification of GASSITY, is another example of a hierarchical data structure imposed on commercial software. SUBSIDE stores and manipulates ground and aerial survey data as part of a research effort to examine alternate means of collecting mining-induced subsidence data. A pattern of 40 to 150 monuments is established on the ground above a longwall coal panel. Before production begins, the horizontal and vertical positions of the monuments, and therefore the ground, are determined by conventional ground surveying techniques. A series of aerial photographs are taken before, during (approximately 10), and after mining under the study area. Photogrammetric techniques are then used to determine the monument positions and ground movements resulting from mining. The data collected include the date and longwall face position at the time of the survey, and the map coordinate and elevation of each surface monument. The three-dimensional coordinates of the monument positions can be transformed for comparison with spatial data from other map projections, and various graphs depicting mine subsidence can be constructed. The structural hierarchy of subdirectories is database, mine, longwall panel, and survey. In addition to the survey subdirectories, the longwall panel subdirectory contains the FACE table. FACE is analogous to DRILL from the previous example and contains the names of all surveys, along with their respective dates and positions. The survey subdirectory contains the various tables of surface monument positions and graphs. Menus, almost identical to those in GASSITY, are used to select input-output options for the data.

SUMMARY

The essence of this system is to impose a dynamic hierarchical data structure on a commercial full-function decision-support software package by using table entries as nodes of a tree that point to nodes at the next level of branching, until the actual data are finally reached. Files are created that contain pointers to other file names, each describing the source of the data found at the next level of the hierarchical data structure. A general purpose procedure, written to produce a multileveled tree structure of sub-directories, greatly simplifies the handling of large amounts of data by allowing an operator to access specific subsets of data without needing detailed knowledge about the organization of the information. Menu-driven utilities for data input, access, error detection, file maintenance, report writing, and graphics are included.

The combination of the menu interface and the hierarchical structure greatly enhances the usefulness of commercially available software for small computers. This technique is being used by PRC to input, reduce, organize, and output data from diverse projects ranging from oil shale and coal core gas desorption analysis to ground and aerial surveys of mining-induced subsidence.

BIBLIOGRAPHY

Anon., 1983, RS/1 User's Guide, BBN Research Systems, Cambridge, books 1-3.

Berztiss, A. T., 1975, "Trees," Data Structures, Theory and Practice, Rheinboldt, W., ed., Academic Press, New York, pp. 213-268.

Finman, J., 1983, "A Computer Software
    System For Use In Mining Research,"
    CIM Bulletin, Vol. 76, No. 858, Oct.,
    pp. 94-95.

Hohner, G., 1984, "Evaluating Database
    Software For PCs," Machine Design,
    Vol. 56, No. 20, Sep., pp. 129-132.

Kroenke, D. M., 1977, Database
    Processing, Science Research
    Associates, Chicago, 408 pp.

Lesk, M., 1984, "Computer Software For
    Information Management," Scientific
    American, Vol. 251, No. 3, Sep.,
    pp. 163-172.

Morgan, C.,ed., 1981, "Data Base
    Management Systems," Byte, Vol. 6,
    No. 11, Nov., 560 pp.

Russell, C. H., 1984, "Today's Integrated
    Scientific Software Adapts To The
    Problem," Research & Development,
    Vol. 26, No. 4, Apr., pp. 94-98.

Wirth, N., 1984, "Data Structures and
    Algorithms," Scientific American,
    Vol. 251, No. 3, Sep., pp. 60-69.

Chapter 8

AN EMPIRICAL APPROACH
TO DETERMINING THE INFORMATION NEEDS
OF A DISPERSED COAL MINING ORGANIZATION

Robert B. Knepp
Manager, Information Systems Planning
Natural Resources Group
Bethlehem Steel Corporation
Bethlehem, Pa. 18016

ABSTRACT

This paper discusses the application of the Information System Study (ISS) methodology in determining the information needs of the Coal Business Unit, Natural Resources Group, of Bethlehem Steel Corporation. The purpose of the study was to determine, at a strategic level, the information needs for developing an Information Systems Plan for this newly formed Business Unit. The ISS methodology was developed by IBM-Belgium and utilized in over 350 studies in Europe. At the time, this study was the third application of the ISS methodology in the United States; two of which were within Bethlehem Steel. This paper will discuss the methodology and its application for: defining the business processes; data gathering - interviewing key employees at five different geographic locations; the encoding of the data for purposes of numerical analysis; determining the quality of existing information, and the information needs in support of the business processes. To date, the results of the study have been used to adopt baseline strategies in the use of information and information systems. These strategies have been prioritized by Management and are in varying stages of development and implementation.

INTRODUCTION

It is best to begin this paper with some mention of the organizational background and changing cultures. Some awareness, in this regard, will be beneficial in understanding where the new organization was coming from and the significance of the timing of the study. Also, the changing culture and management philosophies of the parent company as a positive influence on the structure and business activities of the mining organization, is worth noting.

Historically, the parent company was a highly centralized, operations oriented and managed company. In the late 1970's, the management style, business philosophies and practices began to undergo change. Corporate-wide a participative style of management was being stressed. Within the steel industry, labor-management participation efforts were evolving. With this came the formulation, internally, of business groups and business units. The traditionally centralized functions of Sales, Accounting, Planning, etc. were being merged into the new business units. In mid-1982, the Natural Resources Group was formed and consisted of three business units - Coal, Stone-Anthracite, and Affiliates.

Prior to 1982, the mining organization existed to provide raw materials, in the form of iron ore, coal, and limestone to the parents' steel producing facilities. Mining operated and sold to a captive market. Sales of raw materials to external customers were minimal. As the fortunes of the steel industry went, so did that of the mining operations.

In the last decade, the state of the economy, in general, has challenged business organizations and created pressures which have been a challenging experience for management; in particular, those within or closely allied with the steel industry.

Internally, the formation of the Natural Resources Coal Business Unit, coupled with the Corporate emphasis on adopting and practicing a participative style of management, has been a dramatic change. It has: necessitated the evaluation and redefinition of organizational structures and roles; opened up new markets for selling coal; and, significantly impacted traditional approaches to managing the business. Transition from an operations oriented to a market-driven organization is critical for success.

Such a broadening of scope and responsibility creates needs for more accurate, consistent, and up to date information. Information about the market, customers, and competition are a sudden requirement. The work force and costs of operation now are viewed in a different light. At this point, identifying information needs and adapting to use relative information in the "new environment" is still an evolving process.

An integrated and consistent approach toward establishing and sharing of common information becomes a real necessity, in support of management in the "new environment".

The framework, or architecture, for information must be based on the business lifecycle (plan, acquire, manage, release). It must be independent of existing organization structures and yet be consistent with the strategies and long-range plans of the business. In this manner, the information strategies can be more accurately defined; the decision-making and business activities can be better supported. More than ever, information and data are being considered as a resource and as an aid to management in meeting the goals and objectives of the business.

Prior to the existence of the Coal Business Unit, there had been some use of mechanized information systems. The traditional functions of payroll, supplies inventory, and operating costs were mechanized. Also, there were isolated uses of mechanized production information and equipment maintenance applications. The use of SAS (Statistical Analysis Software, SAS Institute, Inc.) was beginning to evolve.

WHY THE STUDY

Being a newly structured organization, the strategic objectives as a business unit required altering the traditional approaches to managing the business. The business, by design, had become market driven. New markets were now made available for selling the products. Market sensitivity will require more flexibility and elasticity on the part of the production facilities and those who manage them.

With success in the new coal markets, the potential for load-leveling of the production capacity becomes more realistic. The flip-side, however, requires that operating costs-per-ton be reduced and that the quality of the product be maintained or improved, to make a successful transition from a predominantly captive market to that of a more volatile domestic-export marketplace. To support this transition, management will need, more than ever before,

quality information to successfully manage
the business.

The above mission prompted the question:
"What really is our business, and what information supports our business?" Management
needed to determine

o   What information is essential to more
    effectively and efficiently manage the
    business?

o   What information is currently gener-
    ated, based on the past, that may not
    be needed now? Can it be eliminated?

o   What information is needed by the
    organization, but not available?

o   What information do the employees need
    to perform their duties in a more effec-
    tive and efficient manner?

o   What are the opinions of the employees
    regarding the quality of the informa-
    tion they work with?

o   What areas of our business (informa-
    mation) need to be enhanced to provide
    more reliable and timely information?

o   Strategically, what should be the
    direction of future information related
    activities within the organization?
    Are our past and current efforts on
    target with the plans for the business?

o   What information should be mechanized
    to improve the overall efficiency of
    the organization? (By the very nature
    of the times, organizations are
    straining to do more with less. Where
    practical and economically attractive,
    strategic mechanization efforts could
    improve the efficiency and effective-
    ness of the human resources and
    management of the business.)

o   What compatibility exists between
    past/current information mechanization
    efforts, the employees and their
    duties, and the goals of the organiza-
    tion?

Given the above questions and concerns,
it was agreed to conduct a study of the coal
mining organization, its' activities and the
information utilized. Up to this point,
there had been no formal strategy for utilizing information. As mentioned previously,
there had been individual activities of
mechanization, but these had been localized
in their use. Very little had been done to
integrate these developments across the
organization, except for the accounting functions and spare parts inventories.

OBJECTIVES OF THE STUDY

The primary objective of the study was
to establish the basis for satisfying the
information needs of the Coal Business Unit
– a strategic plan. The Information Systems
Strategic Plan was to be consistent with the
strategies and plans for the organization.

Additional objectives were defined by
management, to be accomplished as part of
the study. They included:

o   identifying what information is criti-
    cal to the business, and its quality;

o   development of a more efficiently func-
    tioning system (information network);

o   identification and elimination of
    redundant information;

o   getting necessary information to the
    right people;

o   improving the timeliness of informa-
    tion;

o   the development of a functional flow
    diagram of the information network.

CHOOSING THE METHODOLOGY

At the time, there were three known and
proven methodologies that had been utilized
elsewhere within the parent company for
organizational and information studies.
These were the IBM Corporation's Business
System Planning (BSP) methodology; the IBM
Corporation's Information Systems Study
methodology; and a Functional Analysis methodology that was primarily developed and
applied within the parent company.

The Functional Analysis methodology had
been used to make a more microscopic study
of organization functions and activities,
and the information and personnel utilization in performing the activities. The main
thrust was organizational analysis more than
information analysis. The Functional Analysis methodology had been used successfully
in conducting "Impact Studies" to establish
salaried workforce level requirements and to
determine requirements in Office Automation
studies, to mention a few applications of
the methodology. The process involves use
of interviewing techniques, by a taskforce
or study team, to decompose the organization
and functions to a level of detail desired.

The functional levels are identified as

Gaining the Commitment

Preparing for the Study

Starting the Study

Defining Business Processes

Defining Data Classes

Analyzing Current
Systems Support

Determining the
Executive Perspective

Defining Findings
and Conclusions

Reviewing Information
Resource Management

Defining Info.
Architecture

Determining
Architecture
Priorities

Developing Recommendations
And Action Plan

Reporting Results

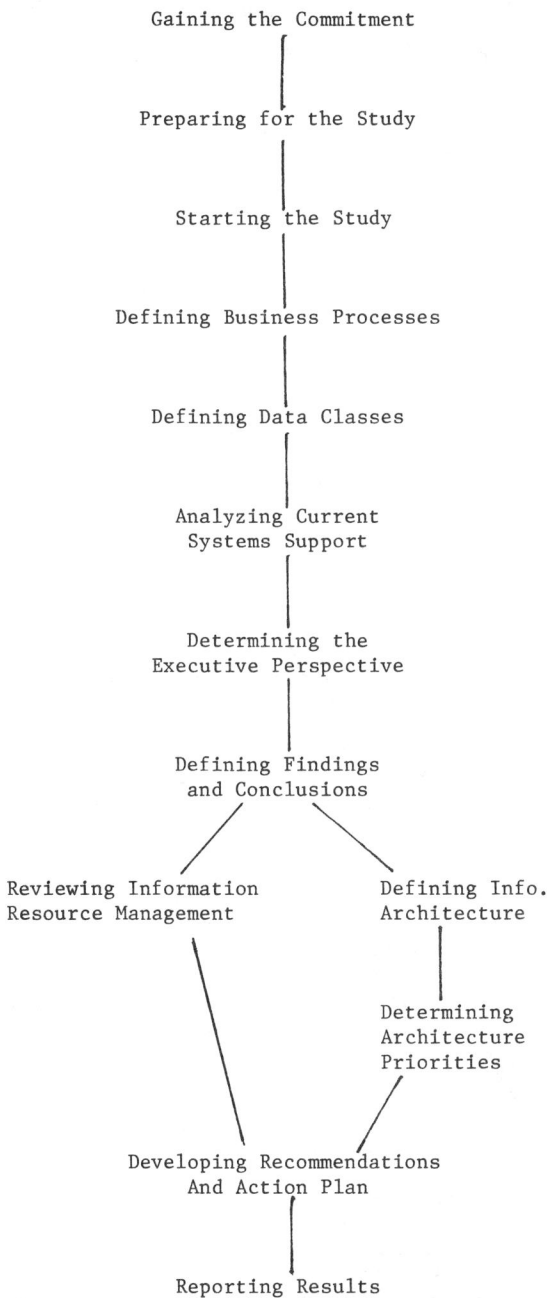

FIGURE 1. Flow of the BSP Study (1)

a subset of the next higher level in a pyramidal process. Graphically, this process appears similar to that of a Bill-of-Mat'l. breakdown for a piece of equipment. The lowest level function could be that of an engineering drawing for an assembly, sub-assembly, or part - depending on the level of detail desired.

The Business Systems Planning (BSP) methodology has been utilized and taught by IBM, quite intensely, over the last decade. This methodology, also referred to as Information Systems Planning, is a structured approach to helping an organization establish an information systems plan, both near- and long-term.

Business Systems Planning is geared to help provide such a plan through:

o A top-down approach to getting people committed and involved, and studying the business (working from the overall to the detail level);

o A bottom-up approach to implementation;

o Use of a structured methodology proven in hundreds of studies; and

o The translation of business objectives into information objectives.

The major activities and flow of the BSP methodology are very similar with that of the ISS methodology, or vice-versa. These activities and flow are illustrated in Figure 1.

The Information Systems Study (ISS) methodology was developed by IBM-Belgium. The ISS methodology had been used successfully in over 350 study applications in Europe prior to being introduced in the United States. The first study in the United States took place at an integrated steel plant in eastern Pennsylvania, which is owned by Bethlehem Steel Corporation.

The activities and flow of the ISS methodology are illustrated in Figure 2. This process, though very similar in format, offered some unique differences from that of the BSP approach. The length of the overall study is considerably shorter in length. This is attributable to: (1) the interviews are done in a group setting (4 - 6 interviewees) versus the single interviewee, one-at-a-time approach of BSP, and (2) the data collection is manually massaged and analyzed in BSP, whereas the ISS approach used a mechanized software tool to aid in analyzing the data and provided mechanized report generation of statistical results.

In the final analysis of the methodologies and fitting them to the constraints of the study and the organization, the Information Systems Study (ISS) methodology was chosen.

As indicated previously, the ISS methodology offered certain advantages not readily available with the other two methodologies. One was the limited duration of the study effort (6-8 weeks); the other was the availability of a computer program to automate the analysis and reporting effort. These two factors compensated for the major study constraint - limited personnel. Another constraint was the limited availability of the members of the study team. In obtaining executive commitment and sponsorship, the duration of their personal availability and involvement was a prime consideration.

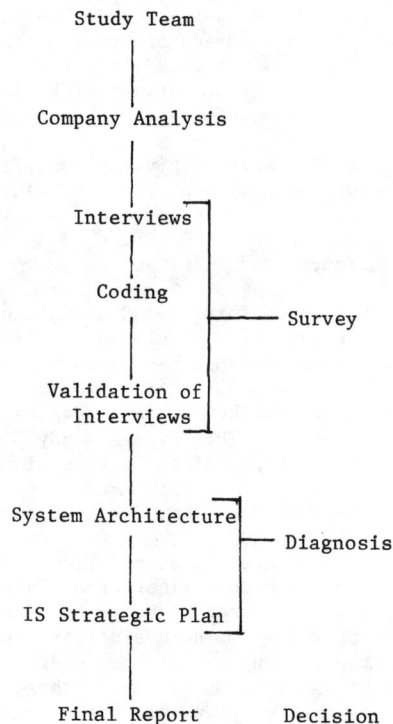

Study Team

Company Analysis

Interviews

Coding   Survey

Validation of
Interviews

System Architecture
                       Diagnosis

IS Strategic Plan

Final Report   Decision

Figure 2. Flow of ISS (2)

THE STUDY

The ISS methodology captures and diagnosis existing information systems, whether they are computerized or not. Based on the data gathered, it further defines the future information architecture and statistically establishes a priority development plan. The results consider end users and management as they are concerned and involved in the study.

The primary phases of the study are 1) Company Analysis, 2) Survey, 3) Diagnosis, and 4) Final Report.

Prior to starting the study executive commitment and sponsorship was obtained. The IBM Information Systems Services organization was contracted to facilitate the study and also provide for the use of the software tool. The study team was formalized and consisted of the top executive, five functional department managers located at Home Office, and the managers of the coal mining operations.

COMPANY ANALYSIS

During this phase (four consecutive days) the study team identified 109 internal business processes and 9 external (origin) processes which originated and/or used information. These processes were grouped functionally: Coal/Facilities, Planning and Production; Coal/Facilities After Production; Equipment and Facilities; Suppliers; Financial; Customer/Marketing; Personnel; Organizations Outside The Coal Business Unit.

The study team then identified key individuals (interviewees) who were involved in the processes. There were 90 individuals identified from 6 different organizations which geographically were located in Eastern Pennsylvania, Western Pennsylvania, West Virginia, and Kentucky. There were 15 interview groups scheduled at four geographical locations over a three week period. Each interview was scheduled for a two day period. Each interview group consisted of six different disciplines (6 employees). This provided for a broader coverage of the processes and created an environment for cross-fertilization of business disciplines within each group. Three dimensions were addressed: Business Processes, Organization, Business Data.

SURVEY

During each interview, participant identification was established (interviewee code, name, and function). The lists of business processes, developed during Company

Analysis, were reviewed. Each interviewee was asked to indicate whether they 1) are actively involved in the process, or 2) know enough about the process to be able to critique the information needed to perform the process. Each interviewee that answered "yes" to these questions was asked to identify all of the data, or information, that was needed to perform the process.

At the end of the interview, each interviewee was asked to evaluate the data they identified - Is the data MISSING?, Is this KEY data?, Is this COMPUTERIZED data? Regarding the data, is the interviewee satisfied in terms of FREQUENCY, DELAY, CONTENT, RELIABILITY, or USEFULNESS?

During each interview, a member of the study team was present to provide continuity from the Company Analysis phase and answer questions regarding the business process definitions. In addition, they began the data encoding process by identifying and assigning data codes and origin process codes to the data that was identified.

Each interviewee also created their interviewee files as they identified the processes and their involvements, along with the data needed to do the processes. This information, their critiques and data characteristics along with the study team member's data, was recorded on documents ready for data entry into the computer program and processing by the mechanized software tool.

After the interviews were completed, the data was coded and the origin process codes assigned (processes that created the data). The information was entered into the computer utilizing the software tool. Interviewee printouts were generated and returned to the interviewee for validation and clarification. Returned changes were then entered into the computer files.

DIAGNOSIS

During this phase, the software tool was used quite heavily to analyze the data. The primary outputs are in the form of statistics, matrices, and simulations. The primary results are statistically represented as a "percent of satisfaction" with the data.

Regarding statistics, the software tool provides the opportunity to select statistics on various criteria. The statistics can be requested in any sequence based on four elements of the data - process, data, origins, interviewee, or in any combination thereof. Additional parameters were also utilized to compare responses relative to weighted vs. non-weighted comparisons;

essential (key) data vs. nonessential data; doers of the processes vs. those who know about the processes, etc. This mix of criteria was applied to statistical selections for subsystems and functional process groupings; such as maintenance, planning, costs, inventory, payroll, marketing, personnel, etc. The statistical results quantified the interviewees satisfaction with the data based on information quality (percent of satisfaction.)

Additionally, based on interviewee data critiques, the sum of and specific data items can be identified as: Essential; Computerized; as having dissatisfaction with Frequency, Delay, Content, Reliability, and Usefulness; Satisfactory and Unsatisfactory; Missing; Problem Data (missing plus unsatisfactory); and Satisfaction Factor. These statistics are computed and displayed based on any or all of the four elements of the data.

The development of statistical matrices is another dimension of the software tool. Any possible combinations of process, data, origins, and interviewees taken two by two can be put in a matrix format. The matrices show more clearly the results of the simulations. The most important, in this study, was the matrix "process vs. data" in that it defined the information system of the business.

In preparing simulations, the software tool assumes that some particular data are made perfect and calculates the influence on this fact on the entire information system. These results identify this influence and aids in prioritizing future plans regarding the information system (network).

During this phase a clustering activity took place. This created a very large matrix (approximately 12 ft. by 16 ft.) which related all of the data to the process. By examining the data-to-process relationships, it was possible to group processes such that the processes create data which are used primarily within the group itself. This is what we refer to as a subsystem within the total information system. Subsystems exchange data with other subsystems. (The subsystems were listed earlier in the Company Analysis section of this paper.)

FINAL REPORT

The final report was handled in two segments. The first segment was the formal preparation of the Diagnosis Phase results in a report to the study team.

The study covered approximately seven weeks, from the initiation of the Company Analysis Phase through the Diagnosis Phase. At that time the study team reviewed the findings and requested that additional criteria be added to the data and that some additional analyses be made based on these criteria.

The two major criteria requested was to revisit the interviewees for clarification of their responses in certain areas, and the addition of coding to the interviewee code to add a position code to provide statistical information by position (job classification).

Following the completion of these activities, the Final Report was modified and the study team then prioritized the recommendations which resulted from the study. These prioritized recommendations became the near term strategic information system plan. Included in the near-term priorities was the finalization of the long-term strategic information systems plan.

CONCLUSION

The study identified the processes of the business and the essential information that is required. The methodology, in itself, enhanced the understanding and awareness of the participants and the organization.

Regarding the data, based on the opinion of those surveyed, we now know:

o What information is considered essential

o What information is currently computerized; what information should be computerized

o The areas where we have problems with our data and, generally, what type of problems

o What information is needed by our business processes and by those performing the processes

o The current information that has limited value and should be reshaped or eliminated

o The perceived quality of existing information within the organization

Most importantly, we have captured on a database a most comprehensive strategic definition of our business, and the information needed in support of the business and management decision-making activities.

REFERENCES

1. IBM Corporation, Third Edition, July 1981, "Business Systems Planning, Information Systems Planning Guide" No. GE20-0527-3

2. VEYS, Michel, IBM-Europe, March 1982, "IS Architecture, I.S.S."

Chapter 9

DEVELOPING AN INTEGRATED, COMPUTERIZED, PURCHASING/INVENTORY/

MAINTENANCE SYSTEM--ITS PROBLEMS AND POSSIBILITIES

James E. Barker

Manager, Business Systems
North River Energy Company

ABSTRACT

Many of the normal tasks associated with maintenance, purchasing, inventory control and accounting involve a logical and consecutive flow of information. The information is usually copied, stored, retrieved, updated, and passed on to the next functional department several times in various formats. Sometimes it is lost. The duplication of this process many times over, with the inherent possibilities for error or additional costs, can be alleviated, to a large extent, through data processing (DP). The computerization of these tasks, with the automatic migration of information between business functions, provides a means to minimize these human errors and duplication of effort. Computerization requires a high degree of commitment by employees--both management and labor--to make it work. However, it represents a potentially significant economic benefit for the company.

This paper outlines the steps one coal mining company used in bringing DP to its operations, the problems encountered, and the solutions that worked.

INTRODUCTION

North River Energy Company (NREC) is a partnership of NRE Corporation, a wholly owned subsidiary of Republic Steel Corporation and now a part of the LTV Corporation, and The Pittsburg & Midway Coal Mining Co., a wholly owned subsidiary of Gulf Oil Corporation. NREC mines coal in Alabama by surface and underground methods and produces over 1 million tons of steam coal annually for the electric utility market.

In its corporate beginning in the early 70's, North River Energy Corporation was part of Republic Steel Corporation. Republic's business plan called for initiating development of DP support for NREC in 1979. Even though the plan was progressing, NREC still had minimal computerized support by 1981. This paper will discuss NREC's development of a DP capability in the first half of the decade of the 1980's. In particular, it will address the development of an integrated system to assist the purchasing, inventory, and maintenance functions. (Keep in mind this development was occurring within the framework of providing complete DP support for NREC as a whole.) Emphasis will be placed on the problems encountered during the developmental effort, the solutions found, and the benefits experienced (or projected).

BACKGROUND

Midway through the planning for Republic's remote DP support from the vicinity of Pittsburgh, Pennsylvania, the partnership was formed and North River Energy Company came into existence, as a company, in July 1981. Until that time, computer support from Republic consisted of payroll, periodic personnel listings, and end of month summaries of the accounts payable and supply activities. With the advent of the partnership, even this minimal support was scheduled to end.

Faced with this problem, NREC had to develop plans for some means of data processing support in order to grow. After extensive study and discussions, the partners decided that a stand alone operation for NREC would provide the greatest flexibility and benefit. NREC began developing a plan for full DP capability complete with staff, hardware, and software, plus an implementation schedule. The plan that evolved basically categorized our computer software support needs into three phases: Those programs that would help us continue our day-to-day activities (called Phase I); programs that would help us reduce costs and increase production (Phase II); and programs to support important, but less urgently needed, functions (Phase III).

Phase I programs were defined as those supporting:

            Payroll
            Accounts Payable
            Fixed Assets
            Human Resource Management

Phase II was to support:

> Maintenance
> Inventory
> Production Tracking
> Job (or Work) Orders
> Safety

Phase III was to support such additional functions as:

> General Ledger
> Purchasing
> Engineering
> Surveying
> Marketing
> Legal/Contractual Matters
> Statistical/Economic Analyses

This plan was developed, analyzed, economically justified and approved by the partnership in 1982. NREC hired two DP professionals and acquired the hardware in 1982. The hardware consisted of the Hewlett-Packard Model 3000 Series 44 Processor, two 404 megabyte disc drives, two tape drives, three printers, and thirteen screens. By early 1984, NREC had purchased three software packages to support Phase I. One program supported the payroll/personnel functions; another, the accounts payable and general ledger functions; and another, the fixed assets. Each vendor was selected after careful analysis and an on-site demonstration to make sure the programs satisfied NREC requirements. Two delays were encountered prior to the on-line use of the programs. The first was arriving at mutually agreeable language with the vendors regarding their purchase and maintenance contracts. The second and significantly greater delay was the programming necessary to make the payroll module comply with the provisions of the company's wage agreement with the UMWA.

In summary, by early 1983, NREC was in the following data processing situation:

A good but ambitious plan to acquire computer support.

Good hardware to support this plan.

Experienced DP professionals on the staff.

A majority of employees who had virtually no appreciation for, or experience with, the capabilities and limitations of computers and computer programs.

DEVELOPING A MIMS PROGRAM

In the Spring of 1983, NREC formed a development team to define requirements for the Phase II functions of a maintenance, inventory, and job order system. The acronym "MIMS" (Maintenance and Inventory Management System) was used. The purpose of this team was to:

Describe by flow chart the functions currently performed by each department.

Develop departmental data handling/information requirements to be performed by a MIMS type program.

Quantify the benefits achievable through this program.

Determine project economics.

Evaluate the commercially available package and select one.

The members of this team were:

From the underground mine--general foreman, underground maintenance; foreman, longwall maintenance; storekeeper; general foreman, surface operations.

From the surface mine--supervisor of personnel and stores, assistant to the superintendent.

From the Business Systems group--manager, Business Systems; supervisor, Management Information System.

From the beginning, the development team identified some cardinal requirements for the MIMS program. First, it must integrate all the applicable business functions; second, it must minimize the clerical functions being performed by management but still have pertinent information readily available for decision making purposes; third, the information in this program would be electronically available to accounts payable and general ledger; and fourth, the program was to be as "user friendly" as possible, thereby allowing the "generators" of the raw data to also be "inputers" as much as possible. In other words, if a foreman has to write an item of operating data on paper, he might just as well type it on the screen of a CRT.

The development team defined "integrating" as being the automating of the administrative (or clerical) processes associated with purchasing, inventory, and maintenance. For example, if an item of operating supplies or a replacement part were to be needed, the MIMS program would interrogate the inventory module to see if an uncommitted part is available. If the part is not, it would cause the purchasing module to print out a purchase requisition to be approved by the mine superintendent and would issue a purchase order for action by the purchasing manager or storekeeper. When the part is available, the program would write a work order listing the maintenance skills required, parts required, and estimated time to complete. When the parts are issued, the appropriate cost and inventory files would be adjusted. When the work order is completed, the equipment history file would be updated. To aid in the planning of this integrated effort, because of the many purchasing functions associated with this process, the manager of purchasing was added to the development team.

PROBLEMS/SOLUTIONS

In the first few meetings, the most prevalent problem was the availability of committee members. The members of the development team were selected because they were the managers of the functional areas to be supported by MIMS. As such, they not only knew the day to day operating conditions and informational needs of their department, but they also were the ones most familiar with future requirements and conditions. Unfortunately, their regular job usually required 8-10 hours a day or more. Because of this, they did not always have time for the committee work. If there was a production stoppage, they were totally unavailable for MIMS development. To help solve this problem, the assistant superintendent of the underground mine was added to the team. He is a progressive, imaginative individual who clearly recognizes the benefits of a MIMS program. Whenever possible, he reassigned priorities of daily work to allow the team members time for MIMS development. He was also able to provide the mine-wide requirements for information from his perspective. While there is no good solution to the availability problem, it is still felt that functional managers are the best members to be on a development team.

The second and perhaps most formidable problem for most of the employees was a lack of experience or familiarity with a computer. The problem manifested itself in a number of different ways. Some of these problems were expressed by the MIMS committee members, while other problems were a concern of the employees in general. Specifically, they were:

Fear that it would cost them their jobs.

Belief that computerization is not suitable for, or of any help to, them in their jobs.

Belief that it will add to their work load, hence something to avoid.

Lack of appreciation for the computer's capabilities and limitations. They were unable to visualize or quantify the benefits upon which the economic justification depends.

Apprehension about being committed to the projected benefits.

Willingness to accept anything the DP professional might recommend without a critical evaluation of how it might affect them.

The solution to these problems can be summed up in one word: education. Just what form or forms the educational process takes is the variable. Some of the means NREC used were:

Using various game programs to introduce the user to the idea of the interactive process.

The Hewlett-Packard users group contributed library, which includes a variety of good programs, was used.

Having the computer staff actively seek ways to support the business through a quick return programming effort.

Visiting similar coal mining operations that are supported by computer software in order to get the benefit of their experience.

Using programs that are easy to use and that can help solve an immediate number crunching need. For example, NREC used two different electronic spread sheets that gained immediate and enthusiastic acceptance.

By no means were the educational efforts confined to having the computer oriented members of the committee introduce and enlighten operating members into the wonders and mysteries of data processing. Of equal importance was to make sure that the DP members understood the problems and constraints placed on the people who actually mine the coal.

A third problem the MIMS Committee addressed was the economic benefits of such a program. It is easy to feel that we need computerized help in this or that area; however, it's quite another problem to be able to state--with numerical or statistical justification--that we will reduce costs by X dollars a day, or increase production by Y tons per year through the use of this program.

To reach an agreement, the committee visited or talked with personnel of other coal mining companies who were using computerized support. Members reviewed demonstrations from software vendors, reviewed and analyzed time study data, and, through use of relatively simple programs already on the equipment, gained a better appreciation of the system's capabilities and limitations. MIMS benefits that the committee expects are:

A 5 percent reduction in equipment downtime through better knowledge of parts availability and costs.

A 5 percent reduction in maintenance overtime labor costs.

A reduction of 58 man-hours/month through automated printing of purchase orders, reports, etc.

Based on other companies' experiences around the country, it is felt these numbers are conservative. Since MIMS is but one program in our overall development effort, it would be difficult to isolate its specific contribution to the economic indicators of discounted cash-flow rate of return, return on investment, or payback. However, benefits from MIMS are estimated to be on the order of $500,000 per year.

While not a serious one, a fourth problem confronted was to address the attempts by the "squeaking wheels" to get their pet projects programmed and on-line. These were projects that were not planned to be in Phases I and II. There was certainly no argument that their projects were worthwhile and would benefit their department, if not the company as a whole. However, the plan covered not only hardware and software purchases, but people acquisitions as well. NREC did not have enough people on the DP staff to do this additional programming and still adhere to the schedule for Phases I and II. On the other hand, the committee was attempting to "sell" a product (computerization) and did not want to alienate future DP "customers" by being unresponsive to their informational needs. Remember, on-line, real-time DP support was new to most NREC employees, and some were apprehensive about it. To solve this and other similar problems, a Data Processing Evaluation Committee (DPEC) was formed.

The DPEC has, as its members, the head of each of the four major departments that make up the NREC organization. These are the Administration, Accounting, Engineering, and Operations departments. In addition, the membership included the Vice President as Chairman, plus the supervisor of the Management Information System. The purpose of the DPEC is two fold. First, it establishes guidelines and priorities for the development of software. Second, it is an educational forum to allow dissemination of information to the individual departments on progress being made, the capabilities and limitations of computerization, as well as acquainting the Management Information System staff with the day-to-day problems encountered by the prospective users.

A fifth problem was to determine the best means to economically and efficiently get the basic raw data into the computer system. As mentioned earlier, a cardinal requirement in the MIMS development was to minimize the clerical effort of the front line managers. The evaluation committee feels that, in general, the job order program drives the majority of the other programs. To get accurate and timely job orders out of such a system, one must have the same type of input. Therefore, two data collection devices are being considered: a mine monitoring system and a small hand-held device that can be carried by the foremen.

In mine monitoring, NREC plans to place sensors in selected areas of the mine to monitor the mechanical, electrical, and environmental conditions. These sensors would be tied to a small monitoring processor. This processor would then be tied via telephone data line to the HP processor. Examples of the information thus obtained would be equipment downtime and number of operating hours of the monitored equipment (e.g., continuous miners, roof bolters, belt drives). From these data the MIMS program could then publish preventive maintenance schedules, job orders, purchase orders, monitor inventory levels, etc. As of this writing (Oct. 1984), NREC has reviewed 11 vendors of monitoring systems and selected two for an in-mine evaluation.

The small hand-held collection device, which would complement the mine monitoring system, is slightly futuristic. Nevertheless, it has potential to aid the section foreman in recording elements of shift operating data as it occurs (e.g., clock numbers of the crew, hours that a crew member performed outside his bid job, number of shuttle cars loaded, and supplies and spare parts used). This could provide the raw data for production information, mining costs, safety analysis, etc. At the end of the shift the data in these devices could be electronically sent to the HP processor. The section foreman could then have his end-of-shift reports automatically printed out while he discusses problems with the mine management, bathes, etc. As of this date, a device has not been found that is small enough to be conveniently carried by the section foreman, rugged enough for the underground mine and powerful enough to record the necessary information. However, NREC is still searching. If one is found, one task of the committee would be to convince the section foreman to use it. He is already carrying many devices on his miner's belt that are legally required. The key to his acceptance would probably be the time saved at the end of his shift.

Other problems that we are now addressing are:

The amount of equipment history we desire to maintain.

The amount of man hours required to load the inventory data such as parts numbers, vendors and prices, usage rates for parts and supplies, and economic order quantities.

Changes in accounting procedures.

SUMMARY

As already mentioned, the development of the integrated, computerized MIMS system is taking place within the larger context of providing overall DP support for the company as a whole. While many industries have computerized inventory control or maintenance management programs, it is felt that a totally integrated system--from work order and purchasing through accounts payable and general ledger--is somewhat unique to the coal industry. The problems that have been discussed are not unique in the development of DP support for a company. After considerable trial and error, NREC did find the solutions that were appropriate to the situation and personalities involved.

If in bringing these problems and solutions before this conference NREC has helped others experiencing a similar development, the author is satisfied. NREC's computer efforts have certainly been helped by other companies.

Chapter 10

MINESTAR - A Revolutionary Approach
to the problems of Mine Scheduling

Andrew J. Lingard[1] and Thomas F. Doherty[2]

ABSTRACT

Existing mine scheduling systems are characterised by a lack of user control over the logic that directs the scheduling process. A large part of the decision-making function of these systems is not available to the mining engineer for evaluation or modification, resulting in the mining engineer having to compromise and approximate his particular mining operation.

MINESTAR has been developed in the recognition that each and every mining operation has its own combination of scheduling problems for the mining engineer to address. MINESTAR is a high level scheduling language, enabling the mining engineer to tailor a general scheduling framework to each specific mining operation. The system is highly interactive, and the mining engineer can also select his own preferred method of interfacing with the system through a workstation.

The mining engineer can alternatively use each element of the workstation - terminal, graphics screen and digitiser - for input and output.

1.   Senior mining engineer, Mincom Pty Ltd., Atlanta, GA.
2.   Analyst/programmer, Mincom Pty Ltd, Australia.

INTRODUCTION

REVIEW OF EXISTING SYSTEMS

Mine scheduling systems are used by mining engineers to schedule production and equipment in operating mines, and to assess alternative production strategies during the evaluation of future plans and/or potential new mines.

Such systems typically simulate the removal (mining) of various materials (overburden, coal) using various items or fleets of mining equipment by applying mathematical formulae (equipment production rates) to a mining block database. Volumes of materials removed are accumulated for reporting.

Whilst there are numerous mine scheduling systems available today, there is no shortage of disillusioned and frustrated mining engineers trying to use these systems to solve problems.

At the very centre of the mine scheduling dilemma are two truisms:

1.   Every mine is DIFFERENT.
2.   Every model or simulation has inherent limitations due to the simplifying assumptions made in constructing the model.

If a mining engineer examines the causes of the shortcomings in the scheduling system he/she is using, he/she will find that those causes can be traced back to one (or both) of the above.

One could propose a third truism - that both mines and mining engineers change over time, the former as a consequence of both market requirements and geological factors, the latter by virtue of increased experience.

Whilst the problem is relatively easy to define and observe, the solution - as always - requires an innovative approach.

The revolutionary approach adopted to develop MINESTAR (Mine Simulation, Targeting and Reporting) relied heavily on recent advances in software engineering, and the evolution of mining software from a "cottage industry" (mining engineers cutting code) to a factory-style manufacturing industry, with highly qualified computer scientists exploiting computer harware advances.

Existing mine scheduling systems fall into two broad categories -

(i)   <u>specific</u> systems - which address the mine scheduling requirements of one mine; and

(ii)   '<u>general</u>' systems - which address a mining method, and which can be applied (with varying degrees of success) to a range of mines.

Specific Systems

These normally originate from mining engineers-turned-programmers in a DP department guided by a mining enginer. Such systems have good potential 'fit' with the mine for which they were designed, and are easy to use for the engineer who specified the system. The user also understands the limitations of the model.

However, such systems suffer a number of disadvantages, including:

(i)   the model is not transportable, either to a different mine or to the same mine over a long period of time.

(ii)   the system is not flexible enough to cope with the increasingly complex problems generated by competitive markets (eg. coal quality).

(iii) for anything other than a simple mining operation, the development costs can be inordinately high.

(iv)   development and modification time can be lengthy as the engineer's perception of the problems changes with increased experience.

(v)   system usability suffers dramatically when the key engineer leaves the company.

In many cases, the end result is that the engineer compromises with an approximation to the mining operation.

General Systems

Such systems normally originate from a software supplier, as a package, or from the corporate office of larger companies. Despite initial aims of wide generality, such systems are often released with a limited range of application, due to the designer's range of experience and the need for a software supplier to remain commercially viable (thus development times must be minimised). The dilemma facing the developer fo a general system can be stated as -

> The more simplifying assumptions incorporated into the model, the shorter the development time, but the less general its application range.

User Flexibility

From the user standpoint, the most limiting feature of existing systems is that, in order to 'drive' the system, the mining engineer has to continually and consciously adjust his/her thinking between the 'visually concrete' (mine plan layout) and the abstract (the mining block database).

Such a situation is clearly inefficient and is prone to error and oversight.

THE OPTIMAL SCHEDULING SYSTEM

The optimal scheduling system is one which combines all the advantages of existing systems and excludes all the disadvantages, as well as incorporating the most recent advances in hardware capability and software engineering, for the lowest cost.

Because of the complexities of most modern mining operations, packaged general solutions will usually represent the lowest cost per user, since the software supplier can spread his development costs over a number of users.

The challenge, therefore, lies in designing a system which is general enough to cover a wide range of applications, yet which can be tailored to be perceived by a given user as a system designed specifically for his/her mine and its inherent problems, to minimise unnecessary compromises and approximations.

Further, this tailorability must be easy and rapid.

An ideal system would also overcome the user difficulties in continually translating mine plans into the abstract terms of the mining block database.

DESIGN CRITERIA FOR MINESTAR

Minestar was designed to be the optimal scheduling system, as defined above.

The user (mining engineer) criteria were -

(i)   the system must be general enough to be applied to any mine in the world (coal or hard rock; open pit or underground).

(ii)   the tailorability from the general to the specific must be simple to use and rapid in its implementation.

(iii) the system must be capable of being used for the most "broad-brush" long term schedule and the most detailed short term schedule and all points in between.

(iv)   the system must enable the engineer to schedule directly from mine plans and to visually (ie. graphically) monitor scheduling progress.

(v)   the system should do all the bookkeeping tasks, including commenting of decisions.

Since the system was developed in the commercial environment of a software house, there was the inevitable marketing criteria - the system should be demonstrably better than existing systems, yet development costs limited to permit pricing at a level not more than the most prominent existing systems. Further, the system must be readily enhanceable to take advantage of rapid advances in hardware technology so as to ensure its continued pre-eminence.

When confronted with these criteria, the programming group concluded that -

(i)   development costs would be minimised and user flexibility maximised if the system was designed around powerful, updatable modules such as generalised database, report writer, screen mapping and data dictionaries.

(ii)   tailorability would be provided by developing a high level language, which the engineer could use to rapidly define his own mine and its scheduling constraints.

(iii)   usability would be provided by a graphics workstation, with the system accepting input from anywhere within the workstation. Further, the utilisation could be tailored by the user to suit individual preferences.

SYSTEM OVERVIEW

The Workstation

A typical workstation comprises a terminal, a digitiser and a graphics screen. Whilst this represents the best configuration, the system can be driven from a terminal alone, or from terminal and one other device.

Both mine plans and user-defined menus can be mounted on the digitiser. Typical menus might include an equipment menu, a material type menu and a menu of commonly used commands and abbreviations.

Furthermore, the buttons on the digitiser puck can be assigned commands and responses. The engineer can therefore conduct his/her entire scheduling operation from the digitiser where he can select blocks from his plans, select equipment and material types from menus, and select commands and responses from a menu or the mapped buttons.

The progress of his schedule can be displayed in numeric form on the terminal screen and in graphics form on the graphics screen (eg. blocks shaded as they are mined, with different patterns for different materials or equipment or time periods).

Alternatively, the mining plans can be displayed on the graphics screen and blocks selected by using the cross-hairs.

On terminals with function keys, common commands and responses can be assigned to these keys, thereby minimising keystrokes for the terminal user.

Each mining engineer is free to choose the workstation configuration that suits his preferences. Workstations can be saved and restored as required.

Graphics Output

As the polygons which represent each mining block can be loaded in to the system, a large range of output graphics can be produced. This includes plots with any user specified values "posted" for each block, and multiple shading options for any user specified criteria. Graphics specifications are defined, where all the attributes of the plot are input and saved. A plot is produced by simply selecting from the specifications defined.

For example, scheduling progress by equipment or period (or both) could be displayed. Different colours and/or patterns could be used to highlight each period or machine.

Online graphics is also provided, enabling the scheduling progress to be updated on the graphics screen. Plot specifications are used to define the characteristics of the displayed plot.

Bar charts, graphs and pie charts of the scheduling results can also be produced. The scheduled status of a block can be readily portrayed as a bar chart showing the amount of each scheduling item mined.

Reports

Continuing the user definability theme, reports can be fully specified in terms of content and format. Report specifications are produced, and Minestar data records are sent to the specification at report generation time.

Reports of any type can be produced, using Minestar selection logic to define what data is sent, and the report specification to detail the report layout. Hence, reports of machine movements, block status, inventories and summaries can be readily generated.

Report output can optionally be directed to the terminal screen, so that reports can be previewed at creation time.

Screens

Two types of screens are used in Minestar. User formatted screen lists are used as the main interface to the commands of Minestar. Screen lists comprise command mnemonics and descriptions, which are selected by moving a selection bar through the displayed screen. List commands are provided for selection bar movement, execution of the current command and displaying the online help for the current command.

The other type of screen is the screen "map". Screen maps are completely definable by the user, in terms of content and format. They are used for all information display during scheduling. User summaries, block data and all stepping results can be displayed using screen maps. Hence, what is displayed during the scheduling process can be changed to best suit the current requirements. Especially in short term detailed scheduling, where constant monitoring of many variables is necessary, screens can be selected to highlight these.

The screens used in the editing of the block records can also be formatted by the user, so that specific subsets of the record can be changed. This provides one method of limiting access to the entire database.

For example, an editing screen can be used to input surveyed block data as the mine actually progresses, to enable comparisons between planned and actual, and to enable updating of the block database prior to the next round of scheduling.

Database

The Minestar database consists of user definable records, containing all the block information necessary for scheduling. The record consists of the "block" name followed by any numeric or character items entered when defining the database format. Each item is given a unique name, which is used within the system for addressing the database values.

Data included in the block record can include seam information, burden information, coal information, quality information and any specialised characteristics or flags. NO assumptions are made about the data types, relationships or units.

Once defined, the database can be upgraded in the prescheduling stage, using user expressions defining the logic required. User expressions can be simple arithmetic operations or programs which incorporate powerful system functions which can return values from table or curve enquiry. For example, the overburden values in a block could be adjusted by the dragline rehandle, by using a rehandle versus overburden depth curve.

Interactive enquiry and update of block records is also provided, using a powerful, interactive screen editor.

Item values can be loaded and/or updated from a sequential file, either selectively or for all the items in the block record. The database can be readily modified to meet the demands of scheduling case studies, for example.

Each item in the database can be categorised into groups, providing a higher level association of values of the same "type". For example, if the database includes a number of coal seams in a block, the average block quality could be retrieved by defining each type of coal quality as a group, and enquiring on the group value for a block.

Scheduling

Machines

Machines are used in Minestar to control the scheduling step. The time taken for a step is calculated from the machine rate, and the associated machine calendar is advanced.

The machine definition involves nominating a production rate and a calendar. The rate is a user expression which can be as simple as a numeric constant or rate selection based on block database item values and table and curve enquiry. For example, shovel productivity at the coal face can be dependent on the seam thickness. The base shovel rate could be adjusted by a productivity factor, which is retrieved from a productivity versus thickness curve.

The calendar is defined in terms of the major scheduling period and the component time intervals in each. Other time intervals can be used by defining a relationship to the currently defined intervals. The time frame being used can be set by the user as the reference calendar for evaluating the schedule status.

Hence, a calendar in terms of operating days per annum could be used for long term scheduling, or shifts (or even hours) per day for short term scheduling.

Breakpoints

A large part of scheduling is the recording and monitoring of the values considered critical in determining the schedule progress. The criteria used can be simple targeting of production quantities required, or more specific tests such as testing whether overburden is about to be spoiled into a strip still containing coal.

Minestar allows the user to define all the conditions to be checked during scheduling, as breakpoints. Breakpoints are user expressions which are tested during a step. Actions resulting from an activated breakpoint can be to abort the step, or warn the user and proceed with the step, or query the user as to whether the step should proceed.

Hence, breakpoints can be simple warnings about coal not being uncovered during coal mining or more powerful conditions such as coal quality specification checking or checking if a drill is too close to the dragline.

Stepping

The material type to be removed from a block is selected by the user during the block sequencing. Operational modes can be defined which specify the activities carried out during a step. Typical modes for short term scheduling in an open pit mine might include topsoil removal, prestripping, overburden drilling, overburden blasting, overburden removal and coal removal.

The quantity of the material to be taken can be selected as an amount, or a percentage of the current value, or a percentage of the original value.

Hence, by selectively choosing the items to modify during the step, and the amount to be removed, one step can be designed that will perform quite complex upgrades to the databases.

An alternative to the single step interactive approach provided in Minestar is the use of sequence files. These are ordered block sequences that can be run in a batch mode. Depending on the requirement, they can also include information on the current machine, mode and material being removed, so that the file can be run to produce a complete schedule. Using this facility in conjunction with the backtracking functions provides a "what-if" capability, where various alternatives in the sequence can be tested.

All information generated during each step is stored in a transaction database. This allows sequence backtracking while still retaining the integrity of the summaries and block information used. Further, all reporting and plotting of the schedule results is produced from this database.

Summaries

Minestar includes a general summary handling capability. A summary is a list of items that are to be updated after a scheduling step, using the accumulation type defined. The items to be summarised may be any of the database items, or user items defined during the summary definition. In addition to these attributes, summaries also allow the user to define an update condition, enabling selective accumulation of the items included, and an initialisation cycle (per machine, period for example), which causes the current values to be saved, and the summary to be cleared when these variables change..

Summaries can be interrogated during scheduling at any time. Summaries which are updated during a step have the associated update record written to the transaction database for the step. This allows the summaries to be correctly updated during backtracking.

Summaries are used to record the quantities of materials going to selected stockpiles (coal or waste), so that stockpile volumes and capacities can be monitored throughout the schedule.

System Applications

Minestar addresses all the scheduling tasks undertaken by a mining engineer:

· short term scheduling - of mining strategies to examine the effects of variations in shipping schedules, equipment availabilities and product qualities.

· long term scheduling - to evaluate different production rates or pit combinations or mining sequences or blending from a range of pits or production faces.

· targeting - to rapidly assess the rate of burden removal required to achieve nominated coal production targets, or, alternatively, the rate at which coal can be uncovered by a nominated waste removal capacity.

· equipment selection - to evaluate the impact of alternative equipment fleets or sizes on a production schedule.

· operating basis assessment - to ascertain the effects on production schedules of adopting different working shift arrangements.

· evaluation of mining criteria - including different mining thicknesses for different equipment, different minimum separable parting thicknesses, and alternative recovery/dilution combinations.

Development History

The conceptual specifications for Minestar were developed by a team of 3 mining engineers meeting occasionally between August and October 1983.

The functional specifications commenced in November 1983 and were completed in January 1984, involving approximately 4 to 5 person-months of effort between mining engineers and analyst/programmers.

Following a review and revision in February 1984, the actual programming commenced in March 1984. A team of 4 programmers and a mining engineer was established.

In late June 1984 - after 14 man months of programming - a prototype system was available for demonstration and testing. The pre-release version came in September 1984 - 9 man months of effort - and the first release was completed in December 1984 - another 8 man months.

Future Developments

The main thrust of the future development will be enhancement of the high level language which permeates the system. The objective is to make the language more powerful by moving it closer to everyday spoken English, whilst retaining the terminology of the mining engineer.

CONCLUSION

It is not the role of the authors to pre-empt the marketplace by declaring that Minestar achieved the design aim of being the optimal scheduling system. The system was developed as a commercial product in a market-driven organisation, and the final judgment on the system's merits rests with the mining engineers of this world.

Chapter 11

SEAMSYS AT PEABODY COAL COMPANY

Richard Lusk

Peabody Coal Company

ABSTRACT

Peabody Coal Company purchased SEAMSYS from Control Data Corporation in March of 1982. This paper will address the following:

- History of SEAMSYS at Peabody
- Enhancements to SEAMSYS
- Gridded Seam Model
- Desired Enhancements to SEAMSYS
- Future Directions

Introduction

Peabody currently utilizes two large IBM compatible mainframe computers in its St. Louis Data Center. Both computers are linked to engineering, environmental, geology, design, financial and other users via an extensive telecommunication network.

The communications network uses IBM's Systems Network Architecture (SNA) to support micro and mini computers, graphics display terminals, digitizers, engineering plotters, low speed impact and laser printers, and approximately seven hundred (700) 3270 terminals. The network services over eighty (80) business, operating and engineering locations throughout the Company.

The operating system is VM/CMS which allows users to sign onto the host computer as though they were on their own micro computer. This analogy is even carried to the extent that the users must format their own mini-disks (i.e., floppies).

Operating over this network from a central site solves all of the problems associated with local area networks. There is one copy of each load module shared by all users; data may at the determination of its owner be shared by any and all users throughout the corporation.

HISTORY OF SEAMSYS AT PEABODY COAL COMPANY

The history of SEAMSYS at Peabody Coal has gone through the following phases:

- Benchmark
- Installation
- User Acceptance and Training
- On Going Support

Benchmark

The initial benchmark of SEAMSYS was started in January 1982 after a two year investigation of software packages. The benchmark was primarily 'results' oriented. A multi-seam operation in a western mining area was used to compare hand generated results, results from another computer system, and the results from SEAMSYS. In all cases the results were within 3% which was considered to be an acceptable level of error.

Installation

SEAMSYS, as sold by Control Data Corporation (CDC) was designed to run on their Cyber series computers. Installation of the product at Peabody required modifications to resolve differences between CDC and IBM implementations of the Fortran compiler, file structure, and terminal access methods. The modifications were completed and the system was running in an alpha test mode within one month. Benchmarks of the western mining area that had been run on CDC were again run to compare the results on the IBM. The errors were within 3% of the hand generated results. Additional benchmarks were run to compare hand results with some of the single seam operations in the east.

User Training & Acceptance

The overall strategy was to develop at least one strong user in each division who would be responsible for SEAMSYS. One week training classes were scheduled at Peabody's St. Louis corporate office and at each of the division offices. About eight trainees attended each session. In addition to SEAMSYS training, most of the users were being introduced to the operating system, VM/CMS, which tended to make the training seem more difficult.

The command language of SEAMSYS is free form input (most of the time); however, there is a specific order to the commands which can cause difficulties and is a frequent source of errors. This, along with the many optional input parameters, makes SEAMSYS not too readily useable to the casual user. Peabody has committed a full time analyst to support the SEAMSYS users.

The majority of the divisions are using SEAMSYS to do their 10 year estimates. The corporate engineering office has used SEAMSYS extensively for evaluation of potential reserve areas. There are currently 60 SEAMSYS models which have been developed at Peabody Coal Company.

On Going Support

The on going support for SEAMSYS is expected to require one full time person for the forseeable future; however, the nature of the support has changed considerably during the last year. Training and phone support has decreased significantly, but there is an on going effort providing program enhancements.

PEABODY ENHANCEMENTS TO SEAMSYS

Major enhancements by Peabody have included the following:

- Drill Hole Data Base
- Full Screen Input
- SEAMCALC
- Report Writer Modifications

Drill Hole Data Base

The Drill Hole Data Base is more than an enhancement. It is a system that maintains all of the Peabody drilling information and related data - quality analysis, geophysical logs, geological logs, traverse, etc. SEAMSYS was not capable of handling this magnitude of data which accounts for 160,000 drill holes and 40,000 quality samples. An interface has been developed to load the data from the Drill Hole Data Base directly into the SEAMSYS composite data base.

Full Screen Input

The SEAMSYS command language presents a somewhat less than formidable but more than trivial barrier to the casual user. A series of input screens was developed that leads the user through the development of a SEAMSYS run by menu prompting.

This program then builds the command files for SEAMSYS. It insures that keywords are correct, that logical relationships are adhered to and that the order of the commands is correct.

SEAMCALC

SEAMCALC was developed to supplement the internal mathematical computations of SEAMSYS (e.g., stripping rations). This is a system that allows the user to define variable names in terms of grid values stored within the model (e.g., seam top, bottom, ash, sulfur. BTU, etc.). The variables are defined by full screen input with defaults allowed, should that particular value not exist at the grid point. The user may then write a normal equation using parentheses arithmetic operators, the functions MAX, MIN and IF statements to mathematically describe the problem.

Report Writer Modifications

The SEAMSYS reports are adequate if the system is being used for reserve evaluation. However, as use proceeded throughout the corporation, it was necessary to modify the report writer. This turned out to be a non-trivial task. The logic for generating the reports is extremely difficult and our efforts in this area were reduced to post processing output records from the SEAMSYS report writer.

SEAMSYS GRIDDED SEAM MODEL

The SEAMSYS gridded seam model is the real strength of SEAMSYS. The outstanding features are:

- Grid Generation
- Grid Manipulation
- Consolidation of Data
- Report Writer

Grid Generation

SEAMSYS allows a great deal of flexibility when calculating gridded data. The user may specify an interpolation radius to determine the value (if there is one) at a particular grid point. Closest points within moving sectors (usually 8) are chosen as interpolation values if inverse square of the distance is being used. Kriging is also offered; however, the results for coal structures and quality values do not seem to be worth the extra cost in computer time. The user may also define an extrapolation radius for those grids that exist on the boundary of a crop or recovery line. SEAMSYS also allows the user to specify a crop line (a digitized polygon) and to generate grid values only within the user defined area.

Grid Manipulation

After the gridded seam model is built SEAMSYS provides post processing of the gridded seam model. One type of processing is "push down" which is a technique that SEAMSYS uses to insure that the geologic realities of coal seam structures are adhered to. That is, coal seams do not cross nor do they extend above topography. Basically seam index values are compared on a grid by grid basis to determine if seam relationships have been violated; if so, then the seam which is defined to be dominant pushes down or nulls the interfering seam. SEAMSYS also offers technique called "windowing" which will null the data for a seam at all grids around a hole, which is deep enough to intersect the seam but in which the seam does not appear (e.g., sand channels).

Consolidation of Data

SEAMSYS is a legitimate data base in the sense that the data is stored in a hierarchical fashion for the gridded seam /Normalizationgrid point owns the seams that exist at that point, each seam owns a list of up to 250 parameters (e.g., top, bottom, ash, sulfur, BTU, etc.). Virtually all of the quantitative data associated with seams in a given mining area can be stored in this central location.

By using this structure it is as easy to model 15 seams as it is to model a single seam.

Report Writer

The report writer has a great deal of flexibility that gives the user different views of the established gridded seam model. The user can vary mining depth, weathering depth, minimum seam thickness, and cut off limits on any of the quality parameters (e.g., ash must be less than 10%) in order to produce a report that will give detailed output for tons, yards and acres. This same report request can be used to show the contours from which the report is derived. The report may also be constrained by user input polygons so that the values are reported within a certain area of interest. An outstanding feature is that many different requests may be issued (the user can apply "what if" logic) against the gridded seam model.

DESIRED ENHANCEMENTS TO SEAMSYS

SEAMSYS has been successful; however, the following are enhancements that we believe would increase its effectiveness:

- Normalization of Coordinates
- Improved Data Base Access
- Topography Generation
- Targeting

Normalization of Coordinates

The normalization of coordinates is a feature that would increase the effectiveness of SEAMSYS in two areas.

The first area involves the manipulation of coordinates within the computer. IBM stores numeric data as single precision decimal with six significant digits (and sometimes 7). When SEAMSYS intersects and unions polygons it must, of necessity check to see if the results are legal. If errors occur, (e.g. the union of two polygons is not a single polygon) then processing stops and precision or tolerance problems can occur. This has been the most outstanding problem in implementing SEAMSYS at Peabody. Mines which are on a local coordinate system have never experienced these problems, but mines on state plane coordinates are plagued with these difficulties. Given that most mines cover an area of twenty miles, one method of solving the problem would be to internally transform the mining area by translating the mines coordinates to an area bounded by the difference of the extremes of the northing and easting.

The second area of improvement and an argument for using normalization or transformation of coordinates, is to handle more efficiently those mines whose length lies diagonally with respect to the coordinate system. The gridded seam model is a rectangle whose dimensions are the maximum northing minus the minimum northing by the maximum easting minus the minimum easting. Significant storage could be saved by internally logically rotating the mine model so that it would be perpendicular to the coordinate system. A solution would be to normalize the coordinates.

Improved Data Base Access

The design of the data base is good; however, the implementation leaves something to be desired. The data base access is written in Fortran by necessity in order to provide portability across many types of computers. The end result is that there is a lot of processing time lost in performing read and write operations. This significantly affects the computer run time.

Topography Generation

SEAMSYS does not offer a method of generating a topographic grid from anything other than drill hole collar elevations.

Since most topography is generated from digitized contour lines another software package must be purchased or written and made to interface with SEAMSYS.

Targeting

Targeting is a SEAMSYS function that allows the engineer to simulate machine operations and indicate the machines position whenever a specified number of tons or yards has been removed. All of this is accomplished by laying cut polygons over the gridded seam model. The cut polygons are generated by SEAMSYS within zones for which a mining direction is given, a cut width is given and opposite sides of the pit are given. This is a method that works well for general estimates; however, there is a need for the user to explicitly specify the cut polygons to indicate the non-uniform mining operations of the equipment.

FUTURE DIRECTIONS

The gridded seam model reflects the geological structure of the seams as accurately as is possible, given current technology, and the coal quality is similarly represented. This gridded seam model then becomes the data base for the mine and becomes the source of data for many other applications.

End of month calculations is a textbook application utilizing this 3-D model. The user can digitize polygons from aerial photographs and reports can be written to give yards, tons and pertinent quality values for each seam and polygon.

Operations planning models already exist that use gridded seam models to simulate the mining process. Equipment and production rates can be given that will produce reports giving the daily status of the equipment.

CONCLUSION

While it is true that computers cannot mine coal, the gridded seam model is one of the first and most significant steps in aiding the engineer to develop a strategy that will let him use the computer as an efficient and effective tool for mine planning.

Chapter 12

DEVELOPMENT AND IMPLEMENTATION OF THE MPFS MINING SYSTEM ON
DRUMMOND COAL COMPANY'S IBM 4341-II COMPUTER SYSTEM

Bradley A. Brasfield

Drummond Coal Company, Inc.
P.O. Box 1549
Jasper, AL 35501

Abstract. Drummond Coal Company has installed an
on-line mine engineering system to evaluate
reserves and mine plans with reporting capabilities by lease. The system was integrated with
existing data collection files to load geological
and analytical data in a streamlined ongoing basis.
The MPFS (Mine Planning Feasibility System) system
was chosen to be the hub from which other related
computer applications would hook to. The most
significant enhancement to the MPFS system
capabilities to date is the facility PRPDFN, which
is a database containing property boundaries and
associated ownerships, both surface and subsurface.
These boundaries can then be imposed on mine areas
or production periods to provide reserve reports
by property, property type and owner.

The hardware selected for the on-line engineering workstation is the standard ASCII type devices
found attached to "super-mini" based configurations. The interface was established with a KMW
3271 emulator which also functions as a protocal
converter. A Calcomp plotter and digitizer,
Tektronix graphics terminals and various other
ASCII terminals are attached to the KMW protocal
converter. The low cost of the ASCII terminals,
support of graphics input, machine portability and
software vendor support were the major reasons
they were chosen.

INTRODUCTION

Drummond Coal Company's mining operations are
located in the Black Warrior Coal basin of Alabama.
Drummond is currently operating 11 mines in the
coal field. Average annual production is 7,000,000
tons which is marketed domestically and worldwide.
Drummond mining operations include 10 draglines and
1 shovel. Drummond also operates 3 coal preparation plants and two barge loading facilities.
Total bucket capacity for the operating mines is
778 cubic yards. Corporate offices are in Jasper
which is located near the center of the coal field.

Mine areas vary extensively in terms of size,
number of seams or splits, geology and number of
drillholes. The operating mines have reserves
containing approximately 5,500 drillholes with
some reserves having up to 10 seams or splits. The
vast amount of drillhole and coal lab data generated along with the extensive reporting requirements
needed by the company were the major factors which
led Drummond to investigate the potential for
computer applications.

Drummond's Mine Engineering Department has
offices at each division location and in Jasper.
The functions of this group are quite diversified
ranging from exploration thru reclamation, however,
the primary function is to evaluate reserves,
produce mine plan projections (short and long term)
and maintain regulatory compliance. Operating
mines utilize a full range of engineering requirements and each field office fulfills these
requirements. These requirements include preparation of short and long term mine projections,
royalty reporting and feasibility studies.

History

Prior to the installation of the MPFS system
all mine planning and reserve analyses was
accomplished by use of hand calculators and paper
files. Retrieval of information could only be
accessed from these paper files. Mine projections,
property evaluations, mapping, and all other
reports were generated "by-hand". These methods
were adequate for quick looks and one-time studies,
but most of the engineering work required was not
of this nature. A slight change in mining plans
often required the whole mine planning procedure to
be repeated.

In January, 1981, a program was initiated to
store all coal lab analyses and associated basic
geologic information on computer file. Initially,
the file was a sequential "flat file" on disk.
The data was keyed in batches by lab personnel
after mine engineering personnel edited the data.
All coal lab data is keyed and stored but only the
core drilling data goes to the engineering file.

Reporting and access from this file was limited
to "Easytrieves" ran on the mainframe and statistical programs written by engineering personnel
executed on an IBM Series I RJE attachment.
Easytrieve is an information retrieval and data

management system designed to simplify computer-programming. Easytrieves were used to compile listings of data to check for input errors. The RJE link proved to be very cumbersome since data could be retrieved in only 80 bit record lengths and the database records were 400 bit record lengths. A multiple pass and reorder of the data files had to be executed to reformat the data. Since existing records could not be updated and VSE/VSAM (Virtual Sequence Access Method) is the only key indexed file supported by IBM, the sequential file was converted to a VSE/VSAM file.

VSE/VSAM is a software product offered by IBM which allows the following choices of data organization:

1. Entry sequenced files
2. Relative record files
3. Key sequenced files

Existing records can be updated and accessed following conversion to VSE/VSAM. The capability is required for an on-line engineering system supporting daily editing and reporting.

The next project initiated was the addition of all noncored seams and drillholes along with the entry of all drillhole and lab data generated prior to January, 1981. This phase was very tedious and time consumming but it required our staff to closely examine their data prior to entry. Close examination of the data proved to be beneficial since past data handling errors or ommissions were found and corrected.

SOFTWARE SELECTION

A review of mine planning software was conducted by a joint committee of engineering and data personnel. The engineers were primarily concerned with system interface and CPU consumption. The review committee met with various groups to:

1. Learn what is available in the marketplace.
2. See how other large coal companies moved into mine planning computer applications.
3. Determine if Drummond's 4341 was adequate and compatible with available software and engineering requirements.
4. Consider time-sharing.
5. Determine other hardware requirements.

It was found that engineering applications were consistently operating on non-IBM computers however, one of the large companies visited had begun evolving toward a total IBM environment. It was therefore concluded that an engineering system could be installed on Drummond's IBM system.

Thus the task was to find an IBM compatible mine planning system that met our criteria listed as follows:

1. The system must be tested and proven with good recommendations from users.
2. The price should be competative with other systems.
3. The system should be user-friendly.
4. The system should be versatile in terms of geologic modeling and mine planning needs.
5. The vendor support should be good.
6. Interface to Drummond's drillhole database and enhancement for future needs is required.

The MPFS system was chosen because it was the only system evaluated compatible with our total requirements. The criteria that an IBM compatible system and versatile mine planning system was required with property reporting capabilities were the key factors influencing the decision.

The MPFS programs were originally developed under MVS/TSO and were batch in nature. The agreement between Drummond and the vendor (Energy Resource Solutions, Inc.) included enhancing the system to provide:

1. Interface to Drummond's drillhole database with load programs.
2. On-line digitizing and plotting.
3. Menu driven user interface.
4. Conversion to VM/CMS.
5. Provide IBM terminal graphic support or non-IBM terminal support.

WORKSTATION CONFIGURATION

Non-IBM or asynchronous terminals were chosen along with a Calcomp 965 plotter and 9000 digitizer. The Tektronix 4100 series color graphics terminals were chosen along with the Televideo terminals. All of these devices are attached to a KMW 3270-II protocol converter.

The KMW provides high-speed EBCDIC binary synchronous data transmission between a remote computer and one or more local ASCII serial asynchronous devices. "Transparent ASCII" mode is used when the devices are in MPFS. This mode allows the IBM system to transmit and receive ASCII information, thus allowing the MPFS programs complete control over the graphics CRT's, plotter and digitizer. Standard protocol conversion occurs when the devices are not in MPFS. Thus the terminals function in IBM 3277 mode or in standard asynchronous mode.

This configuration was chosen for the following reason:

1. Purchase of KMW was required to support an on-line interface with the Calcomp plotter and digitizer and additional ports were available.
2. All peripheral equipment is portable between machines.

3. The current version of IBM graphical
    support interface did not support
    graphical input from the terminal.

APPLICATIONS

The purpose of the paper is to explain how the
MPFS system was implemented on Drummond's IBM
system, however, a brief description of the system
will be presented. The MPFS system consists of
four major subsystems which have their own unique
databases. These subsystems are as follows:

1. Drillhole database
2. Grid database
3. Mine Design database
4. Digitizer database

Each subsystem provides the standard capabilities found in most mine planning systems therefore
a detailed presentation will not be made.

The system is project and user oriented therefore reserve areas can be defined as unique
projects or as mine areas within projects.
Generally, it is expedient to define reserves as
unique projects rather than mine areas with a
project since under VM/CMS only one user can be in
database at the same time.

The priority from the beginning of our computer
applications has been data integrity. Many months
were spent editing and adding data prior to actual
use. However, since August, 1984 the mine engineers from each division have used the system to
successfully produce five-year mine plans for all
mines. As-mined and washed tonnages were produced
showing projected tonnages by ownership. Reserves
reflected any "old works" or faulting encountered.

CONCLUSION

Drummond Coal Company has developed new and
better ways to mine coal efficiently for many
years.   Computer aided mine planning has given
Drummond engineers a tool which has significantly
increased the ability to manage data and evaluate
mine plans.

The system described in this paper involved the
integration of functions from various departments
within Drummond from exploration thru land
acquisition. This integration of effort within
our company is significantly enhancing the
feasibility decision making process.

REFERENCES

Pansophic Systems, Inc., 1984, Easytrieve Plus,
    p. 1-1.

International Business Machines Corporation, 1979,
    1982, VSE/Virtual Storage Access Method, Release
    3 Modification Level 0, p. 1-1.

KMW Systems Corporation, 1981, BAC-3270FS
Bisync/Async Data Converter Reference Manual,
p. 1-1.

Chapter 13

MICROCOMPUTER SOFTWARE FOR ROCK MECHANICS AND VENTILATION : A TOOL FOR MINE OPERATIONS

Dr J. Arcamone, Dr M. Dejean, Dr N. d'Albrand, Dr J.P. Josien

Mining Techniques Division - CERCHAR - GROUPE CHARBONNAGES DE FRANCE
Boîte Postale n° 2 - 60550 VERNEUIL EN HALATTE - FRANCE

Abstract. Coalfields in France are characterized by unequal dimensions and diverse geological conditions. In order to adapt to the special conditions, Charbonnages de France (CdF) has developed or modified different exploitation methods. A large effort has been given to increase mechanization for social and economic reasons. This increase in mechanization entails increased investment costs and imposes appropriate options to be taken at different steps in the planning. CdF's research center has developed numerous computer aids to help in making these decisions.

In this article, we will examine different computer programs that have been developed to solve problems of rock mechanics and ventilation. Specifically, we will look at how subsidence predictions, dimensioning calculations for the working by room and pillar methods can be made rapidly using this software on a microcomputer. Bolting pattern too can be determined automatically with such tools. Once the planning of the working has been completed, it is possible to study the ventilation network on a micro or mini computer depending on the objectives and quality of the output. This approach, concerning different specific problems at the project stage, can be examined overall using CAD tools on which Cerchar has also conducted research. However, when extraction begins and during the working, it is often necessary to verify that the predictions reflect reality and that the materials chosen are well suited to do the job. For this, an electronic installation has been developed that can be placed underground to verify the state of the face. The recorded data, analysed at the surface with appropriate equipment, allows one to establish a diagnosis of the state of the working. In parallel, micro-computer software was developed that was specially conceived for processing continuous measurements made underground in the roadways or workings.

The article concludes with a discussion of the problems that need to be solved to promote the more systematic use of high performance computer aids in mining companies.

Coalfields in France are characterized by unequal dimensions and diverse geological conditions. In order to adapt to these special conditions, Charbonnages de France (CdF) has developed or modified different exploitation methods. A large effort has been made to increase mechanization for social and economic reasons. Mechanization to improve productivity entails higher investment costs and requires the choice of sound options during the different stages of the planning process. Computers facilitate the integration of various kinds of data, technical and economical for example and contribute to an improvement in decision making. For technical choices, to which we address ourselves in this paper, a computer can be used in different ways in the prediction and the monitoring of mine works. It can be installed at a distance, thanks to improved communications systems, or on the spot. A full-sized or mini-computer responds to different objectives according to the conditions encountered, the exploitation methods, importance of the mine, and the degree of technology attained. Essentially these objectives are (Dejean, Raffoux, 1982) :

- the calculation of solutions to specific technical problems (ventilation, ground pressure) using analytic, numeric, or simulation methods

- assembling, storing and analysing, and restoring data that allow the mining operation to be followed in real time, for example, or to follow a borehole reconnaissance campaign.

One chooses either the mini or full-sized computer based on the applications and financial situation of the mining company. Most of the implements currently used by Charbonnages de France were initially developed by its research center for full-sized 32 bit computers. Progressively, these first applications were adapted to the ever improving micro- computers (8-16 bits). The list of computer programs has been growing gradually as a function of the needs related to modifications in the method of exploitation, new advances in micro and mini-computers and more precise approaches for certain technical problems.

ROOM AND PILLAR DIMENSIONNING
dimensionnement de chambres et piliers

MENU

GENERAL METHODS OF ANALYSIS - methodes generales d'analyse

1 - MODEL OF OVERBURDEN WITH SQUEEZED PILLARS
    modele de recouvrement avec piliers deformes
2 - TRIBUTARY AREA METHOD
    methode de l'aire tributaire
3 - MODIFIED TRIBUTARY AREA METHOD (ELASTO-PLASTIC PILLARS)
    methode de l'aire tributaire modifiee (piliers elasto-plastiques)
4 - MODIFIED TRIBUTARY AREA METHOD (INFLUENCE OF HEIGHT TO DIAMETER RATIO)
    methode de l'aire tributaire modifiee (influence de l'elancement du pilier)

STABILITY OF THE PILLARS - stabilite des piliers

5 - CALCUL WITH ELASTO-PLASTIC PILLARS (RECTANGULAR OR SQUARE PILLARS)
    calcul de piliers en elastoplasticite (piliers rectangulaires ou carres)

STABILITY OF THE ROOF - stabilite du toit

6 - MODEL OF BEAM (BENDING BREAKING)
    modele de poutre (rupture en flexion)
7 - MODEL OF BEAM (BUCKING BREAKING)
    modele de poutre (rupture par flambage)
8 - MODEL OF BEAM (COMPRESSION BREAKING)
    modele de poutre (rupture en compression)
9 - MODEL OF SHELL (BENDING BREAKING)
    modele de plaque (rupture en flexion)

DO YOU NEED SOME INFORMATION ABOUT THE METHODS OF CALCUL (Y/N)
Voulez-vous consulter le mode de calcul des modeles (o/n) Y
IF NOT,GIVE THE NUMBER OF THE MODEL
Si non,indiquer le numero du modele 1

Figure 1 : Menu for the computer program "STAB PILLAR"
           (room and pillar working)

* PILIERS ALLONGES *

COEFFICIENT DE SECURITE SUR SURFACE PILIERS :2

	* PILIERS ELASTO-PLASTIQUES			* PILIERS TOUT PLASTIQUE *	
	CONTRAINTE MAXI (LIM)	LARGEUR PLASTIQUE	LARGEUR PILIER	CONTRAINTE AU PIC	LARGEUR PILIER
EPONTES RESIST.	800	6.2	30.4	1473.8	29.8
EPONTES TENDRES	800	11.9	48.6	540.4	49.2
TUN. ELAS.PLAS.	160	6.0	80.2		

CODE DES COMMANDES
  /   DEMANDE UNE LISTE DES PARAMETRES
  RUN   DEMANDE L'EXECUTION DU CALCUL
         (PAS DE COMMANDE) DEMANDE STOP
  X=V   LE PARAMETRE X PREND LA VALEUR V

Figure 2 : Example of the output for calculation performed by
           the stable pillar program, "STAB PILLAR"

In this article, we will examine some tools that were developed at CERCHAR in the mining techniques program. We will make the distinction between software adapted to planning, more precisely to the prediction and resolution of specific problems, and software adapted for monitoring, in workings and in roadways.

This last point is very important in the mining domain where rules for making predictions or forecasts are often founded on statistical analysis where the data base must be up-dated regularly.

SOFTWARE AS A PLANNING TOOL

Improvements in mine planning can be envisioned in the following two stages :

- during the evaluation of the coalfield, in order to predict with an increased dependability, the irregularities presented by the seams

- during the definition of the working strategy for a new panel in the mine.

The first point has been greatly studied and complete and efficient software has been presented by numerous authors at previous conferences (see the preceding conference at Morgantown or the APCOM 84 Conference, London, March 1984). We will pay particular attention to developments made to improve the definition of the working strategy. The assistance brought to bear can be imagined following two very different approaches

- one assumes that the planner has a general view of the problem, or works a relatively simple mine, and he is given the necessary data processing tools to solve very precise problems : to calculate the subsidence above a planned longwall or to study the stability of a pillar or the roof of a room

- the mine is complex, by its geology, by the proximity of the workable seams, by the working method necessary, and, thusly, the planner must take numerous data into account (characteristics and positions of the seams and old works among others). In this case, recourse to a more complete system, analagous to the CAD system widely used in the automotive or aeronautics industry could prove itself to be profitable.

We shall examine these two approaches successively by describing the relevant software.

Resolving specific problems on a microcomputer

We assume that the person in charge of a room and pillar exploitation should prepare the projects of works to be done in a seam with the imperative to protect certain zones on the surface. He will need to choose several types of pillars, collapsible or not according to the zone and to define precisely the barrier pillars that are to remain stable with respect to the caved zones.

In order to dimension these different pillars and associated rooms while verifying the stability of the roof if it is required, one can resort to the program "STAB PILLAR". This program is entirely "user-friendly" and is available for the APPLE II microcomputer or the IBM PCXT with printer 132 characters, in text and graphic modes. It gives the user a menu proposing different calculation modes for pillars or the state of the roof in the room (figure 1). The calculation methods used in the program were developed by several specialists in the matter and amply tested. These methods, gathered together by Ghoreychi (1983) are based on analytic calculations, statistical analysis on application of the strength of the materials. By calling the function "HELP" or simply by reading the notice, substantial information about the methods of calculation are given. This program was used by CERCHAR to dimension several workings. The calculations are based on data that we judge to be relatively simple to obtain and in any case necessary to describe the panel adequately.

Essentially these data are :

- seam thickness

- the depth of the working

- a characteristic of the number of layers of the roof and of the floor

- the uniaxial compressive strength, possibly the elastic strength of the seam, roof and floor, and, if necessary, the elastic moduluses

- the average number of rooms in panel

- room widths.

The results given are :

- the length and width of the stable pillars in the short or long term, the weakened zone on the pillar's fringe for certain calculations (figure 2)

- the state of the roof (stable/instable) in the given configuration.

Complementary details are sometimes required for certain subroutines.

The simultaneous study of the pillars and the roof in the rooms, facilitated by this software, allows one to use the complementary program "PC BOLTING" to define the bolting pattern if it is necessary to sustain the sidewall or to support the roof. The program "PC BOLTING" has been developed on IBM-PC XT microcomputer with graphics devices (graphics display, hard-copy, X-Y plotter). A simplified non graphic version will be distributed for the IBM-PC XT with only a printer.

The following parameters are required :

- nature and type of rock to be bolted (number of layers in the roof, thicknesses ...)

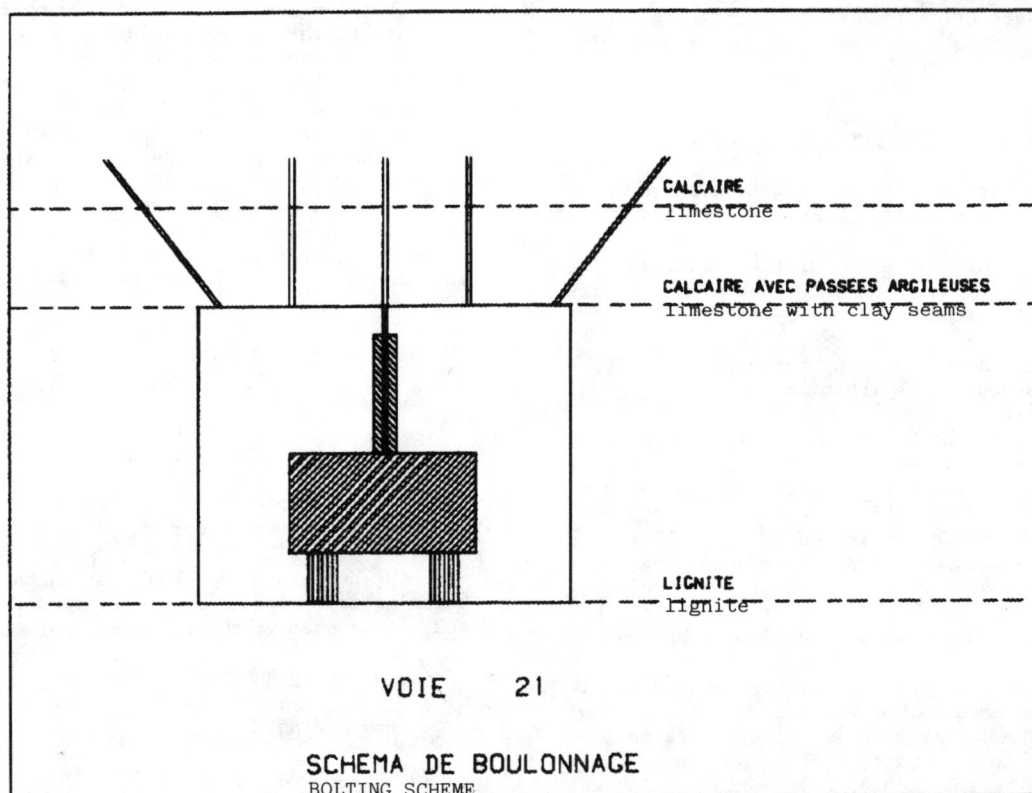

Figure 3 : Example of a bolting pattern

Tailles 26,27,28,29,30   du panneau   ETOILE

REPRESENTATION DU PANNEAU EQUIVALENT

representation of an equivalent longwall

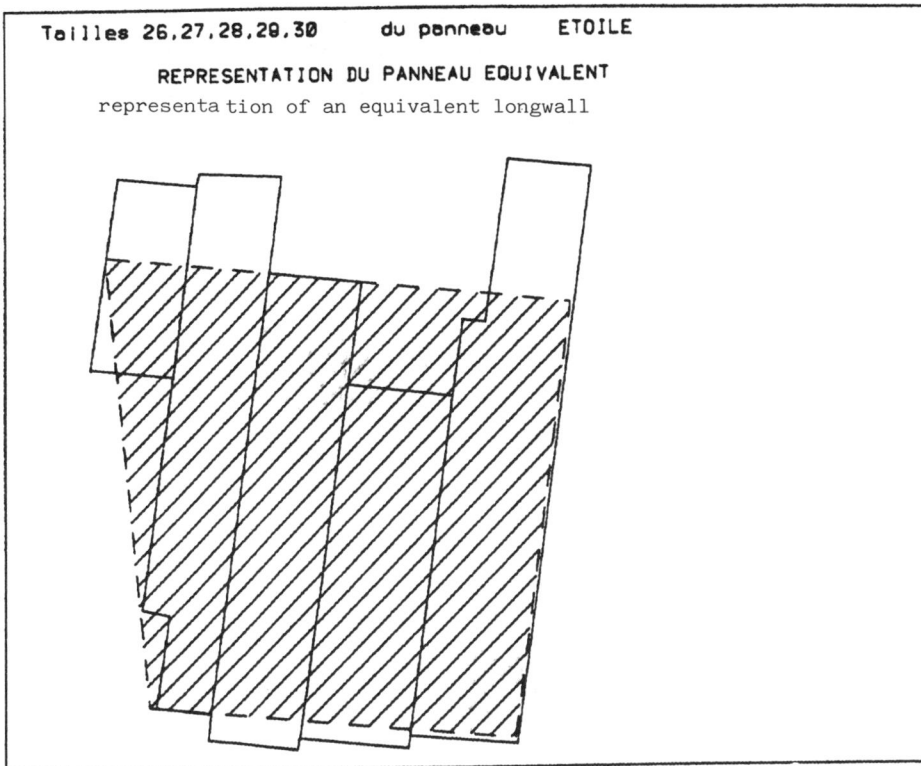

pour continuer-presser la touche [RETURN]

Figure 4 : Definition
of an equivalent long-
wall in the program
SUBSIPRED

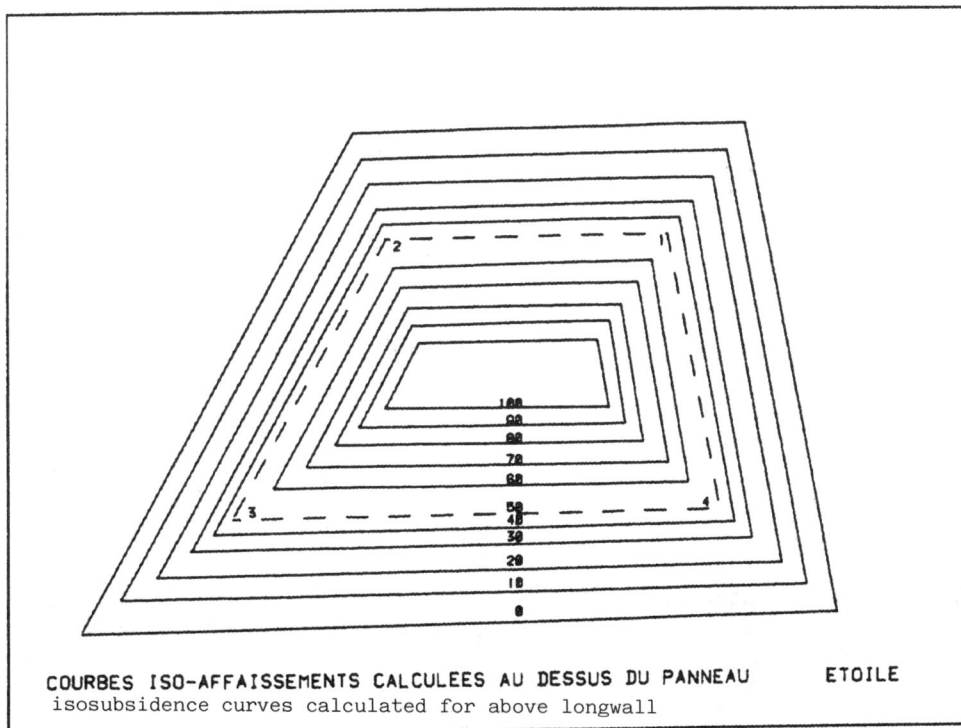

COURBES ISO-AFFAISSEMENTS CALCULEES AU DESSUS DU PANNEAU   ETOILE

isosubsidence curves calculated for above longwall

POUR ARRET PRESSER LA TOUCHE 'A' - POUR UN AUTRE DESSIN PRESSER LA TOUCHE [RETURN]

Figure 5 :Isosubsi-
dence curves - Program
SUBSIPRED

- mechanical properties of these layers (modulus of elasticity, uniaxial compressive strength)

- fractures pattern

- disposition of the rock to weathering

- hydrological pattern and aggressiveness of water

- geometry of the roadways and natural stresses (estimated from the depth if necessary)

- predicted lifetime.

The computer program provides a graphic display of the bolting pattern in the form of a fundamental schematic diagram and dimensioned drawings easily usable underground, completed by a table of the bolt characteristics and their fixing with :

- rod length
- rod type
- diameter
- type of anchoring (point-anchored or resin)
- rod direction
- density by m^2
- lagging (plate, grid).

This computer program is based on the knowledge acquired by the specialists at CERCHAR during many dimensionings in very different kinds of mines. After the feasibility of the prescribed bolting pattern has been examined in light of the financial situation of the mine, it is possible to modify the dimensioning of the working eventually in order to reduce the stresses in the roof.

When a dimensioning of the longwall has been defined taking into account ground pressure problems underground, the program "SUBSIPRED" issued for the IBM-PC XT calculates surface movements above workings with collapsible pillars. Starting from the initial mining layout described by the coordinates of its top, the operator on the program defines an equivalent simplified panel for the calculation. A few complementary parameters are required :

- the depth of the seam
- thickness
- dip angle
- a characteristic of the overburden (overburden plastic, competent or intermediate)
- a characteristic of the ground at the surface
- a characteristic of the overburden tectonic.

In the case where this program is used for longwall faces, one can specify the treatment behind the face front.

In conversational mode, the program gives the simplified plan of the equivalent panel and the lines of isosubsidence of this panel on the graphics display (figures 4 and 5) or on a 132 character printer in graphics mode. The program also gives the subsidence of one point when the operator specifies the plane coordinates. With these results, the operator can verify if the subsidence at different points is compatible with the tolerences fixed for certain works or structures on the surface.

This program is based on works carried out at CERCHAR and by specialists from French coalfields. A synopsis was made as a part of thesis research (Arcamone, 1980).

Once the panel has been defined with respect to the underground and the surface, it is possible to study the ventilation network of the panel in the heart of the mine and to predict, for example, the modifications of forms on certain circuits.

The program PC VENT, available for the IBM-PC or micro PDP-11 performs this analysis. Using the program "VENDIS" on a micro-computer and aided by a console, the network can be displayed and certain parameters can be modified interactively.

This program is described in detail by d'Albrand et. al. 1985. Here, we recall the principal characteristics. Given the description of the ventilation network (coordinates and depths of the nodes, branch resistances and temperatures, fan characteristics), it is possible to simulate the ventilation network and to determine interactively the modification to be made on the general network in order to obtain the desired flowrate at the different points in the mine.

This program takes less than one minute on an IBM PC for a network with about 100 nodes and 200 branches (example of the table of results, figure 6) necessitating 80 iterations to assure a reasonable convergence in the calculation.

As with the programs "SUBSIPRED" and "PC BOLTING", PC VENT is applicable to longwall faces and is based on a large amount of research by CERCHAR (Froger, 1976).

Presentation of an integrated planning system

When the geology of the mine is difficult to schematize with numerous stacked seams, a very complex situation could result with mine workings being conducted in different seams, and neighboring seams that are being worked could have an influence on one another. These phenomena, if not gauged properly, can make the working difficult if not dangerous.

These phenomena, essentially ground pressure, have been studied in France for thirty years and rules have been established for the order in which the seams are worked on rate of advancement of the working ... These rules allow the

SIEGE P E V G A R . M   CALCUL DU 73-78- 0

SIEGE T

```
************************************************************************************************
*        *           *      *            *          *        *          *             *       *            *
* BRANCHE * R.NUM(Mu) * I( C)* R(Murque) * Q(m3/s)  * H(mmCE) * IN(mmCE) * OBSERVATIONS * PU(kW) * PRESSION (mmCE) *
*        *           *      *            *          *        *          *             *       *            *
************************************************************************************************
*  1 -  2 *   0.50 * 17 *   0.49 *  23.72 *   0.25 *    3.95 *             *  0.057 * 11174 * 11178 *
*  1 -  5 *   5.70 * 16 *   5.60 * -16.15 *  -1.46 *  -89.50 *             *  0.231 * 11174 * 11086 *
*  2 -  3 * 300.00 * 17 * 769.78 *   6.97 *  37.44 *  -66.33 * Q IMP.  7.0 *  2.562 * 11178 * 11080 *
*  3 -  4 *   0.70 * 17 *   0.69 *  15.75 *   0.17 *  -39.45 *             *  0.026 * 11178 * 11138 *
*  3 - 24 *   3.00 * 16 *   2.98 *  -0.37 *  -0.00 * -108.21 *             *  0.000 * 11080 * 10972 *
*  4 -  5 *   1.01 * 17 *   1.00 * -21.43 *  -0.46 *   -3.94 *             *  0.096 * 11138 * 11135 *
*  4 - 26 *   8.10 * 15 *   7.96 *  37.17 *  11.00 *  -99.94 *             *  4.014 * 11138 * 11028 *
*  5 -  6 *   1.90 * 15 *   1.86 * -28.00 *  -1.46 *  -50.10 *             *  0.401 * 11135 * 11086 *
*  6 -  7 *   1.20 * 13 *   1.17 * -44.15 *  -2.28 *  -13.24 *             *  0.989 * 11086 * 11075 *
*  7 -  8 *   2.70 * 16 *   2.68 * -23.34 *  -1.46 * -135.38 *             *  0.335 * 11075 * 10941 *
*  7 -  8 *   3.40 * 16 *   3.38 * -20.80 *  -1.46 * -135.38 *             *  0.298 * 11075 * 10941 *
*  8 -  9 *   2.30 * 15 *   2.29 * -51.79 *  -6.14 *   15.59 *             *  3.120 * 10941 * 10963 *
*  9 - 10 *   0.10 * 15 *   0.10 * -51.79 *  -0.27 *  -29.88 *             *  0.136 * 10963 * 10934 *
* 10 - 11 *   0.10 * 16 *   0.10 * -51.79 *  -0.27 * -217.60 *             *  0.138 * 10934 * 10716 *
* 11 - 12 *   0.40 * 17 *   0.42 * -116.75 * -5.67 * -333.25 *             *  6.496 * 10716 * 10389 *
* 11 - 15 * 107.00 * 18 * 110.20 *  19.92 *  43.75 *   -7.53 *             *  8.554 * 10716 * 10665 *
* 11 - 15 *  21.00 * 17 *  21.55 *  45.04 *  43.72 *   -7.56 *             * 19.324 * 10716 * 10665 *
* 12 - 13 *   0.40 * 16 *   0.43 * -118.05 * -5.96 * -320.60 *             *  6.901 * 10389 * 10074 *
* 13 - 14 *   0.10 * 14 *   0.11 * -118.05 * -1.51 *  -80.05 *             *  1.747 * 10074 *  9996 *
* 15 - 16 *  73.00 * 20 *  75.74 *  19.97 *  30.21 *   71.04 *             *  5.921 * 10665 * 10706 *
* 16 - 17 *   4.90 * 19 *   5.07 *  71.92 *  26.24 *  -37.48 *             * 18.517 * 10706 * 10642 *
* 16 - 17 *   1.00 * 20 *   1.03 * -51.95 * -2.79 *   16.25 *             *  1.434 * 10706 * 10725 *
* 17 - 18 *   2.70 * 19 *   2.81 *  43.54 *   5.33 *  -52.17 *             *  2.277 * 10642 * 10585 *
* 17 - 19 *   3.10 * 20 *   3.25 *  28.38 *   2.60 *   -1.24 *             *  0.734 * 10642 * 10658 *
* 20 - 21 *  11.00 * 13 *  11.22 * -48.28 * -26.16 * 182.50 *             * 12.391 * 10725 * 10934 *
* 21 - 22 *  10.00 * 13 *  10.11 * -26.71 * -7.21 *  -25.57 *             *  1.391 * 10934 * 10915 *
* 21 - 25 *   1.00 * 13 *   1.01 * -21.57 * -0.47 *    6.40 *             *  0.099 * 10934 * 10940 *
* 22 - 25 *  17.50 * 18 *  17.44 * -26.71 * -11.44 *   5.13 *             *  3.260 * 10915 * 10933 *
* 23 - 24 *   3.00 * 19 *   3.03 * -17.74 * -0.95 *   30.77 *             *  0.166 * 10940 * 10972 *
* 24 - 27 * 100.00 * 21 * 101.44 * -13.66 * -18.94 *   30.59 *             *  2.540 * 10972 * 10960 *
* 25 - 27 *   2.00 * 18 *   2.01 * -19.06 * -0.73 *   27.00 *             *  0.137 * 10933 * 10960 *
* 25 - 28 * 200.00 * 15 * 594.59 *  -7.64 * -23.08 *  -46.65 * Q IMP.  7.3 *  1.731 * 10933 * 10909 *
* 26 - 27 *   5.60 * 16 *   5.57 *  32.73 *   5.97 *  -61.10 *             *  1.916 * 11028 * 10960 *
*        *           *      *            *          *        *          *             *       *            *
************************************************************************************************
```

Figure 6 : Example of table of results - Program PC-VENT

Figure 7 : Structure of the database

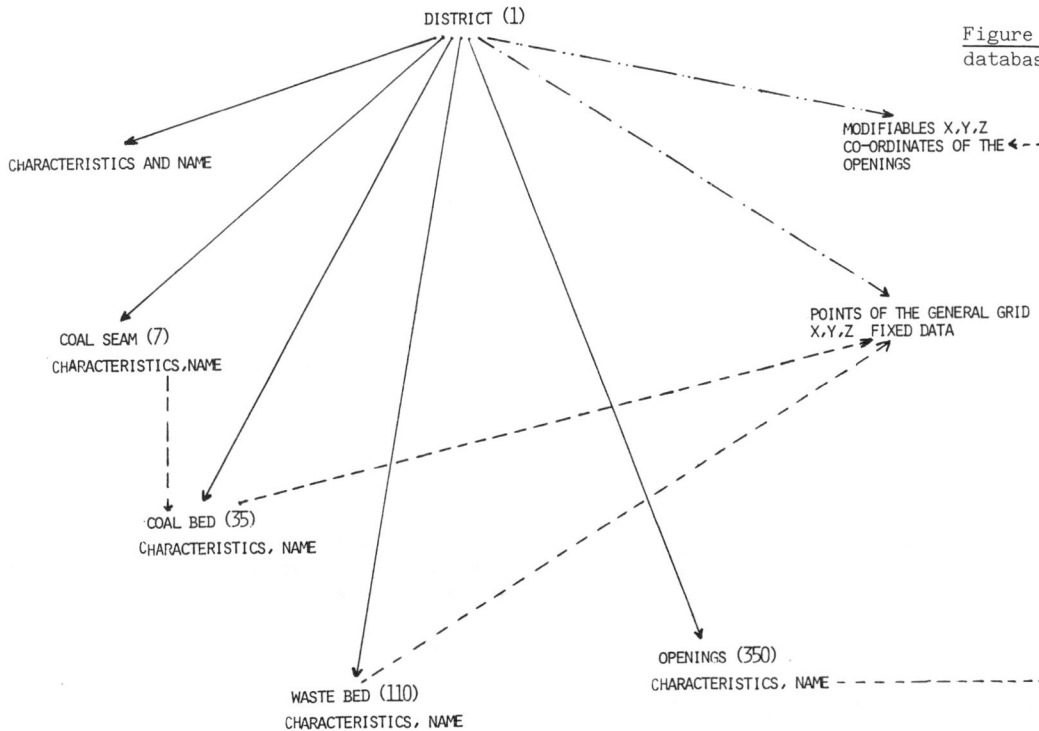

DISTRICT (1)

MODIFIABLES X,Y,Z CO-ORDINATES OF THE OPENINGS

CHARACTERISTICS AND NAME

COAL SEAM (7)
CHARACTERISTICS,NAME

POINTS OF THE GENERAL GRID
X,Y,Z FIXED DATA

COAL BED (35)
CHARACTERISTICS, NAME

WASTE BED (110)
CHARACTERISTICS, NAME

OPENINGS (350)
CHARACTERISTICS, NAME

NOM DE LA COUCHE

ERNA2

ENTREE LE : 17 JANVIER 83

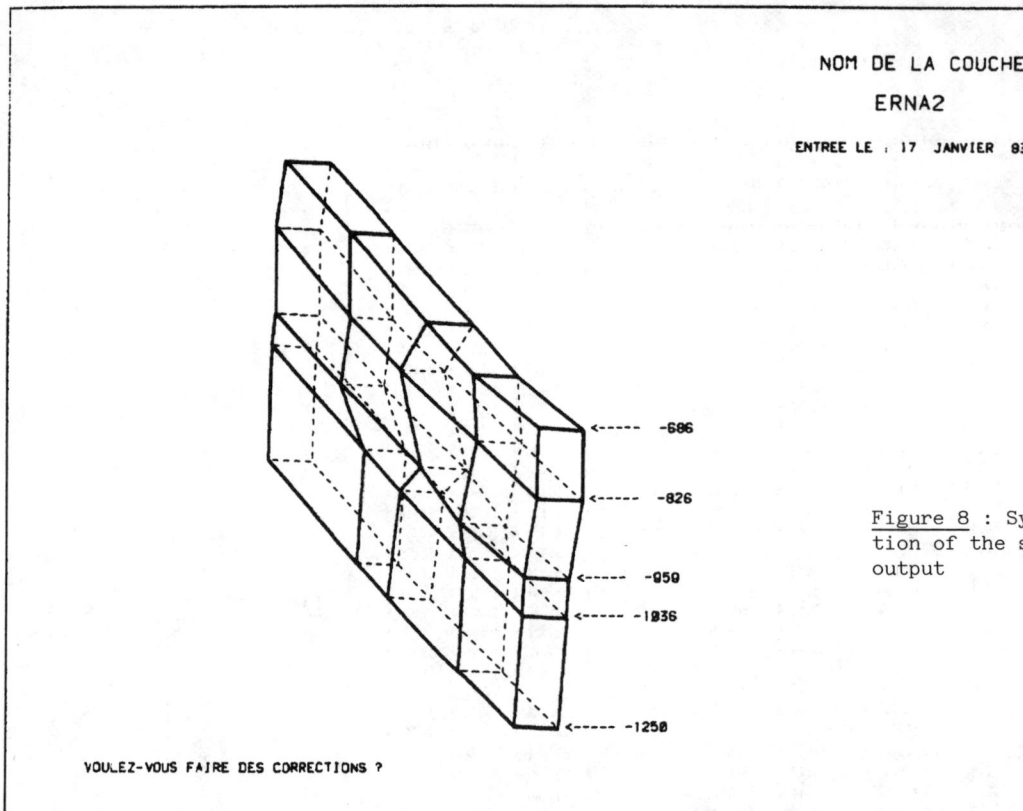

←---- -686

←---- -826

←---- -959

←---- -1936

←---- -1258

VOULEZ-VOUS FAIRE DES CORRECTIONS ?

Figure 8 : Synthetic representation of the seam graphics display output

---

82-MAY-84 11;50;52

PROBLEME 5
TEST

LEGENDE

TAILLES
INFLUENCANTES :

1..CECILE

2..ERNA1

3..ERNA1

OUVRAGE PROJETE :

TEST

OUVRAGE INFLUENCE :

TEST

N PAS   2
PAS   52.50

Figure 9 : Reciprocally inter-acting faces

ECHELLE :

1 cm represente   8.6 m

pour continuer,
pressez (RETURN)

-1845

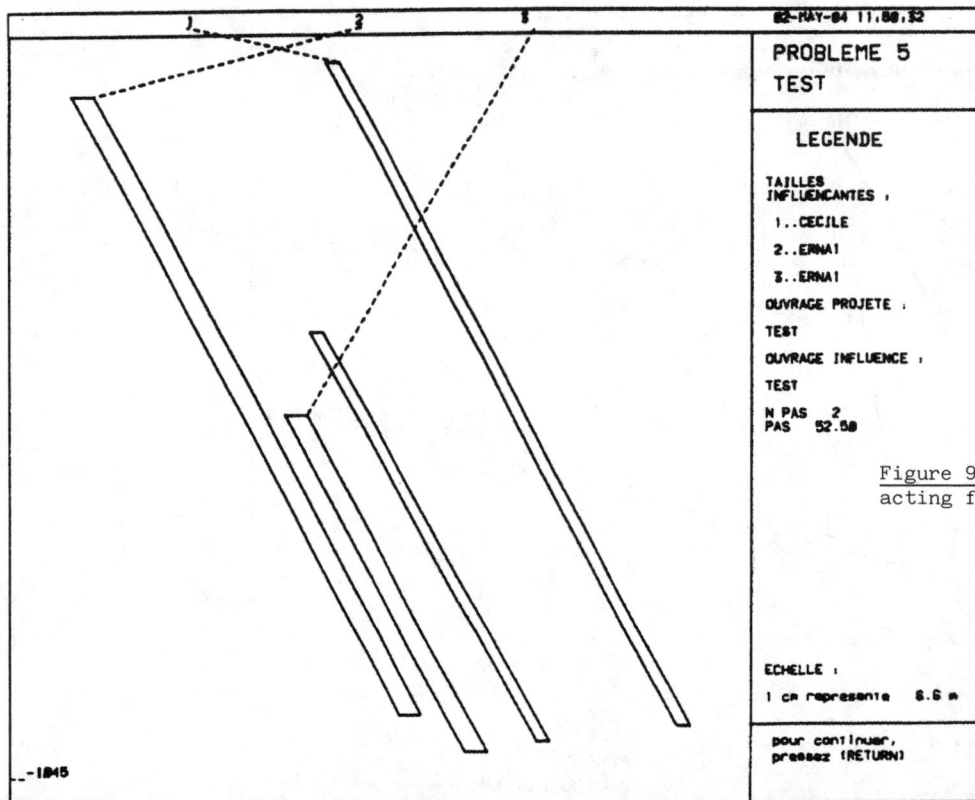

mine operator to choose the optimum working strategy for an ensemble of workings in a sub-ensemble of the mine. The boundary of this sub-ensemble is generally defined by the geology, the homogeneity of the dip angles, the close-ness of groups of seams. Often, it is a compart-ment limited by faults. In any case, good use of these rules is dependant on long calcula-tions which can back off the mine operator who has other exploitation problems. Under these conditions, the problem posed by their systema-tic utilization during the decision making pro-cess leads us to insist upon two points :

- to represent the zones of interinfluence between workings, present and future, by a su-perposition of the plans

- to perform the calculations automatically after having encompassed the consequences of the reciprocal influences.

To respond to this objective, a complete sys-tem has been developed for steep seams. The sys-tem PLAMIMET operates on a Tektronix 4054 A equipped with the following peripherals :

- graphic display
- hard copy thermal printer
- digitizer
- X-Y plotter

This work is integrated in the general stream of a more systematic approach to mining problems and in the development of Computer Aided Design tools.

Having access to the data base for the mine (figure 7) allowing the management of informa-tion (seam characteristics and position) on the geology and old workings (characteristics of the works and position) the operator can ensure the three essential functions of a CAD system :

- loading and updating information in the da-tabase

- restoration of geological sections, plans of works, and plans of superpositions of works in space (figure 8, example of a schematization of a seam)

- calculation of ground pressures at diffe-rent points in the mass submitted to the influ-ence of mining works and choice of the optimum position of future works.

This system, described by Arcamone et.al. 1984, has been improved to predict the produc-tion curve for a group of workings during its lifetime.

This system was developed as part of a pro-gram to provide a global approach to certain mi-ning problems following a progressive approach. Indeed, such an approach provides a much easier integration of very different mining techniques in the planner's decision making process. He be-nefits :

- from a data base unique and complete for the mine

- from the results of calculations, some-times complex, performed in real time, if he wishes, by mini-computer.

The examination of multiple variants for the exploitation project if facilitated.

SOFTWARE AS A MONITORING TOOL

The miner's environment, especially under-ground, is complex because it evolves continu-ously throughout the life of the vein. Among other things, this is due to modifications in the seam and its walls, fluctuations in the cost of marketable products and changes invol-ving the method of exploitation that depend for the most part on the two preceding points. Un-der these conditions, if the forecasting aspect is important, in the short or middle term, the monitoring aspect is even more so between the forecasting stage and the execution stage, it is a common occurence to note a modification of certain predicted constraints.

In particular, this is the case in workings where huge investments are made in the begin-ning for chock shield supports and winning ma-chines. With profitability in mind, one should try to optimize the use of the implements under the actual conditions encountered underground. Thus, the variability of the coal field, roof irregularities or difficulties in controlling goafs often seem to be responsible for bad re-sults obtained during the first advancements of the face. In fact, the origin of these problems may simply be due to certain defects in the hy-draulic system or in an organization not well adapted to the work cycle in the working. In other cases, poor performances can be due to premature wear of the implements creating defor-mations or rupture of supporting elements. In still other cases, it can be due to certain fun-damental parameters for the method of exploita-tion that are poorly adapted, such as the length of the face, the orientation of the face in the coalfield or the treatment of the goafs. The objective of CERCHAR's measuring system at the face "Check up your longwall" is to esta-blish a diagnostic for fundamental parameters that have an effect on the behaviour of working (Poirot, 1984). It gives all the values of the elements that allow one to define immediately the most efficient intervention that can be per-formed on the working to optimize its opera-tion. Thus, for example, in a French mine, a di-agnostic study showed that there was a defect in the regulation of the hydraulic network fee-ding the support. After correction, the produc-tion at the face went from 700t/day to 1.500 t/day.

The implementation of a diagnostic system at the face is composed of three phases.

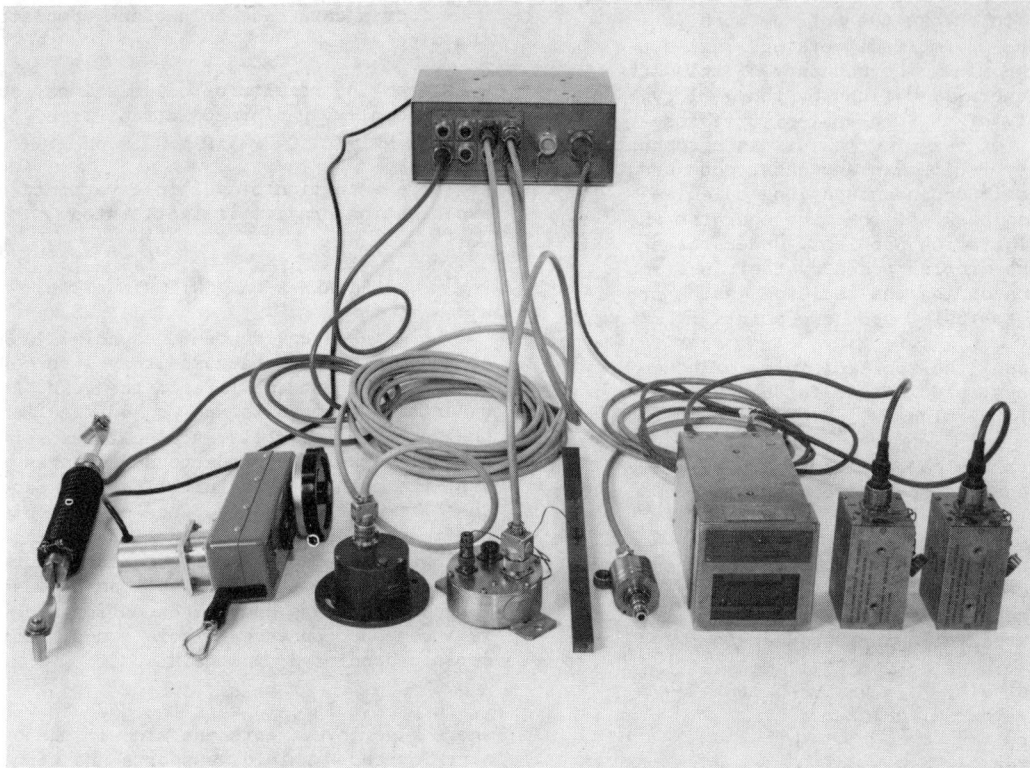

Figure 10 : Devices used by the system "Check up your longwall" (recorders and sensors)

position of the gauges on the upper spacer bar of the long- wall support

position of the gauges on the back of the longwall support

expansion at 4 m in the roof

mm of expansion

Figure 11 : Records on a longwall support in seam 12 B Simon shaft

interval of interrogation

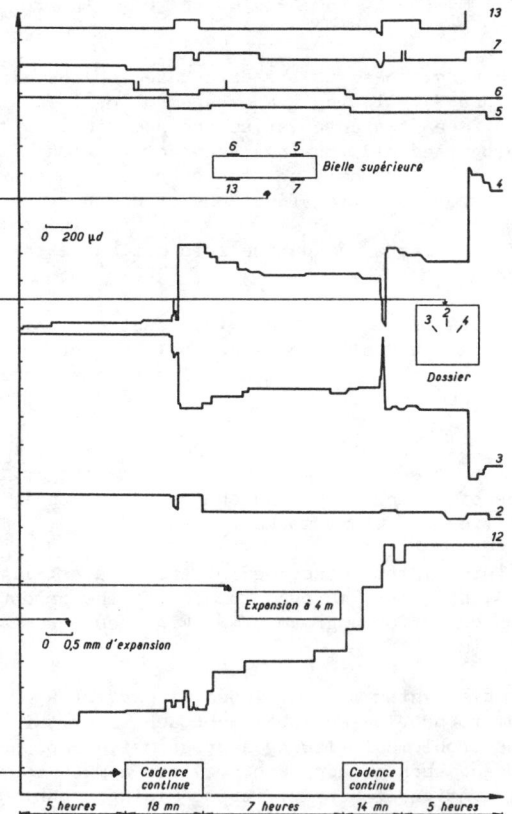

Figure 12 : <u>LOGIC CHAIN OF ROADWAYS DEFORMATION ANALYSIS ON COMPUTERS</u>

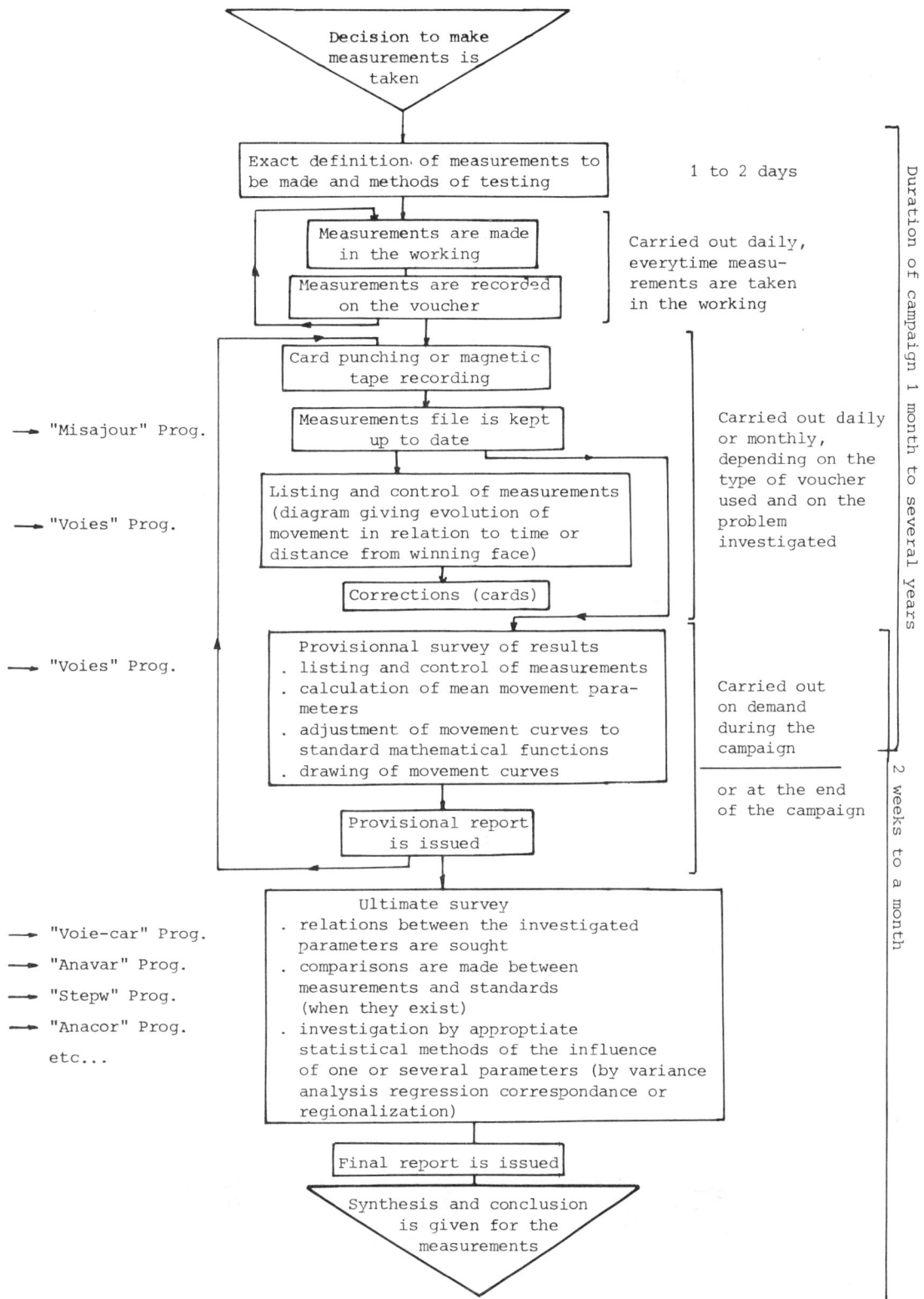

Figure 13 : Flow-chart for the monitoring of the bolted roadways

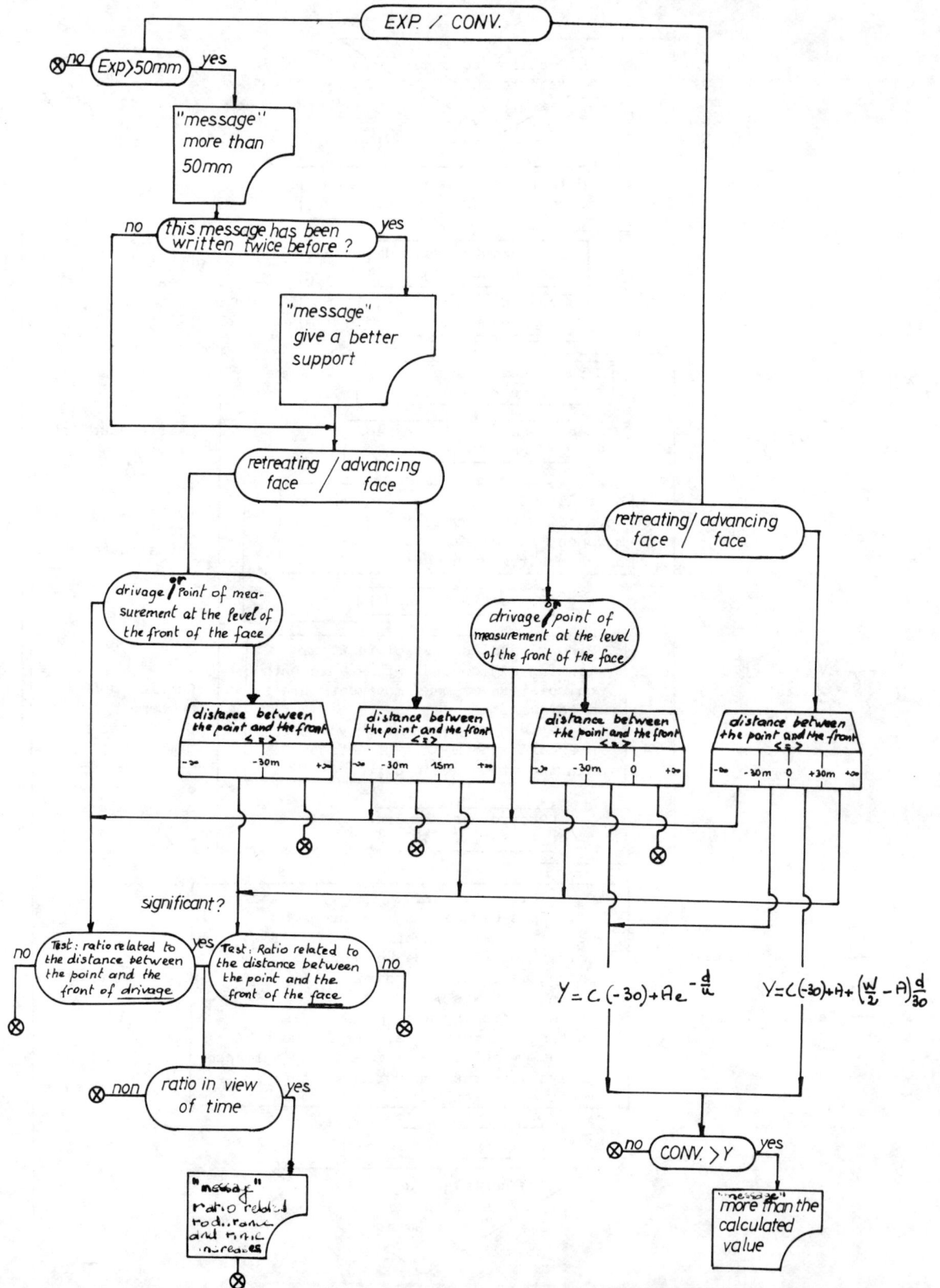

- previous analysis of the problem that goes back to choosing the sensors that will be installed underground and their position on the support and the walls. These sensors are linked to a recording system that stores the information on cassette underground not far from the sensors (figure 10)

- making the measurements which goes back to installing the recording devices and ensuring its operation for the duration of the test campaign

- analyzing the measurements and the diagnostic that is furnished by a chain of programs running on an APPLE II micro-computer or an IBM-PC. This system performs the processing based on the cassettes from underground (an example of the printer output is shown in figure 11).

In many cases, the forecasting tool is based on a group of equations established by statistical analysis.

This analysis corresponds to a sample from several workings and therefore corresponds to a domain of the variation of the parameters studied and a domain of validity for the mode of calculation. This mode gives reliable predictions, in all rigor, only if the parameters fall within the variation domain of the initial sample.

In view of this fact, it is often necessary to verify the good behaviour of the works following a major modification in the method of exploitation that occasionally led to prediction calculations on the limit of validity.

For that, an underground data acquisition and storage "package" was developed on the Bull IRIS 80 computer. The organigram of the different computer programs is shown in figure 12. It permits the restitution of the curves "movement-time" or "movement-distance with respect to a point" for different types of measurements. A tool especially adapted to monitoring and to the diagnostic in real time of bolted roadways initially developed for a Bull IRIS 80 computer now has been adapted to an IBM-PC XT. The diagnostic is based on the organigram presented in figure 13 starting with the case of roadways near the faces.

CONCLUSION

In this article, we have given a summary of the computer programs developed at CERCHAR for personal computers and applied to mining techniques. These programs which have been habitually used at CERCHAR by an advisor to the mine operator for predicting or monitoring, can now be used directly by the mine operator thanks to the great capabilities of current micro-computers. Modifications in the method of exploitation are now possible on these computers in real time. Predictions of their consequences and monitoring the works afterwords, are facilitated by these programs.

The transfer of tools for decision making and for responding to specific technical problems from the research center to the man in the field has started due to micro-computers and associated software but has not yet completely entered into practice at the mines. None the less, the transfer has started if we judge by the papers presented during these latest conferences on "mines and computers" or the appearance of newspapers that take it upon themselves to test commercial softwares destined for the mines in order to help mine operators, who are not specialists, acquire well adapted software. Far from diminishing the jurisdiction of the researcher, this evolution should, on the contrary, allow him to perfect his knowledge if he stays in contact with mine operators equipped with computer implements, thanks to following the consequences of modifications of the method of exploitation better. This improvement in knowledge should bring an improvement in prediction tools.

REFERENCES

Dejean, M. JP., and Raffoux, J.F., January 1982, "Planning and data processing in collieries operations", Revue de l'Industrie Minérale.

Ghoreychi, M., 1983, "Stabilité d'exploitations partielles avec piliers en état post-rupture", thèse docteur-ingénieur INPL, Nancy-France.

Dejean, M. JP., Piguet, J.P., and Raffoux, J.F., September 1983, "Rock bolting in France", International Symposium on Rock Bolting, Abisko-Sweden.

Arcamone, J. A., 1980, "Méthodologie d'étude des affaissements miniers en exploitation totale et partielle", thèse docteur-ingénieur INPL, Nancy-France.

Froger, C., et al., April 1976, "Calcul des réseaux d'aérage", Numéro spécial Aérage, Revue de l'Industrie Minérale.

d'Albrand, N., Gunther, J.P., and Josien, J.P., 1985, "Interactive graphic display of ventilation networks", Tuscalosa-Alabama.

Arcamone, J. A., Dejean, M., and Sayetta, C., March 1984, "PLAMIMET a CAD system to improve mining design with Strata Control", APCOM, 18th Symposium, London-England.

Poirot, R., October 1984, "Check up your longwall, a system developed by CERCHAR", Internal Report.

Chapter 14

COMPUTER-BASED INTEGRATED MINE MANAGEMENT SYSTEMS

- THE SHOP FLOOR APPROACH

by Alan J. McElrea[1] and Des P. Scherman[2]

ABSTRACT

The cost of performing equipment maintenance is an increasingly significant component in the total operating cost of today's capital-intensive, large-scale mining operations. Because of the inter-dependence of equipment maintenance on other functional areas of the minesite, an integrated approach to the monitoring and control of maintenance costs is essential. An important element of the introduction of computer-based systems in a mine is the acceptance of the systems by the users. The development history and major features of a minesite-developed system, which emphasises the user's importance, is described. The system outlined is designed to provide information for decision-making to the place where the company's maintenance money is actually spent - on the shop floor.

1.   Manager, Commercial Systems, Mincom Pty. Ltd., Brisbane, Australia
2.   Mine Management Consultant, Mincom Pty. Ltd., Brisbane, Australia

INTRODUCTION

In recent years, the costs of performing maintenance on mining equipment has become an increasingly significant component in the total cost of operating a mine. With depressed world markets for coal and minerals, the international competitiveness of a mining company is determined by its ability to contain operating costs. Companies are now directing their attention towards close monitoring and control of maintenance and associated operational areas such as materials supply and inventory management, with a view to containing costs.

Because equipment maintenance impacts on several departments in the operating mine - workshops, maintenance planning, purchasing, warehousing, accounting and mine planning - and since it consumes the major resources of labour, materials and equipment, it is apparent that an effective system for managing maintenance requires integration of information from a wide range of functional areas. Such a task is ideally suited to the power of modern computers.

THE ROLE OF EQUIPMENT IN MODERN MINING

In the past 15 years, major increases in the output of coal and minerals have occurred throughout the world. This has been achieved by the use of large-scale equipment in open cut mines and strip mines in countries such as USA, Canada, South Africa and Australia.

For example, the saleable coal output of the State of Queensland, Australia in 1968 comprised 2.7 million tonnes from open cut mines and 2.9 million tonnes from underground mines.[1] By 1983, underground output had risen to 3.82 million tonnes, whereas open cut production leapt to 32.0 million tonnes - a compound growth rate of about 20% per annum. This growth was achieved by the investment of hundreds of millions of dollars in mining equipment - there are currently, for example, more than 40 large walking draglines (40 cubic metres or over) in Queensland.

The introduction of such equipment resulted in major increases in labour productivity - between 1968 and 1970, the output per manshift for open cut mines increased from 17.88 tonnes to 29.15 tonnes! Since then, however, productivity has steadily declined to about 19 tonnes per manshift. A similar trend in labour productivity is apparent elsewhere in the world.

Whilst many factors have contributed to the decline in labour productivity - including labour attitudes and increasing overburden depths - a major factor is the steadily increasing age of the mining equipment. As equipment gets older, its reliability generally decreases and its maintenance costs increase.

The need for companies to maintain and improve their international competitiveness requires that the slide in labour productivity must be halted. There are two avenues open to mining companies - either invest large amounts of capital in new and improved equipment (an unlikely choice in times of depressed markets) or improve the reliability and productivity of existing equipment by improvements in the maintenance of that equipment.

In Australian open cut mines, the maintenace manpower is typically 30% to 40% of the minesite workforce and the operating costs for maintenance labour and materials is typically 50% to 60% of the mine operating costs and rising.

Thus, maintenance is a significant element in the mine's cost structure and, as such, represents an area where management can focus attention on controlling costs and improving labour productivity.

The maintenance function in mines is affected not only by the economic forces of the international market place, but also by the harsh environmental conditions prevalent at many minesites.

The remoteness, temperature extremes, and dust affect the minesite personnel as well as the equipment. It is generally difficult to recruit and retain skilled maintenace personnel. Against this background, it is apparent that management efforts to control maintenance costs must take into account the diverse skill levels of the maintenace personnel and the labour turnover.

THE NEED FOR INTEGRATED SYSTEMS

The maintenance function in a mining operation does not exist in isolation. Indeed, the maintenance department has everyday interactions with the warehouse and purchasing departments, the accounting department, the personnel department and the mining department.

An integrated system enables these interactions to be addressed at the minesite by the people who need the information to make the day-to-day decisions.

The capability of such a broadly-based system is reflected in the range of management information it can provide in response to questions such as:

- What is the availability and productivity of equipment?

- What is the frequency and causes of equipment downtime?

- Is a failure pattern developing in a particular equipment type, or equipment component?

- What planned maintenance activities must be performed in future periods?

- What are the manpower and material requirements for those maintenance tasks?

- Can these requirements be met by the current workforce? Are necessary parts on hand?

- How are costs going against budgets? Against jobs or projects?

- What are the costs of operating a particular piece or class of equipment?

- Should we scrap, replace or upgrade certain pieces of equipment?

- What items of inventory will become obsolete through changes in equipment? What items are already obsolete?

- What is the impact of a planned major shutdown on provisioning and manpower schedules?

- Could we be wasting good oil as a result of our oil changeout schedules?

- What did we mine yesterday? Which equipment did we use to mine it?

- What are current stockpile volumes?

A systematic and integrated computer-based approach can improve equipment maintenance by:

(i) improved utilisation of labour resources (by allocating the right people to the right tasks at the right time; by defining those tasks etc.);

(ii) reducing the inventories of spare parts and materials whilst maintaining or improving the service efficiency of the warehouse;

(iii) reducing shop floor problems - removing the panic operation mode and personnel frustrations, eg. for a 3000 hour service on a coal hauler, when the job order is initiated, the stores inventory for all parts would be checked to see if everything needed was available, the parts and materials would be transferred to a "job bin" and delivered to the shop floor prior to job commencement;

(iv) providing dynamic scheduling for more accurate scheduling and job continuity;

(v) providing history files which can be used to:

- analyse equipment performance.

- design modifications to equipment.

- design and specify new equipment.

- store data for estimating upcoming jobs and for future budget projections.

- provide continuity of equipment experience, maintenance standards and work procedures despite discontinuities in maintenace personnel as a result of labour turnover.

- accurately report the costs of owning and operating a piece of equipment through its life cycle, a key element in the decision of when to replace items.

Such an integrated approach to maintenance and inventory control was identified by the Keynote Speaker[2] at the First Conference on Computers in the Coal Industry as an area with potential high benefit to cost ratios for coal companies.

THE HUMAN ELEMENT

THE MINCOM INTEGRATED MINE MANAGEMENT SYSTEMS

Development History

The major factor in the acceptance of computer-based systems as a management tool is the ease with which the people who need to use the system can obtain the information they desire. Computer systems operate well when people understand how they can use them effectively and then want to use them.

This aim is achieved by having the users heavily involved in the design of the systems. A frequent shortcoming in the design of maintenance systems has been the failure to recognise the importance of the minesite maintenance personnel.

Tomlingson (1983)[3], a US maintenance consultant, states that "the axiom that the experts in maintenance management are in Maintenance, not data-processing, has been overlooked".

Stewart (1983)[4], an Australian personnel manager, also recognises this aspect - "all ability in analysis and synthesis is not encoded in the mid brain of analyst/programmers from computer departments. There is a host of clear thinking people out in operating departments who do have these skills in abundance and have operating experience to back them up".

The people in the workshop and in the warehouse know why specific information is needed, when and in what form. Logically, these people have all the information necessary to design a system so that it produces exactly what is needed to allow the maintenance department to perform more efficiently.

As noted by Dr Weiss[2] in his keynote address to the First Conference on Computers in the Coal Industry - "To implement multisource systems, management of the human side of this business within an organisation framework is at least as important, if not more so, than all the sophisticated hardware and software".

The systems described in this paper are based on systems originally developed by Comalco Aluminium Limited at the Weipa bauxite mine on Queensland Cape York Peninsula, where they have been successfully used for a number of years. The systems have recently been upgraded, expanded and enhanced by Mincom Pty. Ltd. and have been released to mining companies throughout the world. The systems are operating or being implemented on more than 10 sites in Australia.

Melvin (1983)[5] describes the development of the Weipa systems as a series of steps comprising:

(i)   recognition that the existing, manual paper-based system was rarely used as a reference and was becoming more complex as the mine grew;

(ii)   a drive from the maintenance department for computerisation of the system;

(iii)   the appointment of an independent (outside) group to design the system;

(iv)   the (eventual) production of a result;

(v)   recognition by the maintenance department that the result could not solve its problems;

(vi)   abandonment of the project;

(vii) later recognition that a computer-based system was still needed;

(viii) the assignment of systems design responsibility to the maintenance people themselves, and the appointment of computer systems designer to assist the group;

(ix)   the creation of an acceptable system for maintenance scheduling within 3 months.

In the words of Melvin (1983)[5], "This resulted in a system that they understood very well and were extremely keen to use. It has been very successful for the entire time that it has been in operation".

MINCOM　　INTEGRATED MINE MANAGEMENT SYSTEMS

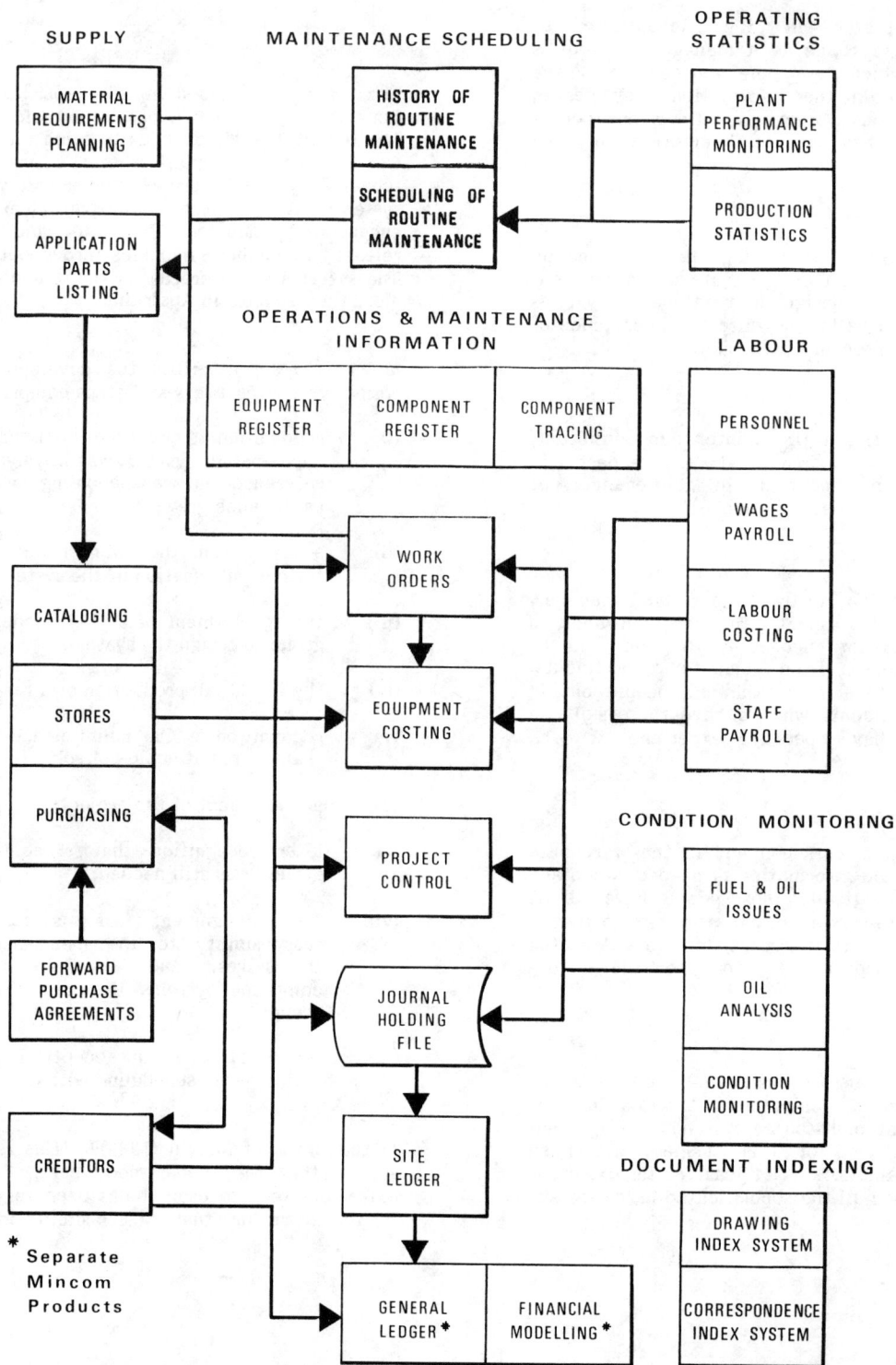

SUPPLY　　　MAINTENANCE SCHEDULING　　OPERATING STATISTICS

MATERIAL REQUIREMENTS PLANNING

HISTORY OF ROUTINE MAINTENANCE

SCHEDULING OF ROUTINE MAINTENANCE

PLANT PERFORMANCE MONITORING

PRODUCTION STATISTICS

APPLICATION PARTS LISTING

OPERATIONS & MAINTENANCE INFORMATION

LABOUR

EQUIPMENT REGISTER

COMPONENT REGISTER

COMPONENT TRACING

PERSONNEL

WAGES PAYROLL

CATALOGING

STORES

PURCHASING

WORK ORDERS

EQUIPMENT COSTING

LABOUR COSTING

STAFF PAYROLL

PROJECT CONTROL

CONDITION MONITORING

FORWARD PURCHASE AGREEMENTS

JOURNAL HOLDING FILE

FUEL & OIL ISSUES

OIL ANALYSIS

CONDITION MONITORING

CREDITORS

SITE LEDGER

DOCUMENT INDEXING

DRAWING INDEX SYSTEM

* Separate Mincom Products

GENERAL LEDGER *

FINANCIAL MODELLING *

CORRESPONDENCE INDEX SYSTEM

FIGURE 1

A similar approach was taken with the existing trouble-plagued supply system and the method of giving responsibility for the design and supervision of computer-based systems to the end-users, and having data processing people assigned as assistants and advisers to the users, became the philosophy for the development of all Weipa's data processing systems.

For example, on the development of the Component Tracing and the Maintenance Scheduling Systems, the project manager was a person who had commenced at the mine as a Mechanical Tradesman.

Melvin (1983)[5] states that this approach "proved to be very successful in that the users get what they want - a system that they understand and are very happy to use".

In 1982, Mincom acquired the rights to expand, develop and market these systems. The systems have been generalised to ensure their applicability to any mining operation. The resultant systems are therefore a combination of years of shop floor experience at a remote, highly mechanised minesite and Mincom's extensive experience in implementation, support and use of computer-based methods in the mining industry.

Systems Overview

The Integrated Mine Management Systems are a set of modular computer packages which together comprise a comprehensive methodology for control of production, maintenance and materials management activities, and for collecting, processing and displaying day-to-day mine management information.

The information is available throughout the workshops, warehouse and office on visual display units, and the system collects data at a high level of detail, allowing useful reporting to all levels of management from the shop floor to the General Manager.

The major modules of the system are:

- Operations and Maintenance Information
- Maintenance Scheduling
- Operating Statistics
- Fuel and Oil
- Supply
- Labour

The modules are integrated, as shown diagrammatically in Figure 1, so that information maintained by any one application area is accessible as required by all other applications. The systems are not only integrated among themselves but also provide for data interface with corporate financial and management information systems.

OPERATIONS AND MAINTENANCE INFORMATION

This system is a collection of sub-systems which together provide comprehensive facilities for recording and reporting information on significant operations and maintenance activity at the minesite. The sub-systems include:

- #### Equipment and Component Registers

  Many applications of the Mine Management Systems are related to the operation and maintenance of mine equipment. Identification of equipment and components is therefore vital to the total system and is the major role of the Equipment and Component Registers.

  The Component Register provides formats for various types of components, so that static information peculiar to that type can be recorded. Only registered components can be subject to component tracing.

- #### Work Orders

  The main purpose of the Work Order sub-system is to record the operational circumstances and costs associated with a job. The facility provides a comprehensive tool useful in both operations and maintenance areas.

  A job is identified to the system as one of several types. For example, work orders can be created for:

  - one-off breakdown maintenance.

  - standing work order to cover many small jobs of the same type, without costing individual jobs.

  - standing work orders for collecting the period costs of scheduled maintenance work.

  - project work orders for jobs that are part of a project in the Project Costing sub-system.

- training work orders to accumulate individuals' time and cost for on-the-job training.

The system provides powerful recording, reporting and analysis mechanisms, using both coded and narrative information.

A series of codes are available to comprehensively categorise a job. In addition, extensive English narrative can be entered by completing the work order to permanently record facts about the job.

· Component Tracing

Information within the Operations & Maintenance Information System is categorised by the combination of equipment number, component code and modifier code. This combination identifies portions of machinery, but does not identify the individual component presently fitted. Component Tracing provides an additional facility for tracing individual rotable spares through their life cycle of purchase, fitment, removal and rebuild.

The information held in the Supply System and Work Order files for each of these documents is used to present a comprehensive picture of the operating and maintenance history of individual components.

· Equipment Costing

An important element of mine-site expenditure is the full cost of owning, operating and maintaining equipment. To provide detailed information on this cost element, the Equipment Costing facility combines work order costs related to an equipment item with costs which have been charged directly to that item to present a comprehensive cost picture for each individual piece of equipment.

Costs are reported based on normal accounting periods, and include:

- routine work orders for each equipment number.

- standing work orders opened to cover multiple occurrences of the same job or a series of small jobs on several equipment numbers (ie. all light bulb replacements on a fleet of trucks). The system segregates costs for the period for each equipment number appearing on these work orders. This feature is convenient for grouping small jobs which are not important enough for individual work orders, while still allowing costs to be assigned to individual equipment numbers for equipment costing purposes.

- fuel and oil costs from transactions supplied by the Fuel & Oil System.

- items charged directly to equipment numbers from individual transactions that are not part of a work order.

· Project Control

This facility is used to control construction and similar projects falling in either capital or operating cost categories. The system supports the engineer in controlling costs against estimates and provides appropriate accumulations for transfer to fixed asset accounts based on taxation units of property. It also supports planning and cost control for large infrequent operating jobs such as major plant shutdowns.

Project Control also provides for reporting committed costs as well as actual costs, when orders for goods and services for projects are placed through the Mine Management Supply System.

· Site Ledger

The Site Ledger facility is used primarily for preparation of minesite operating budgets and reporting of actual costs against budgets for use by maintenance and operating personnel. The provision of a Site Ledger devoted to onsite mine costs separate from the company's general ledger provides a number of distinct advantages. Generally, operations and maintenance personnel require accounts at a detailed level in order to properly classify and control costs. Such information needs to be integrated with job costing information from the work order system, equipment costing information and operating statistics for variance analysis. At this more detailed level, site ledger information is an important segment of the mine manager's total information requirements.

On the other hand, the general ledger of the company is oriented towards statutory and accounting requirements, at a higher summary level. The general ledger also includes additional complexities of foreign exchange and revenue accounting not required by the minesite.

The Mine Management System therefore provides a site ledger geared to the budget and costing information requirements of minesite management, and capable of transmitting any level of summary information to the corporate general ledger.

MAINTENANCE SCHEDULING SYSTEM

The Maintenance Scheduling System is devoted to two functions:

- ### Schedules of Preventative Maintenance

  Provides lists of recurring maintenance tasks to be performed on specific equipment and components by individual maintenance groups working in rostered shifts. The system can schedule tasks based on either elapsed time or operating hours and can be extended to incorporate other scheduling criteria.

- ### History of Preventative Maintenance

  Reports history of scheduled and performed occurrences of selected tasks. This history can include measurements, readings, English comments and references to related work orders opened as a consequence of the routine maintenance.

The objective of the maintenance scheduling facility is to perform those routine clerical tasks necessary to remind maintenance personnel when recurring tasks need to be done. This approach encourages easy acceptance of the system by maintenance staff who receive immediate benefit from system use.

Computerised recording of routine maintenance procedures reduces the impact of personnel turnover, and provides a basis for implementation of more sophisticated planning procedures. The system is designed to be totally controlled and directly operated by maintenance personnel.

OPERATING STATISTICS SYSTEM

- ### Plant Performance Monitoring

  Performance statistics on individual pieces of plant and other productive units are an important segment of the total information requirements of the minesite. Not only do such statistics highlight problem areas of low availability and poor utilisation, they can also indicate areas of excessive availability, possibly caused by over maintenance.

- ### Production Statistics

  A wide range of production statistics can be maintained and displayed by the system, such as:

  - haul truck trips made from particular mine blocks or pits to particular dumping points, the average speed, distance travelled, wet and dry tonnes hauled.

  - loader usage in terms of the number of trucks loaded, wet and dry tonnes loaded.

  - process plant production by number of wet and dry tonnes input and produced, water consumption, primary screen hours.

  - shiploader statistics on wet and dry tonnes of each product loaded per weightometer for each ship, time taken to load, ship draft survey results.

CONDITION MONITORING

- #### Fuel & Oil Issues

  Recording issues of fuel and oil which are costed and passed to the costing sub-systems for permanent reporting of consumption rates and costs. This facility is treated as part of condition monitoring because certain information is germane to the oil analysis module.

- #### Oil Analysis

  Monitoring the condition of machine lubricating oil through the recording and reporting of oil sample laboratory test results.

  Monitoring oil performance has large savings potential, both through prevention of catastrophic failures which are signalled by changes in oil composition, and by facilitating extended oil use.

  The system retains analytical results of oil samples taken from individual compartments. It also maintains additional information to assist in interpretation, including:

  - machine, compartment and oil operating hours.

  - top-up litres.

  - narrative comment or warnings issued by the oil chemist in reviewing analysis results.

  - narrative responses to warnings from maintenance personnel.

- #### Condition Monitoring

  This module provides for holding and reporting all other condition monitoring information. This could include:

  - shock pulse monitoring
  - vibration analysis
  - load box test
  - wear thickness measurements.

SUPPLY SYSTEM

The Mine Management Supply System includes seven interrelated functions:

- #### Cataloguing

  Cataloguing is a mechanism within the Mine Management Supply System which provides for disciplined identification and description of material used in the operation of the mine. Each item of material is identified by a unique stock code assigned by the system, but can also be fully cross-referenced by colloquial names and manufacturers' part numbers, terms more commonly used by maintenance personnel.

  Online system access is provided by:

  - part number
  - colloquial name
  - stock code
  - application part listing (APL) number.

- #### Stores

  The Stores sub-system of the Mine Management Supply System provides for the accounting and control of all items held as inventory.

  The main functional areas supported are:

  - receipt of items into inventory, either at a warehouse or offsite receiving point.

  - controlled issue of material to authorised requestors.

  - control of the quantities of stock on hand and their physical location within the warehouse.

  - performance of cyclical or periodic stock audits, in the form of physical inspection and count, and technical assessment.

  - provision of detailed and summarised information for the management, control and analysis of inventory.

  The system provides for multiple warehouses, and multiple locations for each stock code.

- ### Purchasing

The Purchasing function of the Mine Management Supply System, provides for the ordering of all material required for the operation.

An important feature of the purchasing facility allows inventory controllers and purchasing officers to preselect those items which they wish to review before purchasing, thus allowing the system to automatically provision those items which do not require their attention.

- ### Forward Purchase Agreements

The Forward Purchase Agreement (FPA) facility is designed to support the negotiation of contract for the future purchase of material required for the operation.

The negotiation of such contracts has a number of advantages:

- negotiation based on annual usage rather than single order volumes can effect lower prices.

- purchasing information remains stable for a set period, allowing the system to automatically purchase a large portion of the material required, thus freeing purchasing officers to spend more time in negotiation, expediting and follow-up.

- alternate part numbers and alternate sources are often volunteered by suppliers for review by supply management, providing more complete information on purchasing options.

The system also supports the preparation of tender documents, evaluation of quotes, notification of successful suppliers, and updating of Supply System files with new purchasing information.

- ### Creditors

The Creditors sub-system of the Mine Management Supply System provides for recording and payment of monies owing to creditors.

The system caters for the approval and payment of liabilities incurred both for all ordered goods and services, and for items for which specific orders are not placed (rent, fees, etc.).

- ### Application Parts Listing

The Application Parts Listing (APL) sub-system of the Mine Management Supply System provides for identification of material (ie. parts) required for a particular maintenance job or making up a particular piece of equipment, as the basis for subsequent requisitioning.

The sub-system serves an important role as a major communication link between the maintenance department as users of material and the supply department as providers of that material. APL's are also an integral part of the cataloguing function, providing vital fitment (where used) information and the ability to identify obsolete stock items.

- ### Material Requirements Planning

This module brings together all of the information contained in the maintenance planning system and the application parts listing system to plan material requirements for the Supply system, to ensure material for planned jobs is procured and is available on time.

LABOUR SYSTEM

The Mine Management Labour System provides three basic facilities:

- ### Wages and Staff Payroll

Detailed calculation and reporting of employees' pay. Pay envelopes can include a report of the employees' earnings at the timesheet-entry level of detail.

•   Labour Costing

Extraction of costing data from timesheet entries. Labour costs may be charged to work orders or directly to individual pieces of equipment.

•   Personnel

Maintenance and reporting of employment history, skills inventory, leave entitlements and absences.

DOCUMENT INDEXING

Two systems are provided to maintain access to documents related to the operation.

•   Drawing Index System

This system provides a comprehensive means of indexing drawings and technical information in regard to the plant. Using the same codes as in the other maintenance systems, the operator can retrieve the location references to drawings and other documents for areas of plant, equipment numbers, construction contracts and other criteria.

The system allows for multiple locations for documents, and maintains information on the form (print, operation card, microfilm) and version number maintained at each location. Companies can therefore have a number of drawing offices working on the same base set of drawings and documentation without danger of updating out of date versions.

The documents registered can be cross-referenced to Application Parts Listing in the Supply System to link drawing and bill of material information.

•   Correspondence Index System

This system provides the same sort of multiple accesses to locate letters, telexes and memoranda, held in central archives. It can be linked to microfiche processing to do away with paper files.

CONCLUSION

The rising costs of equipment maintenance can be monitored and controlled by the use of computer-based management systems which address the needs of the users. The rapid availability, upon request, of detailed and accurate information to workshop foremen, stores clerks, tradesmen and purchasing offices is a key element in enabling these people to make informed decisions on what course of action to take in a given set of circumstances.

The information is available to the front line decision-makers at the place the company's money is actually spent - on the shop floor and in the warehouse. It is here that control of maintenance costs can be achieved and the company's competitiveness maintained and improved.

REFERENCES

1.   Queensland Coal Board - Annual Reports 1978-1982.

2.   Weiss, A. (1983), "The Human Element - Key to Profitable Computer Applications in Mining", keynote address to the First Conference on Computers in the Coal Industry, Society of Mining Engineers of A.I.M.E.

3.   Tomlingson, Paul D. (1983), "Computers for Maintenance", Coal Age, July 1983, pp 86-92.

4.   Stewart, Karl W. (1983), "Management of Computer Systems Implementation within an Operating Mine", keynote address to the 'Computers in Mining - 1983' symposium, Aus. I.M.M.

5.   Melvin, G.J. (1983) "Computerised Maintenance Systems at Weipa", in 'Computers in Mining - 1983' symposium, Aus. I.M.M.

MINE MONITORING

Chapter 15

COSTS AND BENEFITS OF MINE MONITORING

Jeffrey H. Welsh

Pittsburgh Research Center
Bureau of Mines
U. S. Department of the Interior
Pittsburgh, Pennsylvania

Abstract. The Bureau of Mines, U.S. Department of the Interior, performed cost/benefit analyses for the computerized monitoring systems installed at two U.S. coal mines. One system monitors and controls the belt conveyors, the second monitors the belt conveyors and environmental parameters. Actual costs and benefits were used in the analysis when available. Both systems proved to be beneficial to the mine.

INTRODUCTION

Computerized monitoring is becoming common in the mining industry. There are currently 42 systems installed in U.S. underground coal mines and 8 in underground metal and nonmetal mines. Twelve monitoring system manufacturers are actively pursuing the market. Even though monitoring systems have been installed in United States mines for several years, records of costs and benefits of these systems are not available. This type of information would be extremely useful to mining companies that may be considering the installation of a system.

The Bureau currently has a cooperative agreement with two mines and has been collecting data from them to determine the in-mine cost-and-benefit performance of monitoring systems. The monitoring system at each mine is different and the mines are different; that is, they have different methods of mining, are at different stages of development, etc. Therefore, both costs and benefits vary for each system. All systems have recently been installed, so data collection is still in progress. Costs and benefits associated with each system are identified. The data collected are presented through a 5-year net present value (NPV) cash flow analysis.

COST AND BENEFITS OF IN-MINE SYSTEMS

In this section, a description of the mines and monitoring systems are given, costs and benefits are identified, then results of an investment analysis are discussed. The investment analysis presents the yearly incremental effects on each of the mines' balance sheets resulting from the acquisition and installation of a monitoring system.[1] Where possible, actual costs and data are used. In some cases, estimates were made since both systems are recent installations and complete data were not available. Each investment is evaluated over a 5-year timeframe based on the assumption that the life span of the monitoring systems is 5 years. It is also assumed that the mines are steady state, i.e., the current level of production will remain the same and there will be a market for the coal produced. For the investment analysis, the discount rate is 20%, investment tax credit 10%, tax bracket 48%, and depletion allowance 10%. Straight line depreciation is used. Specific mine names are not identified but are designated as A and B. A summary of the investment analysis for these two mines is shown in table 1.

<u>Table No. 1</u>

SUMMARY OF INVESTMENT ANALYSIS

Mine	Mine size (Ktons/ year)	Mining method	Parameters monitored	Initial system cost ($)	NVP_{20} ($)
A	500†	Cont.	Belt and environmental	272,900	335,124
B	300	Cont. and longwall	Belt	628,523	1,385,658

† Mine A is 500,000 tons per year but an additional 800,000 tons per year from an adjoining mine is carried by Mine A's belts.

Mine A

This mine is classified at the upper end of the scale for small mines with an annual production of approximately 500,000 tons. It is located in the

[1]Bolt Beranek and Newman, Guidelines for Environmental Monitoring in Underground Mines (contract J0100039). Bureau of Mines OFR 180-82, 1982, NTIS PB 83-147777.

eastern United States. There are five working
sections, but only four operate at any given
time. Maintenance is performed on the idle
section. The mine has three production shifts
per day, 5 days per week. The coal seam is about
3.3 ft thick and ranges from 100 to 250 ft below
the surface. Coal is extracted by the
room-and-pillar continuous miner method and
hauled by conveyor belts. A simplified mine map
is shown in figure 1. There are three sections
off the north mains and two off the south mains.
The coal (800,000 tons per year) from an
adjoining mine enters at the end of the south
mains. This general mine plan is expected to
continue for the next 5 years.

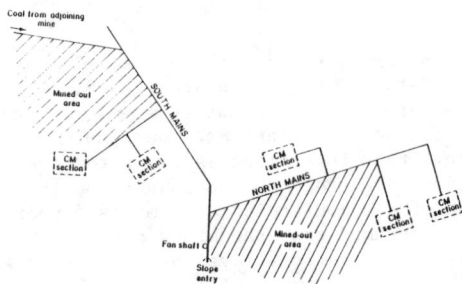

FIGURE 1. - MINE A MAP

The system consists of a central computer,
transducers placed in the mine, and cable
connecting the computer to the transducers.
Transducers monitor CO, CH_4, air velocity, and
conveyor belt status. Currently, 12 each of CO,
CH_4 and air velocity transducers are installed.
The environmental transducers are currently
operating in intake air; however, approval was
recently granted by the State mine regulatory
agency for operation in return air. Five belt
monitoring stations are currently being
monitored, and near future plans call for a total
of 14 belts to be monitored. The analysis is
performed for the monitoring of all 14 belts.
Parameters monitored for each belt are power
on/off, pull cord, fire sensor, overload breaker,
slip detector, plugged chute switch, sequence
switch, controlled stop, and bearing overheat.

Benefits. The benefit produced by this system is a
decrease in downtime of the conveyor belts, which
allows for more coal production. During a
continuous 2-month period at this mine, 27.3
shift-hours were lost because of belt problems.
The maintenance foreman also indicated that 65%
of this downtime was spent in determining the
source of the problems. Monitoring belt
parameters can eliminate the downtime associated
with determining the source of belt failure.
Using a coal sales price of $30 per ton, gross
revenues that result from belt monitoring are

$369,686 per year. Calculations are shown in
appendix A. Since the environmental monitoring
has not been fully utilized to date, its benefits
are not considered. Even when the environmental
transducers are placed in returns, the benefits
associated with environmental monitoring will be
difficult to quantify.

Acquisition Costs. Initial acquisition costs
associated with this system include the
following:

Central station........................	$42,900
Color graphics.........................	25,600
Power backup...........................	6,600
Cable (24,000 ft)......................	43,000
Environmental transducers	
(12 each, CO, CH_4, AV)...............	52,800
Belt stations (14 @ $2000 each)........	28,000
Spares*	8,700
Total	$207,600

*15% spares of outstation circuit boards, CO
and CH_4 sensor elements, and air velocity
transducers.

Installation Costs. Most of the installation
including cable and transducers was done by a
mine service company. Installation costs are
$65,300.

Operating Costs. This mine currently does not
have a dedicated system operator, but to fully
benefit from the system, the system would have to
be attended during production. Using a salary
for a system operator of $15,000 per year and a
multiplication factor of 2 to determine total
costs to the mine for this employee, operation
costs = 3 shifts x $15,000 per year x 2 = $90,000
per year.

Maintenance and Repair. Maintenance and repair
consist of two parts, a maintenance engineer's
time and equipment repair and replacement. A
full-time maintenance engineer is necessary to
keep the system operational. This engineer's
salary is $30,000, and a multiplication factor of
2 is used to determine total costs to the mine.
The mine currently does not have such a person
but the cost was included. Annual parts
replacement involves underground sensors and
circuit boards.

In-mine experience indicates the following
mean life of system components: CO sensor
elements, 2 years; CH_4 sensor elements, 1 year;
air velocity transducers, 2 years; belt station
circuit boards, 5 years. Using these mean lives
and maintaining 15% spares, the following part
replacements are needed:

6 CO sensor elements per year @ $200 each
12 CH$_4$ sensor elements per year @ $75 each
6 Air velocity transducers
                 per year @ $1,950 each
4 Belt station circuit boards
                 per year @ $300 each

Even though the benefit of environmental monitoring is not included in the analysis, maintenance and repair of environmental transducers is included to obtain a total system cost. Total annual maintenance and repair costs are as follows:

Maintenance engineer ($30,000 salary)$60,000
Parts replacement   15,000
                        $75,000

Investment Analysis. The NPV analysis for this system over a 5-year period is presented in table 2. The NPV_{20} is $335,124 with a payback period of less than 1 year. Results clearly indicate that this is a favorable investment. Environmental monitoring also would provide benefits that have not been considered in the analysis. It should be noted that including in the analysis the coal being carried by the belts of mine A from the adjoining mine has the net effect of analyzing a 1.3 MM ton-per-year mine (500,000 tons per year from mine A and 800,000 tons per year from the adjoining mine). Calculations for coal from mine A alone (500,000 tons per year) results in only a marginally favorable investment. The NPV_{20} in this case is $32,684.

Mine B

This underground coal mine produces approximately 300,000 tons of coal per year. It is also located in the eastern United States. There are two continuous miner sections and one longwall. The mine operates two production shifts and one maintenance shift per day, 5 days per week. Coal is transported by belt conveyors from the face to a silo at the preparation plant. Average seam thickness is 4.6 ft, at a depth of 550 ft at the portal. The product is primarily metallurgical coal with steam coal as a byproduct. A simplified mine map is shown in figure 2.

The monitoring system installed at this mine is an English MINOS system. The system consists of a central computer, transducers, outstations, and cable connecting the central computer to the outstations. The only function of the system is to monitor and control the belt conveyors. Twelve conveyors are currently under MINOS control with expansion planned to include a total of 15. The analysis is performed for the monitoring of all 15 conveyors.

Table No. 2

NPV ANALYSIS FOR MINE A

Description	Initial ($)	1985 ($)	1986 ($)	1987 ($)	1988 ($)	1989 ($)
Gross revenue......		369,686	369,686	369,686	369,686	369,686
Operating cost.....		0	0	0	0	0
Depletion allow. ..		-36,969	-36,969	-36,969	-36,969	-36,969
Monitor costs						
Acquisition......	-207,600	0	0	0	0	0
Installation.....	-65,300	0	0	0	0	0
Operating........		-90,000	-90,000	-90,000	-90,000	-90,000
Repair..........		-75,000	-75,000	-75,000	-75,000	-75,000
Depreciation.......		-54,580	-54,580	-54,580	-54,580	-54,580
Taxable income.....	-272,900	113,137	113,137	113,137	113,137	113,137
Nondeductible exp.	0	0	0	0	0	0
Pre-tax profit.....	-272,900	113,137	113,137	113,137	113,137	113,137
Taxes on profit....	-130,992	54,306	54,306	54,306	54,306	54,306
Invest. tax credit.	27,290	0	0	0	0	0
Taxes paid.........	-158,282	54,306	54,306	54,306	54,306	54,306
After tax profits..	-114,618	58,831	58,831	58,831	58,831	58,831
Depreciation.......		54,580	54,580	54,580	54,580	54,580
Depletion allow. ..		36,969	36,969	36,969	36,969	36,969
After tax cash.....	-114,618	150,380	150,380	150,380	150,380	150,380
Net present value..	-114,618	125,312	104,439	87,025	72,528	60,438
Total net present value.....	335,124					

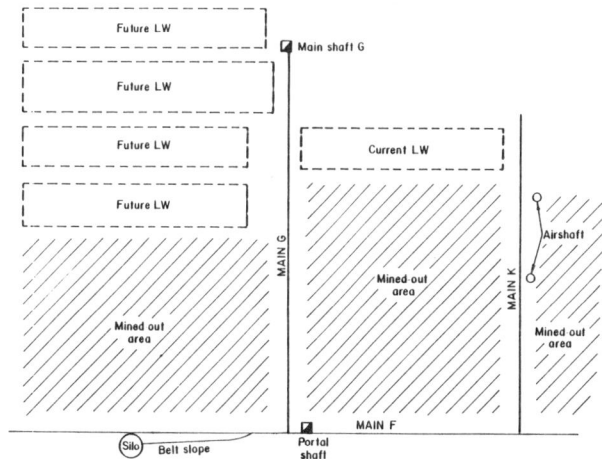

FIGURE 2. MINE B MAP

Parameters monitored at each belt are belt slip, conveyor sequence, pull cord, belt fire sensors, deluge, incomplete start, motor overload, blocked chute, belt alignment, belt tear, and bearing and equipment overheat.

To provide effective use of the monitoring and control system as a management tool, an extensive communication system was installed. It consists of a voice communication system utilizing multiphones and pagerphones.

Benefits. The most important benefit realized so far from this system has been in the location of belt faults and the coordination of belt repair. This has reduced belt downtime, which allows for more production time. Through the use of a chart recorder to record on-off operating status of two unmonitored conveyors inby the monitored belts, the mean-time-to-repair (MTTR) of faults on monitored and unmonitored belts was compared.

Over a noncontinuous interval of 209 hours, 118 faults were experienced on the belts. MTTR was 27.01 minutes per incident on the unmonitored belts and 13.48 minutes per delay incident on the monitored belts, or 50% less per incident. Using coal sales price of $30/ton, gross revenues that result from belt monitoring are as follows:

1985	1986	1987	1988	1989
1,099,767	$1,154,104	$1,333,762	$1,156,708	$1,109,815

Different revenues result due to the changing conveyor arrangements as mine plans change. Calculations are shown in appendix B.

Acquisition Costs. Initial acquisition costs associated with this system include

Central station	$ 87,053
Outstations	91,706
Spares	20,377
Warranty	22,578
Hardware and software engineering	39,635
Other	213,236
Total	$474,585

Costs covered under "other" could not be broken down into specific charges. However, included are costs for transducers, cabling, training, U.S. customs, and miscellaneous materials and services.

Installation Costs. Installation of the system was done by mine personnel. It consisted of the time of two maintenance mechanics plus part time supervision for 40 weeks or 2 person-years. At a salary of $30,000 per person-year, installation labor costs totaled $60,000. To determine the total cost of the employee to the mine, the salary is multiplied by a factor of 2. Another $33,938 was spent by the manufacturer of the system for installation and startup. Total installation costs = $153,938.

Operating Costs. Mine personnel operate the system three 8-hour shifts per day, 5 days per week. Each operator is paid $15,000 per year, and total costs incurred by the mine for the employee equal a factor of 2 times the salary. Total operation costs are 3 shifts x $15,000/year x 2 = $90,000/year.

Maintenance and Repair. Annual maintenance and repair consist of three parts: a maintenance agreement to service the central computers, a maintenance engineer's time (1 person-year at $30,000 salary times a factor of 2) and equipment repair and replacement.

Maintenance agreement	$ 3,000
Maintenance engineer	60,000
Parts replacement	22,100
Total	$85,100

Parts replacement consists of $2,100 for belt sensors and $20,000 for outstation circuit boards.

Investment Analysis. The NPV analysis for this system over a 5-year period is presented in table 3. The NPV_{20} is $1,385,658 with a payback period of less than 1 year.

Table No. 3

NPV ANALYSIS FOR MINE B

Description	Initial ($)	1985 ($)	1986 ($)	1987 ($)	1988 ($)	1989 ($)
Gross revenue		998,350	1,047,377	1,028,707	1,050,937	1,006,475
Operating cost		0	0	0	0	0
Depletion allow		-99,835	-104,738	-102,871	-105,094	-100,648
Monitor costs						
Acquisition	-474,585	0	0	0	0	0
Installation	-153,938	0	0	0	0	0
Operating		-90,000	-90,000	-90,000	-90,000	-90,000
Repair		-85,100	-85,100	-85,100	-85,100	-85,100
Depreciation		-125,705	-125,705	-125,705	-125,705	-125,705
Taxable income	-628,523	597,710	641,835	625,032	645,039	605.023
Nondeductible exp	0	0	0	0	0	0
Pre-tax profit	-628,523	597,710	641,835	625,032	645,039	605,023
Taxes on profit	-301,691	286,901	308,081	300,015	309,619	290,411
Invest. tax credit	62,852	0	0	0	0	0
Taxes paid	-364,543	286,901	308,081	300,015	309,619	290,411
After tax profits	-263,980	310,809	333,754	325,016	335,420	314,612
Depreciation		125,705	125,705	125,705	125,705	125,705
Depletion allow		99,835	104,738	102,871	105,094	100,648
After tax cash	-263,980	536,349	564,196	553,592	566,218	540,964
Net present value	-263,980	446,940	391,834	320,364	273,087	217,413
Total net present value	1,385,658					

SUMMARY

Both monitoring systems have been shown to provide economic benefits to the mines where they are installed. Cost and benefit data will continue to be collected as more operational experience is gained. The analyses will be updated to reflect the new information. Even though the monitoring systems have proved beneficial to the mines discussed, the benefits may vary if the systems were installed in other mines.

APPENDIX A—CALCULATIONS FOR MINE A

(1) Determine production per shift-hour per section (without monitoring):

P — Production (tons per shift-hour per section).
AP — Annual mine production (tons) = 531,458.
BD — Belt downtime (hr/shift): 27.3 hr/2 months = .21.
PT — Time schedule for production (hours per shift): 6.67.
T — Time available for production per shift.
S — Number of production shifts per day: 3.
X — Number active mine sections: 4.

$$T = PT - BD = 6.67 - 0.21 = 6.46 \text{ hr/shift}$$
$$P = AP \div (S \times 5 \text{ days/week} \times 52 \text{ weeks}) \div X \div T =$$
$$531,458 \div (3 \times 5 \times 52) \div 4 \div 6.46 = 26.4 \text{ tons/shift-hr/section.}$$

(2) Determine amount of coal dumped on belts from adjoining mine per shift-hour:

Y — Coal dumped on belts from adjoining mine per day (tons/day): 3,200.
J — Coal dumped on belts from adjoining mine (tons/shift-hr).
J — $Y \div S \div T = 3,200 \div 3 \div 6.46 = 165.2$.

(3) Determine the benefit of belt monitoring. A mine conveyor network analysis is used to account for the sequencing of belts. It is assumed that all belts have an equal probability of failure. Since the mine has five active sections, with only four operating at any given time, the arrangement of active belt conveyors will vary depending on which sections are active; therefore, five conveyor networks are analyzed. The expected benefit (B) for each conveyor network is determined by:

$$B = \left(\sum_i P(F)_i \times D_i\right) \times R,$$

where $P(F)_i$ = Probability that a belt carrying i tons of coal/hours fails,
D_i = Gross revenue lost if a belt carrying i tons of coal/hour fails,
and R = Percentage reduction in belt downtime with monitoring: 65%.

$$D_i = i \times BD \times C$$

where i = Tons per hour of coal transported by the belt
and C = Coal sales price: $30/ton

The following five diagrams show the conveyor arrangements possible for Mine A. Also shown is the calculation of B for each conveyor arrangement.

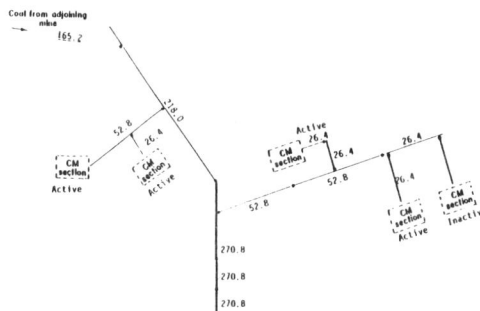

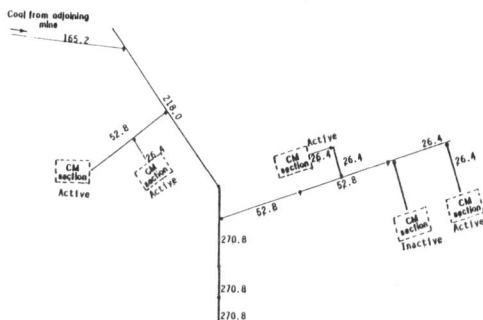

i (tons/hr)	x	BD (hours/yr)	x	C ($/ton)	=	D ($/yr)	P(F)	D x P(F)	
26.4	x	163.8	x	30	=	129,730	5/13	49,896	
52.8	x	163.8	x	30	=	259,459	3/13	59,875	
165.2	x	163.8	x	30	=	811,793	1/13	62,446	Σ x R = B = 561,708 x 65% =
218.0	x	163.8	x	30	=	1,071,252	1/13	82,404	$365,111/yr
270.8	x	163.8	x	30	=	1,330,711	3/13	307,087	

i (tons/hr)	x	BD (hours/yr)	x	C ($/ton)	=	D ($/yr)	P(F)	D x P(F)	
26.4	x	163.8	x	30	=	129,730	5/13	49,896	
52.8	x	163.8	x	30	=	259,459	3/13	59,875	
165.2	x	163.8	x	30	=	811,793	1/13	62,446	Σ x R = B = 561,708 x 65% =
218.0	x	163.8	x	30	=	1,071,252	1/13	82,404	$365,111/yr
270.8	x	163.8	x	30	=	1,330,711	3/13	307,087	

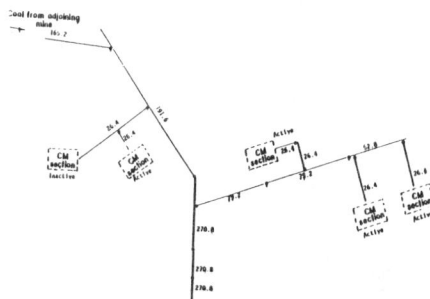

i (tons/hr)	x	BD (hours/yr)	x	C ($/ton)	=	D ($/yr)	P(F)	D x P(F)	
26.4	x	163.8	x	30	=	129,730	6/14	55,598	
52.8	x	163.8	x	30	=	259,459	1/14	18,533	
79.2	x	163.8	x	30	=	389,189	2/14	55,598	Σ x R = B = 540,118 x 65% =
165.2	x	163.8	x	30	=	811,793	1/14	57,985	$351,077/yr
191.6	x	163.8	x	30	=	941,522	1/14	67,252	
270.8	x	163.8	x	30	=	1,330,711	3/14	285,152	

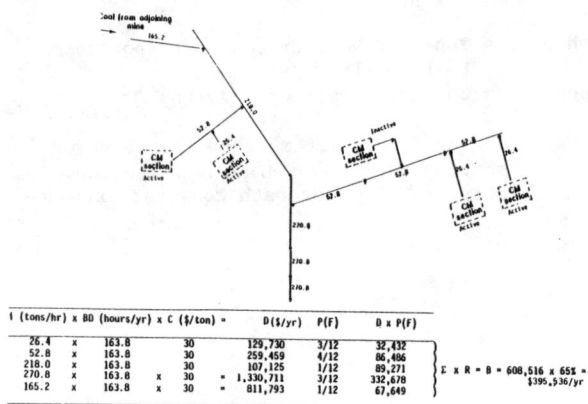

i (tons/hr)	x	BD (hours/yr)	x	C ($/ton)	=	D ($/yr)	P(F)	D x P(F)
26.4	x	163.8		30		129,730	3/12	32,432
52.8	x	163.8		30		259,459	4/12	86,486
218.0	x	163.8		30		107,125	1/12	89,271
270.8	x	163.8	x	30	=	1,330,711	3/12	332,678
165.2	x	163.8	x	30	=	811,793	1/12	67,649

Σ x R = B = 608,516 x 65% = $395,536/yr

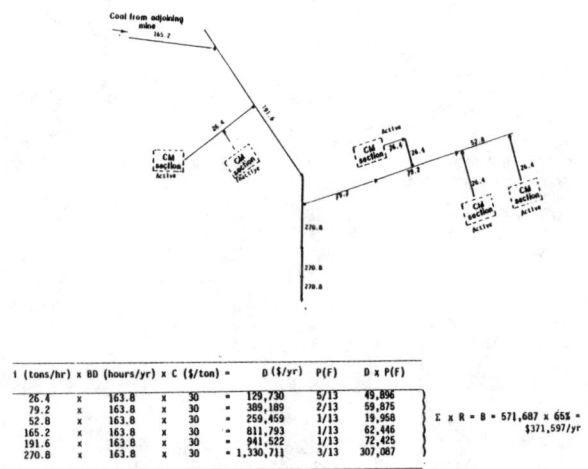

i (tons/hr)	x	BD (hours/yr)	x	C ($/ton)	=	D ($/yr)	P(F)	D x P(F)
26.4	x	163.8	x	30	=	129,730	5/13	49,896
79.2	x	163.8	x	30	=	389,189	2/13	59,875
52.8	x	163.8	x	30	=	259,459	1/13	19,958
165.2	x	163.8	x	30	=	811,793	1/13	62,446
191.6	x	163.8	x	30	=	941,522	1/13	72,425
270.8	x	163.8	x	30	=	1,330,711	3/13	307,087

Σ x R = B = 571,687 x 65% = $371,597/yr

Once the expected benefit from each of the five conveyor arrangements is determined, a total expected benefit from conveyor monitoring at mine A is determined by multiplying the probability that a particular conveyor arrangement is being used times the expected benefit (B) for that conveyor arrangement and summing the product for each conveyor arrangement:

(1/5 x 365,111) + (1/5 x 365,111) + 1/5 x 351,077) + (1/5 x 395,536) + (1/5 x 371,597) = $369,686.

Since the general mine plan (two continuous miner sections off the south mains and three continuous miner sections off the north mains) is expected to remain the same for the next 5 years, the gross revenues expected for each of the 5 years in the investment analysis is $369,686.

APPENDIX B--CALCULATIONS FOR MINE B

(1) Determine production per shift-hour per section (without monitoring): BD (belt downtime (%)] =

118 faults/209 hr x 27.01 min down/fault = 25%.

If 14 mine belts (12 monitored and 2 unmonitored) had 118 faults in 209 hr, the probability of a belt fault is as follows:

118 faults/(209 hr)(14 belts) = 0.040 faults/belt hr.

Continuous miners produce 120 tons/shift. If nine belts in sequence carry coal out of the mine from a continuous miner section, the continuous miner cannot mine coal 1.08 hr/shift.

PT x No. belts x faults/belt hour x downtime/fault = 6.67 hr/shift x 9 belts x .04 faults/belt hour x 27.01 min/fault = 1.08 hr/shift.

Therefore, the production rates of the continuous miners are:

120 tons per shift ÷ (6.67 hr/shift - 1.08 hr/shift) = 21.5 tons/hr.

Longwalls produce 375 tons/shift. If 10 belts in sequence carry coal out of the mine from a longwall section, the longwall cannot mine coal 1.20 hr/shift.

PT x No. belts x faults/belt hr x downtime/fault = 6.67 hr/shift x 10 belts x .04 faults/belt hr x 27.01 min/fault = 1.20 hr/shift

Therefore, the production rate of the longwall is

375 tons per shift ÷ (6.67 hr/shift - 1.20 hr/shift) = 68.6.

(2) Determine the benefit of belt monitoring

A mine conveyor network analysis is used to account for the sequencing of belts. It is assumed that all belts have an equal probability of failure. Since the mine plan, and therefore the belt conveyor arrangement, changes over the 5-year period used for the investment analysis, different revenues result from belt monitoring in each year. A separate conveyor network analysis is performed for each year in the following five diagrams.

1985

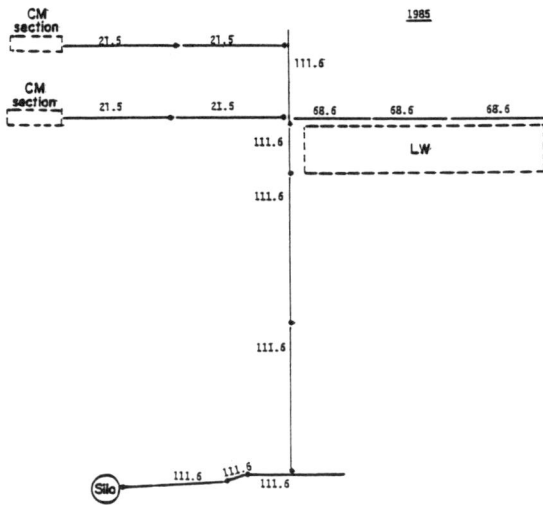

i (tons/hr)	x 80 (hours/yr)	x C ($/ton)	=	O($/yr)	P(F)	O x P(F)	
21.5	x 868.4	x 30	=	560,118	4/14	160,034	
68.6	x 868.4	x 30	=	1,787,167	3/14	382,964	} Σ x R = B
111.6	x 868.4	x 30	=	2,907,403	7/14	1,453,702	

B = 1,996,699 x 50% = $998,350/yr

1987

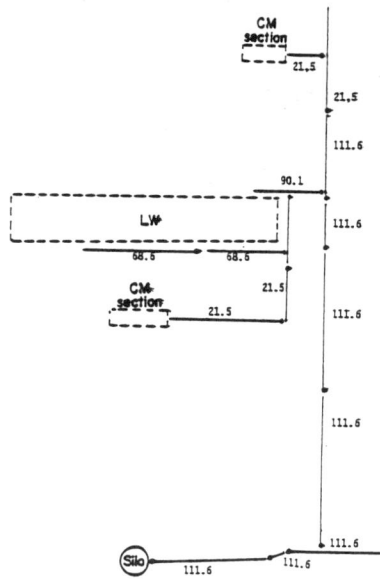

i (tons/hr)	x 80 (hours/yr)	x C ($/ton)	=	O($/yr)	P(F)	O x P(F)	
21.5	x 868.4	x 30	=	560,118	4/15	149,365	
68.6	x 868.4	x 30	=	1,787,167	2/15	238,289	
90.1	x 868.4	x 30	=	2,347,285	2/15	312,971	} Σ x R = B
111.6	x 868.4	x 30	=	2,907,403	7/15	1,356,788	

B = 2,057,413 x 50% = $1,028,707/yr

1986

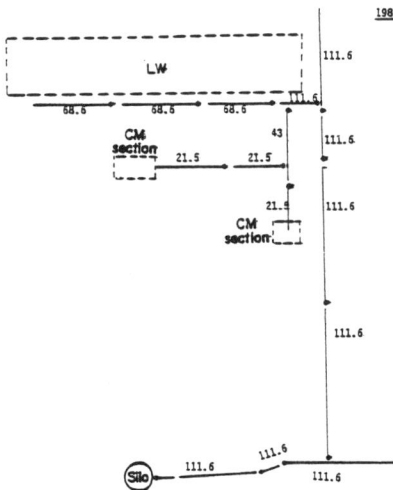

i (tons/hr)	x 80 (hours/yr)	x C ($/ton)	=	O($/yr)	P(F)	O x P(F)	
21.5	x 868.4	x 30	=	560,118	3/15	112,024	
43	x 868.4	x 30	=	1,120,236	1/15	74,682	
68.6	x 868.4	x 30	=	1,787,167	3/15	357,433	} Σ x R = B
111.6	x 868.4	x 30	=	2,907,403	8/15	1,550,615	

B = 2,094,754 x 50% = $1,047,377/yr

1988

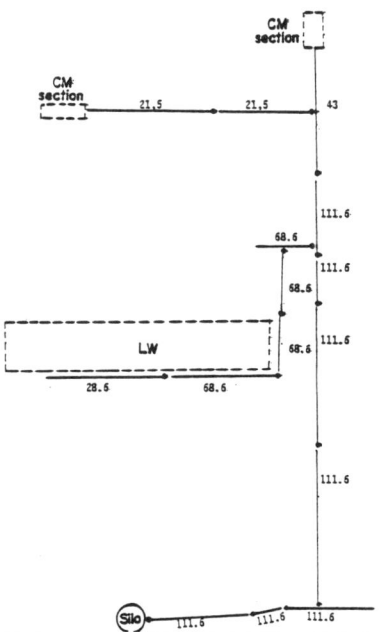

i (tons/hr)	x 80 (hours/yr)	x C ($/ton)	=	O($/yr)	P(F)	O x P(F)	
21.5	x 868.4	x 30	=	560,118	2/15	74,682	
43	x 868.4	x 30	=	1,120,236	1/15	74,682	
68.6	x 868.4	x 30	=	1,787,167	5/15	595,722	} Σ x R = B
111.6	x 868.4	x 30	=	2,907,403	7/15	1,356,788	

B = 2,101,874 x 50% = $1,050,937/yr

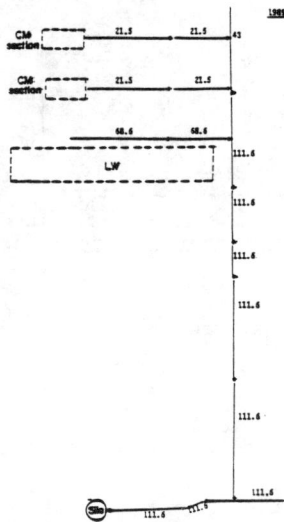

i (tons/hr) x	BD (hours/yr)	x	C ($/ton)	=	D($/yr)	P(F)	D x P(F)	
21.5	x 868.4	x	30	=	560,118	4/15	149,365	)
43	x 868.4	x	30	=	1,120,236	1/15	74,682	)
68.6	x 868.4	x	30	=	1,787,167	2/15	238,284	) Σ x R = B
111.6	x 868.4	x	30	=	2,907,403	8/15	1,550,615	)

$$B = 2,012,951 \times 50\% = \$1,006,475/yr$$

B (expected benefit for each conveyor network) =
$$\left(\sum_i P(F)_i \times D_i \right) \times R,$$

where $P(F)_i$ = Probability that a belt carrying i
          tons of coal/hour fails
     D_i = Gross revenue lost if a belt
          carrying i tons of coal per hour
          fails,
  and R = Percentage reduction in belt
          downtime with monitoring = 50%.

      D_i = i x BD x C,

where i = tons per hour of coal transported by
     the belt,
 and C = coal sales price: $30/ton.

Chapter 16

A GUIDE TO MINIMIZE MINE-WIDE
MONITORING SYSTEM START-UP PROBLEMS

RAYMOND C. VOIGE

INSTRUMENT DIVISION
MINE SAFETY APPLIANCES COMPANY
PITTSBURGH, PENNSYLVANIA

Abstract. The installation of a computer-based
Mine-Wide Monitoring System is more than just a
routine job at most underground coal mines.
Problems that show up on the day the system is
started-up can be minimized by careful planning
and installation of both central and remote
stations. The experience of Mine Safety Appliances
Company in assisting with the installation of more
than 10 Mine-Wide Monitoring Systems in the
United States, is the basis for this planning and
installation guide.

INTRODUCTION

Starting up a Mine-Wide Monitoring System can
be a routine task or a difficult one. Each
installation has its unique requirements, but
following a guideline based on extensive experience will improve the chances of an easy start-
up.

Mine Safety Appliances Company (MSA) has sold
and assisted in installing more than 10 Mine-Wide
Monitoring Systems (MWMS) in the United States.
They are located in coal mines in Pennsylvania,
West Virginia, Kentucky, Tennessee, Illinois,
Colorado, and Virginia. System size ranges from
several remote stations and 2-3 thousand feet of
wire to 14 remote stations spread out over 15
miles on three data lines.

This guideline will start with system planning,
review installation suggestions, and end with a
start-up check list.

PLANNING

Central Station Planning

The location of the central station portion of
the Mine-Wide Monitoring System is important and
will be governed by several factors; convenience,
access to communication systems, and proper environment. Some mines will be able to give full
priority to each of these, others will have only
one location available and compromises will be
made.

Convenience is an important factor in making a
Mine-Wide Monitoring System a part of daily operations. Mine operating and maintenance personnel
will be making decisions based on information
provided by the MWMS. The mine office area is
usually centrally located and an ideal place for
the monitoring system to provide timely information on hazardous conditions, mine operations, or
maintenance problems. Communication system and
environmental requirements can usually be met at
the mine office location.

A long range consideration is system expansion.
The small carbon monoxide (CO) belt fire detection
system may grow into a system monitoring belts,
power centers, mining machines, ventilation, and
methane drainage. Provision for a larger system
which might include one or more graphic displays
should also be considered.

Finally, the location within a room should take
into account, operator use of the keyboard, video
screens, and printer. This requirement plus the
need for simple cabling and easy access for maintenance suggests a table top or console arrangement of equipment.

The communication system is a vital link in a
MWMS. Local alarms for high CO or methane build-
up need to be confirmed and action taken. Maintenance men must be informed about belt problems
and where they are located. If both pager and
vehicular communication (i.e. trolley) systems are
used in the mine, units should be located near the
central computer station or combined into a console arrangement. An outside telephone unit
should also be available for routine, emergency,
or computer suggested calls (i.e. call Joe Smith,
Mine Superintendent if a CO sensor signals an
alarm or if the main belt is down for more than 30
minutes). The communication systems are also
useful when trouble-shooting the data line(s) and
calibrating sensors.

Environmental conditions are important for all
computers, but especially high speed units with
large memory capacity. High temperatures, above
40°C cannot be tolerated for more than a few minutes or a loss of data memory may result. Most

115

mines locate their central station computers in the mine office where basic air conditioning is provided for the building. If the computer is located in a small room, however, supplementary air conditioning may be required.

Dust is also a problem. If the room cannot be kept to commercial office standards, supplementary dust filters must be used on both the computer and disk drive units.

NOTE: The use of floppy disks for on-line computer memory is usually not recommended for mining locations because of the dust problem. Floppy disks should only be used for system start up. Solid state memory is more reliable for on-line data storage.

A separate 110 vac power line from the building circuit breaker panel will provide a reliable power source and prevent unexpected shutdowns from overloaded circuits. Further protection can be provided by an uninterruptible power supply (UPS). A minimum of 10 minutes protection is recommended to bridge minor power outages. If the monitoring system is used to comply with MSHA requirements for a belt fire detection system a 4 hour UPS and remote station battery back-up may be required for the complete system.

Remote Station Planning

The remote stations and associated wiring require as much planning as the central station. For example, the CO Belt Fire Detection System is currently the largest use for a MWMS. The remote stations for this application must be located at a point where several CO sensors can be wired to it to reduce equipment costs. This same location must have 110 vac available for powering the remote unit. Future system expansion to belt monitoring would also suggest the remote be located near a belt drive to make the required electrical connections a simple job.

The data line from the central station must be connected to each of the remote stations. In a large mine several lines may be required to service different areas of the mine. A layout on a mine map is essential to review alternate wiring schemes to optimize the installation.

The vehicular entry installation is usually the easiest to make, but if remote stations are located in belt entries, extra wire will be needed to run down the appropriate crosscuts. Checking the data line for breaks and shorts is also easier in the vehicular entry, but trolley type vehicles can create a lot of electrical noise that must be filtered out by the monitoring system. The belt entry on the other hand may be quiet from an electrical noise stand point, however it may have other problems such as poor access for installation and trouble shooting.

Figure 1.   Central Station Installation

Figure 2.   Remote Station

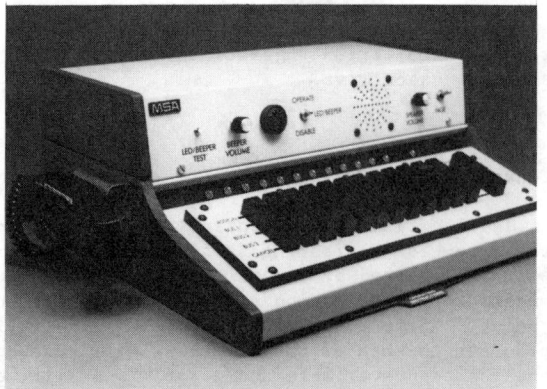

Figure 3.   Communication System Console

INSTALLATION

Central Station Installation

If a reasonable amount of planning was done for the location and environment of the central station, the installation should be very easy. Console type equipment will take longer to wire up because of wiring constraints, but power and data cables are exposed to more hazards in a table top installation. Excess wiring should be coiled up and taped to prevent the operator from stepping on it and possibly causing a bad connection. The use of anchor screws will prevent data cable connectors from backing out of receptacles and causing intermittent operation.

Lightning arrestors must be used on data lines that are run between buildings and enter the mine portal. Gas discharge type arrestors are recommended in conjunction with waterproof enclosures. This procedure will eliminate loss of signal on the data line due to possible carbon arcing or water paths to ground.

When data lines are run on the surface on power line poles or near roadways, shielded type cable is recommended to reduce electro-magnetic interference (EMI) from high voltage power lines and automobile ignition systems.

Underground Installation

The installation of remote stations, sensors, and wiring will also be accomplished faster and with fewer problems if adequate planning is included in the job. Mine maintenance personnel have a great deal of experience in hanging cable and mounting pager and other electronic equipment properly. Sometimes things can be overlooked though.

Data line cable can be specified to be identical to pager wire; black and white single twisted pair No. 14 gauge. Confusion may result however, when both cables are run in the same entry. Another color combination or method of differentiation will help prevent splicing the two systems together.

Cable splices must be made carefully with attention to good connections and waterproof taping. A bad splice can shut a MWMS down until a maintenance crew can locate and repair it. The central station terminal on the MSA DAN™ System, alarms if the data line is broken or is not responding. Information from all or part of the system is not being received however.

The data line is usually electrically balanced to ground. That means stray signals are picked up equally and cancel out. If one side of the data line becomes grounded on a metal support such as a conveyor chain the balance is lost. EMI and signals from equipment like trolley radios are now able to get on the data line and may cause loss of data throughput (the data security system discards messages with errors).

Figure 4.   Reseating Computer Circuit Boards

START-UP

Start-up Check List

Writing out a list of items to check prior to starting-up a MWMS may seem time-consuming, but it may prevent major problems and unnecessary expense. The following list contains some items that are basic to many types of computer-based systems and others that may be unique to the MSA DAN systems.

Central Station

1.   Is the computer and its accessory equipment on a separate fuse or breaker at the building load center? (Check to see what equipment goes off when the power is purposely shut off.) Are the disk drive and terminal shutoff by the computer reset switch? (This makes system rebooting easier.) Make sure the equipment powered by the uninterruptable power supply does not exceed its power rating.

2.   Is the environment for the central station holding to planned specifications? What happens to the temperature in the area of the computer after 24-48 continuous hours of operation? Check seasonal variations. If standard dust filters are loading up in a few days or less, special filters are probably required. Are there any problems with viewing the terminal screen; glare?

3.   Are all the data cables routed properly and securely fastened?

4. Were all the circuit boards in the
computer reseated to insure good contact?
Was the cardboard head protector removed
from the disk drive? Don't forget to put
it back in any time the disk drive is
moved. Are extra disks available with the
program loaded on it? Dirt and
unintentional exposure to magnetic fields
can wipe out a disk. (See Figures 4 and 5).

5. Does a remote station temporarily
connected up near the computer as a mini
system check out okay? With all the
remote stations deactivated, is there
any dc voltage on either side of the
balanced data line to ground? Readings
higher than 5 volts dc should be
investigated. The ac voltage across
the data line should also be minimal and
values greater than 1v p-p also require
investigation. Check for proper
lightning arrestor installation (gas
discharge type in waterproof enclosure).

Figure 5. Inserting A Head Protector
Into a Floppy Disk Drive.

Remote Station

1. Is the remote station mounted securely
(i.e., cement block stopping) and
protected from dripping water? Are
sensors and ac power connected properly?
Is the remote station easily found by
the fire boss when the backup battery
must be reconnected?

2. Were sensors such as CO installed as
required by the manufacturer? Can the
sensors be calibrated easily when it's
required? Is the wiring from sensor
to remote station installed to
prevent accidental breaks or shorts?

Additional items can be added for each
installation. MSHA classified systems have
specific installation instructions for both
red and blue type underground remote stations.

SUMMARY

A Mine-Wide Monitoring System is capable of
excellent operation if the installation is
well planned and the conditions of the
environment are taken into account. As more
systems are installed, additional expertise
will be gained and incorporated in new
installations. Also, most systems are
purchased with expansion expected over
several years. This time frame is more
than adequate for normal evaluation and
modification to take place.

Chapter 17

THE EVOLUTION TO COMPLETE MINE MONITORING

by

Mr. J. D. Armstrong
Jim Walter Resources, Inc., Mining Division
Sr. Maintenance Engineer
Brookwood, Alabama

Mr. T. E. McNider
Jim Walter Resources, Inc., Mining Division
Deputy Manager, Ventilation Department
Brookwood, Alabama

ABSTRACT

Jim Walter Resources' first experience with data acquisition from an underground source to a central receiving station on the surface was in 1979. The first system was of simple design with a carrier frequency unique to that one input. With the vast improvements in mine monitoring equipment systems are now capable of supporting up to four floppy disk drives with a storage capacity of 1.2 megabytes each plus four hard disk drives with up to 58 megabytes each. The data transfer rates are in the order of 4800 bits/sec. With the sophistication of the present computer system for mine monitoring, a new tool is available to help the Mine Manager operate his mine more efficiently. This paper details the evolution from the first system to the present system used today.

INTRODUCTION

Jim Walter Resources, Inc., operates five underground coal mines in Tuscaloosa and Jefferson counties in Central Alabama. All five mines are extracting coal from the Mary Lee group of seams in the Warrior Coal Basin. The four newer mines are in the deeper part of the basin with overburden thickness ranging from 1300 feet at No. 3 Mine to 2100 feet at No. 5 Mine. Coal is mined with continuous and longwall type mining equipment, conveyed by belt to a central location and hoisted out of the mine by a pair of skips.

Due to difficult geologic conditions which limit ventilation capabilities, Jim Walter Resources in 1976 decided to explore the possibility of using beltlines as additional intakes. Section 75.326 of Title 30 of the Code of Federal Regulations governs beltline ventilation and it states in part:
"In any coal mine opened after March 10, 1970, the intake and return air courses shall be separated from belt haulage entries and each operator of such mine shall limit the velocity of the air coursed through belt haulage entries to the amount necessary to provide an adequate supply of oxygen in such entries, and to insure that the air therein shall contain less than 1.0 per centum of methane, and shall not be used to ventilate active working places".

In order to use belt air to ventilate a working face, a petition for modification of the Code of the Federal Regulations was required, additional safe guards were needed, and the company has had to install low level carbon monoxide monitors along the beltlines with a communication system capable of collecting the data and transmitting it to a central location. With the decision to utilize intake air on the beltlines, this prompted the first carbon monoxide monitoring system and subsequent evolution into the central computer system used today.

Designing The First System

Jim Walter Resources worked with Giangarlo Scientific Company, Inc., to design a system which would perform this task. The final design was installed at No. 4 Mine, and consisted of the Ecolyzer 4000 made by Energetic Science, Inc., as the carbon monoxide monitor, the series 6405 variable frequency Telemeter System made by RFL Industries, Inc., as the display and alarming unit. This system was purchased and installed with six underground monitoring stations in September 1979.

The Ecolyzer 4000 is an instrument designed for our application to measure carbon monoxide in the 0-50 ppm range. The instrument can be powered from either 120 volts AC or 12 volts DC and has both a meter display and a 0-1 volt DC recorder output.

The RFL system used in the No. 4 Mine project operates on the telemeter type of line transmission. In the transmitter, the DC voltage input is changed to a frequency and superimposed on a carrier frequency unique to that one input. At the receiver the reverse occurs, and the end product is a DC voltage corresponding to the input voltage. The Datalogger 2000 continuously scans the output voltage from the telemetry receivers, displays them and initiates an alarm if the preset level is exceeded.

JWR No. 3 Mine

Niagra Scientific Microprocessor Based System
Soon after the approval of the first petition at No. 4 Mine, a similar petition was filed for No. 3 Mine. This petition was granted in August 1980. From experience gained from the installation at No. 4 Mine and advances in technology in this field, the system purchased for installation at No. 3 Mine contained a vastly improved communication and data acquisition system. Giangarlo Scientific Company had contracted Niagra Scientific Company to design and build a microprocessor based data acquisition system to increase the flexibility of the communication network. As before the Ecolyzer 4000 was the carbon monoxide monitor purchased.

The communications network purchased for No. 3 Mine is based around the Datkon 700 central microprocessor unit. This unit controls the polling of each underground station, the alarming criteria and data output through programmable computer software. The underground link to the central computer is the Dardak 700, a limited memory minicomputer capable of accepting and storing up to sixty-four analog or digital (contact closure) inputs. In operation, the Dardak accepts and stores the input data from the carbon monoxide monitor until the central asks it to transmit that data to the surface. The central computer continuously polls each remote station and asks it to dump its data into the central memory. The central memory is continuously updating, retaining only the most current output from each remote station. This data is converted back to engineering units (ppm for carbon monoxide, percent of methane, etc.) and printed out on the accompanying data terminal at preset intervals. Communication is via the same type of dedicated cable as in use at No. 4 Mine. Identification of each remote and data transfer is accomplished by a unique digital bit structure superimposed on a common carrier frequency, eliminating the need for additional hardware to transmit and receive a number of frequencies.

Up-Grade of The System

This system has been expanded to approximately forty remote units underground. It was then decided to up-grade the Datkon 700 central microprocessor to a Giangarlo Scientific Company Model 2002 central computer. The up-grade to the central computer enabled the mine to atuomatically monitor analogue and digital signals for 128 remotes (approximately 1500 points) expandable. The new central computer also provided auto and manual control, RGB color monitor, keyboard entry with 32K RAM, trending and plotting, 30-day history, dual disc mine mapping, multi-level alarming, operator instruction messages, real time clock, color graphic, printer with graphic capabilities, I/0 interface for multi-tasking applications, and interfaceable to a host computer.

Ultimate Goal of Complete Mine Monitoring

The use of intake air on the beltline was a major accomplishment for Jim Walter Resources providing added ventilation to the working sections and a positive flow of air on the beltline. Even with this accomplishment, the goal of Jim Walter Resources from the conception of underground monitoring was to completely monitor the mine in its entirety. The systems discussed did a good job for what they were intended, monitoring of carbon monoxide on beltlines, but technology was limited in real time control and memory storage to do complete mine monitoring.

Since the first system was installed at No. 4 Mine, technology has improved to the point where complete mine monitoring is possible. This led to the two systems that are presently being installed at No. 5 and No. 7 mines. The system selected for each mine are both capable of complete monitoring and control. The different types of equipment selected was based on mine managements preference.

JWR No. 5 Mine

Minetics 8000 System At No. 5 Mine the M/C-8000 system is being installed. The M/C-8000 System is a micro-computer with the following components – primary computer, color video display, character generator, printer keyboard, cartridge drive, communication controller, and outstation. The primary computer, see Figure 2, is the standard Minetics 96K RAM System complete with multi-tasking, real-time software for digital and analog monitoring and remote control. Color video displays of mimic diagrams of equipment status, alarm surveillance, analog graphs and bar charts, alarm logs, clock/calendar, downtime analysis, and complete password setup capabilities are included.

Because the system uses RAM memory instead of magnetic data storage, video displays are generated fast, usually within a few seconds. This is important in the real-time operating mode of the system. A printer provides hard copy to help management study problem areas and evaluate production and safety performance.

The central station is installed in an ordinary mine office. There was no requirements to provide cleanroom conditions. Except for a cartridge unit used infrequently for recording setup information, there are no moving parts involved in the day-to-day operation of the system. All data is stored in RAM memory, thus avoiding the delay and wear problems normally associated with motor driven magnetic recording equipment (e.g. floppy disks) in the dusty mine office environment.

A 600 VA, 120VAC Sentinel-600 uninterruptable power supply maintains power to the primary and peripherals during power failures of 2 to 3 hours duration. Two 12V, 50 HA lead-calcium batteries are automatically recharged and maintained in ready state by the two stage charger. The UPS also protects the electronic components from line transients.

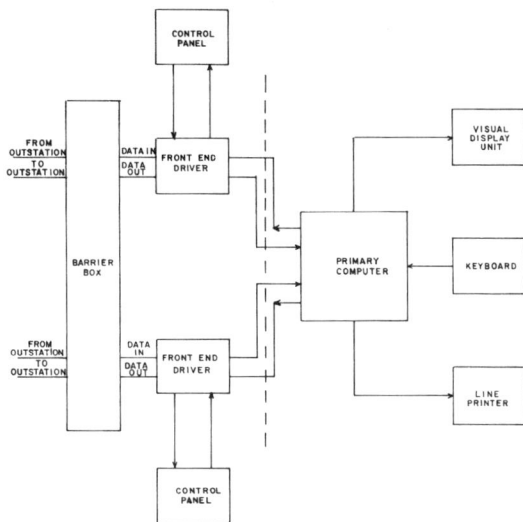

FIGURE 2 TYPICAL MINETICS SYSTEM

A communication controller maintains continuous communication with all outstations using two independent 4-wire circuits. Each outstation is interrogated at least once every second. The system is initially equipped for handling 28 outstations but can be expanded by adding additional communication modules, 28 outstations per block.

Communications is 600 baud, full duplex, (100% duty cycle both ways) over 2 twisted pairs of AWG 16 cable. Transformer isolation is provided by barriers on the surface and at each outstation. This provides a current loop with low DC impedence to reject ambient noise. Transmission distances can be up to 10 miles. Data reliability is assured by various electronic checks.

The outstation is a single card design with 8 digital inputs, 8 analog inputs, and 8 digital outputs, see Figure 3. Battery backup provides power to the outstation and sensors for at least 4 hours after loss of main power. A power "down module" is included to allow the line power and battery to be disconnected by keyboard command in the event of loss of ventilation.

The Fireboss *100 carbon monoxide fire sensors (manufactured by Rel-Tek Corp.) are connected to analog input channels of the outstations. Units are spaced along the conveyor belts at the MSHA designated spacings (not exceeding 2000 feet).

Section alarms alert personnel in the working section if a fire is detected that requires evacuation. Each section alarm has a loud horn, a xenon flasher and electronics to interface with the outstations. When a high CO level is detected, usually 10 ppm above ambient, a warning is sounded at the central control station. The operator can then display a graph of this sensor vs time on the video screen. A comparison can

also be displayed of other CO sensors inbye the alarming unit, since the products of combustion will drift in the direction of air flow. This will help confirm the fire.

When the CO level reaches the alarm level, usually 15 ppm above ambient, a horn sounds at the control console. The operator can then activate the appropriate section alarms by simple keyboard command.

In the event of loss of ventilation, as sensed by a designated channel in the fan outstation, a timer is activated in the computer which automatically issues a shut-down command to all outstations after a designated time delay, presently 5 minutes.

JWR No. 7 Mine

Conspec Senturion 200 System The Senturion 200 System is being installed at No. 7 Mine. The Senturion 200 is a main frame mini-computer with the following components - color graphics processor, processor and co-processor, color terminals, and printers.

The color graphics processor is a multi-tasking multi-user computer which can support up to four color terminals simultaneously. It continuously accesses the main processors for monitoring and control in addition to supporting multi-users and multi-tasking functions at the same time. The color graphics processor has one 1.2 megabyte floppy disk and one 10 megabyte hard disk with expansion capabilities of 16 disk drives in configuration of either hard or floppy disk. Random access memory is 208 kilobytes expandable to one megabyte.

Communication between the central console and the underground field device, ie. accessors, is handled through a 4-conductor wire referred to as the trunk line. Communication is routed around the central processor to a local modem (on the surface) where the RS232C level communication is converted to signal level PSK and then transmitted to the remote modem and trunk driver. It is then converted once again to RS232C levels and placed on the underground trunk line. The above is repeated in reverse for communication from underground to the central console. As this system is full duplex, both modes of communication can take place simultaneously. The communication data rate (or band rate) is 4800 bits per second.

An accessor is the communication link between all field devices (CO monitors, contacts, motor starters, etc...) and the main processor. The accessor is a single or multiple point data transmitter/receiver processor. It is powered around the data trunk. Therefore, the need for local power supplies is eliminated. All accessors are wired in parallel and can be connected at any point on the trunk. By installing the accessor directly beside the equipment to be monitored and/ or controlled, the cost and potential interference of long inter-connect cables is eliminated. Due to the complexity of both the Minetics and

and Conspec Systems a more detailed explanation
of the Minetics System will follow.

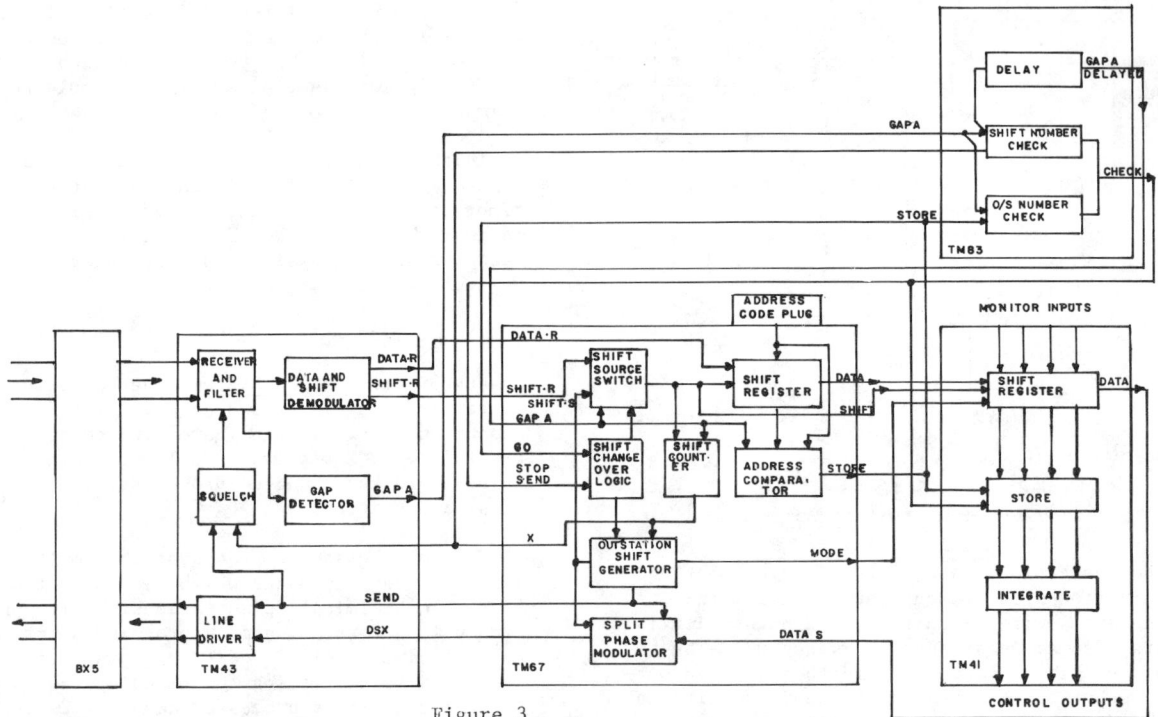

Figure 3

In the Minetics M/C-8000 System a device
known as a Front End Driver is used. The FED is
essentially a small computer which is dedicated
to the control of telemetery rings. Each ring
contains up to 14 outstations. Figure 1 shows
the components necessary to make up the FED.

The MC15 processor card uses an Intel 8085
which controls the operation of the bus structure.
The MC15 has a serial interface which communicates
with the primary computer and an interrupt controller. The interrupt controller indicates when
one of the interface cards has completed a task
and is ready for the next one.

The memory which is the MC32 can store program information, data or a combination of both.
For program information the MC32 stores it in a
EPROM so the MC15 can read the contents but is
unable to change them while RAM is used to store
data so that it can be read and changed as
necessary. The memory of the MC32 is organized
into four blocks of 4096 address locations for a
total of 16384. It is necessary to include
memory in all the blocks but each block must be
either all RAM or all EPROM.

Various types of control panels are available.
Meters, leds, switches, etc., can be included on
the control panel therefore a range of interface
cards are available. If meters are included on
the panel the digital information from the micro-

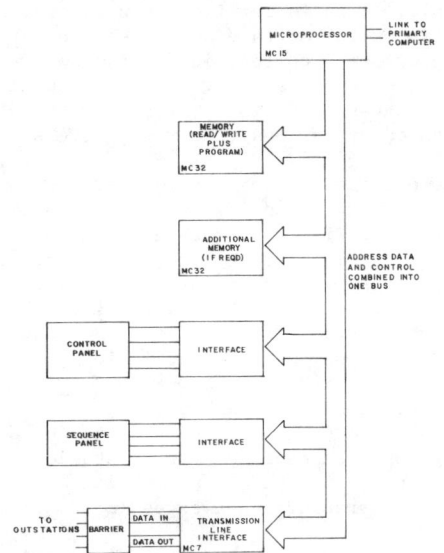

FIGURE I FRONT END DRIVER - GENERAL CIRCUIT

computer must be converted to an analogue value.

Each FED is supplied with a sequence panel to
specify the outstation types and their physical

relationship. The operation is best described in two parts. First, during the initialization program the linking arrangement is examined and supplies to the FED information on the current configuration of the outstations. During initialization the FED generates a sequence file. Then under normal operating conditions the sequence file can initiate for instance, the start or stop of conveyors.

The signals to the outstation are outputs from the processor as a series of bytes. The MC7 translates these parallel signals into a serial format. In the opposite direction the serial responses from the outstations have to be formed into eight parallel signals before they are input to the microcomputer. One byte of information is output via a port on the MC7 to an eight bit shift register operating in the parallel mode. After loading the information, the register returns to a serial mode and the shift generator clocks the data through the register to the split phase modulator. In the split phase modulator the data and shift signal are coded and transmitted to the outstations. The replies from the outstations are first demodulated to give data and shift. These two signals from the serial data to the second shift register. When all eight bits have been received the output byte is input to the microcomputer.

An outstation is made up of two groups of circuit cards. The first is concerned with the telemetry link, called the transmission block. The second group is the control circuit. The transmission block is concerned with the processing of information to and from the control station. The basic shift register board has four stages and is capable of accepting four parallel inputs and providing an equal number of outputs to the control circuits. Figure 3 illustrates a simple outstation transmission block. When receiving, the split phase modulated signals from the central station are routed through the BX-5 barrier to the receiver and filter where any electrical noise is removed. At the data and shift demodulator the combined signal is separated. The data signal goes straight to the shift register while the shift signal is fed through a shift source switch before clocking the data through the registers. At the same time the shift counter records the number of received pulses and when a preset number is exceeded the output goes high preventing any further serial transfers. Throughout the serial transfers of data a timer within the gap detector has been continually reset by the incoming signals. If no signals are received for 1-1/2 shift periods the transmission is considered complete. If the six bits held in the address register are not identical with those of the address code no store signal is produced and the outstation is ready to receive. If both addresses match, the contents of the register are stored. The shift generator outputs a signal called mode and within it a shift-S pulse. The mode is fed direct to the TM67 and TM41, whereas shift-S is routed via the source switch. The combination initiates a parallel transfer of information to the registers. After mode has

returned to zero the shift generator outputs a send signal enabling the split phase modulator and line driver. The outstation can now transmit to the central station. As can be seen, the outstation can not transmit and receive data at the same time. The control circuit cards of an outstation are numerous due to the different types of control and monitoring situations.

SUMMARY

With the purchase of the last two systems, Jim Walter Resources now has sufficient computer capacity and speed to completely mine monitor and do control functions.

Jim Walter Resources is in the process of completely automating the belts by monitoring slip switches, blocked chutes, pull cords, power on and off. This will be a direct cost savings by limiting down time. Coal flows and storage capacity of coal can be monitored to maintain a steady feed of coal and limit surges. Complicating this is the overall plan now underway to tie No. 4 and No. 5 Mines together underground to transfer coal from one mine to the other. This will enable better use of the lift capacity of the portals of the two mines, keeping both facilities at maximum capacity.

The logic for controlling the flow of coal between mines will reside in the primary computers, one for each facility. Information exchanged between computers is used to control the flow to optimize production. This allows the most critical sections to run during surge times and more evenly distribute the flow of coal which will in turn lower peak power demands. Mine face equipment can be monitored for location, number of face cuts, power usage and downtime. Mine fans are being monitored for negative pressure induced by the fan, vibration, bearing temperature, and motor amperage. ther ventilation parameters rather than carbon monoxide are now being monitored. Coal washer plants are being investigated to completely automate to yield a higher percentage product recovery. All of these things will help the Mine Manager better manage his mine and therefore be reflected in cost savings to the mine.

COAL PREPARATION

Chapter 18

USING A MICRO-COMPUTER TO PREDICT THE PERFORMANCE OF COAL PREPARATION

Lu, Maixi

China Institute of Mining, People's Republic of China

Zhang, Shenggui

Anshan Ferrous Metal Mining Design and Research

Institute, People's Republic of China

ABSTRACT

Two computer packages -FIT and PRD- have been systematically developed to predict coal preparation performance. Package FIT is used to simulate raw coal washability curves and the coal preparation distribution curve, while package PRD is for predicting the performance of specific gravity separation, floatation, screening and blending.

FIT uses 6 models to fit specific gravity separation distribution curves, 3 models of which are also used to fit raw coal washability curves. Calculating 81 sets of distribution curve data and 40 sets of washability curve data indicates that these models are suitable and flexible for representing distribution curves and washability curves.

PRD employs the network flow method to predict the performance of various flowsheets. For specific gravity separation, raw coal and distribution curve data can be input through either model parameters or float-sink test data. For floatation, PRD predicts the performance according to either float-sink test results of raw coal and efficiency or pilot plant floatation test data. A case study predicting a flowsheet separation performance is described in this paper.

INTRODUCTION

In order to predict the performance of specific gravity separation, it is necessary to know both raw coal washability curves and the separation distribution curve of the devices. It is well-known that these curves result from float-sink test data. The traditional method to calculate the specific gravity separation performance is shown on Table 1.

Table 1   PREDICTION OF PRACTICAL YIELD AND ASH USING RAW
COAL WASHABILITY DATA AND DISTRIBUTION FACTORS

Specific gravity	Raw coal (%) Yield	Raw coal (%) Ash	Partition (%)	Clean coal (%) Yield	Clean coal (%) Ash	Midding (%) Yield	Midding (%) Ash
1	2	3	4	5=2*4	6=3	7=2-5	8=3
-1.30	12.9	5.1	99.6	12.8	5.1	0.1	5.1
1.30-1.35	10.0	7.1	99.5	9.9	7.1	0.1	7.1
1.35-1.40	17.6	8.6	99.0	17.4	8.6	0.2	8.6
1.40-1.45	18.5	11.6	98.5	18.2	11.6	0.3	11.6
1.45-1.50	13.9	15.5	88.7	12.3	15.5	1.6	15.5
1.50-1.55	9.7	20.1	54.0	5.2	20.1	4.5	20.1
1.55-1.60	5.4	25.3	15.6	0.8	25.3	4.6	25.3
1.60-1.65	3.2	30.1	6.3	0.2	30.1	3.0	30.1
1.65-1.70	1.9	33.8	3.0	0.1	33.8	1.8	33.8
+1.70	6.9	48.5	0.0	0.0	48.5	6.9	48.5
Total	100.0	15.45		76.9	10.68	23.1	31.3

The basic principle of this method is:

1) Assume the material of each interval of specific gravity is of the same property, i.e., ash, sulphur, and specific gravity, etc.

2) Assume the partition within a specific gravity interval is the same.

Obviously, these two assumptions will be well satisfied only when the specific gravity interval is small enough.

In fact, the specific gravity interval can not be designed small enough in a float-sink test because of both experiment cost and test error.

It is natural, using interpolation, to get additional data other than the experiment points. However, this method neither smooths the test error nor generates satisfactory curves (1).

Recently, using mathematical models to fit distribution curve has found its way in coal preparation. In this study, the authors have not only introduced more models to fit the distribution curve, but also extended them to fitting washability curves.

After the washability curves and distribution curve are fitted, the coal preparation performance for a given flowsheet can be predicted. However, the amount of calculation for both fitting curves and predicting separation performance is tremendous, needless to say, optimizing the flowsheet and the combination of device parameters. The authors have developed another personal computer package using the network flow method to calculate the separation performance after working out washability curves and a distribution curve fitting package. What is more, the authors use the orthognal design method to figure out the optimum combination of operation conditions for a given flowsheet.

FITTING DISTRIBUTION CURVE AND RAW COAL WASHABILITY CURVES

Distribution Curve Models

The distribution curve eventually gives a graphics representation of the partition y, as a function of specific gravity, x. In packages FIT, six models have been employed to fit distribution curve. They are Erasmus, Quasi-Normal Integral, Modified Normal Integral, Hyperbolic Tangent, Modified Hyperbolic Tangent, and Weibull models. The following is a brief introduction to them.

Erasmus model:

Erasmus (2) proposed the following formula

$$Arctan(k(x-c))=t2-(t2-t1)y/100 \qquad (1)$$

as distribution curve model where k, c, t1, t2 are model parameters.

Gottfried et al. (3) proposed using Weibull function to fit the distribution curve, the equation for which is as follows:

$$y=100(y0+a*exp(-(x-x0)**b/c)) \qquad (2)$$

with model parameters y0, a, x0, b, and c.

The normal integral function can be used as a distribution curve model and was recognized as such by Tromp and Terra many years ago. In package FIT, the normal integral function is represented by the Quasi Normal Integral (NI) equation:

$$y=100(a+c\int_{1.2}^{x} exp(-k(x-x0)**2)\ dx) \qquad (3)$$

Where a, c, k and x0 are model parameters.

To improve the fit at the extremes of the distribution curve, a linear term is added to (3), which now becomes:

$$y=100(a+b(x-1.2)+c\int_{1.2}^{x} exp(-k(x-x0)**2)\ dx) \qquad (4)$$

with the additional parameter b. (4) is called Modified Normal Integral model (MNI).

In this package Hyperbolic Tangent model (HT) is defined as:

$$y=100(a+c*th(k(x-x0))) \qquad (5)$$

where a, c, k and x0 are model parameters.

The Modified Hyperbolic Tangent model (MHT) is written

$$y=100(a+bx+c*th(k(x-x0))) \qquad (6)$$

by introducing a linear term bx with another parameter b.

The authors also suggest using the damped least square method to calculate the model parameters. Practice has shown the damped least square method is a fast, effective and easy-to-use method.

Washability Curve Models

Five curves are customrily used for representing washability data, (see figure 1), but the key curves are only two, the yield-specific gravity curve (Y-SG) and the cumulative float-ash curve (CF-A). Since all the necessary data are available from these two curves and the other three curves can be derived from them, only these two curves are studied in this paper.

A great deal of calculation has shown that only Erasmus, Hyperbolic Tangent and

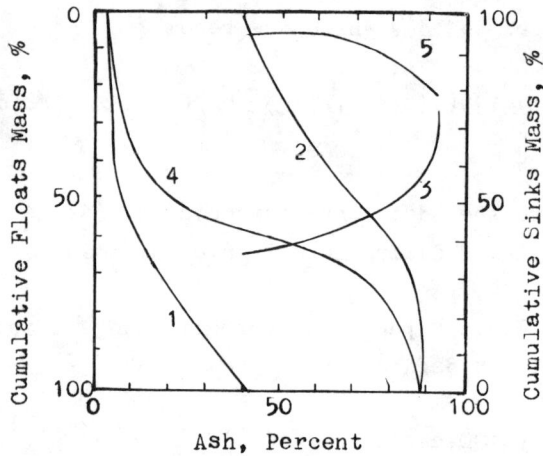

1. Cum. Float Coal - Ash Curve
2. Cum. Sink Coal - Ash Curve
3. Yield - Specific Gravity Curve
4. Elementary Ash Curve
5. ±0.10 Near Gravity Material Curve

Figure 1 Washability Curves

Modified Hyperbolic Tangent Models of the six distribution curve models are well suitable to fitting washability curves.

It should be pointed out that the ordinate of the washability curve is cumulative recovery and is conventionally plotted from 100 to 0. This is the inverse of the practice with the distribution curve which is plotted from 0 to 100. The ash content of the cumulative float coal varies from less than 5% to more than 40%. In order to use the same programs as were used for distribution curve fitting, corrdinate transformation was used before and after fitting.

Fitting Result and Analysis

A comparison of jig distribution curves from the six models is shown in figure 2. A comparison of washability curves from

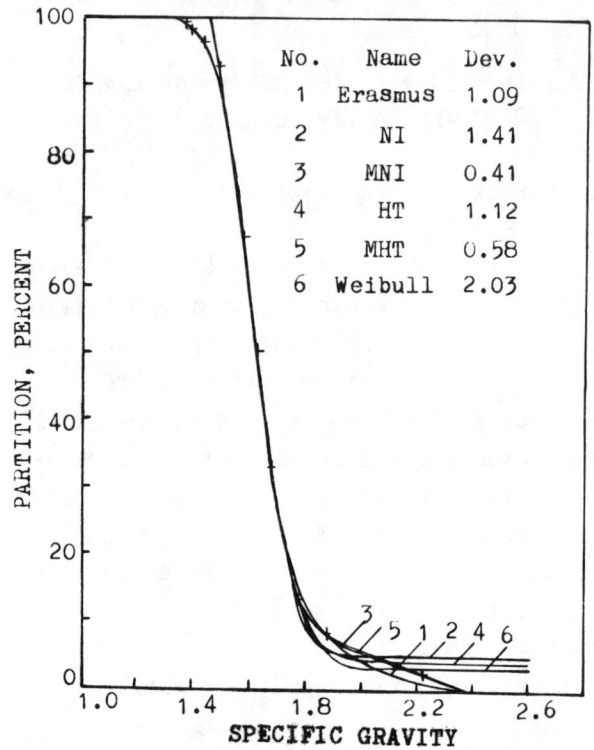

Figure 2 Fitted distribution curves for a jig

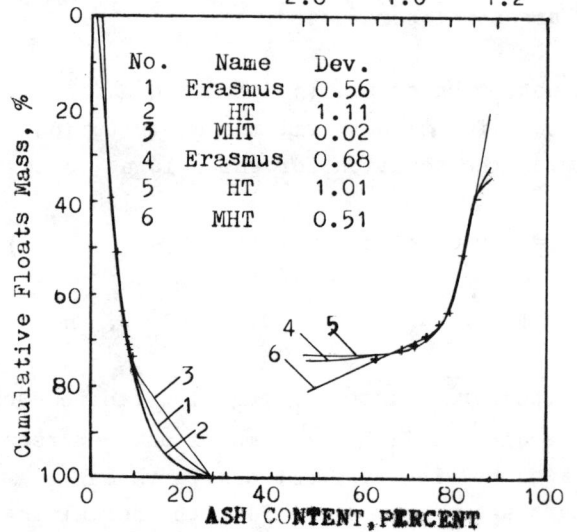

Figure 3 Fitted Washability Curves

the three models is shown in figure 3. The deviations for fitting 81 sets of distribution curves and more than 40 sets of washability curves are shown on Table 2

Table 2 FITTING DEVIATIONS FOR VARIOUS SEPERATION METHOD

	NO. of runs	fitting model					
		Weibull	Erasmus	HT	MHT	NI	MNI
Jig	34	2.14	1.30	1.70	1.01	1.97	1.05
Heavy Medium	15	2.03	1.01	0.82	0.60	0.92	0.79
Chance Cone	9	3.12	1.19	1.77	0.92	1.98	1.08
Clean Table	12	1.79	1.71	1.29	1.04	1.29	0.94
Hydraulic Cyclone	6	1.43	1.37	1.45	1.05	1.45	1.24
Spiral Seperation	5	1.08	1.26	1.55	1.08	1.71	1.03
Total	81	2.08	1.31	1.45	0.94	1.62	1.01

Table 3 THE SUMMARY OF FITTING DEVIATION FOR WASHABILITY CURVES

	Yield-Specific Gravity			Cumulative Float-Ash		
	Erasmus	HT	MHT	Erasmus	HT	MHT
Total Number	44	43	44	42	41	45
Mean Dev.	0.82	1.30	0.36	1.35	1.82	0.47
% of dev.<1	66	39	96	50	27	89
% of dev. 1-2	25	49	4	29	49	7
% of dev.>2	9	12	0	21	24	4
% of bad shape	5	0	5	0	0	4

and Table 3. The results illustrate that MHT is the best model according to the fitting criteria for both distribution curve and washability curves. Although MNI also generates a satisfactory curve, it takes a much longer time than others. It is a good idea to choose another model to refit when one model does not fit well. A considerable amount of calculation implies that fitting deviation of less than one for distribution curve means good fitting while the value above 2 is unsatisfactory, which maybe results from either bad fitting or abnormal test data (4). The conclusion is also suitable to washability curves with good fitting, resulting in the order of 0.5 of fitting deviation.

PREDICTING THE COAL PREPARATION PERFORMANCE

Generally speaking, the coal cleaning flowsheets are different with different plants. Obviously, a generalized computation procedure is very useful and convenient for the user to evaluate the performance of various flowsheets. The authors found that the network flow method is the best way to do so. In package PRD, each node of the network represents an operation of the flowsheet while the flow between two nodes depicts the material flow between the operations. Since the individual operation is the basis of the flowsheet, an introduction to various separation operations is first given in the rest of this paper.

Specific Gravity Separation

The washability curves and distribution curve discussed above are of fundamental importance to quantitatively describe the performance of specific gravity separation. This package predicts the specific gravity separation according to the same principle with manual method as showed in Table 1. The advantage of this package is that the specific gravity interval can be divided into very small parts, for example 0.01; thus it greatly increases the calculating accuracy as mentioned in introduction.

It is very clear that a distribution curve is directly associated with the specific gravity of the separation. Therefore, when the specific gravity of separation changes, the distribution curve varies correspondingly. Three methods for altering the distribution curve are used in package PRD.

To Keep Probable Error Unchanged The probable error is defined as

$$Ep=|(d25-d75)|/2 \qquad (7)$$

where d25 and d27 are the specific gravities corresponding to partitions 25% and 75%. In essence, this method simply translates a distribution curve along x axis (specific gravity) without changing the curve shape.

Keeping Imperfection of Jig Unchanged The imperfection for jig, water cyclone, and concentrating tables is defined as

$$I=Ep/(d50-1) \qquad (8)$$

where d50 is specific gravity of separation, i.e., the specific gravity corresponding to partition 50%, Ep is the probable error.

Keeping Imperfection of Dense Medium Separation Unchanged The imperfection for

dense medium separation is defined as

$$I=Ep/d50 \qquad (9)$$

where Ep and d50 are the same as above.

It should be noted that the above two methods change the fatness of the distribution curve.

Floatation

There are two ways to calculate floatation results in package PRD. One is using pilot or coal preparation plant data to express floatation results, while another uses float-sink data.

Using Pilot or Plant Data After getting the ash of clean coal and tailing each cell, the user can use the following formula to compute the yield for each cell:

$$Rci=Rri*(Ati-Ari)/(Ati-Aci) \qquad (10)$$

where Rci is clean coal yield for cell i, Ati, Ari and Aci are the ash of tailing, raw coal and clean coal respectively for cell i. Rri is the yield of raw coal entering cell i. Figure 4 shows floatation washability curves derived from Table 4 which is obtained from the pilot plant data using this formula.

Using Float-Sink Test Data Through float-sink test, a set of data of recovery and ash vs. specific gravity can be obtained. Sometimes, float-sink test data is used to predict floatation results when it is difficult for the floatation data to be obtained directly. The reason is that the characteristics of the surface chemistry of raw coal pertain to specific gravity to a certain extent. When float-sink test data is used to calculate the recovery and ash,

Table 4 FLOTATION TEST RESULT

	cell	direct		cumulative	
		W %	Ash %	W %	Ash %
	1	20.23	7.20	20.23	7.21
	2	8.48	7.78	28.71	7.39
	3	10.48	8.36	39.19	7.65
clean	4	10.32	8.59	50.01	7.85
coal	5	7.37	10.58	57.38	8.20
	6	14.39	15.03	71.77	9.58
	7	5.20	16.83	77.07	10.03
	8	0.83	18.29	77.90	10.12
tailing		22.10	62.27	100.00	21.65
total		100.00	21.65		

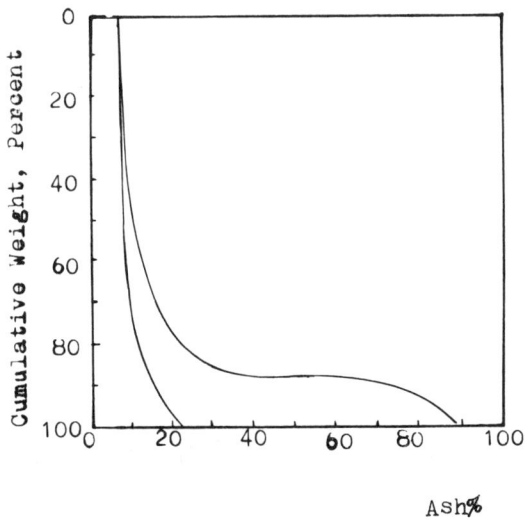

Figure 4   Washability Curves of a
            Floatation

package PRD takes into account the separation efficiency to modify the recovery in order to get the real floatation recovery.

Screen, Classification and Splitter

Screen and classification separate material according to the material sizes while splitter simply separates the material into two parts and does not change

the material property at all. In package PRD, the authors provide several schemes for the user to select.

Blending

When several materials enter the same operation or the same stockpile, material blending takes place. The blended material is either the 'raw coal' for an operation which it enters or the final product when it enters a final stockpile. In package PRD, blending is automatically performed by a subroutine according to the network flow. Therefore, the user need no longer consider material blending.

Network Flow Model

As mentioned above, the flow in a network flow represents a real material flow in a flowsheet of coal preparation. There are two kinds of flows--inflow and outflow-- for each operation. However, an outflow of one node is at the same time an inflow of another node which is called the destination node of the outflow. Therefore, if the outflow information ('raw coal' characteristics and destination nodes) of each node is known, the inflow information of each node can be obtained. In package PRD, each operation can produce only two products (a complicated operation producing more than two products may be decomposed into several operations, each of which produces only two products); therefore, each node has only two outflows, and only two destination nodes need to be inputted for each node. This technique reduces the network flow input data to minimum. Figure 5 describes a coal preparation flowsheet. Figure 6 is the network flowchart of this flowsheet. There are eight operations and six final products. After the input of

these data, the computer printed out the flowsheet as showed on Table 5.

Table 5   Coal Preparation Flowsheet

		Destination Nodes	
node	sort	fine or clean	coarse or Refuse
1	screening	2	3
2	screening	8	5
3	SG sep	10	4
4	SG sep	9	13
5	SG sep	6	13
6	SG sep	12	7
7	SG sep	12	11
8	flotation	12	14

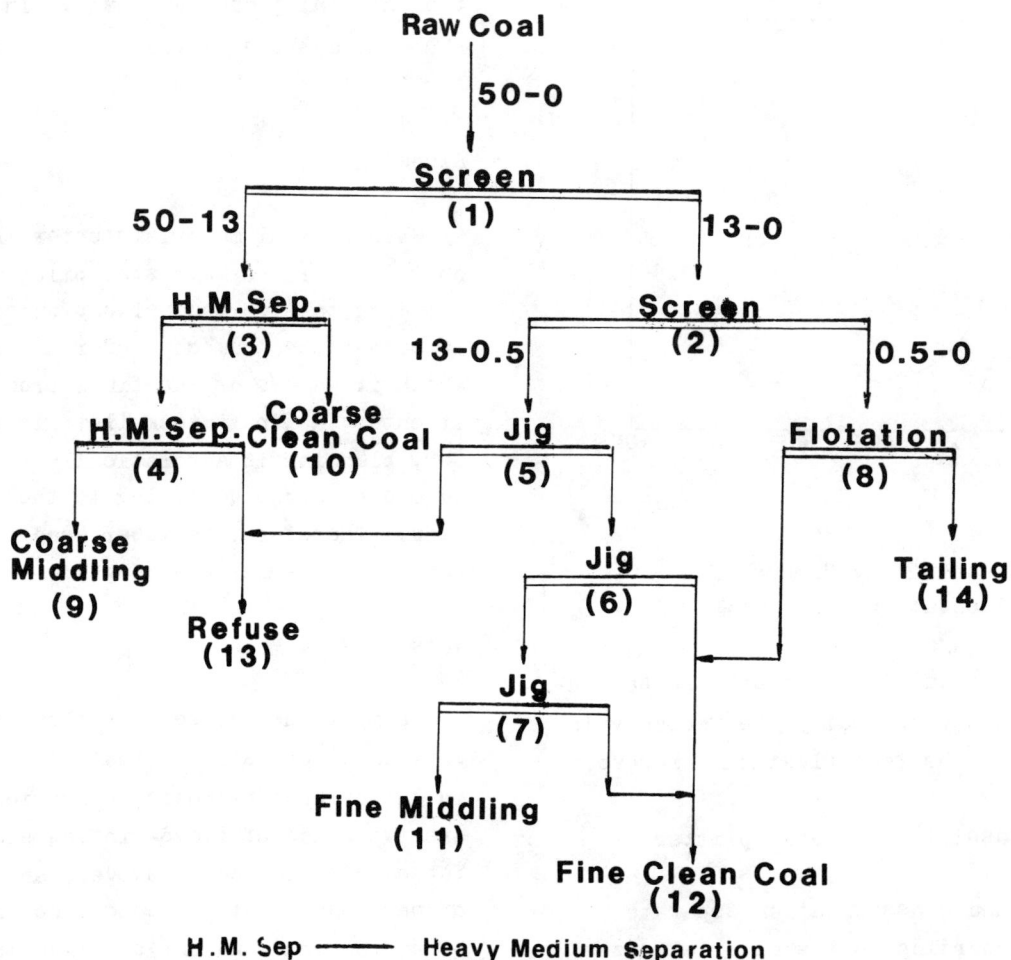

Figure 5   Coal Preparation Flowsheet

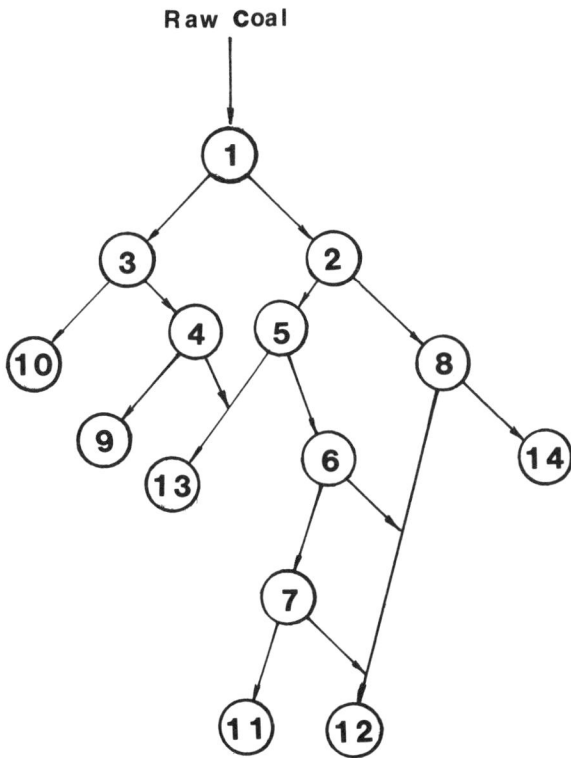

Raw Coal

Figure 6   Network Flowchart of a Coal
Preparation Flowsheet

It should be noted that the first column of destination nodes is connected with
clean or fine coal flow while the second
with coarse coal or refuse flow.

Another outstanding feature of package
PRD is that the user can arbitrarily
number the operations and need not indicate
the sequence over which the operations will
be performed. In fact, there is a subroutine, JUDGE, which automatically judges the
computation sequence of each operation. All
the considerations make this package easy
to use, even if the user has no network
flow knowledge.

Table 6 is the computer calculating

result for every node. It includes all the
information for a node. Column 2, 3, 8, 9,
10, 11 are recovery and ash for raw coal
and products. Column 4 is the sort of node.
For example, node 1 is Screen, node 3 is
specific gravity separation, node 8 is
floatation. Column 5 is specific gravity
of separation for SG separation node. Table
7 is the final product's quantity and quality. It should be mentioned that the
three Tables (Table 5, 6, 7) are all printed out by computer.

Table 7   Final Product Quantity and Quality

product	recovery %	ash %
9	2.40	28.89
10	12.86	11.19
11	4.61	56.16
12	63.84	10.79
13	11.96	77.94
14	4.33	59.92

The Optimum Design of Operation Parameters

It is obvious that package PRD can be
used to calculate the separation performance for various flowsheets. For a given
flowsheet, it is common practice to find
out the optimum operation strategy, i.e.,
the optimal combination of operation parameters. Considering so many parameters in
a real flowsheet, this package uses the
orthogonal design method to find out the
best scheme. Of course, the so-called best
scheme is approximately optimal, but not
too far from the real optimum. The user is
required to input an appropriate orthogonal
test table according to the number of the
parameters and their value ranges.

Table 6   Computer Calculating Result for Every Node

Node	Raw Coal		Separation Condition				Fine or Clean Coal		Coarse or Refuse	
	R%	A%	sort	SGS	Q	K	R%	A%	R%	A%
1	100.00	22.66	screen				79.77	20.54	20.23	31.02
2	79.77	20.54	screen				20.00	15.41	59.77	22.26
3	20.23	31.02	SG sep	1.55	4	1	12.86	11.19	7.37	65.64
4	7.37	65.64	SG sep	2.10	5	1	2.40	28.89	4.96	83.43
5	59.77	22.26	SG sep	2.10	1	3	52.77	15.39	7.00	74.04
6	52.77	15.39	SG sep	1.60	5	3	45.94	11.18	6.83	43.71
7	6.83	43.71	SG sep	1.55	5	2	2.23	17.94	4.61	56.16
8	20.00	15.41	flot				15.67	8.65	4.33	39.92

CONCLUSION

These two packages have been successfully run on an IBM personal computer. For distribution curve or washability curves fitting, package FIT takes several minutes. Computation time for predicting the performance of a given flowsheet with ten operations is less than ten minutes with package PRD. Obviously, they are cost-effective. In addition, they are also user-friendly packages. These packages are run in question-and-answer form so that the user can make full use of his experience and needs neither computer knowledge nor mathematics.

Another distinguishing feature of package PRD is that using network flow to represent a flowsheet makes the package very flexible for the user to evaluate various flowsheets, and the amount of imput data minimum. All the considerations make them convenient, effective and easy-to-use.

ACKNOWLEDGEMENT

Special acknowledgement is given to Dr. K.J.Reid, Professor and Director of Mineral Resources Research Center, University of Minnesota, for providing the research conditions and kind assistance during the time we studied at the University of Minnesota, USA.

REFERENCES

1. Reid, K.J., Lu, Maixi and Zhang, Shenggui, 1984, "Computer Reliability Analysis of Coal Distribution Curves," International Journal of Mineral Processing, (in press).

2. Erasmus, T.C., 1976, "Predicting the Performance of a Coal Washer With the Aid of a Mathematical Model," Proceedings, 7th International Coal Preparation Congress, G1, Sydney.

3. Gottfried, B.S., 1978, "A Generalization of Distribution Data for Characterizing the Performance of Float-Sink Coal Cleaning Devices," International Journal of Mineral Processing, 5:1-20.

4. Reid, K.J., Lu, Maixi and Zhang, Shenggui 1983, "Simulating Coal Preparation Distribution Curves Using a Microcomputer," International Coal Preparation. (in press).

Chapter 19

"COALBLND"
A Computer Program For Blending Coal
by

Andrew Sos III
Geologist/Hydrologist

BCM Inc.
127 Goff Mountain Road
Cross Lanes, WV 25313

Abstract. "COALBLND" is a user friendly coal blending computer program used to calculate mixing proportions needed to produce a blend of a desired quality. The need and application for such a program is discussed. An example run of the program is included to demonstrate its operation and user friendliness.

In addition to the determination of the correct blending proportions, the program calculates the correct proportions under varying moisture conditions and will perform an economic analysis on the blend if desired.

INTRODUCTION

Coal blending is becoming a very important function in today's coal marketing and processing world. It is increasingly being recognized as a viable economical and conservative method for more efficient utilization of one of the most important energy resources available today.

THE NEED FOR BLENDING

The market, public, and governmental agencies set the quality requirements that coal producers must meet with their product. The acid rain controversy has sparked both the public and the Environmental Protection Agency to demand more stringent quality control on coal consumers. These stiffer quality requirements have, in turn, created many problems for the coal industry to solve. Although sulphur is the most environmentally damaging component of coal, it is not the only parameter of importance to the consumer. Other parameters of concern are commonly moisture, ash and BTU. Suppose the market requires the coal to contain two percent sulfur, twelve percent ash and thirteen thousand BTU's. The producer may have several coals that meet one parameter requirement, but none that meet all of them.

Failure of the producer in meeting the quality requirement results in a penalty or reduction in the price received for his coal. In order for coal to be a viable and competitive source of

energy today it is necessary for it to meet these new standards.

How can coal operators meet the specifications that the marketplace, government and public impose on coal users? In some cases it may be impossible for a producer to market his coal at a fair price because it does not meet the quality standards of the marketplace. The operator has three choices: 1) shut down his operation, 2) sell his product at a very small profit to someone who has compatable coals that can be blended to produce a marketable product, 3) or he may purchase a compatable coal and blend it himself. Blending has been going on for years, but the practice has not reached its full potential and, as a result, operators are losing money needlessly.

Blending two different coals to produce a mixture that meets a certain specification for one parameter is not very difficult. For example, if one ton of coal X with a sulfur content of three percent is mixed with one ton of coal Z at one percent sulphur, two tons of a blend with two percent sulphur is produced.

However, it is a very complicated task to control more than one parameter during a blending operation. Most specifications require that at a minimum, sulphur, ash, and BTU meet specific values. Not only does the sulphur have to be controlled, but the ash content must also be held in check. In order to control or manipulate two parameters in a blend, three sources of coal are necessary.

Assume coal X has three percent sulphur and eleven percent ash, coal Y has one percent sulphur and fourteen percent ash and the specifications call for two percent sulphur and twelve percent ash (Table 1). If the same blend is used as before—one ton of coal X and one ton of coal Y—a blend is produced with two percent sulphur and twelve and one half percent ash. This obviously does not meet the specifications of twelve percent ash and will result in a penalty. These two coals cannot be blended to meet this specification. If less coal Y is used with one percent sulphur, the ash content

decreases, but the sulphur is over the
specification.

In order to meet the criteria another coal
compatible to the blend must be introduced. The
blending problem becomes more difficult when
working with three coals, all with different
parameter values. Assume coal Z is available and
has point five percent sulphur and thirteen
percent ash. Table 1 lists the coals to be
blended and shows their parameters and the target
specifications which are to be met.

	SUL	ASH	BTU
COAL X	3%	11%	13000
COAL Y	1%	14%	12000
COAL Z	0.5%	13%	14000
SPECS.	2%	12%	13000

Table 1

How can the operator calculate the correct
proportions to best utilize the coal he has on
hand and meet the specifications without wasting
parameters? Wasting parameters means selling
coal with a higher quality than the market
specifies when it could be blended with a coal of
lesser quality, thus creating a blend that meets
the specifications and conserves the high quality
coal. This allows the operator to sell his
lesser quality coal at a higher price and
increase his profits. For example, selling coal
with one percent sulphur when the specification
calls for two percent sulphur, is wasting
sulphur. This wasted sulphur could be blended
with a coal containing a higher sulphur
percentage and possibly make an unmarketable coal
marketable or a marketable coal more profitable.

THE BLENDING PROGRAM

Getting back to the question above, what
proportions of the coals must be mixed in order
to meet the specifications? Does the operator
use trial and error or does he use the computer
program "COALBLND"? "COALBLND" is a quick and
inexpensive way to solve coal blending problems.
It is user friendly (the program prompts for
input) and a rapid method for determining the
blending proportions necessary to meet specified
parameters.

In the above example, coal X is out of
specification on sulphur, coals Y & Z are out of
specification on ash. Here are three sources of
coal, all of which are out of specification for
our particular market. By blending these coals
the specification will be met and all of the
coals can be marketed. Figure 1 is an example
run of "COALBLND" using the inputs from Table 1.
If a blend is impossible, that can be determined
right' away too. Obviously all conditions cannot
be met with a limited number of coals. In other
words, a given set of parameters cannot always be
achieved by the available coals. The coals being
blended must be compatible for the specified

perameters.

The program also calculates the approximate
BTU of the blend provided the BTU of each coal in
the blend is entered at the beginning of the
program. BTU is one of the important parameters
that must be considered when blending. The BTU
cannot be handled in the same manner as sulphur
and ash because BTU is a property of the coal and
not a physical part of its composition like
sulphur and ash. For this reason "COALBLND" does
not proportion the blend based on BTU, but rather
calculates the resultant BTU based on the various
proportions of the coals necessary to produce the
blend.

Moisture correction is another feature of
"COALBLND". "COALBLND" allows the operator to
make a moisture correction prior to blending in
order to produce the proper blend under varying
moisture conditions. This is especially
important if blending is being done using a
weight rather than a volume method.

The blend is calculated on a dry basis and
then the amount of moisture in each coal blended
is entered and a new ratio is calculated based on
wet weight. As already mentioned the blend can
now be accomplished on a weight basis regardless
of the coal's moisture content.

"COALBLND" can also perform an economic
analysis of the blend. In some cases it may not
be most profitable to blend the coal, thus it can
be sold unblended.

The program prompts or asks if the user wants
a cost analysis. A "yes" answer initiates
additional prompts which include: the production
cost of each coal, the market price of each coal,
and the market price of the blend. The program
calculates the percent profit for each coal which
can be compared with the percent profit for the
blend.

CONCLUSION

The producers who will profit most from the
undiminishing completion in the mining industry
will be those who take advantage of beneficial
computer oriented methods geared to solving
production problems.

"COALBLND" is an easy to use computer program
designed to help achieve this goal. Those with

limited or no computer experience need not be
apprehensive about its usage. The program has
its limitations, but where applicable will prove
a most important tool for the coal industry.

RUN

      COALBLEND

   A PRODUCT OF B-C-M

==
THIS PROGRAM IS A COALBLENDING PROBLEM SOLVER

==

DO YOU WANT COST ANALYSIS Y/N ? N

==
DO YOU WANT MOISTURE CORRECTION Y/N ? N

==
HOW MANY COAL SEAMS ARE TO BE BLENDED? 3

==
NAME THE COAL SOURCES TO BE BLENDED? X

NAME THE COAL SOURCES TO BE BLENDED? Y

NAME THE COAL SOURCES TO BE BLENDED? Z

==
 ENTER THE BTU OF X COAL
? 13000
 ENTER THE BTU OF Y COAL
? 12000
 ENTER THE BTU OF Z COAL
? 14000

==
BY BLENDING 3 COALS THIS PROGRAM ENABLES YOU TO
CONTROL 2 PARAMETERS

==
ENTER THE PARAMETERS ? SUL
ENTER THE PARAMETERS ? ASH

==

ENTER THE VALUES FOR THE SUL CONTENT OF EACH
COAL TO BE BLENDED:

 1) X SUL CONTENT = ? 3
 2) Y SUL CONTENT = ? 1
 3) Z SUL CONTENT = ? .5

 4) DESIRED SUL CONTENT =? 3

==
-- TO MAKE A CHANGE ENTER NUMBER --? 4
NEW VALUE ? 2
 1 : SUL CONT OF X COAL IS 3
 2 : SUL CONT OF Y COAL IS 1
 3 : SUL CONT OF Z COAL IS .5
 4 : DESIRED SUL CONTENT IS 2

==
-- TO MAKE A CHANGE ENTER NUMBER --?

==
ENTER THE VALUES FOR THE ASH CONTENT OF EACH
COAL TO BE BLENDED:

 1) X ASH CONTENT = ? 11
 2) Y ASH CONTENT = ? 14
 3) Z ASH CONTENT = ? 13

 4) DESIRED ASH CONTENT =? 12

==
-- TO MAKE A CHANGE ENTER NUMBER --?

==
THE CORRECT PROPORTIONS NEEDED TO PRODUCE THE
DESIRED BLEND ARE

    57 PERCENT X COAL WITH A BTU OF   13000
    14 PERCENT Y COAL WITH A BTU OF   12000
    29 PERCENT Z COAL WITH A BTU OF   14000

       THE BTU OF THE BLEND IS   13143

==
DO YOU WANT TO RERUN Y/N? N
DO YOU WANT A PRINTOUT Y/N ? N

==

FIGURE 1. AN EXAMPLE RUN OF "COALBLND" USING DATA FROM TABLE 1

Chapter 20

A MODEL FOR DETERMINING THE COST-BENEFIT OF MODIFYING
CLEAN COAL SAMPLING PROGRAMS

by

E.L. Gillenwater
Department of Management
University of Kentucky

G.T. Lineberry
Department of Mining Engineering
University of Kentucky

Abstract. The primary objective of this study is to determine if altering the sampling program, by differing sampling decision rules, can result in tighter quality control of a preparation plant's clean coal output, and if so, to determine if it is economically desirable to do so. The basic trade-off that must be investigated is the product gain possible by altering the sampling program and the increased cost of sampling and analysis.

To aid in the analysis of this trade-off between product gain and sampling and analytical costs, a computer model has been developed. Given the sampling history of a plant under study, the program analyzes the plant's current sampling scheme. Applying systematic decisions on sampling frequency based on results of previous sample analysis by the use of the Markovian Chain Theory and other statistical and mathematical techniques, new sampling schemes can be modeled and compared using economic considerations. Through this method, it may be found that for a given plant, overall economics can be improved by altering the sampling program, most likely by requiring additional sampling. Effects of premiums and penalties, as well as sampling and analytical costs on plant economics are also included. An example of the model's use will conclude the paper.

INTRODUCTION

Coal sales agreements invariably contain a quality description of the coal being purchased. This description states specific requirements for ash, heat content, moisture, and other characteristics, which are largely dependent on the ultimate use of the coal. The contractual range within which the quality of coal must fall is known as the "deadband," or, "area of acceptance." In this study, ash content is used for illustrative purposes, although the same principles apply to all other quality descriptors.

From a buyer's viewpoint, coal not meeting the contractual requirements can be rejected or assessed penalties, which has the net effect of reducing the selling price. Furthermore, the buyer desires a uniform, homogeneous fuel product with known qualities. This is of particular im-

portance to electric utilities. From the seller's point of view, potential penalties and shipment rejection provide incentive to keep coal quality within specified ranges, while attempting to maximize revenue.

In order to meet contract specifications, a large percentage of U.S. coal production is processed in cleaning facilities. The primary objective of the coal preparation plant is not to produce coal of the lowest possible ash and sulfur content, as has been suggested (Anderson, 1981), but to produce a near-uniform product which maximizes plant economics. One method to achieve a near-uniform product is to closely control the ash content of the clean coal output so that high-ash, low-cost material can be blended with low-ash, high-cost material, effectively narrowing the deadband.

If, for example, a coal sales agreement calls for an 8% ash coal, but in ensuring acceptance of the shipment, a 6.5% ash coal is being shipped, an important source of revenue is being lost. If the plant has the option of replacing only 20 lb/T of $25/T coal for 20 lb/T of $30/T coal in a raw coal blend in a 400 TPH plant (assuming 250 operating days at 16 hours per day), a yearly gain of approximately $80,000 could be realized, exclusive of additional sampling. In addition, increased recovery of clean coal would be expected at the increased ash, resulting in additional cost savings.

The current model considers the trade-off between product gain and sampling and analytical costs, as well as the impact of premiums and penalties. It is expected that the model will find greatest usage by operators of high-volume coal preparation plants and of electric power plants. To make full use of the model requires extensive historical plant performance data, but approximations can be made with available plant data. Historical plant data is used to model the effect of systematic decisions on sampling frequency, and the overall economics is considered for each feasible sampling scheme. The optimum sampling schemes for a given plant can then be offered to the plant operator for consideration and the improvement in plant economics for each sampling program can be estimated.

BACKGROUND

To date, clean coal sampling for quality control has been little more than an "autopsy" of the previous day's throughput. The time-consuming methods of ASTM coal sample analysis have in part been responsible for this situation. Results of samples taken one day were normally not available until the following day. As a result sampling, to some extent, has been considered as a "necessary evil" and has not served as a preparation plant control procedure. This view of sampling is now beginning to change, as more major coal producers and processors are investing in coal analysis laboratories near their operations, allowing feedback of plant performance in less than one shift. In addition, development of rapid off-line and on-line sample analysis equipment has given preparation plant operators the capability to sample their output and receive the results of that sample within minutes or hours. This has added a new dimension to coal preparation plants, permitting adaptive capability and the feasibility of meaningful process feedback and control.

In the course of operation, the quality level of clean coal output fluctuates, not only from day to day, but also hourly. Many factors, both controllable and uncontrollable, contribute to this fluctuation, including variable input quality, variable input rate, and operating characteristics of the plant. For any coal preparation plant, there is some target level of output quality that plant operators attempt to achieve. Actual output quality will vary around this target level. Figure 1 illustrates this fluctuation.

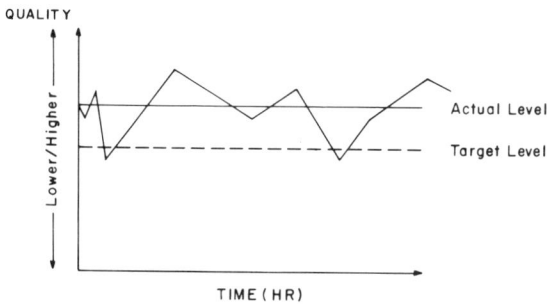

Figure 1. Output Quality Fluctuation

All contracts contain some description of coal quality. Specifications may vary among contracts for different suppliers and consumers, with some requiring tighter quality standards than others. These quality specifications are generally stated in ranges of values around an expected quality level. This range of acceptable values is known as the "deadband" (see earlier). If actual clean coal quality does not fall within the deadband range, then penalties or premiums are incurred on the per ton sales price. These penalties or premiums are assessed against (or attributed to) the entire shipment for which the samples of unacceptable (or highly acceptable) quality were found.

A hypothetical deadband for ash content is shown in Figure 2. In this example the expected value of ash content for any sample is 7%; however, samples drawn within the range of 6% to 8% are acceptable. Samples drawn with ash content outside these ranges cause the shipment to be assessed a penalty (or premium). Many contracts also contain rejection limits, a severe penalty for shipment of an "out-of-spec" coal. Rejection limits represent quality levels below which entire shipments may be rejected by the buyer. This limit is shown as a dashed line in Figure 2.

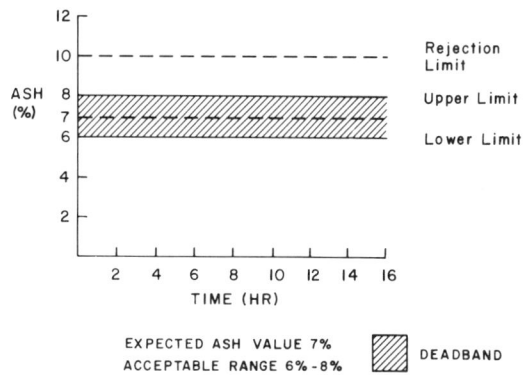

Figure 2. Deadband Range

Ash was chosen as the main control variable for this study since it is an acceptable indicator of overall coal quality. In addition, if it is shown that additional sampling of clean coal output can improve plant economics, there is available technology to implement rapid ash analysis.

SIGNIFICANCE

As noted in Figure 1, the quality level of output varies with time. In order to maintain clean coal quality within the limits imposed by the deadband, sample results must be used as adjustment signals so that control can be exerted over the preparation process. The following two procedures illustrate the use of increased sampling frequency, and plant adjustments implied therein, to move the level of coal quality from an unacceptable one to an acceptable one. An illustration of these two sampling procedures is given in Table 1 and figures 3A and 3B.

TABLE 1. Example of results of two different sampling procedures.

Procedure 1

EXPECTED ASH LEVEL, 7% Deadband range, 6%-8%
Sampling Rule - Once per shift and at plant shutdown.

Plant operates 2 shifts per day (16 hours).

Shift	Hours Into Shift	Hours Into Day	Sample Result (% Ash)
1	4	4	9.0
2	1	9	8.5
2	8	16	7.25

WEIGHTED AVERAGE ASH for day is 8.43.

Procedure 2

EXPECTED ASH LEVEL, 7% Deadband range, 6%-8%
Sampling Rule - Sample one hour into 1st shift of the day, if previous sample acceptable ($\leq$ 8%) sample in 4 hours. If previous sample unacceptable, (> 8%) sample in 1 hour, then sample at end of shift.

Shift	Hours Into Shift	Hours Into Day	Sample Result (% Ash)
1	1	1	9.0
1	2	2	8.5
1	3	3	7.25
1	7	7	7.0
2	3	11	7.5
2	7	15	8.0
2	8	16	7.75

WEIGHTED AVERAGE ASH 7.62.

Procedure 1 represents a situation in which clean coal output is sampled once per shift and at plant shutdown. Three samples are taken, the first, four hours into the shift, the second at the beginning of the next shift, and the third at the end of the day. The sample results were 9.0%, 8.5%, and 7.25%, respectively, with a resultant daily weighted average ash of 8.43%. The daily average ash of Procedure 1 is outside the acceptable range, making it very probable that a shipment sample drawn from this coal would be assessed a penalty.

Procedure 2 represents a situation in which clean coal output is sampled one hour into the first shift of the day, and then depending on whether the results are acceptable ($\leq$ 8% ash) or unacceptable (> 8% ash), another sample is taken on four or one hour intervals, respectively. (Full explanation of model theory will be given in the next section.) Seven samples were taken over a sixteen hour period (the same time frame as Procedure 1). This procedure resulted in daily weighted average ash 7.62%. By examining Figures 3A and 3B, it can be seen that the quality level of the clean coal, based on percent ash, can be more quickly brought back into the acceptable range by more frequent sampling.

It is vital that a systematic procedure be developed to optimize the sampling program for any given coal preparation plant. Such a procedure can minimize the risk of shipment rejection and can prevent the shipment of material of significantly better quality than contracted.

Literature Review

Most approaches to coal sampling have taken one of four directions:

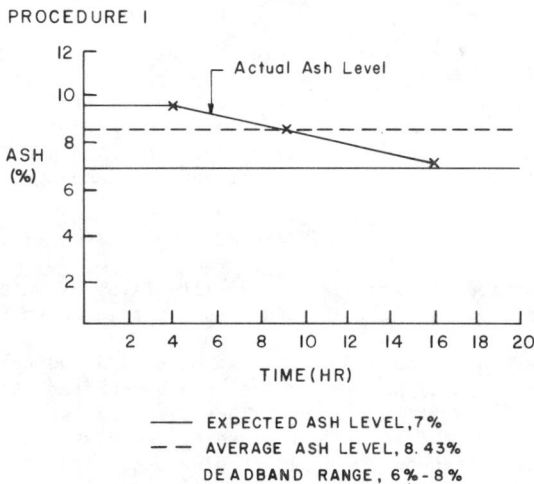

PROCEDURE 1

— EXPECTED ASH LEVEL, 7%
– – AVERAGE ASH LEVEL, 8.43%
DEADBAND RANGE, 6%-8%

Figure 3A

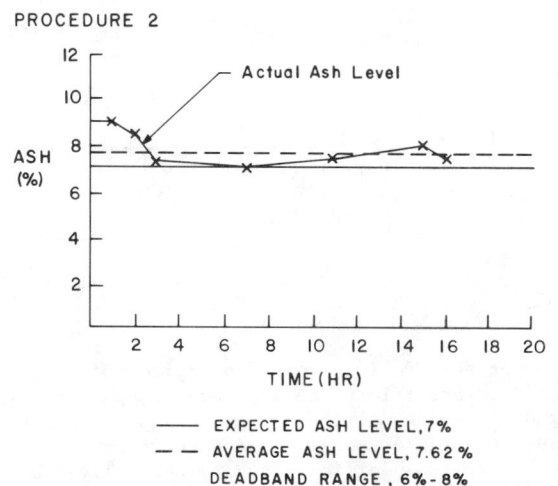

PROCEDURE 2

— EXPECTED ASH LEVEL, 7%
– – AVERAGE ASH LEVEL, 7.62%
DEADBAND RANGE, 6%-8%

Figure 3B

1) Problems associated with the actual sampling process (representativeness, etc.),
2) Problems connected with sampling for end-users of coal (e.g., utilities, steel, etc.),
3) Purely technical treatises on sampling,
4) Process monitoring and control for coal preparation plants.

Most of the pertinent literature examines sampling from the first three perspectives.

Work in the area of sampling methodology (e.g., Aresco, 1965; Keller, 1965; Symonds and Charmbury, 1978; Gy, 1981; Gould, 1982) has primarily dealt with sample representativeness, as well as sample analysis.

Quinlan and Venkatesan (1976) examined the possibility of producing higher quality clean coal for power generation. The likelihood of additional revenue for the producer was mentioned but not fully explored. It was suggested that further work be done in this area. Studies that examined sampling as a tool to increase the revenues of preparation plant operators or to provide a homogeneous clean coal product were not found.

A new technique of interpreting washability data, using three dimensional graphics, has been developed (Reeves, 1984), but the author has warned "that proper sampling and laboratory techniques are essential for accurate evaluation." New on- and off-line sampling devices have created the ability to more closely monitor the quality of plant output (Cooper, 1983; Gozani and Stenftenagel, 1983; Cooper, 1984), but little has been done to fully investigate the economic feasibility of their use. Nevertheless, advancements in process monitoring and control of coal preparation plants are receiving increasing attention (Kemp, 1984; Penny, Longden, and Lidgate, 1979; Cooper, 1984). Kemp (1984), in his discussion of this technology, notes that the mining industry in general, and the coal industry in particular, has been slow to participate in the technological advances in process monitoring and control. If additional sampling is prescribed by the results of this study beyond the current capabilities of plant operators, more serious attention should be given to this available technology, which has already "contributed greatly to improve the productivity of virtually all manufacturing industries throughout the world" (Kemp, 1984).

MODELING TECHNIQUE

A computer model, using Markov Processes (Anderson, Sweeney and Williams, 1982), was developed to simulate differing sampling procedures for a coal preparation plant. Markov models are useful in studying the evolution of certain systems over repeated trials. These repeated trials are often successive time periods, where the state or outcome of the system in any one time period cannot be determined with certainty. A set of transition probabilities describes the manner in which the system makes the transition from one period to the next.

The model has a finite number of states, the transition probabilities remain constant over time, and the probability of being in a particular state at any time period is dependent only upon the state of the process in the immediately preceding time period.

The current model has been restricted to only two possible states:

State 1: Coal quality as measured by percent ash is within acceptable ranges
State 2: Coal quality as measured by percent ash is not within acceptable ranges

From historical data, a matrix of transition probabilities can be developed. Questions about the probability of being in a particular state in a future time period can then be answered. This matrix of transition probabilities is shown in Table 2. The probability that the system is in state i at period n is referred to as a state probability.

TABLE 2. Matrix of transition probabilities.

Current Time Period	Next Time Period	
	State 1	State 2
State 1	P_{11}	P_{12}
State 2	P_{21}	P_{22}

P_{ij} = probability of making a transition from state i in a given time period n, to state j in the next time period, n + 1

Program Development

A computer program was developed that simulates the decision-making process that could occur as a result of a given sample outcome. The program is based on the assertion that if a preparation plant is not operating in an acceptable range, some corrective action should be taken, after which another sample should be drawn to determine the effect of the corrective action on current output. The program simulates this procedure, examining all feasible sampling decisions and finally listing the twenty most desirable in decreasing order of potential cost savings, or, net revenue gain.

In order to develop this program for a specific coal preparation plant, a data base of sample results in chronological order is needed. The larger the data base, the more reliable the calculation of the state probabilities. To calculate the state probabilities, the chronologically ordered data must be classified into four groups:

I) Acceptable sample results that occurred in period t in which sample results in period t-1 were also acceptable.
II) Acceptable sample results that occurred in period t in which sample results in period t-1 were unacceptable.
III) Unacceptable sample results that occurred

in period t in which sample results in period t-1 were acceptable.

IV) Unacceptable sample results that occurred in period t in which sample results in period t-1 were also unacceptable.

State 1 is defined as the occurrence of an acceptable sample (the plant is operating in an acceptable range). State 2 is defined as the occurrence of an unacceptable sample (the plant is not operating in an acceptable range).

A transition matrix can then be set up and the matrix of transition probabilities calculated, as shown in Table 3. The samples are classified into one of the four groups and entered into the transition matrix. The transition probabilities for each transition cell are then calculated by:

$$P_{ij} = \frac{\text{the number of occurrences in cell ij}}{\text{the total number of occurrences of state i}}$$

where the number of occurrences in cell ij is the number of transitions from state i to state j. These probabilities are then used in the simulation to determine the occurrence of a specific sample result.

The program also allows other relevant information to be input, regarding key contractual limits and values and productivity parameters. Table 4 lists allowable inputs and sample values. These sample values will be used in an example of the model's use.

TABLE 3. Transition matrix and calculation of transition probabilities.

Sample Number	Sample Result	Sample Number	Sample Result	Sample Number	Sample Result
1	.08	12	.092	23	.077
2	.09	13	.085	24	.070
3	.075	14	.078	25	.076
4	.078	15	.069	26	.079
5	.076	16	.078	27	.082
6	.082	17	.082	28	.088
7	.076	18	.079	29	.082
8	.065	19	.078	30	.075
9	.069	20	.082	31	.070
10	.078	21	.081	32	.065
11	.081	22	.076	33	.055

	Transition Matrix			Transition Probabilities Matrix		
	State 1	State 2		State 1	State 2	
State 1	15	6	21	0.714*	0.2857	1.0
State 2	6	5	11	0.5455	0.4545	1.0

$*P_{ij} = 15/21 = 0.7143$

TABLE 4. Allowable inputs and sample values.

1. Number of operating days to be simulated: 22 days (month's operations)
2. Sampling cost on a per sample basis: $50
3. Clean coal output in tons per hour: 500
4. Number of plant production hours per day: 16
5. Selling price per ton of coal: $30
6. Penalty in $ per ton on unacceptable coal shipments: $3
7. Premium in $ per ton on shipments above a specified quality level: $2
8. Present average ash of current coal shipments: .06
9. Contract specification for ash: .08
10. Upper limit on ash content: .095
11. Lower limit on ash content: .04
12. % ash below which shipments will receive a premium: .05
13. % ash above which shipments will incur a penalty: .085

Example of Model's Use

Numerous runs were made with a variety of different inputs to validate the program. The example listed here was chosen simply to illustrate the expected results and not to indicate a specific optimum sampling decision rule.

The input values for the sample run are listed in Table 4. The run examines all sampling combinations from one to eight hours between samples, in one hour increments for acceptable samples, and one-half hour increments for unacceptable samples. This generated 72 different sampling combinations. An example of the output associated with each combination of decision times is shown in Table 5. The simulation results for all sampling combinations are sorted by net revenue gain with the top twenty combinations listed in Table 6.

TABLE 5. Example of output of one selected sampling decision rule.

Scenario information summary 8

Decision Rule: 3 hours between acceptable samples
                1 hour between unacceptable samples

Number of operational days:	22
Average ash for the 22 operational days:	.072
Total product savings (or loss):	1317.6
Value of product savings:	39528
Total sampling cost:	8350

Total Premium Received At $2 per ton	Total Penalty Incurred At $3 per ton	Difference
0	0	0

Value of product savings less total sampling cost @ $50 per sample	31178
Total gain or loss (-)	31178

TABLE 6. Top twenty feasible decision rules, sorted by decreasing order of estimated net revenue gain.

Time Between Acceptable Samples (h)	Time Between Unacceptable Samples (h)	Estimated Net Revenue Gain ($/mo.)
7	4	75,244
8	1.5	72,278
6	.5	68,366
7	.5	68,020
8	3	64,950
8	1	60,304
1	1	54,304
8	3.5	53,798
5	1	52,394
5	3	52,110
3	3	51,216
7	5	50,632
6	1	50,358
4	1.5	49,682
8	2	49,448
8	.5	46,484
6	1.5	44,234
5	2	43,970
2	2	42,348
8	2.5	41,842

A simple regression analysis was performed on the 72 feasible decision rules, using revenue as the dependent variable and time between acceptable and unacceptable samples as independent variables. With only time between unacceptable samples considered:

$$REVENUE = 44579.2 - 7816.2 \, T_1$$

where, REVENUE = total net gain or loss ($)
and T_1 = time between unacceptable samples (h).

This model defines the relationship between revenue and time between unacceptable samples. The model has an R^2 value of 0.215, indicating that the variable accounts for 21.5% of the variance in revenues.

With both variables (time between acceptable samples and time between unacceptable samples) considered, an overall negative relationship still exists:

$$REVENUE = 38522.9 - 8534.4 \, T_1 + 1459.6 \, T_2$$

where, T_2 = time between acceptable samples.

The R^2 value in the full model is 0.220, indicating that 22.0% of the variance in revenues is explained by the model. This indicates that virtually all variance is explained by time between unacceptable samples. Correlations for time between samples and revenues were calculated and both showed negative correlations, −0.16 for acceptable samples and −0.46 for unacceptable samples. This accentuates the relationship between decrease in time between samples and revenues. In general, the simulation program supports an increase in the number of samples taken per day.

If these results had been output from actual sample data and operating characteristics of a preparation plant, a rational course of action would be to consider some of the sampling combinations in Table 6. Not all recommended combinations will be acceptable to a specific operator for a variety of reasons, including time for sampling and analysis, manpower shortages, and plant operating schedule. In this example, however, since even the adoption of the twentieth best sampling decision rule projects a gain in net revenue of $503,104 ($41,842/mo. x 12 mos/yr), there is adequate justification for questioning the current sampling procedure.

Only one example run has been examined here; however, similar results were obtained for runs over a wide range of plant operating characteristics. Revenue gains varied but remained positive over all sample runs.

Conclusions and Recommendations

The current study provides justification for increased frequency of sampling when a preparation plant is not producing to specification, but does not rule out decreased frequency of sampling to effect overall plant economics. The program also allows rapid prediction of plant economics, which result from variable contractual limits and operating characteristics.

The program can be easily modified to accommodate any transition matrix, developed for any preparation plant. In order for this to be accomplished, adequate records of previous sample results must be maintained. Because no operators were found with sufficient historical data, or with a willingness to allow the data to be used, actual sample data was not used in the program development. More thorough recordkeeping is recommended for full benefit of the proposed model.

The results of this preliminary study also justify further study of the economic feasibility of the use of on- and off-line rapid sample analysis equipment. If the sampling frequency is to be increased, then reliable results must be more rapidly obtained to permit practical diagnostic process feedback and control. The reliability of rapid analyzers can be tested in laboratory settings, but the relationship between speed and reliability should only be verified by plant practice, with actual economic conditions being evaluated in conjunction with equipment measures. Results of this study seem to support on- and off-line analysis, but more extensive investigation is recommended.

BIBLIOGRAPHY

Anderson, D.R., Sweeney, D.J., and Williams, T.A., 1982, "Markov Processes," in Management Science, Third Ed. West, St. Paul, pp. 603–606.

Arthur Anderson & Company, 1981, The Business of Coal, (New York: McGraw Hill), pp. 53-54.

Aresco, S.J. and Orning, A.A., 1965, "A Study of the Precision of Coal Sampling, Sample Preparation and Analysis," SME Transactions, Vol. 232, Sept., pp. 258-264.

Cooper, H.R., 1983, "On-line Coal Analysis Promises to Improve Profitably," Coal Mining and Processing, Vol. 20, No. 9, Sept., pp. 53-56.

Cooper, H.R., 1984, "Progress in On-line Coal Quality Measurement," Journal of Coal Quality Vol. 3, No. 1, Jan., pp. 16-23.

Gozani, T.G. and Stenftenagel, J.S., 1983, "On-line Rapid Ash Meter," Paper No. A1/D2000-27, American Mining Congress Coal Convention, St. Louis, Missouri.

Gould, G., 1982, "Coal Sample Representativeness," The Journal of Coal Quality, Vol. 1, No. 3, Summer, pp. 23-26.

Gy, P.M., 1981, "Sampling or Gambling?" Coal Mining and Processing, Vol. 18, No. 9, Sept., pp. 62-67.

Huck, W., 1983, "Instrument Maintenance Ensures Coal Quality," Coal Mining and Processing, Vol. 20, No. 2, Feb., pp. 32-34.

Keller, G.E., 1965, "Determination of Quantities Needed in Coal Sample Preparation and Analysis," SME Transactions, Vol. 232, Sept., pp. 218-226.

Kim, Y.C. and Barwa, S.L., 1984, "Need for Closely Spaced Samples for Short-Term Coal Mine Planning," Mining Engineering, Vol. 36, No. 3, March, pp. 276-280.

McGraw, M., 1982, "Legal Principles of Sampling and Testing," The Journal of Coal Quality, Vol. 1, No. 4, Fall, pp. 24-26.

Penny, D.F., Longden, C., and Lidgate, P.D., 1979, "Process Systems Control for Coal Preparation Plants," Symposium Proceedings, The Association of Mining, Electrical and Mechanical Engineers.

Phillips, P.J. and Cole, R.M., 1980, "Economic Penalties Attributable to Ash Content of Steam Coals," Mining Engineering, Vol. 32, No. 3, March, pp. 297-302.

Quinlan, R.M. and Venkatesan, S., 1976, "Economics Count in Coal Preparation Planning," Coal Mining and Processing, Vol. 13, No. 9, Sept., pp. 80-88.

Reeves, R., 1984, "Washability Data at a Glance," Coal Mining, Vol. 21, No. 8, Aug., pp. 52-56.

Symonds, D.F. and Charmbury, H.B., 1978, "Coal Prep Sampling: An Update," Coal Mining and Processing, Vol. 15, No. 9, Sept., pp. 88-92.

Vaninetti, G.E., 1982, "Mineral Analysis of Ash Data: A Utility Perspective," The Journal of Coal Quality, Vol. 1, No. 2, Spring, pp. 22-31.

Chapter 21

COAL STOCKPILE MANAGEMENT SOFTWARE FOR MICROCOMPUTER

SUKHENDU L. BARUA AND YOUNG C. KIM

DEPARTMENT OF MINING AND GEOLOGICAL ENGINEERING
THE UNIVERSITY OF ARIZONA, TUCSON, AZ 85721

ABSTRACT

A Fortran based software to aid in dealing with coal stockpile management problems is discussed here. The objective of coal stockpile management is to control coal flows from different sources to a series of stockpiles and subsequently from stockpiles to the power plant in order to maintain a smooth and proper feed to the boiler.

The developed software has two functionally distinctive, but interconnected, parts. The first part consists of a set of routines to create and update coal stockpile information in the computer. Here, the user can interactively determine the disposition of the run-of-mine coal or coal from other sources to any existing stockpiles or to a new stockpile depending on its quality and quantity. The user can also determine when and how much to remove from each stockpile using the second part of the software. The program displays the stockpile status before and after the coal movement and generates a hardcopy output, if so desired. The stockpile file becomes input to the second part.

In the second part, multi-objective goal programming technique is utilized to solve the feed blending problems using varying qualities of stockpiled coal. The program can be run either interactively or in a batch mode. Goal programming determines the amount of coal needed from potential stockpiles to make up the desired blend of coal. This blended coal is either sent to the cleaning plant for coal washing or to the boiler for coal burning. Whenever there is a change in the stockpile status, this change requires updating the stockpile file using the first part of the program.

This coal management software is simple, easy to follow, and a very useful tool for planning engineers at power plants.

INTRODUCTION

Coal stockpiling is an important part of an overall feed control strategy available for a coal burning power plant. Power plants are designed to burn coal of a certain quality. Design specifications on parameters such as sulfur, ash, heating value, and moisture contents of the feed should be met both to ensure efficient power plant performance and to meet emission regulations.

Often it is difficult to obtain coal from a single source that provides all the desired qualities and quantity. To ensure desired quality end product and continuity of supply, power plants must rely on consortium of suppliers where each one generally supplies different quality coal. When coal arrives at the power plant, from a mine or from an external supplier, it can be sent to 1) coal silos (and/or stockpiles) for mixing with other coals, 2) coal stockpiles for future use, 3) coal cleaning facility, or 4) directly to boiler. This possible movement of coal is shown in Figure 1.

There are three main factors that dictate as to which stockpile the coal should be sent. These are: 1) quality and quantity of existing coal in a stockpile, 2) the overall storage capacity of each stockpile, and 3) stockpile availability. Once coal is temporarily stored in a stockpile, the coal industry does not allow stockpiled coal to exceed a 45-day tonnage requirement of a utility plant (Baafi 1983, p. 85). This is due to possible oxidation and spontaneous combustion of stockpiled coal.

To maintain a proper level of coal inventory in terms of both quality and quantity in various stockpiles and to solve multi-objective coal blending problem, a dynamic coal stockpile management system is needed. This paper describes an interactive, microcomputer-based program written for this purpose. The developed system is designed specifically to aid planning engineers in arriving at daily decisions involving; 1) the movement of daily run-of-mine (ROM) or externally pur-

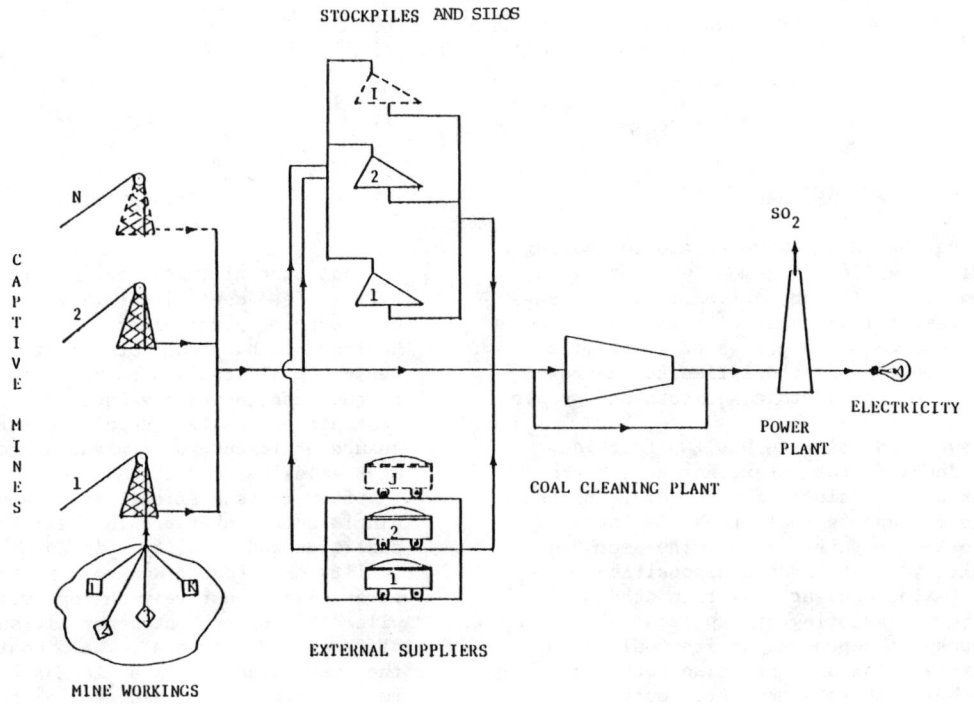

FIGURE 1. A Schematic Diagram of Coal Flow from Mines to Power Plant.

chased coal to the stockpile and 2) the movement of stockpiled coal to the power plant or to the cleaning plant. For example, when coal arrives at the power plant, should it be sent to one stockpile or a multiple of stockpiles? Or, should it be blended with externally supplied coal? If so, how should the blend be? A careful look at Figure 1 clearly shows the many possible permutations of material flows in the overall stockpile management system.

THE COAL STOCKPILE MANAGEMENT PROGRAM (CSTOCK)

The program CSTOCK has two distinctive, but interconnected parts. The first part consists of a set of routines to create and/or update coal stockpile files. Before updating, the user can interactively review the current status of each stockpile. Based on this information, the user can guide the ROM coal or coal from other sources to any existing stockpiles or create a new stockpile depending on quality and quantity of the coal in question with respect to the stockpile status of coal.

The second part of the program is used to determine the removal amounts from each or any combination of stockpiles to satisfy the need of the power plant. In the second part, multi-objective goal programming technique is utilized to interactively solve the feed blending problems, subject to varying quantity and quality of available stockpiled coal.

The interconnected aspect of these two programs lies in the sharing of the common stockpile file information. The program CSTOCK is a menu driven software, written in MICROSOFT FORTRAN 77, Version 3.13. It consists of nearly 1,600 lines of commented codes. The source code is divided into 19 subroutines, each of which performs a specific task. Table 1 contains a list of the major subroutines and their functions.

TABLE 1

List of Major Subroutines in CSTOCK

STATUS: Checks the status of stockpile file.

UPDATE: Updates or creates stockpile by calling subroutine ADD.

ADD: Adds coal to a stockpile.

TAKOUT: Removes coal from a stockpile.

CRITER: Checks stockpile criteria file and logically selects stockpile/s suitable for incoming coal.

DATE20: Obtains current date for file updating. It needs an assembly language program ACSV20, which is a part of FORLIB.PLUS package.

BLENDC: Accesses the goal programming part.

BATCHM: Provides batch mode operation for goal programming.

INACTM: Provides interactive mode operation for goal programming.

GOAL: Solves goal programming problem using modified simplex algorithm.

CHKOUT: Displays output on the terminal screen.

CHECK2: Checks user input before accepting it.

CSTOCK FILE STRUCTURE

To initiate the programming logic of CSTOCK, it (CSTOCK) needs a stockpile information file. The user must create this file prior to running CSTOCK. This file should contain stockpile information of at least one stockpile. Information of other stockpiles can be created during the execution of CSTOCK.

The program is currently limited to four (4) coal variables that are stored in stockpile information file in a fixed format. The stockpile information include serial I-D number, four coal variables, stockpile description (i.e., SILO-1, SILO-2, etc.), and the stockpile creation date or the last updated date.

The program assumes that the first variable in the file record is the total tonnage of coal currently stored in each stockpile. The remaining three variables of coal quality (i.e., such as ash, sulfur, BTU) can be stored in any order. Except the first variable name, the remaining names are user specified. The program then uses these input variable names during subsequent runs.

CSTOCK sequentially reads the user supplied stockpile file and stores stockpile information in a new direct access file where one record of 70 characters long stores all the necessary information for one stockpile. Therefore, the number of stockpiles that can be stored in a single direct access file is unlimited. CSTOCK keeps track of a particular stockpile in the direct access file using the serial ID number of that stockpile. No special link-list structure is used here for file management.

During updating information of a stockpile, the user must supply the serial ID number of that stockpile. CSTOCK then copies the corresponding record of that stockpile onto a new file, and modifies the stored information only on the new file, as per user's request. Program also changes the previous date to the current date in the new file.

If the updated file information is correct and if the user is satisfied with this new file, then the original file can now be deleted. As the incoming coal is added to an existing stockpiled coal, the program assumes that the entire coal (incoming coal plus stockpiled coal) becomes homogeneous in quality.

The resulting grades are calculated by simply taking the weighted (by tons) average of incoming coal and stockpiled coal.

WHY NOT USE AVAILABLE DBMS SOFTWARES

Since the stockpile information file is nothing but a data base management system (DBMS) software, the question naturally arises as to whether or not it would be advantageous to use currently available DBMS softwares. According to a recent article in PC Magazine (June 12, 1984), there are about 66 DBMS programs available for microcomputers and every week another program appears on the market, promising to surpass every current and future microcomputer DBMSs. If there are so many DBMS programs available then why develop another one?

There are several reasons for this. First, the data management program developed here is a part of the overall stockpile management program, specifically designed to solve coal stockpiling and coal blending problems of a power plant. As mentioned before, incoming coal is added to a stockpile if it meets the requirements of that stockpile. Each stockpile is specifically designed to accept certain quantity and qualities of coal. It is, therefore, desirable to use a DBMS program that has the above decision making power in it. Unfortunately, no marketed DBMS program has such capability. This is because, coal stockpiling is a specified operation of a power plant and the marketed DBMS programs are general business oriented softwares.

Secondly, aside from solving coal stockpiling problems through simple file-handling system, CSTOCK solves goal programming problems as well. For an interactive goal programming software to solve coal blending problems using stockpiled coal, it must have internal access to the stockpiled coal information. Internal access means that, while executing the program, the user must have easy access to the stockpile file without having to exit from the program. This is because mixing varying qualities of stockpiled coal in order to find a desired power plant feed is an interactive process. The first part of CSTOCK provides an easy access to the stockpile file that the goal programming (the second part of CSTOCK) needs. This is one of the unique features of CSTOCK.

Thirdly, if one of the DBMS programs out of 66 is considered then it will make the blending program dependent on that DBMS program. In other words, the user has to get two different and disjointed softwares instead of one, not to mention the software commands that user must learn in order to use them.

GUIDING COAL TO-AND-FROM STOCKPILE USING CSTOCK

CSTOCK is driven by the main menu which has four items as shown in Figure 2 below:

```
MENU:   SELECT ONE BY TYPING AN INTEGER #

  1.  Check stockpile status
  2.  Update stockpile file
  3.  Blend coal
  4.  Quit

  Your selection ==>  1
```

Figure 2 Items on the Main Menu of CSTOCK

Menus 1 and 2 access the first part of the program, whereas menu 3 acesses the second part. Table 2 shows the example run of the program in which the existing stockpile file (OLD5.DAT) is being updated to a new file called NEW5.DAT, after entering the information "R-STOK-3", i.e., the raw coal to be stored on stockpile #3.

In the previous example given in Table 2, the user selected the stockpile #5 which did not exist. Therefore, a new stockpile (I.D.#4) was created during the session. Should the user let the program select the ID of existing stockpile depending on the stockpiling criteria, then CSTOCK requires the stockpile criteria file. This criteria file is simply a sequential access file that contains maximum and minimum limitations on quantities and qualities pertaining to each stockpile. A sample stockpile criteria is shown in Table 3.

During coal transfer from the stockpile to either the boiler or cleaning plant the user can remove any amount of coal from any existing stockpiles. However, if the chosen amount is more than what the stockpile can provide then CSTOCK warns the user of this apparent miscue.

After the initial updating of stockpiles, the user is sent back to main menu (see Figure 2). Now the blending part of CSTOCK can be assessed by selecting item 3 from this menu.

COAL BLENDING USING GOAL PROGRAMMING

At the power plant, it is imperative in the short term that the blend resulting from the stockpile coals be calculated regardless of whether or not it satisfies the plant specifications. In order for the planner to recommend a course of action to take, the magnitude of the difference between the target blend and the expected blend over the short term must be known to the planner. Goal programming provides this information which can be used to rank various options available such as; 1) purchasing coal on spot market, 2) coal cleaning, 3) further stockpiling, and 4) blending.

```
TODAY`S DATE:   12/27/84


WELCOME TO THE COAL STOCKPILE MANAGEMENT PROGRAM



MENU:  SELECT ONE BY TYPING AN INTEGER #

    1.   CHECK STOCKPILE STATUS
    2.   UPDATE STOCKPILE FILE
    3.   BLEND COAL
    4.   QUIT

    YOUR SELECTION ==> 1


    DO YOU HAVE A STOCKPILE FILE?  ENTER Y OR N ==>      Y


        NAME OF ORIGINAL STOCKPILE FILE
        (NO MORE THAN 11 CHARACTERS LONG) ==> OLD5.DAT

        NAME OF NEW STOCKPILE FILE
        (NO MORE THAN 11 CHARACTERS LONG) ==> NEW5.DAT

        NAMES READ ARE:
                ORIGINAL FILE:  OLD5.DAT
                NEW FILE:       NEW5.DAT
        ARE THESE NAMES CORRECT?  ENTER Y OR N ==> Y

        THE ORIGINAL STOCKPILE FILE

    ID  K-TON   SULFUR   ASH    K-BTU   DESCRP       DATE
    1.  14.00    2.50   18.00   11.00   R-STOK-1   12/17/84
    2.  10.00    2.60   19.00   11.00   R-STOK-2   12/19/84
    3.   5.00    2.90   20.00   10.00   C-SILO-1   12/20/84

    MENU:  SELECT ONE BY TYPING AN INTEGER #

        1.   CHECK STOCKPILE STATUS
        2.   UPDATE STOCKPILE FILE
        3.   BLEND COAL
        4.   QUIT

        YOUR SELECTION ==> 2

        SELECT ONE OF THE FOLLOWING

        1   ADD COAL TO A STOCKPILE
        2   TAKE OUT COAL FROM A STOCKPILE
        3   DO NOTHING

        YOUR SELECTION ==> 1

        ENTER AMOUNTS OF K-TON , SULFUR , ASH  , K-BTU ,
        WITH DECIMAL AND SEPARATED BY A COMMA OR A BLANK
        ==> 3.,2.9,20.,10.
```

Table 2 Sample Run of Program CSTOCK

```
VALUES READ ARE:

          K-TON  =       3.00
          SULFUR =       2.90
          ASH    =      20.00
          K-BTU  =      10.00

ARE THESE VALUES CORRECT?  ENTER Y OR N ==> Y

SELECT ONE
1    LET THE PROGRAM SELECT THE STOCKPILE DEPENDING
          ON THE STOCKPILE CRITERIA
2    USER SELECTS THE STOCKPILE REGARDLESS OF THE
          STOCKPILING CRITERIA
3    CHANGE OF HEART - DO NOT ADD COAL

YOUR SELECTION ==> 2

TO WHICH STOCKPILE YOU WANT TO ADD? ==> 5

THIS STOCKPILE DOES NOT EXIST.  WOULD YOU WISH TO
CREATE A NEW STOCKPILE ?   ENTER Y OR N ==> Y

THE NEW STOCKPILE NUMBER IS  4

DECRIPTION OF THIS NEW STOCKPILE?
NO MORE THAN 10 CHARACTERS LONG ==> R-STOK-3

SELECT ONE
1    SEE STOCKPILE FILE
2    SEE STOCKPILE CRITERIA FILE
3    GO BACK TO MAIN MENU

YOUR SELECTION ==> 1

WHICH ONE ?
1   ORIGINAL STOCKPILE FILE
2   UPDATED STOCKPILE FILE
3   BOTH FILES

YOUR SELECTION ==> 3
THE ORIGINAL STOCKPILE FILE

ID  K-TON    SULFUR    ASH    K-BTU     DESCRP       DATE
1.  14.00     2.50    18.00   11.00     R-STOK-1    12/17/84
2.  10.00     2.60    19.00   11.00     R-STOK-2    12/19/84
3.   5.00     2.90    20.00   10.00     C-SILO-1    12/20/84

THE UPDATED STOCKPILE FILE

ID  K-TON    SULFUR    ASH    K-BTU     DESCRP       DATE
1.  14.00     2.50    18.00   11.00     R-STOK-1    12/17/84
2.  10.00     2.60    19.00   11.00     R-STOK-2    12/19/84
3.   5.00     2.90    20.00   10.00     C-SILO-1    12/20/84
4.   3.00     2.90    20.00   10.00     R-STOK-3    12/27/84

MENU:  SELECT ONE BY TYPING AN INTEGER #

       1.   CHECK STOCKPILE STATUS
       2.   UPDATE STOCKPILE FILE
       3.   BLEND COAL
       4.   QUIT

YOUR SELECTION ==> 4
```

Table 2   Sample Run of Program CSTOCK
Continued

```
TYPE CRT2.DAT
ID  MX-KTN MN-KTN MX-SUL MN-SUL MX-ASH MN-ASH MX-KBT MN-KBT DESCRIPT
1.   15.0   1.0   2.45   2.55   30.0   15.0   20.0    0.0   R-STOCK-1
2.   15.0   1.0   2.75   2.56   25.0   15.0   20.0    0.0   R-STOCK-2
3.   15.0   1.0   3.50   2.76   20.0   15.0   20.0    0.0   R-STOCK-3
4.   14.0   0.0   2.44   1.95   15.0    0.0   20.0    7.0   C-SILO-1
5.   14.0   0.0   1.94   1.73   15.0    0.0   20.0    7.0   C-SILO-2
6.   10.0   0.0   1.72   1.50   15.0    0.0   20.0   10.0   E-SILO-3
7.   10.0   0.0   1.49   0.00   15.0    0.0   20.0   10.0   E-SILO-4
```

Table 3 Sample Stockpile Criteria File

To follow this second part of CSTOCK, one should have some basic knowledge of goal programming and how to formulate a goal programming problem. In Table 1, subroutine GOAL, START, and FINISH were taken from S. M. Lee's text (1984) and modified to suite CSTOCK. A detailed discussion on coal blending using goal programming appears on a paper by J. E. Lonergan (1983).

Once the blending problem is formulated, it can be solved by the second part of CSTOCK. In running the second part, the user can choose either batch execution or interactive mode of execution. For batch mode execution, CSTOCK requires an input file (sequential) which must contain the goal programming formulation. In case of interactive mode, no such input file is needed. Instead, CSTOCK creates a formulation file which used in batch mode operation is for program execution.

If the planner is satisfied with the blending solution and wants to make it final then he/she can enter the first part of CTSOCK and carry out the necessary updating to reflect the decision. On the other hand if the blended solution is found unsatisfactory, then the planner can reformulate the problem and run it till desired goals are met. Reformulation involves one or more of the following:
1. modify the priority structure
2. change the goal targets (RHS)
3. add or drop a goal
4. add or drop a stockpile
CSTOCK allows an easy access to carry out these modifications.

NOTEWORTHY FEATURES OF CSTOCK

This coal stockpile management program brings the basic data management (part one) and an operations research tool together in a single software for microcomputer.

Provisions are made in CSTOCK for both batch mode and interactive mode of operations. The batch mode operation becomes specially handy for those users who are familiar with the file editing in a microcomputer. With file editing knowledge, one can easily modify the goal programming formulation and obtain solution via batch mode. Also, one can take the advantage of CSTOCK and solve any kind of linear goal programming problem, not necessarily coal blending only, which is the case in interactive mode.

Special care was taken to make this program as "easy to follow" as possible by providing: 1) brief comments wherever appropriate, 2) "do nothing" type of easy exit from any part of the program, and 3) "hand shake" attitude, i.e., input check before the program accepts it.

CONCLUSION

In order to provide the flexibility needed to analyze the stockpiling and blending problem faced by coal fired power plant owners dependent on multiple sources of coal supply, a coal stockpile management program was developed utilizing linear goal programming. Goal programming was considered over the more common linear programming because all constraints in linear programming are considered as additional goals with some target values. The solution that comes closest to satisfying these target values is the "optimal" solution; thus, avoiding the possible impasse of "infeasible" soltuion in linear programming.

Microcomputer usage of CSTOCK enhances its practicality and makes it extremely economical to the power plant owners for its employment. Up-to-date results and experiences prove that today's microcomputers can economically and efficiently perform some tasks that were the premise of minicomputers.

REFERENCES

Baafi, E.Y., 1983, Application of Mathematical Programming Models to Coal Quality Control, Ph.D. Dissertation, The University of Arizona, Tucson, AZ, 106 pp.

Lee, S.M., 1984, Goal Programming for Decision Analysis, Auerbach Publishers, Philadelphia, 2nd Edition.

Longergan, J.E., 1983, Computerized Solutions to Mine Planning and Blending Problems, Ph.D. Dissertation, The University of Arizona, Tucson, AZ 117 pp.

UNDERGROUND MINING APPLICATIONS

Chapter 22

FORECASTING UNDERGROUND COAL MINE PRODUCTIVITY WITH A MICROCOMPUTER

Robert L. Grayson and Y. J. Wang

Department of Mining Engineering
West Virginia University
Morgantown, WV 26506

Abstract. The development and use of a productivity forecast model for an underground coal mine is presented. With the aid of a microcomputer, an extensive, interactive, multiple linear regression analysis was performed on a set of mine operating data, and 93% of the variability in section productivity is accounted for by the fitted linear model. Productivity is forecast as a function of seven variables which characterize physical mining conditions, expected levels of downtime, and the quality of production crew performance.

The model can be used as a tool in developing productivity forecasts for 5-year and annual operating plans or for short-term (e.g., quarterly) revisions. Productivity forecasts for new sections and for operating sections with changing conditions or personnel are made for illustration. The model will allow the mine manager to eliminate historical subjectiveness from predicting productivity or temper his subjectivity with much needed objectivity.

Introduction

Given a production demand from the corporate office, from individual buyers on the spot market, and/or from a buyer under contract, the coal mine manager must respond by meeting the required demand in a timely manner. One ingredient of planning for meeting demand is the prediction of the performance of the separate production centers and the projected output of the entire operation. In any production system, productivity is a random variable subject to a complicated cause system. Cause systems change frequently and, at times, rapidly; therefore, it is necessary for the forecast model to be responsive to significant changes in the cause system immediately without overreacting to random fluctuations (Mize, White and Brooks, 1971).

Some components of the coal mine productivity cause system are:

1. type of mining,
2. roof condition,
3. floor condition,
4. mining height,
5. mechanical downtime,
6. systematic downtime,
7. effect of water influx,
8. seam inclination,
9. hardness of coal,
10. geographical location,
11. crew quality,
12. supervisor effectiveness,
13. affiliation of labor force,
14. incentive plan, and
15. labor relations.

Mutmansky (1966), Wilson (1973), Malhotra (1975), Suboleski (1978), and Stefanko (1983) address the effect on productivity of many of these variables. Statistical procedures have been used (with aid of a computer) in developing forecast models for coal mine productivity. Due to the extreme variation in the cause system of coal mines in general, and even for specific operations, the approach appears sound. A very well-performing coal producing section can turn into a very poor performer practically over night in some coal seams. Therefore, a forecast model which can encompass the cause system in predicting productivity will be an invaluable tool in the coal mining industry. A closer look, however, needs to be taken at including human performance factors.

In assessing the cause system which affects coal mine productivity, one must consider very important components which embrace the labor force (human elements). It is well documented that the United Mine Workers of America play an active role in negotiating with coal industry management competitively and that "labor-management relations in the coal industry are still in the dark ages," as suggested by Lawson (1983). Risser and Malhotra (1976) remind the reader of the historical effect of the cyclic demand for coal on work force and community relations. This is obviously a contributing factor to the labor union position of opposition today. The single-mindedness of the labor union can be highlighted by the fact that the six contract expirations prior to 1984 resulted in over 300 lost production days. Coal operators have indeed avoided opening new mines unless they could be ensured of a non-union operation. A survey by Gordon (1981) of coal industry executives reveals this discontent with labor-management relations. Therefore, the evaluation of the labor force is a critical step in implementing any productivity forecast model.

Productivity forecasting for a specific operation need not include a long list of cause system variables. Many factors such as inclination of seam, some forms of systematic downtime, hardness of coal, geographical location, affiliation of labor force, incentive plan, and labor relations will drop out because of homogeneity for the operation in time or consistency throughout the mine. Also, some factors when combined are tantamount to their individual effects while some of the forecast error is eliminated. This is especially apparent in a regression model. Variables such as roof or floor condition and the effect of water influx on either of them could be combined in such a way. Little would appear to be lost when crew quality and supervisor effectiveness factors are combined.

Budnick, Mojena and Vollman (1977) suggest that a subjective forecast which results in small errors is superior to a more sophisticated one which also yields accurate forecasts but is expensive. Adam and Ebert (1982) suggest evaluating the cost of forecast error in terms of the costs of overproduction and underproduction. The variability of coal mine productivity and the extensiveness of the cause system lead to the conclusion, as does experience, that large errors do occur in predicting productivity and production. Indeed, an adequate, affordable forecast model which incorporates cause system variables is needed. Time series or qualitative (e.g., Delphi, Adam and Ebert, 1982) models which perform technological forecasting without a quantitative input of cause system factors do not appear effective in predicting coal mine productivity. Production simulators such as those developed by Virginia Polytechnic Institute (D.O.E., 1981) and Pennsylvania State University (Manula et al., 1975) may be too sophisticated and require too much time and expense for incorporation into the managerial framework of a mining operation of moderate size.

Multiple regression, whether linear or not, appears to be the most useful tool for predicting multivariate, real-world coal mine productivity. It is capable of incorporating the cause system variables which can quickly and drastically affect productivity. As stated by Walpole and Myers (1978), multiple regression gives "an adequate representation ... in the ability of the entire function to predict the true response in the range of the variables considered." The regression model can be used interactively on a microcomputer to perform a detailed sensitivity analysis. It also appears to be the proper selection when considering sophistication and cost factors. Figure 1 depicts a generalized graph of the tradeoff between cost and accuracy in selecting a forecast model (Chambers et al., 1971). However, the system operating costs due to inaccurate forecasts could be much higher when considering the absolute necessity of meeting supply contract demands in the coal industry—a very competitive one. Also, with the advent of the microcomputer, the cost of utilizing a regression-correlation model has substantially decreased.

Description of Operating Mine Sections and Conditions

Pursuant to the foregoing discussion, productivity data covering an 8-month period was obtained from an operating underground coal mine. Coal producing sections in the mine utilized the room-and-pillar system of mining. Continuous mining machines, roof bolting machines, shuttle cars, and auxiliary fans were used in the production cycle of each section. There existed five operating sections, one of which was primarily used as a spare location. The four regularly operated sections were:

1. A "super" development section which consisted of a 10-man crew, 2 continuous mining machines (CM's), 2 roof bolting machines, 4 shuttle cars, and 2 auxiliary fans.

2. A development section of standard manpower and machinery.

3. A full pillar-extraction section which consisted of an 8-man crew and standard machinery except for auxiliary fans.

4. A partial pillar-extraction section which consisted of a 6-man crew and standard machinery except for auxiliary fans. Roof bolting was not normally done.

Physical conditions varied considerably throughout the mine. A steady, substantial influx of water persisted in one of the sections. Intermittent wet conditions were encountered in another of the sections. The other two sections experienced very rare occurrences of wetness.

Roof conditions also varied significantly throughout the mine. Conventional 6-foot bolts were used in three sections, while 5-foot resin grouted rods were used in the other one. One section was able to hold roof coal which protected a 2- to 3-foot layer of drawslate from weathering and kept the coal product from being substantially diluted by inorganic material. It also sped up roof bolting of individual cuts. The regular development section had an immediate roof layer of from 8 to 10 feet of soft shale which could not be bolted in place. By leaving some coal bottom, the mining height was reduced to approximately 12 to 14 feet. Occasionally, although unpredictably, somewhat better roof conditions occurred and permitted a normal mining height of 7.5 feet to be achieved.

Development of the Forecast Model

An analysis of the operating data was made in order to select variables which appeared significant in affecting section productivity. Many of the variables mentioned previously are not needed for a specific mine model. Two such variables, which are consistent throughout the mine, are seam inclination and hardness of coal. Variables initially selected to be incorporated into the multiple regression model included:

V1 - mining height (feet)
V2 - roof type (rank ordering)
V3 - floor condition (rank ordering)
V4 - mechanical delay percentage
V5 - transportation delay percentage

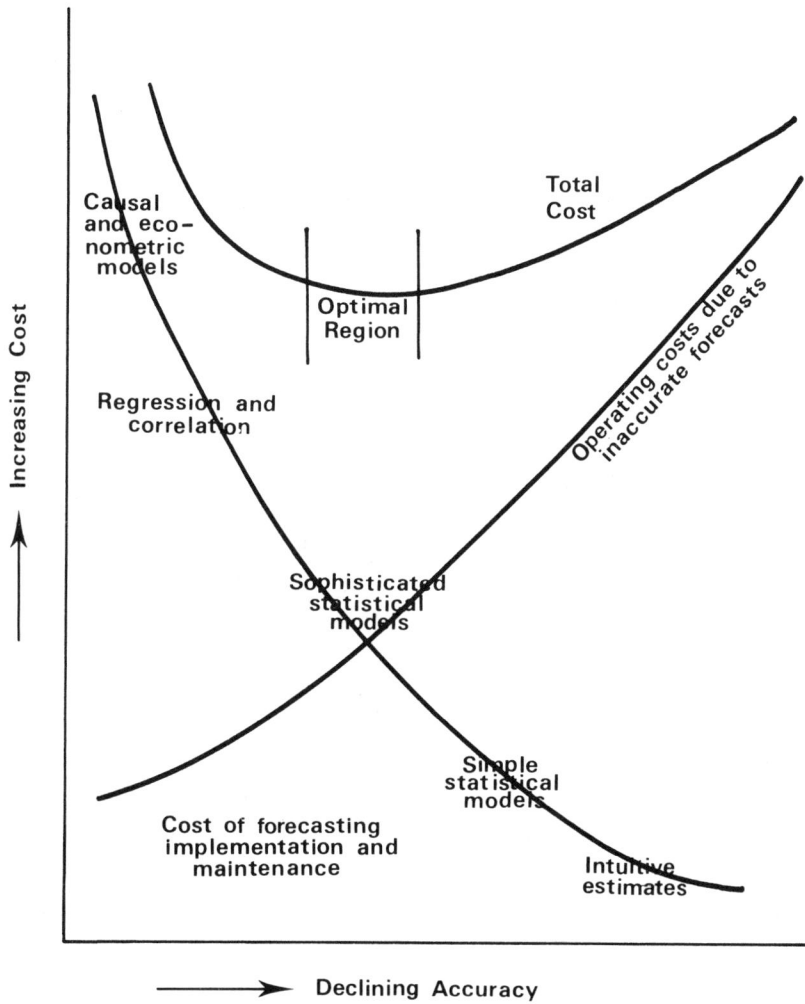

Figure 1.   Cost versus accuracy tradeoffs in forecasting.

V6 – supervisor quality (rank ordering)
V7 – crew quality (rank ordering)
V8 – unusual delay percentage
V9 – grade on section (%)
V10 – number of places mined
V11 – type of mining (rank ordering)

The multiple regression analysis of 8-month data relating to these parameters yielded the following formulation for prediction of section productivity:

$$\text{Productivity} = 429.20 - 15.59V1 + 12.51V2$$
$$+ 17.21V3 - 9.83V4 - 4.99V5$$
$$- 9.93V6 + 30.68V7 - 4.59V8$$
$$+ 14.66V9 - 18.78V10 + 45.38V11$$

with

Coefficient of multiple correlation, $r = 0.967$

Coefficient of determination, $r^2 = 0.935$
Standard error of estimate, $SE = 33.282$

Actual productivity during the period ranged from 229 raw tons per shift (RTPS) to 578 RTPS among the sections. Although the response of this model must be considered adequate, the high constant term indicates that the effect on productivity by the cause system variables is small. It appeared that some redundancies were involved or that some variables were missing in explaining productivity.

Realizing that the existence of redundancies was the most likely answer, further analysis was made. Neither stepwise regression software nor software for principal component analysis is normally available at a mine site. Thus, extensive reapplication of regression analysis was used while eliminating redundant variables from a practical point of view.

First of all, it was clear that the variable "number of places mined" depended largely on the method of mining employed on a section, e.g., two places were mined on pillar extraction whereas 6 places were mined on a "super" section. Elimination of this variable, and reapplication of multiple regression yielded the next formulation:

$$\text{Productivity} = 209.85 - 9.99V1 + 15.19V2$$
$$+ 24.68V3 - 5.07V4 - 2.37V5$$
$$- 4.51V6 + 31.29V7 + 0.54V8$$
$$+ 8.43V9 + 48.55V10$$

where all variables are the same except for V10 becoming the old V11, and with

$r = 0.967$
$r^2 = 0.934$
$SE = 32.779$

Now the constant term is low compared to the range of actual productivity and reflects a base level of expected productivity which could be

obtained in spite of poor conditions. However, the manager can "intuitively feel" that the contribution of V8 (unusual delay %) is very small and that the grade variation influence in the mine is low.

Making variable changes to reflect these thoughts led to the new model:

$$\text{Productivity} = 209.84 - 7.28V1 + 17.09V2$$
$$+ 19.94V3 - 5.10V4 - 2.30V5$$
$$- 6.01V6 + 32.43V7 + 47.81V8$$

where now
V1 – mining height (feet)
V2 – roof type (ranked)
V3 – floor condition (ranked)
V4 – mechanical delay (%)
V5 – transportation delay (%)
V6 – supervisor quality (ranked)
V7 – crew quality (ranked)
V8 – type mining (ranked)

and with

$r = 0.966$
$r^2 = 0.933$
$SE = 31.563$

Note that the constant term remained relatively low and that the standard error of the estimate decreased significantly compared to the concomitant small decrease in correlation. Worthwhile changes were made.

All of the remaining variables are known by a mine manager to be significant (without principal component analysis) except for one small problem. Coefficients of variables included in the model should give a logical interpretation of results from calculations made in incremental analysis. For example,

$-7.28V1$ indicates that as the mining height increases an additional foot above normal height, then productivity decreases by 7.28 raw tons (i.e., bad roof is encountered and/or longer bolting times are required).

$-5.10V4$ indicates that as the % mechanical delay increases by 1 unit, then a reduction in productivity of 5.10 raw tons occurs.

A problem, however, develops regarding the prediction of the effect of supervisor (V6) and crew (V7) quality on productivity. Obviously, the two variables are strongly interrelated. Many mine managers believe that a good foreman will bring out the best in an average crew who are willing to do their work properly. But what are the effects of a good crew coupled with a poor supervisor? Which factor dominates? What is the effect when a good supervisor is placed with an unwilling (poor) crew?

This particular model reflects the uncertain relationship by giving

$-6.01\,V6 + 32.43\,V7$

as the combined effect. It suggests that an

increase of 1 unit in supervisor quality decreases productivity by 6.01 RTPS, whereas an increase of 1 unit in crew quality increases productivity by 32.43 RTPS. A contradiction exists regarding supervisor quality and productivity.

Combining these two factors into a single variable "crew quality" yields the final model:

$$\text{Productivity} = 204.94 - 2.98V1 + 20.01V2$$
$$+ 22.11V3 - 5.24V4 - 1.95V5$$
$$+ 4.77V6 + 49.75V7$$

where
V1 - mining height (feet)
V2 - roof type (ranked)
V3 - floor condition (ranked)
V4 - mechanical delay (%)
V5 - transportation delay (%)
V6 - crew quality (ranked)
V7 - type mining (ranked)
and with

$r = 0.966$
$r^2 = 0.933$
$SE = 31.03$

Note that an improvement of the model has again been obtained by reduction of the standard error and by reduction of the constant term. No loss in correlation resulted. Also, at this point, the manager desires not to further reduce the cause system variables, and he is interested in discerning the results of variation of each parameter successively.

Use of the Model

Now that an adequate model has been developed for predicting section productivity based on operating data, variables can be manipulated to discover "what happens if situations are changed?" The first analysis is made on the final model itself, paying particular attention to coefficients to variables which will be used in incremental analysis. As the mining height increases, 2.98 RTPS are lost per foot. As the roof type ranking increases, 20.01 RTPS are gained per unit rank. An increase by one unit in rank of the floor condition results in an increase of 22.11 RTPS in productivity. As the mechanical delay percentage increases one point, a 5.24 RTPS reduction in productivity results. Similarly, an increase of one point in transportation delay percentage causes a decrease of 1.95 RTPS in productivity. As the quality of the crew (including supervisor) increases one unit in rank, an increase of 4.77 RTPS results. And finally, as the method of mining increases one unit in rank, productivity increases by 49.75 RTPS. It must be noted that this model applies only to the data for this mine, and a separate regression analysis must be made for any other mine unless similar production results have occurred. The ranking systems used for certain variables are discussed next.

The ranking system used for rating the roof type was:
5 - coal top (good condition)
4 - good sandstone/shale roof
3 - competent sandstone/shale roof
2 - irregular sandstone/shale roof
1 - poor sandstone/shale/drawslate roof.
The ranking system used for describing the condition of the mine bottom was:
5 - hard, dry, no ruts
4 - firm, dry, few ruts
3 - firm, damp, some ruts
2 - soft, damp, some holes
1 - soft, wet, many holes
The ranking system used for assessing crew/ supervisor quality was:
14 - very good crew/very good supervisor
13 - very good crew/good supervisor
12 - very good crew/average supervisor
11 - good crew/good supervisor
10 - good crew/average supervisor
9 - good crew/below average supervisor
8 - average crew/good supervisor
7 - average crew/average supervisor
6 - average crew/below average supervisor
5 - below average crew/good supervisor
4 - below average crew/average supervisor
3 - below average crew/below average supervisor
2 - poor crew/good supervisor
1 - poor crew/average supervisor
The ranking system used for describing the type of mining was:
4 - partial retreat mining
3 - development mining (2 CM's)
2 - full retreat mining
1 - development mining (1 CM)
A full range of productivity projections can now be explored for each type of mining to be utilized. Table 1 shows the results of productivity prediction for the standard development section by using the model. Some actual conditions are included in order to compare the forecast with the actual productivity obtained. In the first three columns, it is clear that the model's response closely approximates the actual productivity under varying conditions. The fourth column tells the manager what the predicted productivity is if good mining conditions were encountered in a standard development section and average levels of delays and an average crew are maintained. Other scenarios could be developed if different conditions are anticipated. The results of the fourth column give a prediction for a new section located in an area known to have good conditions initially.

Table 2 shows the results of productivity prediction for the "super" section. The super section was typically located in areas where good conditions existed (at least initially). The first three columns of Table 2 show how productivity varied with changing conditions and the close approximation of actual productivity by the model. Delays to production were never high during the period of operation; therefore, column 4 explores the possibility of a substantial increase in delay time on productivity. This could be done incrementally also. With an analysis such as this, a range of expected

Table 1. Prediction of Productivity for tne Standard Development Section

Variable	Situation	Bad Conditions Poor Delays < Avg. Crew	Bad Conditions Avg. Delays < Avg. Crew	Avg. Conditions Poor Delays < Avg. Crew	Good Conditions Avg. Delays Avg. Crew
V1	MHT	10.5	10.5	8.0	7.5
V2	RFTYP	2.0	2.0	3.5	4.0
V3	FLCND	1.25	1.25	2.5	4.0
V4	MDLYS	14.0	10.0	14.7	10.0
V5	TDLYS	5.6	2.0	5.6	2.0
V6	CREW	6.0	6.0	6.0	7.0
V7	TYPMNG	1.0	1.0	1.0	1.0
Predicted Production		235	263	297	378
Actual Production Same Conditions		229–251	262	274–278	Never Occurred

Table 2. Prediction of Productivity for the "Super" Section

Variable	Situation	Good Conditions Good Delays Good Crew	Bad Conditions Good Delays Good Crew	> Avg. Conditions Avg. Delays Avg. Crew	Bad Conditions Bad Delays Good Crew
V1	MHT	7.0	11.2	7.5	12.5
V2	RFTYP	5.0	1.5	4.0	1.5
V3	FLCND	4.5	1.5	3.0	1.25
V4	MDLYS	7.0	5.7	12.6	17.0
V5	TDLYS	1.5	2.1	1.9	9.0
V6	CREW	13	13	7	13
V7	TYPMNG	3	3	3	3
Predicted Production		555	412	442	330
Actual Production Same Conditions		551–566	387	421	Never Occurred

productivity levels can be easily developed to analyze the sensitivity of mine cost to productivity changes. A high level of delays, of course, would never be expected.

Table 3 reveals the results from a similar analysis for a partial pillar extraction section. The data of the second column relates to the actual results obtained by a section which worked half a month as a "super" section and half a month on partial retreat. Column 3 shows what would occur if improved mining conditions and reduced delays were realized. Column 4 shows the effect of bad conditions on this type (non-cycle) mining. The 451 RTPS predicted for the mining conditions of column 4 can be attributed to the facts that not as much tramming of the continuous mining machine is done, no roof bolting is done, and a pretty well maintained shuttle car can handle the poor quality bottom even though problems will be substantial.

Table 4 presents the productivity forecast for a full pillar extraction section in subject mine under varying conditions. The first 3 columns reveal forecast versus actual productivities. The fourth column shows probable results whenever improved mining conditions would be realized.

With the results already discussed, a forecast for any time period can be made. Adjustments to the model can be made on a continuing basis by using additional, newer data to augment the original data or by using a 6-month, rolling data base. The driving goal is to get and maintain a reliable forecast tool.

Summary

A method has been presented for making objective productivity forecasts at a mine site by using a multiple, linear regression model. The forecast model consists of variables which embrace the cause system that directly influences mining productivity. It provides a tool that can be used interactively on a microcomputer to investigate the impact of changing conditions on section productivity or to forecast production from a new section given information on probable conditions.

A specific model developed for an underground coal mine explains 93% of the variation in operating section productivity by using seven variables. The variables included mining height, roof type, floor condition, mechanical delay percentage, transportation delay percentage, crew quality, and type of mining. A set of operating data covering an 8-month period was used to obtain the model. Use of the model was then demonstrated in a series of calculations and the results were compared with actual performances. The model is seen to perform well and thus it provides a reliable tool for the mine manager to infuse objectivity into his productivity forecast. The model provides a flexible tool which can be adjusted readily by using further regression analysis on an updated data set.

References

Adam, E. E., Jr. and Ebert, R. J., 1982, Production and Operations Management, 2nd Ed., Prentice-Hall, Englewood Cliffs, NJ.

Budnick, F. S., Mojena, R. and Vollman, T. E., 1977, Principles of Operations Research for Management, Richard D. Irwin, Homewood, IL.

Chambers, J. S. et al., 1971, "How to Choose the Right Forecasting Technique," Harvard Business Review, July-August.

D.O.E., 1981, "Design Optimization in Underground Coal Systems," Report No. FE-1231-27, Vol. IX, NTIS, Springfield, VA.

Gordon, R. L., 1981, "Coal Industry Problems," Final Report, EPRI EA-1746, NTIS, Springfield, VA.

Lawson, J. W. R., II, 1983, "Preparing for 1984 UMWA Negotiations," Coal Mining and Processing, August, pp. 40-42.

Malhotra, R., 1975, "Factors Responsible for Variation in Productivity of Illinois Coal Mines," Illinois Minerals Note 60, Illinois State Geological Survey.

Manula, C. B. et al., 1975, "A Master Environmental Control and Mine Systems Design Simulator for Underground Coal Mining," Final Report, USBM Grant G0111808, NTIS, Springfield, VA.

Mize, J. H., White, C. R. and Brooks, G. H., 1971, Operations Planning and Control, Prentice-Hall, Englewood Cliffs, NJ.

Mutmansky, J. M., 1966, "A Statistical Method for Investigating Variables Affecting Underground Mining," M.S. Thesis, The Pennsylvania State University.

Risser, H. E. and Malhotra, R., 1976, "Coal," Economics of the Mineral Industries, 3rd Ed., ed. W. A. Vogely, Part 3.13A, AIME, New York.

Stefanko, R., 1983, Coal Mining Technology Theory and Practice, AIME, New York.

Suboleski, S. C., 1978, "Evaluation of Operational Constraints in Continuous Mining Systems, Predicting Productivity," Final Report, Volume II, USBM Grant G0166028, NTIS, Springfield, VA.

Walpole, R. E. and Myers, R. M., 1978, Probability and Statistics for Engineers and Scientists, 2nd Ed., Macmillan, NY.

Wilson, T. E., 1973, "A Rate Determining Model for a Continuous Mining Machine," M.S. Thesis, The Pennsylvania State University.

Table 3. Prediction of Productivity for the Partial Pillar Extraction Section

Situation Variable	> Avg. Conditions Avg. Delays Good Crew	Avg. Conditions Avg. Delays > Avg. Crew	Good Conditions Good Delays Good Crew	Bad Conditions Avg. Delays Good Crew
V1 MHT	7.5	8.1	7.0	12.0
V2 RFTYP	4.0	4.0	5.0	2.0
V3 FLCND	3.0	2.5	4.0	1.5
V4 MDLYS	10.0	7.8	8.0	10.0
V5 TDLYS	2.0	5.4	2.0	2.0
V6 CREW	14	12	14	14
V7 TYPMNG	4	3.5	4	4
Predicted Production	538	496	592	451
Actual Production				
Same Conditions	578	458	Never Occurred	Never Occurred

Table 4. Prediction of Productivity for the Full Pillar Extraction Section

Situation Variable	> Avg. Conditions Avg. Delays Good Crew	Bad Conditions Poor Delays Avg. Crew	> Avg. Conditions Poor Delays Avg. Crew	Good Conditions Avg. Delays Good Crew
V1 MHT	7.5	8.1	7.7	7.5
V2 RFTYP	4.0	2.0	4.0	5.0
V3 FLCND	3.5	2.0	3.5	4.0
V4 MDLYS	8.1	12.5	15.6	8.0
V5 TDLYS	5.8	7.7	2.3	2.0
V6 CREW	14.0	7.0	7.0	14.0
V7 TYPMNG	2.0	2.0	1.95	2.0
Predicted Production	453	318	384	492
Actual Production				
Same Conditions	413	311	409	Never Occurred

Chapter 23

A SIMULATION MODEL FOR LONGWALL SYSTEMS UTILIZING SHEARERS

by
Changwoo Lee
and
Jan M. Mutmansky

Department of Mineral Engineering
The Pennsylvania State University
University Park, Pennsylvania

Abstract. Most simulation models oriented toward the longwall mining process are designed to be production predictors and are used as a tool of the mining engineers in evaluating the parameters of the longwall mining operation. The simulator described in this paper has two goals: to optimize the design parameters of the major longwall subsystems and to predict the production potential of the longwall operation. The simulator is designed for use in analyzing longwalls using shearers and contains individual subroutines to handle longwall production by the unidirectional, bidirectional, half-face, and full-face mining methods. Further, this method is used to study the effects of changes in the web depth, face length, and availability of each longwall subsystem on the overall productivity of the longwall mining system. Sensitivity analyses are performed in order to obtain optimal values for these parameters and to provide information concerning the order of importance of each subsystem on the total production.

INTRODUCTION

The increasing number of longwall mining units in the United States reflects intrinsic advantages of the longwall mining method over other underground mining methods. Higher recovery and production rates plus less hazardous mining have resulted from the longwall mining process. Even though problems associated with maintenance and subsidence are not uncommon in the longwall mining system, the relatively mechanized and simplified mining processes seem to have potential for better control over those factors.

To compensate for the capital-intensive characteristic of the longwall mining system, the initial system design at the early stage of development and the mining operation must be optimized. This task must be accomplished in order to maximize the positive aspects of the longwall mining system as well as the production potential.

The objective of this paper is to develop a simulation model of the longwall mining system so that it can help engineers and management optimize the design and operating parameters of the longwall system. In addition, the model is designed to predict the longwall production potential by stochastically simulating the manner in which all the longwall subsystems interact with each other. In the last section, several problems to which this simulator can be applied are illustrated: determination of the optimal web depth and face length plus the order of relative importance of each longwall subsystem to total production and system availability.

LONGWALL MINING SYSTEM

Basically, the longwall mining system is composed of several subsystems: the shearer subsystem, the face conveyor subsystem, the roof support subsystem, and the external conveyor subsystem. Each subsystem plays its own unique role in the mining process as follows:

· Shearer subsystem - extracts coal from the face and conveys it onto a face conveyor.

· Face conveyor subsystem - carries the coal mined to the end of the face and over a stage loader onto a belt conveyor.

· Roof support subsystem - provides temporary support over the immediate face area created by the extraction of coal.

· External conveyor subsystem - transports the coal out of the mine.

All these subsystems are connected in series, so that a delay and/or failure of any one subsystem results in a total system shutdown. Even though the shearer is the major entity in the shearer subsystem, some of its operating parameters are determined by constraints set by the other subsystems including the face conveyor and roof support subsystems. The major interactions within these three subsystems are:

· the interaction between the shearer haulage speed and the face conveyor speed due to the limit of face conveyor loading capacity, and

· the interaction between the shearer haulage speed and the roof support advancement speed caused by the allowable length of unsupported face exposed when removing the coal seam.

However, among the four principal subsystems, the external conveyor subsystem does not have a direct effect on the others unless it encounters a delay and/or failure. In this paper, the panel entry development process is excluded from the simulation model even though, in most cases, the longwall face operation cannot proceed without panel development.

The behavior of a longwall coal production system was simulated by Manula et al. (1979) as a part of the study of operational constraints in continuous mining systems. This simulator uses a time incrementing method and has three subsystems: the shearer, face conveyor and roof support subsystems.

The simulation output includes optimal values for several design and operational parameters as well as the production potential. Even though there exists significant variation of the availabilities between different subsystems (Jet Propulsion Lab, 1977; Herhal et al., 1978), the system availability in this simulator is assumed to be fixed at 55% for all subsystems. Therefore, specific delays associated with each subsystem are not treated separately.

In a subsequent simulator outlined by Haycocks et al. (1981), an attempt was made to optimize longwall panel dimensions and mine ground control using both analytical procedures and the simulation method. In order to find the optimal face length, an analytical formula was developed to quantify the relationship between the face length and the shift production. Here, the production potential is deterministically calculated as a function of coal cutting dimensions, density of coal and face length under the assumption that the cycle time at the face is fixed. However, this approach does not take into account stochastic components associated with the cycle time and the production potential.

Kuroda et al. (1982) developed a longwall production simulator based on performances of the shearer and roof support systems. All operating parameters are assumed to follow the beta distribution which has a fixed range. However, from a practical standpoint, it seems to be impossible to quantify the minimum and maximum operating times for different equipment and subsystems.

In another production simulator developed by Xin and Topuz (1983), the shearer cutting and flitting speeds are assumed to be constant variables. The exponential distribution is used to approximate the shearer continuous working and downtimes, and times required for roof support advancement. However, several studies (Jet Propulsion Lab, 1977; Herhal et al., 1978; Pimental et al., 1981; Lee, 1983) show that exponential distributions are not a good model of operating and downtimes of some longwall subsystems. In addition, exclusion of the stochastic components in the shearer speed seems to reduce the applicability of this simulator.

SIMULATION MODELING

As in most other mining systems, the longwall mining system possesses deterministic as well as stochastic components. The longwall mining model can be deterministic if that is the way in which the system is designed but variable factors associated with human and geological variables as well as system failures must be modeled stochastically. The means of handling each of the elements of the longwall process will be explained in the next few sections.

Design Simulator

The first half of the simulator is designed to provide optimal values for the design and operating parameters associated with the shearer, face conveyor, and the roof support subsystems. Optimization of all parameters in question is guided by a common objective of maximization of the production rate.

The basic structure of this simulator is to maximize the production under the constraints of panel description, equipment specifications, and geological conditions, so that any changes in those constraints can be tested through this part of the simulator. Therefore, at the stage of panel development or at the time when modification of the system is desired, this simulator can provide important information concerning the optimal description of the equipment and the operating conditions of the total system.

The design simulator consists of three subsystems: the shearer, face conveyor, and roof support subsystems. The interactions between parameters in each subsystem and those between the three principal subsystems are quantified in order to reduce the longwall operation into simplified mathematical models. The major parameters quantified in each subsystem are listed below.

(1) Shearer subsystem

- interactions between coal properties and coal extracting performance,

- interactions between cutting mechanism and shearer cutting speed, and

- interactions between drum specifications and shearer operation.

(2) Face conveyor subsystem

- relationship between conveyor specifications and conveyor loading capacity,

- interaction between shearer cutting speed and conveyor speed, and

- interaction between power requirement and the conveyor specifications and operation.

(3) Roof support subsystem

· relationship between face design and behavior of the immediate roof, and

· relationship between the immediate roof and support requirement.

The dynamic relationships among the above parameters in each subsystem are illustrated in Figures 1 through 3.

Shearer Subsystem. The main entity in this subsystem is a shearer which employs one or more rotating cutting drums. Around the periphery of the drum are mounted cutter picks. Since the purpose of this subsystem is to extract coal from the face and convey it onto a face conveyor, the emphasis of the model is placed on quantification of the relationship between the physical properties of the coal seam and the coal cutting mechanism.

The following parameters are quantified in a deterministic manner in this subsystem:

(1) Specific energy. This variable is defined as the energy required to remove a unit volume of coal. The specific energy depends primarily on the web depth and the cleat orientation of the coal (Osterman, 1966; Evans and Pomeroy, 1966).

(2) Depth of cut. For a given capacity of the shearer, drum speed is inversely proportional to web depth for a given pick penetration and a constant specific energy of cutting. Therefore, the depth of cut is defined as a function of specific energy, available shearer power, number of picks per cutting line, drum speed, web depth, and height of extraction (Guay and Ludlow, 1981).

(3) Cutting force acting on each bit. The cutting force can be quantified as a function of the depth of cut, shear strength of coal, stress distribution factor, rake angle of picks, angle of internal friction, and angle of friction of cutting (Nishimatsu, 1972). In addition, the cleat orientation indirectly affects the cutting force (Evans and Pomeroy, 1966).

(4) Shearer cutting speed. Geometrically, the linear distance travelled by the drum can be described as a function of the depth of cut, the number of picks per cutting line, and the drum speed (COMINEC, 1976). This distance is equal to the distance travelled by the shearer.

(5) Power requirement. The shearer power is composed of two components - cutting power and helixing power. The cutting power is determined by the volume of coal extracted per unit time and the specific energy. The helixing power depends on loading rate, width of drum, and material factor which is a function of the

characteristics of the coal (Guay, 1980).

(6) Torque delivered to drum. The torque is defined as a function of the power requirement of the shearer for cutting and helixing plus the drum speed (Guay and Ludlow, 1981).

As shown in Figures 1 and 2, the most critical output from this subsystem is shearer cutting speed which can be maximized subject to the following constraints:

· available power of shearer,

· allowable range of the bit speed,

· limit on the pick penetration,

· allowable torque delivered to the drum,

· face conveyor loading capacity, and

· limit on the length of the unsupported face.

Face Conveyor Subsystem. On longwalls, the coal extracted by a shearer is loaded continuously onto a flexible chain conveyor that carries it to the end of the face and over a stage loader onto a belt conveyor, which transports it from the mine. Once the shearer cutting speed is determined in the shearer subsystem, the conveyor speed and power requirement can be calculated on the basis of the production rate set by that particular cutting speed.

The following parameters are modeled in this conveyor subsystem:

(1) Loading capacity of the face conveyor. This capacity is geometrically determined by the density of coal and width of conveyor. In addition, if a spill plate is attached, its height also affects the loading capacity (Guppy and Wittaker, 1970).

(2) Conveyor speed. Generally, all coal cutting machines are operated at cutting speeds less than the conveyor chain speed. In other words, the production rate determined by the shearer cutting speed has direct effects on the loading capacity and speed of the conveyor, while the conveyor specifications affect the shearer haulage speed. The conveyor speed at which all coal cut by the shearer is loaded onto the conveyor can be calculated from extraction height, web depth, swelling factor of coal, and shearer cutting speed (Teale, 1970; Guppy and Wittaker, 1970).

(3) Power requirement. The power requirements at the headgate and tailgate drives are determined by the pulling force in the upper and lower chain strands. Pulling force in the upper chain strand has a maximum value when the shearer is at the tailgate after its cutting trip, or when the shearer is on its trip to the headgate and

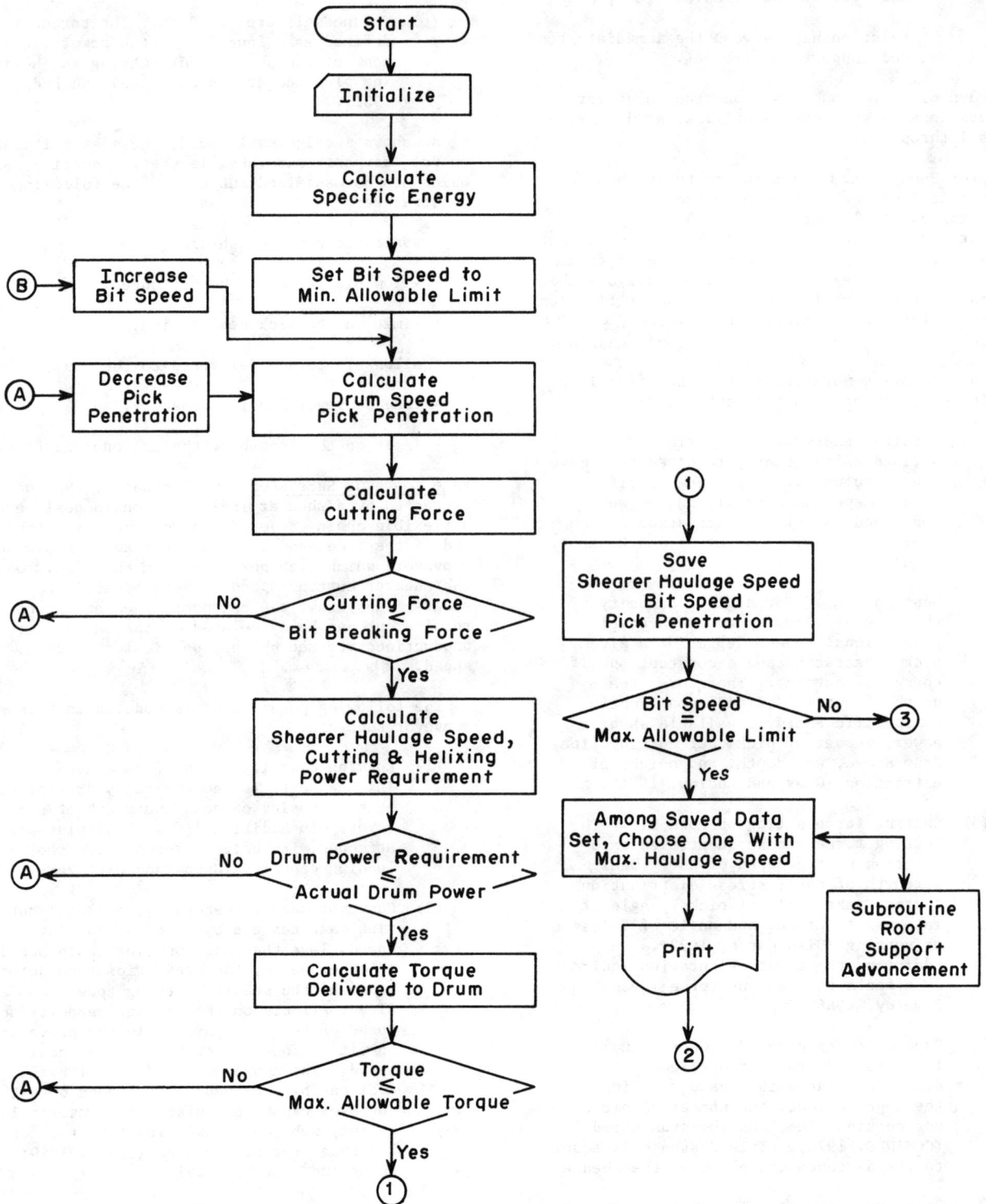

Figure 1. The Shearer Subsystem.

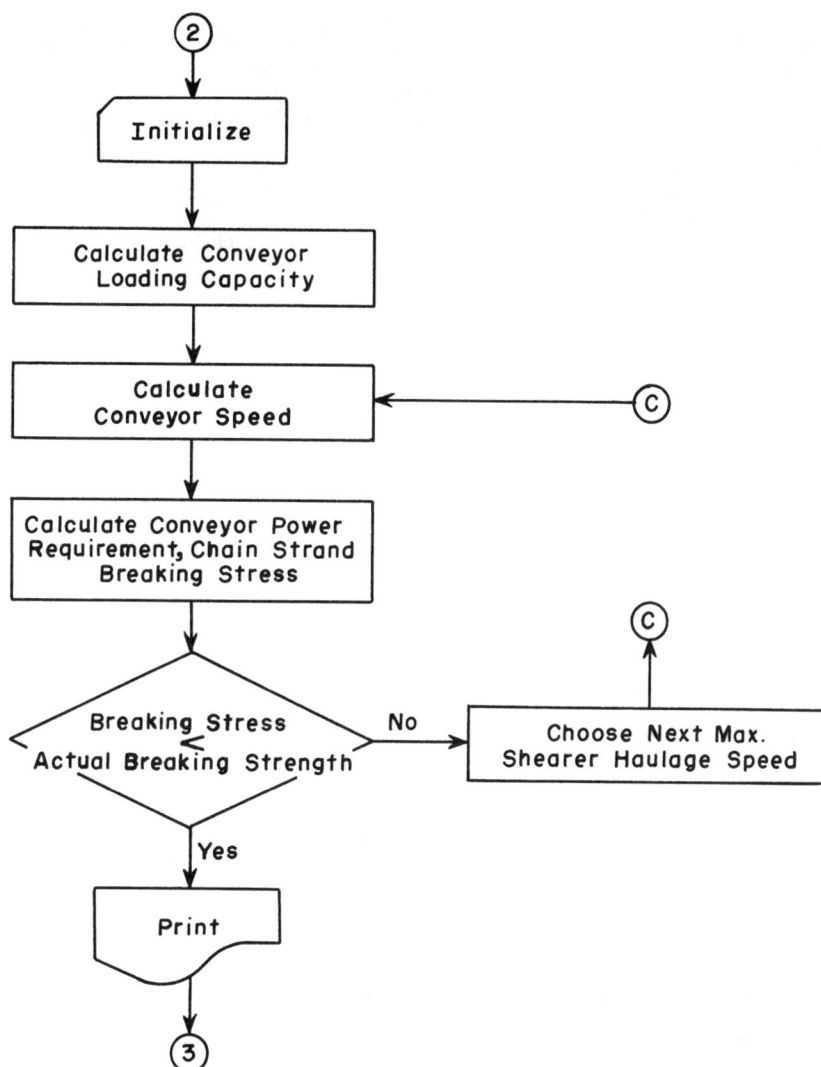

Figure 2. The Face Conveyor Subsystem.

the first loaded section of the conveyor reaches the headgate. Therefore, the power of the headgate drive depends on the maximum force of these two cases. That of the tailgate drive depends on the pulling force in the lower chain strand which can be obtained from the weight of chain and the face length (COMINEC, 1976).

(4) Breaking load of the chain strand. Calculation of the maximum allowable load of the chain strand is based on the assumption that the pulling force of all drive units is required at one point on the chain. In addition, the design of the chain strand must include a safety factor (COMINEC, 1976).

Roof Support Subsystem. Temporary support is required over the immediate face area created by the extraction of coal. This is accomplished by hydraulically-powered, self-advancing roof supports. As the face advances, the unsupported roof behind the roof support should fall harmlessly into the space occupied originally by the coal bed. Therefore, to achieve this purpose, the relationships between the roof materials and roof support specifications must be identified.

(1) Height of the immediate roof. This dimension is known to be proportional to the extraction height and inversely proportional to the bulking factor of the roof materials (Wilson, 1975).

(2) Roof beam length. It is assumed that the roof is fractured ahead of the face at a distance equal to the height of the face. Therefore, the roof beam length is a summation of overhang length, canopy length, distance to the new face, and extraction height. In this calculation, the maximum overhang length among those for all layers in the immediate roof is considered as the overhang length of the entire immediate roof.

(3) Weight of the immediate roof. The weight of the immediate roof can be represented by a freed block of strata which is defined by the overhang length, face length, and immediate roof height.

(4) Prop-free front. The distance between the front prop and the coal face can be calculated on the basis of the worst condition for weak roof, that is, vertical caving at the rear of the support. It is a function of the total thrust of the support and the geometry between the immediate roof, support, and coal face (Wilson, 1981).

(5) Mean support load density. Support load density depends on the yield load which should be at least equal to the weight of immediate roof with an additional safety factor taken into consideration.

The calculation of these parameters is performed as shown in the flowchart illustrated in Figure 3.

Production Simulator

Prediction of the production potential of a longwall system is possible through simple parametric analysis if all variables associated with the system are deterministic. For example, let's suppose that the availability of all subsystems and the total system are known. Then the production rate is just a function of extraction height, web depth, shearer haulage speed, and total system availability. However, this is not the case in a real situation; different types of delays or failures which occur in each subsystem determine the continuous running time of each subsystem, while the level of maintenance and other factors have direct effect on the duration time of those delays and failures.

In this simulator, all delay times related to each subsystem are lumped into one category rather than differentiated by their causes. Then they are approximated by Weibull distributions with three parameters. However, in the case where the Weibull distribution model fails to fit the delay data, empirical models are substituted to generate times-to-delay and delay duration times. The procedure used for simulation time generation within the simulation program is the event-oriented technique. This procedure was selected mainly due to its convenience for master clock control using statistical distribution models.

On the basis of the type of longwall mining method, the production simulator is divided into four subprograms for the unidirectional, bidirectional, full-face and half-face methods. The basic differences between these four subprograms lie in the differences in the mining method itself:

· flitting of the shearer on its return trip in the unidirectional method,

· flitting of the shearer along half of the face in the half-face method, and

· the sumping procedure at the end of the face in the full-face method.

As shown in Figure 4, the basic structure of the production simulator consists of a main program which simulates the four different methods and nine subroutines which have the following tasks:

(1) providing a constraint on the shearer cutting speed calculated in the design simulator in order to keep the length of unsupported face within given limits,

(2) generating pseudo-uniform random numbers,

(3) generating different seeds for random number generation,

(4) initializing necessary variables at the beginning of each production shift,

(5) generating times-to-delay,

(6) generating a delay duration time for a subsystem which has experienced a delay,

(7) calculating the cumulative production in each shift,

(8) preparing a shift report which contains information concerning shift production and delay data for each subsystem, and

(9) preparing a summary report of the entire simulation.

The input data for the production simulator are transferred from the design simulator except for several parameters that are dependent on the particular mining method. The most important data for the production simulator is the shearer cutting speed which is optimized in the design simulator. However, the production simulator can be run without running the design simulator as long as the shearer cutting speed is known.

APPLICATION

Other than optimization of the operating parameters and prediction of the production potential, one of the most distinctive features of this computer simulation model is its capability to estimate effects of changes in a variable on others. Therefore, whenever changes occur in

```
                    (3)

              ┌──────────────┐
              │  Initialize  │
              └──────────────┘
                     │
                     ▼
          ┌────────────────────┐
          │  Calculate Height  │
          │ of Immeadiate Roof │
          └────────────────────┘
                     │
                     ▼
          ┌────────────────────┐
          │     Calculate      │
          │  Average Density   │
          │  Adjusted Density  │
          └────────────────────┘
                     │
                     ▼
          ┌────────────────────┐
          │     Calculate      │
          │  Overhang Distance │
          └────────────────────┘
                     │
                     ▼
          ┌────────────────────┐
          │ Calculate Weight   │
          │ of Immeadiate Roof │
          └────────────────────┘
                     │                                    (A)
                     ▼                                     ▲
                 ◇─────────◇      No    ┌──────────────────────────┐
                 │Dip Angle │─────────▶ │ Calculate Yield Load Using│
                 │ = Zero   │           │ Wilson's Ordinary Equations│
                 ◇─────────◇            └──────────────────────────┘
                     │ Yes
                     ▼
                 ◇─────────◇      Yes   ┌──────────────────┐
                 │Support is│─────────▶ │    Calculate     │
                 │ Shield ? │           │   Yield Load     │
                 ◇─────────◇            └──────────────────┘
                     │ No                        │
                     ▼                          (A)
          ┌────────────────────────┐
          │ Calculate Yield Load   │
          │Using Wilson's Moment   │
          │      Equations         │
          └────────────────────────┘
                     │◀───────────────────────────────(A)
                     ▼
          ┌────────────────────────────┐
          │ Calculate Mean Load Density,│
          │ Max. Prop-Free-Front Distance│
          └────────────────────────────┘
                     │
                     ▼
              ┌──────────────┐
              │    Print     │
              └──────────────┘
                     │
                     ▼
                    (4)
```

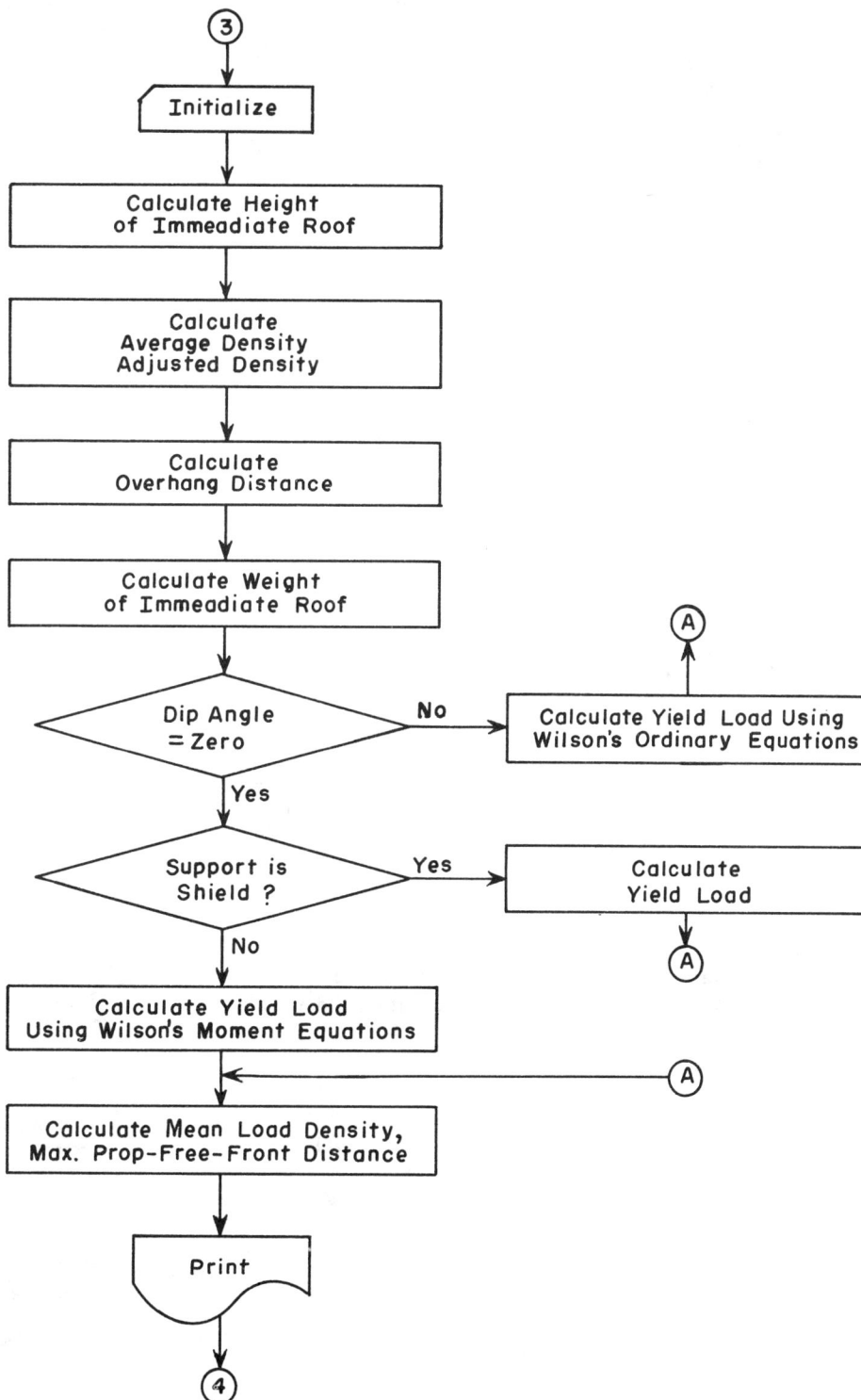

Figure 3. The Roof Support Subsystem.

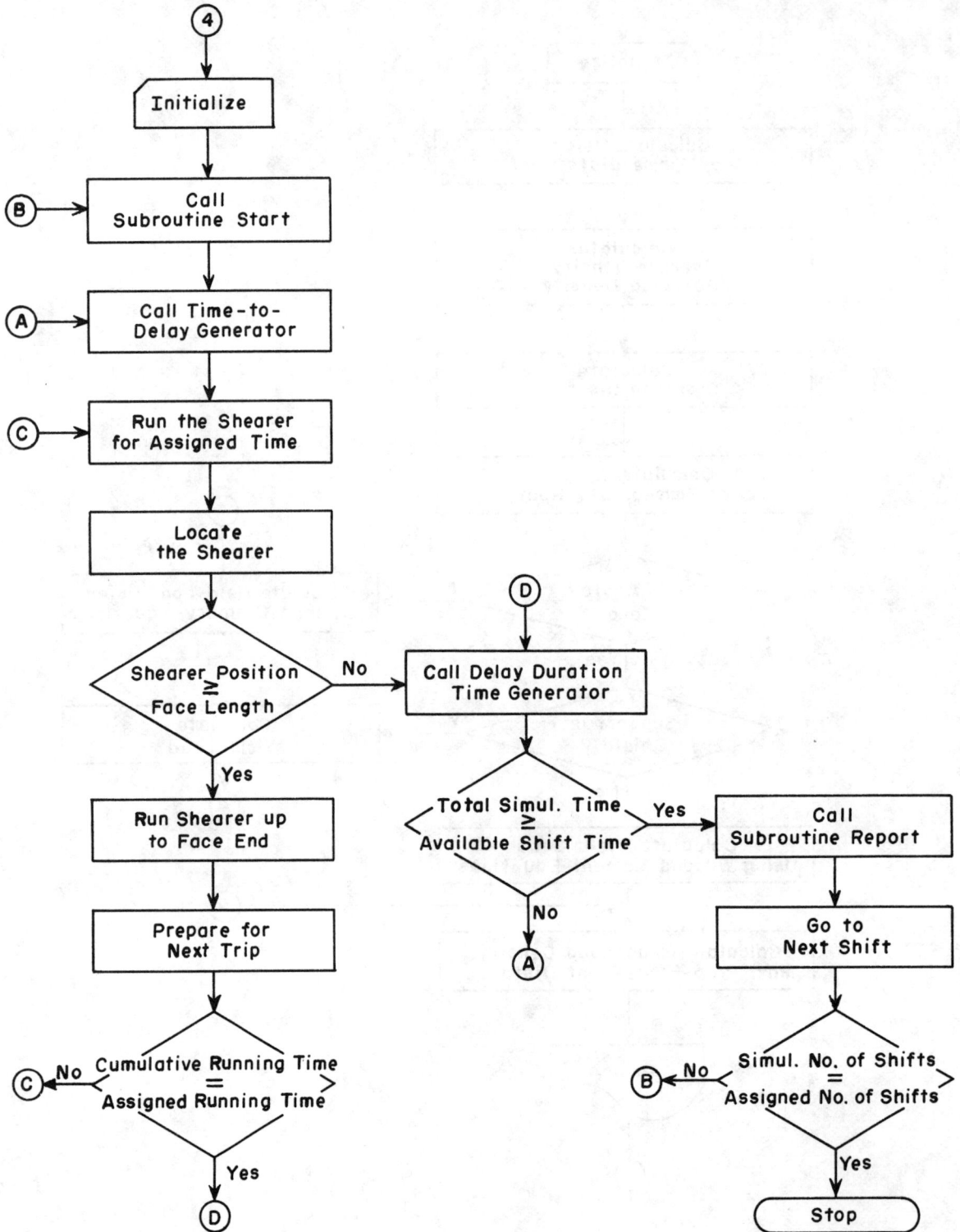

Figure 4.   The Production Simulator.

operating parameters and/or non-operating parameters including panel description, the resulting
effects on production can be approximated through
the simulator. In this paper, three parameters
(web depth, face length and availability of each
subsystem) are analyzed. All data used in this
paper were collected from several published
reports on a longwall demonstration project in the
southern part of the United States (Gentry, 1976;
Jet Propulsion Lab, 1977; Lawrence and King, 1980;
Lawrence and Hackett, 1981).

Web Depth

   The web depth cannot be increased without modification of several other variables associated
with the shearer and face conveyor subsystems.
This is due to the limitation set by the specifications of those two subsystems: that is, drum
design and capacity of the shearer, loading capacity of the face conveyor, and power of the head
and tailgate drives. The relationship between
those parameters results in changes in the shearer
cutting speed and production rate.   Table 1 shows
results of the sensitivity analysis of web depth
using the simulator.

Table 1. Web Depth, Cutting Speed and Production.

Web Depth (inches)	Cutting Speed (ft/minute)	Total Delay Time (minutes/shift)	Shift Production (tons)
15	15.61	147.38	820.10
18	15.61	147.39	984.13
20	15.61	148.92	1083.74
21	15.61	147.39	1148.15
24	15.04	151.04	1269.34
27	12.72	138.08	1384.20
30	10.85	148.24	1368.08

Face Length

   At the mine planning stage or whenever any
modification of mine layout is required, the optimal face length is one of the major concerns. If
it is defined as a length which maximizes the
total production with the maximum availability of
the system, face length can be optimized using a
sensitivity analysis.

   As shown in Table 2, shift production increases
at a smaller rate as face length increases.
Therefore, the face length which generates the
maximum marginal increase in the total production
is a possible candidate for the optimal face
length if the economic terms other than production
are not taken into account.

Availability of the System

   Variation of the availability of each subsystem
affects the total system availability as well as
the production rate. Given that all the other
subsystems have fixed availabilities, an increase
or decrease in the production resulting from
changes in the availability of one subsystem can
be simulated. This implies that the results of

Table 2. Face Length and Production.

Face Length (feet)	Total Delay Time (minutes)	Shift Production (tons)
300	149.01	913.24
350	144.33	976.16
400	145.89	1010.35
450	145.02	1051.08
500	144.85	1076.66
550	148.92	1083.74
600	147.24	1111.47
650	145.79	1138.90
700	145.71	1156.30
750	146.42	1168.10
800	139.78	1204.93
850	139.77	1218.54
900	148.93	1189.34
950	146.72	1210.58
1000	144.57	1227.10
1050	144.54	1235.80
1100	148.45	1226.51

the sensitivity analyses of the system provide the
relative importance of each subsystem with respect
to the production. An analysis of this type can
be applied to allocate limited resources in longwall mines.

   The procedure adopted for this purpose was:
(1) increase or decrease times-to-delay or delay
duration times of each subsystem by a certain percentage, (2) calculate a modified set of parameters in Weibull distributions, and (3)
using new distribution models, run the simulator
and predict the production potential.

   As shown in Figures 5 and 6, the output from
the data set mentioned in the previous section
indicates that change in the delay duration time
of the external conveyor has the greatest effect
on the total system availability and the production. As far as the times-to-delay are concerned,
the total system availability and the production
can be most effectively increased by an increase
in the shearer availability.

CONCLUSIONS

   The principal objective of this project was to
develop a computer simulator that will aid
engineers and management in choosing optimal
design and operating parameters for a longwall
utilizing a shearer and in predicting production
potential of the longwall system.

   Using the longwall simulator, the following
operating parameters which are critical to the
longwall operation are optimized under given
specifications of equipment:

   · shearer cutting speed,

   · drum rotary speed,

   · face conveyor speed,

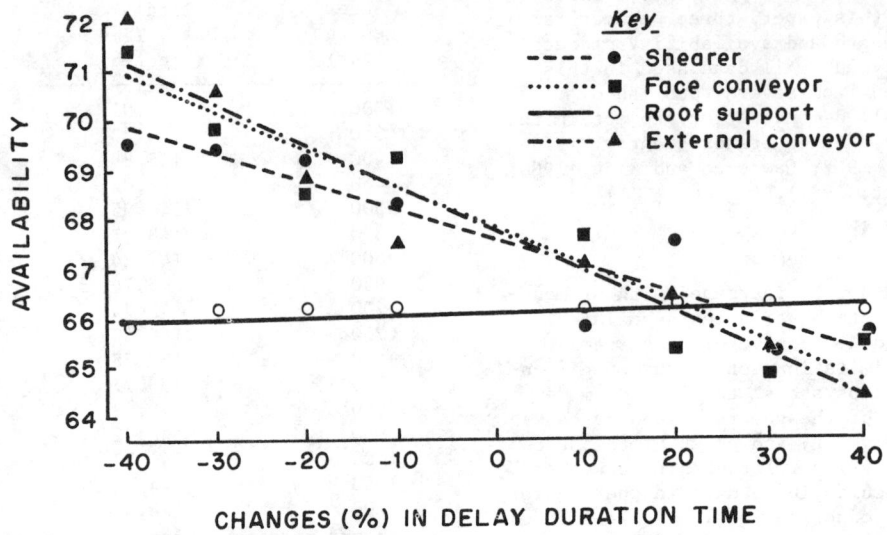

Figure 5. Variation of the Delay Duration Time
and Total System Availability.

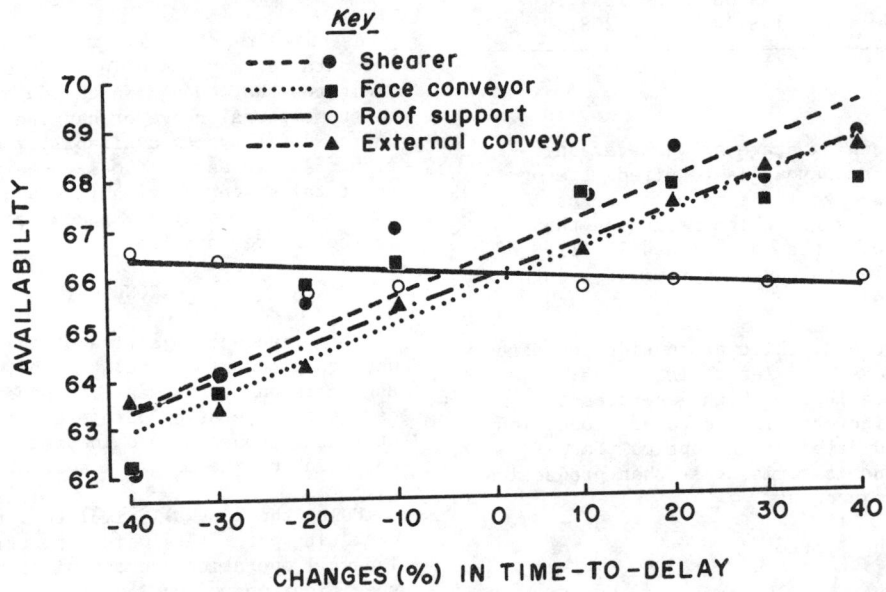

Figure 6. Variation of the Time-to-Delay
and Total System Availability.

· chain strand breaking strength, and

· mean load density and yield load of the roof support.

Further, power requirements for the shearer and face conveyor subsystems at the optimal operating condition or under the maximum load are estimated. Finally, the simulator was used to determine the following items through sensitivity analysis:

· optimal web depth which maximizes production,

· optimal face length which maximizes the system availability and the production, and

· maximum increase in the total production and the system availability with limited resources.

The major feature that distinguishes this simulator from others is its orientation towards both design problems and the prediction of productivity. This feature may be of interest to engineers who are charged with both design and planning functions for longwall operations.

REFERENCES

COMINEC, Inc., 1976, Conceptual Design of an Automated Longwall Mining System, Report, U.S. Bureau of Mines Contract No. S0241051, Tulsa, Oklahoma.

Evans, I. and C.D. Pomeroy, 1966, Strength, Fracture and Workability of Coal, Pergamon Press, London.

Gentry, D.W., 1976, Rock Mechanics Instrumentation Program for Kaiser Steel Corporation's Demonstration of Shield-Type Longwall Supports at York Canyon Mine, Raton, New Mexico, U.S. Department of Energy Report FE 9067-1, Golden, Colorado, July.

Guay, P.J., 1980, Evaluation of the Kloswall Longwall Mining System, U.S. Bureau of Mines Report, BuMines-OFR-102-80.

Guay, P.J. and J. Ludlow, 1981, "The Effect of Mining Wider Webs on a Longwall Face," Ch. 28 in Longwall-Shortwall Mining, State-of-the-Art, R.V. Ramani, Ed., SME-AIME, New York, New York, pp. 223-234.

Guppy, G.A. and B.N. Wittaker, 1970. "Relationship between Machine Output and Face Conveyor Capacity," The Mining Engineer, Vol. 129, No. 329, July, pp. 598-615.

Haycocks, et al., 1981, Department of Mining and Mineral Engineering. Virginia Polytechnic Institute and State University, Design Optimization in Underground Coal Systems - Vol. X - Underground Longwall Ground Control Simulator Report, U.S. Department of Energy, Contract No. DE-AC01-76-ET-10722, Virginia Polytechnic Institute and State University, Blacksburg, VA.

Herhal, A.J., Pimental, R.A., and W.J. Douglas, 1978, "Longwall Conveyor System Study," Report, U.S. Department of Energy, Contract No. ET-77-C-01-8915(2). KETRON, Inc. Wayne, Pennsylvania.

Jet Propulsion Laboratory, 1977, Longwall/Shortwall Mine Equipment Availability and Delay Analysis, U.S. Department of Interior, JPL Report No. 5030-62, Pasadena, California.

Kuroda, N., Uchida, S. and T. Kozuchi, 1982, "Application of Simulation Model in Design and Evaluation of Longwall System," SME-AIME Preprint No. 82-375, The First International SME-AIME Meeting, Honolulu, Hawaii.

Lawrence, R. and T. Hackett, 1981, "Operating Experience in Thick Coal Longwall Mining, York Canyon Mine, Raton, New Mexico, Ch. 23 in Longwall/Shortwall Mining, State-of-the-Art, R.V. Ramani, Ed., SME-AIME, New York, New York, pp. 183-190.

Lawrence, R. and R. King, 1980, Performance of Shield Longwall Demonstration at Kaiser Steel Corporation's York Canyon Mine, U.S. Department of Energy Report No. DOE/ET/12530-T1, Washington, D.C.

Lee, C., 1983, Statistical Analysis of Longwall Delay Data and Simulation of Longwall Mining Systems, Unpublished M.S. Thesis, Pennsylvania State University.

Manula, C.B., Mukherjee, S. and L. Dooley, 1979, Evaluation of Operational Constraints in Continuous Mining System - Vol. IV - A Longwall Simulation Model, Report, U.S. Department of Energy Contract No. ET-76-6-01-8982, The Pennsylvania State University, University Park, Pennsylvania.

Nishimatsu, Y., 1972, "The Mechanics of Rock Cutting," International Journal of Rock Mechanics and Mining Science, Vol. 9, pp. 261-270.

Osterman, W., 1966, "Cutting Performance of a Drum Shearer-Loader as a Function of Web Depth and Drum Speed," Gluckauf, Vol. 102, No. 4, February, pp. 125-136.

Pimental, R.A., Urie, J.T., and W.J. Douglas, 1981, "Evaluation of Longwall Industrial Engineering Data," Report, U.S. Department of Energy Contract No. ET-77-C-01-8915(11), KETRON, Inc., Wayne, Pennsylvania.

Teale, R., 1970, "Some Factors Governing the Rate of Output on a Longwall Face," The Mining Engineer, Vol. 129, No. 118, July, pp. 616-634.

Wilson, A.H., 1975, "Support Load Requirement on Longwall Faces," The Mining Engineer, Vol. 134, No. 173, June, pp. 479-491.

Wilson, A.H., 1981, "An Alternative to 2 m Prop-Free-Front Limit," Colliery Guardian, Vol. 229, No. 5, May, pp. 148-150.

Xin, J. and E. Topuz, 1983, "A Simulation Model for Longwall Mining Operations," SME-AIME Preprint No. 83-387, Fall Meeting and Exhibit, Salt Lake City, Utah, October.

Chapter 24

AN INTERACTIVE MICRO-COMPUTER SIMULATOR FOR LONGWALL OPERATIONS

E. Topuz*, E. Nasuf**, and N. Michalopoulos***

*Associate Professor, **Visiting Professor, and *** Former Graduate Student

Department of Mining & Minerals Engineering
Virginia Polytechnic Institute & State University
Blacksburg, Virginia 24061

ABSTRACT

This paper presents an interactive computer program that simulates the operations of longwall mining systems. Both deterministic and probabilistic methods are used to estimate the production potential of the system. Among the factors that have been considered in performing the analysis were the characteristics of the equipment and the coal seams, the support requirements, the equipment failures and various other delays.

In order to determine the relations between the various factors that affect the production of the system, both analytical and statistical models were employed. The statistical basis for the latter type of models were the data obtained from longwall mines in the United States. A case study that was used to validate the model, has been included in the paper.

INTRODUCTION

Longwall mining has become an increasingly popular method in underground coal mining due to its potential for improving safety and productivity, better ground control and higher levels of recovery.

The production capacity of longwall panels depends on several external and internal factors. The external factors, which are set by the environment, include the depth of extraction, the seam thickness, physical characteristics of the coal, and the geological conditions. The internal factors, which can be manipulated by the operator, include the characteristics of the equipment, the dimensions of the panel, the quality of labor and the operational policies of the management.

This paper introduces an interactive simulation model for longwall mining systems (LONGWALL. SIM) that can be used in estimating the production potential of a longwall system using either shearers or plows, and in identifying the effect that a change in the parameters of the system would have on its production.

Most of the computer simulation models in use are designed for computers with large memory and high speed of execution. The introduction of micro-computers and their highly interactive capabilities, however, diminished the necessities of learning the large computer systems, simulation modelling and extensive data preparation. It has become much easier now to enter and change the data values and evaluate the effects of the changes. Therefore, it was decided to develop a simple and user-friendly computer program for Longwall mining operations that can be used in micro-computers, interactively.

The program is written in FORTRAN-77 and runs on the IBM PC and PC XT under PC-DOS releases 1.1 and 2.0. Minimum system configuration includes 128k bytes main memory, and a printer or CRT.

SIMULATION MODEL

Simulation consists of imitating the performance of a system step-by-step in order to determine the outcomes at critical points in the process. Among the variables of the system, only those that have a substantial effect upon the performance measures should be represented in the model.

The production process in a longwall face essentially consists of the following operations:
1. Cutting – the coal from the face;
2. Loading – the broken coal onto face conveyor;
3. Conveying – the broken coal out of the face;
4. Supporting – the newly exposed roof; and
5. Controlling – the exposed roof behind the advancing supports.

The shearer or plow performs the cutting and loading operations. The conveying operation works closely with the cutting and loading operations. Therefore, all three of these operations can be incorporated into one "coal extraction" operation. The supporting and controlling operations are performed by the support units. Hence, these two operations can be considered as one "support-advancement" operation. Consequently, the operations in a longwall face can be combined into two main activities: coal extraction and support advancement.

The regular sequence of these operations is interrupted by predictable and unpredictable delays. Predictable delays include travel time, lunch breaks, safety meetings, fire drills, and preventive maintenance. Unpredictable delays can be caused by men, by the geological environment or by the equipment failures. The failures related to the roof control and ventilation can also be considered in this category.

The simulation model is event-oriented, that is, the system's main operations are considered to be "events" which are sorted in a record according to their time of occurrences.

In a longwall mining system the following activities may be considered as events:

- Coal extraction (cutting, loading, transporting)
- Support advancement (supporting, controlling)
- Equipment failures,
- End of a shift or end of a simulation run.

The current activity of the coal cutting machine is determined by the control variables within the program. Time variables that record

the time of occurrence of each activity are established. The above activities are kept in a record along with the corresponding time of occurrence. Each time the record is examined, the activity with the minimum occurrence time is selected. If this event is the end of shift, the simulation results are printed.

A macro flow-chart of the longwall simulation program (LONGWALL.SIM) is given in Figure 1.

INPUT DATA

The input data for longwall simulation model can be classified as variable input data and fixed input data.

Variable Inputs

The variable input data for the model include the following:

- Speed of cutting machine (backward and forward, when sumping, and tramming empty),
- Rate of advance of support units,
- Head and tail gate delays,
- Time required to reverse drums,
- Time between failures (cutting machine, face conveyor, and outby haulage system),
- Repair times (cutting machine, face conveyor, and outby haulage system).

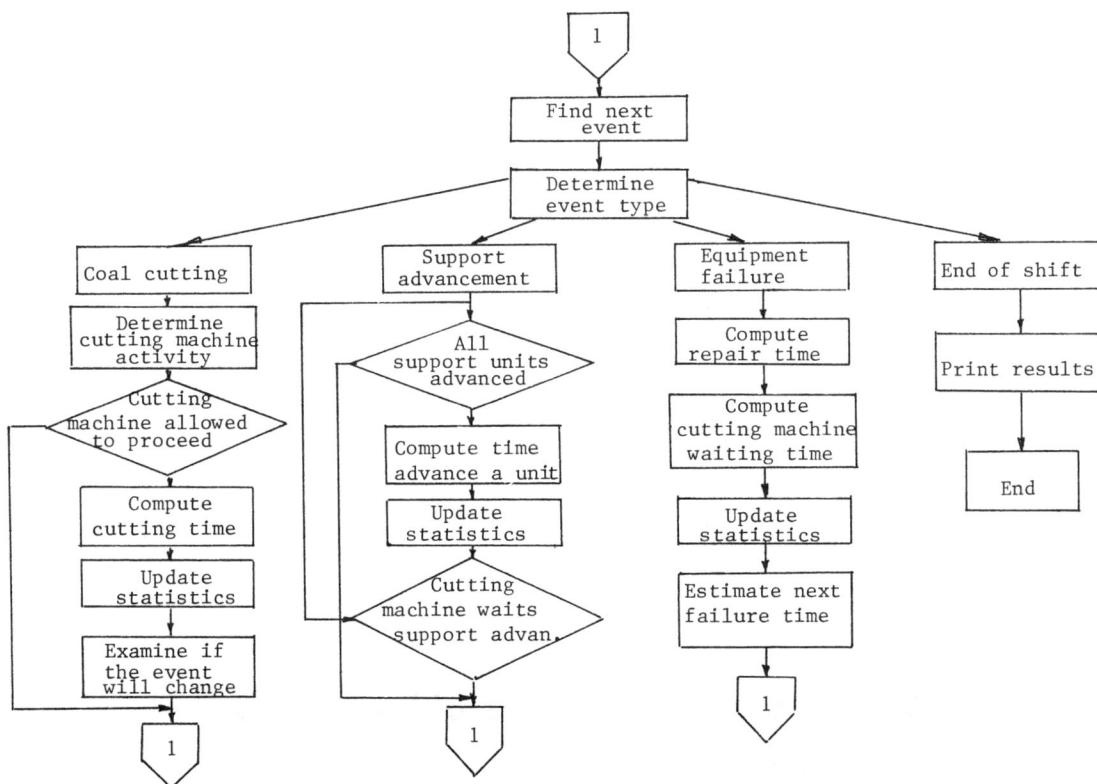

FIGURE 1 - MACRO FLOW-CHART FOR LONGWALL.SIM

Fixed Inputs

The fixed input data for the model include the following:

- System parameters (face length, web thickness, seam height, shift duration, etc.),
- Seam characteristics (strength of coal, density, swell factor, angle of repose),
- Equipment characteristics (cutting machine: dimensions, HP, efficiency, dimensions and strength of cutting bits; face conveyor: length, width, the height of the pans and spill plates, the weight of the chain, HP, efficiency, friction factors, etc.).

Input data for LONGWALL.SIM is directly entered by the user by running the interactive data program. The data is divided into six groups and displayed on the screen as a menu at the beginning of the program. The menu as it appears on the screen is shown in Figure 2.

MENU

1 : OPTIONS OF THE PROGRAM

2 : PARAMETERS OF THE SYSTEM

3 : CHARACTERISTICS OF THE COAL

4 : CHARACTERISTICS OF THE SHEARER/PLOW

5 : PARAMETERS OF THE FACE CONVEYOR

6 : RANDOM VARIABLES

7 : QUIT

Select a group of variables you want to change ?

FIGURE 2 - MENU FOR INPUT DATA

After choosing the group of data which the user desires to change, the entry of new data and the updating of the existing data are possible through a series of menus which guide the user. Any errors found during the data entry and updating process are highlighted on the screen, which permits the user to make immediate corrections. A typical data entry screen is illustrated in Figure 3.

For the variable inputs (Group 6 in the Menu), their average values or their probability distributions obtained by time studies can be used. The program also includes a statistical data processing module which consists of three parts. The first part generates the random values of the variables, given the type of distribution and its parameters. The second part estimates the distribution parameters, if they are not known, from

observations that are provided as input. The third part classifies the observations in histograms and may conduct a chi-square test in order to find whether the data follow the specified distribution. The program permits the user to select random variables interactively from uniform, empirical, exponential, normal, weibull and triangular distributions.

Assumptions

The main assumptions underlying the development of the model are given below:

a. The mean time between the failures and the repair time of the equipment do not change with the parameters of the system.
b. It is desirable to maintain the maximum cutting speed without violating certain contraints.
c. The height and depth of extraction are constant.
d. No multiple cuts are allowed.
e. At the start of simulation, all the equipment is in working conditions.
f. The program does not calculate the shift time lost because of excessive methane liberation and for roof control problems.

Options

The simulation model offers the following options:

a. It can be used in main-frame and in micro-computers,
b. It simulates the operation of either a plow or a shearer,
c. The shearer may be single or double drum and may operate either uni- or bi-directionally,
d. The support units can be advanced as normal system or one-web-back system,
e. The cutting speed of the shearer or plow can be estimated deterministically or entered as a random variable,
f. The variable inputs can be entered either by their average values or by their probability distributions,
g. The user may specify the output options.

Output

The LONGWALL.SIM program permits the user to see the simulation results on the screen or on the printer. The output of the program, at the option of the user, may include the following information:

i. The estimation of the cutting speed and the power requirements of the shearer or plow,
ii. The estimation of the transportation speed and power requirements of the face conveyor,
iii. A report of the time spend in each of the face activities.

The output is printed at the end of the simulation. It can be printed after each pass and/or after each shift depending on the option specified in the program. The program also offers a graphical illustration of the outputs which may help the user

```
                     OPTIONS OF THE PROGRAM
                     ----------------------


          IS CUTTING SPEED A RANDOM VARIABLE (Y/N) ?  n

          DO YOU WANT TO USE A PLOW  (Y/N) ?  n


                MODE OF OPERATIONS ARE :

          1: SINGLE DRUM SHEARER UNIDIRECTIONAL CUTTING
          2: SINGLE DRUM SHEARER BIDIRECTIONAL CUTTING
          3: DOUBLE DRUM SHEARER HALF FACE METHOD
          4: DOUBLE DRUM SHEARER FULL FACE METHOD

          Sellect the mode of operation (ENTER 1,2,3 OR 4) ?4

          DO YOU WANT THE CONVEYOR SPEED TO BE ESTIMATED FOR
          THE FORWARD OR THE COUNTER CUT  (ENTER 1 OR 2) ?1

          WHEN DO YOU WANT THE RESULTS TO BE PRINTED ?

          0: RESULTS WILL BE PRINTED AFTER EVERY CUT AND AFTER EVERY SHIFT
          1: RESULTS WILL BE PRINTED AFTER EVERY SHIFT
          2: RESULTS WILL BE PRINTED AFTER EVERY CUT
          3: RESULTS WILL BE PRINTED ONLY AT THE END OF THE SIMULATION

          (ENTER 0,1,2 OR 3) ?3
```

FIGURE 3 - A TYPICAL SCREEN FOR INPUT DATA

to see and analyze the results easily. Graphical display of shift activities are given in the form of bar charts as shown in Figure 4. Another graphical routine displays the variations in shift production throughout the simulation time as a line graph which may help the user to determine the steady state as shown in Figure 5.

CASE STUDY

The data for the case study were obtained from a coal mine in West Virginia. The average thickness of the seam was 1.5 to 2.0 m. and it was lying almost flat. The depth of cover at the portal point was about 150 m. Both immediate roof and floor were strong. It was a moderately gassy seam and the amount of methane liberated was never excessive enough to stop mining. The mining face was retreating and had a length of 114 m. with an average production of 2,130 t. (2,350 st) per shift.

The face was equipped with a Sagem DTS-300 double drum ranging shearer designed to operate in seams 140 to 300 cm. thick using the full face method. The shearer had a 400 HP motor; the drums rotated at 60 rpm and were 1.5 m. in diameter. The web depth was 61 cm. and the mining height was 2.1 m. The normal operating speed of the shearer was 7.6 meters per minute.

A double inboard chain conveyor of 76 cm. wide was used. It was equipped with two 300 HP electric motors. The operating speed was 79 meters per minute.

The input data values used in the case study are shown in Table I.

TABLE I - INPUT DATA FOR THE CASE STUDY

System parameter	Value
Face length	114 m.
Height of extraction	210 cm.
Web thickness	61 cm.
Density of coal	1.285 kg/m^3
Shift time (excluding travel,etc.)	390 min
Mode of operation	Full face
Cutting speed of shearer	7.6 m/min
Conveyor width	76 cm.
Gate delays	2 min

Time studies were used to determine the delay time at the head and tail gates, the time required to reverse drums, the speed of the shearer when sumping and tramming empty, and the time required to advance the support units. Exponential distributions were assumed for the time between failures and triangular distribution for the repair times.

The program was run for 100 shifts and the variations in shift production were plotted using one of the graphics routines shown in Figure 5. A sample output of a shift result is also shown in Figure 6.

The output of a simulation is a time series with a transient and steady state phases. A system is said to have reached steady state when succes-

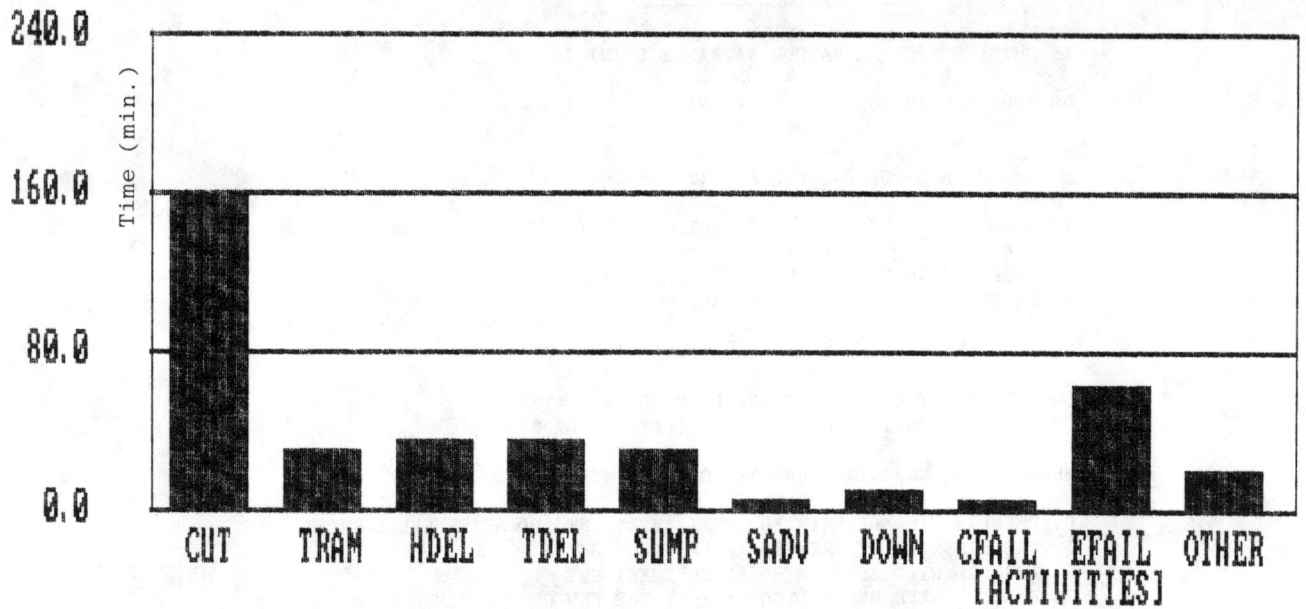

FIGURE 4 - GRAPHICAL ILLUSTRATION OF LONGWALL ACTIVITIES (Average results of 100 shifts)

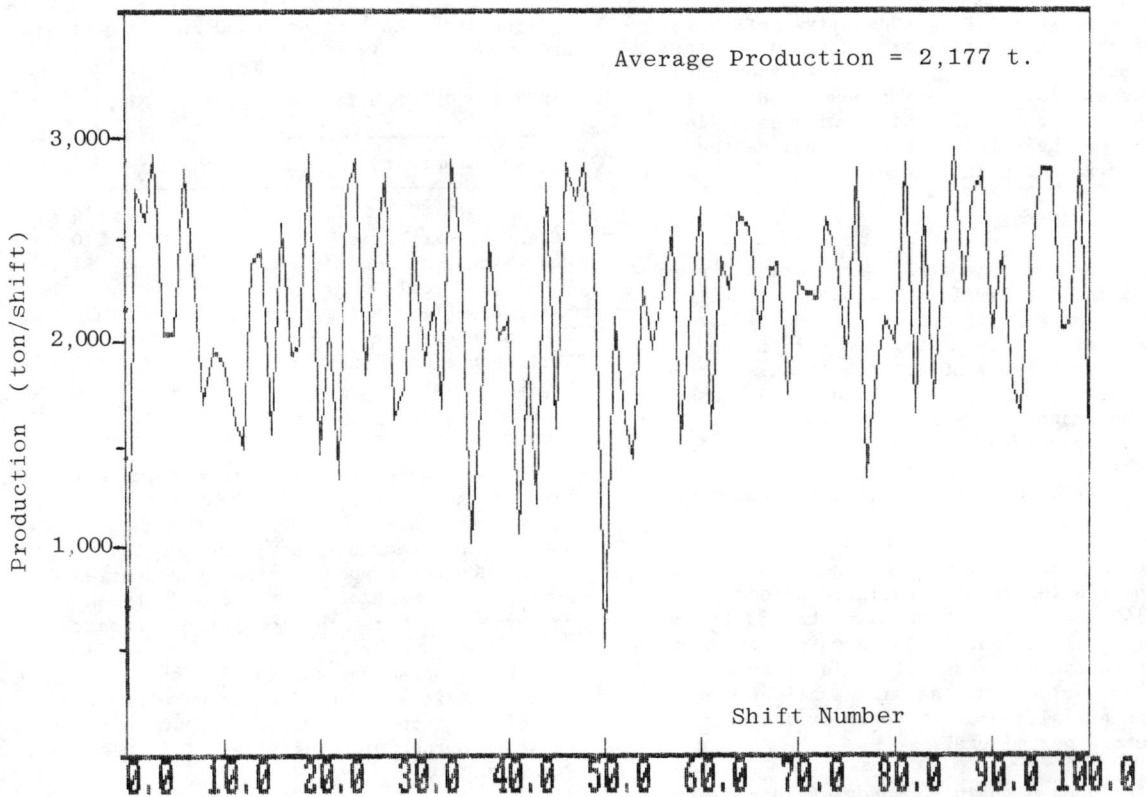

FIGURE 5 - VARIATIONS IN SHIFT PRODUCTION

sive observations of the system's performance are statistically indistinguishable. Steady state implies that the system's performance is independent from the initial conditions. Transient phase, however, occurs because of the simulation output is influenced by the initial conditions and is not representative of the average performance of the system. Even though the transient phase is not observed in Figure 5, in some applications, the user may observe considerable fluctuations in the early shifts. It is advisable to discard the performance of these early shifts and to consider the remaining shifts as a representative performance of the system. Moving average of the shift outputs may be useful to identify the transient phase. A graphical routine included in the program may be used to analyze the remaining shifts corresponding with the steady state phase.

Average production of 100 shifts is 2,177 tons (2,403 st) and reported average production from that section was 2,130 tons (2,350 st) per shift. The difference of 47 tons which corresponds to 2.2% of the average production is considered reasonable for a simulation model.

Once the longwall simulation model is validated, it can be used to study the effects of various parameters on the performance of the system. This way a sensitivity analysis can be

SHIFT #   1

PRODUCTION =   2740 t (3028 st)

ACTIVITY	TIME (MIN)	PERCENT
CUTTING	201.67	51.71
TRAMMING EMPTY	38.59	9.90
HEADGATE DELAY	42.02	10.77
TAILGATE DELAY	43.99	11.28
SUMPING	36.02	9.24
WAITING FOR SUPPORT ADVANCE	4.66	1.20
DOWN - SHEARER/PLOW FAILURE	23.04	5.91
IDLE - FACE CONVEYOR FAILURE	.00	.00
IDLE - OUTBY HAULAGE FAILURE	.00	.00
OTHER DELAY	.00	.00

FIGURE 6 - OUTPUT OF A SHIFT

performed, an optimum combination of parameters can be found and longwall equipment can be selected without costly trial and error methods. Figure 7 illustrates the effects of changes in

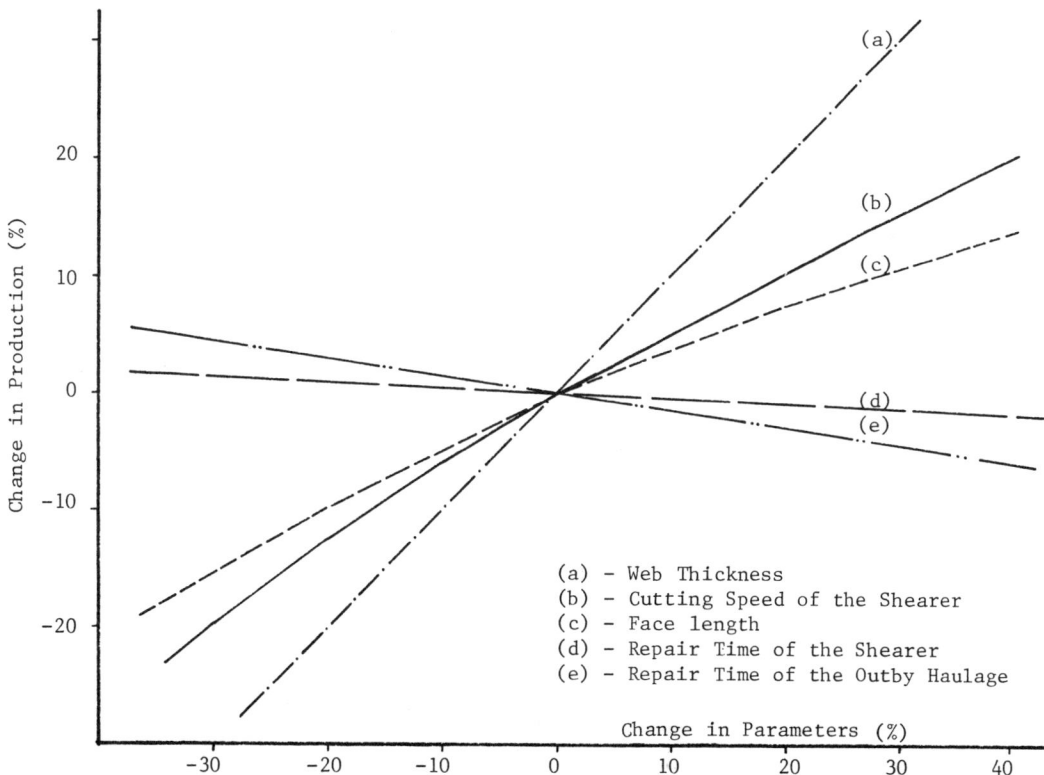

(a) - Web Thickness
(b) - Cutting Speed of the Shearer
(c) - Face length
(d) - Repair Time of the Shearer
(e) - Repair Time of the Outby Haulage

FIGURE 7 - SENSITIVITY ANALYSES OF SOME PARAMETERS

the web thickness, cutting speed of the shearer,
face length, repair time of the shearer and repair
time of the outby haulage on the production
potential of the longwall operation.

Concluding Remarks

    The interactive longwall mining simulation
model is simple and versatile. It provides the user
a tool to predict system performance and productivity. It eliminates the need for costly trial
and error methods in selecting longwall equipment.
The model can be used to study the effects of
changes in equipment design, panel dimensions,
operating methods and equipment reliability to
improve the performance of the system.

Selected Literature

Michalopoulos, N.G., "Simulation of Longwall
      Mining Systems," MS Thesis, August 1983,
      Virginia Polytechnic Institute and State
      University.

Michalopoulos N.G., and E. Topuz, "Simulation of
      Longwall Mining Systems," Proceedings, The
      International Sy,posium on the Application of
      Computers and Mathematics in the Mineral
      Industries, March 26-30, 1984, London, U.K.

Nasuf, E., E. Topuz, and N. Michalopoulos, " A
      Simulation Model For Longwall Mining Systems,"
      Proceedings, Second Annual Workshop, Generic
      Mineral Technology Center- Systems Design and
      Ground Control, November 10-11, 1984, Reno,
      Nevada, (in press).

Topuz, E., E. Nasuf, and N. Michalopoulos, " User
      Manual - LONGWALL.SIM : An Interactive Micro-
      Computer Program For Longwall Mining Systems,"
      Virginia Polytechnic Institute and State Uni-
      versity, (in press).

Xin, J., and E. Topuz, "A Simulation Model for
      Longwall Operations," Preprint, SME-AIME Fall
      Meeting, October 19-21, 1983, Salt Lake City,
      Utah.

Chapter 25

-CONSIM-
AN INTERACTIVE MICRO-COMPUTER PROGRAM FOR CONTINUOUS MINING SYSTEMS

E. Topuz* and E. Nasuf**
*Associate Professor and **Visiting Professor

Department of Mining and Minerals Engineering
Virginia Polytechnic Institute and State University
Blacksburg, Virginia

ABSTRACT

CONSIM, An Interactive Micro-Computer Program for Continuous Mining Systems, is developed for simulation of face operations in continuous mining systems.

CONSIM is a discrete, event-oriented simulation model for the face operations. The events in CONSIM include cutting and loading, hauling, and roof bolting operations, equipment breakdowns, completion of a cut, and the end of the job. CONSIM requires a minimum amount of data to be input by the user. These data are essentially the dimensions of workings and the operational characteristics of the face equipment. The model permits the inclusion of both deterministic and probabilistic interaction among the equipment. The output may be obtained by cut, by shift, and over simulation time needed to mine a number of cuts specified by the user.

INTRODUCTION

The ease in modelling of mine systems on computers and the ability of computers to store, retrieve and manipulate large amounts of data have made computer simulation the most widely recognized operations research technique in the mining industry. As a result, computer simulations have been increasingly used in mine planning and analysis of mine operations and design alternatives.

Most of the computer simulation models in use are designed for computers with a large memory and a high speed of execution. These models necessitate a user experienced in large computers, simulation modelling and extensive data preparation and require large computer memory and long execution time. Therefore, their use is rather limited. In order to minimize these difficulties, thought was given to development of a simple, user-friendly simulation model that can be used on micro- and mini-computers.

FACESIM (Prelaz, et al., 1964), a computer simulation model developed at Virginia

Polytechnic Institute and State University, has been widely used in simulation of face operations in conventional room-and-pillar mining. Because of this widespread acceptance, it was decided to develop a simulator for face operations in continuous mining systems by modifying the original FACESIM. This modified program, CONSIM, that can be used on micro-computers, minimizes the input data requirements and also augments the capability of the program to simulate equipment breakdowns without greatly adding to its complexity.

The program is written in FORTRAN-77 and runs on the IBM PC with two disk drives under PC-DOS releases 1.1 and 2.0. Minimum system configuration includes 128k bytes main memory, and a printer or CRT.

THE MODEL

A simulation model is a mathematical and logical representation of a real system and it permits experimentation without disturbing the operations in the real system. In simulation, the operations in the real system are imitated to determine the outcomes of interest.

CONSIM is a discrete, event-oriented simulation model; i.e., the status of the system and the simulation clocks are updated when an event takes place. The following are the events in CONSIM: (1) cutting and loading, (2) haulage, (3) roof-bolting (4) equipment breakdown, (5) completion of cut, and (6) end of job. The macro-flowchart for CONSIM is shown in Figure 1. The functions and subroutines in CONSIM, with their purposes, are given in Table I.

The Cycle Time

In CONSIM, the miner cycle time, i.e., the time between the start of work in consecutive cuts, satisfies the following equation:

$$CCT = CLT + WP + WC + CT + BT + TT$$

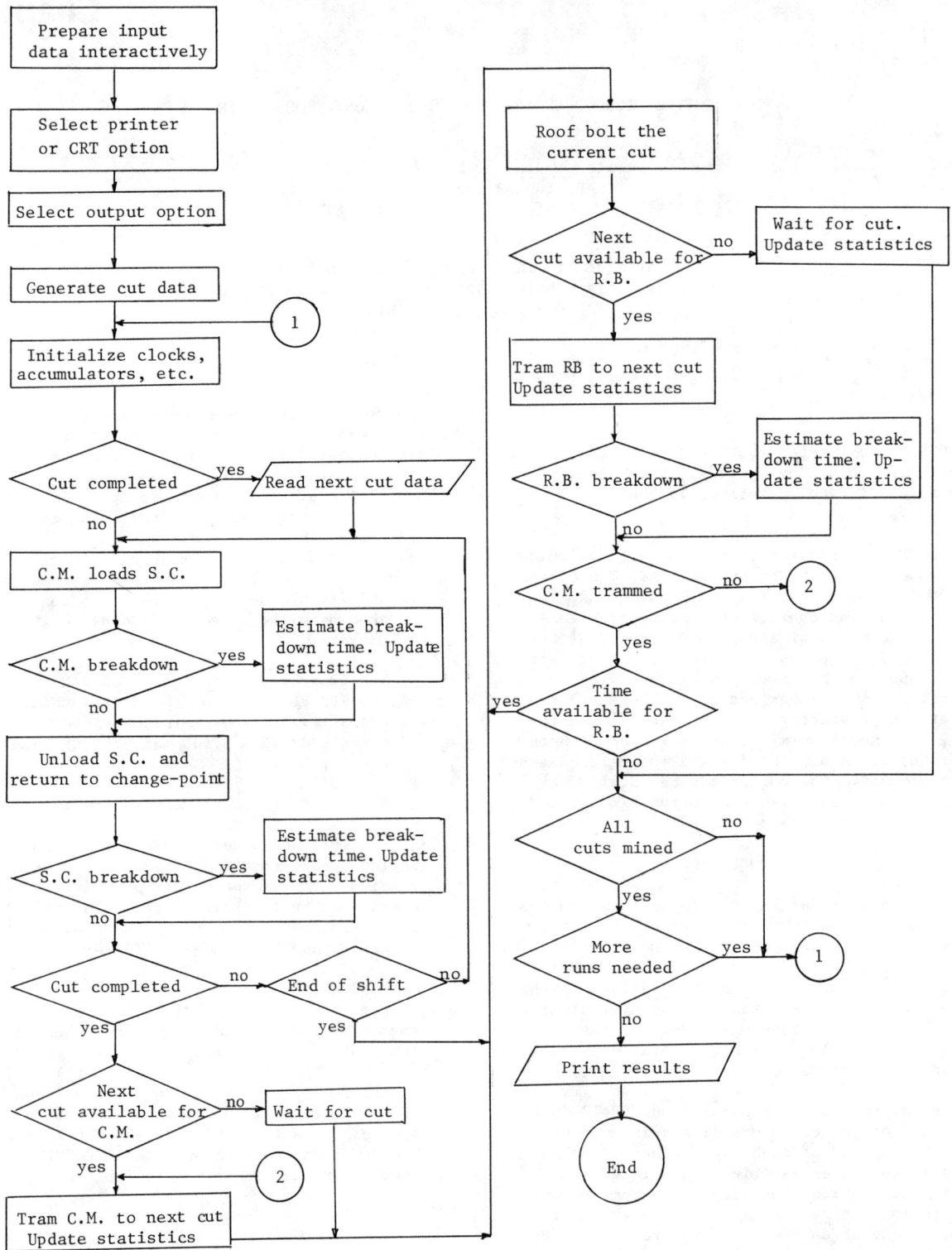

FIGURE 1 - MACRO FLOW-CHART FOR CONSIM

TABLE I - SUBPROGRAMS IN CONSIM AND THEIR PURPOSES

SUBPROGRAM	PURPOSE
RANDG	Subroutine to generate continuous random variates from empirical distributions defined by observed values of loading rate, shuttle car payload, haul rates and discharge rate.
RANDOM(L)	Function which generates uniform (0,1) random numbers.
EXPON(XM)	Function to generate an exponential random variate with mean XM.
GCARD	Subroutine which generates input data for cuts made in development. See Ref. No.3 for some examples of cut numbering and coding for full and partial cuts in single or multiple cut sequencing generated by subroutine GCARD.
DIST	Subroutine called in subroutine GCARD to calculate distances.

where:

CCT - the continuous miner cycle time,
CLT - the time the continuous miner loads in a cycle,
WP - the time the continuous miner waits for coal, i.e., for entry into the next cut,
WC - the time the continuous miner waits for shuttle car during the cycle,
CT - change-out time in the cycle,
BT - the time the continuous miner is under breakdown during the cycle,
TT - the time spent on tramming the continuous miner to the next cut.

This relationship between the miner's cycle time and the time it spends on different activities also holds good for shift time and the total time required to mine all cuts.

Assumptions

The main assumptions made in CONSIM are summarized below. A complete list is given by Ramachandran (1983), and Topuz, et.al. (1984).

- The change-out time is the time spent by shuttle cars hauling over the change-out distance and the time required for switch-out. This time is considered to be zero if it occurs while the miner is under breakdown. The change-out also excludes the time spent hauling the last shuttle car over the change-out distance.

- All the coal in the cut will be loaded into shuttle cars even if the last load is negligibly small.

- Any maneuver of a continuous miner in the cut is considered part of production work.

- The breakdown time of any equipment is the production time lost due to the breakdown.

Time between breakdowns of equipment is the production time between consecutive breakdowns of the equipment, excluding the breakdown time. Shift time refers to the duration of the shift from the time when the face work is started at the beginning of the shift to the time when work is stopped at the end of the shift.

- Breakdown of a continuous miner may occur only after the loading of a shuttle car is completed.

- Breakdown of a roof bolter may occur only after moving it into the cut to be roof-bolted.

- At the start of simulation all the equipment at the face is in working condition and the shuttle car numbered one is ready to be loaded. The continuous miner is initially in cut number one which may be in the first or last heading.

- The loading rate, shuttle car payload, shuttle car haul rates and discharge time are either deterministic values or continuous random variables. If they are random variables, their cumulative distribution functions are defined empirically, i.e., by the data collected by time studies.

Options

The following options are incorporated in the model:

- Face operations up to 200 cuts can be analyzed. This limitation may be relaxed by making necessary changes in dimension statements.

- Any number of headings can be driven.

- Any of the headings can be used as the transportation heading.

- It can be used to generate single or multiple cuts in breakthroughs and headings.

- The discharge point may not be advanced or may be advanced any specified length of belt segments.

- Deterministic or probabilistic data may be used for the analysis.

- Cut data may be generated automatically by using subroutine GCARD or may be input by the user.

- Ten different types of codes may be used for cuts resulting in varying section layouts and cut sequencing.

- The output may be obtained for cuts, for shifts, or for a specified number of cuts.

- The program can be run by a user a specified number of times in order to obtain a statistically more reliable performance of the system.

- The output may be seen on the CRT or may be printed.

Limitations of the Program

The program has several limitations. Some of these limitations are already mentioned under the assumptions. Additional limitations include the following:

- It cannot simulate continuous mining operations in more than one section simultaneously.

- It requires that all shuttle cars travel the same length of haul distances. Therefore, it may not be used for more than two shuttle cars.

- It cannot be used for pillar mining operations.

- Pillars should have regular sizes if subroutine GCARD is used for generation of cut data. This subroutine permits use of 10 different types of cuts in single and multiple cut sequencing operations. The cut numbers and codes are assigned according to a certain repetitive procedure.

- The program permits the use of both deterministic or stochastic data. The simulation done by using deterministic data does not consider the probabilistic nature of face operations. The stochastic simulation, however, needs time study data.

RUNNING THE PROGRAM

The computer program consists of two separate diskettes: Data Diskette and Program Diskette. The Data Diskette includes two interactive data programs, DATA and GCARD, with the total disk space requirements of 150k. The Program Diskette includes the two versions of CONSIM, CONSIMP and CONSIMS, with the total disk space requirements of 150k. The compiled version of the programs can be easily run in a computer having 128k RAM (random access memory).

Input data for CONSIM can be created or changed by running the program DATA in the data diskette. It is also possible to create the user's own cut distances and roof bolter sequence data by running the program GCARD in the same diskette.

CONSIM has two versions, printer version (CONSIMP) and screen version (CONSIMS), in the program diskette. Running these programs separately provides the user with the opportunity to see results on the screen and obtain a printout later if it is needed. Only the printouts are obtainable by running CONSIMP.

Preparation of the Data Files

CONSIM requires a minimum amount of data to be input by the user. These data are essentially the dimensions of workings and the operational and maintenance characteristics of the equipment.

The output of the system obtained from a simulation model are dependent upon the values of the input variables. Since the user is interested in evaluating the changes in performance measures as a result of the changes in input variables, it is important for a successful simulation study that the user prepares the input data very carefully so that possible errors are eliminated and accuracy in estimation is improved as much as possible.

The nine different types of input data cards are needed for CONSIM. These cards are named Card Type A through Card Type I, in alphabetical order. These data cards can be created or changed by running the DATA program. When the program is run, the following menu will be displayed on the screen:

```
*************************************************
*                                               *
* THIS PROGRAM CREATES DATA FOR THE CONTINOUS   *
*                                               *
*   MINING SYSTEM SIMULATOR PROGRAM (CONSIM)    *
*                                               *
*************************************************


                    1.Create a new file
                    2.Change the existing file
                    3.Quit


Enter the option number ?
```

If the user desires to create a new data file then the option number "1" should be entered. In this case, the tables of Card Types A through I will be displayed on the screen as the user continues to enter the data. Each of these tables includes a brief definition of the variables involved and the format that should be used while entering these variables. A typical table is shown below:

```
                        Card Type A
------------------------------------------------------------------

<VARIABLE>                    DEFINITION
------------------------------------------------------------------

<KT> The first cut to be worked by the roof bolter
<NTSC> If all shuttle cars have the same breakdown characteristics
       input (1) ,if different input the number of shuttle cars
<NSC> Number of shuttle cars
<NDUMP> Type of dump , (1):one-way ,(2):two way
<NA> Max.number of input cards in card type H
<NB> Max.number of input cards in card type I
------------------------------------------------------------------
```

Enter the variables of card type A using the following format

```
   KT   NTSC   NSC  NDUMP    NA     NB
 ----- -----  ----- ----- ----- -----
##### ##### ##### ##### ##### #####
    5     2     2     1    20     4
```

The data files so created can be changed either by running the program DATA again or by using DOS system commands.

Sample Input Data

To provide an example for the preparation of input data for CONSIM, a hypothetical five-heading continuous mining section is considered (Figure 2). The pillars are 30.4 by 30.4 m. (100 by 100 ft) centers. The widths of headings and breakthroughs are 6.1 m (20 ft). The depth of a cut is 6.1 m (20 ft). The dump is located at heading number 3 and it is 9.1 m (30 ft) apart from the intersection of the heading and the breakthrough. As the mining progresses, the belt is advanced forward by adding 38 m (125 ft) belt segments. Two shuttle cars of 5 t (4.5 st) capacity and having different failure characteristics are used to haul coal from the miner to the dump point. It was specified to make 41 single cuts. The distance of the first cut from the center-line of the breakthrough is 15.2 m (50 ft). Initially the miner is in the first and the bolter in the fifth heading. The production time in a shift is 360 minutes.

Output

When one of the programs is run, the following menu and questions will appear on the screen:

```
****************************************************
*                                                  *
*                  CONSIMS                         *
*          ( SCREEN VERSION OF CONSIM )            *
*     A SIMULATOR FOR CONTINUOUS MINING SYSTEMS    *
*                                                  *
****************************************************

                  MENU


          1.Analysis by Cut

          2.Analysis by Shift

          3.Analysis Over Simulation time

          4.Quit CONSIMS


Enter the option number ?1


How many times you want to run the program ?10


[ Answering YES to the following question means
you have created your own GCARD data,NO means
you want it to be created by CONSIMS    ]


Are you using your own GCARD data  (Y/N) ? n
```

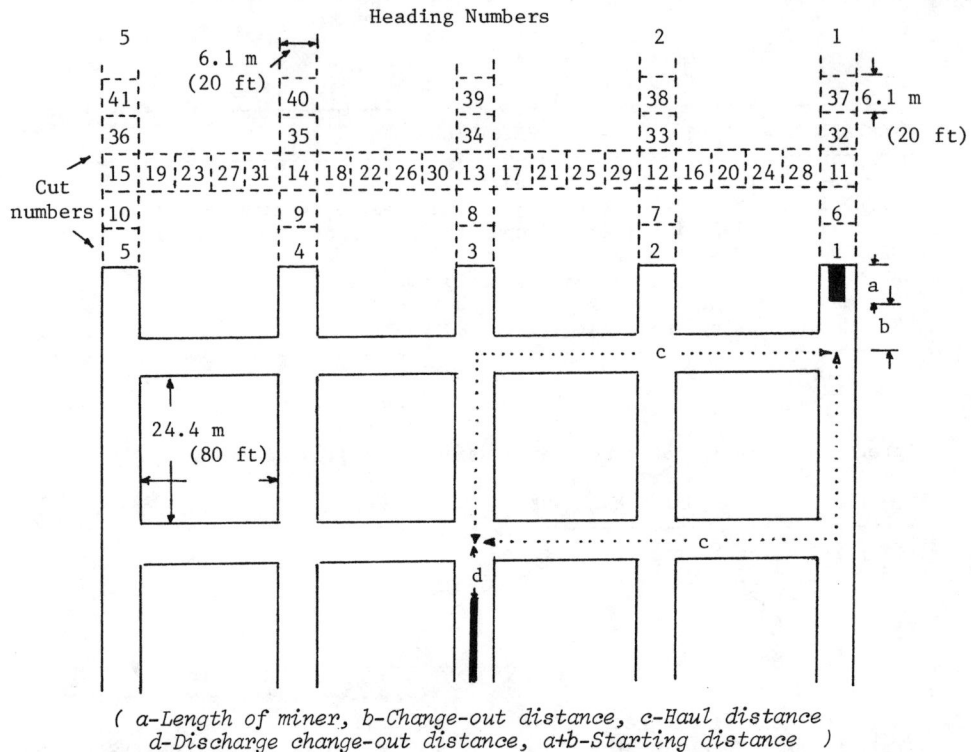

(a-Length of miner, b-Change-out distance, c-Haul distance
d-Discharge change-out distance, a+b-Starting distance)

FIGURE 2 - SECTION LAYOUT SHOWING CUT SEQUENCING AND SOME SYSTEM VARIABLES

The user may choose the type of analysis by entering the number given in the menu. The second question is important for statistical analysis. For reliable statistical results, the program should run about 10 times. It is suggested that during statistical analysis, Analysis Over Simulation Time is used, since the analysis by cut or by shift are time-consuming and not very informative.

The answer to the third question will be YES (Y) if the user has different mine layout, cut dimensions and roof bolter sequence data which necessitate the use of his own GCARD data created by running the GCARD data program and then transferring it to the program diskette. Otherwise, the answer will be NO (N), since the GCARD data will be created by the subroutine GCARD within the program.

The same questions are asked in both CONSIMP and CONSIMS programs. After answering these questions, the results will either appear on the screen or on the printer, depending on the program chosen.

The output of the program may be obtained by cut, by shift, and by over simulation time, at the choice of the user. These analyses are shown schematically in Figure 3.

The output for the sample input data is given in Tables II through VII. Table II shows a sample cut output. The cut cycle time and mining rate are given at the top of the table. The activities performed and their corresponding times are given in the left side of the table. The location of roof bolter and the next cut which will become available for the miner are shown at the right side. At the bottom of the right side, the cumulative times for the miner and bolter are given. These times indicate the time at the change point of their next cut location.

Table III shows a sample shift output. The time spent, the tons produced, and the mining rate during a shift are given at the top of the shift output. The activities performed and their corresponding times together with their percentages of continuous miner's time are given on the left side of the shift output. Times related with roof bolter and shuttle cars are also given on the bottom of the output. The positions of roof bolter and continuous miner at the end of this shift are given on the right side of the output.

Table IV shows a sample output of system performance over simulation time. Total simulation

TABLE II – OUTPUT OF A CUT

CUT NO.   1

```
----------------------------------------------------------------
| CYCLE=  37.0  TONNAGE =  91.4 t    MIN.RATE = 2.24 t/min     |
|                         101.0 st              2.72 st/min    |
|--------------------------------------------------------------|
|   ACTIVITY          TIME  |  CUT    AVAILABLE   |  R.B.  |
|   --------          ----  |  NO.  FOR C.M.[MIN] |  AT    |
|   LOADING           20.0  |  -----  ------------ |--------|
|   CHANGE OUT         3.4  |   5  -->   52.7     |   5    |
|   WAITING FOR S.C.  10.4  |--------------------------------|
|   WAITING FOR COAL    .0  |TIME (C.M.) =   37.0            |
|   TRAMMING           3.3  |TIME (R.B.) =   57.2            |
|   REPAIRING C.M.      .0  |HAUL DISTANCE [MILES]  =    2.4 |
----------------------------------------------------------------
```

TABLE III – OUTPUT OF A SHIFT

SHIFT NO.   1

```
----------------------------------------------------------------
| TIME =  360.4  TONNAGE = 885.1 t   MIN.RATE =  2.46 t/min    |
|                          977.0 st              2.71 st/min   |
|--------------------------------------------------------------|
|   ACTIVITY          TIME    PERCENT  |  POSITION            |
|   --------          ----    -------  |------------------|
|   LOADING          180.4     50.1    |  C.M  |  R.B  |
|   CHANGE OUT        41.9     11.6    |  AT   |  AT   |
|   WAITING FOR S.C.  63.6     17.7    |------------------|
|   WAITING FOR COAL  24.4      6.8    |  10   |   9   |
|   TRAMMING          36.9     10.2    |------------------|
|   REPAIRING C.M.    13.2      3.7    |                  |
|------------------------------------------|  TOTAL  S.C.  |
|   R.B. WAITING FOR CUT   .0             |  HAUL DISTANCE |
|   REPAIRING R.B.         .0             |    [MILES]     |
|------------------------------------------|----------------|
|   REPAIRING S.C.(1)      .0             |    17.8        |
|   REPAIRING S.C.(2)      .0             |                |
----------------------------------------------------------------
```

TABLE III – OUTPUT OF A RUN

SYSTEM PERFORMANCE OVER SIMULATION TIME

```
****************************************************************
* T.TIME = 2320.7  T.TONNAGE =  3744 t   MIN.RATE = 1.61 t/min *
*                               4133 st             1.78 st/min*
*--------------------------------------------------------------*
*          ACTIVITY          TIME    PERCENT                  *
*          --------          ----    -------                  *
*          LOADING          754.0     32.5                    *
*          CHANGE OUT       314.8     13.6                    *
*          WAITING FOR S.C. 241.7     10.4                    *
*          WAITING FOR COAL 718.8     31.0                    *
*          TRAMMING         212.2      9.1                    *
*          REPAIRING C.M.    79.1      3.4                    *
*--------------------------------------------------------------*
*          R.B.WAITING FOR CUT    .0                          *
*          REPAIRING R.B.        11.2                         *
*          TIME (R.B.)         2479.8                         *
*--------------------------------------------------------------*
*          REPAIRING S.C.(1)     30.3                         *
*          REPAIRING S.C.(2)     43.4                         *
*          TOTAL HAUL DISTANCE   80.2                         *
*          TONS PER MILE OF S.C. 51.5                         *
****************************************************************
```

Analysis
By Cut

Analysis
By Shift

Analysis over
Simulation Time

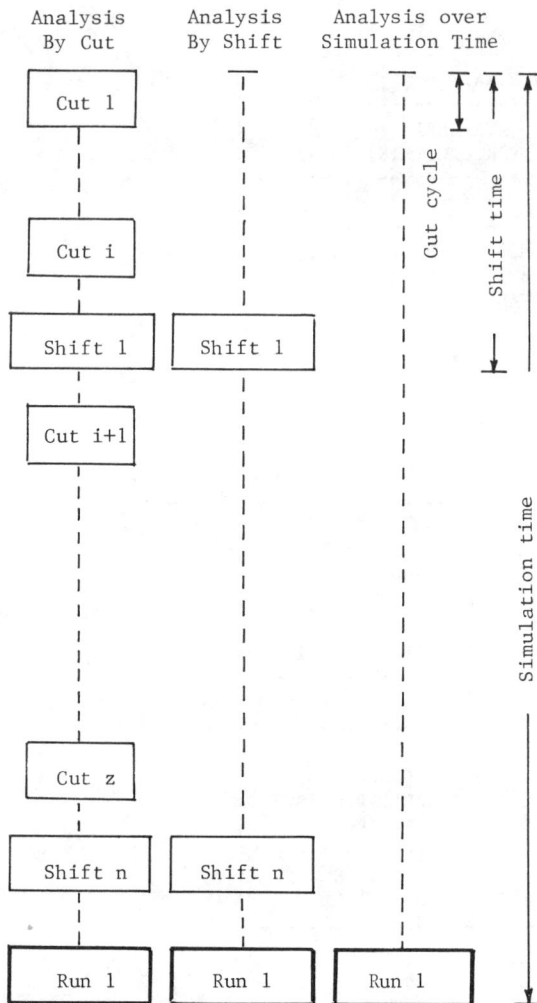

FIGURE 3 - ILLUSTRATION OF OUTPUT OPTIONS

standard deviation of some activities by including all the simulation runs specified by the user. The statistical output over simulation time is given in Table VI.

POSSIBLE APPLICATIONS OF THE PROGRAM

CONSIM can be applied for the evaluation of several production problems in continuous mining sections. Some of its application areas are listed below:

Designing Section Layouts: An efficient design of a section layout is fundamental and most important. The significant factors that are under the control of mine management are the number of entries, widths of entries and breakthroughs, center-line distances between entries and breakthroughs, depth of cut, and location of auxiliary facilities. The program can be used to evaluate any number of section layouts to determine the most favorable layout under a given set of conditions.

Evaluating Equipment Performances: CONSIM can be used to evaluate the production potential of equipment. For example, the output of the sample study indicates that a considerable amount of face time is spent for "waiting coal" and "waiting shuttle car." "Waiting for coal" results from the operations of the bolter. This time can be shortened by using different bolting method or machine. Similarly, "waiting for shuttle car" time can be decreased by increasing shuttle car haulage speeds or their payload capacities, or by decreasing their haulage distances. Figure 4 shows the effects of shortening bolting cycle time and increasing shuttle car payloads on the production rate of a section. The figure illustrates that the production rate is very sensitive to changes in roof bolting cycle time, but it is not sensitive to shuttle car payloads. Also, availability of equipment can be increased by preventive maintenance or by using more reliable equipment. The program can easily be used to evaluate the effects of these changes.

Evaluating New Equipment: The performance potential of several new equipment such as the continuous bridge conveyor, the automated miner bolter, and the automated temporary support units can easily be evaluated.

If bridge conveyors are used to haul coal from the face, the haul rates of shuttle cars, the discharge rates, and the switch-in and switch-out times should be zero. For the empirical distribution of the payload of a shuttle car, enter the average hauling rate of the conveyor with probability one.

To simulate the operation of an automated miner bolter, the values of the input variables for the roof bolter should be zero and the miner's loading rate should be adjusted to include the time required for roof bolting.

The effects of the automated temporary support (ATS) units can be simulated by making necessary

time, total tonnage produced, and mining rate are given at the top of the output. Activities done by the continuous miner and the times together with their percentages over simulation time, and also the times related with roof bolting and shuttle cars are given in the output.

For statistical results over simulation time, the program should be run more than once and also for shift statistics the simulation time should be long enough to have more than one shift.

Shift statistics calculate the mean and the standard deviation of some activities by including only the number of completed shifts in that simulation run, as shown in Table V. As for the statistics for the system performance over simulation time, it gives the mean and the

TABLE IV - STATISTICAL OUTPUT BY SHIFT

STATISTICS FOR SYSTEM PERFORMANCE BY SHIFT
****

		MEAN	STANDARD DEVIATION
TONNAGE (st)	-->	644.9	231.7
MINING RATE (st/min)	-->	1.79	.64
LOADING	-->	118.0	42.9
CHANGE OUT	-->	51.9	30.5
WAITING FOR S.C.	-->	37.5	14.4
WAITING FOR COAL	-->	110.2	89.2
REPAIRING C.M.	-->	9.1	16.8
R.B. WAITING FOR CUT	-->	.0	.0
REPAIRING R.B.	-->	1.9	4.6
HAUL DISTANCE	-->	12.7	4.6

TABLE V - STATISTICAL OUTPUT BY RUN

STATISTICS FOR OVER SIMULATION TIME
****

		MEAN	STANDARD DEVIATION
TOTAL TONNAGE (st)	-->	4132.8	
TOTAL TIME	-->	2198.5	93.1
MINING RATE (st/min)	-->	1.88	.08
LOADING	-->	752.6	4.2
CHANGE OUT	-->	312.5	2.5
WAITING FOR S.C.	-->	276.2	26.4
WAITING FOR COAL	-->	545.5	100.2
TRAMMING	-->	209.1	
REPAIRING C.M.	-->	102.6	79.5
R.B. WAITING FOR CUT	-->	2.7	8.5
REPAIRING R.B.	-->	33.4	44.4
TIME (R.B.)	-->	2377.7	91.5
TOTAL HAUL DISTANCE	-->	79.8	.3
TONS PER MILE OF S.C.	-->	51.8	.2

Stop - Program terminated.

increases in the depth of a cut by adjusting the loading rate of the miner to include the time required for the ATS and other auxiliary operations.

Justifying Capital Expenditures: The program can be utilized to evaluate the proposed changes in mining systems which require capital expenditures. The simulation provides the mine operator with reliable estimates concerning the effects of the proposed changes. These estimates may serve as a guide for justification of capital expenditures.

Developing Production Standards: By the careful use of the programs, production standards can be established in face mining operations. This can be done under a variety of operating conditions, if the conditions are consistent. Once a base standard is established by simulation and validated by field data, any changes, either by choice or necessity, can be quickly evaluated, and the base standard readjusted to conform to these changes. By this approach, all the components of the system can be evaluated and an accurate means of measuring the performance of the operations can be established.

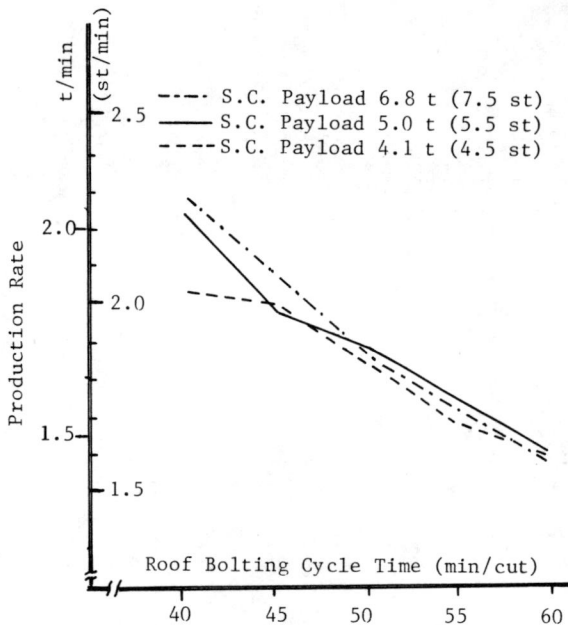

FIGURE 4 - CHANGE IN PRODUCTION RATE VS R. BOLTER
            CYCLE TIME AND SHUTTLE CAR PAYLOAD
            (Average of 10 simulation runs)

REFERENCES

Prelaz, L. J., Sironko, P. T., Bucklen, E. P. and
    Lucas, J. R. (1964), 'Optimization of Under-
    ground Mining', Report No. 6, Vol. 1-3, Office
    of Coal Research, U. S. Department of the
    Interior.

Ramachandran, D. (1983), A Computer Simulation
    Model for Continuous Mining Systems," MS
    Thesis, Virginia Polytechnic Institute &
    State University.

Topuz, E., E. Nasuf, and D. Ramachandran," User
    Manual-CONSIM: An Interactive Micro-Computer
    Program for Continuous Mining Systems,"
    October 1984, Department of Mining & Minerals
    Engineering, Virginia Polytechnic Institute
    & State University, Blacksburg, Virginia.

Chapter 26

INTERACTIVE UNDERGROUND COAL MINE PLANNING USING A MACINTOSH

P.K. Chatterjee*, D.E. Scheck*, and B.S. Li**

* Professor, Industrial and Systems Engineering, Ohio University
** Visiting Scholar, Industrial and Systems Engineering, Ohio University

ABSTRACT

Effective mine planning is generally regarded as the key to increased profitability in underground mining. The potential rewards of superior planning can be assessed by increased productivity, reduction of non-productive (downtime) time and generally greater extraction levels.

This paper outlines the development of a computer aided mine planning program using a Macintosh microcomputer. The program developed examines exploratory bore hole information and produces a structured database of the sampled data. Information from this database is utilized to determine the optimum productivity levels for any given combination of geology, face length, seam thickness, rate of advance, etc., evaluations being made for a range of face machine configurations. The design is conducted on an interactive basis which allows the planning engineer to be fully involved in all design activities. Once a design layout has been completed, the evaluation routine computes the total tons of raw coal, recovered tons, refuse tonnage and other relevant information.

The mine planning and design system includes a simulation routine that provides a basis for predicting the productivity, tonnage, etc., over a specified number of shifts. An alternative approach is to start with a target tonnage, and use the simulation routine to determine the time to achieve this target.

1. INTRODUCTION

Increased productivity can be achieved in underground coal mining through efficient mine planning. Currently, most mine planning is performed manually with perhaps three to four people assigned solely to the task of tracing and drafting mine plans. Because of the time consuming nature of this task, very few alternative plans are considered. Indeed, alternatives are usually not even generated. As a result, the optimum mine plan is seldom implented.

With the recent developments in interactive graphics, much if not all of the mine planning can be performed by computers. Little work, however, is being done in the area of interactive graphics for mine planning. This is evidenced by the relatively few literature reports on the subject. In the recent APCOM conference (18th. APCOM, 1984), the use and benefits of computer aided mine design were discussed in several sessions, but the number of papers with actual interactive graphics applications was few. Haycocks (1984) outlined the use of the program FACESIM. Hartley et.al. (1984) reported the progress in the development of application programs with graphics capabilities for surveying, geology, planning and engineering. There were a few more papers with interactive mine planning applications, but the general trend seemed to indicate that unintegrated almost batch type analysis was still much in vogue.

The emphasis of this paper is on two-dimensional computer aided interactive graphics coal mine design. The programs discussed and other programs related to geostatistical analysis (see Figure 1) have been developed for a main-frame computer. Generally the costs associated with running programs on a main-frame can be large and it was decided to investigate possible ways of reducing these costs without compromising the capability and efficiency of the package. The powerful graphics capabilities of the Macintosh computer together with its fast execution speed was considered a suitable alternative to a main-frame computer.

2. OBJECTIVE

The objective of this project was to develop an interactive computer-aided underground coal mine design program for the Macintosh computer that will help maximize coal recovery at greatly reduced costs. A suite of programs capable of performing geostatistical coal reserve estimation has been developed for a main-frame IBM computer but is not included here because of the limited memory of the Macintosh. The two programs that have been developed for the Macintosh are for

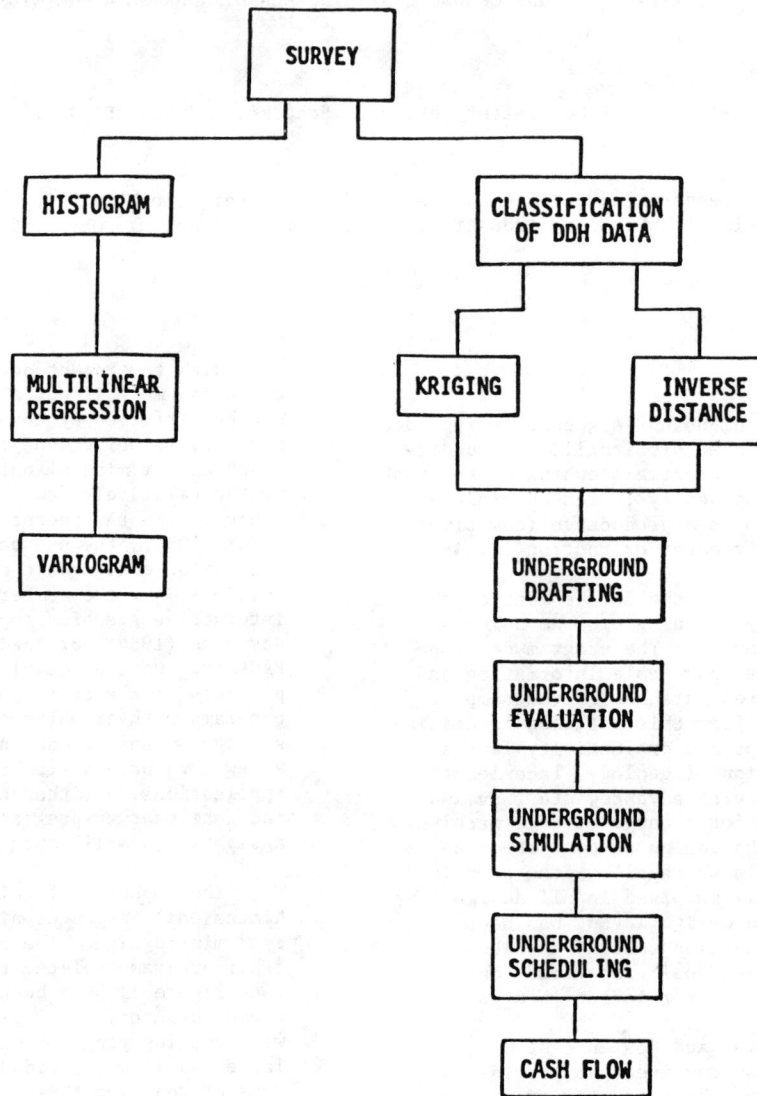

Fig. 1   Underground Coal Mine Program - Stages of Program Development

mine planning and simulation. In these programs underground mine layouts can be drawn to user specifications and a probabilistic simulation performed to determine production levels of different extraction strategies. A comparative evaluation of the alternative stratgies enables the optimum design to be selected thus permitting coal recovery to be maximized at lower costs.

3. STAGES OF PROGRAM DEVELOPMENT

There are three phases in the programs developed. Phase 1 consists of a geostatistical package capable of computing ore reserves, variogram models and kriging. Phase 2 consists of the design and drafting of a mine layout while phase 3 performs a probabilistic simulation of the area and layout selected to be mined. Each phase is discussed in detail below.

3-1. Phase 1 - GEOSTATISTICAL ANALYSIS

A suite of geostatistical programs have been developed for a main-frame computer which perform exploratory drill hole classification and various statistical analyses. Assay values of coal samples are used as input to a program which computes such statistics concerning the deposit as means, variances, experimental and cumulative histograms. These statistics provide information regarding the type of distribution associated with each variable.

A multilinear regression analysis is performed to determine means, variances, partial and multiple correlation coefficients associated with the variables and a final regression equation that relates the first variable in the data base to the rest of the variables. Regression and correlation reveal underlying trends of data, i.e. the degree of association between variables, and can be useful in determing extraction sequences and production scheduling.

Finally a set of programs was developed to perform geostatistical analyses. For each variable a semi-variogram and variance statistics are calculated along with an option to compute directional semi-variograms. The object of variogram analysis is to detect the major characteristics of the variables under investigation. For example, geological studies may indicate a direction of preferred variability of ash. Block variances can be calculated thus providing information on tonnage in a block and its quality parameters, i.e. ash, sulphur, etc.

The next geostatistical routine performs kriging on the input data. Kriging, defined as the best linear unbiased estimator, is now considered the most superior estimation technique available. The technique is to find a set of weights that will minimize the estimation variance. Kriging assigns low weights to distant samples and vice-versa and takes into account the relative spatial location of the samples with respect to the block and each other.

The objective of this geostatistical program is to calculate the kriging variances for each block. The kriging variance permits a computation of the confidence interval of the estimates. This confidence interval provides a measure of the precision of the predicted tonnage and quality parameters such as sulphur and ash. Kriged values can be compared with actual values for accuracy. With this information a coal deposit can be precisely defined and subdivided into smaller blocks whose tonnage and quality are known with some degree of confidence, thus making the task of optimizing a mine design relatively simple.

It was not possible to include the geostatistical programs in the package developed for the Macintosh because of its limited core memory. However, these programs will be available on the Macintosh when its memory is upgraded to 512 Kbytes.

3-2. Phase 2 - DESIGNING A MINE LAYOUT

Mine design consists of drawing a series of design layouts and performing evaluations of the technical feasibility of these layouts. Technical evaluations can be made for a range of face machine configurations, thus enabling a determination of optimum productivity levels for any given combination of geology, face length, seam thickness, rate of advance, etc. The design is conducted on an interactive basis allowing the planning engineer to be fully involved in all design activities.

The programs have been designed to be highly interactive and in this section the user is taken step by step through the drafting process of a mine layout. The user has the option of repeating any step within the drafting sequence through the use of an intelligent menu system. Initially the user selects a grid size and scales it according to the real world coordinates of the land area being mined. When this is completed boundary coordinates are specified for the area to be mined. These points can be entered in one of two ways, either by drawing the area on the screen using the mouse or by entering real world x,y coordinate pairs that define the boundary. The total area enclosed within the boundary is computed and displayed on the screen before the next step is initiated.

Faults that may traverse the mine property are then drawn using the mouse or real coordinates. The program numbers the faults according to the order in which they are entered. Main or staple shafts and other entries are subsequently drawn using either the mouse or by giving their respective x,y coordinates.

Once the physical attributes of the mine (boundary, faults, shafts) have been defined the drafting and designing of the mine layout begins. The first step in this process is to draw the main airways or roadways of the mine. When the first main airway/roadway has been

drawn the program computes the actual length
and azimuth of each straight line segement.
Subsequent roadways can be drawn parallel to or
at any angle to the first roadway and is generally dependent on user prompts. Base lines
for panel sections are drawn next. The base
lines, faults and main entries conveniently
divide the property into mining areas as shown
in Figure 2.

The last step in the drafting phase is
to select the panel sizes and the direction
of advance in each mining area. The user selects
a starting point for the panels in a given area.
The panels are drawn with reference to the base
line and the angle between each panel and the
base line is dependent on user imput. The ending
point for the panels in any area are pre-
determined; this is usually either a base line,
fault or the boundary line of the property. The
program first computes the number of complete
panels it can allocate to the specified area
and estimates the distance between the last
panel and the edge of the property. This distance
is an indication of how much coal would be left
unmined if the panel specifications are unaltered.
If this distance is too large the user can select
an alternative design or another face length.
This cycle can be repeated until the maximum
recovery of coal is achieved in that area.

When the face and panel layouts for the
entire property have been drawn the program
stores the completed drawing on the disk.
Because the amount of data necessary to describe
the picture is large its storage can be achieved
only by dividing it into several segments (8-10)
and storing each segment as a separate file
on disk. The drafting process is complete at
this point and the user can terminate the program
or continue onto the next phase of operation,
the simulation phase. This layout can be recalled to repeat any of the drafting steps outlined above.

3-3. Phase 3 - SIMULATION

Once a design layout has been completed
the total in situ tonnage of coal, production
tonnage, non-coal tonnage and other relevent
information can be computed. A simulation is
then run to provide a basis for prediction of
productivity, tonnage, etc., over a finite number
of shifts. Alternatively, given a target tonnage,
the simulation enables the planner to determine
the length of time needed to achieve this target.
The simulation can be extended to cover other
face areas in the mine and once all areas of
a mine are investigated, a production schedule
of the mine is attempted. Judicious manipulation
of productivity values from different areas
allow maximization of productivity for the entire
mine.

Phase 3 begins by restoring the mine plan
from disk memory to main memory and displaying
this result on the screen. The user has a choice
of six mining strategies to select for the

simulation. The strategies available are shown
in Table 1. For example, the user may want
to know when the mining of several areas will
be completed given the starting date. To simulate
this the user would select strategy 2.

In the simulation phase the user must specify
production variables based on historical data,
past mining experience and knowledge. The
information requested is shown in Table 2.
Some of these variables, e.g., mineable thickness
of the coal seam and the specific gravity of the
coal are dependent on the actual geology of the
deposit. Other variables are dependent on the
particular management procedures followed in the
mine such as the average production per shift and
the number of working days in a month etc.

Information that is computed by the program
is shown in Tables 3 and 4. The user is furnished
with an evaluation, a detailed panel and summary
reports. The detailed panel report provides
monthly figures for total tons produced, average
production per day and the number of idle shifts
per month. Both the panel and summary reports
give a description of the attributes of each
panel that is being mined, the date the mining
begins and finishes on each panel and final
summary at the end of the simulation.

The user can perform as many simulations as
he desires and compare the final evaluations to
help him select the optimum mine plan and
schedule.

4. DISCUSSION OF RESULTS

The main emphasis of this paper is on an
interactive graphics approach to planning and
though an outline of borehole data analysis and
geostatistical applications have been made, the
results that are discussed here deal with outputs
generated in phases 2 and 3. The programs
developed on the Macintosh are essentially
iterative and interactive software routines that
enable different design strategies to be examined
on a display screen. A manual process of elimination is pursued in selecting the optimum design.

It is possible to generate seam contour lines
from the information obtained from exploratory
boreholes. Seam topography is particularly
useful in laying out roadways and face panels.
Once seam contours are drawn it is possible to
draw on the screen any system of roadways and/or
working panels that the design engineer desires.

An underground coal property with an irregular shape is identified in Figure 3. The
property is intersected by several faults. As a
first design exercise, the longwall panels for
this property were arranged East-West and the main
roadways were also allowed to run East-West. The
face width, shearer speed, shift, geological
and other relevant information were read in to
simulate a "first" design layout. Figure 4 shows
how the simulation results appear on the display
screen.

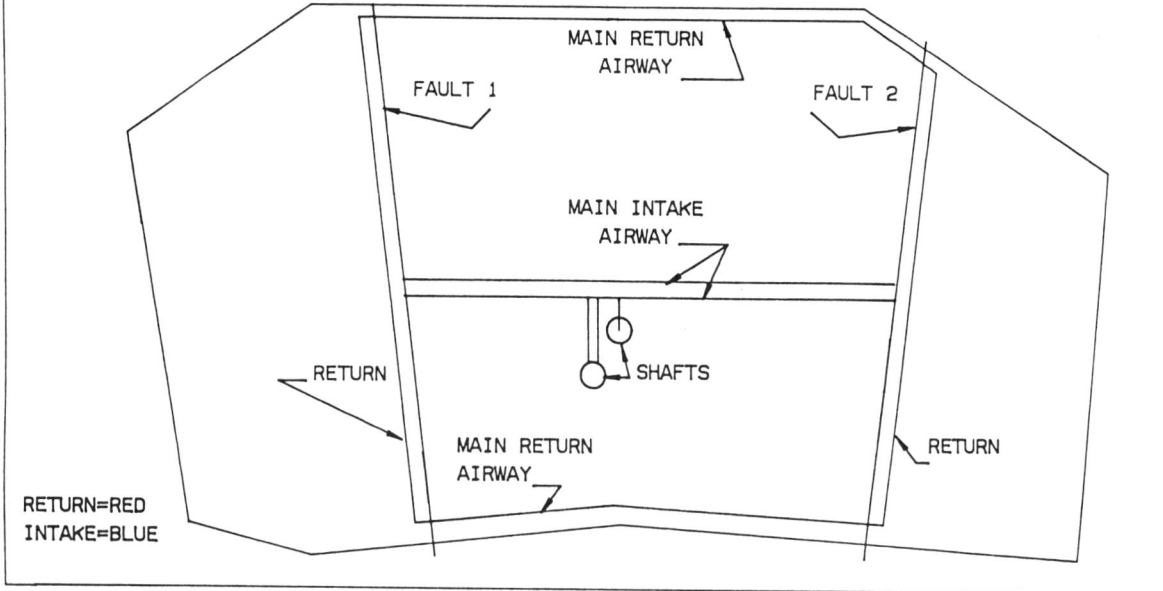

FIG. 2 DIAGRAM SHOWING LOCATION OF FAULTS, SHAFTS, MAIN INTAKE & RETURN AIRWAYS

MAIN RETURN AIRWAY

FAULT 1

FAULT 2

MAIN INTAKE AIRWAY

RETURN

SHAFTS

MAIN RETURN AIRWAY

RETURN

RETURN=RED
INTAKE=BLUE

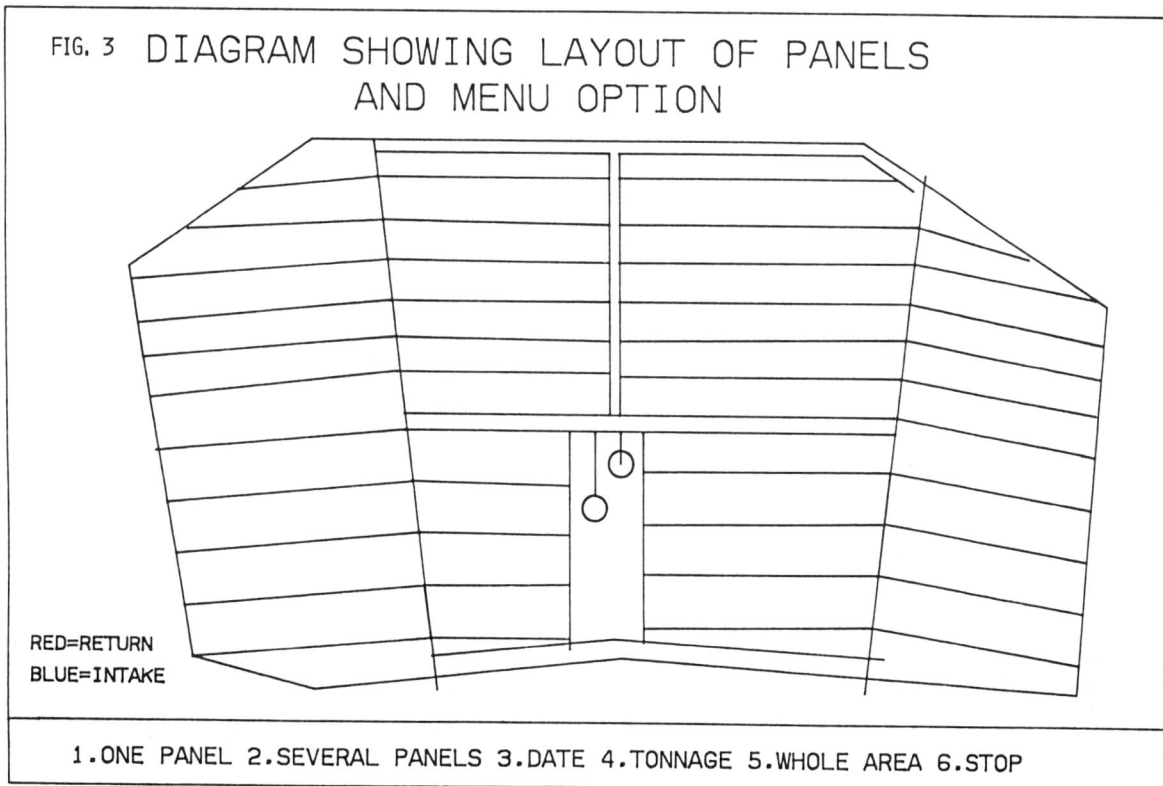

FIG. 3 DIAGRAM SHOWING LAYOUT OF PANELS AND MENU OPTION

RED=RETURN
BLUE=INTAKE

1.ONE PANEL 2.SEVERAL PANELS 3.DATE 4.TONNAGE 5.WHOLE AREA 6.STOP

TABLE 1

MENU FOR SIMULATION STRATEGIES

Number	Strategy	Description
1	Determine the ending date for a given panel.	This option allows the user to specify a panel and the date on which mining will begin on that panel. The program performs simulation and computes the date by which the panel will be completely mined out.
2	Determine the ending date for several panels.	This strategy is similar to the first but allows the user to specify several panels to be mined in sequence. The completion date for mining all the panels is computed and displayed.
3	Determine the working position of a panel on a given date.	In this option the user selects a panel to be mined. The program simulates mining the panel from the starting date until the specified date to illustrate the panel advance that will have been made in that period of time.
4	Determine the date for mining given tonnage.	This option will simulate the mining of a sequence of panels until a given tonnage of coal has been removed.
5	Determine the date for mining several areas	This option allows several areas in a property to be mined. The information required in each case is tabulated to include all parameters of interest.

TABLE 2

USER SPECIFIED PRODUCTION PARAMETERS

1.	The specific gravity of the coal	84.0	7.	The mineable thickness of the coal in each panel	6.4 ft.
2.	The number of working days in a month	20 days	8.	The average production per shift	1500 tons
3.	The number of productive shifts per day	2	9.	The standard deviation of the normal production rate	100 tons
4.	The starting date for the simulation	YY, MM, DD 85 07 14	10.	The probability of having normal production	.90
5.	Which panels are to be mined	1,2,3, 6,8	11.	The face recovery of each panel	.90
6.	The height of the coal seam in each panel	6.5 ft.	12.	The number of days required to move the face equipment from one panel to the next	24 days

TABLE 3

EVAULATION REPORT

PANEL FT.	PANEL LENGTH FT.	FACE LENGTH FT.	HEIGHT OF SEAM FT.	MINEABLE THICKNESS FT.	PANEL RESERVE TONS	ACCEPT. TONNAGE TONS	TONS/ SHIFT TONS	S.D.	PROB.	RECOVERY
1	2562.6	642.4	6.5	6	412.6	380.8	2	.2	.9	.95
2	2842.1	642.4	6.5	6	457.6	422.4	2	.2	.9	.95
3	2699.1	642.4	6.5	6	434.6	401.1	2	.2	.9	.95
4	2556.2	642.4	6.5	6	411.6	379.9	2	.2	.9	.95
5	2413.2	642.4	6.5	6	388.5	358.6	2	.2	.9	.95
6	2270.2	642.4	6.5	6	365.5	337.4	2	.2	.9	.95

TONS IN THOUSANDS

TABLE 4a

DETAIL PANEL REPORT

YEAR	MONTH	WORK DAYS	TOTAL TONS	AVERAGE PER DAY	IDLE SHIFTS
85	8	13	43.2	3.32	3
85	9	20	65.8	3.29	5
85	10	20	66.7	3.34	4
85	11	20	69.2	3.46	5
85	12	20	65.4	3.27	6
86	1	16	51.5	3.54	2

PANEL 1 BEGINS ON   85 8 12
PANEL FINISHED ON   86 1 25   TOTAL TONS PRODUCED = 361.8

TONS IN THOUSANDS

TABLE 4b

FACE SIMULATION REPORT

AREA	PANEL NO.	PANEL LENGTH	FACE LENGTH	PANEL RESERVE	ACCEPT. TONNAGE	START YY	MM	DD	FINISH YY	MM	DD	PRODUCED TONS
2	1	3337.1	562.1	470.1	434.0	84	10	16	85	05	04	412.3
2	2	3377.4	562.1	475.8	439.2	85	05	25	85	12	13	417.2
2	3	3417.8	562.1	481.5	444.4	86	01	04	86	07	27	422.2
2	4	3458.1	562.1	487.2	449.7	86	08	18	87	03	13	427.2
2	5	3498.4	562.1	492.9	454.9	87	04	04	87	10	25	432.2
2	6	3538.8	562.1	498.5	460.2	87	11	16	88	06	15	437.2

TOTAL = 2548.3

Evaluation of the mine design can be obtained in either detailed or summary form. Tables 3 and 4 show the detailed report and summary evaluation reports respectively. The detailed report produces monthly statistics of production variables such as, total tons mined average daily production, and the number of idle shifts. Included in the detailed report is a summary of the production status after each panel is completely mined. The total tonnage produced as a result of the simulated mining is also shown in the detailed report.

The summary report contains production information on each panel mined, including the date a panel was started, date completed and total tons produced from that panel. It also reports the production summary for the entire simulation period.

This layout can be further subjected to design and cost calculations, and a final Net Present Value obtained for the entire extraction process. A second layout can be drawn similarly with the longwall panels running North-South (see Figure 5) and another simulation and evaluation obtained for this layout. A compartive analysis of several of these designs will produce the optimum.

4-1. LIMITATIONS OF THE MACINTOSH

The major limitation of the Macintosh is its insufficient user accessible memory which is about 20K of RAM. Due to this restriction much time and effort was spent overcoming programming requirements.

In order to circumvent the memory problem the programs were written in small modules. The drafting phase is contained in four modules and the simulation phase is contained in three. These modules are then accessed sequentially by an intricate system of chaining.

The lack of sufficient available memory not only hindered programming speed but it also made it impossible to integrate the geo-statistical routines into the new mine planning package. Each mine plan was required to be divided into 8-10 segments for storage purposes. To restore the plan from memory the pertinent data files were recalled sequentially. The segmenting, sequential recalling and storing of the plan are very machine intensive and time consuming.

4-2. INDUSTRIAL SIGNIFICANCE AND BENEFITS

It is widely acknowledged that current coal reserves and quality estimation procedures fail to account for the complexities of a deposit. Reserve calculations are generally over-simplified, leading to incorrect approximation of the actual coal and quality to be recovered. The programs and techniques described in the preceding paragraphs will be an extremely useful tool to apply to any coal reserve. These methods are equally applicable to surface as well as underground mining.

The underground mine planning programs will be invaluable to the operation of both small and large underground coal mines. Small mine owners are generally unaware of the benefits associated with optimum design layouts. These owners will now be able to employ novel and proven technology to achieve lower costs, higher recovery levels and most importantly proper resource utilization.

The use of these programs will offer medium and large coal producers significant gains. In the larger mines, computer aided design will provide an economy of scale which is not achievable by conventional methods. One of the most important benefits of using these programs is the ability to schedule extractions from different working areas to maximize the quality of run-of-mine coal. As the number of working areas increase, manual scheduling of extraction to meet quality and tonnage targets becomes more difficult.

Another obvious benefit of this program is the elimination of tedious manual drafting methods and repetitive computations. These programs offer a genuine reduction in the actual manhours spent in drafting and generating mine plans, thus greatly reducing the costs associated with such tasks. Other benefits specific to the Macintosh package include:

. Updates and changes can be made quickly and easily with instant feedback on the screen.

. The program is extremely user friendly and requires virtually no computing skills from the user.

. It has a powerful computational base which reduces manual computations to a minimum thereby providing longterm and shortterm mine planning projections quickly.

. Many plans and layouts can be generated giving the user an opportunity to evaluate different mine plans and to select the optimum.

. The program keeps a running record of coal inventory (both mined and in situ coal), calender and scheduling dates of each coal area mined or to be mined. This information is particularily useful when more than one seam is being mined.

. Ventilation planning can be easily integrated into the design and shortterm and longterm ventilation requirements evaluated.

. Each plan can be custom designed based on user specifications thus providing an opportunity to test ideas on paper that otherwise would not have been possible.

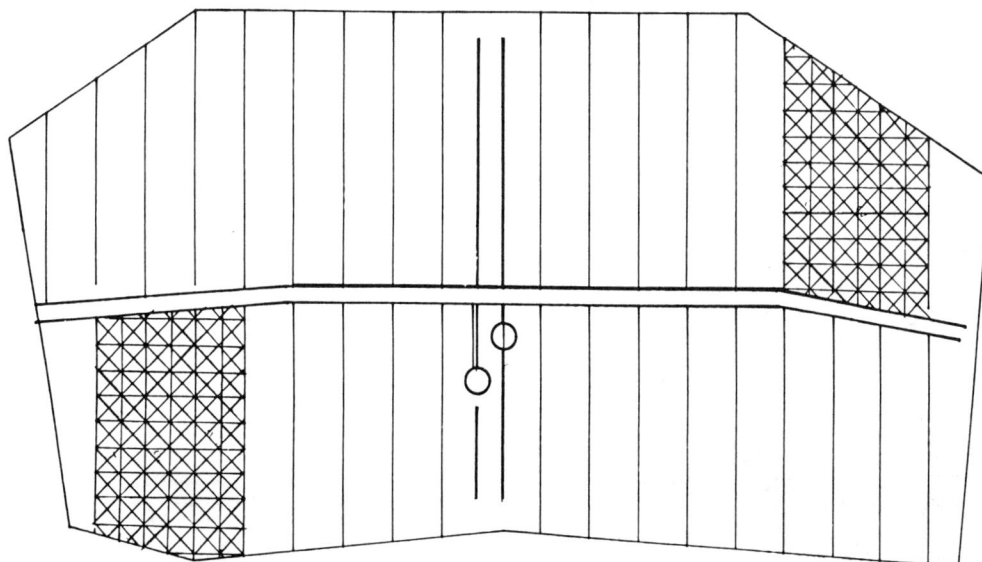

FIG. 4 DIAGRAM SHOWING PANELS
EXTRACTED BY SIMULATION

⊠ AFTER MINING
▨ BEFORE MINING

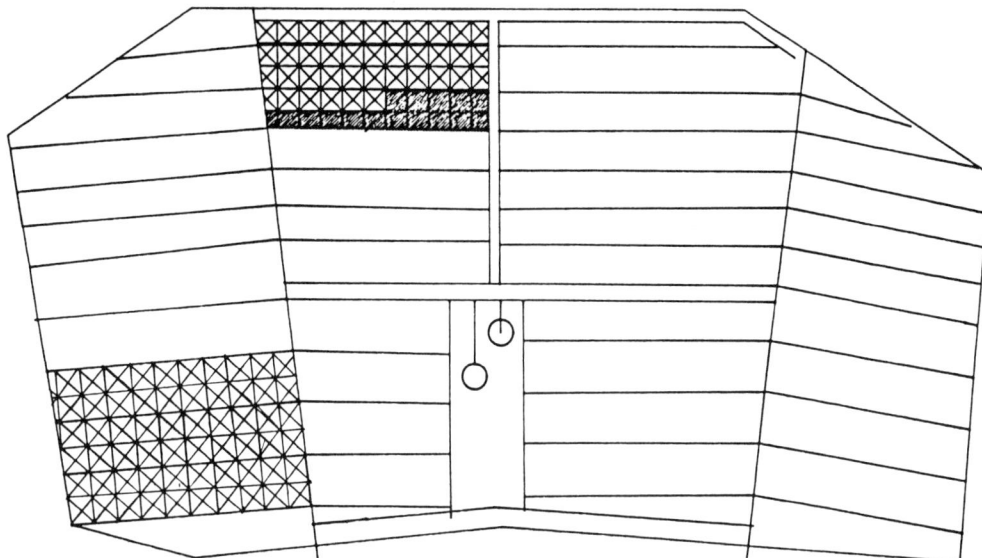

FIG. 5 LONGWALL PANELS
MINING NORTH-SOUTH

⊠ PANELS EXTRACTED BY SIMULATION

4-3. FUTURE ENHANCEMENTS

Although the current program, is powerful and useful, it is still under development. In the near future, it is intended to include several other features that will greatly enhance its current capabilities. A few of these, that are presently being examined are discussed below.

The selection of face and panel lengths is guided largely by economic and technical considerations. Generally the cost of equipment ownership increases with face length, while the cost of panel development per unit of mined coal decreases (Guay et al., 1981). It is intended that eventually the simulation will attempt to quantify the precise relationships between the cost per ton and face length, and the cost per ton and panel length.

The program retains a historical schedule of the entire operation. It is possible to determine at any stage of working the progress of each area that is being mined or developed. This is extremely useful particularily when mining more than one seam at a time. For example, a schedule can be generated that will show the location of longwall faces, mains and submains and other development for each seam being mined. It is intended to utilize the historical data to compute various production schedules and to employ optimization techniques to maximize the coal production of any coal property.

5. CONCLUSIONS

The proposed computer aided mine planning system will obviously reduce the manual effort required to draw and trace mine plans but the greatest benefit will come from the information provided to management. During the pre-mining stage the system can be used to evaluate many alternative designs, a task that is not practical to attempt using manual methods. While mining is in progress the mine plans can be updated easily to reflect the actual progress against planned progress and to provide management with graphical outputs that enhance the visibility of separate but intrinsically linked production areas and help identify potential problems. When unforseeable conditions require a change of plans, the ability to quickly evaluate alternative designs will allow management to make informed decisions more effectively.

REFERENCES:

Haycocks, C., Kumar, A., Unal, A., and Osei-Agyemana, S., 1984, "Interactive Simulation for room and Pillar Mines, Proceedings 18th. APCOM, Inst. Min. Met., London, p. 591-597

Hartely, D., 1984, "Development of Interactive Graphics Within the National Coal Board", IBID, p. 201-210

Guay, P.J. and Ludlow, J., 1981, "The Effect of Mining Wider Webs on a Longwall Face", Longwall-Shortwall Mining, State of the Art, ed. R.V. Ramani, SME-AIME p. 223-234

Chapter 27

GRAPHICAL MINE PRODUCTION SCHEDULING
AND FINANCIAL PLANNING SYSTEM

WILLIAM J. DOUGLAS
ROY F. WESTON, INC.

KATHLEEN Y. KNOEBEL
KETRON, INC.

INTRODUCTION

Mining is an industrial process which employs batch production methods to extract earth minerals from their natural locations in order to obtain usable ores. A mine may be characterized by a process diagram which is a model that describes the mining operations in their logical sequence, as dictated by the mine plan and its associated geometry. Each mining operation is a process step, and the logically sequenced totality of these operations constitutes the mining process or mine plan.

In planning an underground coal mine many geotechnical factors must be considered, and these will affect decisions concerning the design and location of mine adits, ventilation shafts, main entries, submains, and panels. The result is a mine layout which is a geometric description of the locations of the mine process elements. The completion of this layout sets the stage for production scheduling, during which decisions are made concerning the scheduling of mining operations in order to determine a sequence which is efficient in its utilization of mining equipment, geotechnically sound, and capable of achieving targeted production levels.

Once a production plan has been established, economic considerations move into the forefront. The financial planner is concerned with the time schedule of capital and operating costs which will be required to sustain mining operations and with the cost per ton of ROM material and saleable product. The confidence which can be placed in these cost estimates is directly related to the methods used in developing the production schedule and to the consistency of the procedures used when making cost computations based on the engineering variables which are inherent to the mine plan.

In many mine planning activities the production and cost are conducted by two different groups. The transfer of information between these groups can be cumbersome and inconsistent unless systematic procedures are developed to structure the analyses in a logical and orderly manner. Furthermore, if a basic assumption is changed or the mine plan is modified, the extensive re-calculations which are required can often consume a quantity of time and effort comparable to the first analysis. Since assumptions and production schedules are likely to change many times during the planning process, it is evident that there are advantages to having an automated procedure which can address variations to the mine plan or to economic variables with a minimum of computational effort.

APPROACH

In order to address the above stated issues, a computerized process modeling system has been developed to assist mine planners in analyzing production profiles and their associated costs in a manner which is consistent, repeatable, and responsive to the uncertainties which are common during the planning period. The system consists of a series of three models which can be used to analyze production schedules, compute the costs for any selected schedule, and apply computer graphics to illustrate the resulting production and cost profiles.

Figure 1 describes the computational flow of the three models. The MPROD model is the production scheduling system which accepts a mining process diagram in computer-coded form as its principal input and determines the dates during which each process step is completed and the associated production levels during each time period of the plan. The MCOM model accepts the production schedule from MPROD and uses it to develop a cost spread sheet which contains a number of cost line items for each time period and an overall summary of the project in the form of graphical bar charts. The MPLOT model uses a geometric file which describes the mine process steps to produce a mine layout and the MPROD process step completions to produce a mine map which designates the areas which are mined during each time period in a color-coded format.

PROCESS DIAGRAM

A mining process consists of elements which are standardized process types occurring repeatedly to a greater or lesser degree during the course of the mine life. Typical process types are: shaft sinking; driving a main entry; driving a submain; room-and-pillar advance mining; pillaring; and longwall mining. A process type is characterized by its geometric configuration and the associated production rate, and the process step may be repeated many times throughout the mine plan. For example, a seven-entry main or a five entry room-and-pillar panel mining sequence each having definable face production rates would represent two process types. A series of panels of the same process type are called process steps, which are distinct segments of the mine.

The construction of a mining process diagram consists of several steps. The first is to identify each process type and to give it a designated name.

The second is to define the production rate (tons/shift) and the corresponding advance rate (feet/shift) for each process type. These values may be obtained by using face production models, industrial engineering studies, or historical production figures under similar mining conditions. They may also be based on experienced engineering judgment and related computations

affected by the cut plan geometry and seam height.

The starting point for developing the process diagram is a mine layout which describes the location of each process step in the mine plan.

The process diagram defines the precedence relationships among process steps, and thus defines the principal constraints to mining. Figure 2 illustrates a sample coal mine layout which consists of a set of entries and room-and-pillar panels representing a portion of a coal reserve which is to be extracted. It is evident that the production profile will depend on how the entries and panels are scheduled and on the number of producing sections which can operate simultaneously within the mining area.

Five process types are included in the process diagram which represents the extraction plan:

a. Submain
b. North Panel
c. South Panel
d. North Panel Belt Return
e. South Panel Belt Return

Types a-c represent standard room-and-pillar mining operations while d and e are special process types. These latter processes consume time and labor resources but do not produce coal; they represent the removal of belt from one panel so that it can be used in another. Process types such as these are useful in constructing a process diagram which is representative of the operations which occur during the extraction sequence.

The mine process diagram is developed by defining process steps and then linking them in a sequence which describes the extraction plan. In Figure 3, the mine is divided into segments and the process steps associated with each segment are defined. The submain has been separated into five distinct segments. Each of these is a process step of the "submain" process type. These steps represent decision points in the process at which submain or panel steps may be initiated, depending on the structure of the process diagram and the availability of mining equipment.

Figures 4 and 5 are process diagrams representing two distinct extraction sequences. Figure 4

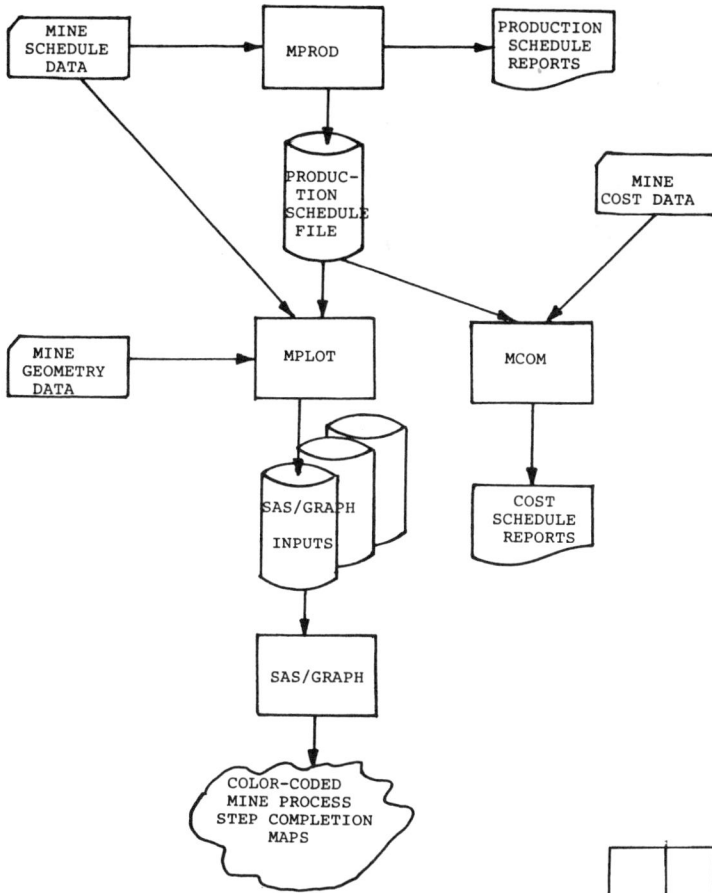

SYSTEM ORGANIZATION
FIGURE 1

NORTH PANELS

SUBMAIN

SOUTH PANELS

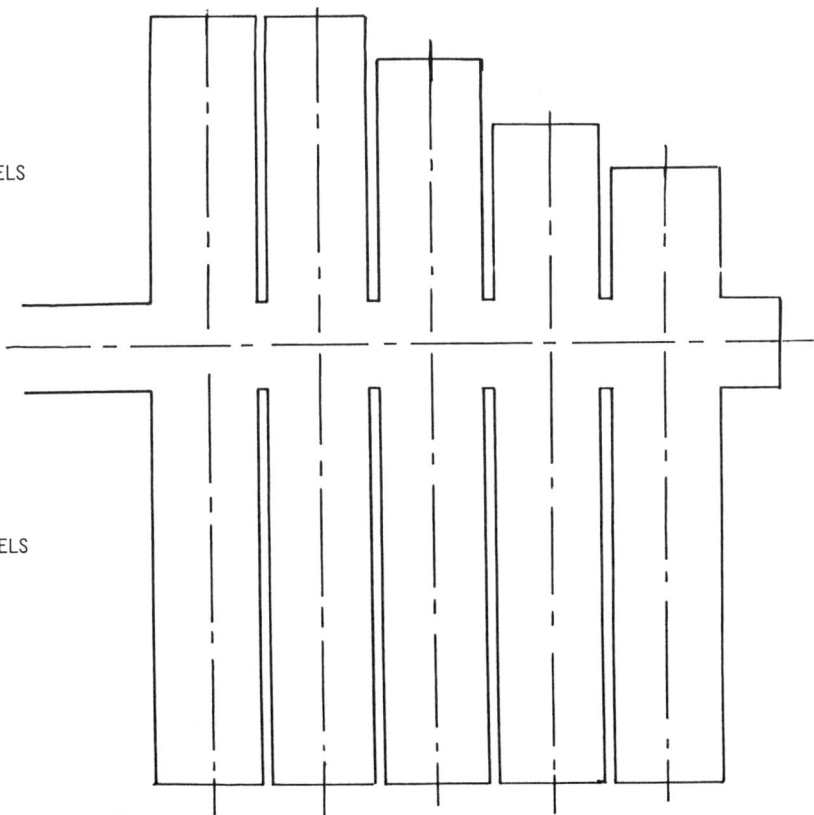

MINE PLAN
FIGURE 2

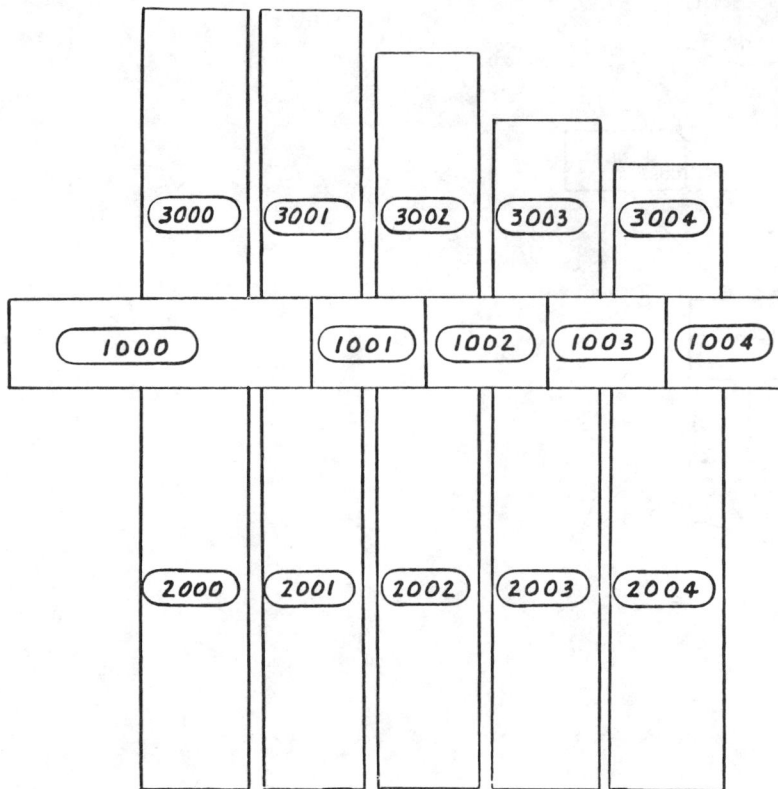

MINING PROCESS STEPS
FIGURE 3

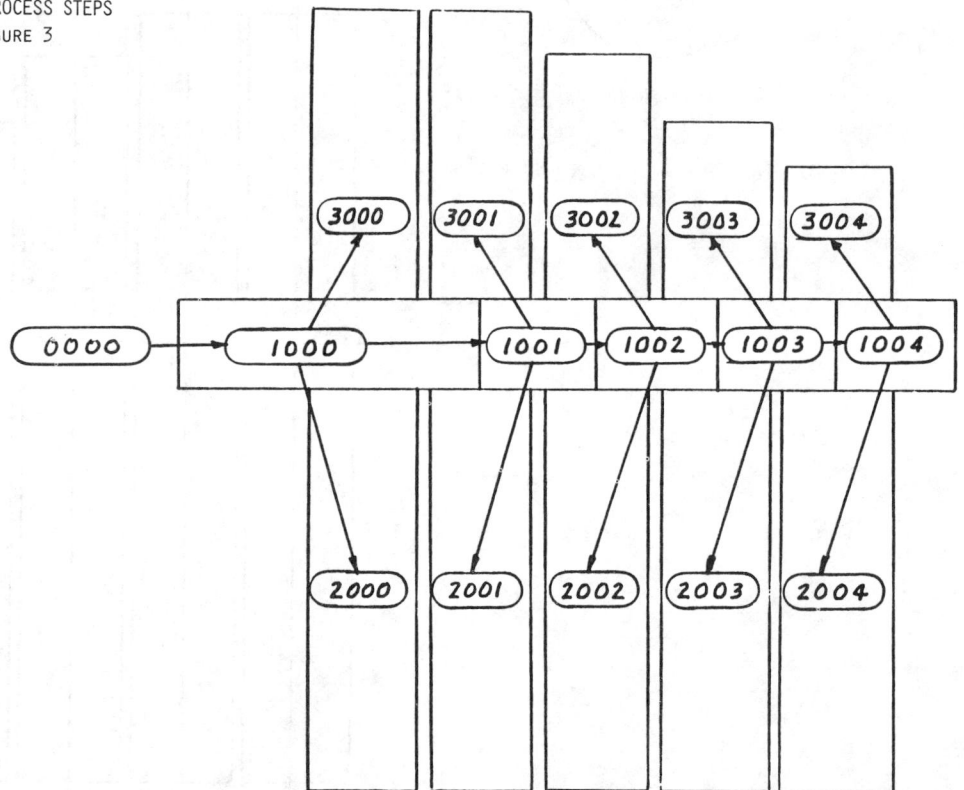

MINING PROCESS #2 DRIVE SUBMAIN
MINE NORTH AND SOUTH PANELS ON ADVANCE
FIGURE 4

describes the mining of both north and south panels on advance, since each panel is mineable as soon as the submain development for the panel has been completed (it is accessible). In Figure 5, only the south panels are mined in advance, while the north panels are mined after the submain has been completed (on retreat). It is evident that the production profiles based on these plans will differ.

The MPROD model is an event processor which determines the start and end date of each process step. The principal input to MPROD is a description of the connectivity of the diagram, as illustrated in Figure 6 for the extraction plan represented by Figure 4. This data file describes each process step as a "son" and its predecessor step as a "father." A step may require several predecessors before it can be initiated. For example, panel 2001 can only be started when its submain has been developed (step 1001) and the belt from the previous panel is available (step 4000), so that it has two fathers. Such multi-fathered steps are referred to as "and" steps since the completion of fathers A and B and C etc. are required before they can be initiated.

In addition to the connectivity, the data file contains the length of each process step. Once mining operations are started for the step, the duration of mining activities (number of shifts) is determined from this length and the advance rate (feet/shift) for the process type. This is the basis for determining the completion date for the step.

MPROD processes each step in the extraction plan to determine its start and end date. When a step is completed it achieves completion status, and when all fathers for a given son have achieved completion, the son becomes available for mining operations by entering the eligibility queue. The mining equipment which has been used in the step which has been completed is returned to the equipment pool, and MPROD processes the eligibility queue to select the next step to be started, based on its priority in the queue and the availability of equipment in the pool. Eligibility is determined by priority number, and when priorities are identical, on the basis of queueing time (first in/first out).

Figure 7 describes the remaining input data to MPROD. The equipment types and quantities which are initially available and the additional equipment purchases are defined, as well as the advance rates, production rates, and equipment requirements for each process type. MPROD processes the network diagram to produce annual or monthly (user-selected) production reports, and step completion dates are determined from the number of shifts per day, the step length, and its advance rate.

Figure 8 illustrates the production report for the sample problem which defines the ROM tonnage, feet of advance, and shifts consumed during each time period (month) for each process type. Cumulative values for each time period and for the complete mining project (35 months) are also illustrated. Figure 9 describes the process step completion schedule, defining the month, day, and shift of each start and completion and other data relating to the tonnage mined, feet of advance, and equipment utilization.

MPLOT

In order to provide additional visibility into the time sequencing of the extraction plan, a graphical plotting procedure called MPLOT has been developed to illustrate the relationship of planned production to mine geometry. Each mine segment is described as a rectangle. The length of the segment is obtained from the MPROD input file for the process step associated with mining the segment. Four additional variables are required to describe each segment: 1) width; 2), 3) starting location (x,y); and 4) angle of orientation of length to the x-axis.

The input file for MPLOT is illustrated in Figure 10, and the resulting mine layout in Figure 11. Using the MPROD process step completion file, a color-coded diagram is produced by MPLOT and the SAS/GRAPH plotting software which illustrates the portion of each segment which is scheduled to be mined during each time period. Figure 12 is a graphical description of the mining schedule produced by the MPLOT/SAS/GRAPH system. It is evident that the visibility provided by such graphics is effective in supporting mine planning decisions. Moreover, the MPROD and MPLOT software can be exercised readily over a range of

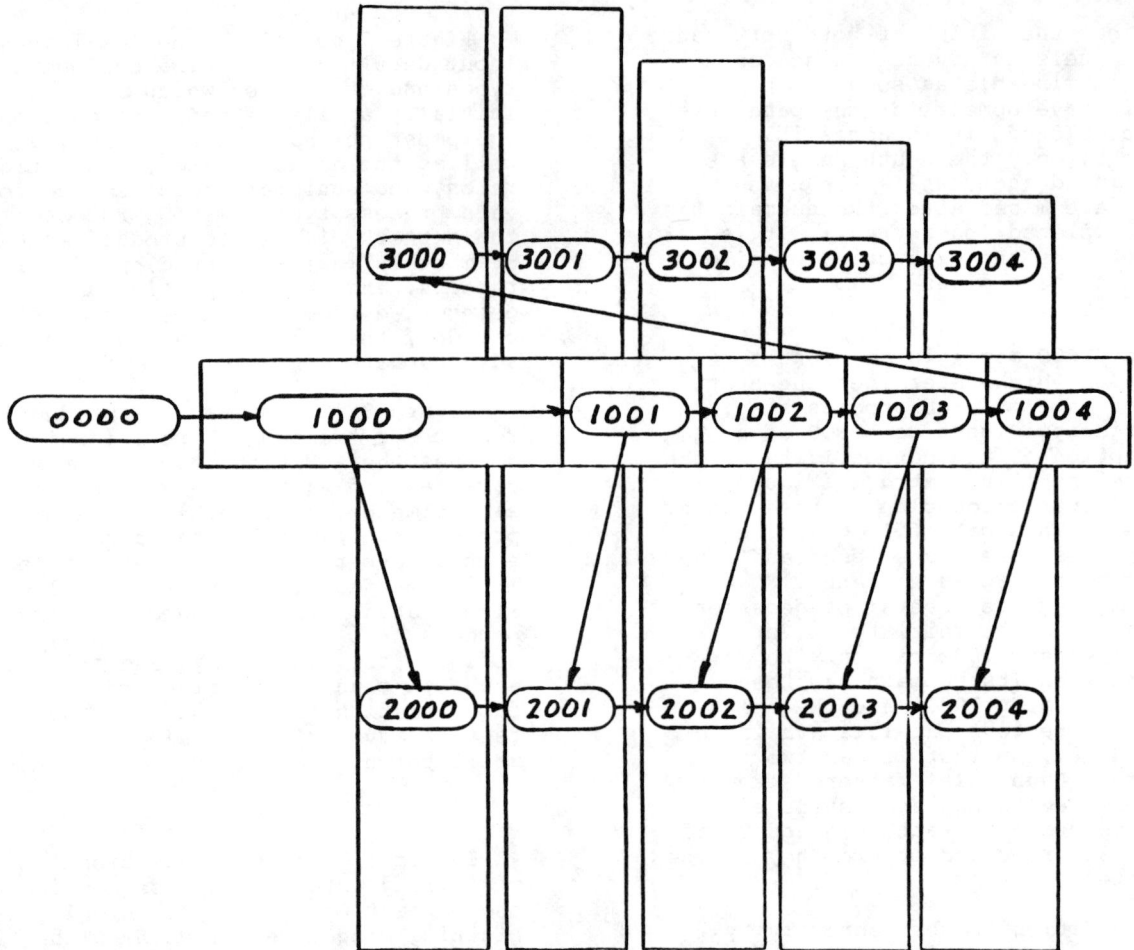

MINE PROCESS #3 DRIVE SUBMAIN MINE SOUTH PANELS ON ADVANCE
IN SEQUENCE. MINE NORTH PANELS IN SEQUENCE AFTER DEVELOPMENT
FIGURE 5

INPUTS TO TREE PROCESSOR FOR   ARCH MINERAL TRAINING EXAMPLE

SEGMENT NUMBER	TYPE SONS	# OF SONS	SON NUMBER	LENGTH (FEET)	SON NAME (OPTIONAL)	SON NUMBER	LENGTH (FEET)	SON NAME (OPTIONAL)	SON NUMBER	LENGTH (FEET)	SON NAME (OPTIONAL)
60000	0	1	11000	1380	ESUBMAIN	0	0		0	0	
11000	0	2	11001	540	ESUBMAIN	22000	1980	EPANELS1	0	0	
11001	0	2	11002	540	ESUBMAIN	22001	1980	EPANELS2	0	0	
11002	0	2	11003	540	ESUBMAIN	22002	1980	EPANELS3	0	0	
11003	0	2	11004	540	ESUBMAIN	22003	1980	EPANELS4	0	0	
11004	0	2	33000	1320	EPANELN1	22004	1980	EPANELS5	0	0	
22000	0	1	44000	9	EPANELBR	0	0		0	0	
44000	0	1	22001	1980	EPANELS2	0	0		0	0	
22001	0	1	44001	9	EPANELBR	0	0		0	0	
44001	0	1	22002	1980	EPANELS3	0	0		0	0	
22002	0	1	44002	9	EPANELBR	0	0		0	0	
44002	0	1	22003	1980	EPANELS4	0	0		0	0	
22003	0	1	44003	9	EPANELBR	0	0		0	0	
44003	0	1	22004	1980	EPANELS5	0	0		0	0	
22004	1	1	44004	9	EPANELBR	0	0		0	0	
33000	0	1	55000	6	EPANELBR	0	0		0	0	
55000	0	1	33001	1320	EPANELN2	0	0		0	0	
33001	0	1	55001	6	EPANELBR	0	0		0	0	
55001	0	1	33002	1100	EPANELN3	0	0		0	0	
33002	0	1	55002	5	EPANELBR	0	0		0	0	
55002	0	1	33003	880	EPANELN4	0	0		0	0	
33003	0	1	55003	4	EPANELBR	0	0		0	0	
55003	0	1	33004	660	EPANELN5	0	0		0	0	
33004	1	1	55004	3	EPANELBR	0	0		0	0	

FIGURE 6

A N D - N O D E   I N P U T S

DEPENDENT #FATHERS INDEPENDENT NODE SEGMENT NUMBERS

2001	2	4000	1001	0	0	0	0	0	0	0	0
2002	2	4001	1002	0	0	0	0	0	0	0	0
2003	2	4002	1003	0	0	0	0	0	0	0	0
2004	2	4003	1004	0	0	0	0	0	0	0	0

S E G M E N T   I N P U T S

SEG. TYPE	FEET PER SHIFT	TONS PER SHIFT	B T	BELT FR/FA	R T	RAIL FR/FA	C T	CABLE FR/FA	SCALE FACTOR		EQUIPMENT TYPES AND QUANTITIES REQUIRED 1 2 3 4 5 6 7 8 9 10 11 12 13 14 15 16 17 18 19 20
1 EAST SUBMAIN	4.300	300.000	1	1.00	0	0.0	2	2.00	1.000	MAX MIN	1 1 2 0 0 0 0 0 0 0 0 0 0 0 0 0 0 0 0 0 1 1 1 0 0 0 0 0 0 0 0 0 0 0 0 0 0 0 0 0
2 EAST PANEL SOUTH	10.000	400.000	2	1.00	0	0.0	1	1.00	1.000	MAX MIN	1 1 2 0 0 0 0 0 0 0 0 0 0 0 0 0 0 0 0 0 1 1 1 0 0 0 0 0 0 0 0 0 0 0 0 0 0 0 0 0
3 EAST PANEL NORTH	10.000	400.000	2	1.00	0	0.0	1	1.00	1.000	MAX MIN	1 1 2 0 0 0 0 0 0 0 0 0 0 0 0 0 0 0 0 0 1 1 1 0 0 0 0 0 0 0 0 0 0 0 0 0 0 0 0 0
4 EAST PANEL BRS	1.000	0.0	2-220.00		0	0.0	1-220.00		1.000	MAX MIN	0 0 0 0 0 0 0 0 0 0 0 0 0 0 0 0 0 0 0 0 0 0 0 0 0 0 0 0 0 0 0 0 0 0 0 0 0 0 0 0
5 EAST PANEL BRS	1.000	0.0	2-220.00		0	0.0	1-220.00		1.000	MAX MIN	0 0 0 0 0 0 0 0 0 0 0 0 0 0 0 0 0 0 0 0 0 0 0 0 0 0 0 0 0 0 0 0 0 0 0 0 0 0 0 0
6 DUMMY	100.000	0.0	0	0.0	0	0.0	0	0.0	1.000	MAX MIN	0 0 0 0 0 0 0 0 0 0 0 0 0 0 0 0 0 0 0 0 0 0 0 0 0 0 0 0 0 0 0 0 0 0 0 0 0 0 0 0

E Q U I P M E N T   I N P U T S

EQUIPMENT TYPE	EQUIPMENT NAME	PURCHASED IN TP 0
1	CONTINUOUS MINER	2
2	ROOF BOLTER	2
3	SHUTTLE CAR	4

G E N E R A L   T R E E   P R O C E S S O R   I N P U T S

TIME PERIOD OPTION (0=YEARLY, 1=MONTHLY)   1
NUMBER OF DAYS PER TIME PERIOD   20
NUMBER OF SHIFTS PER DAY   2
MAXIMUM NUMBER OF TIME PERIODS   50
NUMBER OF SHIFTS PER TIME PERIOD   40
NUMBER OF ADDITIONAL EQUIPMENT PURCHASES/RETIREMENTS   0

S P E C I A L   E Q U I P M E N T   I N P U T S

MAXIMUM FEET TO ACQUIRE (=0 --> NO MAXIMUM)

	TYPE 1	TYPE 2	TYPE 3	TYPE 4	TYPE 5
BELT	0.0	0.0	0.0	0.0	0.0
CABLE	0.0	0.0	0.0	0.0	0.0

FIGURE 7

ARCH MINERAL TRAINING EXAMPLE

TOTAL REPORT OF TONS MINED, FEET ADVANCED AND NUMBER OF SHIFTS
(TONS MINED IN THOUSANDS)

KEY - "K TONS" = KILO TONS, "FT AD" = FEET ADVANCED, " # SH" = NUMBER OF SHIFTS

MON.	TYPE REPORT	SEGMENT TYPE 1	SEGMENT TYPE 2	SEGMENT TYPE 3	SEGMENT TYPE 4	SEGMENT TYPE 5	SEGMENT TYPE 6	TOTAL MONTH	CUMULATE TOTAL
1	K TON	11.7	0.0	0.0	0.0	0.0	0.0	11.7	11.7
	FT AD	167.7	0.0	0.0	0.0	0.0	0.0	167.7	167.7
	# SH	39.0	0.0	0.0	0.0	0.0	0.0	39.0	
2	K TON	12.0	0.0	0.0	0.0	0.0	0.0	12.0	23.7
	FT AD	172.0	0.0	0.0	0.0	0.0	0.0	172.0	339.7
	# SH	40.0	0.0	0.0	0.0	0.0	0.0	40.0	
3	K TON	12.0	0.0	0.0	0.0	0.0	0.0	12.0	35.7
	FT AD	172.0	0.0	0.0	0.0	0.0	0.0	172.0	511.7
	# SH	40.0	0.0	0.0	0.0	0.0	0.0	40.0	
4	K TON	12.0	0.0	0.0	0.0	0.0	0.0	12.0	47.7
	FT AD	172.0	0.0	0.0	0.0	0.0	0.0	172.0	683.7
	# SH	40.0	0.0	0.0	0.0	0.0	0.0	40.0	
5	K TON	12.0	0.0	0.0	0.0	0.0	0.0	12.0	59.7
	FT AD	172.0	0.0	0.0	0.0	0.0	0.0	172.0	855.7
	# SH	40.0	0.0	0.0	0.0	0.0	0.0	40.0	
6	K TON	12.0	0.0	0.0	0.0	0.0	0.0	12.0	71.7
	FT AD	172.0	0.0	0.0	0.0	0.0	0.0	172.0	1027.7
	# SH	40.0	0.0	0.0	0.0	0.0	0.0	40.0	

FIGURE 8

S E G M E N T   C O M P L E T I O N   R E P O R T

SEG#	SEGMENT NAME	*SEGMENT BEGUN* MON.	DAY	SHIFT	*SEGMENT COMPLETED* MON.	DAY	SHIFT	KTONS MINED	FEET ADV.	TOTAL SHIFT	1	2	3	4	5	6	7	8	9	10	11	12	13	14	15	16	17	18	19	20
1000	ESUBMAIN	1	1	1	9	1	1	96.3	1380.	320.9	1	1	2	0	0	0	0	0	0	0	0	0	0	0	0	0	0	0	0	0
1001	ESUBMAIN	9	1	1	12	4	1	37.7	540.	125.6	1	1	2	0	0	0	0	0	0	0	0	0	0	0	0	0	0	0	0	0
2000	EPANELS1	9	1	1	13	20	1	79.2	1980.	198.0	0	0	0	0	0	0	0	0	0	0	0	0	0	0	0	0	0	0	0	0
4000	EPANELBR	13	20	1	14	4	2	0.0	9.	9.0	1	1	2	0	0	0	0	0	0	0	0	0	0	0	0	0	0	0	0	0
1002	ESUBMAIN	12	4	1	15	7	1	37.7	540.	125.6	0	0	0	0	0	0	0	0	0	0	0	0	0	0	0	0	0	0	0	0
1003	ESUBMAIN	15	7	1	18	9	2	37.7	540.	125.6	0	0	0	0	0	0	0	0	0	0	0	0	0	0	0	0	0	0	0	0
2001	EPANELS2	14	4	2	19	3	2	79.2	1980.	198.0	0	0	0	0	0	0	0	0	0	0	0	0	0	0	0	0	0	0	0	0
4001	EPANELBR	19	3	2	19	8	1	0.0	9.	9.0	1	1	2	0	0	0	0	0	0	0	0	0	0	0	0	0	0	0	0	0
1004	ESUBMAIN	18	9	2	21	12	2	37.7	540.	125.6	0	0	0	0	0	0	0	0	0	0	0	0	0	0	0	0	0	0	0	0
2002	EPANELS3	19	8	1	24	7	1	79.2	1980.	198.0	0	0	0	0	0	0	0	0	0	0	0	0	0	0	0	0	0	0	0	0
4002	EPANELBR	24	7	1	24	11	2	0.0	9.	9.0	1	1	2	0	0	0	0	0	0	0	0	0	0	0	0	0	0	0	0	0
3000	EPANELN1	21	12	2	24	18	2	52.8	1320.	132.0	0	0	0	0	0	0	0	0	0	0	0	0	0	0	0	0	0	0	0	0
5000	EPANELBR	24	18	2	25	1	2	0.0	6.	6.0	1	1	2	0	0	0	0	0	0	0	0	0	0	0	0	0	0	0	0	0
3001	EPANELN2	25	1	2	28	7	2	52.8	1320.	132.0	0	0	0	0	0	0	0	0	0	0	0	0	0	0	0	0	0	0	0	0
5001	EPANELBR	28	7	2	28	10	2	0.0	6.	6.0	1	1	2	0	0	0	0	0	0	0	0	0	0	0	0	0	0	0	0	0
2003	EPANELS4	24	11	2	29	10	2	79.2	1980.	198.0	0	0	0	0	0	0	0	0	0	0	0	0	0	0	0	0	0	4	0	0
4003	EPANELBR	29	10	2	29	15	1	0.0	9.	9.0	1	1	2	0	0	0	0	0	0	0	0	0	0	0	0	0	0	0	0	0
3002	EPANELN3	28	10	2	31	5	2	44.0	1100.	110.0	0	0	0	0	0	0	0	0	0	0	0	0	0	0	0	0	0	0	0	0
5002	EPANELBR	31	5	2	31	8	1	0.0	5.	5.0	1	1	2	0	0	0	0	0	0	0	0	0	0	0	0	0	0	0	0	0
3003	EPANELN4	31	8	1	33	12	1	35.2	880.	88.0	0	0	0	0	0	0	0	0	0	0	0	0	0	0	0	0	0	0	0	0
5003	EPANELBR	33	12	1	33	14	1	0.0	4.	4.0	1	1	2	0	0	0	0	0	0	0	0	0	0	0	0	0	0	0	0	0
2004	EPANELS5	29	15	1	34	14	1	79.2	1980.	198.0	0	0	0	0	0	0	0	0	0	0	0	0	0	0	0	0	0	0	0	0
4004	EPANELBR	34	14	1	34	18	2	0.0	9.	9.0	1	1	2	0	0	0	0	0	0	0	0	0	0	0	0	0	0	0	0	0
3004	EPANELN5	33	14	1	35	7	1	26.4	660.	66.0	0	0	0	0	0	0	0	0	0	0	0	0	0	0	0	0	0	0	0	0
5004	EPANELBR	35	7	1	35	8	2	0.0	3.	3.0	2	2	4	0	0	0	0	0	0	0	0	0	0	0	0	0	0	0	0	0

(Equipment columns 1–20: EQUIPMENT AVAILABLE AFTER SEGMENT BEGUN (BY EQUIPMENT TYPE))

FIGURE 9

MINE MAP TO BE PREPARED FOR MAP OPTION 1 - INPUT DATA FOLLOWS:

```
INPUT FOR SEGMENT 1000 X-START   3740 Y-START   2180 WIDTH   320 ANGLE     0
INPUT FOR SEGMENT 1001 X-START   4280 Y-START   2180 WIDTH   320 ANGLE     0
INPUT FOR SEGMENT 1002 X-START   4880 Y-START   2180 WIDTH   320 ANGLE     0
INPUT FOR SEGMENT 1003 X-START   5480 Y-START   2180 WIDTH   320 ANGLE     0
INPUT FOR SEGMENT 1004 X-START   6080 Y-START   2180 WIDTH   320 ANGLE     0
INPUT FOR SEGMENT 1005 X-START   6680 Y-START   2180 WIDTH   320 ANGLE     0
INPUT FOR SEGMENT 2000 X-START   3740 Y-START   2180 WIDTH   320 ANGLE   180
INPUT FOR SEGMENT 2001 X-START   3200 Y-START   2180 WIDTH   320 ANGLE   180
INPUT FOR SEGMENT 2002 X-START   2600 Y-START   2180 WIDTH   320 ANGLE   180
INPUT FOR SEGMENT 2003 X-START   2000 Y-START   2180 WIDTH   320 ANGLE   180
INPUT FOR SEGMENT 2004 X-START   1400 Y-START   2180 WIDTH   320 ANGLE   180
INPUT FOR SEGMENT 2005 X-START    800 Y-START   2180 WIDTH   320 ANGLE   180
INPUT FOR SEGMENT 3000 X-START   4280 Y-START   2340 WIDTH   500 ANGLE    90
INPUT FOR SEGMENT 3002 X-START   4880 Y-START   2340 WIDTH   500 ANGLE    90
INPUT FOR SEGMENT 3004 X-START   5480 Y-START   2340 WIDTH   500 ANGLE    90
INPUT FOR SEGMENT 3006 X-START   6080 Y-START   2340 WIDTH   500 ANGLE    90
INPUT FOR SEGMENT 3008 X-START   6680 Y-START   2340 WIDTH   500 ANGLE    90
INPUT FOR SEGMENT 3010 X-START   7280 Y-START   2340 WIDTH   500 ANGLE    90
INPUT FOR SEGMENT 3500 X-START   4280 Y-START   2020 WIDTH   500 ANGLE   270
INPUT FOR SEGMENT 3502 X-START   4880 Y-START   2020 WIDTH   500 ANGLE   270
INPUT FOR SEGMENT 3504 X-START   5480 Y-START   2020 WIDTH   500 ANGLE   270
INPUT FOR SEGMENT 3506 X-START   6080 Y-START   2020 WIDTH   500 ANGLE   270
INPUT FOR SEGMENT 3508 X-START   6680 Y-START   2020 WIDTH   500 ANGLE   270
INPUT FOR SEGMENT 3510 X-START   7280 Y-START   2020 WIDTH   500 ANGLE   270
INPUT FOR SEGMENT 4500 X-START   3200 Y-START   2340 WIDTH   500 ANGLE    90
INPUT FOR SEGMENT 4502 X-START   2600 Y-START   2340 WIDTH   500 ANGLE    90
INPUT FOR SEGMENT 4504 X-START   2000 Y-START   2340 WIDTH   500 ANGLE    90
INPUT FOR SEGMENT 4506 X-START   1400 Y-START   2340 WIDTH   500 ANGLE    90
INPUT FOR SEGMENT 4508 X-START    800 Y-START   2340 WIDTH   500 ANGLE    90
INPUT FOR SEGMENT 4510 X-START    200 Y-START   2340 WIDTH   400 ANGLE    90
INPUT FOR SEGMENT 4000 X-START   3200 Y-START   2020 WIDTH   500 ANGLE   270
INPUT FOR SEGMENT 4002 X-START   2600 Y-START   2020 WIDTH   500 ANGLE   270
INPUT FOR SEGMENT 4004 X-START   2000 Y-START   2020 WIDTH   500 ANGLE   270
INPUT FOR SEGMENT 4006 X-START   1400 Y-START   2020 WIDTH   500 ANGLE   270
INPUT FOR SEGMENT 4008 X-START    800 Y-START   2020 WIDTH   500 ANGLE   270
INPUT FOR SEGMENT 4010 X-START    200 Y-START   2020 WIDTH   400 ANGLE   270
```

FIGURE 10

mining alternatives to establish a variety of extraction plans to provide decision makers with a thorough assessment of the feasible alternatives. The time needed to achieve the same level of visibility using manual techniques is often not available to planners, and the corresponding personnel resources which would be required usually exceed available budgets.

MCOM

The economic performance of a mining project is of principal concern to company management and to the financial analysts who are responsible for determining the costs and cash flows required to support mining operations.

Financial studies are based on the mine plan and the associated production profile, from which equipment, personnel, and material quantities are established. These are converted to costs through the application of usage factors from the mine plan and their related unit costs.

A financial plan can assume various levels of detail depending on the extent of available information, management objectives, and the allowable time for conducting financial studies. Mine economics will be affected by technical factors inherent to the mine plan and by financial factors which affect items such as unit costs, interest and inflation rates.

Assumptions must be made to account for the uncertainties in other technical and financial variables. As a consequence, it is often desirable to evaluate a number of planning scenarios in order to ascertain the impact of different assumptions, and to obtain a spectrum of financial plans. The computational effort required to achieve equivalent emphasis on each plan is often prohibitive, so that additional assumptions or simplifications in analytical procedures are used to approximate alternative scenarios as variations to the base plan.

The MCOM model has been developed to provide financial analysts with a means for obtaining consistent economic evaluations of each proposed mine plan in a timely manner, and with a maximum degree of computational efficiency. Engineering variables relating to the mine plan, to the process step completions, and to the production schedule are passed from MPROD to MCOM for application to cost computations. Additional data elements which do not affect the mine plan are entered directly into MCOM. The financial analyst can obtain reports describing a variety of cost categories in spread sheet format by exercising MCOM in conjunction with the results of a selected production scenario.

Chapter 28

COMPUTER SIMULATION AND ITS APPLICATION
IN THE EXTRACTION, CONVEYANCE AND HOISTING SYSTEM OF COAL MINES

by

Mr. LIN ZAIKANG

Master Degree of Mining Engineering
China Institute of Mining Technology
The People's Republic of China

INTRODUCTION

A software package, which simulates the main production process in a coal mine, is discussed in the paper. The simulation is mainly concerned with a transportation system that transfers coal from longwall faces to surface bunkers. The package allows us to study overall factors affecting productivity and to analyse interactive effect between these factors. The study leads to a possible optimization design of a coal mine.

MODELLING PRINCIPLES

A typical system as shown in Fig.1 is simulated based on the features of working area (WA) and transportation system in the most mines in China.

All longwall faces in different WAs may be regarded as a "coal source" where the coal is exploited. The mined coal is conveyed from longwall faces towards working bunkers and then through locomotive (or belt) haulage to the underground bunker, to surface finaly. The general system consists of three major parts: WA, main haulage and hoisting. In order to simplify the model, some restrictions have been made as follows:

----- The number of WA at the same working level in a mine must be less than six, and the number of longwall faces in a WA must be less than four, including full machenized mining (FM), generally mechanized mining (GM), and shot-firing mining (SM).

----- The production capacity of the mine may be less than 4 MIL.TPY, while locomotives are used. No capacity restriction is made for main belt.

----- The working system may be three

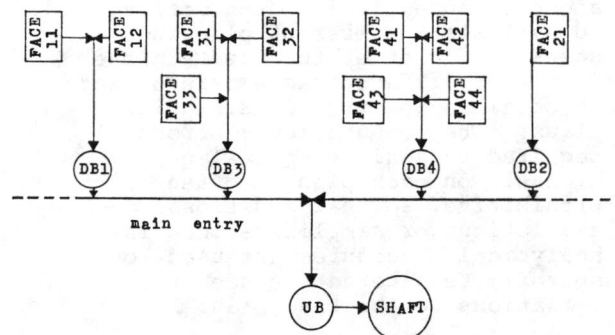

Fig.1 Flow diagram of the transportation system in a coal mine

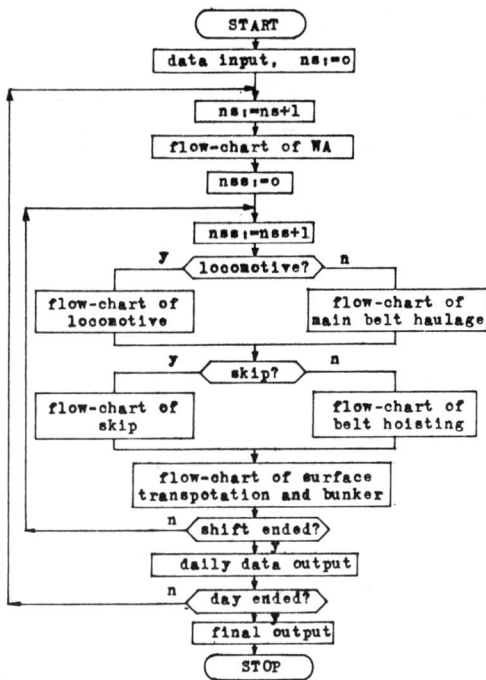

Fig.2 The flow-chart of the overall
software package.

shifts or four shifts per day.

-----Double-track locomotives including
wagon or trap-bottom car are concerneed.

-----The number of the bunkers to one
main belt must be less than six.

-----The quantity of coal and the
hoisting time per skip are treated as
constants.

-----The delay time by transportation
fault is less than two hours.

SIMULATION PACKAGE

The simulation package (LY811) developed based on the discussion above

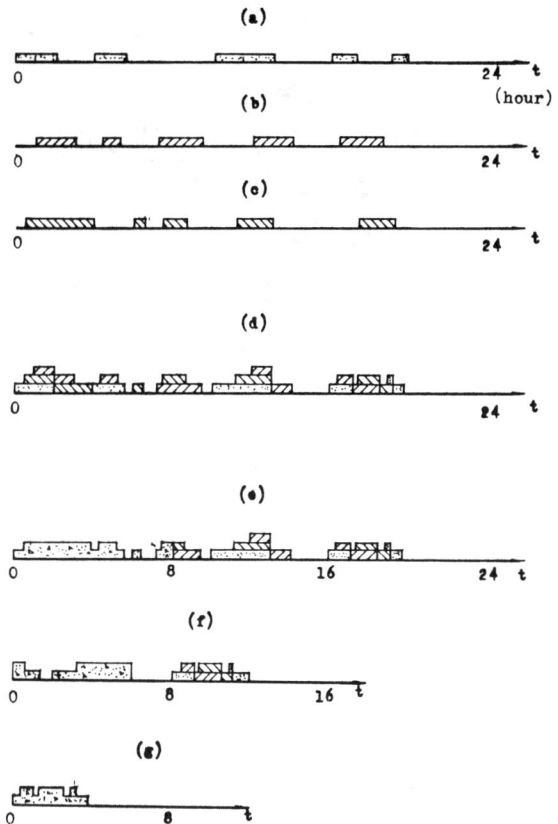

Fig.3 Coal flow collection in a WA and
regulation to a bunker during
each shift.

includes six subroutines which are WA,
locomotive, main belt haulage, skip,
belt hoisting and surface transpotation and
bunkers. A block-diagram of the package
is shown in Fig.2.

WA: The three major kinds of longwall
faces in China are FM, GM and SM. However,
the different number of drums of a shearer
and the mode of start-cut may contribute
to a variety of operation modes. Ten operation modes, such as mode of "FM start-cut
bidirectional mining" and "GM middle-cut
unidirectional mining", are considered in
the package which are widely used in the
most of mines.

Fig.4 flow-chart (left):

START

generation of random, negative exponential and normal distributions

lo working models of longwall faces

data input, ns:=o

ns:=ns+1

working model selection

calculation of shift output, daily output and machine running ratio

coal flow collection from faces

coal flow regulation in a WA

to "main entry"

H: from "day ended?"

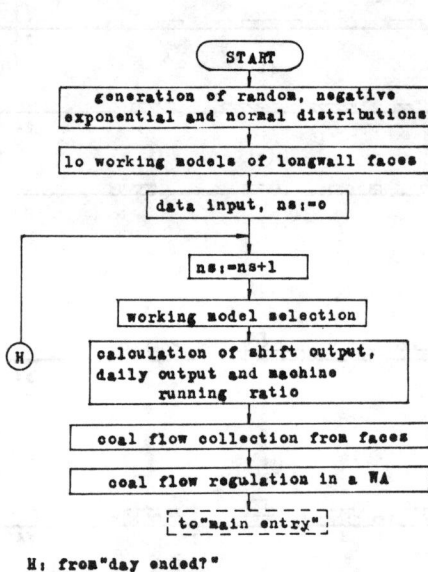

Fig.4 The flow-chart of working areas

Fig.5 flow-chart (right):

input data from "WA"

locomotive? y / n

initial status

selection minimum current time's locomotive

shift ended? n / y

loading? y → loading chart / n

loaded and running? y → loaded and running chart / n

un-loading? y → un-loading chart / n

empty and running chart

all back to pit bottom

calculate locomotive time and the stored coal

main belt chart

to "skip"

Fig.5 The flow-chart of locomotive.

Since there are usually one to four faces in a WA and the coal from the faces is collectively transported to the working bunker, therefore a model is used which simulates both the collection of coal flow in a WA and coal-flow regulation to the working bunker. The model is illustrated in Fig.3, where (a) (b) (c) are discontinous coal-flow from three different faces, (d) is the collected coal-flow from these three faces on the collection conveyer, (e) (f) (g) are the regulated coal-flow to the working bunker for each shift.

Based on the fact of discontinuity of the coal-flow from a longwall face, the package is programmed using "event-oriented" as shown in Fig.4.

LOCOMOTIVE: A locomotive train may be at one of four basic status, i.e., loading, loaded and running, unloading, empty and running, which are of cycling.

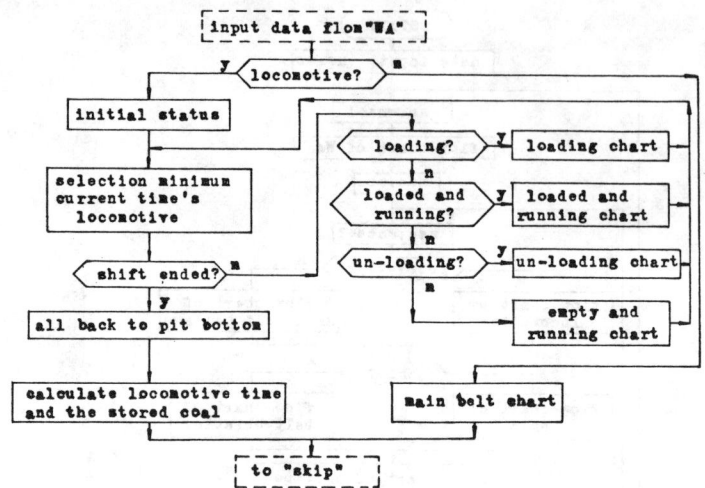

The models for such transpotation system may be constructed using "min-event-oriented" as shown in Fig.5.

SKIP: The "interval-oriented" is used for skip as shown in Fig.6.

The production capacity of Mine 7 in Huaibei is 2 MIL.TPY and the coal there is mined at three different seams, denoted as $3^{\#}$, $5^{\#}$, $6^{\#}$. The second production level are divided by 15 WAs. The eight WA at the level are in seam $5^{\#}$ and the rest are in seam $6^{\#}$. The layout is shown in Fig.7.

Train car in the mine is 3-ton trap-bottom car and the capacity of the skip pair is 7.5 ton. LY811 package is used to simulate the product process of the mine during early and later production periods. The main purposes of the simulation are:

----- to study production situation at the faces.

----- to choose proper locomotives.

----- to design proper production capacity of the mine and WAS bunkers.

During early product period, there are three WAS and seven faces in the mine as shown in Table 1. The simulated resultes are 15-20 tables including following four tables (See Table 2, 3, 4, 5).

In order to choose a proper number of locomotives (or trains), the case from 4 to 10 locomotives are simulated individually and the results are shown in Table 6 and Fig.8.

It can be seen from Fig.8 that K22 is decreasing with the increasment of the number of locomotives and seven locomotives appear to be a reasonable choice. Therefore seven locomotives are used in the design.

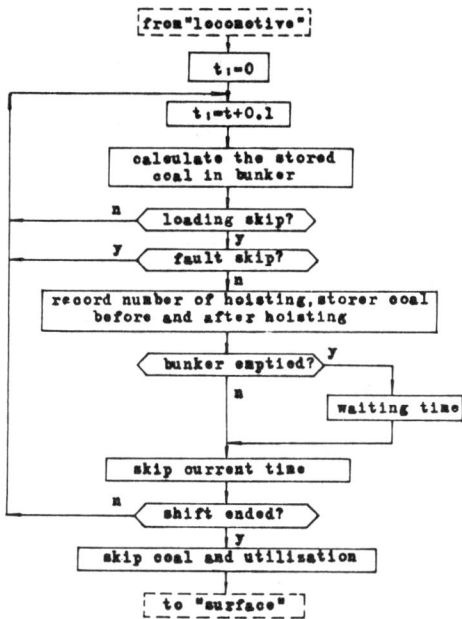

Fig.6 The flow-chart of skip (from "locomotive").

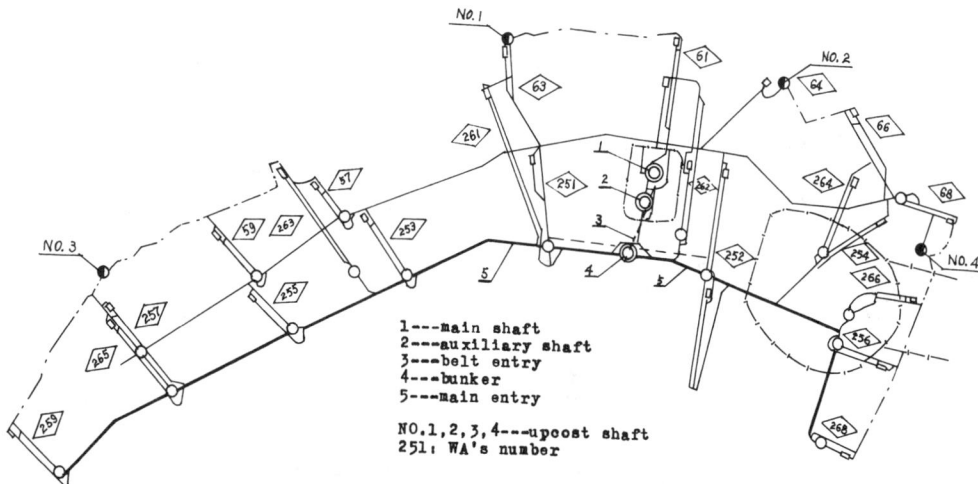

Fig.7 The lay-out diagram of the second level entry in Huaibei No.7 Mine.

Table 1. The design of capacity of WA and faces
during early production period

diagram	DB1 DB3 UB DB2 main entry					
computer number (WA)	1		3			2	
computer number (face)	11	12	31	32	33	21	22
extraction model	3#GM	5#FM	3#GM	5#SM	5#SM	5#SM	5#SM
design of capacity (ten-thousand ton)	15.57	56.56	15.57	17.2	14.08	17.20	17.20
total(ten-thousand ton)	153.40(72.13, 46.87, 34.40)						

Table 2. The simulation daily production
data.

OUTPUT(1)

```
QSHIFT=  4723.54
QSHIFT1=  2093.88      QSHIFT2=  2384.50     QSHIFT3=  245.16
   QDIS=  2271.84       1073.67    1378.03
  QDIS1=   944.07        548.05     601.76
  QDIS2=  1152.48        468.38     763.64
  QDIS3=   175.29         57.23      12.63
     WD=        76           59          92
```

NO	W	QFACE	QFACE1	QFACE2	QFACE3	NFACE	PR
11	23	444.88	167.26	277.62	0.00	239.07	19.92
12	53	1826.96	776.81	874.86	175.29	297.60	27.56
13	0	0.00	0.00	0.00	0.00	0.00	0.00
21	27	491.77	245.96	245.82	0.00	369.23	34.19
22	32	581.89	302.09	222.57	57.23	438.43	40.60
23	0	0.00	0.00	0.00	0.00	0.00	0.00
31	36	444.88	148.29	283.96	12.63	239.07	19.92
32	36	522.16	252.28	269.88	0.00	389.03	36.02
33	20	410.99	201.19	209.80	0.00	328.96	30.46

SYMBOL

(in Table 2, 3, 4, 5)

QSHIFT---daily output (ton)

QSHIFI1 (or2, or3)---shift 1 (or2,or3)
output (ton)

QDIS---WA daily output (ton)

QDIS1 (or2, or3)---WA shift 1 (or2, or3)
output (ton)

WD---shearer started times in each WA

NO---face number

W---shearer started times in each face

QFACE---face daily output (ton)

QFACE1 (or2, or3)---face shift1 (or2,or3)
output (ton)

NFACE---shearer running time in faces (min)

PR---shearer utilization ratio (%)

DIS---working area number

FACE---face number

SCH---shearer operation scheme (10
options)

REG---face regulation (31 mean: 3 shifts
working and 1 shift preparation a
day)

DIS	FACE	SCH	REG	W	F	RT	RQ	RP	RQP	S
1	2	3	31	1	0.00	20.00	21.45	2.80	7.66	1
1	2	3	31	2	35.10	57.90	33.71	4.40	7.66	1
1	2	3	31	3	13.90	76.20	7.66	1.00	7.66	1
1	2	3	31	4	2.00	79.20	47.50	6.20	7.66	1
1	2	3	31	5	4.00	89.40	9.19	1.20	7.66	1
1	2	3	31	6	49.10	139.70	41.37	9.90	4.18	2
1	2	3	31	7	1.00	150.60	6.37	1.90	3.35	2
1	2	3	31	8	7.30	159.80	130.57	17.10	7.64	3
1	2	3	31	9	18.70	195.60	18.58	4.90	3.79	4
1	2	3	31	10	26.20	226.70	3.35	1.00	3.35	4
1	2	3	31	11	16.60	244.30	8.38	2.50	3.35	4
1	2	3	31	12	20.40	267.20	49.17	7.60	6.47	1
					NCUT= 1					
1	2	3	31	13	13.50	288.30	121.09	20.70	5.85	2
1	2	3	31	14	27.80	336.80	5.12	1.40	3.66	3
1	2	3	31	15	1.40	339.60	26.81	3.50	7.66	3
1	2	3	31	16	3.80	346.90	7.66	1.00	7.66	3
1	2	3	31	17	18.10	366.00	32.18	4.20	7.66	3
1	2	3	31	18	12.00	382.20	65.12	8.50	7.66	3
1	2	3	31	19	23.60	414.30	109.36	19.90	5.50	1
					NCUT= 2					
1	2	3	31	20	25.60	459.80	7.66	1.00	7.66	1
1	2	3	31	21	9.90	470.70	24.51	3.20	7.66	1
1	2	3	31	22	1.50	475.40	29.59	4.20	7.05	2
1	2	3	31	23	11.00	490.60	18.77	5.60	3.35	2
1	2	3	31	24	1.00	497.20	7.04	2.10	3.35	2
1	2	3	31	25	10.70	510.00	30.21	4.90	6.17	3
1	2	3	31	26	26.10	541.00	7.66	1.00	7.66	3
1	2	3	31	27	14.20	556.20	7.66	1.00	7.66	3
1	2	3	31	28	6.10	563.30	52.09	6.80	7.66	3
1	2	3	31	29	7.60	577.70	190.56	30.50	6.25	1
					NCUT= 3					
1	2	3	31	30	52.20	660.40	48.31	11.20	4.31	2
1	2	3	31	31	3.90	675.50	3.35	1.00	3.35	2
1	2	3	31	32	24.40	700.90	6.37	1.00	6.37	3
1	2	3	31	33	1.60	703.50	32.94	4.30	7.66	3
1	2	3	31	34	13.00	720.80	100.79	14.00	7.20	4
1	2	3	31	35	10.90	745.70	18.10	5.40	3.35	4
1	2	3	31	36	64.50	815.60	109.98	16.10	6.38	1
					NCUT= 4					
1	2	3	31	37	16.20	847.90	16.85	2.20	7.66	1
1	2	3	31	38	11.80	861.90	36.05	7.80	4.62	2
1	2	3	31	39	15.30	885.00	4.36	1.30	3.35	2
1	2	3	31	40	1.00	887.30	3.35	1.00	3.35	2
1	2	3	31	41	5.60	893.90	27.29	4.80	5.69	3
1	2	3	31	42	23.30	922.00	123.53	17.70	6.98	4
1	2	3	31	43	20.20	959.90	24.90	7.30	3.41	1
					NCUT= 5					
1	2	3	31	44	11.00	978.20	42.13	5.50	7.66	1
1	2	3	31	45	1.00	984.70	7.66	1.00	7.66	1
1	2	3	31	46	15.00	1000.70	7.66	1.00	7.66	1
1	2	3	31	47	24.80	1026.50	7.66	1.00	7.66	1
1	2	3	31	48	3.60	1031.10	7.66	1.00	7.66	1
1	2	3	31	49	9.00	1041.10	13.79	1.80	7.66	1
1	2	3	31	50	5.30	1048.20	30.64	4.00	7.66	1
1	2	3	31	51	8.70	1060.90	10.73	1.40	7.66	1
1	2	3	31	52	1.40	1063.70	15.08	3.60	4.19	2
1	2	3	31	53	4.50	1071.80	7.37	2.20	3.35	2

QFACE= 1826.96   NCYCLE= 5

Table 3. The simulation result of a shift at FM's face 12.

W---shearer running times

F---shearer stop time (negtive exponential distribution, AV=15 min)

RT---shearer started time (min)

RQ---coal output during each run (ton)

RP---shearer running time (engtive exponential distribution, AV=6   min)

RQP---coal output per time unit (ton/min)

S---shearer current status (four-status circular)

NFACE---shearer running time in faces (min)

PR---shearer utilization ratio (%)

NO---the order of unloading

RT---unloading started time (min)

RQ---unloading coal output (ton)

RP---unloading time (min)

Q/P---unloading output per time unit (ton/min)

J2---locomotive number

G---WA bunker number

NS---number of days simulated

QSHIFT---daily output (ton)

QD6---daily entry transpotation output (ton)

QSS6---daily skip hoisting output (ton)

K6---QD6/QSHIFT (%)

K7---QSS6/QD6 (%)

CLOCK6---daily transpotation operation time (hour)

CLOCK7---daily skip operation time (hour)

AV---mean volue

SD---standard deviation

SUM---sum

NSS---number of shifts simulated

QSHIFT I ---shift output (ton)

K22---locomotive relative utilization ratio (%)

K44---skip relative utilization ratio (%)

Table 4. The locomotive un-loading status table of a shift.

Table 5. The table of final simulation results (a part).

RESULT(L.T)

NO.	RT(MIN)	RQ(T)	RP(MIN)	Q/P	J2	G
1	73.76	53.71	1.00	53.71	1	1
2	95.28	48.79	1.00	48.79	2	1
3	98.47	55.00	1.00	55.00	4	3
4	102.75	49.66	1.00	49.66	5	2
5	115.83	55.00	1.00	55.00	7	3
6	121.80	41.73	1.00	41.73	3	1
7	126.02	55.00	1.00	55.00	1	2
8	129.74	50.64	1.00	50.64	6	1
9	149.84	46.04	1.00	46.04	2	2
10	152.22	49.22	1.00	49.22	4	3
11	181.49	44.61	1.00	44.61	5	1
12	187.60	53.84	1.00	53.84	6	3
13	190.10	38.12	1.00	38.12	1	2
14	202.24	51.55	1.00	51.55	7	1
15	206.43	53.78	1.00	53.78	4	3
16	208.92	50.07	1.00	50.07	3	1
17	220.25	44.58	1.00	44.58	2	1
18	231.29	53.83	1.00	53.83	6	2
19	241.78	51.15	1.00	51.15	1	3
20	245.71	40.73	1.00	40.73	5	1
21	261.89	54.33	1.00	54.33	4	2
22	267.18	47.66	1.00	47.66	7	1
23	268.18	48.17	1.00	48.17	3	3
24	279.46	50.89	1.00	50.89	2	3
25	289.97	42.70	1.00	42.70	1	2
26	308.85	39.92	1.00	39.92	6	1
27	316.35	55.00	1.00	55.00	4	2
28	320.66	54.67	1.00	54.67	3	3
29	331.18	41.05	1.00	41.05	5	1
30	342.67	50.56	1.00	50.56	7	1
31	345.43	51.58	1.00	51.58	1	2
32	348.95	39.39	1.00	39.39	2	1
33	359.99	45.46	1.00	45.46	4	3
34	380.54	52.56	1.00	52.56	3	2
35	384.39	42.70	1.00	42.70	6	1
36	388.53	53.64	1.00	53.64	7	3
37	411.46	43.55	1.00	43.55	5	1

WA---fault times

TA---delay time due to fault (min)

CLOCK3---shift transpotation operation time (min)

CLOCK4---shift skip operation time (min)

The coal stored in the underground bunker is variable during a shift and its dynamic diagram is shown in Fig.9.

The capacity of underground bunker may be calculated statistically from the

THE TOTAL OUTPUT DATA

NS	QSHIFT	QD6	QSS6	K6	K7	CLOCK6	CLOCK7
1	4719.17	4787.20	4837.20	101.44	101.04	16.14	17.55
2	4699.67	4692.23	4692.23	99.84	100.00	16.11	17.93
3	5030.02	4999.52	4999.52	99.39	100.00	16.73	17.59
4	5262.91	5297.69	5297.69	100.66	100.00	16.23	17.54
5	5135.11	5149.79	5149.79	100.29	100.00	16.22	17.54
6	4330.08	4299.73	4299.73	99.30	100.00	16.18	17.55
7	4440.99	4462.99	4462.99	100.50	100.00	16.07	17.67
8	4707.59	4683.44	4683.44	99.49	100.00	16.25	17.54
9	4909.58	4961.70	4961.70	101.06	100.00	16.14	17.54
10	4738.24	4722.50	4722.50	99.67	100.00	16.07	17.54
AV	4797.34	4805.68	4810.68	100.16	100.10	16.21	17.60
SD	292.52	303.85	303.92	0.74	0.33	0.19	0.12
SUM	47973.36	48056.79	48106.79				

Nss	QSHIFT(I)	K22	K44	WA	TA	CLOCK3	CLOCK4
1	2019.32	70.56	76.47	0	0.00	366.40	380.80
2	2481.91	96.43	92.81	0	0.00	372.81	381.00
3	217.94	53.22	65.64	0	0.00	229.19	291.30
1	2079.01	73.64	74.08	0	0.00	359.03	403.60
2	2293.79	88.62	88.05	0	0.00	378.01	380.90
3	326.88	51.45	60.73	0	0.00	229.78	291.30
1	2550.30	78.73	79.30	2	81.80	398.67	383.60
2	2272.26	94.53	98.00	0	0.00	375.91	380.70
3	207.46	48.39	64.30	1	1.80	229.13	291.30
1	2628.09	92.88	84.84	0	0.00	375.24	380.00
2	2202.83	92.56	100.00	1	1.00	369.12	380.90
3	431.99	57.28	73.22	0	0.00	229.58	291.30
1	2705.13	93.73	83.39	0	0.00	364.46	380.40
2	1832.73	87.95	98.29	1	6.50	371.01	380.90
3	597.24	55.61	68.31	2	188.30	237.98	291.30
1	2186.77	66.81	68.67	0	0.00	378.02	380.50
2	1848.58	84.88	91.74	0	0.00	362.83	381.20
3	294.73	40.11	45.55	0	0.00	229.90	291.30
1	2182.61	70.82	72.95	1	46.10	363.58	388.50
2	1917.20	82.81	83.73	0	0.00	370.66	380.40
3	341.17	51.10	58.50	1	1.00	230.18	291.30
1	2222.28	69.79	73.42	0	0.00	374.50	380.70
2	2277.41	94.44	93.59	0	0.00	370.64	380.60
3	207.89	46.77	59.84	1	35.60	229.60	291.30
1	2528.81	92.49	83.11	0	0.00	366.38	380.10
2	2174.21	95.80	99.92	0	0.00	373.11	381.20
3	206.55	32.20	55.37	0	0.00	228.68	291.30
1	2293.20	77.11	78.32	0	0.00	363.49	380.10
2	2203.62	85.96	90.74	0	0.00	371.30	381.10
3	241.41	55.30	59.39	0	0.00	229.63	291.30
AV		72.73	77.41	10	362.10		
SD		19.30	14.79				

formula below:

$$Xp = X + K \cdot b$$

here: Xp---bunker's capacity

X ---mean value

b ---standard deviation

K ---constant

Table 6. The comparison data table of varity
for different number of locomotives

NUMBER of LOC	QSHIFT	QD6 (ton)	QSS6	K6	K7	K22 (%)	K44	UB	DB1	DB2	DB3	ΣB[I] (ton)
4	4922.39	3596.92	3606.92	73.31	100.28	98.45	58.71	148	2805	2661	2705	8319
5	4832.98	4560.12	4569.55	94.69	100.21	98.72	74.88	214	912	727	825	2678
6	4848.21	4858.30	4868.30	100.20	100.17	87.82	79.55	270	526	281	355	1431
7	4651.51	4660.94	4670.94	100.20	100.21	71.32	75.61	339	504	179	276	1298
8	4880.83	4899.68	4909.68	100.39	100.19	67.31	79.59	362	460	222	282	1326
9	4699.47	4716.71	4726.71	100.36	100.21	57.02	75.95	413	460	218	263	1354
10	4916.75	4933.50	4943.50	100.33	100.20	55.37	80.22	457	510	191	216	1419

Table 7. The table of choice the bunker
capacity.

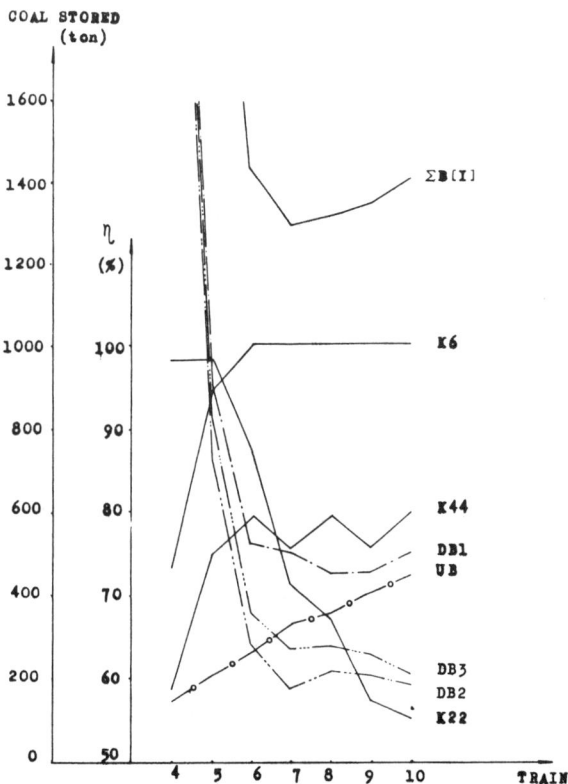

Fig.8 The "CS-TRAIN" and "η-TRAIN"
curves of different scheme.

	UB	DB1	DB2	DB3
AV	294	308	137	164
SD	99	77	51	43
MAX	476	503	227	262
CB1	513	465	241	252
CB2	661	539	291	294
CHOICE	550	500	250	260

SYMBOL (in Table 7)

CB1---lower limit of the bunker capacity
    calculated using $\alpha=0.05$
CB2---upper limit of the bunker capacity
    calculated using $\alpha=0.01$
CHOICE---the properly chosen value between
    CB1 and CB2

Fig.9 The dynamic diagram of the coal
stored in the underground bunker
during a shift

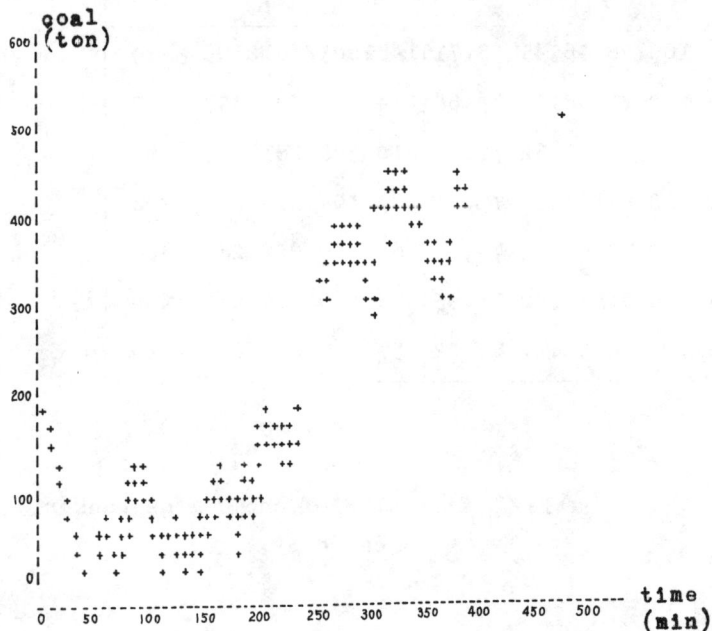

K is obtained from a required confidence limit. In practice, the capacity
of a WA bunker is normally distributed
and the one of a underground bunker obeys
a Gumbel distribution (Zhang, 1982).

A design of bunker capacity of Mine 7
based on the simulation is shown in
Table 7.

In the addition, LY811 software package is used to design for the skips, main
belts and mine bunkers of Mine 2, Shandong
provice, 4 MIL.TPY.

REFERENCES

1. Xu Guozhi, and Guo Shaoxi, 1980,
   The Applications of Computer
   Simulation in Management

2. Lin Shaogong, 1978, The Fundamentals
   of Prosibility and statistics

3. Zhang Xianchen, and Lin Zaikang,
   1982, "Simulation of the System of
   the Extraction and Conveyance"
   Journal of China Institute of
   Mining Technology 1982 No.4

Chapter 29

AN UNDERGROUND COAL MINING MODEL
FOR OPTIMIZING WORKING AREA PLAN AND ITS APPLICATION
by

Wang Yu-Jun   Xu Yong-Qi
Dept. of Mining Engr.
China Institute of Mining Technology
The People's Republic of China

INTRODUCTION

For underground coal mining in China, the mine field are usually divided into working areas. Recently there are about 1500 nation-controlled working areas in operation. In order to keep their normal production, it is always necessary to have a certain number of new working areas in preparation. The working area plan is important for new coal mines as well.

The content of a working area plan includes: the division of the mine into working areas, roadways layout, determination of the main parameters and mining technology. Their selection is related to the results, such as: if the working areas and faces could be in operation effectively, if the equipment could be fully used, how much the volume for driving roadways and the cost for drivage and maintenance of the roadways are. Once the plan has been decided and the construction finishes, it is very difficult to change the roadways system in production period. Hence, when making the working area plan, varied alternatives and parameters should be considered and compared, an optimum decision could be selected by applying system engineering technology. The widely-used computer has become a powerful tool for realizing this aspiration.

This economic-mathematical model for optimizing the working area plan has been successfully used in several coal mines. After continually modified, this model has been more perfect for common use.

THE ECONOMIC-MATHEMATIC MODEL

Actually, the selection of working area plan alternatives is a multi-objectives decision problem. Its criterion consists of technical feasibility and economical profitability. The main index chosen for evaluating different alternatives is the cost per ton coal mined from the working area as a comprehensive index reflecting the consumption of man-power and material resources. By means of mathematic programming method the model is formed to optimize the roadways system and main parameters.

Working area roadways layout systems can be divided into two groups: long-wall face mining along strike and long-wall face mining along pitch. Each group includes many sub-alternatives. Depending on different roadways layout, single or duble wings mining, the location of the rise and the connection among each coal seam, there are 50 alternatives for long-wall face mining along strike. As to long-wall face mining along pitch, 36 alternatives can be composited according to: dividing into working areas or not, advancing to the dip or to the rise, single face or double faces and the layout of inclined roadways. For a specific alternative there could be different parameters to be chosen. The main optimized parameters are as follows: for long-wall face mining along strike --- the working area length along strike (s), the number of blocks (district) (N), the length of the extraction face (L), the number of extraction faces in production (n_0), accordingly the inclined length and the productively of the working area can be abtained; For long-wall mining along pitch, the number of strips (N), the inclined length of the strip (H), the length of the extraction face (L), the number of extraction faces in production (n_0), accordingly the length of the working area along strike and the productively as well. The total number of alternatives are 50×1280 = 64000 and 36×1280 = 46080, correspondingly. As to long-wall mining along strike, it's shown in Fig.1, Fig.2.

The cost per ton coal mined from the working area consists of eight items: working area roadways drivage, maintenance, coal haulage, ventilation, station and chamber drivage, auxiliary haulage, extraction faces movement and extraction face operation cost. The general expressions are shown in Table 1, 2. For different alternatives, a series of factors can be changed by means of longical algebra with (0, 1) matrix and the specific formulas can be obtained.

DEFINITION OF PARAMETERS

Controlled Variables:

- S --- the working area length along strike, S=600 -- 2000m, step 200m;
- N --- the number of blocks (districts), N=2--6 ;
- L --- the length of the extraction face L=100--240m, step 20m;
- n_0 --- the number of the extraction faces in production, n0=1--4;
- H --- the working area length along inclination, m, H=N×(L+LO);
- V_0 --- annual advance speed of the extraction face, m/yr. VO=1/(c+d×L);
- A --- annual output of the working area, t/yr. A=n_0∑m·r·k_0·L·V_0·k_7/n
- K --- recovery of the working area, % K=L·(S-s1)·K_0/S(L+L_0)

Geological Condition Parameters:

- α --- seams inclination, degree;
- ∑m --- total thickness of workable coal seams, m ,∑m=m1+m2+......+mn;
- n --- the number of workable coal seams;
- ∑h --- total vertical distance of coal seams, m , ∑h=h_{12} +h_{23}+......+ $h_{(n-1)n}$
- r --- volume weight of coal, t/m³ ;

Mining Technique Condition Parameters:

- L_0 --- coal pillar width between blocks (districts), m (inclusion subentry width);
- S1 --- coal pillar width of rise, m
- S_0 --- distance between connection roadways, m;
- h_0 --- vertical distance between centralized rock entry and coal seam (mn), m;

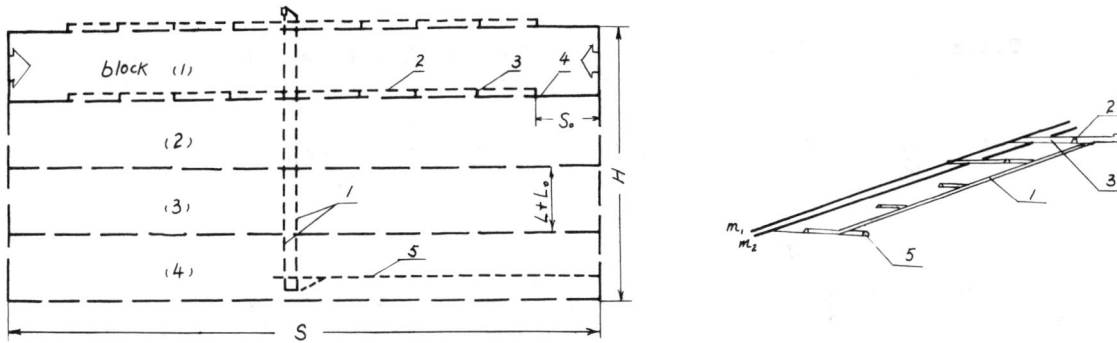

Fig.1 The alternative of long-wall mining along strike
1-rises; 2-centralized rock entry; 3-connection roadway;
4-extraction entry;   5-main entry;

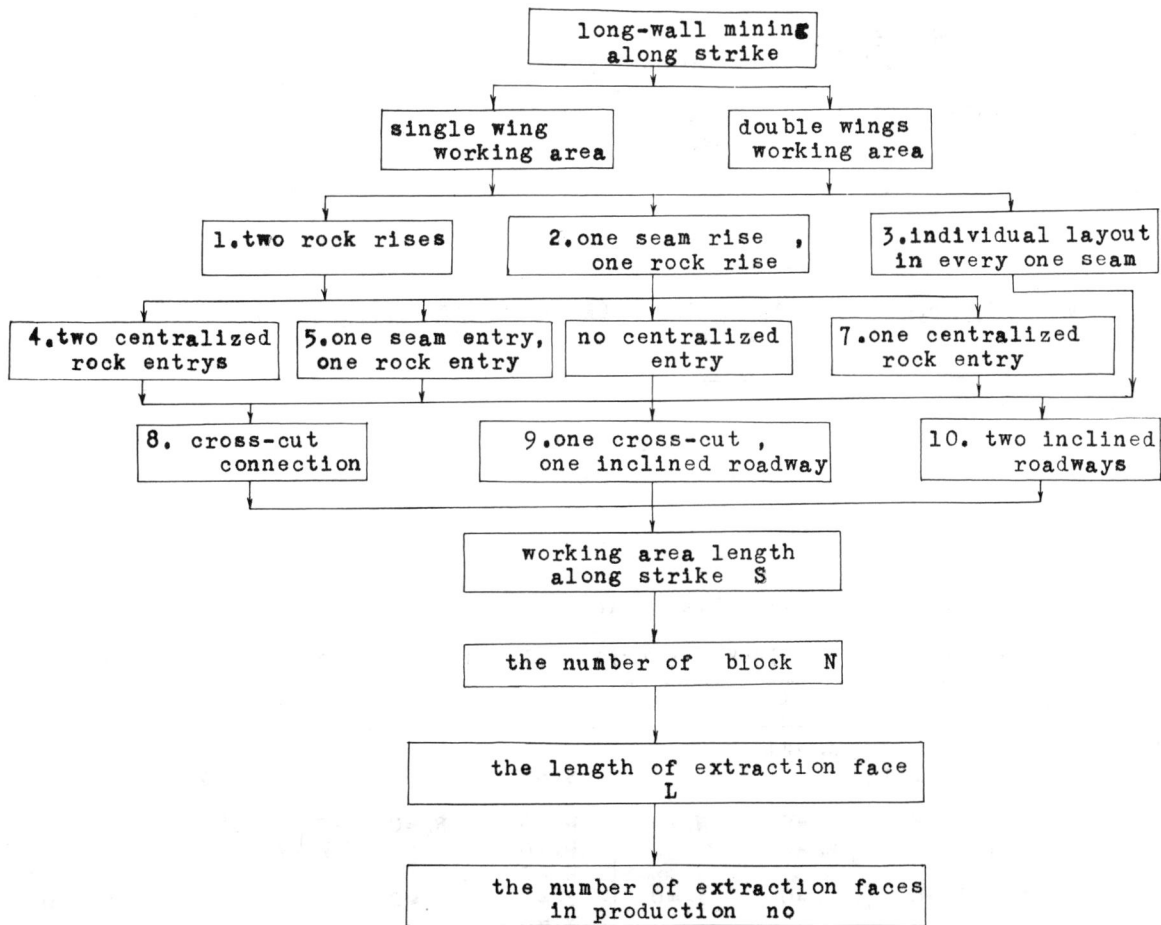

Fig. 2. The number of alternatives

Table 1 Cost of per ton coal for long-wall mining along strike

cost of	general formulas
roadways drivage Z_1	$\dfrac{J_1 \sin\alpha/\sin\gamma + J_2}{RS} + \dfrac{(N-1)(S-a_0 S_0)J_3 + 2nNSJ_4}{RSN(L+L_0)} + \dfrac{2(\Sigma m + \Sigma h + h_0)J_5}{R_0 S_0 L\sin(\alpha+\beta)}$ $+ \dfrac{(N-1)(L+L_0)tg(\alpha-\gamma)J_7}{2R_0 SL}$
roadways maintenance Z_2	$\dfrac{N(L+L_0)(W_1 \sin\alpha/\sin\gamma + W_2)}{A} + \dfrac{2K_0 L(S-a_0 S_0)(W_3+W_4)}{AK(L+L_0)} + \dfrac{2nS_0 W_5}{R_0 LV_0} + \dfrac{nSW_6(1/V_7 + 1/V_0)}{a_0 R_0 L}$ $+ \dfrac{S(\Sigma m + \Sigma h + h_0)(W_7 + W_8)}{S_0 A\sin(\alpha+\beta)} + \dfrac{(N-1)(L+L_0)tg(\alpha-\gamma)W_9}{2A}$
coal haulage Z_3	$\dfrac{K_0 L(N-1)(Y_1+Y_2)}{2K} + \dfrac{(S-a_0 S_0)Y_3}{2a_0} + \dfrac{S_0 Y_4}{2} + \dfrac{SY_5}{2a_0}$ $+ \dfrac{(\Sigma m + \Sigma h + h_0)Y_6}{2\sin\alpha}$
ventilation Z_4	$\dfrac{10.45\times10^{-6}(n_0^3 N(L+L_0) + S/a_0 + 2(\Sigma m + \Sigma h + h_0)/\sin(\alpha+\beta))C_F^3 A^2 a_3}{n_0^3 F_0^5}$
station drivage Z_5	$\dfrac{(Q_0 + Q_c + 550N)J_0}{RSN(L+L_0)}, \quad Q_c = 0.6n_0 R_0 L/n \leqslant 500$
auxiliary haulage Z_6	$\dfrac{N(L+L_0)(K_0+K_c)Y_7}{2} + \dfrac{SK_c Y_8}{2a_0}$
faces movement Z_7	$\dfrac{a_0 n(J_6 + Y_0)}{R_0 S}$
face operation Z_8	$\dfrac{350nb(a_1 + a_2/L)}{R_0 V_0} + \dfrac{350nK_f GTa_3}{R_0 LV_0} + \dfrac{n(E_1/T_1 + E_2 L/T_2 + E_3 L/T_3)}{R_0 LV_0} + C_r$

Table 2 Change the variable value

rises layout			block entrys layout				connection roadways		
1	2	3	4	5	6	7	8	9	10
$J_2 = J_1$		$J_1 = (2n-1)J_2$ $\gamma = \alpha$	$J_3 = 2J_3$		$J_3 = 0$ $S_0 = S$		$\beta = (\alpha+\beta)/6.5$	$\beta = 0$	
$W_2 = W_1$		$W_1 = W_2$	$W_4 = W_3$ $W_6 = 0$	$W_6 = 0$	$W_3 = 0$ $W_4 = 0$ $W_5 = 0$	$W_4 = 0$	$W_7 = W_7/2$ $W_8 = W_8/2$	$W_6 = 0$	
$Y_2 = 0$	$Y_2 = 0$	$Y_1 = 0$	$Y_5 = 0$	$Y_5 = 0$	$Y_3 = 0$ $Y_4 = 0$	$Y_5 = 0$	$Y_6 = 0$		$Y_6 = 0$

β --- inclination of connection
     roadways, degree;

K_o --- recovery of extraction face, %;

K_j --- output coefficient in driving
     coal roadway;

K_g --- waste coefficient in working
     area;

K_c --- auxiliary haulage coefficient in
     working area ;

K_f --- load coefficient of the ex-
     traction equipment ;

V_j --- annual advance speed of the
     drivage face, m/yr.

T --- getting coal time, Hr./Day

γ --- inclination of the conveyer rise,
     degree ;

G --- total electric power of the ex-
     traction face dequipment, KW.

C_f --- air intake, $(m^3)/(min-t)$

F --- average net section of roadways,
     m^2 ;

c, d --- coefficient with relaton to
     extraction face length

C --- constant,
     single wing working area, C =1
     double wings working area, C =2

$R_o = \Sigma m \cdot r \cdot K_o$

$R = \Sigma m \cdot r \cdot K$

a1 --- the number of extraction workers
     with relation to extraction
     face length, man/(day-m)

a2 --- the number of extraction workers
     no with relation to extraction
     face length, man/day

T1 --- miner amortization time, yr.

T2 --- face conveyer amortization time,
     yr.

T3 --- face support amortization time,
     yr.

Unit Costs:

J1 --- drivage cost of rock rise, yuan/m

J2 --- drivage cost of seam rise, yuan/m

J3 --- drivage cost of block centra-
     lized rock entry, yuan/m

J4 --- drivage cost of block seam entry,
     yuan/m

J5 --- drivage cost of connection
     roadway, yuan/m

J6 --- drivage cost of open-off cut,
     yuan/m

J7 --- drivage cost of block chute,
     yuan/m

J0 --- drivage cost of station and
     chamber, $yuan/m^3$

W1 --- maintenance cost of rock rise,
     yuan/(yr.-m)

W2 --- maintenance cost of seam rise,
     yuan/(yu.-m)

W3 --- maintenance cost of block cen-
     tralized rock entry, yuan/(yr.-m)

W4 --- maintenance cost of block cen-
     tralized seam entry, yuan/yr.-m)

W5 --- maintenance cost of advance
     seam entry, yuan/(yr.-m)

W6 --- maintenance cost of block
     seam entry. yuan/(yu.-m)

W7 --- maintenance cost of connection
     cross-cut, yuan/(yur.-m)

W8 --- maintenance cost of inclined
     connection roadway, yuan/(yr.-m)

W9 --- maintenance cost of block chute,
     yuan/(yr.-m)

Y1 --- haulage cost of centralized con-
     veyer rise, yuan/(t-m)

Y2 --- haulage cost of seam conveyer
     rise, yuan/(t-m)

Y3 --- haulage cost of block centralized
     entry, yuan/(t-m)

Y4 --- haulage cost of advance seam
     entry, yuan/(t-m)

Y5 --- haulage cost of seam entry,
     yuan/(t-m)

Y6 --- haulage cost of connection
     cross-cut, yuan/(t-m)

Y7 --- auxiliary haulage cost of track
     rise, yuan/(t-m)

Y8 --- auxiliary haulage cost of
        entry, yuan/(t-m)

YO --- equipment movement cost of
        extraction faces, yuan/(t-m)

Ct --- material cost of the extraction
        face, yuan/t

a3 --- electric cost, yuan/(KW-Hr).

E1 --- miner equipment and overhual
        cost, yuan

E2 --- face conveyer equipment and
        overhual cost, yuan

E3 --- face support equipment and
        overhual cost, yuan

Taking long-wall mining along strike
as an example, the flow chart for cost
analysis program is shown in Fig.3. The
outputs of the program are: the costs per
ton coal mined from the working area for
different alternatives and parameters
(ΣZi), the minimum cost among them (Zmin)
and their relevant parameter values.

For comparing different extraction
technology, different initial conditions
and cost values are used as input data.

There are is only one alternative which
has the minimum cost obtained from the
computer as output. However, we take
those who have nearly minmum costs as
economically feasible alternatives by
making graphic analysis. The final decision will be made by considering other
factors. Sensitivety alalysis can also
be taking by means of graph.

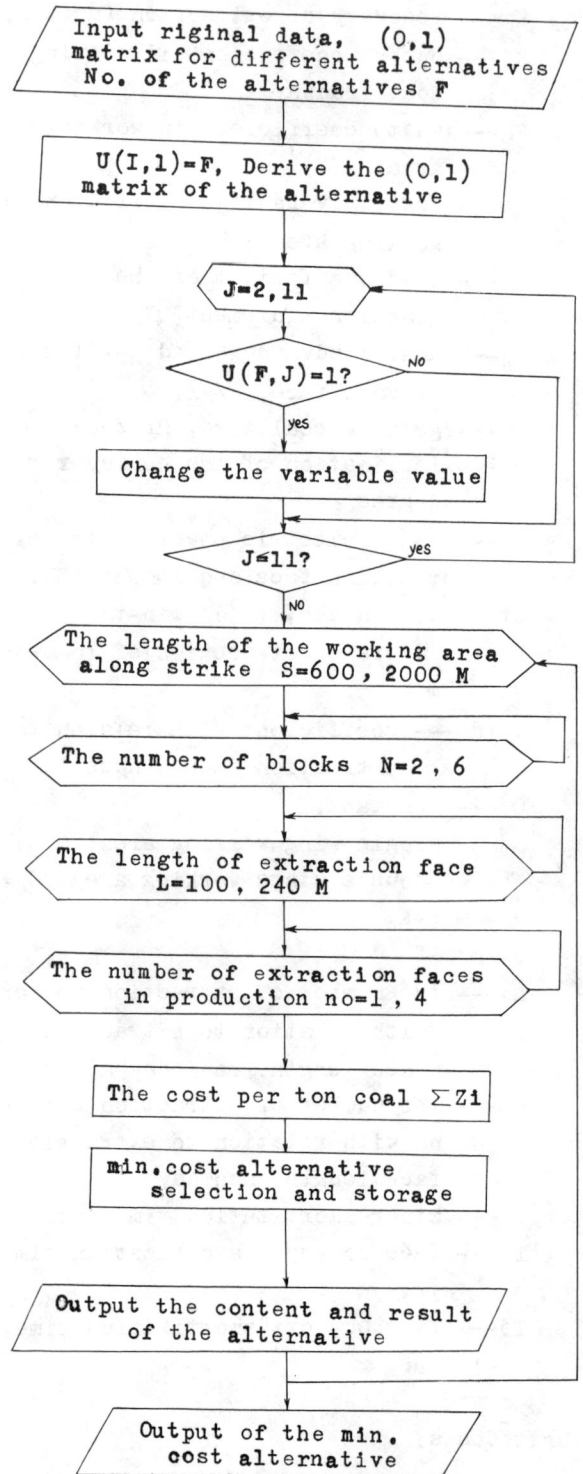

Fig.3 The flow chart for cost
      analysis program

CASE STUDIES

This model has been used to optimize the working area plan in six coal mines with different geological and mining conditions. The main results for long-wall mining along strike are shown in Fig.4. Herein the relationships of the extraction face length (L), working area length along strike (S), the number of extraction faces in production (n_d) and the productivety of the working area VS. the cost per ton coal mined from the working area are reflected. The main conclusions drawn from these case studies are as follows:

1. Different alternatives can be compared and the minimum cost or relatively low cost alternatives can be obtained by mean of this model. The comparision could be made between different roadways, different parameters or different mining technology.

2. Taking the working area as a complicated system, the model gives the comprehensive optimum result reflecting the interaction of varied factors instead of the conventional mathematical method with isolated factor analysis.

3. From the case studied, the different of the cost per ton coal between different alternatives is obvious, usually more than 10%, sometime over 50%. This proves the necessity of optimizing the working area plan. The economic effect cannot be ignored. If 0.1 Yuan/t more profit could be reached, the overall benefit would be significant.

4. The model is a general-purpose one. The applied cases are diversified: single and multiple coal seams with the dip from 8° to 30°, alternatives with single wing and double wings mining, roadway layout in single seam and combined layout, etc. For different geological and technological conditions, the relevant cost can be drawn once the specified initial data has been input. Selected variables and their range can be decided by analyzing the practical situation. For instance, the inclined length of the working area may have been defined for a mine in production, thus the relation between the block (district) number and the length of the extraction face is fixed.

5. The working area length along strike and the productivety usually have their economically feasible values. On the contrary, the extraction face length does not have a corresponding valley cost within the technical feasible range in most cases and is taken one of the boundary value of the technical feasible range as the economically feasible value. The effect of roadways layout mainly depends on the operation cost, especially drivage and maintenance cost of the roadways.

6. Different mining conditions cause different results. It is recommanded to do calculation and analysis with the model to fit the specified condition instead of giving a general solution of some parameters.

ANALYSIS OF SOME KEY FACTORS

1. The relationship between the advance speed (VO) and the length of extraction face gives strong influence upon the running result. By making regression analysis of the practical data from a number of underground coal mines, an expression is derived:

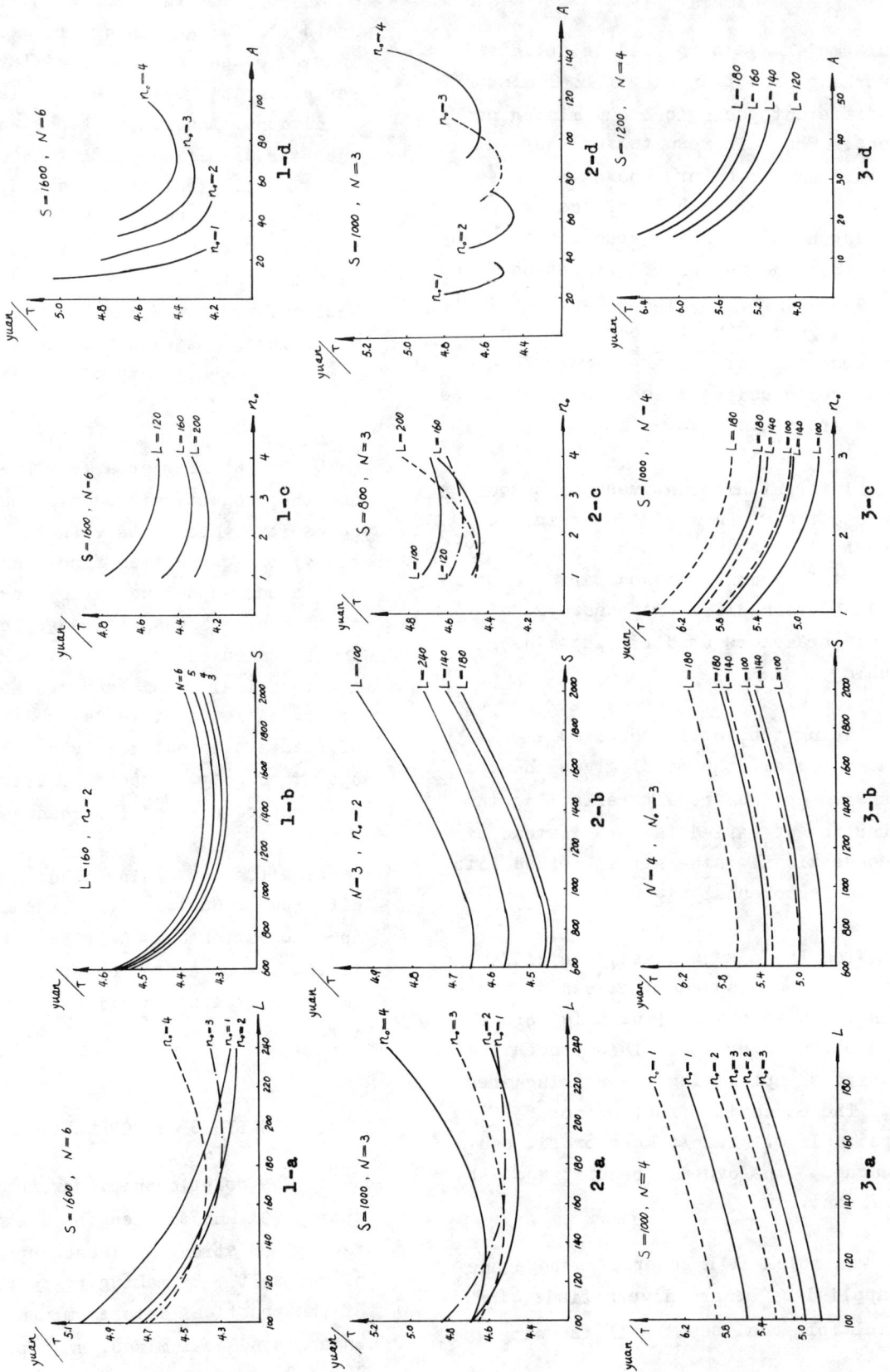

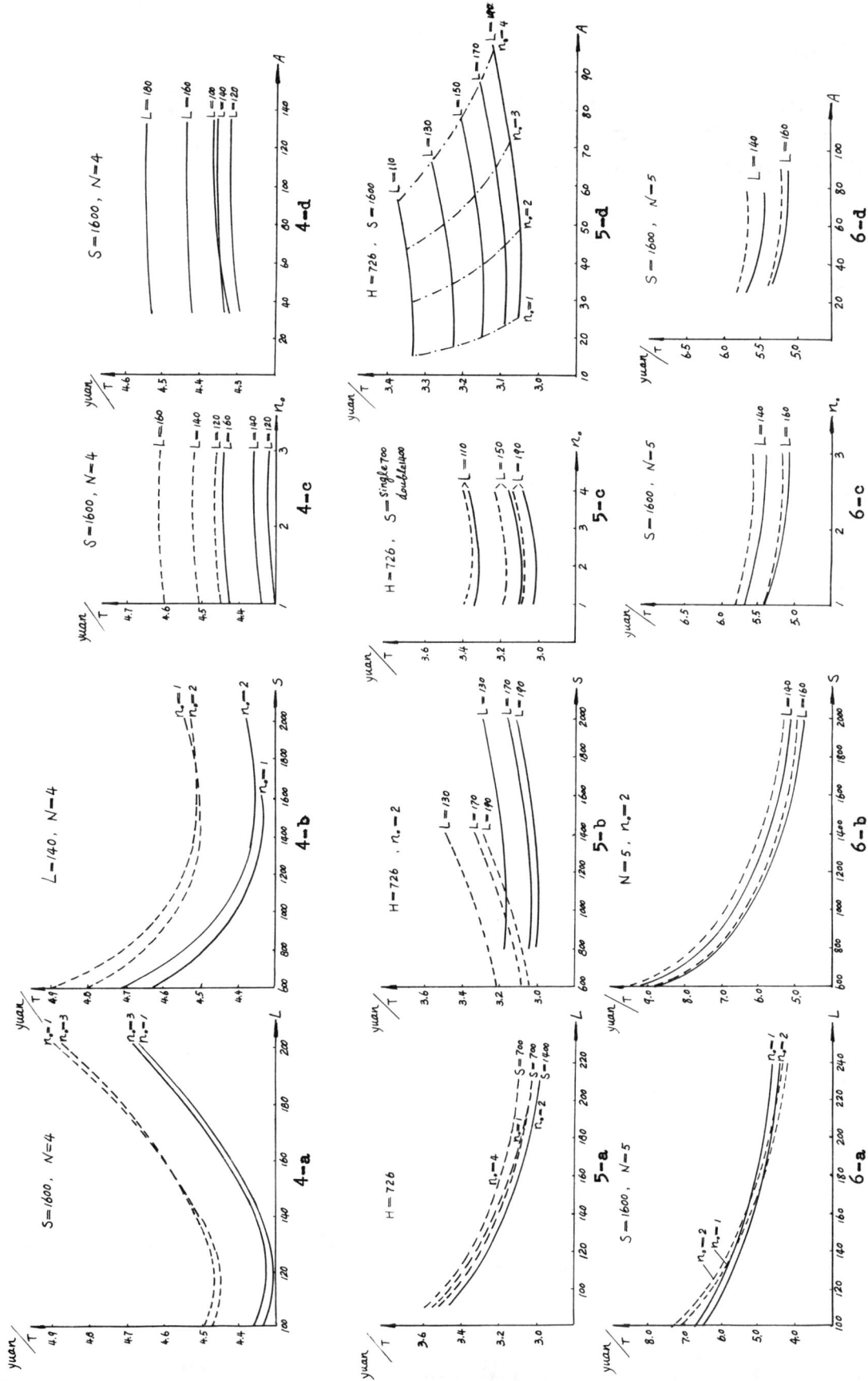

Fig. 4   1-Huaibei Yangzhuang Mine; 2-Yanzhou Nantun Mine; 3-Huainan Xinzhuangzi Mine;
4-Xuzhou Quantai Mine;   5-Xinwen Suncun Mine;   6-Jixi Donghai Mine;

$$VO=f(L)=1/(c+d \cdot L)$$

The constants c and d are various for different coal deposit condition and different extraction technology. However, further study should be engaged in.

2. The accuracy of the unit costs greatly influence the exactitude of the optimization result and the practical value of the model. The more accurate the data of basic parameters, the more reliable the running result.

3. As mentioned above, graph analysis makes the optimization more convenient. Rapid developing computer aided graphic system could support this optimization model to be more efficient.

SURFACE MINING APPLICATIONS

Chapter 30

STRIPSIM: AN INNOVATIVE SURFACE MINE PLANNING AND RECONTOURING SYSTEM

Jonathan R. Baker, Graham A. Mathieson, and Robert J. White

Pincock, Allen & Holt, Inc., Denver, Colorado*

Abstract. The recontouring and reclamation of mined overburden spoil to acceptable standards represents areas of significant activity and major expense to any surface coal mine today. Pincock, Allen and Holt, Inc.'s Strip Mine Simulation and Recontouring Analysis System (STRIPSIM) represents an innovative computerized mine planning system which aids the engineer in finding practical, economical solutions to these problems. The program lays out the overburden bench and spoil lift geometries for accurate volumetrics and spoil placement. The user can vary spoil topographies, pit geometries, and mining methods when simulating the mining of the property. STRIPSIM provides volumetric, equipment and cost output which allows the engineer to determine the least-cost mining approach to achieve a given recontoured topography.

With the increased sensitivity to post-mine reclamation required today, backfill planning for surface coal mines demands a high degree of emphasis in order to minimize adverse economic impacts. STRIPSIM allows the engineer not only to check the feasibility of a particular recontouring plan, but also to estimate the expected costs in achieving the recontouring plan. STRIPSIM's fundamental concepts are applicable to every major form of continuous surface coal mining.

INTRODUCTION

With the increased pace of computerization in mining brought about at least partially by the personal or micro-computer, a relatively large number of software packages targetted at solving specific problems in the industry are becoming available. The purpose of this paper is twofold: first, to describe the development of one such software package named STRIPSIM; and second, to illustrate the need for engineers and software designers to stick to the basics when developing specialized software for the mining industry. We believe that the primary reason for STRIPSIM's success and acceptance by the surface coal mining industry is that the package was designed upon basic mining principles to solve complex mining

*Formerly Golder Associates, Inc., Denver, Colorado

problems. Moreover, it is a tool which, when properly applied, can and already has resulted in considerable cost savings to coal mine operators.

Powder River Basin Mining Practices

STRIPSIM began its life more than two years ago as the product of a group of Denver-based consulting engineers working closely with a client who was intimately familiar with the specific problems to be analyzed. The client operates a large surface mine in the eastern portion of the Powder River Basin of Wyoming, a mine which currently produces close to 10 million tons of coal per year from a coal seam which averages 40 to 60 feet in thickness. The property contains several hundred million tons of coal over many square miles and is quite similar to most of the other dozen mines in the area. Overburden depths on the property range from about 20 feet to almost 300 feet, and most mining is done by large electric shovels and 170-ton trucks.

Most of the mines employ some form of terraced haulback mining consisting of continuous removal of overburden and coal by trucks and shovels, with the overburden being hauled back into the previously mined void as backfill (see Fig. 1). Since a typical property might exceed 10,000 ft. by 20,000 ft. in dimension, numerous mining passes are usually made through a property. Although the geology and the basic mining equipment at most of the mines in the area are similar, the specific mining techniques employed are not. Mining face lengths vary from 1500 feet to over 6000 feet. The placement of overburden haul roads in the pits ranges from one-sided haulage to double-sided haulage with or without additional ramps across the middle of the pit. The practice of leaving coal wedges or "fenders" to reduce the amount of rehandled overburden material ranges from no coal being left to almost full seam-height wedges being left, etc.

While some basis for the great disparity in operating practices can be attached to local geologic or other constraints, an historical review of mining in the Basin tends to suggest that the past experience of company personnel has

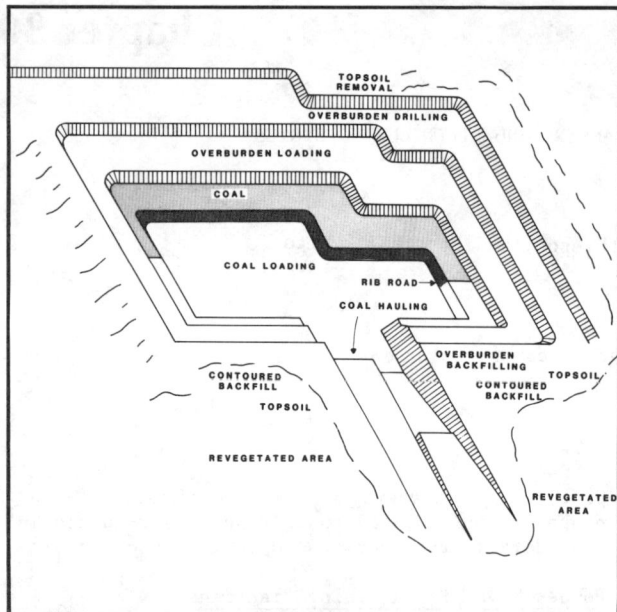

FIGURE 1. PERSPECTIVE VIEW OF TERRACED PIT

played a greater role. For example, former midwest dragline miners, when faced with the great amount of latitude available in the design of new mines in the Basin, developed long pits similar to the dragline mines of the Illinois Basin. Former hard rock engineers, on the other hand, accustomed to mining large porphyry deposits in the southwest, designed narrow pits due to their work which required selectivity and relatively small working areas. Based upon this large variety of pit designs existing in the Basin today, one could conclude that insufficient analysis has been made on the question - what is the best pit design?

Mining reclamation practices in the Basin are not quite as varied. Most operations have been in production at least five years and have consequently developed sound reclamation practices which include stripping of topsoil, segregation of toxic waste material, reestablishment of the mined land to Approximate Original Contour (AOC), spreading of topsoil and revegetation. The reclamation area of most concern to operations and planning personnel is the recontouring of the mined land to AOC at minimum cost (including the reestablishment of proper drainage). Typically, the mine operator must place mined overburden precisely in accordance with a permitted reclamation plan which, when it was developed (usually by environmental engineers), may have had little resemblance to what could be practically achieved in reality. This situation has resulted in numerous instances of an operator becoming "spoil bound," or not being able to mine efficiently because of a lack of space in which to place the mined overburden.

The Problem

Our client, having operated for several years in low cover, was facing deeper cover and a more competitive market. Motivation, therefore, existed toward fine-tuning the operation through a determination of the best or "optimum" pit design and long-range mine plan. Our client had two basic computing needs: First, they desired a computer tool with the ability to quickly assess the impact of any change within the pit design parameters, especially with respect to volumetrics, equipment needs, and cost. Second, they wanted to have a system which could accurately simulate the placement of overburden into backfill for a given mine plan, in order to select a plan which would best match an existing permitted reclamation plan. In addition, such a system should be able to provide realistic input into a future reclamation plan so that the rehandle of mine spoils would be minimized. The obvious intent here was, and still is, to minimize the relatively significant mining costs which are associated with achieving AOC. These two needs were initially considered mutually exclusive of each other.

DESIGN CONCEPTS

Given our client's wishes concerning the capabilities of the system, the design team undertook, with the help of our client, a thorough study of mining in the Powder River Basin in order to identify basic mining constraints and objectives which would have to be modeled. Some of the more significant concepts which were identified along with their origins are discussed in the following sections.

Pit Design

Some of the more significant objectives in pit design are:

o to ensure that a proper balance is maintained between exposed coal and overburden stripping such that coal is always available for loading.

o to minimize operating costs through "optimizing" haulage distances and reducing rehandle of overburden, allowing the equipment to be utilized in the most efficient and productive manner.

o to allow flexibility in overcoming localized problems such as building or moving roads, undulations in the coal seam, etc.

Some of the key design parameters which must be addressed in meeting these objectives are illustrated in Figs. 2 and 3. They are as follows:

o Mining Face Length
o Depth of Overburden/Number of Benches
o Thickness of Coal
o Amount of Exposed Coal in Inventory Within the Pit

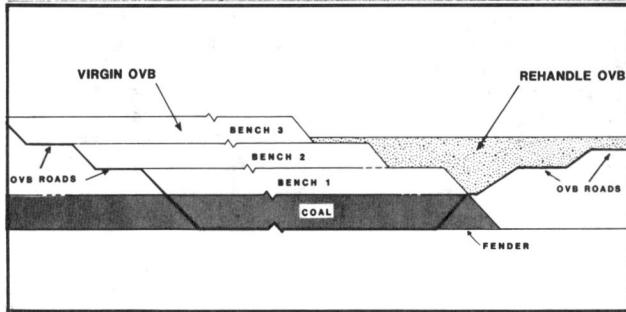

FIGURE 2. VERTICAL CROSS-SECTION OF MINING FACE

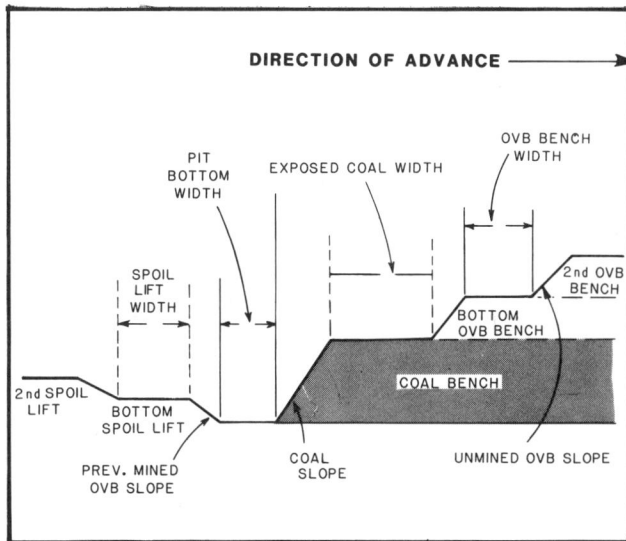

FIGURE 3. LONGITUDINAL SECTION OF TERRACED PIT

o Working Room
o Road Widths and Grades
o Spoil Lift Heights
o Slope Angles

Through study of these basic pit design concepts, it does not take long to appreciate the high level of dependence between individual pit design parameters. For the purpose of illustrating this point, let us examine the following question: "Given that everything else is equal in a particular pit, how long should the face be?"

Obviously, the longer the face length, the longer the truck haul to move material from the face around the sides and into the backfill dumps and, hence, the greater the haulage costs. However, not all things are equal! For example, the longer the mining face, the smaller the width of uncovered coal required to maintain a given inventory of mineable coal and, therefore, the shorter the haulback distance for the trucks. Also, the longer the face, the smaller the percentage of overstrip (which is associated with

haul roads and benches on the pit sidewalls) relative to total overburden mined, thus effectively lowering the amount of rehandled overburden and thereby reducing the actual stripping ratio over the life of the property as well as lowering the cost per ton of coal.

If this is the case, what then would be the tradeoff between:

o A pit with a short face and single-sided haulage versus a pit with a long face and single-sided haulage?

o What about a pit with a long face and double-sided haulage (with or without haulage through the middle of the pit)?

o What about the impact of placing coal roads in the center of a pit versus on the side?

o What about steep coal ramps versus flatter ramps?

o What impact would various pit design parameters have on spoil backfill room and spoil height?

o What amount of a coal wedge can be economically recovered?

o What would be the cost of expanding the in-pit coal inventory?

o How much could overstripping be reduced by only hauling overburden back on every other bench instead of every bench?

o What design parameters should be changed in low overburden versus high cover situations?

It is apparent that to accurately answer all of the above questions manually would require large amounts of manpower and would be extremely tedious and time-consuming, not to mention error-prone. It is not hard to understand why our client wished to automate the pit design process in order to quickly assess the economic impacts of varying design parameters.

Long-Range Planning

Some of the more significant long-range planning objectives are:

o to ensure the removal of sufficient overburden in order to sustain production of a specified tonnage of coal which meets specific quality constraints.

o to identify the best (or at least a better) mine plan for the exploitation of the property throughout its life.

o to identify a recontouring plan which allows the coal to be extracted with a minimum amount of spoil rehandle or additional haulage.

Traditionally, major emphasis has been placed on coal extraction planning and economics versus backfill planning. Substantial effort has usually been expended in the development of a path or sequence through the property which will best allow the producer to meet the given set of coal contracts at the lowest production cost. Typically, given the relatively constant coal thickness and quality in the Powder River Basin, these coal-oriented plans have started in shallow cover along the coal subcrop and progressed gradually into the deeper cover areas.

Only after completion of a long-range mining plan are backfill plans usually addressed. Typically, a simplistic "cookie-cutter" approach has been utilized which amounts to "lifting" the overburden which vertically overlays the coal being mined, removing this coal, swelling the overburden, and then placing the swelled material directly back into the void left by the coal. Many Powder River mines have utilized this technique in developing reclamation plans as this "cookie-cutter" approach is accepted practice in many dragline operations and is relatively easy to use. To date, most of the Powder River mines which have employed this technique have not fared too badly, as their mining has not proceeded beyond shallow cover areas. However, as mining progresses into deeper cover on these properties, significant volumes of overburden materials will need to be rehandled or at least hauled substantial distances in order to achieve reclamation plans developed from the "cookie-cutter" post-mine topography.

An example which illustrates why this can happen follows. A theoretical mining path is developed for a property (see Fig. 4). Fig. 5 contains the unmined surface topography. Mining boundaries of panels 1 and 2 are indicated with the direction of mining marked. Fig. 6 shows the same area as a post-mined topography generated by the "cookie-cutter" technique. Notice that other than a general lowering of elevation, the topography including drainage appears almost unchanged. Figs. 7 and 8 depict the same tract following a more realistic simulation of each mining pass. Notice how portions of the hills and valleys have been displaced in opposite directions. One should also note the significant impact that adjacent panels can have on the current mining pass due to the sometimes large amount of overstrip on the pit sides.

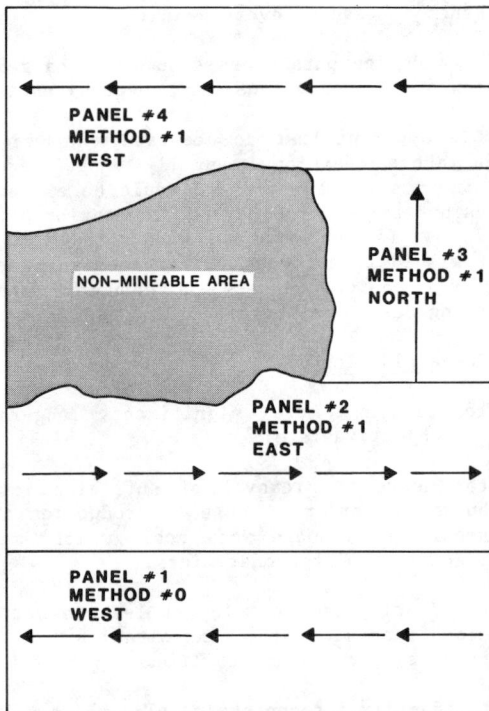

FIGURE 4. TYPICAL PANEL LAYOUT

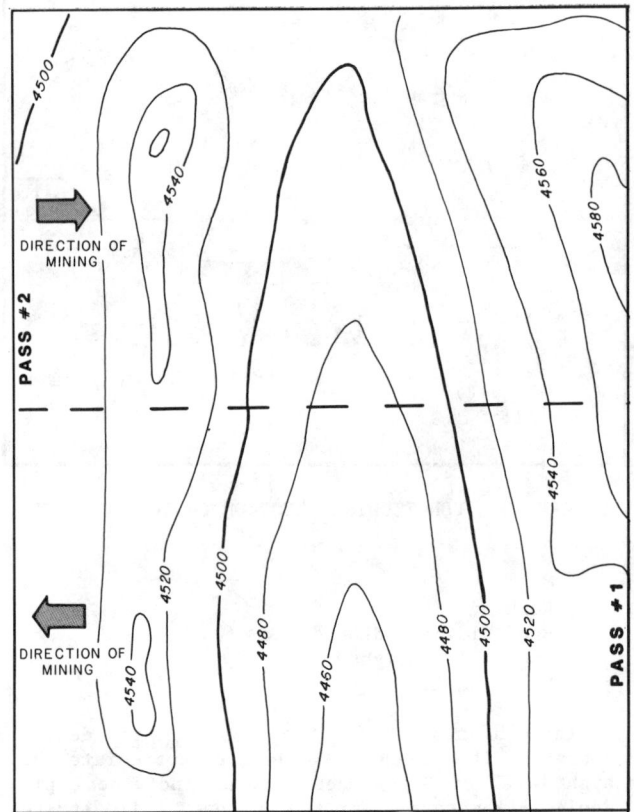

FIGURE 5. UNMINED TOPOGRAPHY

One can readily observe the significant differences which exist between the two post-mine topographies shown in Figs. 6 and 8. If a mine operator were forced to backfill and recontour according to Fig. 6, substantial volumes of material could be out of place relative to where the material should be placed under a minimum cost haulback plan. This could result in a significant increase in overburden haulage costs measurable in cents-per-ton. On the other hand, one should also note how much more complicated

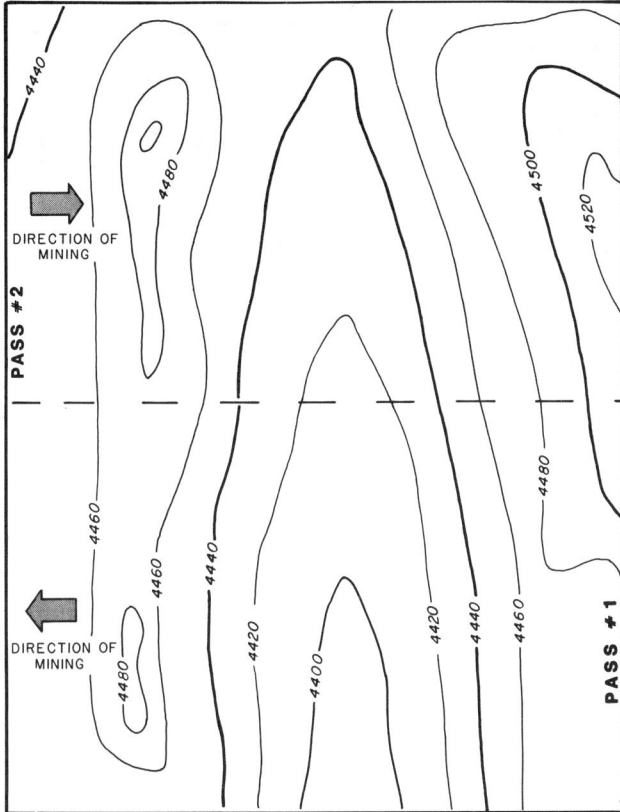

FIGURE 6. "COOKIE-CUTTER" POST-MINE TOPOGRAPHY

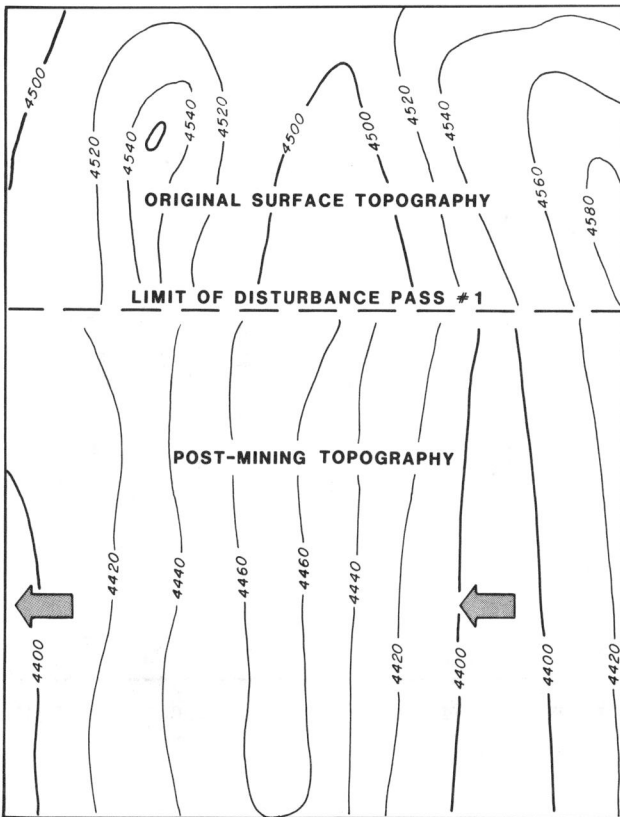

FIGURE 8. PANEL #2 POST-MINE TOPOGRAPHY

the task is of generating a realistic post-mine backfill topography and then designing a proper drainage system while achieving AOC. Three to six months is usually required to manually generate a realistic backfill map for just one particular mine plan. Usually, once an engineer has completed one backfill map, he is extremely reluctant to make any changes or to consider other alternatives due to the amount of work involved. Obviously, this is not an ideal situation as conditions constantly change which affect mining and backfill plans. An opportunity emerges to utilize computer methods for simulation of the mining process, allowing quick evaluation of the results, "optimization" of the mine plan, and minimization of operating costs.

The mining-related problems associated with recontouring and long-range mine planning are:

o How would one actually generate an accurate post-mine topography in a reasonable amount of time?

o How should one select an optimal recontouring plan for a given mining plan?

o How significant is the impact of adjacent panels on the panel of interest?

o Would the "optimum" backfill plan or the "optimum" mine plan actually guarantee an "optimal average plan?"

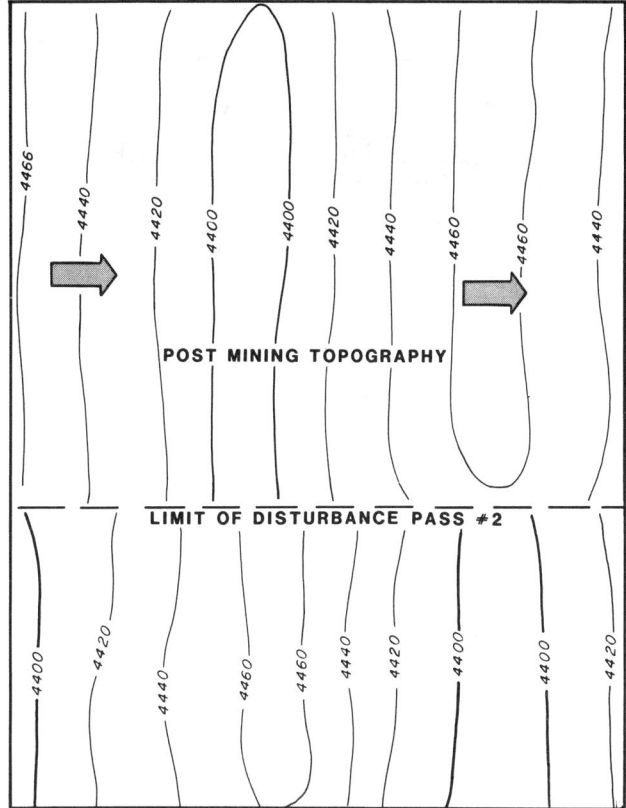

FIGURE 7. PANEL #1 POST-MINE TOPOGRAPHY

o What impact would a turn or change of direction have on a backfill plan?

o Should mining be oriented parallel to existing hills and valleys to overcome the reclamation problems associated with mining through these topographic features?

o Given a specific reclamation plan which is permitted and therefore "cast in concrete," what is the cheapest way to mine the deposit while maintaining coal specifications?

In attempting to identify a solution to this problem, our client (who had manually generated one backfill plan previously) first proposed a system involving the use of relationships such as overburden height versus displacement distance. They believed the problem to be too complex to accurately simulate with a computer. However, a much more advanced simulation tool in due course emerged.

STRIPSIM

The results of our work with the client was the development of the so-called "Surface Coal Mine Planning and Recontouring Analysis System" (STRIPSIM). STRIPSIM is an interactive computer software system incorporating modular design features. The system contains two basic modules designed to meet the short-range or pit design needs as well as the long-range planning needs, respectively. These two modules are referred to as PITPLAN and DISPLACE. The design and capabilities of these two modules are discussed in the following sections.

PITPLAN

PITPLAN was designed to allow the engineer to rapidly assess the effects of changing pit slopes, road widths, road grades, bench widths, face lengths, coal inventories, haulage strategies, etc. This is accomplished by placing almost every pit design parameter under user control. Fig. 9 contains a sample of PITPLAN's interactive input screen which clearly shows how everything from face length to slope angles is user-controlled. One should also note that PITPLAN allows the user to evaluate various haulage methods (currently six). These methods were selected so as to simulate all of the actual haulage methods currently being used in the Powder River Basin, as well as a few innovative variations of those methods. An illustration of one of these methods--the "so-called" Center Incline Method--is illustrated in Fig. 10; a brief description follows:

The Center Incline Method is a double-sided haulback technique with all overburden being hauled back on in-pit roads placed on both ends of the pit. The coal haul road is modeled as a double ramp system located in the center of the pit extending back through the spoil. A portion of the top overburden bench is allocated for periodic "plating" or filling of the coal ramp to allow it to move forward. For contrast, another

technique known as "Advance/Retreat" is shown in Fig. 11.

FIGURE 9. PITPLAN INPUT SCREEN

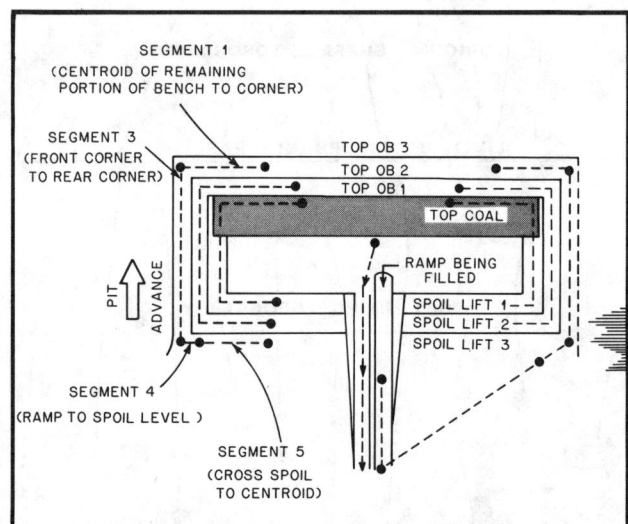

FIGURE 10. PLAN VIEW OF "CENTER INCLINE" METHOD

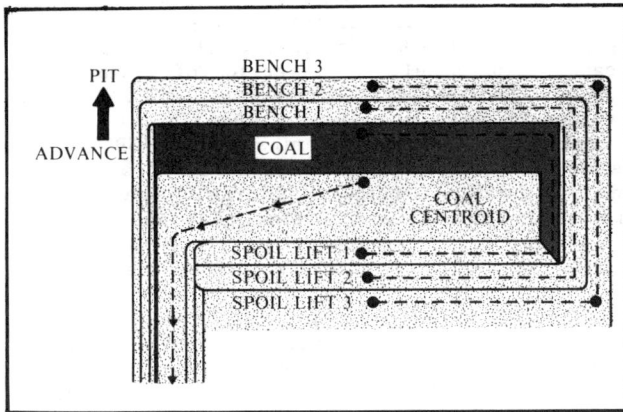

FIGURE 11. PLAN VIEW OF "ADVANCE" METHOD

The major differences between the six alternate haulage methods is in the number and placement of haul roads. PITPLAN automatically lays out pit geometry by taking the input parameters and accessing the appropriate haulback model. Then, based on simplified planar geometry, and the road placement criteria contained within the given haulback model; PITPLAN constructs the requested pit geometries, calculates volumes, and generates haulage profiles by bench.

PITPLAN first lays out in vertical cross-section the virgin and rehandle portions of the overburden face (see Fig. 2). The road and catch bench placement as well as the virgin/rehandle contact are controlled by the haulback model; all other parameters are specified by the user. Cross-sectional areas are calculated for the trapezoids representing the coal seam and the overburden bench segments. These areas are later converted to volumes based on the feet of advance of the particular pit necessary to achieve a required coal production.

Upon completion of cross-sectional area calculations for the advancing face, PITPLAN lays out the vertical cross-section of the spoil or backfill side geometry. The overburden areas for each bench are then multiplied by the swell factor, and the appropriate quantity of swelled overburden is used by PITPLAN to calculate the height of each spoil lift, such that the area in the lift is just equal to the swelled overburden area.

Once all the geometries, volumes, and lift heights are calculated, PITPLAN generates overburden haul profiles and assigns the appropriate volumes to each profile. Every haul profile within the pit is broken into a maximum of five haul segments. These haul segments are described as follows (see Fig. 10).

o Segment 1 - Flat haul from the centroid of the overburden face to the center of the haul road on the side of the pit.

o Segment 2 - Face ramp located in the overburden face extending from the center of the haul road down to the bench toe at specified grade. If face ramping is used, segment one represents the haul from the centroid of the face with respect to the toe of the face ramp.

o Segment 3 - Flat haul extending back from the pit measured from toe of overburden face to the crest of the spoil lift.

o Segment 4 - "Adjustment" ramp located in spoil to allow trucks to either climb up or drop down onto their respective spoil lifts from the corresponding bench. For simplicity of calculation, the ramps start from the center of haul road and use the specified slope.

o Segment 5 - Flat haul from the end of the adjustment ramp to the centroid of the spoil lift with respect to the end of the ramp.

Face ramps are used to haul overburden from one level up to the next highest level. This eliminates the need for haul roads on the bench being ramped from and creates a narrower pit.

The standard PITPLAN report has two pages: one with the input parameters used in the run and one with the resulting yardages and distances. This second page (see Fig. 12) starts by listing every haul simulated with the bank cubic yards of virgin material, the bank cubic yards of rehandle material, and the length in feet and grade of each segment of that haul. Below this is an explanation of any special hauls used in the design. The report then lists the total overburden height, total coal height, overburden stripped, coal uncovered pit advance, stripping ratio, and spoil height.

FIGURE 12. PITPLAN REPORT

While the information presented in the PITPLAN report is important, it is not too meaningful as presented. We therefore developed an interface routine which would automatically take the PITPLAN-generated information and arrange it in a form suitable for input into a haulage simulation routine. The latter was designed to estimate the performance of haulage units under various conditions. The haulage simulator utilized by PITPLAN is the Euclid simulator, developed and owned by Euclid, Inc. The PITPLAN/Euclid interface allows the engineer to quickly assess the impact of yardage or haulage changes associated with a given pit design. Fig. 13 contains a modified final summary report generated by the Euclid program which itemizes haul times, productivities, and costs for each haul within the given pit.

EUCS --SUMMARY COMPARISONS-- FORMAT IN U.S. UNITS --EUCLID VEHICLE STUDY PERFORMED FOR

BENCH NO.	HAUL TIME IN MIN.	RETURN TIME IN MIN.	FIXED TIME IN MIN.	CYCLE TIME IN MIN.	CYCLES PER 50.MIN HOUR	UNIT PROD/ 50.MIN HOUR	REQD PROD IN BCY	FLEET SIZE	UNIT O&O COST $/H	TOTAL UNITS @ 85% AVAIL.	FLEET COST/BCY @ REQ. PROD.	FUEL CONSUM IN GAL/H	MAX CALC. TIRE TMPH	OVERBURDEN HAULAGE COST PER TON OF EXPOSED COAL
1	1.20	1.20	4.00	6.40	7.81	693.9	10590.	0.00	130.28		0.103	14.5	142.4	0.000
2	1.54	1.46	4.00	7.00	7.14	635.0	4563220.	1.80	132.31		0.270	16.1	168.0	0.123
3	1.79	1.63	4.00	7.42	6.74	599.1	9134170.	3.81	133.94		0.254	17.4	184.8	0.232
4	2.19	1.91	4.00	8.10	6.17	548.7	10120760.	4.61	135.82		0.280	18.9	202.8	0.283
5	2.60	2.19	4.00	8.79	5.69	505.7	11107340.	5.49	137.32		0.303	20.1	218.4	0.337
6	2.96	2.83	4.00	9.79	5.11	453.9	3770920.	2.08	137.65		0.361	20.4	248.7	0.136
7	4.42	4.43	4.00	12.85	3.89	345.9	1526180.	1.10	140.23		0.544	22.5	281.6	0.083
PIT COMPOSITE					5.99	532.3	40233180.	18.89	136.00	23	0.297	19.1	210.0	1.195

**** NEITHER PINCOCK, ALLEN & HOLT NOR EUCLID CAN OR DOES REPRESENT OR WARRANT, EXPRESSLY OR IMPLICITLY, EITHER THE ACCURACY OF THE RESULTS OF THIS PROGRAM OR THAT EUCLID OR COMPETITIVE EQUIPMENT WILL IN FACT ACHIEVE THE PERFORMANCE INDICATED ON ANY JOB TO WHICH THE RESULTS OF THIS PROGRAM MIGHT APPLY.

FIGURE 13. EUCLID REPORT

Through the use of PITPLAN then, an engineer can quickly assess the impact of a given design change in terms of such common denominators as stripping ratio, haulage costs per ton of exposed coal, and equipment requirements. The engineer is also able to rank various alternate pit designs by use of the same criteria.

DISPLACE

The DISPLACE module within STRIPSIM was designed as a tool in to address long-range coal mining planning needs. In order to properly handle these needs, substantial amounts of data are required as input for DISPLACE. This input data can be classified into the following three groups - geologic/topographic data, basic mine plans, and a desired final post-mine topography.

With regard to the initial geologic data required, DISPLACE expects to receive three gridded files in a northing, easting, and value format for the specified area of interest. These three files contain the existing surface topography, the base of coal elevations and a cumulative coal thickness or isopach. DISPLACE has built-in interactive interfaces to facilitate the transfer of this gridded data and can accommodate most gridded models generated by most coal deposit modeling software.

In order to input a basic mine plan, the user must supply a file containing digitized shapes which define the base of coal width and length of each mining panel, the direction of mining, the desired haulback method (i.e., per PITPLAN), and the panel sequence or number (see Fig. 4). Upon receipt of the digitized mine plan, DISPLACE in essence overlays the mine panels on top of the deposit model matrices. DISPLACE then automatically generates a series of mining "cuts" (each cut width being equal to the gridded cell width) through each of the panels and calculates the coordinates of each cut boundary point. DISPLACE then searches for all coal base elevation and coal thickness point values lying within each cut and averages those values respectively. An internal file is then generated containing the panel, method, and cut numbers for each cut as well as the average coal face length, thickness, and base elevation. In addition, indice coordinate boundaries of the cut are also generated. This file can be manually checked by the user to ensure that DISPLACE has been given the proper data. If the user wishes to run more than one mine plan or iterate on the first mine plan, all that is required is to digitize in the new or revised mine plan and rerun DISPLACE.

Additional information regarding file maintenance and mine planning specifics are input interactively through the main DISPLACE input screen (see Fig 14). The key item of importance in this screen is that pit design parameters such as slope angles and spoil stacking heights can be modified for various panels. This allows the user the necessary flexibility in addressing local geologic or environmental problems associated with a given area of the property.

Also of importance is that a DISPLACE simulation can be stopped or started in any part of the property. This provides the user with the ability to simulate multiple pits or to hone in on problem spots without wasting time or computer resources. Other important features include a boxcut option for opening up new pits as well as the ability to input coal production targets.

Once all of the necessary data is input, DISPLACE proceeds to simulate the mining of coal and overburden as well as the bench-by-bench displacement of that overburden into the backfill void. The procedure utilized by DISPLACE for displacing overburden into backfill is not a set of curves, or rule of thumb as originally proposed. On the contrary, DISPLACE, utilizing similar subroutines to PITPLAN, reads a portion of the pit design parameters from a parameter file and part from the deposit/mine plan model's information file, and then performs the pit layout, volumetric, backfill, and some haulage calculations similar to PITPLAN. However, DISPLACE does not just look at one cut at a time as in PITPLAN, nor does it ignore how that cut is associated with the rest of the deposit. Instead, DISPLACE has a sophisticated system for tracking the progress of backfill cut after cut after cut. This system enables DISPLACE to remember the precise shape of the void to be

```
  RUN DISPLACE
IS THIS A DEBUGGING RUN? (Y OR N)
N
INPUT A TITLE FOR THIS RUN.
DISPLACE RUN FOR TESTING
WHAT IS THE FILE NAME FOR THE PANEL REPORT?
PANREPT
************************ WARNING ************************
THIS TOPOGRAPHY FILE WILL BE WRITTEN OVER WITH THE
SPOIL TOPOGRAPHY, DESTROYING ITS PRESENT CONTENTS.
************************ WARNING ************************
WHAT IS THE NAME OF THE TOPOGRAPHY/SPOIL FILE?
TOPOFLR.DAT
WHAT IS THE NAME OF THE INFORMATION FILE?
INFOFI
ARE THE METHOD PARAMETERS COMING FROM A FILE OR FROM
THE TERMINAL? (F OR T)
F
WHAT IS THE PARAMETER FILE NAME?
DISPARAM
WHAT IS THE MAXIMUM BENCH HEIGHT ALLOWABLE?
55
DO YOU WANT TO SPECIFY PARAMETERS FOR INDIVIDUAL PANELS
OTHER THAN THE STANDARD ALREADY INPUT? (Y OR N)
N
THE FIRST BENCH WILL FLOAT WITH RESPECT TO A PLANE
DEFINED BY THE NEXT THREE QUESTIONS.  YOU NEED TO
SPECIFY THE STRIKE ( AN ANGLE CLOCKWISE FROM NORTH )
THE DIP ( AN ANGLE FROM THE HORIZONTAL DOWN ), AND THE
PLANE ELEVATION AT THE NORTHWEST CORNER OF THE MODEL.

WHAT IS THE STRIKE OF THE BASE PLANE?
0
WHAT IS THE DIP OF THE BASE PLANE?
0
WHAT IS THE ELEVATION OF THE BASE PLANE AT THE
NORTHWEST CORNER OF THE MODEL?
5000
WHAT IS THE MAXIMUM HEIGHT ABOVE ORGINAL TOPO YOU WANT
TO STACK SPOIL TO?
100
DO YOU WANT TO SPECIFY A HEIGHT FOR INDIVIDUAL PANELS
OTHER THAN THE STANDARD ALREADY INPUT? (Y OR N)
N
WHAT IS THE STARTING PANEL NUMBER?
2
WHAT IS THE FIRST CUT MINED IN THE STARTING PANEL?
1
WHAT IS THE FINAL PANEL NUMBER TO BE MINED?
4
WHAT IS THE FINAL CUT MINED IN THE ENDING PANEL?
5
IS THIS A BOXCUT START? (Y OR N)
Y
IS FINAL PIT WALL TO BE VERTICAL? (Y OR N)
Y
WHAT IS THE LONGEST CROSSPIT HAUL EXPECTED?
10000
DO YOU WANT TO USE TARGETING? (Y OR N)
N
DO YOU WISH TO USE THE TOPSOILER FEATURE? (Y OR N)
```

FIGURE 14. DISPLACE SCREEN

```
DISPLACE RUN FOR TESTING


   FOR PANEL    2  USING METHOD    5 :
   **********************************

   NUMBER OF CUTS MINED          =      9.

   TOTAL  ADVANCE                =   1395.00 FEET

              YARDAGES  OF  THE  OVERBURDEN  BENCHES   ARE
   ********************************************************************

     BENCH  1 PRODUCED    7.395 MILLION BCY WITH AVERAGE HEIGHT 15.54
     BENCH  2 PRODUCED   31.336 MILLION BCY WITH AVERAGE HEIGHT 45.00
     BENCH  3 PRODUCED   18.908 MILLION BCY WITH AVERAGE HEIGHT 45.93
     BENCH  4 PRODUCED    3.691 MILLION BCY WITH AVERAGE HEIGHT 20.16
   --------------------------------------------------------------------

   TOTAL OVERBURDEN -    61.330 MILLION BCY


   AVERAGE  COAL  FACE  LENGTH  =   4469.01 FEET

   AVERAGE  COAL  HEIGHT        =     60.20 FEET

   AVERAGE  OVERBURDEN  HEIGHT  =    121.76 FEET


   TONS OF COAL MINED           =     15.028 MILLION

   STRIPPING  RATIO             =      4.08
```

FIGURE 15. PANEL REPORT

filled, the amount of fill already in place, as well as the coal road configuration in existence at the backfill location. In short, DISPLACE performs an extremely realistic simulation of material movement and displacement that is more accurate than would be practical manually. DISPLACE also updates the topography matrix with a user-controlled lag (maximum haulback distance). This update allows a more accurate and realistic final result. DISPLACE also allows the user to make directional changes (turns), but does not attempt to precisely model the complicated mining which occurs in a turn.

DISPLACE generates several types of output consisting of panel-by-panel reports containing a summary of method parameters for the haulback technique used; a summary containing the feet of advance, haulback technique, overburden yardage by bench, coal tonnage, average face length, coal height, and overburden heights (see Fig. 15); and a report on any hauls longer than the minimum haul distance due to backfill constraints (no additional space available for backfilling at that specific cut location). DISPLACE also produces an additional report if production targeting is used. This report gives the tons of coal, feet of advance, overburden yardage removed, etc. for each targeted amount of coal, similar to the panel report. Targeting can be utilized for production scheduling. The output reports described above contain sufficient information to allow the user to accurately estimate mining costs including variable haulage costs for each panel or target.

In addition to the printed reports, a gridded surface output file is generated which can then be contoured and plotted. This surface file depicts post-mine backfill (minimum haulback placement). The topography created by this minimum haulback spoil placement is representative of a series of long "stair steps," which environmentally is unacceptable. As the least-cost placement of the spoil (from a mining viewpoint), this "stair step" topography is probably what would be left by cost conscious engineers lacking environmental awareness and/or regulations. An example of a contoured post-mine backfill topography of a portion of a deposit is shown in Fig. 16.

From the output data, especially the backfill map, one should be able to develop several post-mine recontouring options which would sufficiently address drainage, boundary, and other environmental constraints to be acceptable to the regulatory authorities. Obviously, the main goal of this design effort should be to identify an acceptable recontoured surface which would minimize additional haulages and/or rehandle of backfill material.

The approach taken by STRIPSIM in addressing this problem is to provide the user with the means by which to compare the backfill plan against different alternate recontouring plans. In this way, he is able to determine the amount

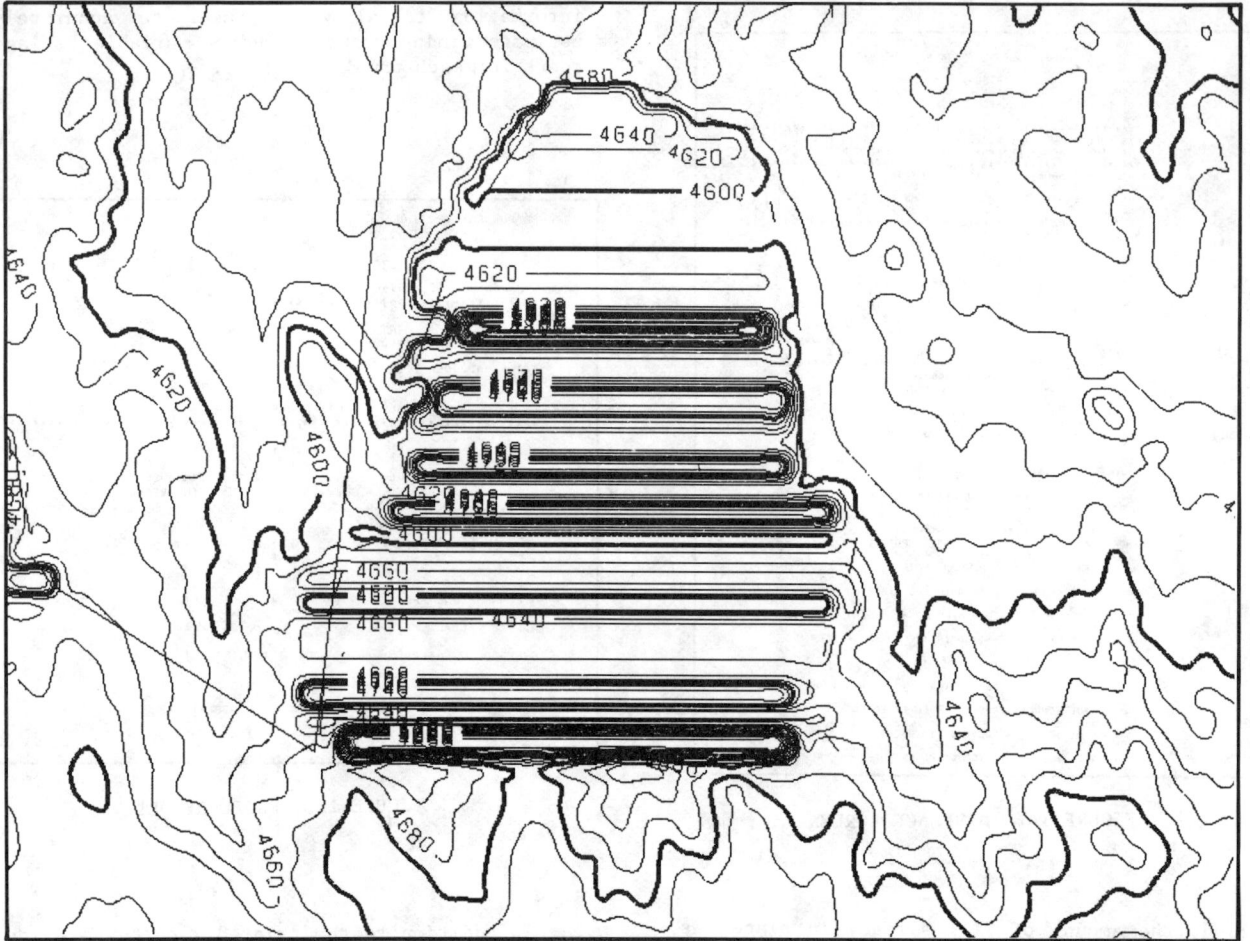

FIGURE 16.   STRIPSIM GENERATED BACKFILL MAP

of material which would have to be cut and/or
filled in order to transform the post-mine
backfill into a final AOC surface. A report
containing cut/fill volumes by panel for the two
surfaces compared is also output by DISPLACE.
While this cut and fill would probably never
occur as a discrete process at the end of the
mine's life, the cut/fill volumes serve as an
indicator of how well the recontoured surface
would blend with the hopefully least-cost mine
plan.

The cut/fill map is also useful in identifying
problem areas in mining and recontouring plans.
Once the major problem areas have been identified, one can modify either the recontoured
surface or the mine plan and repeat the DISPLACE
runs until the most desirable truly integrated
mining and recontouring plan is achieved. An
additional feature only recently added to
DISPLACE, is a topsoil tracking and planning
system which draws heavily from basic DISPLACE
logic.

STRIPSIM APPLICATIONS

Below is a list of currently identified applications for STRIPSIM.

Pit Design
o Identification of best haulback method for a
  given situation.
o Selection of most economical pit dimensions.
o Identification of relationships and inter-
  relationships between pit design parameters.
o Determination of economic coal mining
  limits.

Mine Planning
o Generation of accurate volumetric estimates
  by area or period.
o Development of realistic post-mine spoil
  topographies.
o Good estimation of variable stripping costs
  (i.e., haulage costs) over the property
  life.
o Comparison of various long-range mining
  strategies.
o Determination of economic benching limits
  for truck/shovel pre-benching at a dragline
  property.

Reclamation
- o Identification of the most suitable re-contouring plan.
- o Comparison of various recontouring and mining plans.
- o Generation of topsoil movement plans and volumes.

CONCLUSIONS

In this paper we have attempted to review Powder River Basin coal mining practices in general; to comment on the "optimization" of these operations; and, finally, to present an overview of a computing system specifically designed to be an invaluable tool in this "optimization" process, namely STRIPSIM.

The development of this package has represented a technically challenging and rewarding experience which has led us to conclude:

- o That the successful development of mining-related software fundamentally depends on a team of engineers gaining a thorough understanding of the problem to be solved at the outset.

- o That, by breaking complex mining problems into their lowest common denominator (through the use of basic principles), good algorithms and sound practical solutions can be developed which lead to a relatively straightforward exercise in code writing.

- o That STRIPSIM represents a major advancement in the development of practical mine planning software, especially in the area of mobile equipment placement of overburden into backfill.

- o That other coal mining situations such as Appalachia-type mobile equipment contour and mountaintop surface mines could also benefit from the type of applied research and software development on which the STRIPSIM package was founded.

The authors gratefully acknowledge the management support of their former employer, Golder Associates, with whom the STRIPSIM package was originally developed. In addition, we are deeply indebted to Jim Francis and Dean Larson, without whose inspiration and technical guidance in the development of such innovative software, STRIPSIM would not exist today.

Chapter 31

CALCULATING STRIPPING RATIOS IN STEEPLY DIPPING DEPOSITS

M. Beras and L. Adler

Department of Mining Engineering
West Virginia University
Morgantown, WV 26506

Abstract. Calculating stripping ratios is basic
to open pit mine design. A new approach is developed based on an existing method employing
length rather than area measurements. The existing
method has been digitalized which highly simplifies
computations.

Introduction

This paper is concerned with volume calculations
in steeply dipping deposits. The stripping ratio
is defined as the volume of excavated rock (waste)
per unit volume of excavated mineral. Stripping
ratio calculations are of primary importance in
open pit mine design since they determine pit
limits, mining system selection, haulroad layout
and other design factors. The stripping ratio is
used in the control and management of the pit
operations throughout the life of the mine.

The future of open pit mining in steeply
dipping deposits will depend upon 1) an increase
in working depth and 2) the use of more powerful
machinery. These factors will lead to an increase
in capital investment and operating cost. The
handling of the volume distributions is therefore
of primary concern.

Background

In 1957, Rzhevskii (1) proposed a graphical-
analytical method for volume calculations using
vertical geological cross-sections. This proposed
method allowed the user to measure lengths instead
of areas or volumes. The advantages of the method
are clear, it made obsolete planimetring or
counting squares. Since then, the method has been
widely adopted in Europe, but there has not been
any published attempt to digitalize it. The
proposed method allows the simulation of a large
number of mining and geological parameters that
will lead to better decision-making.

For initial simplicity the analysis is limited
to a single vertical cross-section having a zone
of influence of one unit.

Algorithm Description

Assume that the deposit illustrated in Figure 1
can be represented by a digital type model (2).
The model of the deposit is constructed by defining
the coordinates x, y and z, see Table 1, of the
points of contact of the deposit and the center
line of the horizontal layers Δz, see Figure 2.
The block qualities such as rock type, grade, etc.,
see Table 2, are defined also. Assume that the
direction of mining is given by the position of
the opening trench at each horizon. The direction
of mining is determined by the points k, p, 1, 2,
3, and 4 (see Figure 2). The preliminary boundaries of the pit are defined by the closed line
a-h-s-r-a and are determined by the user. Also
determined are the angle of the working edges of
the pit ϕ, and the pit slopes γ_1 and γ_2.

In the first horizon (z_1) the opening trench
will be at point k. The ore volume will be represented by the area d-e-k and the stripping ratio
will be equal to zero. At the second horizon the
opening trench will be at point p. The waste
volume will be represented by the area of the
triangles b-c-j and f-g-t. The mineral volume
will be represented by the area of the polygon
c-d-k-e-f-t-p-j.

When the opening trench is at the z_2 horizon,
the waste volume will be w_{z2} and the mineral
volume will be m_{z2}. Then, the current stripping
ratio between horizon 1 and 2 is:

$$SR_{1-2} = \frac{w_{z2}}{m_{z2}} \qquad (1)$$

To calculate the stripping ratio SR_{1-2} by the
Rzhevskii algorithm, it is necessary to project
to the z_2 horizon, the points of contact between
the working edges of the pit (points j, b, t, and
g in Figure 2) with the ore body and the pit
limits. These points are projected by tracing
lines parallel to the line k-p. The length of the
segments ℓ-m and o-q will represent the volume of
waste between the z_1 and the z_2 horizons, and the
length of the segment m-o will represent the amount
of ore volume for the same interval. The waste
volume will therefore be:

$$w_{1-2} = \overline{\ell\text{-}m} + \overline{o\text{-}q} \qquad (2)$$

and the ore volume will be:

$$m_{1-2} = \overline{m\text{-}o} \qquad (3)$$

where $\overline{\ell\text{-}m}$, $\overline{o\text{-}q}$ and $\overline{m\text{-}o}$ are the length of segments
ℓ-m, o-q, and m-o respectively.

Proposed Algorithm

1. Input the model of the deposit.
2. Set the tentative pit boundaries.

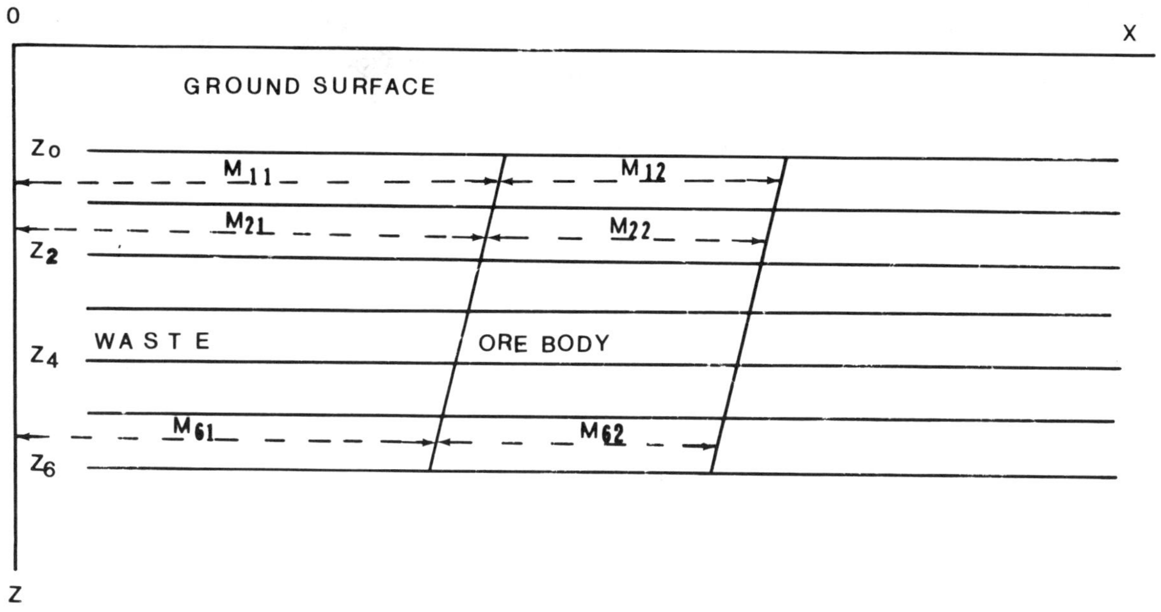

Fig.1 MODEL CONSTRUCTION

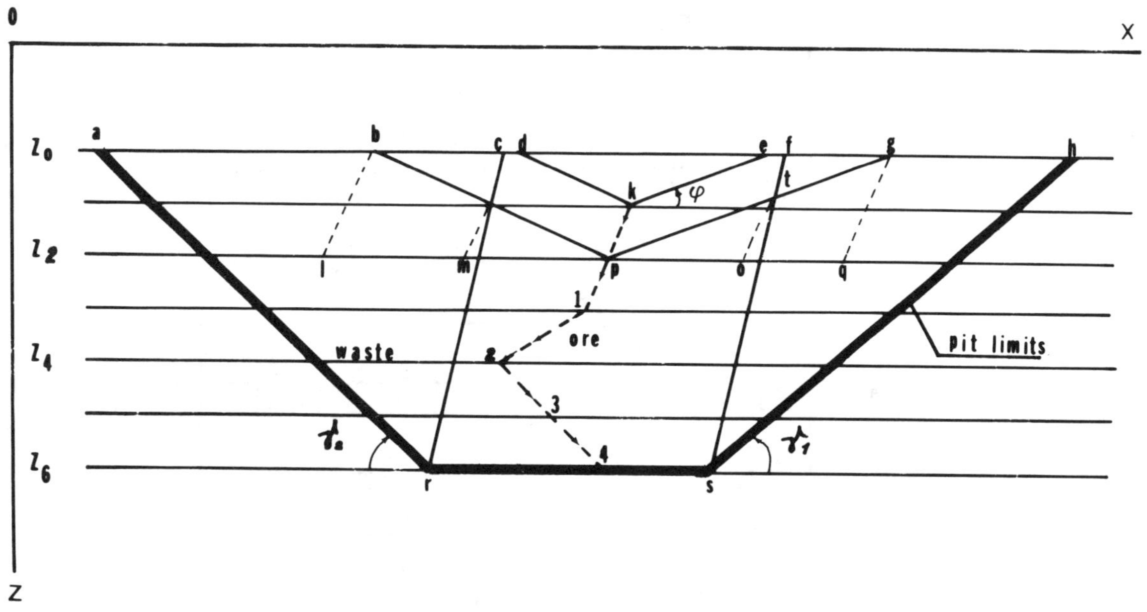

Fig.2 GEOMETRIC RELATIONSHIPS

Table 1. Ordinates of Points of Contact of Rock
Mass

Mark of Horizon	Block Number	
	1	2
z_1	M_{11}	M_{12}
z_2	M_{21}	M_{22}
z_3	.	.
z_4	.	.
z_5	.	.
z_6	M_{61}	M_{62}

Table 2. Content of Useful Component

Mark of Horizon	Block Number	
	1	2
z_1	0	α_1
z_2	0	α_1
z_3	0	α_1
z_4	0	α_1
z_5	0	α_1
z_6	0	α_1

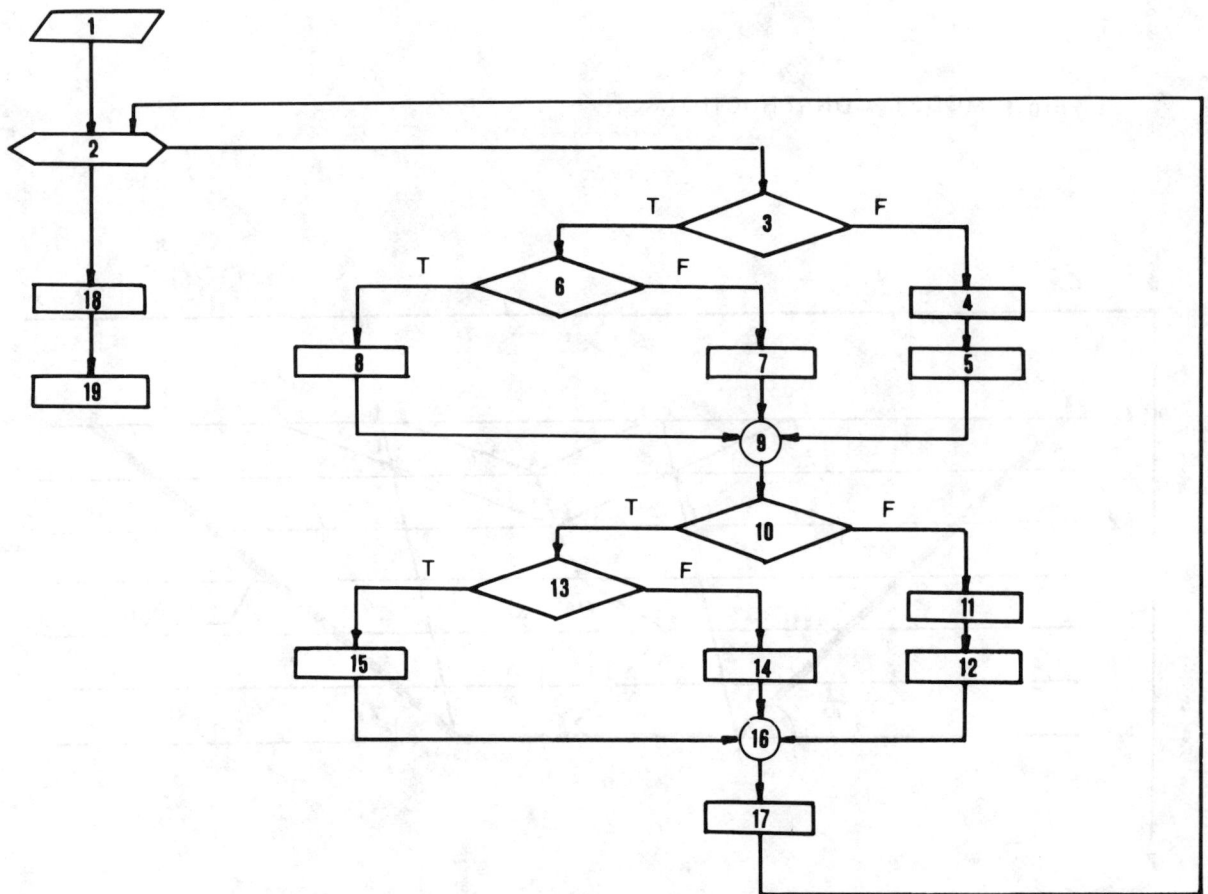

Fig. 3 FLOWCHART FOR STRIPPING RATIO CALCULATIONS

3. Input the angle of the working edge of the pit (ϕ).
4. Input the direction of sinking (or the co-ordinates of points k, p, 1, 2, 3, 4).
5. Find the coordinates of points b, g, j and t.
6. Project b, j, t, and g to the z_2 horizon, parallel to k-p.
7. Find the length of the segments ℓ-m, m-o, and o-q.
8. Calculate the stripping ratio, the ore volume and the waste volume by formulae (1), (2), and (3) respectively.
9. Move down to horizon z_3 and define point 1 as point p. At the z_2 horizon replace point p, by point k.
10. Calculate the coordinates of all points as well as the length of all segments.
11. Repeat the process until the last horizon has been reached, and calculate the total ore and waste volumes as well as the average stripping ratio.
12. Print result.

Discussion

For the deposit (see Figure 1) a flowchart of the algorithm solution has been developed, see Figure 3. The flowchart is described below. The flowchart solution is intended only for explanatory purposes and is not a general solution. Even though the example does not provide a general solution it can be used for deposits approximated by two parallel lines. For more complex shapes it is necessary to find the points of contact of the working edges with a polygon representing the ore body perimeter.

In a real situation every representative vertical cross-section is analyzed and the zones of influence for each cross-section determined, then the volumes are found for each zone of influence.

Flowchart Description

1. Input a, h, r, c, s, f, k, p, ϕ, γ_1, and γ_2.
2. Do I = 1, N.
3. Does the left side of the working edge intercept the surface?
4. Find the intersection between the left side of the working edge (p-b) and a-r.
5. Find the coordinates of b.
6. Does the left side of the working edge intercept a-r?
7. Find the intersection between p-b and the surface.
8. b = a.
9. End if.
10. Does the right side of the working edge intercept the surface?
11. Find the intersection between the right side of the working edge and the surface.
12. Find the coordinates of g.
13. Does the right side of the working edge intercept h-s?
14. Find the intersection between the right side of the working edge and h-s.
15. g = h.
16. Connector.
17. a) Find the coordinates of j, t, ℓ, m, o, and q.

b) Calculate the current stripping ratio.
c) Go to next horizon.
18. Calculate the cumulative volumes of waste and mineral as well as the average stripping ratio.
19. End of program, print results.

Conclusions

The present technique greatly simplifies calculating stripping ratios for steeply dipping deposits while improving the accuracy. Still greater accuracy can be achieved by reducing the distance between horizons. The advantages of the method are: 1) programming the algorithm requires a small amount of machine time, 2) memory storage is vastly reduced, and 3) as a result programming skill is minimized and analytical flexibility is increased.

References

1. Rzhevskii, V. V., 1957, "The Scheduling of the Open Pit Mining of Coal and Ores," (in Russian), UGLIEKHIZDAT, pp. 66-88.

2. Yasil'ev, E. V. et al., 1974, "Computer Storage of Mining-Geological Information," Soviet Mining Science, Volume 10, No. 4.

3. Lemieux, M., 1979, "Moving Cone Optimizing Algorithm," Computer Methods for the 80's, SME, pp. 329-345.

Chapter 32

COMPUTER SIMULATION OF STRIPPING OPERATIONS
IN OPEN PIT LIGNITE MINES

BY

KOSTAS G. FYTAS,
Graduate Student
PETER N. CALDER,
Head of Department
Department of Mining Engineering
Queen's University
Kingston, Ontario, Canada K7L 3N6
DIMITRIOS STEFANIDIS,
National Technical University of Athens
Athens, Greece

ABSTRACT

This paper describes the development and application of a generalized computer simulation model, PITSIM-II, to the stripping operations, involving front-end loaders and trucks, in a Greek open pit lignite mine.

PITSIM-II is a fixed time incrementing computer simulation model, capable of simulating any combination of cyclic loading machines (front-end loaders, backhoes or shovels) and trucks working together in an open pit mine. With proper manipulation of the input data, such as route profile characteristics, truck assignments and loaded material classification, the haulage system under consideration can be analyzed and if necessary improved.

The model can assist the mine operator in selecting the size and number of loading and hauling machines and operating them in the most efficient way in order to achieve certain production and blending targets. It can also aid management in both short and long range production scheduling in terms of forecasting the expected production rates and efficiencies of the loading and hauling equipment.

INTRODUCTION

With the depletion of rich underground deposits and the ever increasing demand for metals and industrial minerals it became apparent that large scale open pit mining offers the unique solution towards mining close to surface deposits. On the other hand the energy crisis has created the need for mining coal deposits of low calorific value (e.g. lignite deposits) that lie under thick layers of overburden. Therefore, the complexity of open pit mining operations has been increased during the last two decades.

The materials handling costs in an open pit mine usually range between 40 and 70 percent of the direct mining cost. Any inefficiencies in materials handling systems therefore have a considerable effect on net cash flow for the mine. One of the most common materials handling method in open pit mining operations consists of cyclic loading (e.g. shovels, front-end loaders, backhoes) and hauling machines (e.g. off-highway trucks). The basic problem in such systems is the selection of the right size and number of these machines and their scheduling in order to maximize their utilization. These tasks have been simplified with the development of computer simulation techniques.

Many computer simulation programs have been developed and used in the past for the design and analysis of open pit haulage systems (O' Neil, Ramani, Chick, Fytas). PITSIM-II is such an interactive computer simulation model of open pit haulage systems that enables the user to simulate and analyze any loading and hauling system. The application of this model is illustrated in two case studies involving the simulation of stripping operations in a Greek open pit lignite mine.

PITSIM-II MODEL

PITSIM-II is an interactive, fixed time incrementing computer simulation model written in FORTRAN-IV. It consists of several interrelated subroutines simulating the various activities in an open pit haulage system. The computer model is the result of a long development process which has resulted in the computer package, whose flow chart is shown in Figure 1.

Within the PITSIM-II model itself, two submodels can be distinguished :
- The Conventional Dispatching Submodel, which is a standard simulation model in-

volving a conventional dispatching subroutine (DISPAT) that dispatches the trucks of various truck types, to shovels for maximum utilization of equipment. It can operate in both an automatic mode or in an interactive one, which allows the operator to play the role of a truck dispatcher.
- The Linear Programming Dispatching Submodel, which is a simulation model involving a linear programming dispatching subroutine (LINDIS) that dispatches various trucks of a mixed fleet size to shovels for achieving certain production and blending targets using the minimum fleet size. Dispatching of trucks in this submodel is carried out according to the optimum truck flows of each section of the haulage network, calculated by a linear programming subroutine.

The main features of the model are:
- It is a fast, generalized and flexible simulator.
- There is no need of an advanced programming knowledge in order to simulate a specific open pit haulage system, since the model is completely interactive.
- It can simulate ANY open pit hauling and loading system, with unlimited simulation capabilities as far as the number of trucks, shovels, waste dumps, crushers and processing plants is concerned.
- The creation and any subsequent modification of the simulation and linear programming data bases is also interactive, allowing the operator to face any "what if" situation such as shovel and truck breakdowns or delays, shovel moves, travel time increases or decreases, shovel digging condition changes, dump and crusher downtime, route hazards and changes in material qualities.

- There are two options, in terms of shovel loading cycle probability distributions :
a. Using a lognormal probability distribution which is the most common cycle time distribution of cyclic loading machines (shovels, front-end loaders, backhoes, e.t.c.)
b. Using empirical shovel cycle time data, taken from field time studies.
-There are also two options in terms of truck haul-return cycle time definition :
a. By means of the performance charts of each truck type (speed-rimpull curves) using basic motion equations.
b. By means of empirical haul-return cycle time data derived from time studies.
- The model allows for four options of truck dispatching :
a. Manual dispatching, when the mine operator (model user), based on information about the shovel availability prompted by the computer, dispatches trucks to the various shovels according to his own judgement.
b. Automatic dispatching, when the program itself based on built-in dispatching criteria, dispatches trucks automatically to shovels. The flow chart of subroutine DISPAT that carries out automatic dispatching is illustrated in Figure 2.
The above two dispatching schemes are striving towards achieving the best utilization of equipment taking into account no production or blending targets.
c. Non-dispatching, when the system operates with original fixed shovel-truck assignments (static system).
d. Linear Programming dispatching system, when the program, based on optimal truck flows of each route, dispatches trucks to shovels in order to meet certain production and blending targets using the minimum fleet

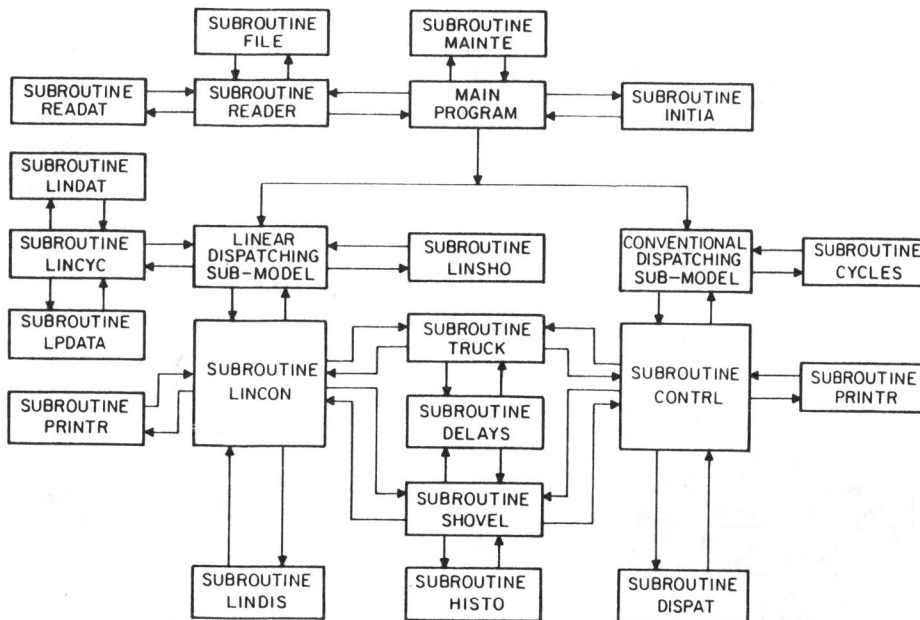

FIGURE 1. GENERAL FLOWCHART OF
PITSIM-II PROGRAM.

size. The flowchart of subroutine LINDIS that carries out linear programming dispatching in the model is illustrated in Figure 3.

PITSIM-II COMPUTER PACKAGE

As illustrated in Figure 1 PITSIM-II computer package consists of twenty subroutines and one main program. The package is designed on the IBM 4341 mainframe computer system at Queen's University. This system operates in a VM/CMS (Virtual Machine/Conversational Monitor System) environment and supports an interactive machine of 960 kbytes.

PITSIM-II package uses standard IMSL (International Mathematics and Statistics Library) routines for random number generation from uniform and normal probability distributions. This IMSL library is a set of subroutines written in FORTRAN-IV.

EQUIPMENT SUBROUTINES

Subroutine TRUCK

Subroutine TRUCK controls and calculates the motion of each truck on the haul-return road. The general flow chart of this subroutine is given in Figure 4. It can be based on either speed-rimpull performance curves of each truck type or empirical time study data.

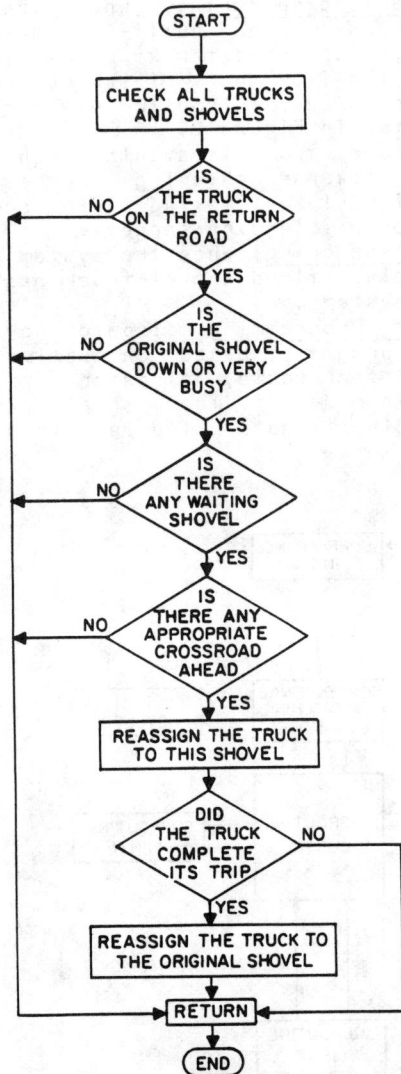

FIGURE 2. CONVENTIONAL TRUCK DISPATCHER FLOW CHART.

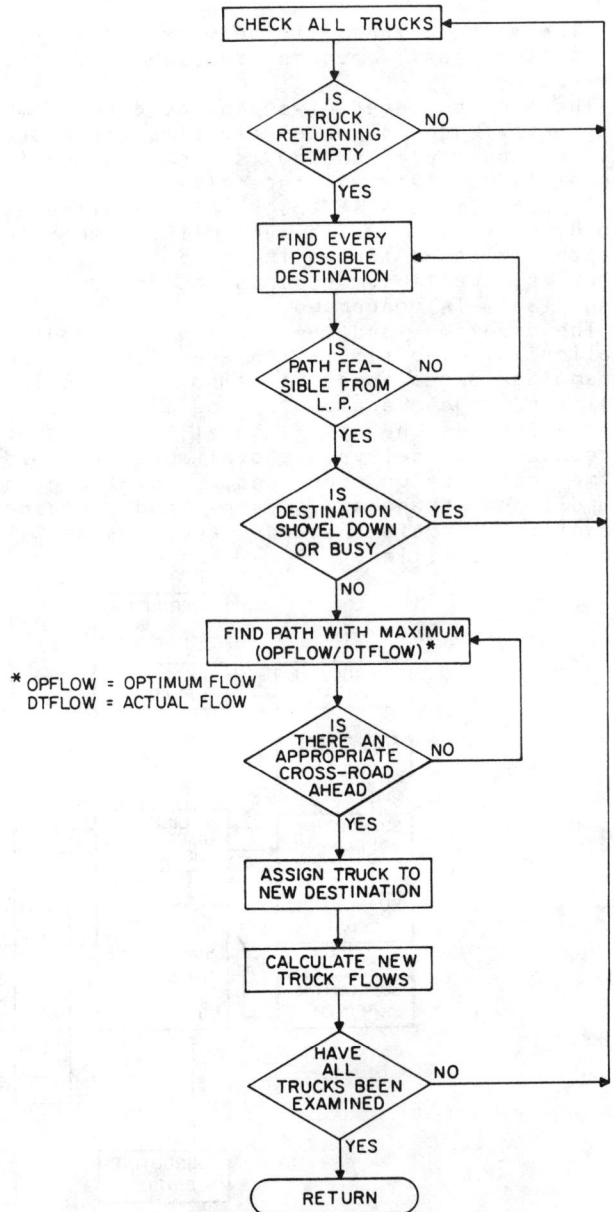

FIGURE 3. LINEAR PROGRAMMING TRUCK DISPATCHER FLOW CHART.

In the first case the subroutine calculates the available rimpull from the truck engine, which corresponds to current speed and then based on the usable rimpull between the wheels and the ground calculates the truck acceleration, new speed and new position at the end of each time increment.

During these calculations the subroutine takes into account the speed limits imposed by safety or operational reasons for each segment of the road and checks for trucks moving ahead and if necessary it readjusts the truck speed. In calculating the truck speeds it also takes into account a speed factor for each truck describing the mechanical condition and a pass status for every section of the haul-return road that allows or prohibits passing situations between two trucks.

Subroutine SHOVEL

Subroutine SHOVEL analyzes the loading cycle of each shovel. In calculating the shovel loading cycles, it can take into account either a log-normal distribution adjusted for each particular shovel (Figure 5) or an empirical cumulative probability distribution of loading cycle times obtained in field.

This subroutine also keeps track of the quantity and classification of the material being handled (ore or waste) along with operational or down status of every shovel. The capacity of the shovel dipper is represented in the model by a normal distribution.

The various delays of every piece of equipment in the model are calculated based on a uniform probability distribution. On a long term simulation basis the proportion of total operating time to the total down time is defined by the mechanical availability of each piece of equipment, which is an input parameter.

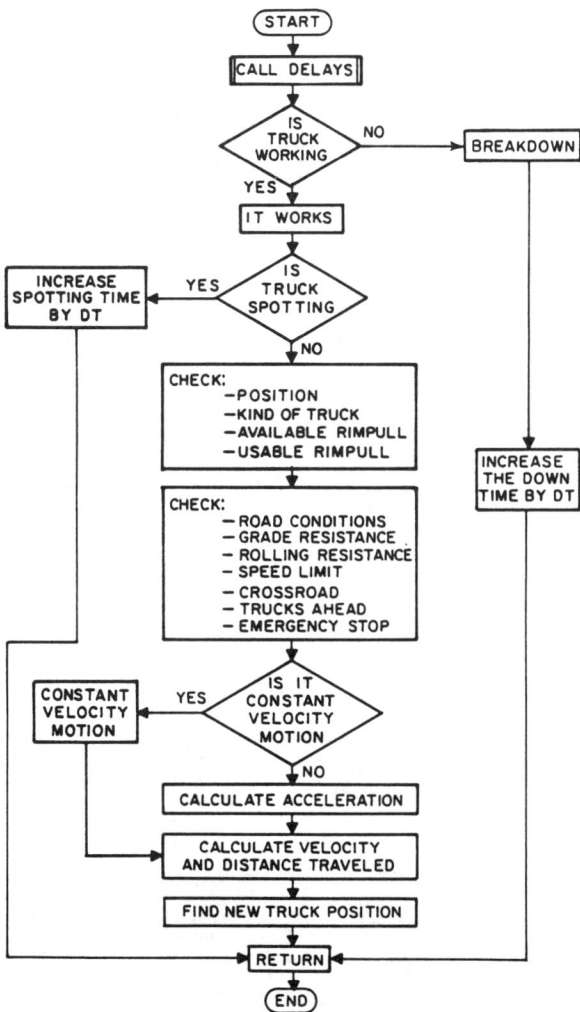

APPLICATION PROCEDURES

PITSIM-II simulation package can simulate every loading and hauling system that employs cyclic loading and hauling equipment.

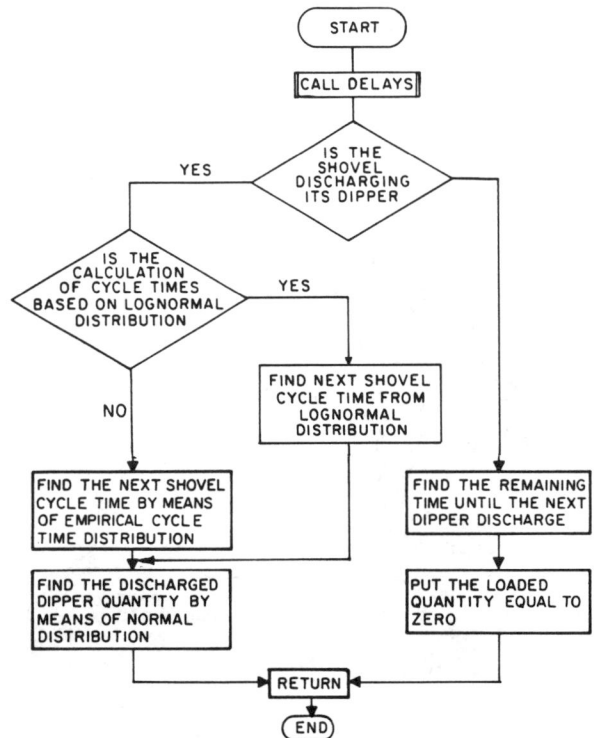

FIGURE 4. FLOW CHART OF SUBROUTINE TRUCK

FIGURE 5. FLOW CHART OF SUBROUTINE SHOVEL

The basic procedures that should be followed in each case study are the following :
- Preparation of a layout of the haul-return road indicating all loading and dumping places (nodes) as well as all possible ore and waste routes.
- Numbering of all nodes and segments using the following terminology :
   1 for every ore loading machine
   2 for every waste dumping place
   3 for every ore dumping place (crusher or stockpile)
   4 for every crushed ore loading place
   5 for the processing (or milling) plant
   Any negative number for the crossroads
   Any other positive number for the haul return segments
- Establishment of loading and dumping cycle time probability distributions for every loading and hauling machine.
- Creation of the simulation data base using the interactive editing facility.
- Subsequent corrections of the data base are carried out using the interactive modification option of the package.

As soon as the simulation data base is created the user can start the simulation procedure for the desired number of shifts. At the beginning of each shift the user specifies the preventive maintenance policy, as well as the initial start-up truck positions.

CASE STUDIES

PITSIM-II computer simulation model (in one of its original forms) was used in the simulation of the loading and hauling operations involved in removing the hard over-

burden formations during stripping in the Kardia open pit mine in Northern Greece. The layout of the open pit lignite mines in the Ptolemais region is illustrated in Figure 6. Both lignite and overburden are mined using a continuous mining system consisting of bucket wheel excavators, coveyor belts and stackers (Panagiotou).

These hard formations are mainly extended inclusions or banks of coarse clastic sediments, such as clays, marls, gravels and conglomerates consolidated together. The existing bucket wheel excavators, which are used for the continuous mining of lignite seams and soft overburden cannot mine these hard formations. Therefore, in the continuous phase of bucket wheel exacavator conveyor-stacker mining intervenes the discontinuous phase of hard overburden mining using explosives for fragmentation and shovel-truck systems for removal (Georgen).

The hard formations constitute the the most serious problem to the open pit lignite mines of the Ptolemais region. The problem is more intense in the Kardia and South

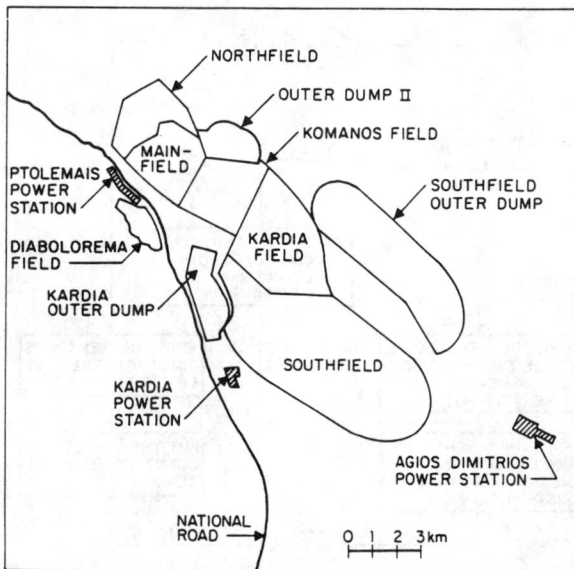

FIGURE 7. LOG-NORMAL PROBABILITY DISTRIBUTION OF SHOVEL CYCLE TIMES.

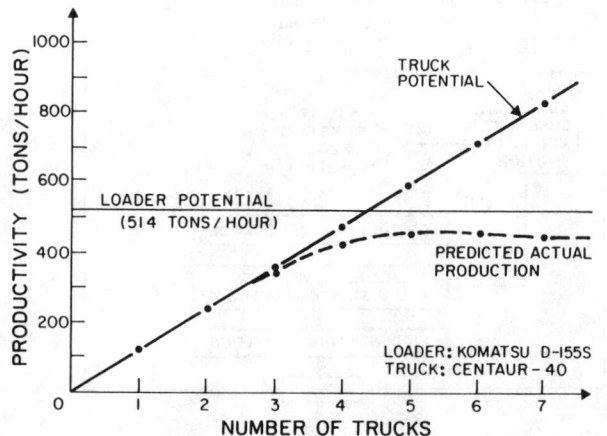

FIGURE 6. LAYOUT OF PTOLEMAIS OPEN PIT LIGNITE MINES.

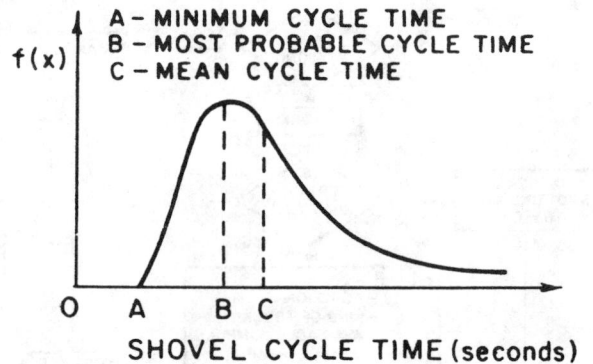

FIGURE 8. RELATIONSHIP BETWEEN PRODUCTIVITY AND NUMBER OF TRUCKS IN THE FIRST CASE STUDY.

Fields, where the hard formations have a negative effect on the efficiencies of bucket wheel excavators, due to frequent interruptions of their operation.

In the Kardia Field the loading and hauling operations of hard overburden were supported by the following equipment during the simulation studies :
- Three (3) wheel loaders (two CAT-988B and one 175-B Michigan.
- One (1) track type loader (KOMATSU D155S)
- Ten (10) off-highway trucks CENTAUR-40.

CASE STUDY NO. 1

The first case study simulated the loading and hauling operations in a system consisting of a track type loader (KOMATSU D155S) and three off-highway trucks (CENTAUR-40) which is a common loader-truck combination for the hard overburden removal. The material consists of hard overburden formations of the second overburden bench, mainly marly limestone (specific weight :1.0 tonnes/cu.m). The objectives of this case study were :
a. Model validation
b. System analysis
c. Sensitivity analysis.

Loading cycle time distribution : The chi-square (X^2) test revealed that the load cycle times of the D-155S loader, as observed in field, fit to the log-normal probability distribution (Deshmukh) (Figure 7) with a 0.05 level of significance.

Dumping cycle time distribution : The chi-square (X^2) test carried out on field observations of the truck (CENTAUR-40) dumping cycle times revealed that they fit to a log-normal probability distribution with a 0.01 level of significance.

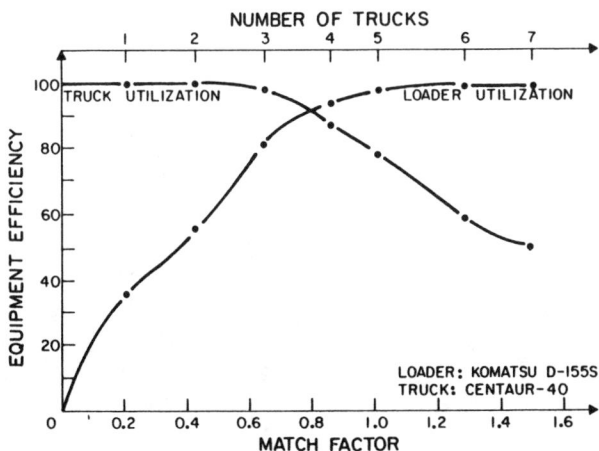

FIGURE 9. RELATIONSHIP BETWEEN EQUIPMENT EFFICIENCY AND MATCH FACTOR IN THE FIRST CASE STUDY.

Equipment breakdown distribution : Due to lack of extensive data in terms of equipment breakdowns, it was assumed that the various equipment delays due to breakdowns follow a uniform probability distribution.

The input data related to this case study are shown in Table 1. Analysis of the model's output revealed that the production figures given by the program are very close to the real ones with less than 5% error.

A sensitivity analysis was carried out using various number of trucks in combination to the same loader. The results are depicted on Figure 8 and Figure 9, where the equipment efficiency has been plotted against the number of trucks in the system or the match factor (Morgan). The match factor of a loading and hauling system is defined as following :

$$\text{Match Factor} = \frac{NT \times LCT}{NL \times HRCT}$$

where:
NT : is the number of trucks
LCT : is the loading time of each truck
NL : is the number of loaders
HRCT: is the haul-return cycle time

A loading and hauling system is working on maximum equipment efficiency when the match factor becomes 1 (perfect match). In our case the perfect match point is at five trucks, which means that the system needs two more trucks for maximum equipment efficiency (minimum waiting time).

CASE STUDY NO.2

The second case study simulated the loading and hauling operations in a system consisting of a wheel loader (CAT 988B) and three off-highway trucks (CENTAUR-40).

The material consists of hard overburden formations of the first overburden bench, mainly marly limestone with limestone intercalations (specific weight: 1.8 tonnes/ cubic meter)

Loading cycle time distribution : The chi-square (X^2) test revealed that the load cycle times of the CAT -988 B loader as observed in field, fit to the log-normal probability distribution with a 0.05 level of significance.

Dumping cycle time distribution : The chi-square (X^2) test revealed that the truck (CENTAUR-40) dumping cycle times fit very well to a log-normal probability distribution with a 0.01 level of significance.

Equipment breakdown distribution : Due to lack of extensive data related to equipment breakdowns, it was assumed that the various equipment delays due to breakdowns follow a uniform probability distribution.

A sensitivity analysis was carried out using various number of trucks in combination to the same loader. A similar analysis using the match factor revealed that the perfect match point is at three trucks, which means that the system operates in the best way as it is (maximum equipment efficiency and minimum waiting time).

CONCLUSIONS

PITSIM-II is a general computer simulation model for open pit haulage systems. It has been used extensively as a teaching tool at the Department of Mining Engineering at Queen's University and as a valuable analysis tool in industrial applications in the following areas.
1. Planning a new open pit haulage system :
   - Design the loading and hauling system.
   - Selection of loading and hauling units.
   - Efficient combination of loading and hauling units in order to satisfy certain production and blending targets.
2. Analysis of an existing open pit haulage system :
   - Analysis and improvements of the existing system.

- Short and long range production scheduling of an open pit.
- Evaluation of alternative loading, hauling and truck dispatching schemes.

PITSIM-II can be a valuable tool in training mine operators, since it provides the user with insight into the feasibility of implementing associated operational policies in the actual mine. Moreover the interactive capability allows the user to act as a shift boss or truck dispatcher and use his own judgement in redeploying trucks as conditions change or breakdowns occur.

REFERENCES

O'Neil, T.J., 1968, "Computer Simulation of Materials Handling Systems in Open Pit Mining", M.Sc. Thesis, Pennsylvania State University.

Ramani, R.V., Manula, C.B., 1978, "Application of a Computer Simulation System to a Contour Area Mine", S.M.E. of A.I.M.E. Preprint No.78-AO-329.

Chick, P.D.D., 1980, "The Interactive Simulation of Open Pit Haulage Systems",M.Sc. Thesis, Queen's University.

TABLE 1 . Input data of the first case study

Number of simulated shifts	80	
Number of waste loading places	1	
Number of waste dumping places	1	
Time Increment	3	seconds
Duration of one shift	360	minutes
Number of loading machines (loaders)	1	
Mean loader (CAT 988B) cycle time	47	seconds
Standard deviation of loader cycle time	12	seconds
Maximum loader cycle time	72	seconds
Minimum loader cycle time	30	seconds
Mean bucket capacity of loader	6	tonnes
Standard deviation of bucket capacity	1.2	tonnes
Maximum bucket capacity	7.3	tonnes
Minimum bucket capacity	5	tonnes
Material classification	Waste	
Loader Mechanical Availability	80 %	
Loader Mean down time	1	minute
Truck Capacity	36	tonnes
Empty truck Weight	28	tonnes
Spotting Time	35	seconds
Mean truck dumping time	35	seconds
Standard deviation of truck dumping time	8	seconds
Maximum truck dumping time	50	seconds
Minimum truck dumping time	25	seconds
Truck type	CENTAUR-40	
Number of trucks	3	
Truck mechanical availability	95 %	
Mean truck down time	1	minute
Number of segments of haul-return road	12	
Rolling resistance on the haul return road	2 %	
Maximum truck speed	15	m/sec

Fytas, K.G., Stefanidis, D.S., 1981, "Simulation of Open Pit Loading and Hauling Systems",B.Sc. Thesis, National Technical University of Athens.

Fytas, K.G., 1983, "An Interactive Computer Simulation Model of Open Pit Haulage Systems", M.Sc. Thesis, Queen's University.

Georgen, H., Fiebig, W., 1980, "Lignite Mining in the Ptolemais Region, Greece", Mining Magazine, February 1980, pp.126-139.

Panagiotou, G.N., 1979, "Computer Simulation of the Mining Operations in an Open cast Lignite Mine", M.Sc. Thesis, University of Newcastle upon Tyne.

Deshmukh, S.S., 1970, "Sizing of Fleets in Open Pits", Mining Engineering, December, 1970, pp. 41-45.

Morgan, W.C., Peterson, L.L.,1968, "Determining Shovel and Truck Productivity", Mining Engineering, December, 1968, pp. 76-80.

Chapter 33

A COST MODEL FOR OPEN PIT LIGNITE MINES

Neş'e Çelebi, A. Günhan Paşamehmetoğlu

Mining Engineering Department, Middle East Technical University, Ankara, Turkey

Abstract. In this paper, a cost model developed to facilitate rapid decision making in determining the economic viability of new coal fields in Turkey is presented. The model has two parts, the first part being the capital investment model and the second part being the operating cost model. Cost of various items in previously available feasibility reports of various open pit lignite mines are used as data for the proposed model. The cost model, its limitations and fields of application are discussed. An example is given.

INTRODUCTION

In recent years, especially after the energy crisis, the fastly growing consumption of coal for energy generation, domestic and industrial use, necessitated a considerable increase in coal production. Therefore the priority was given to search for new coal fields. Considering the limited reserves of coal in Turkey, optimum use of these new coal fields is of vital importance. Hence, the economic feasibilty studies, after the completion of geological studies, are a necessity. Although detailed feasibility studies give the most accurate results, this type of study takes time and requires more information and is expensive. On the other hand, a rapid estimation method can provide sufficient data in the prefeasibility stage.

Rapid cost estimation methods have been used successfully in the chemical industry (Holland, et al., 1974) and have found some applications in the mining industry. Toth and Stinnett (1976) have developed a simple model for open pit coal mines with an accuracy in the range of ±30%. Jarpa (1977) has developed another model for open pit copper mines, which is more practical.

A well-known approach in the preliminary estimates is the Lang Factor approach. This approach assumes that the total investment cost is proportional to the cost of equipment which is necessary for the production:

$$\text{Total investment cost} = K \times \text{Equipment cost} \quad (1)$$

The Lang Factor (K) is obviously greater than unity since it embodies all the other investments for the mine and it can be estimated accurately for a definite range of production capacity.

Another approach is the cost-capacity approach. The following relationship is valid between the investment cost of an equipment and the capacity of that equipment:

$$\text{Equipment cost} = a(\text{capacity})^b \quad (2)$$

where a and b are constants.

It is clear that a similar relationship between the total investment in equipment and production capacity can be established:

$$\text{Total investment in equipment} = A \left(\frac{\text{Production}}{\text{capacity}}\right)^B \quad (3)$$

where A and B are constants. This relationship shows that the total investment cost in equipment is directly proportional to the production capacity raised to an exponential power. This is logical, since an increase in production can only be possible by an increase in the number or the capacities of equipments used. The same exponential relationship is also valid for the operating cost and this type of relationship has already been used in the detailed feasibility studies in mining industry (Clement, 1980), (Mular, 1978).

In this study, using the above relationships, a cost model for a prefeasibility stage is developed. The model has two parts; the first part being the capital investment model and the second part being the operating cost model. Since most of the coal mines are open pit in Turkey, the model is developed for open pit coal mining only.

MODEL DESCRIPTION

Capital Investment Model

The total capital investment cost consists of two basic parts; the fixed investment cost and the working capital:

$$I = F + W \qquad (4)$$

where I is the total investment, F is the fixed investment and W is the working capital.

The fixed investment cost is the capital required for legal fees, land improvement, construction and transportation works, major and auxiliary equipment and shipping, unloading and set up costs of these equipments. Besides, contingency allowances are also included in the fixed investment cost.

Equipments, specified as the major equipment, are those which are used in the overburden removal, coal excavation, haulage and coal preparation. On the other hand, equipments, classified as auxiliary are those which are not directly related with the production such as equipment used for further exploration and laboratory investigations.

Since the major equipment is directly engaged in production, fixed investment can be expressed by using relationship (1):

$$F = K \cdot E \qquad (5)$$

where E is the major equipment cost.

Generally, in an open pit coal mining operation, overburden removal is the most significant cost component. Therefore, the above equation can further be improved to;

$$F = K_1 E_D + K_2 E_C \qquad (6)$$

where E_D is the cost of major overburden removal equipment, E_C is the cost of major equipment for coal production and K_1 and K_2 are individual factors for each cost component, obtained by breaking down the Lang Factor (K).

Refering to relationship (3), the following relations between cost of equipment and capacity can be established:

$$E_D = a_1 D^{b_1} \qquad (7)$$

$$E_C = a_2 C^{b_2} \qquad (8)$$

In the above equations a's and b's are constants, D is the annual overburden removal capacity in m^3/year and C is the production capacity in tons per year. Annual overburden capacity can be expressed in terms of annual production capacity by the use of the stripping ratio (R):

$$D = R \cdot C \qquad (9)$$

Variations in the stripping ratio affect the systems and equipments used in the overburden removal operation, for the same amount of coal production. This fact makes the stripping ratio an important parameter which also has to be considered in the model.

Working capital includes the money required for the inventory, accounts receivable and ready cash necessary to operate the business without being interrupted. Since the inventory requirement of overburden removal operations is very high, working capital can be divided into two parts as:

$$W = W_D + W_C \qquad (10)$$

where W_D is the inventory necessary for overburden removal and W_C is for the other working capital items. As being related with the capacity, these parts can also be expressed in terms of capacities, as:

$$W_D = a_3 D^{b_3} \qquad (11)$$

$$W_C = a_4 C^{b_4} \qquad (12)$$

Combining all the relationships from (4) through (12), the final model can be constructed where the total investment is a function of the annual production capacity and the stripping ratio:

$$I = K_1(a_1(R.C)^{b_1}) + K_2(a_2 C^{b_2}$$
$$+ a_3(R.C)^{b_3} + a_4 C^{b_4} \qquad (13)$$

The above model facilitates the estimation of the total investment for a coal field when the stripping ratio is known and the annual production capacity is determined beforehand.

Operating Cost Model

The items included in the annual operating cost are the cost of parts and supplies, energy cost, labor cost, saleries for administrative and clerical staff, maintenance and repair costs, depreciation, general administrative costs, state royalty and interest payments.

The notations used for these items are as follows:

PASP : cost of parts and supplies
ENRG : energy cost
LABR : labor cost
SALR : saleries for administrative and clerical staff
MNRP : maintenance and repair costs
GADM : general administrative costs
SRTY : state royalty
DEPR : depreciation
INTP : annual interest payment

As in the case of investment model, considering the influence of overburden removal and using the appropriate relationship for each item, operating cost model can be developed step by step and combining these items, the final model can be constructed.

Each of the following costs -cost of parts and supplies, energy cost and labor cost- can be broken down into two components; one relating to the overburden removal, the other relating to all the other operations. Applying cost-capacity relationship for each component, the above operating cost items can be determined as:

$$PASP = e_1 D^{f_1} + e_2 C^{f_2} \qquad (14)$$

$$ENRG = e_3 D^{f_3} + e_4 C^{f_4} \qquad (15)$$

$$LABR = e_5 D^{f_5} + e_6 C^{f_6} \qquad (16)$$

where e's and f's are constants.

Considering the purpose of the model, it was found suitable to relate the saleries for administrative and clerical staff only to the production capacity:

$$SALR = e_7 C^{f_7} \qquad (17)$$

Assuming that the maintenance and repair costs are proportional to the major equipment costs, it is possible to express these costs as:

$$MNRP = g_1 (E_D + E_C) \qquad (18)$$

where g_1 represents proportionality.

According to the general administrative cost structure of Turkish coal mines, some administrative costs are directly proportional to the production capacity. Besides this, using the relationship (1) which is suitable for the remaining part of the general administrative cost, the following equation can be established:

$$GADM = C.H + g_2(E_D + E_C) \qquad (19)$$

where H represents special administrative cost rate per ton of production, and g_2 is proportionality constant.

In Turkey, State Royalty is paid on yearly production and its rate is determined by the royalty schedules. Then,

$$SRTY = C.S \qquad (20)$$

where S is the royalty rate.

Non cash items of operating cost are depletion and depreciation. Since depletion is not allowed in Turkey, only depreciation is considered. In order to put the depreciation into the model, the replacement investments should also be determined. Replacement investments are usually done in five year periods, ten year periods and fifteen year periods according to specified life categories of the replaceable items. As in the case of the fixed investment, relationship (1) can be used for replacement investments:

$$P_5 = K_3(E_D + E_C) \qquad (21)$$

$$P_{10} = K_4(E_D + E_C) \qquad (22)$$

$$P_{15} = K_5(E_D + E_C) \qquad (23)$$

where P_5, P_{10}, P_{15} are used to denote the replacements in five, ten and fifteen year periods respectively. K's are the Lang Factors.

Assuming thirty years operating life and using the straight line method of depreciation which is the only allowable method, the depreciation can be simply put into the model as:

$$DEPR = (F + 5P_5 + 2P_{10} + P_{15})/30 \qquad (24)$$

One of the important item of the operating cost is the interest payments. Considering the financial situations in Turkey, it is assumed that all the fixed investment will be debt financed with repayment of the debt in equal amounts. Then the average annual interest payments, for n years payment plan, will be

$$INTP = F.i \frac{n+1}{2} \cdot \frac{1}{30} \qquad (25)$$

where i is the interest rate for the debt.

Then, the operating cost model is obtained by combining all the relationships from (14) through (25). Constructing the model on the basis of operating cost items provides the opportunity of examining those items separately:

Annual operating costs:

Cost of parts and supplies $= e_1 D^{f_1} + e_2 C^{f_2}$

Energy cost $= e_3 D^{f_3} + e_4 C^{f_4}$

Labor cost including salaries $= e_5 D^{f_5} + e_6 C^{f_6} + e_7 C^{f_7}$

Maintenance and repair costs $= g_1 (E_D + E_C)$

General adm. costs $= C.H + g_2 (E_D + E_C)$

State royalty $= C.S$

Depreciation $= \dfrac{K_1 E_D + K_2 E_C + (5K_3 + 2K_4 + K_5)(E_D + E_C)}{30}$

Interest $= (K_1 E_D + K_2 E_C) \cdot i \cdot \dfrac{n+1}{2} \cdot \dfrac{1}{30}$

DETERMINATION OF MODEL PARAMETERS

In order to determine the parameters of the proposed model, detailed feasibility reports of eight open pit lignite mines were used. Since these feasibility reports have been conducted in different years, they were updated to March 1983, using appropriate price indices.

After obtaining all the data, least square method of estimation was used, in order to determine the regression coefficients of the relations.

As a first step, parameters K_1 and K_2 for the relationship (6) were estimated using the data for F, E_C, E_D:

$K_1 = 1.31$

and

$K_2 = 2.07$

This regression equation has no constant term and there will be no mean corrections. As a result of this, the most popular criteria in determining the significance of regression, r^2, which measures the proportion of the total variation in the dependent variable explained by the regression can not be computed correctly. However, when F-test was conducted, the regression was found significant.

The parameters K_1 and K_2 indicates that invest-

ment in equipment comprises a great portion of the fixed investment and the coal production equipment has more influence upon the fixed cost.

The other relationships for which the parameters should be estimated are relationships (7), (8), (11) and (12). Making logarithmic transformations of $\ln y = \ln a + b \ln x$, linear equations were derived for these exponential relationships and then the parameters are estimated as follows:

				r^2
E_D :	$a_1 = 1\ 621\ 087$	$b_1 = 0.55$	0.61
E_C :	$a_2 = 6\ 296\ 931$	$b_2 = 0.42$	0.34
W_D :	$a_3 = \quad\quad 3\ 787$	$b_3 = 0.77$	0.81
W_C :	$a_4 = \quad\quad 44\ 945$	$b_4 = 0.69$	0.77

F-tests were conducted for all of the four equations and the regression equations were found significant. r^2's are high enough for three of the equations. Although r^2 for equation (8) is lower than the others, it can be considered as sufficient enough for the application, since in the model this equation will be treated together with the others.

When the estimated parameters are inserted, the capital investment model is determined as:

$$I = 2\ 127\ 987\ (R.C)^{0.55} + 13\ 025\ 495\ C^{0.42}$$

$$+\ 44\ 945\ C^{0.69} + 3\ 787\ (R.C)^{0.77}$$

A percentage error of 25% - 30% is acceptable for this type of cost estimation methods. The accuracy of this model was checked by comparing the results with that of the detailed feasibility reports. The maximum percentage error of the model being 13%, it averages to 8%. This indicates the precision of the investment cost model.

Operating cost model parameters were also estimated by using the same procedure discussed above. Estimated parameters for equations (14) - (17) are given below:

				r^2
PASP :	$e_1 = \quad 3\ 000$	$f_1 = 0.78$	0.72
	$e_2 = 34\ 468$	$f_2 = 0.62$	0.64
ENRG :	$e_3 = \quad\quad 1.12$	$f_3 = 1.23$	0.77
	$e_4 = \quad\quad 8.31$	$f_4 = 1.19$	0.87
LABR :	$e_5 = \quad\quad 0.09$	$f_5 = 1.36$	0.66
	$e_6 = 67\ 029$	$f_6 = 0.63$	0.45
SALR :	$e_7 = \quad\quad 0.04$	$f_7 = 1.45$	0.41

The f parameters of energy costs, labor cost for overburden removal and salaries of staff are greater than unity indicating an increasing slope.

There are no constant terms for the other equations. Below estimated values of parameters are given:

MNRP   : $g_1 = 0.034$

GADM   : $g_2 = 0.018$

P_5 : $K_3 = 0.300$

P_{10} : $K_4 = 0.330$

P_{15} : $K_5 = 0.170$

March 1983 values were used for the constant parameters H and S in relationships (19) and (20) respectively. Values of i and n in relationship (25) were taken so to reflect the current financial conditions.

Using all the above parameters, the operating cost model was obtained. The percentage error of the model is in the range of 2%-14%, the average being 7%.

DISCUSSION OF THE MODEL

The data used for model construction were obtained from the mines where the method of coal production systems are similar, namely shovel-truck system. This is also true for overburden removal and in these coal mines dragline-shovel-truck or shovel-truck systems are adopted for stripping operations. Therefore, the valididity of the model is tested for this type of coal mines and the model is applicable to the coal fields which are suitable for the above systems.

The model is valid for a definite range of capacities (650 000 - 4 200 000 tons per year) and stripping ratios (2.89 - 10.06 m^3/ton) since the parameters are estimated using a definite set of data.

As it is stated previously, in developing the model, March 1983 rates (in Turkish Liras) were used and to use the model dynamically, for each item in the model suitable cost indices were added. The model also permits to update the constant parameters when they are changed.

In using the model the required inputs are only the desired annual production capacity and the stripping ratio at the coal field considered. As an output the following can be obtained: - total investment, fixed investment, working capital, annual operating cost and the unit costs. Also, when the price of the coal is inputed to the model, by calculating the internal rate of return, it is possible to estimate the desirability of investing to this field under the given price condition. Therefore, the model can be used manyfold according to the policy followed.

AN ILLUSTRATIVE EXAMPLE

To demonstrate the use of the model, a hypothetical coal field is assumed with a desired production capacity (C) of 3 000 000 tons per year and a stripping ratio (R) of 5.00 m^3/ton. A typical output for this example is as follows:

NAME : SAMPLE

ANNUAL PRODUCTION (000 TONS)	3000.
STRIPPING RATIO (CU M PER TON)	5.00

xxxxxxxxxxxxxxxxxxxxxxxxxxxxxxxxxxxxxxxxxxxxxxx

BASE YEAR	JULY 1984

xxxxxxxxxxxxxxxxxxxxxxxxxxxxxxxxxxxxxxxxxxxxxxx
xxxxxxxxxxxxxxxxxxxxxxxxxxxxxxxxxxxxxxxxxxxxxxx

FIXED INVESTMENT (TL)	46226019000
WORKING CAPITAL (TL)	4488855000
TOTAL CAPITAL INVESTMENT (TL)	50714874000
CAPITAL INVESTMENT PER TON (TL/TON)	16905

xxxxxxxxxxxxxxxxxxxxxxxxxxxxxxxxxxxxxxxxxxxxxxx

ANNUAL OPERATING COSTS

	TOTAL(000 TL)	UNIT(TL/TON)
PARTS-SUPPLIES	2269853.	756.62
ENERGY	1876933.	625.64
LABOR	2051267.	683.76
MAINTENANCE-REPAIR	1075130.	358.38
GENERAL ADM. COSTS	616673.	205.56
STATE ROYALTY	300000.	100.00
DEPRECIATION	3992586.	1330.86
INTEREST	1864449.	621.48

xxxxxxxxxxxxxxxxxxxxxxxxxxxxxxxxxxxxxxxxxxxxxxx

TOTAL ANNUAL OPERATING COST (000 TL)	14046891.
UNIT OPERATING COST (TL/TON)	4682.30

xxxxxxxxxxxxxxxxxxxxxxxxxxxxxxxxxxxxxxxxxxxxxxx

CONCLUSION

The developed model provides the rapid estimation of the total capital investment of a coal field and the unit cost of coal for predetermined production capacity and stripping ratio.

This model is in accordance with the conditions in Turkey and it is valid only for open pit coal mining operations. Although the model uses only yearly capacity and stripping ratio as input and does not consider the effect of other factors such as geological or geotechnical, it is sufficient and suitable for a prefeasibility stage. For more detailed studies, it is possibble to improve the presented model.

REFERENCES

Clement, G.K., et al., 1980, Capital and Operating Cost Estimating System Manual for Mining and Beneficiation of Metallic and Nonmetallic Minerals Except Fossil Fuels in the U.S. and Canada, U.S. Bureau of Mines, 149 p.

Holland, F.A., et al., 1974, "How to Estimate Capital Costs", Chemical Engineering, Vol. 81, Pt. 2, pp. 71-76.

Jarpa, S.G., 1976, "Capital Investment and Operating Cost Estimation in Open Pit Mining", Proceedings of the 14 th APCOM Symposium, AIME, pp. 920-931.

Mular, A.L., 1978, Mineral Processing Equipment Costs and Preliminary Capital Cost Estimations, Special Vol. 18, Can. Ins. of Min. and Metall., 166 p.

Toth, G.W. and Stinnett, L.A., 1976, "Techniques for Rapid Preliminary Estimates of Coal Mining Costs", Mining Convention of the American Mining Congress, Denver, Co., pp.22-41.

Chapter 34

TRACKING MINED LAND REVEGETATION RESULTS
WITH A MICROCOMPUTER

Warren Keammerer, Consultant
and
Betty Gibbs, Consultant
Boulder, Colorado

ABSTRACT

The results of reseeding mined land must be evaluated and proven to be successful so mining companies may qualify for bond release. The amount of data collected which must be evaluated may be extensive and can be managed effectively with a microcomputer. A program has been developed for the IBM PC and related computers which makes the data handling much easier than hand methods and also aids with evaluation of the success of the seeding methods used.

The following capabilities are provided:

● Historical information is maintained, such as seeding methods, types of seed mixes used, fertilizers used, and costs.

● Cover percentages and vegetation productiion data (g/m^2) are saved along with data for species composition.

● Statistical calculations, including t-tests, one-way ANOVA as well as descriptive statistics, can be made on the cover and production data.

● Reports are produced which provide readily accessible information about the success of the revegetation process.

● Graphs can be plotted for a visual representation of the data collected data.

The system will help mine operators determine when the revegetation is not successful so appropriate management techniques can be instituted to accomplish the desired revegetation goals. Mine operators need to determine the continuing success of the revegetation methods so that bond release may be obtained within the least amount of time. Historical information can provide valuable data about what techniques have worked under which soil, slope and other conditions. Random access is provided to stored historical, cover and production data.

The programs are interactive and menu-driven. Information is provided in each program about how to enter data or what the user can expect from the program.

INTRODUCTION

The amount of information and data for reclaimed lands increases directly as the age and size of a mining project increase. It becomes more and more difficult to keep track of reclamation activities and results as the volume of data increases. Tracking this information is especially important for projects where annual reports to regulators are required, and requests for bond release are eventually prepared.

Without a mechanism for rapid, accurate and easy retrieval of reclamation data, several different problems may occur. Data from reclaimed areas may be lost. Area data may be available, but may be stored in a variety of difficult-to-access media.

Because of the time required, detailed analyses of revegetation techniques are not likely to be conducted. Yearly reports to regulatory agencies may be incomplete or inaccurate as a result of lost or improperly recalled data.

The revegetation information monitoring and analysis program was developed to automate the process of collecting, storing and analyzing data for environmental and reclamation managers. The program provides a comprehensive means for storage, analysis, and retrieval of vegetation data from reclaimed lands. A complete historical record of reclamation and revegetation treatments is also maintained.

Individual reclaimed parcels or native vegetation types can be tracked quickly and reports of pertinentat data can be printed. Base data from native vegetation types are used to evaluate the success of revegetation on reclaimed lands.

PROGRAM DESCRIPTION

The computer program designed to manage and analyze the revegetation data includes data handling, report generation, statistical analysis and simple graphics. Figure 1 shows the relationship of the individual modules.

The main menu is shown in Figure 2. Two major types of functions are provided: 1) data entry, display and maintenance, and 2) analyses of the data entered. The data entry options include the following:

- History Information Maintenance
- Species List Maintenance
- Cover Data Maintenance
- Production Data Maintenance

The analysis options include:

- Summarization reporting of cover and production data
- Line graphs and bar charts of selected data
- Statistical analysis
- Similarity index
- Search of history or sample data

All program options are accessed through a series of menus.

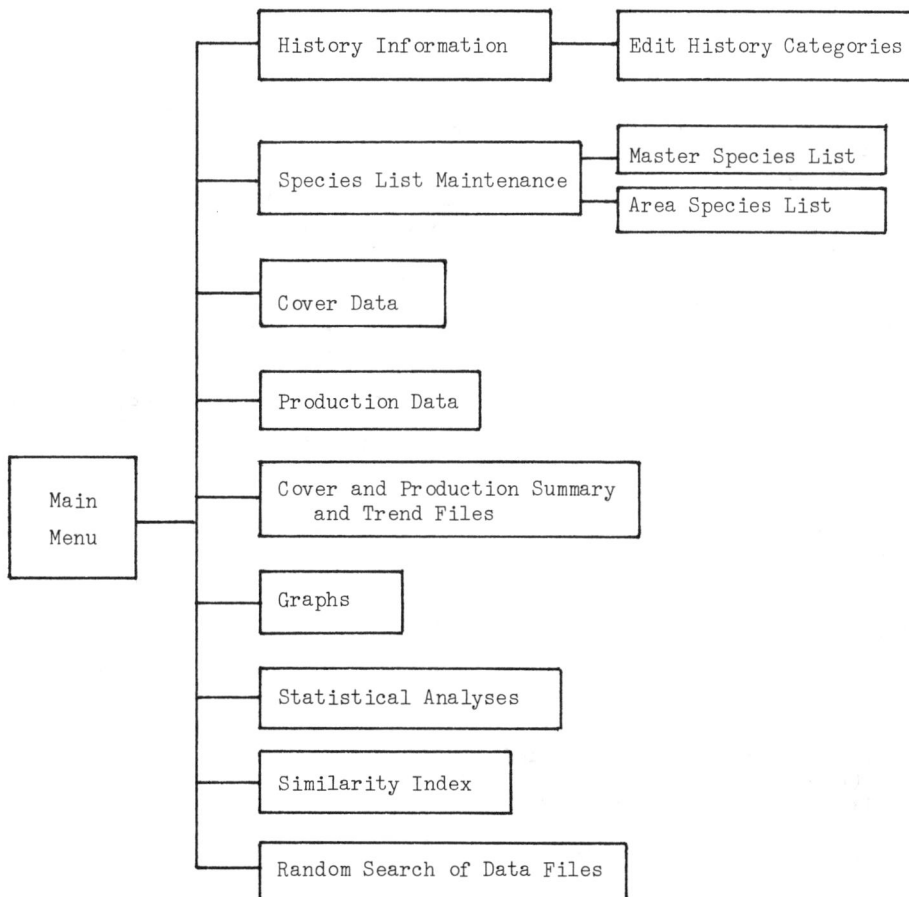

Figure 1. Flow chart showing the relationships of program modules.

REVEGETATION INFORMATION MONITORING AND ANALYSIS

==

OPTIONS:
1. History Information Files
2. Species List Maintenance
3. Cover Data Files
4. Production Data Files
5. Cover and Production Summary and Trend Files
6. Graphs
7. Statistical Analyses
8. Similarity Index
9. Random Search of Data Files
99. Exit From Program

==

Enter selection?

Figure 2. Main menu.

History Information Maintenance

All general historical information related to an area is stored in a file. Nearly 40 information categories are supplied with the program. The categories include items such as the name of the area, the reclaimed vegetation type, area in acres, slope, aspect, seeding month and year, seedbed prepartion, topsoil, fertilizer, seed mix, and costs as well as additional information. An example of a portion of the history file is shown in Figure 3 along with typical data for the categories.

The categories which are supplied with the program can be changed, although it is best to change the categories only at the beginning of use of the program. When historical information has been stored, the category names should not be changed, although new categories can be added.

Sixty lines of text can be included as part of the historical information. The text space can be used to enter progress information about an area, such as changes in seed mix, notes about any work done in the area and any special events which occurred in the area.

When files are created, the program lists the categories to the screen and the data related to each category can be entered. The user enters

HISTORICAL DATA ENTRY AND EDIT   Page   1
==
Name of area   : 80-1
Reclaimed Vegetation Type   : Upland Grassland
Area(acres)   : 0.03
Slope(%)   : 1
Aspect   : nw
Reclamation Type(T or P)   : p
Seeding Month   : Nov
Seeding Year   : 1980
Seedbed Preparation   : Chiseled
Topsoil(Y or N)   : y
   Replacement date   : Oct 80
   Amount replaced   :
Seed Mix   : Upland No. 1
Seeding Method   : Drill

*** Push esc to return to historical menu ***********
*** Push > to go to next screen, < to return to previous screen *****
***** If you need explanation of fields and the data to be entered,
      move to the field and push the question mark (?)

Figure 3. Example of a history file screen
           showing some of the categories and
           related data.

data by moving the cursor to the desired location on the screen and typing in the information. Information is added or changed in the same manner. Information stored for an area can be edited easily at any time.

Categories can be edited with simple commands. The categories are listed on the screen and the user enters either a C, I, A or D beside a category to be changed, inserted, added or deleted. Categories can be added only to the end of the list when history information files have been established. At the beginning of use of the program, categories can be inserted or deleted. Category names can be changed at any time because a change in the name does not affect the structure of the history file.

A printer report of the historical information for an area can be produced. The report includes the category name and the values for the selected area.

Species List Maintenance

A species list is included with the program and includes approximately 250 species of grasses, forbs, shrubs, cacti, trees, and other plant names. Each plant is identified by scientific name, common name and family name. In addition, a numerical life form code and a six character species code is assigned to each plant. A list of the life form codes is provided and spaces in the numerical sequence are left between major groups of plant types for expansion if needed.

The only time the user must assign a life form code is during the creation of a new item in the Species List. The life form code appears in some of the screen reports, but the program handles assignment of the code to the species entered in data files. The numeric life form code is used internally by the system for quick identification of life form groups.

The six-character species code is a shorthand method of identifying a species entered in a data file. The user only needs to enter six characters to identify the plant species, rather than the much longer scientific name or common name. The species code is composed of the first three characters of the first name (genus) and the first three characters of the second name (specific epithet) of the scientific name.

Two types of species lists are used. The primary list is the Master Species List which contains all the species names existing in a particular region. The other list is a list of species by area. The Area Species List is for information only and is not used in any other programs. Part of a report of the Master Species List is shown in Figure 4.

The Area Species List provides a short list of the species found in any particular area. Area Species lists can be maintained at the user's discretion and hard copy printer reports can be produced. The only data which needs to be entered for this list is the six-character species code and the year first observed. When a report of the list is requested, the full scientific name, common name, and family name of the species is listed.

Cover and Production Data Entry

Cover and production data are treated similarly. Cover values are entered as a percent of cover a species provides in the plot. Production values are also entered by species and are the weight in grams per square meter for the particular species in the plot. Information can be entered individually for each plot in the area, or the plots can be combined and the total percent cover or production weight can be entered. The number of plots in which a species occurred can be stored.

The six-character code for each species is entered. This species code entered is checked for a match in the Master Species list. If the entered species code does not match any species code already stored, a message is printed informing the user that the code does not exist.

Life Form	Species Code	Scientific Name	Common Name	Family Name
3	Agrsmi	Agropyron smithii	Western Wheatgrass	Gramineae
3	Agrrip	Agropyron riparium	Streambank Wheatgrass	Gramineae
6	Brotec	Bromus tectorum	Cheatgrass	Gramineae
3	Oryhym	Oryzopsis hymenoides	Indian Ricegrass	Gramineae
9	Chepra	Chenopodium pratericola	Goosefoot	Chenopodiaceae
9	Salkal	Salsola kali	Russian Thistle	Chenopodiaceae
4	Bougra	Bouteloua gracilis	Blue Grama	Gramineae
3	Stivir	Stipa viridula	Green Needlegrass	Gramineae
9	Polavi	Polygonum aviculare	Devil's Shoestrings	Polygonaceae
3	Danuni	Danthonia unispicata	One-spike Poverty Oatgr	Gramineae
9	Kocsco	Kochia scoparia	Summer Cypress	Chenopodiaceae
9	Cleser	Cleome serrulata	Rocky Mountain Bee Plan	Capparidaceae
9	Amaret	Amaranthus retroflexus	Pigweed	Amaranthaceae

Figure 4. A portion of the Master Species List.

If an error is made in entry of the species code, the correct code can be entered before the record is written to the file. If the species code is not already in the Master Species List, the record can still be written to the cover or production data file and the new species code can be added to the Master Species List at a later time.

A report of information stored in the cover and production files can be routed to a printer. The report to the printer or the screen displays the life form numeric code, the full scientific name, and the numerical data entered into the selected file. An example of the information entered in a cover file is shown in Figure 5.

Summary and Trends

Selected summarized cover and production parameters may be stored in summary files. Information stored in cover and production files is summarized and stored in a composite file.

For plant cover in any given area, mean cover values for each species, mean cover for life form groups, and mean total vegetation cover values are stored. For production data, mean production by each species, mean production for life form groups, and mean production are summarized and stored.

The summaries contain data for the above parameters for as many as five different sampling years. The summary data may be viewed on the CRT or sent as a report to a printer. The summary files are later used for generating graphs and serve as the data base for conducting random searches of cover and production data.

Graphics

The graphics module generates line graphs and bar charts. Line graphs are used to portray changes in plant cover and production over time. Any of the cover and production data stored in the summary and trends files may be used to generate line graphs. Several years of data must be available to produce a meaningful line graph.

Cover or production data from any given year which has been summarized may be portrayed on a bar chart. Graphical display of the data can provide greater insight into successional changes in the structure and composition of reclaimed areas.

The graphs can be used to aid in determining the success or failure of revegetation methods.

Statistical Analyses

Descriptive statistics (mean, standard deviation, and variance) are calculated for selected cover and production parameters. The module also provides options for hypothesis testing using a Student's t-test and a one-way analysis of variance.

The two tests permit evaluation of results using different revegetation techniques such as different planting dates, different fertilization rates, different seed mixes, or different mulching rates). The tests also allow for evaluating differences in cover and production between reclaimed areas and native vegetation types.

Reports are produced which are organized by life form group and with each species listed individually. The reports can be viewed on the CRT and also output to a printer.

Similarity Index

In plant ecology, similarity indices are used to evaluate differences in structure and composition between two different areas. Comparisons can be made between any two sampled areas.

The similarity index module calculates similarities of selected areas on the basis of cover by species or cover by various groups of life forms. Comparisons can be made on as many as 30 different data sets at a time. The results are

COVER Data List   File:80-3IF82 Sample Size: 5

Life Form	Species	Cover %	Range	No. of plots
3	Agropyron smithii	79	7 - 21	5
3	Agropyron riparium	21	0 - 8	4
3	Agrtra	34	2 - 11	5
3	Oryzopsis hymenoides	14	2 - 4	5
9	Kochia scoparia	46	2 - 22	5
9	Salsola kali	6	0 - 5	4
11	Atriplex canescens	6	0 - 6	1
10	Artemisia frigida	0.01	0 - .01	2
5	Poa pratensis	8	0 - 8	1
3	Stipa viridula	1	0 - 1	1

Figure 5. Portion of a cover file report as seen on the CRT

DESCRIPTIVE STATISTICS FOR AREA 80-3IF82

==

Sample size = 5

	Mean	Standard Deviation	% Biomass
Grasses	331.14	113.95	77.50
Forbs	93.40	176.15	21.86
Semi-Shrubs or Half-Shrubs	0.06	0.13	0.01
Shrubs	2.66	5.95	0.62
Total Production	427.26	200.52	

Figure 6. Example of descriptive statistics
for a single area.

shown in a half matrix and the mean similarity is calculated for all comparisons. The mean similarity is also calculated for each data set.

The data sets are entered using the cover data entry module. The data sets are then selected individually or as a group for calculation of the similarity index.

Search Cover, Production and History Data

All information in the cover, production and history data files can be searched. Complex search criteria may be entered. An example of a search may be to find all areas which were seeded with seed mix Upland No. 1 after 1980, had percent cover of greater than 9% and production of greater than 3 grams per square meter.

All items of data in the history and summary files are identified with a descriptive code. The code is used to define the search along with the search limits. Before the search is actually performed, all the search criteria are printed to the screen and verification is requested. If the criteria are correct, the search begins.

The result of the search is a list of areas which meet the criteria. The results can be displayed on the screen or routed to a hard copy printer.

HARDWARE REQUIREMENTS

The Revegetation Analysis program runs on an IBM PC or compatible computer. A memory size of 256K is recommended because of the large number of species which may be needed for a complete analysis.

Two floppy disk drives are a minimum requirement, but a hard disk allows the programs to run faster. A printer is needed for hard copy reports and a 132-column printer is recommended because many of the reports are wider than 80 columns.

CONCLUSIONS

The Revegetation Information Monitoring and Analysis program provides an aid to environmental and reclamation managers to produce timely and representative annual reports. The printer reports produced by the modules are designed to be included in a formal report. Table numbers and titles can be entered to customize the tabular information to fit the report being produced.

The system provides a means of organizing the massive amount of information collected to monitor the progress of reclamation. The graphs and reports can be used as a tool to assist with proof of successful revegetation as a condition for bond release.

Chapter 35

THE ROLE OF HANDHELD COMPUTERS/CALCULATORS
IN THE DEVELOPMENT OF THE PRAIRIE HILL MINE
IN NORTH CENTRAL MISSOURI

Mark R. Congiardo, P.E.

Associated Electric Cooperative, Inc.
Senior Mining Engineer
Springfield, Missouri

Abstract. This paper discusses the role played
by handheld computer/calculators in the development of a major strip mining operation in North
Central Missouri. The handhelds have been and
currently are being used as the primary mine
planning tools at this operation. This paper
addresses how these tools are being utilized for
mine planning, budgeting, range diagrams, and
reclamation. The advantages, as well as the limitations and problems of such small systems, are
discussed and an analysis of the accuracy of the
computed data as compared to the actual field
data will also be discussed.

INTRODUCTION AND BACKGROUND

Introduction

Small handheld computers and programmable calculators may be considered by some to be merely
"toys" and not serious mine planning tools.
However, the mine planning associated with the
$250,000,000 expansion of the Prairie Hill Mine
to supply the coal demands of Associated Electric's new 630 MW unit at the Thomas Hill Power
Plant was based largely upon a handheld system
costing less than $1200. This fact, in my mind,
should cause us to take a closer and more serious
look at the handheld computer as a mine planning
tool.

Background

History of the Prairie Hill Mine. The Prairie
Hill Mine is located in North Central Missouri,
approximately 80 km (50 miles) northwest of
Columbia, Missouri in Randolph and Chariton
Counties. The mine was originally owned and
operated by the Peabody Coal Company to supply
two coal fired generating units at Associated
Electric's Thomas Hill Power Plant. In order to
meet those needs, Peabody was using a 20.6 cu m
(27 cu yd) walking dragline and a 10.7 cu m (14 cu
yd) dragline as their primary stripping equipment.
In the early to mid 70's, Associated decided to
expand the Thomas Hill Power Station's generating
capacity by adding a 630 MW coal fired unit. In
its effort to provide electrical power to its
customers at the lowest possible cost, Associated
purchased the Prairie Hill Mine from Peabody and

assumed operational control January 1, 1980. In
1979, Associated formed the Fuels and Mining Division and began hiring its staff. Until this
staffing was completed, Associated commissioned
Morrison-Knudsen of Boise, Idaho as consulting
engineers to aid in the expansion of the mine to
full capacity prior to the final takeover. Their
scope of work included the construction of three
Bucyrus-Erie 2570-W walking draglines and development of the first Five Year Mine Plan.

Draglines: Initially, six draglines were to be
purchased from Bucyrus-Erie; three 2570-W class,
a 1470-W class machine and two 1300-W class
machines. Later, this combination was changed to
four 2570's and one 1570 class machine. In 1980,
after the first 2570 was put into operation, the
original assumptions, regarding operating time and
productivity, were found to be too conservative
and the number of machines was cut to three 2570's
and a 1570. Budgetary restrictions, plus the
decision to supplement the coal supply with outside coal sources prompted Associated to drop the
1570. This led to the final number of machines
now in production, three 2570's.

Five Year Mine Plan: The initial Five Year
Mine Plan was completed in June of 1980 by M-K.
The primary computational tools used were a
Hewlett-Packard 41C handheld computer, a magnetic
card reader, and a thermal printer. The reasons
for this selection were multiple.

- The selection of software system,
  compatible with our mainframe computer
  was to be left to Associated's staff.

- The known geology of the region
  suggested a flat lying seam with
  few anomalies.

- The exploration drilling program had
  not been completed in areas beyond
  the scope of the five year plan.

- The purpose of the plan was to provide
  a starting point and the required effort
  to load a mainframe seemed inappropriate.

Production Data. The Prairie Hill Mine was
designed to produce approximately 1.8 million

tonnes (2.0 million short-tons) of raw coal per year. Table 1 summarizes the other basic data regarding overburden and coal thicknesses.

Table 1

1. Overburden Thickness   34 m (110 ft)
2. Coal Thickness   1.10 m (3.75 ft)
3. Virgin Ratio (cu m/tonne) 23:1

There are insufficient low ratio reserves to supply the required tonnage for the 30 year life expectancy of the mine. Therefore, in order to maintain a constant annual production, we must carefully balance our low and high ratio reserves. This requires accurate and equally careful scheduling of the draglines. When one machine is in high ratio coal, then one machine must be scheduled to be in low ratio and the third scheduled to be at the field ratio in order to maximize the productivity of the mine as a whole. We cannot use a cutoff ratio for each individual machine as is commonly done throughout the industry. The cutoff must be established mine wide, which makes the mine planning engineer's job much tougher. Table 2 summarizes the production data for all three machines for the years 1982, 1983, and the first 3 quarters of 1984.

Table 2

Production Data

	1982	1983	1984*
Virgin cu m (x 1000)	32523.3	36401.8	30975.9
Total cu m (x 1000)	41261.5	43091.7	34622.3
Tonnes (x 1000)	1628.6	1722.5	1511.0
Virgin Ratio (cu m/tonne)	20.0	21.1	20.5
Total Ratio (cu m/tonne)	25.3	25.0	22.9

* First 3 quarters only

Computer History. With the completion of the first Five Year Mine Plan in June of 1980, we began searching for a software package compatible with our PRIME 750 mainframe system. In the interim, the HP-41C system was completely re-written and expanded. It was to be used for the mine planning until a system was located, installed, and the drill hole information inputted. In 1981, we succeeded in finding a system which we believed would be completely compatible with our hardware. The system was originally designed to run on a PRIME 550 and we were able to load it into our system almost immediately. The time consuming ritual of inputting all available drillhole information was now underway. During the course of this process we found some problems with our mapping and because of these problems, we were unable to utilize the system to its fullest. Therefore, we began to rely more

and more upon the HP-41 system for mine planning, reclamation planning, range diagrams, and budgeting. For the past three years, these functions have been performed, almost exclusively, using this system, along with many other functions, such as swing recorder reports, haul road construction estimates, and time study work. In the Spring of 1984, we began switching to Energy Resource Solution's Mine Planning and Feasibility System, which, in our estimation, will allow us to take full advantage of our mainframe capabilities. However, we are sharing the system with 4 other divisions at Associated (accounting, power production, power distribution, and engineering) and the effect of the increased CPU time required to operate this system may not allow us to use it at 100% efficiency, until we add more hardware. Until the system is completely operational and all programs checked out thoroughly, the handheld system will play an important role in all mine planning functions.

CURRENT USES AND PRACTICES

Mine Planning

Mine planning at the Prairie Hill Mine relies upon the structure and thickness maps that were created for the initial Five Year Mine Plan. Using these maps, grids are established for the entire field on 120 m (400 ft) grid intervals. We normally use the same scale for our mapping, thus we have values for the coal, overburden, and parting thicknesses for each square 2.5 cm (1 in) of the map. These maps then become the basis for all mine plans. Once a mining area has been evaluated and a general mining sequence established, the pit layout work is normally done, taking into consideration all shearout, chop-in, and boxcut locations. The layout is then overlayed onto the grid maps and the volumetrics for each pit are computed in inputting the grid values along each 120 m (400 ft) of pit length. Using this method, we are, in essence, "mining" the pit on paper using the known values from the grid data maps, the given dragline parameters, and the current pit configuration (See Fig. 1 for sample output from the Pit Volumes program).

From this data, using the HP-41C system, we are able to check the depth of each 120 m (400 ft) of pit length against the computed critical depth. If the total depth of the interval exceeds critical depth, then the required extended bench re-handle is added to the normal pit cleaning and road material rehandle. After we finish running the volumetrics for each pit, we then begin timing out the pits. By using the historical, or estimated, production rate per day, we are able to compute the total target yardage for a day, month, quarter or year, whatever is required, and the approximate face location at the end of the time period. The associated values for the virgin overburden waste, tonnes uncovered, and ratio are output for later use in the development of the final reports (See Fig. 2 for sample output from the timing routine).

At this point, we are basically done, except for getting someone to type up all of the tables and reports. However, as we all know, revisions are inevitable with any mine plan and it may take several revisions (15 were required in the Summer of 1984) to finally satisfy all those concerned and evaluate all options. If there are major changes in the pit sequencing or mining areas, then the pits must be reconfigured and recomputed. However, we have found that, in most cases, these revisions are minor and can easily be corrected by editing the pit files or by changing the timing and the pit sequencing slightly. By using these methods, and the ever increasing historical data base for dragline productivity, we have been able to forecast pit locations and production values with a good deal of accuracy. This aspect of the system will be discussed, in detail, later in this paper.

```
                                            JUN84

  ---------------------       DAYS =             30.00
                              VIR YDS =     1,646,025.48
  PIT NO.        2            R/H YDS =       198,974.52
  ---------------------       TOT YDS =     1,845,000.00
                              TONS UNC =       62,377.82
  VIR YDS =    557,407.41     VIR RAT =            26.4
  R/H YDS =     55,740.74     EFF RAT =            29.6
  TOT YDS =    613,148.15     FT OF PIT =       4,334.2
  TONS =        19,685.49     PCT PIT USED =       79.0
  VIR RAT =         28.32     ACTIVE PIT =          2.0
  EFF RAT =         31.15     ------------------------
  PIT LENGTH =     800.00
```

Fig. 1-Pit Volumes   Fig. 2-Timing Program
Sample Output   Sample Output

Range Diagrams

At any strip mining operation, the generation of range diagrams for the primary stripping machines, whether they are draglines, shovels, bucketwheel excavators, or hydraulic excavator, are an extremely important, and time consuming, part of a mine engineer's duties. They become the basis for all mine planning, reclamation planning, and other major decisions regarding the operation of the mine. Range diagram studies usually involve the creation of sets of drawings depicting varying pit depths, pit widths, highwall angles, spoil angles, and swell factors. If one considers all of the possible combinations, there would be thousands of range diagrams that should be generated in order to look at all of the possibilities. Considering that each diagram may take 30 minutes to an hour, a mine engineer would have little time to devote to his other duties. The most time consuming range diagram, for a dragline operation such as ours, is the "critical depth" diagram. Critical depth is defined as the total depth of overburden at which the dragline must begin using the extended bench method to spoil all of the waste material. This diagram may take several iterations, and an hour or two, to finally approximate the critical depth value. This is good for one situation only and several cases must be evaluated.

At Prairie Hill, we are stripping so close to

the critical depth of these machines, that it became obvious that the critical depth diagrams were going to be the most important for us. We began by using the time honored methods of drawing and planimetering; of redrawing and replanimetering. After several of these diagrams, it became apparent that we needed a general range diagram in order to quickly compute the required data and some sort of program to make these changes quickly and easily. The output data had to be complete and concise, in order to make the generation of any diagram quick and easy. The general range diagram and the necessary equations were created and checked against existing range diagrams. There were several errors found. However, these errors were not in the computations, but in our drafting.

The next step was to program the handheld system with necessary steps and amenities to make the computations quick and easy. Since we were using the HP-41C system to make these computations by hand, it was a relatively simple matter to put the system in program mode and repeat the steps already being used. Once all of these necessities and "niceties" were programmed in the system, we had a program which would allow us to compute any range diagram, including the "critical depth" diagram, within seconds (See Fig. 3 for sample output from the range diagram program). These same computations are an integral part of the mine planning system, discussed earlier, and in the reclamation planning system.

```
  2570-W                        OPTION A
                                TD-CRIT=      137.02
        BASE DATA               DUMP HGT=      63.36
                                DP=           117.02
  BENCH DEPTH=       20.00
  H/W SLOPE=          0.50
  SPOIL SLOPE=        1.30       OPTION B
  SWELL=              1.10       TOTAL DEPTH=   150.00
  W-OLD PIT=        175.00       DUMP HGT=       64.65
  W-NEW PIT=        175.00       DSP=           318.05
  R-FACTOR=         293.00       WEB=            25.05
  T-COAL=             4.00       % R/H=          25.96
  MAX DP=           160.00
  MAX HD=           126.00
```

Fig. 3 Sample Output
of Range Diagram Program

Reclamation Planning

The reclamation planning system is a direct offshoot of the range diagram program. It was developed to predict the spoil peak and valley elevations, as well as predicting an approximate rough grading or "strike" elevation. Our method of reclamation is unique in two ways.

- We use a "terracing" technique of grading.

- We use the draglines to place glacial till on top of the spoil peaks to aid reclamation.

At Prairie Hill, the topography is a hilly and somewhat harder to farm than other areas of the country. Therefore, instead of simply grading the spoils in a manner which translates and elevates the original topography, we have opted to create "plateaus" or "terraces." These terraces should allow better crop production, help to lower our reclamation liability, and improve our revegetation performance. A great deal of planning and estimating must be done by the engineering and reclamation departments in order to predict the best locations and configurations for the terraces.

The second technique is called "glacial till substitution." This technique utilizes the long boom characteristics of the 2570 draglines to place the clay or "glacial till" on top of the spoil pile. This eliminates the need for the scrapers to haul the required 1.2 m (4 ft) of clay. This helps to minimize costs and the size of our scraper fleet. Again, a great deal of planning is required to maximize this savings.

```
PROGRAM
RECLAMATION
PIT NO.=        1
AVE SUR. EL.=   750.00
AVE COAL EL.=   650.00
PEAK STA.=      2+18.45
PEAK ELEV.=     789.65
VALLEY STA.=    1+30.95
VALLEY ELEV.=   722.35
GRADE STA.=     1+74.70
GRADE ELEV.=    756.00
```

Fig. 4 - Sample Output
of Reclamation Program

The method of using a structural cross section and then creating consecutive range diagrams to estimate the spoil elevations proved to be a tiresome, tedious, and time consuming task. The reclamation programs were developed to help alleviate this situation. Using the relationships defined by the general range diagram and by modifying some of the subroutines in the range diagram program, we were able to create a system which allowed the mine engineer to run a structural cross section in plan view. The necessary input data is obtained from the pit layout maps and the structure maps. The information required, to run the program, is the mean topo elevation and the mean coal elevation. From these computations, we are able to approximate the configuration of the rough graded spoil (See Fig. 4 for sample output from the reclamation program). A discussion as to the accuracy of these methods will be discussed later.

Budget Planning

Budgeting normally requires many of the elements already discussed (a mine plan and a reclamation plan in particular) in order to prepare a realistic budget. Another important element is the ability to perform accurate labor cost estimates and, since accountants seem to have an affinity for long, complex forms, this usually requires a great deal of "number crunching." Such was the case at Prairie Hill. The form developed by accounting to compute the UMWA labor cost was very long and, although not terribly complex, required many

repetitive and redundant computations. The system was set up to compute the labor costs by "function group" on a single form. A function group is defined as those individuals assigned to a particular mining function (i.e. Overburden Removal or Coal Loading). We currently have more than 20 such function groups at the Mine. Each form would take 30 minutes to an hour to complete and 3 to 4 days to complete the labor costs for the entire mine. After the second complete revision, we realized that we needed some help to quickly compute these revisions.

```
PROGRAM
LABOR

OB REMOVAL          112
DRAGLINE OPERATOR                    PROGRAM
PAY GRADE=           5               LABRAT
MEN-1ST=          3.00
MEN-2ND=          3.00               GRADE=       5.000
MEN-3RD=          3.00               DAY-STR=     125.024
MEN-IDLE SAT=     9.00               AFTER-STR=   127.537
MEN-SUN=          9.00               MID-STR=     128.374
                                     DAY-1.5=     184.737
%ABS=             8.00               AFTER-1.5=   188.450
CALL OUT=Y                           MID-1.5=     189.687
NORM DAY                             DAY-2=       238.852
$/DAY=         1,111.62              AFTER-2=     243.652
ADJ $/DAY=     1,022.69              MID-2=       245.252
DAYS=            250.00
ADJ $/YR=    255,671.83
ADJ HRS=      16,560.00
```

Fig. 5 Sample Output from
UMWA Cost Program

Again, we turned to the handheld system to provide this necessary help. The labor programs developed did not eliminate the need to complete the forms by hand. However, it did reduce the computational time drastically, which therefore reduced the total time required to complete each sheet (See Fig. 5 for sample output). The same sheet, which had originally taken up to an hour, could now be completed and checked in about 15 minutes and, in a day's time, a complete revision could be completed.

ADVANTAGES AND LIMITATIONS

Advantages

It would seem unlikely that a system with less than 6K of RAM and a 7 bit microprocessor would have any advantages over much larger systems. Other than initial cost, I can think of at least three things that I would consider advantages. They are:

- Direct transfer of Engineering Methods

- Immediate Help

- Micro-analysis of Mining Methods

Direct Transfer of Engineering Methods. This is a definite advantage to the mining engineer. Basically, to program the handheld computer, all the engineer has to do is switch to the program mode and insert variables for the constants that he would normally key in. This is somewhat of an oversimplification of the process used to program more complex problems. However, for everyday number crunching this is a reasonably accurate analysis. For more complex problems it is necessary to understand some of the basics of programming such as controlled looping and subroutines. Using these techniques, an engineer can program some of the most complex problems and tailor them to suit his individual needs. He can cut to the heart of a problem without going through a lot of unnecessary busy work. And, since he has done the programming, debugging, and necessary checks himself, he should understand the program, what it is supposed to do, how to correct a problem if it occurs, and how to change the programming as his needs change. Thus, he has the advantage of custom built software without the expense.

Immediate Help. There is a great deal of similarity between today's large computer systems and the stripping machines being analyzed. They are usually very expensive and may take up to 2 years to install and debug. In such cases, the handheld system can be used as an interim measure in order to produce the required reports and mine plans. Normally a great deal of time and effort is spent in the selection of software and hardware, due to the size of the expenditure. However, the engineer in the field usually needs help immediately; not 6 months to 2 years in the future and for a relatively small investment (usually less than $1200), the engineer can be provided with the power to do his job quickly and efficiently. In other cases, particular small operations, the handheld system may not only provide immediate help, but may fulfill all of the requirements of the operation.

Micro-analysis of Mining Methods. Most of the large systems are designed to give a "macro" or very broad point of view and they seem to break down when focusing on "micro" parameters that are second nature to most engineers. Rehandle is one of these areas that most systems have a hard time handling on a "macro" basis, especially in a dragline operation. Rehandle is usually given little emphasis and is simply treated as an add-on percentage of the total overburden volume. Rehandle at Prairie Hill is comprised of the road material, pit cleanings, and extended bench material, and cannot be simply treated as an add-on percentage. Our road material thicknesses vary greatly through th highs and lows of the rolling terrain and these effects must be calculated. The overburden may exceed the critical depth of the machine and the effects of this extended bench material must also be anticipated.

As discussed earlier, in the mine planning section, we analyze 120 m (400 ft) sections of each pit independently and apply varying amounts of road rehandle, pit rehandle, and, if critical depth is exceeded, the appropriate amount of extended bench rehandle. A 5% bust in our rehandle estimate can mean a 2.3 million cu m (3.0 million cu yd) difference in our production estimates and, since this type of error would be accumulative, a 5 year mining plan could be off as much as 11.5

million cu m (15.0 million cu yd). This is the equivalent of 9 months production for one of our 2570 draglines. Rehandle is a very real factor at our mine and must be calculated to the best of our ability.

Other "micro" parameters, that larger systems have a hard time handling, would be the additional overburden material and rehandle incurred with "chop-ins" and boxcuts. We are currently chopping-in along a highway, in order to maximize coal recovery, and since a chop-in is very similar to a boxcut, allowances must be made for the additional material. Another parameter would be the "drop-face" technique that we use to swing our pits. This technique allows us to turn our pits much quicker than the fanning technique. We mine about two-thirds of the previous pit and start to shear out but instead of shearing out, we drop half of the face and leave it square. In the next pit we will mine about one-third of the previous pit and do the same thing. The next cut is layed out so that it cuts across both of these faces and is cut the full length and width of the original pit. With the handheld system, handling these techniques is a simple matter and since we can handle these unusual methods, we are able to make our predictions real and acceptable to the operations staff.

Limitations

There are several limitations inherent to the handheld system. The three most apparent limitations are:

- Lack of Random Access Memory (RAM)
- Lack of Speed
- Lack of Structure Modeling Capabilities

Lack of Random Access Memory. The maximum RAM capabilities of the HP-41C that we use is approximately 6K. This is extremely small when compared to other computers on the market today. However, we use external mass storage in the form of a digital cassette drive which helps to cut down on the required on-line memory (See Fig. 6 for a photo of the entire system). Most of the time, we do not have much of a problem with the RAM limitations. The only program in which the storage problems became a real factor was the Labor Cost program which was discussed earlier. Basic spreadsheet operations, such as this one, that other computers handle quite easily, are so memory intensive that the handheld system's memory can be wiped out in a very short time. The program had to be rewritten several times to trim the memory usage in order to have it do all of the required spreadsheet operations.

Speed Limitations. Earlier, in the mine planning section, I discussed how we establish the grid data base maps by hand. It is the speed limitations of the handheld system that forces us to do this. Most large systems, that I am familiar with, use a sequential search of the drill hole data base to establish a uniform grid over the project area. Even on a mainframe computer, this method may take up to several hours of computer time to complete. We tried applying the same search and grid techniques to the handheld system not too long ago and found that it could be done but it would take several days to complete the same area. In this same period of time the grids could be established by hand. We have not given up on using the handheld system to compute the grid data base, but we must find a quicker, direct access method to the drill data base instead of a sequential search of the data base from top to bottom. Computational speed when using other applications seems to be somewhat slower than other systems, although I have not made any comparative tests and, there appears to be no buffering capabilities to allow the user to type ahead and increase data entry speed.

Lack of Structure Modeling. As previously mentioned, the grids are generated from already existing structure maps. This is just the opposite of the techniques used by larger systems. Larger systems usually generate the structure maps from the computed grids. The creation of these maps must be done by hand, using the handheld as an aid for this process. This could be a serious limitation for larger, multi-seam operations. At our operation, this requires us to generate a top-of-seam structure, coal thickness isopach, and consolidated overburden isopach. Using these maps and the existing topo maps, we are able to create the grid maps used for mine planning. As stated in the previous paragraph, we are working on a direct access method to speed up this process.

Fig. 6 Photo of handheld system
From left to right: digital cassette drive,
HP-41C handheld computer,
and thermal printer.

ACCURACY ANALYSIS

Mine Planning and Production

Since the first machine began production in 1980, we have maintained historical records of its productivity. We did this in order to eventually base our mine planning production rates solely upon this accumulated data. In 1982, after the third machine was put into production, we began checking the actual productivities with the forecasted data. During this period, we have had our share of problems with unscheduled downtime due to mechancial failures and a whole gamut of other things which can blow a mine plan's accuracy sky high. The accuracies at Prairie Hill have varied widely. We have had months when the actual production was only 67% of the forecasted values and we have had months when the actual production was 133% of the forecasted production. However, the actual values, on the average, have fallen in the 90 to 95% range. We would like to improve on this in the future. However, considering all of the factors that can influence a mine plan's accuracy, we feel that we are doing pretty well.

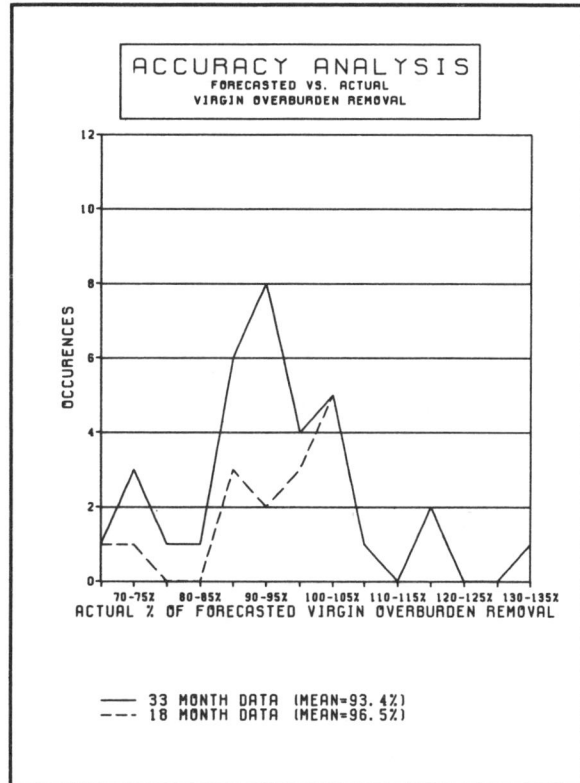

Chart 1

Chart 1 illustrates the accuracy of the forecasted overburden removal versus the actual overburden removed. All forecast data was obtained from the handheld mine planning system. The number of occurences on the y-axis are the number of months in which the percentage accuracy specified on the x-axis was achieved. The mean value for 33 months data was 93.4% of the forecasted data. The 18 months data was included to check on improvements in accuracy. As may be seen, the mean has improved 3.1% to a value to 96.5%. This improvement is largely due to the increasing historical data base and a large decrease in unscheduled downtime.

Chart 2 illustrates the accuracy of the forecasted tonnage versus the actual tonnage produced and Chart 3 checks the accuracy of the forecasted versus actual virgin ratio. On both the 33 month data and the 18 month data the tonnage accuracy is 3.3% less than the virgin overburden accuracy. The mean values for this analysis are 90.1% and

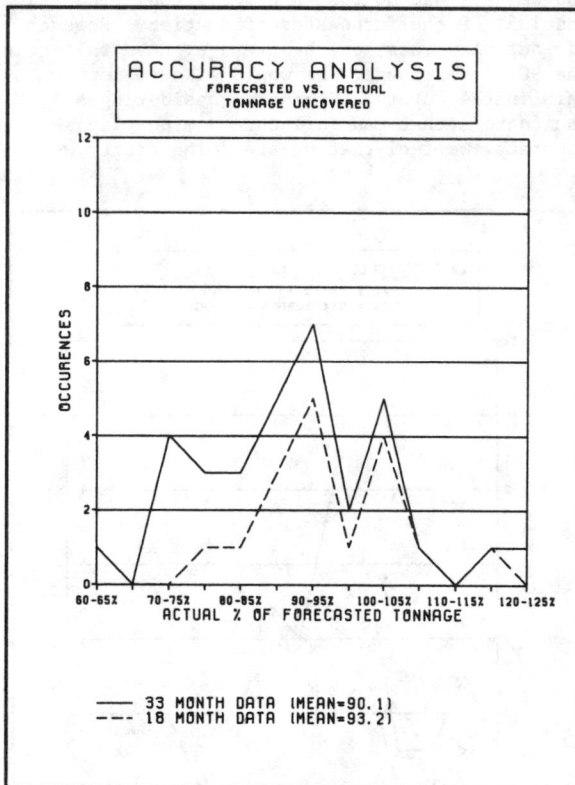

Chart 2

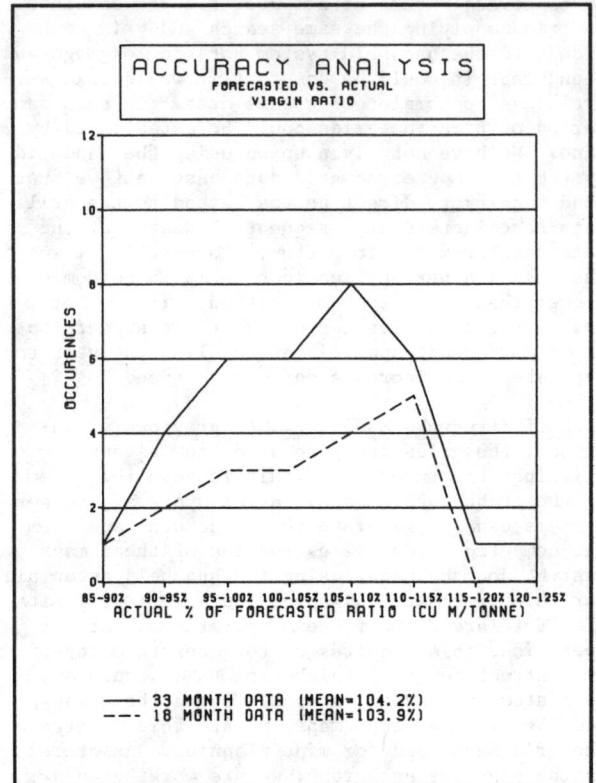

Chart 3

93.2%, respectively. The ratio analysis seems
to reinforce this point since it varies by approximately the same amount. The mean values for the
ratio analysis are 104.2% and 103.9%. This
appears to reflect an overstatement of the recovery
of coal in the pit and amounts to 0.05 m (2 in) of
coal. In future mine plans we will allow for this
difference.

Reclamation Planning

    Unfortunately, at present, the reclamation
program cannot be any more accurate than the structure range diagram technique. Both methods assume
that the area in the pit is spoiled laterally along
the x-axis into the old pit. Neither technique
allows for the translation effect which takes
place along y-axis (the x-axis is assumed to be
parallel to the cross section line and the y-axis
is assumed to perpendicular). This appears to
be the reason why the actual values, when compared
to the forecasted elevations, seem to vary so
greatly (See Chart 4). Because of this translation error, a fairly large area was sampled (305
m x 365 m) to obtain a realistic mean value. This
value was found to be -0.17 m (-7 in). In the
future, we hope to expand the general range diagram to allow for this translation, and this
should bring these variances into line.

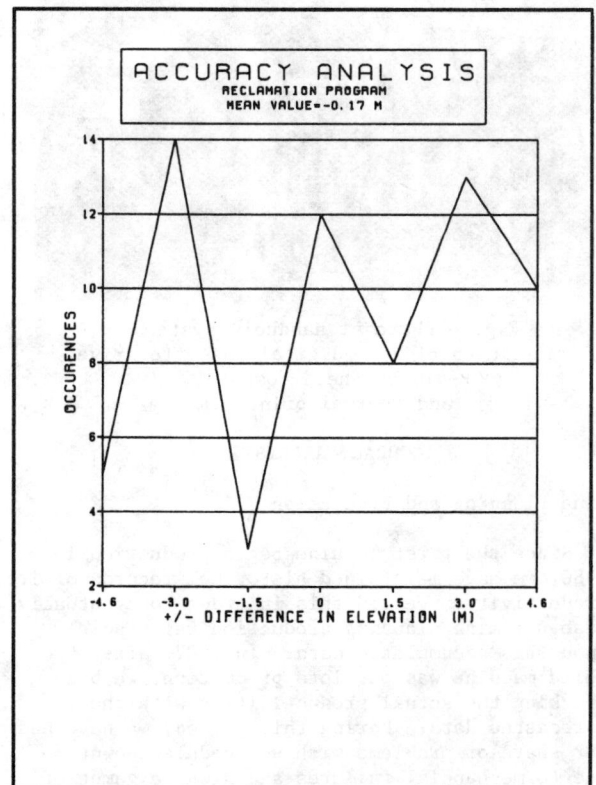

Chart 4

CONCLUSIONS

For the past 5 years the handheld computer
system has served us well as the primary mine
planning tool for the development and operation of
the Prairie Hill Mine. It has been used as an
interim measure to do the required mine planning,
reclamation planning, and budgeting. Thus,
bridging the gap between the initial purchase and
the final development of our mainframe system.
This may be the handheld computer's niche; to provide immediate help for the mine engineer and to
serve him until a larger, faster, and more efficient system can be developed. It is sincerely
hoped that this paper effectively illustrates this
point and that it will induce a greater respect
for the handhelds as effective mine planning
"tools" and not just "toys" with little capacity
for serious work.

Chapter 36

A DYNAMIC MODEL OF A SURFACE COAL OPERATOR

by

Anil B. Jambekar
Michigan Technological University
Houghton, MI 49931

and

Patrick A. Kinek
U.S. Bureau of Mines
P.O. Box 18070
Pittsburgh, PA 15236

ABSTRACT

A computer simulation model has been constructed as a framework to study the consequences of government regulations, and the delays inherent in these regulations, on the functional areas of small surface coal operators. This paper specifically looks at the effect of the Surface Mining and Reclamation Act promulgated by Congress in 1977. The act has altered basic operating decisions of coal operators in land management. Lengthy and costly permit preparation procedures, lengthy local and state review of permits and lands, increased bond fees and costly reclamation requirements are some of the critical changes the regulations have brought about. Policies frequently considered by the surface mining industry and the government to alleviate the hardship caused by the regulations are mechanisms to offset increased bond fees. It is a purpose of this paper to demonstrate the utility of computer based simulation as an effective methodology to study the long term effects of such policies.

INTRODUCTION

This paper reports an application of compute modeling to study the long term consequences of government processing delays on small surface coal operators. The study is done with the aid of a computer model developed at Michigan Technological University. Funding for the development of the Small Coal Operator Model (SCOM) was provided by the Office of Surface Mining (OSM) to simulate the dynamic impact of government regulatory policies on small surface coal operators (Jambekar and Kinek, 1981).

The objective of this paper is to bring to the attention of the academic and non-academic community interested in coal, a computer based formal policy model that can be used to address some operational, regulatory and organizational policy model. The interested readers may also refer to other related papers (Kinek and Jambekar, 1984a, 1984b).

SCOM is the first system dynamics model to look at the operational aspects of a mining company in the United States. It includes a relatively detailed representation of the functional areas of a small coal operator and an aggregate representation of the coal market, lending institutions and the government (Figure 1). The model structure describes land management, the physical flow of coal, equipment capacity management and money management. Inherent in this structure are the policies utilized by the small operator in monitoring and controlling coal production and distribution, land resources, equipment utilization and money flows. The external entities in the small operator's influencing environment are the coal market, lending institutions and government. Coal markets influence the flow of coal shipments and revenue. Lending institutions influence the flow of money by controlling debt, interest rates and bond payments. Government has traditionally influenced money flows via taxes. Recently though, government extended its monitoring and control capabilities by promulgating the Surface Mining Control and Reclamation Act of 1977 (Public Law 95-87). This law mandates that coal operators include permitting and bonding policies that modify their traditional land management practices.

The SCOM was built on information gathered during a two-month trip through the midwest and eastern coal regions (Small Surface Coal Operators, 1980, private communication). Agreements were reached with small coal operators in six states to visit their operations and study mining techniques, local permitting and bonding process requirements, their relationship with banks and bonding companies and, most importantly, their decision-making rules. Specifically, these operators were located in western Pennsylvania, northern West Virginia, eastern Kentucky, southern Ohio, western Missouri and eastern Oklahoma. The

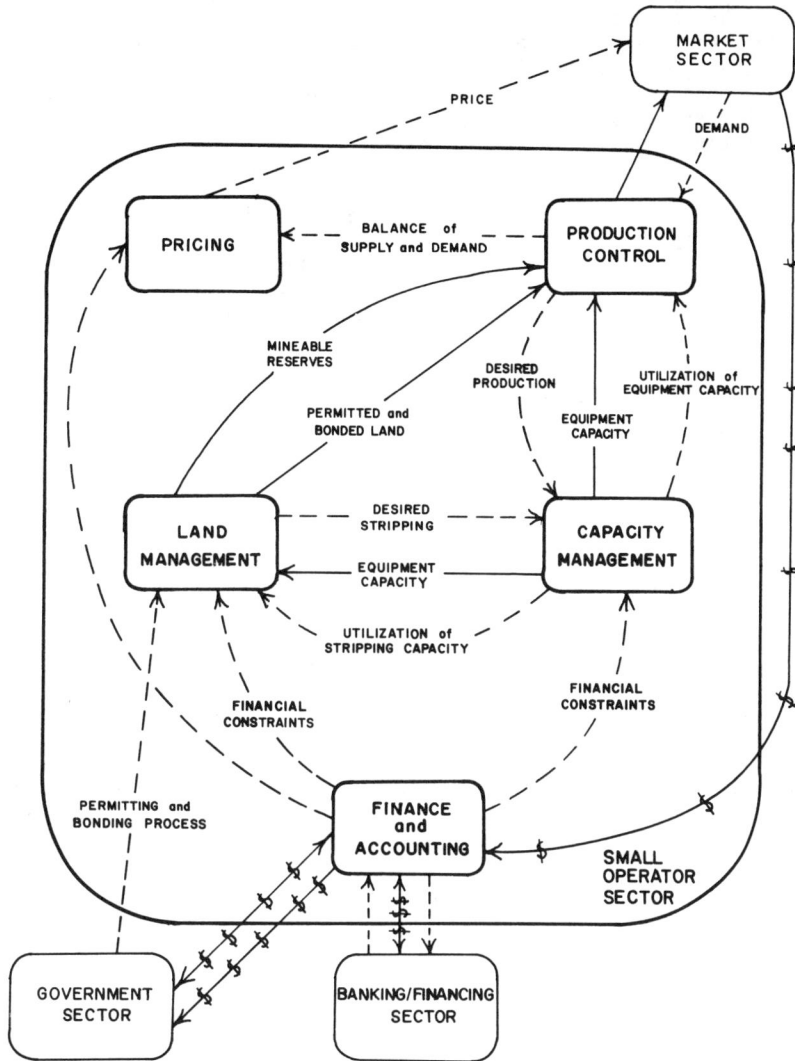

Figure 1. Subsystem Diagram of Small Coal Operator

Pennsylvania operator agreed to our use of his operation in forming the model's basic structure. This was an exceptional sacrifice for them because construction of the model's basic structure entailed a month-long interviewing process. Information gathered at the other mines was used to verify the adaptiveness of this basic structure to each operation.

Other contributors to the formulation of this model were a coal broker, a strip mining and engineering consultant and two banks, (1980), all located in the western Pennsylvania region. The coal broker described the relationship and interactions between small coal operators and the coal market. The consultant gave a timetable and specific costs associated with the permitting process. Both banks outlined their lending policies towards small coal operators seeking equipment loans or working capital loans. Follow-up interviews with the

Pennsylvania operator and one bank were conducted six months later after the model was completed for validation purposes.

The main thrust of the interviewing process was drawing out the mental models each operator used to make decisions. The receptiveness of the operators to the concept of mental models was encouraging and extremely useful in concentrating their thoughts on how and why they made particular decisions. The information captured during these interviews clarified each operator's policies, their unique reactions to similar circumstances and degree of reaction.

For the purposes of this paper, a small operator of size 50,000-100,000 tons/year is selected.

This paper analyzes three different bonding and permitting-related policies. These are 1) reduce the state bonding fee from $4,000 per acre to $1,000 per acre, 2) increase the bond retirement rate by reducing the time to retire a bond from sixty months to thirty months and 3) decrease the permitting and bonding delays gradually while retaining the state bonding fee at $4,000 and the time to retire a bond at sixty months.

METHOD OF ANALYSIS

The vehicle for analysis is a simulation model of a small coal operator's organization. The model can approximately be viewed as an organizational process model (Cyert, R. M. and March, J. G., 1963) and should be distinguished from normative models. An organizational process model is concerned with the interactions between existing decision rules that might account for inefficiencies. Typically an organizational process model will span several functional areas. By contrast, most normative models do not pay great attention to the existing decision rules and are more specifically focused on designing optimal decision rules within a single functional area.

The model was constructed using the structuring principles of system dynamics as defined by Forrester (1961, 1968). System dynamics is particularly appropriate since it readily lends itself to portraying the breadth of organization structure (encompassing several interrelated functional areas) that forms the operating environment for a typical small coal operator. Furthermore, the formulation of a system dynamics model clearly identifies the goal-seeking aspect of decision-making in which management traditions and philosophies are reflected.

A system dynamics model differs from most other simulation models in that it generates its own history from relationships built into it, rather than merely computing the financial results to be expected from a given sales forecast using the firm's financial ratio based on past performance (see Schrieber, A. N., 1970).

Although it does not supply optimal solutions to problems, system dynamics can provide the necessary understanding of the complex operations of an organization that helps policy makers find, at least, good solutions. It focuses on the processes causing the search for more sensible policies for stability and survival.

System dynamics is the application of feedback control theory to the dynamic phenomena arising in economic, industrial and corporate systems. System dynamics integrates the functional areas of a system into a single framework, creating and "organizational laboratory" for the purpose of designing and testing more effective policies. Within the small coal operator sector, the functional areas or "subsystems" are production control, capacity management, land management, pricing and finance and accounting control. Analysis of the dynamic implications of the interactions between these functional areas is possible under a system dynamics framework.

System dynamics uses computer simulation as a means to generate the time-varying behavior of model variables. The computer has the capacity to handle the complex interactions between variables and recreate the nonlinear behavior characteristic of real-world systems. Computer simulation enables policy designers to bring their system into the "laboratory" for analysis and redesign.

System dynamics requires the close involvement and input of the client (coal operator, manager, etc.) during and after model construction. System dynamics models are constructed using terminology the client is familiar with and are based on the client's mental models of how the company operates. The close involvement necessitated by system dynamics philosophy enables the client and the model developer to acquire the knowledge of system interaction. With this knowledge, understanding of the dynamic implications of decisions can be greatly improved.

The results cannot be validated in any rigorous sense because of the nature of the problems and in any case, time was of the essence. There are some philosophical considerations when evaluating the usefulness of model validity. On the one hand, consideration for the more traditional purposes of prediction and optimization, and on the other hand consideration for the purpose of applying science to management for a path of discovery into the policy areas of an organization.

There is some question as to the benefit of a more rigorous validation attempt in this instance. The intention at the time was to take the fragmented cause-effect maps of various actors in the system - upon which they base their decisions anyway - and weave these into a whole map of the operational domain. The assumptions implicit in these cause-effect relationships were examined for plausibility and vaguely defined or missing parts were supplied by further inquiry among the participants. If the resultant map depicting cause-effect relationships is found to be a satisfactory representation of reality by the original participants for the purpose in hand, then it is from the organization's point of view valid. Furthermore, if it results in insights and understandings that permit more complexity to be understood by participants and hence, introduced into their decisions where crude but forceful arguments based on simplistic assumptions prevailed before, then, at the least, it is a useful model. The model, therefore, is not an optimizing or accurately predicting device, but a tool for learning, change and intervention during a time of policy formation.

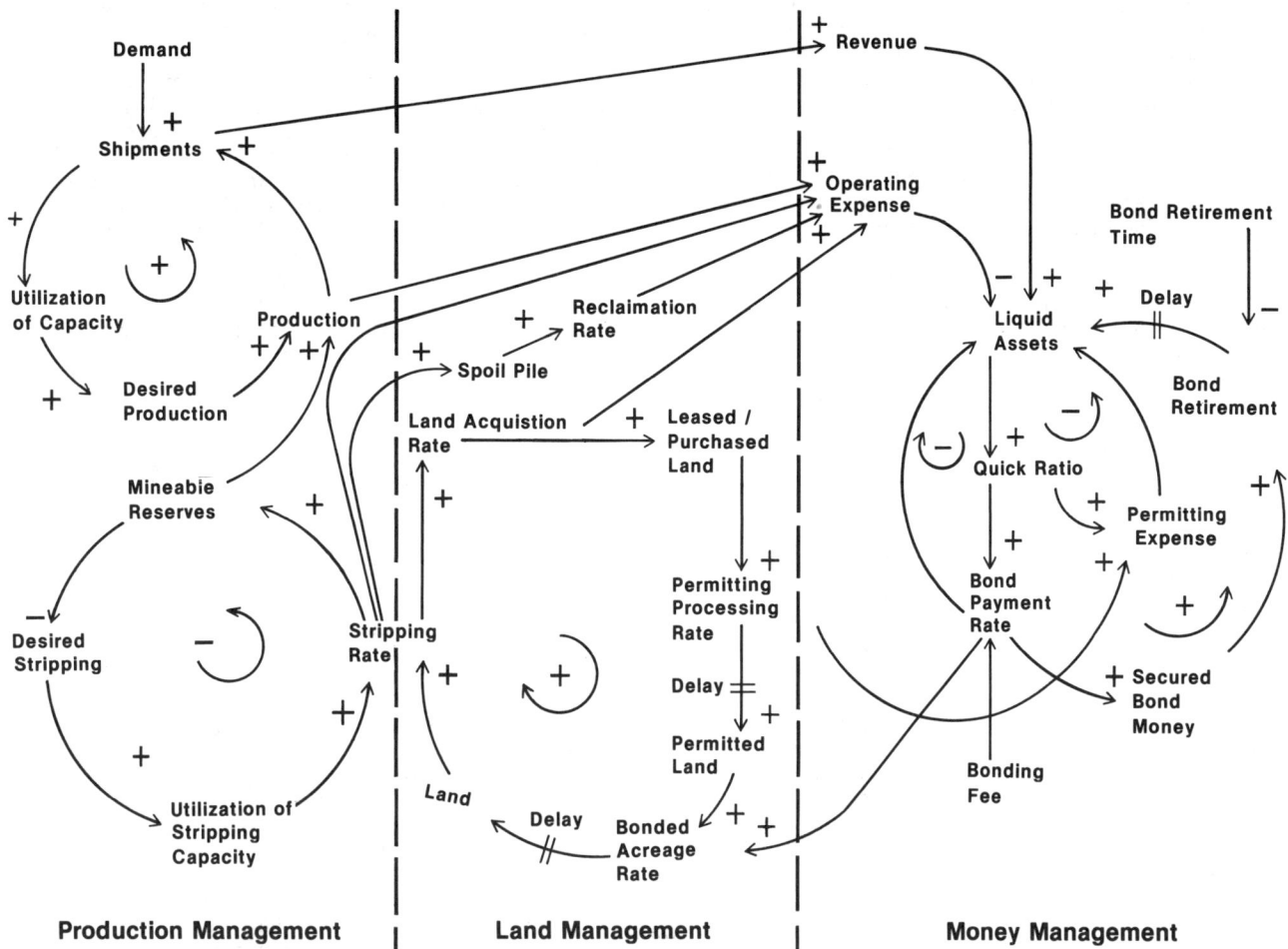

Figure 2. Causal Structure

MODEL STRUCTURE

Land management is the area within the small surface coal operator system that government has felt the greatest need to monitor and regulate. The government's objective is to insure that the operators refurbish the land to its premining condition. The government's method of regulation is based on issuance of permits and bonding requirements. This introduced a disciplined approach to land management. However, the effect of the decisions made in land management in response to government regulations radiated throughout the production and finance areas of small operator.

The essential causal structure of various variables of interest is displayed in Figure 2. The basic elements of causal diagram are very simple. The policy variables, performance variables and concept variables are represented as points on a sheet of paper, and the causal influence are shown by arrows between these points. A positive sign denotes that an increase in the variable at the tail of the arrow will lead to an increase in the variable at the head and a decrease leads to a decrease. A negative sign indicates the opposite movement of the head variable from that of the tail one.

Land Management

Changing the form of the land is a physical process; changing the status of the land is a required regulatory process. Land status must change after the land is acquired and before stripping begins in order to comply with PL 95-87. After a tract of land is acquired, it must achieve "permitted" and then "bonded" status before stripping can commence. This entails completing mining and drainage permit applications for review and subsequent approval by local and state agencies. Attainment of permitted status is followed by procedures to bond a portion of the newly permitted land. The useful life of a permitted section of land is much longer than a bonded section. Due to the large monetary outlays associated with bonding ($2,000-$4,000 per acre), only a section within the permitted area is bonded at any one time.

Acreage equivalent to the small operator's yearly stripping rate is the minimum bonding requirement. Bonded status is also preceded by the review and subsequent approval of the bond by local and state agencies.

The foundation of sustained growth for small surface coal operators is the acquisition of sufficient land and reserves. The stripping rate determines what is a sufficient rate of acquisition. Once the coal underlying a parcel of land has been exposed by stripping and mined, that parcel of land has no future potential in regards to coal production. In order to compensate for this continuous depletion of reserves, land must be acquired at a rate comparable to the stripping rate. If not, the life of the mine and the company will continuously decline. The ability to acquire sufficient land and reserves is expected to become a problem for small surface coal operators in the East within ten years. Figure 2 illustrates the feedback loop used in this model to produce growth within the land management sector along with the delays associated with changing the permitted and bonded status of the land. The delays inherent in the permitting and bonding process are the most critical variables in this model. As the "paperwork and processing time" or the "bonding process time" is increased, the permitting process rate and the bonding process rate, respectively, decrease. Once this occurs, the gain previously observed in this positive loop becomes a degenerative loss. If the behavior is not controlled, the level of bonded land approaches zero, eventually halting the stripping rate and production.

Production Management

Overburden must be stripped off before the coal underlying it can be mined. If the production rate exceeds the stripping rate, the level of mineable reserves (exposed coal) will eventually decrease to a point where production must be halted until more overburden can be stripped away. At this point, since there is no production, coal stockpiles begin to diminish. Shipments will eventually decrease if production cannot resume, decreasing revenues from coal sales. This squeeze on production resulting from an inadequate stripping rate can occur if the production rate exceeds the stripping capacity, or, if the level of bonded land is inadequately maintained due to processing delays or excessive costs. If land is not bonded, a coal operator can not legally begin stripping.

The decision to adjust the stripping rate based on the level of mineable reserves is illustrated in Figure 2. The negative feedback loop captures the process of matching the actual stripping rate with the desired stripping rate. The rate of exposing reserves and the stripping rate are identical variables with different units. The stripping rate is in acres per month while the rate of exposing reserves is in tons

of coal per month. There is an acceptable conversion between acres and tons of coal based on the seam height and number of acres. The conversion states there is approximately 1,742 tons of coal per acre-foot. Multiplying this figure by the seam height and the number of acres, a conversion from acres to tons of coal is made. Thus as the stripping rate increased, the rate of exposing reserves also increases. This results in an increase in the level of mineable reserves. The level of mineable reserves is compared to a constant goal, desired mineable reserves. If the level of mineable reserves is below the desired level, a position adjustment in the stripping rate is made. This is accomplished via the correction for mineable reserves. If the level of mineable reserves is too low, a positive correction for mineable reserves is made. A positive correction results in a positive increase in desired stripping, increasing the utilization of stripping capacity. Utilizing more stripping capacity increases the stripping rate, eventually creating an increase in the level of mineable reserves. Once mineable reserves coincides with desired mineable reserves, adjustments in the stripping rate will cease.

Although the level of mineable reserves has a strong effect on the stripping rate, the level of bonded land has an even stronger impact. Even though we may need to adjust the stripping rate to increase the level of mineable reserves and maintain the production rate, if the level of bonded land is low or nonexistant, the stripping rate will have to decrease or stop regardless of the internal pressures to increase it. Figure 2 demonstrates how land management's positive feedback loop is competing with the goal-seeking negative feedback loop for dominance and control of the stripping rate.

Impact on Cash Flow

The functional area most closely monitored in any business is the financial control sector. Most decisions eminate from this sector and radiate out to all areas of a business. Financial decisions are made for the purpose of hedging against conditions outside the control of the decision-maker that influence financial behavior. For instance, the small coal operator has little or no control over inflation, interest rates, demand for coal, bank loan policies or cost-bearing government regulations. While interest rates and costs are rising (or were), prices are falling and bank loan decisions varying, what effect will different government regulatory policies have on the financial behavior of the small coal operator? The causal diagram in Figure 2 illustrates where government policies intercede in the small operator's finances, creating behavior that radiates throughout his entire system.

The government regulation that creates the largest drain on the small operator's cash flows is bonding. Although the money securing the bond is eventually returned to the small operator, the problem lies in the length of time between purchasing the bond and retiring the bond. It currently takes five years before a bond is retired and the money is returned to the small operator; that is assuming there have been no citizen's complaints contending the operator has not adequately reclaimed the land and water. If such complaints exist and are substantiated by an OSM inspector, the five-year retirement period starts over again. Twenty acres of land is what an average-sized small operator mines yearly. The cost of a bond for one year for an operator of this size would range between $40,000 and $80,000, depending on which state the land exists in. Consider $80,000 being taken out of the cash flows of a highly leveraged business yearly and not recovering the money for a minimum of five years. At the end of the five years, a newly established operation will have put up $400,000 before its first $80,000 is returned. The bond payment rate depends on the image the operator projects to a bonding company. If a bonding company will back a particular operator's bond, the operator may then only pay five percent of every $1,000. On the other hand, if the operator is inexperienced, has never dealt with the bonding company before, has an unstable financial position or has a bad reputation in the local mining community, the operator will probably be denied backing by the bonding company. The operator must then assume the total financial burden himself.

Money expended for the purpose of preparing a permit can average between $50,000 and $60,000 per year for a small operator. Permit expenses are a burden placed by the government on the operator to insure that proper mining techniques are employed and safeguards against spoiling the land and water have been taken. Payments for preparation of permits are never reimbursed.

Counteracting the drains on liquid assets through permitting and bonding are the revenues realized through coal sales (Figure 2). Oftentimes these revenues are insufficient to justify the purchase of a new bond due to increasing operating expenses or a depressed coal market price. If a new bond cannot be purchased, stripping will have to be curtailed. Following a declining stripping rate will be a decline in production and shipments, resulting in decreased revenues. By not purchasing a bond, the operator is wagering that the coal underlying his declining level of bonded land will last long enough to bring in the necessary revenues to afford a new bond in order that mining can continue. This strategy or predicament is extremely risky during a period of faltering demand, rising cost and stagnate prices.

Another strategy frequently observed when a small operator's expenses are outpacing revenues involves cutting back on reclamation and its associated expenses long enough to regain solvency before OSM inspectors order him to resume normal reclamation activities. As shown in Figure 2, reclamation monetarily contributes only to operating expenses. The long term contribution of reclamation is the preserved beauty and productivity of the land and water. To the small operator who is in a short term financial squeeze, reclamation is the one activity in the mining cycle which can be curtailed without decreasing production. Two forces act on the reclamation rate; one is internal and the other external. As explained if the financial performance of the operation is waning, the tendency is for the operator to cut back reclamation in order to regain solvency. Opposing this internal force are government regulations and government inspectors. If the level of spoilpile acreage becomes excessively high, government inspectors will release an order to resume reclamation regardless of the operator's financial condition. To a point then, the internal force dominates the external force and spoilpile acreage is allowed to increase. Once the spoilpile acreage level exceeds a particular inspector's criteria, the pressure to reclaim begins to dominate, requiring reclamation and its drain on liquid assets to resume.

APPLICATION OF THE MODEL

Permit and Bond Reviews Extended or Backlogged

One of the advantages of a system dynamics model is its ability to simulate the open-ended question, "What would happen if...?" Two of those questions were asked here in order to design and test policies under what seem to be realistic assumptions. The just assumption or set of conditions is that demand for coal, operating expenses and equipment costs will all rise in the future. Over the twenty years of simulation time, demand for coal will increase five percent per year (The Wall Street, September 25, 1979), or in other words, a doubling by twenty years end. Based on trends in the seventies, increases in operating expenses and equipment costs were also made over the twenty year period. The trend in expenses was dependent on this study's desire to simulate the effect government regulations have on the marginal small operator. In this case marginal translates into approximately a ten percent profit margin during the entire twenty years. As demand pulls price upwards, the operating expense and equipment cost trends increase accordingly to maintain a ten percent profit margin. The question may be asked, why not simply maintain constant demand, price and expenses and achieve the same ten percent profit margin objective? Statistically, this observation has merit. Dynamically, even though a relatively consistent profit margin is maintained, pressures created by a desire to expand in response to a growing demand build up in the system, producing behavior that otherwise would not have appeared under a constant demand scenario. This study's desire to simulate marginal operators stems from the vulnerability of this type of operator to increased government regulations and an inflationary economy.

The second assumption is based on current patterns in the government permitting and bonding review agencies and also considerations by the agencies to extend the reviewing process. During a sluggish coal market in 1979-80, the processing times for reviewing a permit and a bond stood at twelve months and one month, respectively. These processing times were already excessive and caused problems according to the small operators interviewed. These were also considerations emerging from the review agencies of manually extending the reviewing process (one such consideration during 1979-80 proposed extending the bond review process to one year). These patterns resulted in the question being asked, if the reviewing process during a sluggish coal market is already lengthy, what will happen if or when the demand for coal begins increasing steadily? A continuous increase in the review process times simulates a scenario whereby there is a continual increase in the requests for permits and bonds by large and small operators to government review agencies with already lengthy review procedures and with limited manpower and capacity. This limited capacity results in an increased request backlog and inevitable delays in processing the requests. This scenario is a likely possibility if the government fails to reduce regulatory "red tape" but does reduce a limit government manpower during a period of increasing demand for coal. For the purposes of this paper we have steadily increased the permit review process from one year to two years and the bond review process from one month to six months. Both delays increase steadily over a ten year period for purposes of graphics clarity only. None of the assumptions in this study are predictions or forecasts, especially in regards to the timing of events.

The first model runs in Figure 3 indicates the impact of an increasing permit and bond review process on the small operators. The dynamic behavior of the small coal operator system proves to be undesirable as the processing delays are increased. The degree of undesirable behavior becomes more severe as the size of the small operator decreases. As the larger small operator's stock of bonded land begins decreasing in Figure 3, the bond payment rate is forced to increase rapidly in order to sustain desired rates of production in the face of an increasing demand for coal. The increased bond payment rate depresses cash flows to a critical level. At this point the small operator must backoff the rate at which he is acquiring bonded land to replenish his cash flows. This action allows stocks of bonded land to fall to a level whereby stripping rates and subsequently production rates are also forced downward. When cash flows recover, the bond payment rate begins another rapid increase in order to increase production to meet demand. A large payment is made so that a sufficiently large parcel of bonded land is acquired to

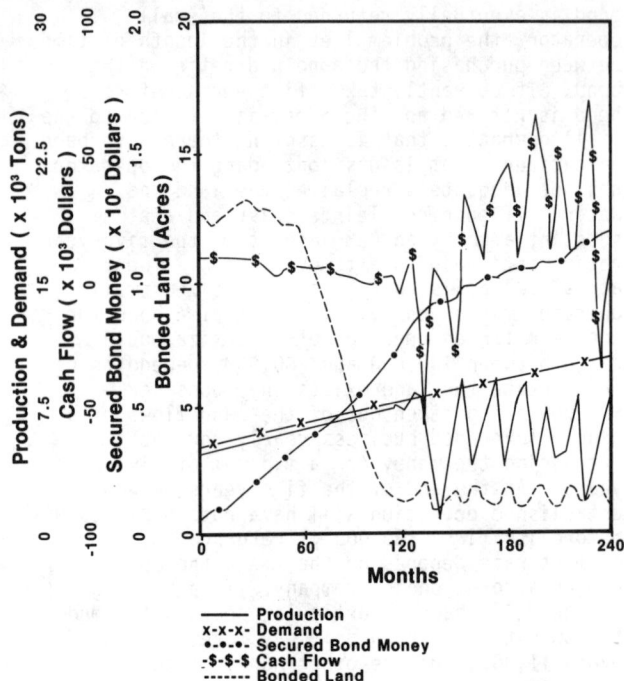

Figure 3. Permit and Bond Reviews Backlogged

sustain the operator until his new bond requests are processed, which at this time in the model takes six months. The increase in bonded land starts the stripping rate upward along with the production rate. The large bond payments, though, and increased system activity drains the cash flows again, causing the cycle to start over.

The smaller operator in Figure 3 begins this oscillatory behavior before time period "120" because he is more sensitive to increased processing times. This sensitivity arises from the fact that the resource base decreases as the size of the coal operator decreases. As a result, the large swings in bond payments, cash flows and production are particularly harmful to an operator with a small resource base. The smaller operator begins exhibiting oscillatory behavior in time period "60" and by time period "120" is almost out of business.

State Bonding Fee Policy

One suggested solution to alleviate the financial burden being carried by the small coal operator is to decrease the state bonding fee. The model runs shown in Figure 3 used a state bonding fee of $4,000 per acre. This will be reduced to $1,000 per acre to determine if decreasing the fee will improve the previously observed behavior. Our assumption that the local and state permitting and bonding agencies have a fixed processing capacity will now come into play. At some point the rate of requests for permitted and bonded land exceeds the rate at which requests can be processed within a reasonable time. When this point is reached, the capacity of the permitting agencies will be fully utilized and a backlog will begin. The backlog will continue to expand to a level where it would be illogical for a coal operator to increase his requests or compensate for the increased processing time by investing more and more money into bonded land per request. Increasing the rate of requests becomes illogical when in order to keep pace with demand permitted and bonded land is needed at time period "120", but the request cannot be processed until time period "130". Regardless of how much money is thrown at the permitting offices, the request would not be processed in time. This scenario would force the bond payment rate, the stock of bonded land, and the stock of permitted land to level off at a point where request can be processed fast enough to sustain a lower than desired rate of production. If demand exceeds this production rate, increasing the price of coal to induce more production will not decrease the delays in processing permit and bond requests. The system is very resistant to change despite the large quantities of money thrown at the problem.

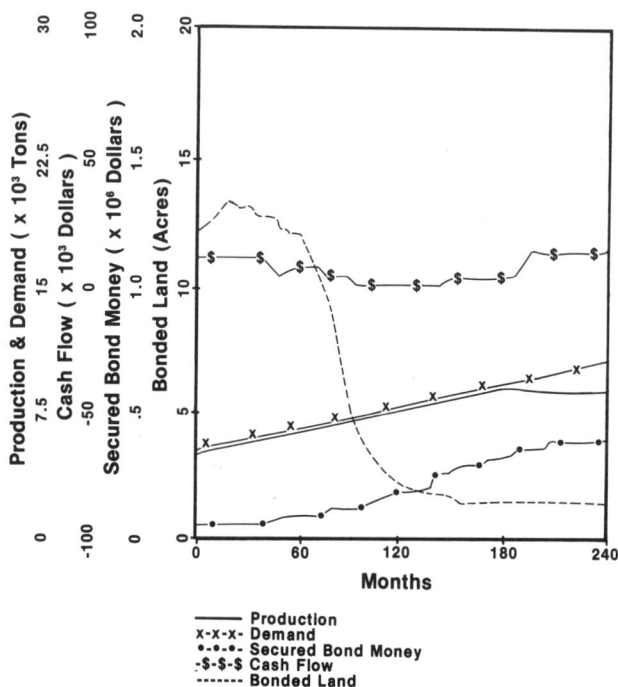

Figure 4. State Bonding Fee Policy

Observing Figure 4 reveals that a reduced bonding fee stabilizes the small operator at a threateningly low level of cash flow and allows the operator to remain operating, albeit, in a somewhat tenuous oscillatory manner. The leveling-off point is where permit requests are being processed in two years and bonds in six months. Attempts to increase the bond payment rate beyond this point would still not aid in matching production with demand given that the permit and bond review agencies have a limited

processing capacity. The policy to reduce the state bonding fee lowers the bond payment rate and the level of secured bond money.

Although the lower bonding fees address the immediate problem of alleviating the small operator's financial burdens, it does not address the primary cause of the financial pressure - the delays created by increased permitting and bonding process times. A government policy to decrease state bonding fees could control financial pressure, but it does nothing to relieve the cause of the pressure. Whether controlled or not, the pressure still remains within the system. Another popular policy often expressed by industry as a possible solution is an increased bond retirement rate.

Bond Retirement Policy

The small coal operators who were interviewed during the course of this study felt very strongly about the excessive length of time bond money is secured. The current length of time to retire a bond is five years. With an increasing state bonding fee, the quantity of funds secured is considerable. A large outflow of money coupled with a much slower inflow is very damaging to the sensitive cash flows of small operators. A long bond retirement time also discourages bonding companies from actively participating in the bonding process. The risk involved as the bond retirement time increases becomes too great for them. The chances of losing all or part of the bond money they secured for the operator in the course of five years for improper reclamation is very likely under the current regulatory framework. Bonding companies, the surface mining industry and banking institutions who loan money to small operators would all desire a shorter bond retirement time.

An increased bond retirement rate was formulated by decreasing the time to retire a bond from sixty months to thirty months. The state bonding fee has been increased to $4,000 per acre again for this test. Thus the bond payment rate is still very high, but by lowering the bond retirement time the secured bond money is returned to the operators much faster. Figure 5 exhibits behavior very similar to that produced by lowering the state bonding fee. The small operator in Figure 5 is still struggling to remain in business despite a quicker bond reimbursement. The small operator system treats any new source of money in a similar manner. As long as excessive delays in the permitting and bonding process exist, the system will remain very resistant to change.

The final policy formulation entails reducing the processing delays to their original values. The policy is formulated by allowing the permitting and bonding process times to constantly increase up to time period "120" as before, then quickly reducing them to their original values by time period "160".

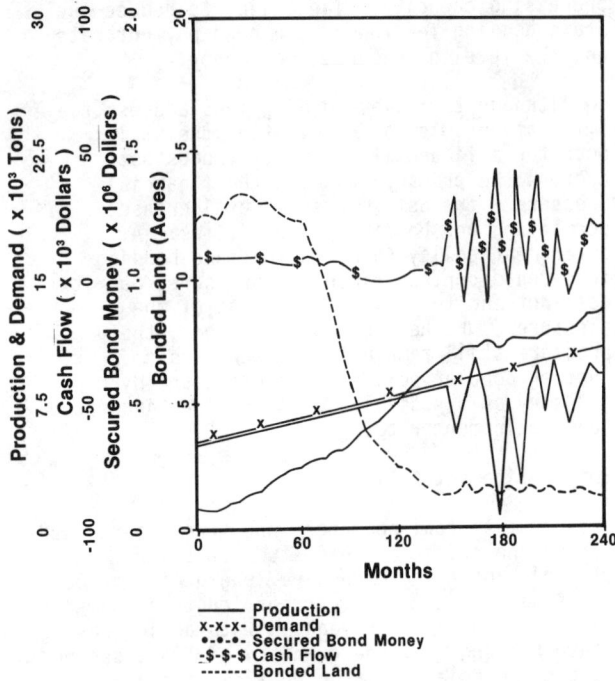

Figure 5.   Bond Retirement Policy

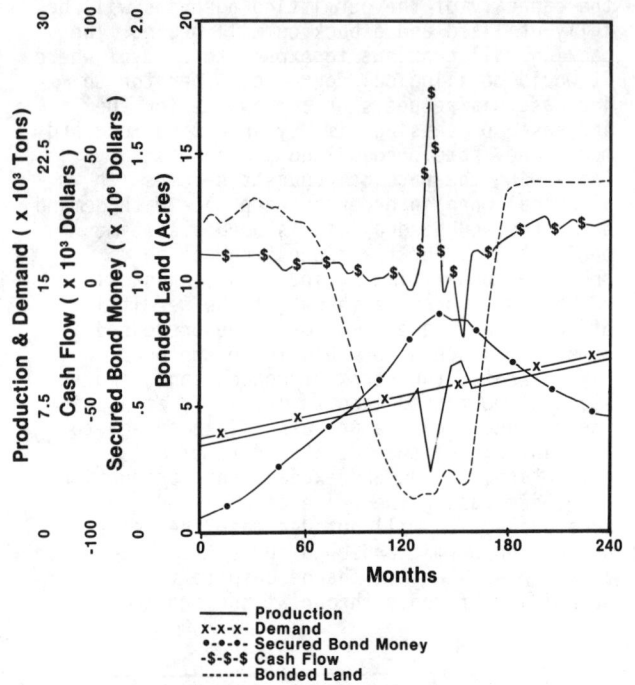

Figure 6.   Review Process Policy

Review Process Policy

To this point there has been no policy
directed toward reduction of the processing
times involved in the permitting and bonding
process. The processing times have been allowed
to increase and stabilize at a very high level.
The behavior of the small operator system when
an increasing, then decreasing processing time
is simulated is shown in Figure 6. After time
period "120", as the processing times begin to
decrease, the production rate starts an
oscillatory ascent. By time period "190" small
operators attain a stable growth pattern
commensurate with demand. By reducing the
processing times, local and state permitting
agencies are capable of processing many more
requests with the same processing capacity.

The bond payment rates for the operator
falls to a much lower rate even though the state
bonding fee remains at $4,000 per acre. The
level of secured bond money is also
significantly reduced even though the bond
retirement time is still five years. Looking
back, as the processing delays increased the
bond payment rates also increased in order to
maintain the stock of bonded land. In other
words, the small operators use up more bonded
land between processing periods as the periods
increased. If the operators do not increase
their request for more bonded land per
application, then their stock of bonded land
will continue to decrease as the processing
delays increase. With the request for more
bonded land increasing, the bond payment

accompanying the request also must increase.
Reversing this logic, as the processing delays
are decreased, the bond payment rate per bond
request can also decrease without jeopardizing
stocks of bonded land. The bond payment rate
now has much more freedom to react without
pressuring the entire system.

CONCLUSION

This paper described the simulation model
SCOM and its uses in understanding the impact of
government processing delays on small surface
coal operators. SCOM captures the physical and
regulatory dynamics of land transformation
within the surface mining cycle.

It was found that policies frequently
considered by the surface coal mining industry
and the government to alleviate the financial
burdens of small coal operators are mechanisms
for channeling money to the operator in an
effort to offset increased bond payment rates.
By formulating policies that merely channel
money to the small operator, financial pressures
are controlled but not relieved. These types of
policies addresssed the symptoms of increased
processing delays, not the causes. Under an
increasing demand scenario, reducing the
permitting and bonding process delays is the key
to improving the behavior of all operating
ranges of small operators. More research into
the structure and procedures of various mining
regulatory agencies is needed to detail how and
where suggested policies could be implemented.

The most unique feature of SCOM is that it is a systems-oriented policy model looking at the operational dynamics of the U.S. mining industry today. There are many areas in the mining industry that need formal policy models to address operational, regulatory and organizational policy problems.

REFERENCES

Cyert, K. M. and March, J. C., A Behavioral Theory of the Firm, Prentice Hall, Englewood Cliffs, N.J., 1963.

Forrester, Jay W., Industrial Dynamics, MIT Press, Cambridge, Mass., 1961.

Forrester, Jay W., Principles of Systems, Wright-Allen Press, 1968.

Jambekar, Anil B. and Kinek, Patrick A., "Dynamic Study of the Impact of Government Regulations on Small Surface Coal Operators," Final Report prepared for Office of Surface Mining, December, 1981.

Kinek, Patrick A. and Jambekar, Anil B., "The Effect of Government's Permitting and Bond Review Process on Small Surface Coal Operator," Materials and Society, Volume 8, No. 1, 1984a, pp. 71-82.

_____ and _____, "Simulation of A Small Surface Coal Operator," Simulation, Volume 42, No. 3, March 1984b, pp. 125-131.

Mining Information Services, Key Stone, Coal Industry Manual, McGraw Hill Mining Publications, 1979.

Public Law 95-87, United States Statutes at Large, 95th Congress, First Session, 91 Stat. 445, Volume 91, 1977.

Schrieber, A. N., Corporate Simulation Models, University of Washington, Seattle, Washington, 1970.

U.S. Bureau of Mines, Mineral Facts and Problems, Bicentennial Edition, U.S. Government Printing Office, 1975.

The Wall Street Journal, "Overcapacity in Coal Industry is Blamed for Weak Prices: Shakeout is Underway," September 25, 1979.

Communication with banking officials, August 20-October 19, 1980.

Communication with engineering consultant, August 20-October 19, 1980.

Communication with small surface coal operators, August 20-October 19, 1980.

Chapter 37

OPTIMIZATION OF SURFACE MINING SYSTEMS UTILIZING INTERACTIVE COMPUTER SOFTWARE

G. V. Goodman, M. Karmis, Z. Agioutantis

Department of Mining and Minerals Engineering
Virginia Polytechnic Institute and State University
Blacksburg, Virginia

Abstract. The economic conditions of most of the mining industry in the recent years has resulted in increasing competition in the market-place among coal producers. To remain competitive in the current market, there is an urgent need for increased efficiency in these operations. To investigate the influence and the interdependence of the many variables that affect a surface mine operation, the MINEPACK model, developed by Mathtech, Inc., was utilized. This interactive model is capable of analyzing several surface operations such as truck haulage, front-end-loader or shovel operation, drilling and blasting, and overburden removal using dozers, scrapers, or draglines.

The Mathtech model was utilized to investigate several surface mining operations in the Appalachian coalfield. Input data was collected from an on-site investigation of each operation and analyzed with this model. To minimize any discrepancies between the model output and the actual field conditions, the input data was slightly adjusted. With the model and the actual operation in close agreement, selective changes were then made to the input data. By carefully examining the output information, the cause-effect relationship of certain changes was determined.

INTRODUCTION

The deteriorating economic condition affecting much of the mining industry has resulted in increasing competition in the market-place among coal producers. Nowhere has this been more apparent than with the small surface mine operators located in the Appalachian coalfield, where increased owning and operating costs threaten their existence. To remain competitive in the existing market, these operators need to consider means of increasing their operating efficiency while maintaining or increasing current production levels. To investigate the influence and the interdependence of the many variables affecting a surface mine operation, computer modeling has become an important and widely used tool.

Many software packages are currently available for the analysis and the design of surface mine systems (Butler and Fouts, 1975; Dzuibek and Weil, 1983; Kirk et al., 1983; and Manula et al., 1975). The practical applications of these programs range from the initial reserve calculations to final production scheduling and total mine systems simulation. Although all models simulate surface mining activities with some degree of accuracy, the research concluded that the MINEPACK surface mine model developed by Mathtech, Inc., was easier to implement on the IBM Personal Computer. This interactive model is capable of analyzing various surface operations such as truck haulage, front-end-loader or shovel operations, drilling and blasting, and overburden removal using dozers, scrapers, or draglines.

THE MINEPACK MODEL

The MINEPACK model utilizes a dual-menu format that transfers user control to the appropriate section of the program. The menu listing is self-explanatory and user selection of the desired option is made by simply striking the appropriate numeric key. Audio and visual error messages are included in the menu coding to prevent an option selection out of the displayed range. A modified program flow chart is given in Figure 1.

The first, or "main option," menu displays those options available to the user for data file construction or data manipulation and analysis. Selection of the data analysis options transfers the user to the second, or "productivity analysis" menu. This screen displays those mine systems available for analysis within the program. Depending on the option chosen, program control will be transferred to the appropriate section and the user prompted for the necessary information.

This menu format was not present in the original MINEPACK model, but was later added to provide a more efficient means of maintaining user control within the program.

The primary unit of analysis is termed a lift and is defined as the volume of overburden removed by one or more machines. Coal need not be present under the lift. The necessary user input required for lift analysis consists of lift dimensions, coal

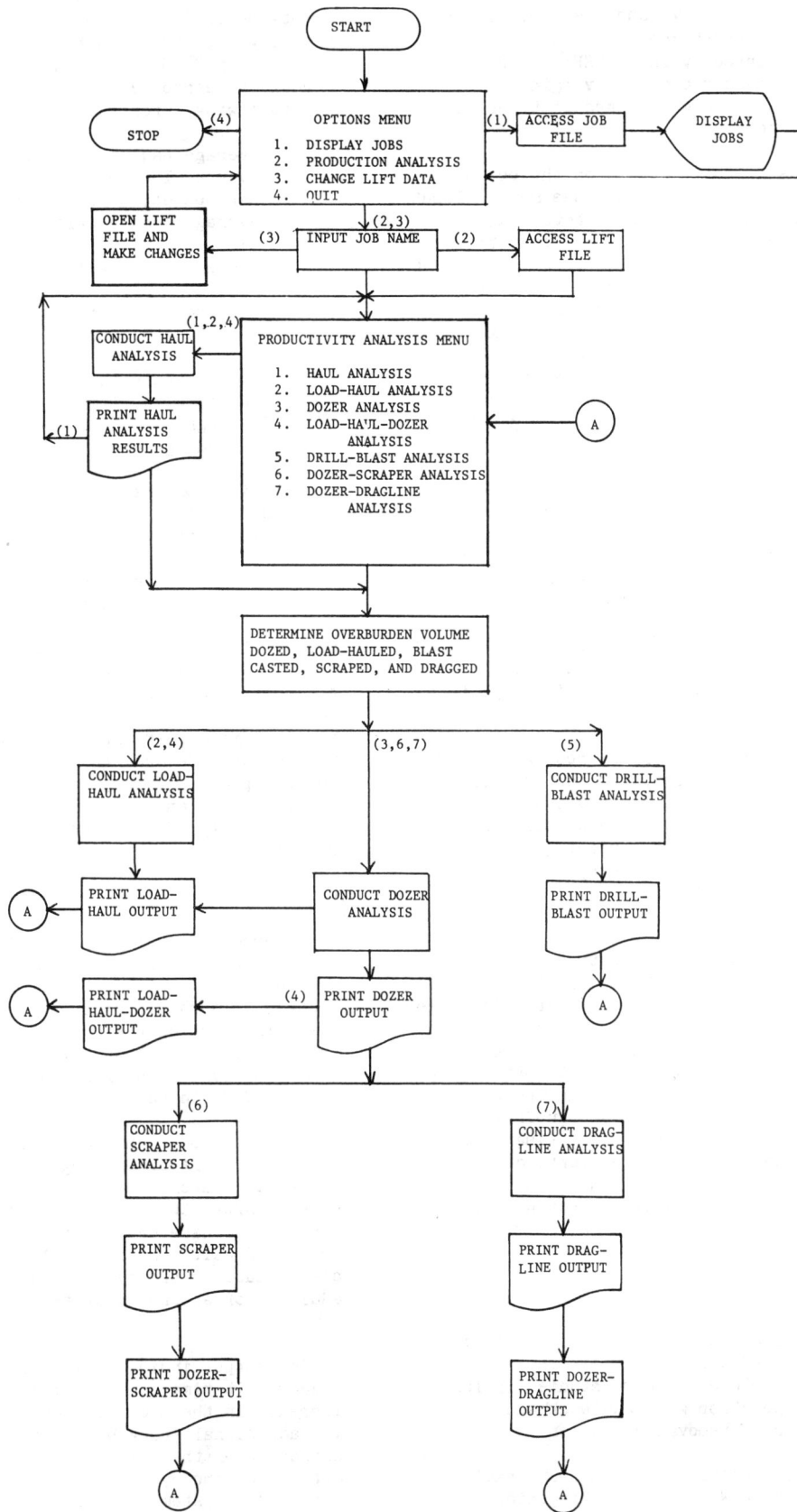

Figure 1: Modified Flowchart for MATHTECH Program

thickness, in-pit recovery, and the swell factor and density of the overburden. The output information calculated by the MINEPACK model includes coal tonnage uncovered, virgin strip ratio, cost per ton uncovered, and cost per bank cubic yard removed.

Additional user input depends on the option selected in the productivity analyses menu. Input for haulroad analysis consists of both hauler (truck or scraper) performance characteristics such as load hauled, empty weight, number of gears, and rimpull-speed data; and haulroad characteristics such as rolling resistance, grade, and length. The haulroad is assumed to be divided into segments to assure that constant conditions are maintained over each length. Output information gives the average load hauled by each truck or scraper, in addition to the haul and return times for each segment of the haul road.

Required input data for the load and haul analysis consists of the number of trucks per loader or shovel, their types, their haul and return times over the known haulage road, spot time, dump time, job efficiency, and hourly owning and operating costs. For the loader or shovel, the input consists of bucket capacity, fill factor, cycle time, job efficiency, and hourly owning and operating costs. Computed output information includes productivity, operating efficiency, total load time, and total cost.

Input data for dozer stripping requires field values of push and back-up speeds for the dozer, dozer type, blade capacity, job and operating efficiencies, and hourly owning and operating costs. The output for dozer stripping consists of average push distance, productivity, total cost, cost per bank cubic yard, and cost per ton uncovered.

Finally, the drilling and blasting routine require user input of drill hole diameter, drilling rate, powder ratio, primer ratio, fraction of holes wet, primer cost, ANFO cost, slurry cost, detonating cord cost, miscellaneous blasting cost, hourly owning and operating cost, bit cost. Output information consists of drilling time, number of holes; depth of hole, stemming, and subdrilling; costs of primer, ANFO, slurry, detonating cord, total drilling cost, and cost per vertical foot. The burden and spacing may be inputted by the user or calculated within the model (Mathtech, Inc., 1983).

MODEL VALIDATION

The MINEPACK model was utilized to investigate several surface mining operations in the Appalachian coalfield. The case studies included a contour mining operation with haulback and a modified mountaintop removal operation.

The first site analyzed was a contour-haulback operation located in Wise County, Virginia. Overburden was removed with a Caterpillar 992C

front-end-loader with a 10.5 cy (cubic yard) bucket capacity and hauled to the backfill with two Caterpillar 773B off-highway trucks with a 50T (34 struck cy) capacity. From on-site investigations, the following information was determined for this job:

1) average hauler load of 30 cubic yards
2) average loader cycle time of 0.75 minutes per bucket; 3 buckets per truck.
3) average loader wait (no truck) of 1.1 minutes
4) average truck wait (no loader) of 1.1 minutes
5) average spot time, haul time, and dump time of 0.1, 3.3, and 0.2 minutes, respectively.

An initial analysis of this case study was conducted using the load and haul option of the MINEPACK program. The output for this job (Figure 2) shows good agreement between the calculated and field values of truck wait time; yet poor correlation for the values of loader wait time, total cycle times of the trucks and loader, and haul time of the trucks.

The research found that this discrepancy was due to the haulage procedure used at this particular operation. After leaving the loader and approaching the dump site, each truck would stop and back the remaining distance to the dump. Although this distance was small compared to the total haul distance, the actions of stopping and backing did significantly increase the haul time of the trucks. Second, the model indicates a maximum loaded speed of over 30 mph, while the field readings show an actual maximum speed of approximately 7 mph. These differences are responsible for the model's underestimation of the haul times, total cycle times, and the loader wait time.

In order to realistically reflect the haulage conditions at this operation another analysis was conducted by inputting the average haul time of the trucks. This was in contrast to the initial analysis in which the haul time was calculated by the model. The output revealed that total cycle times and wait times were well matched with the collected field data.

With the model and the field data in good agreement, selective changes were made to the load-haul system. Despite the good correlation of results after the second run, the loader efficiency was seen to have dropped significantly due to the increased wait time of the loader. The initial change made to the system, therefore, was the addition of another haul truck to reduce this wait time.

The output indicated an obvious reduction in the loader wait time coupled with a significant increase in the operating efficiency of the loader. The additional truck had little effect on the operation of the original truck fleet, in that efficiency and cycle times were only slightly affected. System production on the other hand, increased substantially.

HAULAGE SIMULATION FOR JOB CASE STUDY #1

HAULER TYPE: CAT773B
AVERAGE LOAD: 29.925 CUBIC YARDS
MATERIAL DENSITY: 1.275 LOOSE TONS/CY
ROLLING RESISTANCE: 65 #/TON

SEGMENT	GRADE (%)	DISTANCE (feet)	SPEED LOADED (mph)	SPEED RETURN (mph)	HAUL TIME (min)	RETURN TIME (min)	TOTAL TIME (min)
1	-5	390	23.0	15.3	0.19	0.29	0.48
2	0	50	30.2	36.6	0.02	0.02	0.03
3	9	300	4.7	19.2	0.72	0.18	0.90

TOTAL HAUL TIME FOR ROAD: 1.41 MINUTES

TRUCK DATA

TRUCK TYPE: CAT773B
TRUCK LOAD (CY):   29.9
OF TRUCKS USED: 2
JOB EFFICIENCY: .9
AVERAGE # OF TRUCKS UP: 1.8

SPOT TIME (min):   .1
HAUL TIME (min):   1.4
DUMP TIME (min):   .2
LOAD TIME (min):   1.7
OPERATING EFFICIENCY: .81
WAIT TIME/CYCLE (min): .8
TOTAL CYCLE TIME (min): 4.2

LOADER DATA

LOADER TYPE: CAT992C
JOB EFFICIENCY: .9
BUCKET CAPACITY
STRUCK CUBIC YDS.:   10.5
FILL FACTOR: .95
CYCLE TIME (min):   .57
OPERATING EFFICIENCY: .85
BUCKETS/TRUCK: 3
WAIT TIME/CYCLE (min): .3
CYCLE TIME/TRUCK (min): 2.01

Figure 2: Initial MINEPACK Analysis for Case Study #1

Several more runs were made to study the effects of increased hauler presence on machine efficiency and system productivity. The results are illustrated in Figure 3 and show an increase in truck performance characteristics, such as wait time and efficiency, as well as a decrease in these similar loader characteristics. System productivity also increases as the number of haulers increase; yet above the four-hauler level, the change in productivity with the addition of one hauler is minimal. In fact, the figure shows productivity reaching an asymptotic value at 830 lcy/hr., thus indicating a possible upper bound on productivity for this system.

A second analysis was conducted with the haulers carrying 40 cubic yards or 4 buckets per truckload. Due to the resulting increase in truck payload, a representative haul time and load time were calculated for this situation and determined to be 3.7 minutes and 2.3 minutes, respectively. Spot and dump times were assumed to remain constant in this case. A similar analysis was then conducted by varying the number of haulers present in the load-haul system. The results are shown in Figure 4 and indicate a general improvement in the system until the four-hauler level is reached. At this point both machine and system outputs deteriorate due to the increased truck waiting time. This causes a corresponding increase in the truck cycle time, thus increasing the loader wait time.

Analysis of the model output under these two conditions indicates that system optimization depends on the parameter specified. For Figure 3, a fleet size of two trucks will maximize the operating efficiency of the haulers while minimizing their average wait time. In this respect, the system is operating optimally, despite low productivity. To optimize these operating characteristics for the loader, in addition to increasing the productivity, a larger fleet size is appropriate.

Similar conclusions are possible by analyzing Figure 4. Again, the operating parameters of the haulers are optimized with a fleet size of two trucks. At this point, system productivity is minimal. However, to optimize the operating characteristics of the loader, as well as the productivity, a fleet size of four trucks is recommended. An increase in fleet size beyond this point causes a decline in the system's performance.

The second case study analyzed was a modified mountaintop removal operation located in West Virginia. Overburden removal is conducted using a Caterpillar 992C front-end-loader with a 13 cy bucket and two Caterpillar 777 off-highway trucks with 85 ton capacities. To supplement the load-haul system, 2 Caterpillar D9H dozers are utilized as part of the overburden removal process.

The current mining plan at this site dictates that 40 percent of the overburden is removed with the load-haul system, roughly 37 percent removed or casted during the blasting process, and the remaining 23 percent removed by the dozer system. By varying the amount of overburden handled by the dozer and load-haul systems, the changes in per ton operating costs were investigated.

An initial run was made using the load-haul-strip option of the MINEPACK program and was found to be well matched with the actual operating characteristics. Successive applications of this program were made by varying the overburden percentages handled by the dozer and load-haul systems. The casted volume was assumed constant for all conditions. To further delineate the interactions between the load-haul and dozer systems, the analyses were also conducted by varying the size of the dozer fleet.

As illustrated in Figure 5, a fleet size of one dozer results in an increasing cost per ton value with an increasing percentage of dozer utilization. This was due to the inability of the one dozer to effectively remove the spoil. With a fleet size of two and three dozers, however, the cost per ton value shows an initial decrease followed by a subsequent increase. The research attributed this behavior to the presence of an upper bound on the percentage dozed, beyond which a particular fleet size would operate inefficiently. Examination of Figure 5 shows this percentage to be roughly 13 percent and 53 percent for fleet sizes of two and three dozers, respectively. This figure also indicates that for the conditions investigated, using three dozers with the present load-haul system resulted in greater economy than using two dozers. Furthermore, these savings improved for dozing percentages between 13 and 53 percent. Given that the existing operation is based on 23 percent dozer removal, a savings of approximately $.06/ton could be realized by the addition of another dozer to the overburden removal process. Concerning the dozer percentage, however, the operation appears to be optimal; for any increase in this value will cause an increase in the cost per ton.

CONCLUSION

The MINEPACK model has proved to be a versatile tool for analyzing various surface mining systems. Although a more detailed approach must be utilized to provide complete operations analysis, this model is sufficient for investigating the smaller, individual system components. This model has been applied to several case studies and has provided options to improve both the load-haul and dozer operations at these sites.

As competition grows between coal producers, efficiency and economy of operation will become more important to all operators. This, coupled with the increasing complexity of many mining operations indicates that computer analysis will become the dominant method utilized in the future.

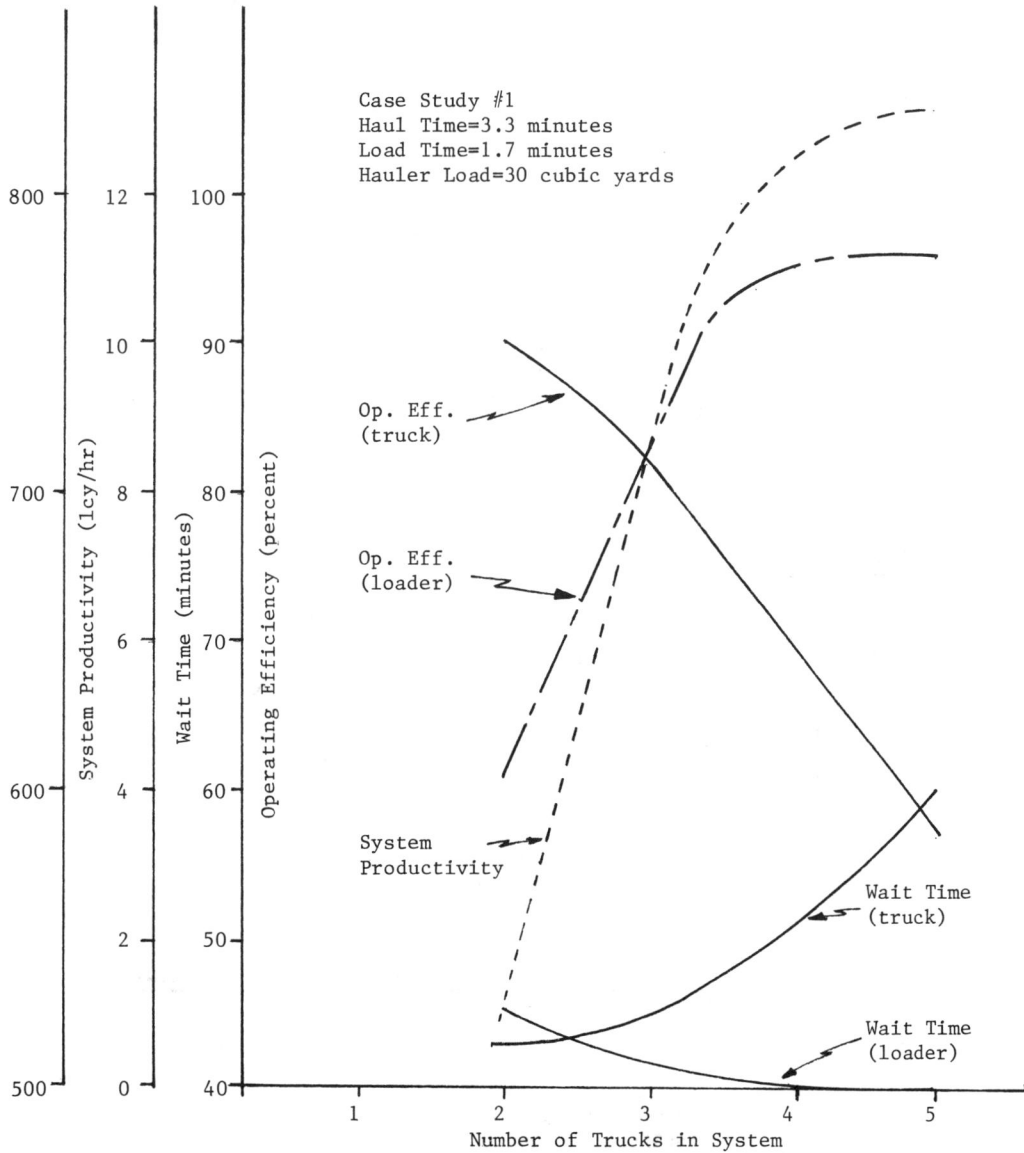

Figure 3: System Productivity and Machine Performance
vs Hauler Fleet Size

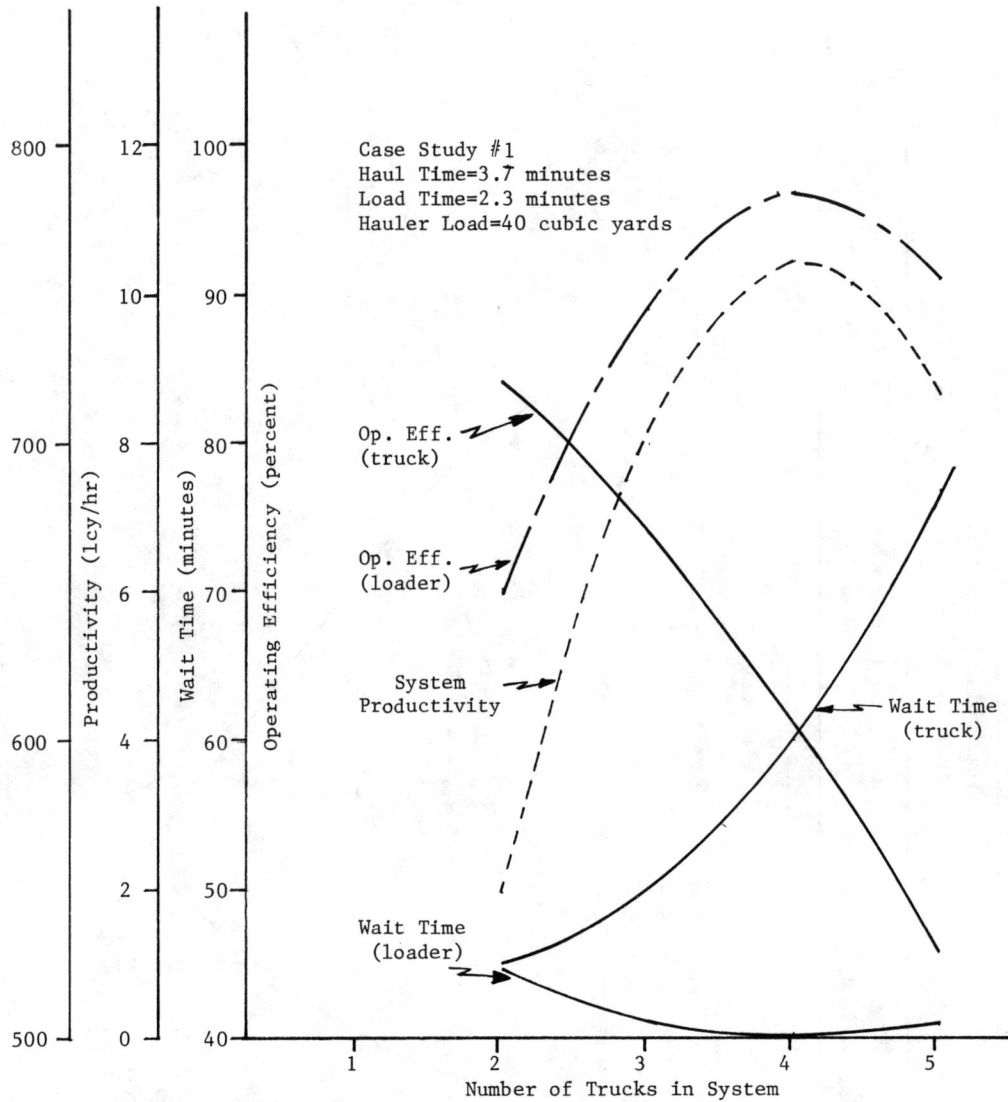

Figure 4: System Productivity and Machine Performance
           vs Hauler Fleet Size

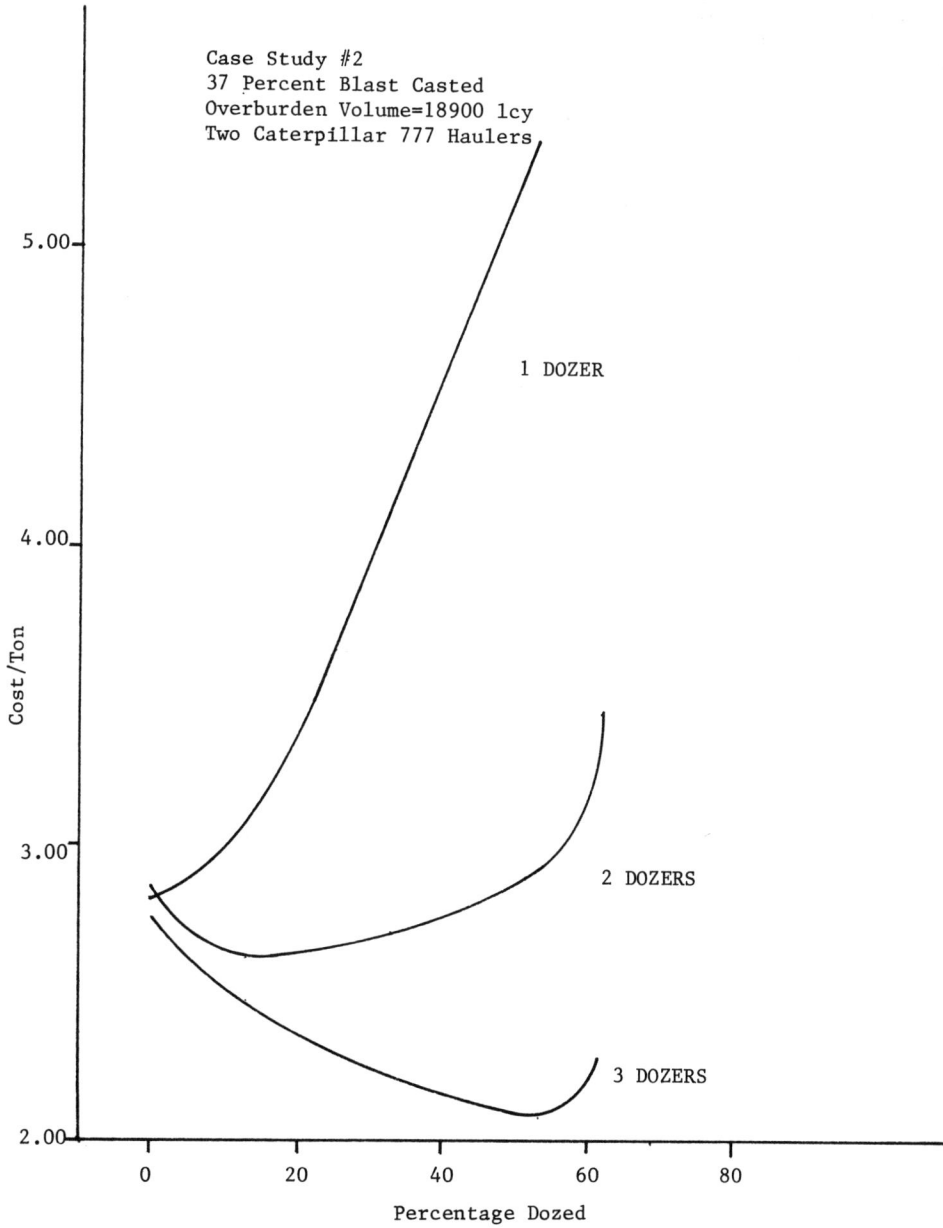

Figure 5: Cost/Ton vs Percentage Dozed for
Various Fleet Sizes

REFERENCES

Butler, J. M. and R. K. Fouts, 1975, "Optimizing
    Open Pit Mining with Computer Simulation," Min.
    Cong. Jour., Vol. 61, No. 3, March.

Dziubek, J. A. and R. R. Weil, 1983, "Computer
    Applications for Surface Mining in the
    Appalachian Coal Basin," Proceedings, First
    Conf. on Use of Computers in Coal Industry, West
    Va. Univ., August, pp. 91-103.

Kirk, K. E., L. E. Welborn and P. A. Ley, 1983,
    "Computer Assisted Mine Planning and Mine Plan
    Review," Proceedings, An Overview of Computer
    Aided Design Models Applicable to the Surface
    Mining Industry, Inst. for Mining and Minerals
    Res., Univ. of Kentucky, Lexington, June.

Manula, C. B., R. V. Ramani and T. V Falkie, 1975,
    "A General Purpose Systems Simulator for Coal
    Mining," Min. Cong. Jour., Vol. 61, No. 3, March.

Mathtech, Inc., 1983, "MINEPACK-Mining and
    Excavation Simulation," Versailles, KY.

Chapter 38

COMPARISON BETWEEN VARIOUS OVERBURDEN STRIPPING METHODS
.FOR TEXAS LIGNITE MINES

JAN K. WOLSKI

DERECK R. PRINCE

NEW MEXICO INSTITUTE OF MINING AND
TECHNOLOGY, SOCORRO, NEW MEXICO

ARCH MINERAL CORPORATION
ST.LOUIS, MISSOURI

INTRODUCTION

Lignite production in Texas has increased from two million tons per year in 1970 to 38 million tons per year in 1983 -- an average annual growth rate of over 25%. The latest forecasts indicate that the total production of lignite in Texas may reach 160 million tons/year by the year 2000 (Miercort, 1984). This trend will result in a high demand for large stripping machines.

The overburden-stripping machine most commonly used in Texas lignite mines has been the large walking dragline. But when draglines are used at pit depths greater than 30 m, considerable material rehandling is required and the stripping process becomes complex, with two or three passes of the dragline needed for each cut.

The maximum economical pit depth for dragline operation is about 45 m. Beyond that depth, other stripping methods have to be considered. The Texas lignite mining industry is becoming more and more interested in bucket-wheel excavator (BWE) technology, which can be successfully applied at the greater depths. Although BWEs have been used in similar situations abroad for decades, they are relatively new in the U.S. and have never been used in U.S. lignite mines. Limited knowledge about the system makes it difficult to select the proper equipment and to design suitable stripping methods. BWEs and the associated hauling systems are expensive, and the results of that investment can be disappointing if the equipment and stripping methods are not properly chosen for the conditions at each particular mine.

This paper is the result of a study intended to find the optimum parameters for various stripping methods and equipment for use in Texas lignite mines, given the types of deposits and the geotechnical conditions that are found there. The computer programs described here were written for fast and efficient evaluation of the various available stripping methods. By enabling engineers to select the most efficient lignite mining systems, these programs will help lower production costs, ultimately leading to cheaper electricity.

Five stripping systems are described and analyzed here: dragline; scraper-dozer; shovel-truck; BWE-cross-pit conveyor; and BWE-around-pit conveyor. A comparison of these methods concludes this paper.

MINE MODEL

A theoretical model of a lignite mine was developed to analyze and compare various kinds of overburden stripping methods and equipment. The parameters of this imaginary mine are typical of Texas lignite mines. The mine has two lignite seams, each 2 m thick, separated by 12 m of interburden. The overburden thickness can be varied from a minimum of 12 or 15 m to a maximum of about 50 m. The maximum total pit depth is 66 m. The pit length can be varied from 1000 to 3000 m.

The overburden is composed of sand, gravel, clay, and soft shale. Some hard streaks less than 30 cm thick are present. The average material density is 1.85 t/m^3. A high water table and high annual rainfall combine to produce wet, unstable spoil piles. The relationship between the spoil pile slope angle and the spoil pile height is

$$BETA = BA + \frac{BB}{HS} \quad (degrees),$$

where HS is the spoil pile height (m), and BA and BB are parameters that can be found from the slope measurements. The values of BA (21.6) and BB (300) used here are taken from a graph presented by Bhattacharyya and Chi (1982). This well-defined formula is valid for spoil pile heights of 20 to 60 m. As the spoil height increases from 20 to 60 m, the spoil angle decreases from 36.6 to 26.6 degrees.

The safe overburden slope angle is given by

$$SALFA = AA - AB * H \quad (degrees),$$

where H is the total highwall height and AA and AB are determined from geotechnical data. Although assuming a linear relationship between the safe overburden slope angle and overburden height is a simplification of the real situation, it is good enough for the purposes of this study. The model assumes that AA = 65.0 and that AB = 0.33. As the highwall height increases from 15 to 60 m, the

safe slope angle decreases from 60 to 45 degrees.
If the highwall working angle (ALFA) is greater
than the safe angle (SALFA), a safety bench must
be cut at an intermediate level. The width of this
bench can be determined from the geometry of the
highwall by the following formula (Fig. 1):

$$SB = HP * \left(\frac{1}{\tan(SALFA)} - \frac{1}{\tan(ALFA)} \right)$$

where HP is the the total highwall height.

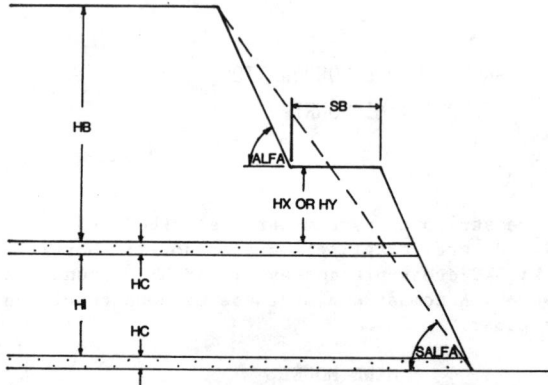

Figure 1. Highwall working slope angle ALFA vs.
           safe slope angle SALFA.

General Description of Computer Programs

Emphasis was placed on ease in making and modifying the input files. The use of a "NAMELIST"-
type file allows the input data to be entered in
the form "variable = value" and does away with
rigid formatting requirements. This allows engineers with very little computer experience to use
the programs. The value of each variable can be
changed easily, which helps in performing the
sensitivity analyses. Other parameters can be
modified directly in the program, if desired. All
of the variables are described in comment lines at
the beginning of each program. To make the programs
easy to understand and to modify, each line of the
program is described.

The output data are stored in tabulated form in
the output files. The output is limited to the
most important technical and economic information,
but other data can be retrieved easily if desired.

Ownership Cost Subroutine. A subroutine called
EQAN (Wolski, 1984) is used in all of the programs
to calculate the average annual ownership cost of
the equipment. This cost is adjusted for federal
and state taxes and for the investment tax credit.
The Accelerated Cost Recovery System (ACRS) or any
other depreciation schedules can be used. The
equipment salvage value is handled in accordance
with current tax regulations. A cost escalation
can be considered in relation to a reference year
(NYEAR), and the analysis may be started in any
selected year (NYEAR0). Any desired percentage of
the capital cost may be paid from the company's
own equity; the rest will automatically be assumed
as a loan for any chosen period. Either constant
total payments or constant principal payments may
be selected.

The present ownership cost is calculated using
the minimum required return on investment, and the
average uniform ownership cost is found. The
results are given in the estimated dollar value
for the year NYEAR. This ownership cost can be
added to the average operating cost and then
subtracted from the income to give the effective
profit. Of course, it is not necessary to calculate
income and profit to compare the costs of various
types of equipment.

The following values for the economic factors
were used in this study:
- Federal/state tax: 50%
- Investment tax credit: 8%
- 5-year ACRS depreciation system
- Cost escalation rate: 5%
- 100% of cost payed from company's equity
- Minimum required return on investment: 10%
- All costs in 1984 dollars.

The resulting unit cost is the total direct
cost. It does not include any overhead and/or
contractors' profits.

DRAGLINE STRIPPING MODEL

General

A typical two-phase dragline stripping system,
as shown in Figs. 2-5, is used. One possible mode
of operation, in which the dragline operates on the
elevated spoil bench, is disregarded in this
model because the high water saturation of the
spoil material may make it unfeasible (Kahle,
Moseley, 1982).

In the first phase of the stripping process,
the dragline operates on the original ground and
excavates to the top of the upper seam. In the
second phase, the dragline moves down via a ramp
at the end of the pit and excavates the material
above the lower seam. The need for movement up
and down a ramp is avoided if two draglines are
used in the same pit.

Stripping Modes

Mode 0. Simple side casting. This mode is used when
the horizontal and vertical dragline reaches are
sufficient to place all excavated material on the
final spoil pile. No material rehandling by the
dragline is involved. The maximum dragline reach in
phase 1 (DR1) and phase 2 (DR2), as well as the
required dragline vertical reach in phase 2 (HD2),
are used to calculate the dragline boom length. If
the assumed maximum boom length is exceeded, Modes
1 and 2, both of which involve material rehandling,
must be considered.

Mode 1. In phase 1, the dragline backfills the
space above the pit floor, forms a pad for itself
that will be used in phase 2, and spoils the
excess material onto the final spoil pile. A
safety bench of width SB (m) is left in order to
stay below the maximum safe highwall slope angle.
The slope of excavation of undisturbed material is
held constant at 60 degrees. For overburden heights
of 25 m or less, it is not necessary to raise the
dragline pad above the level of the upper seam,

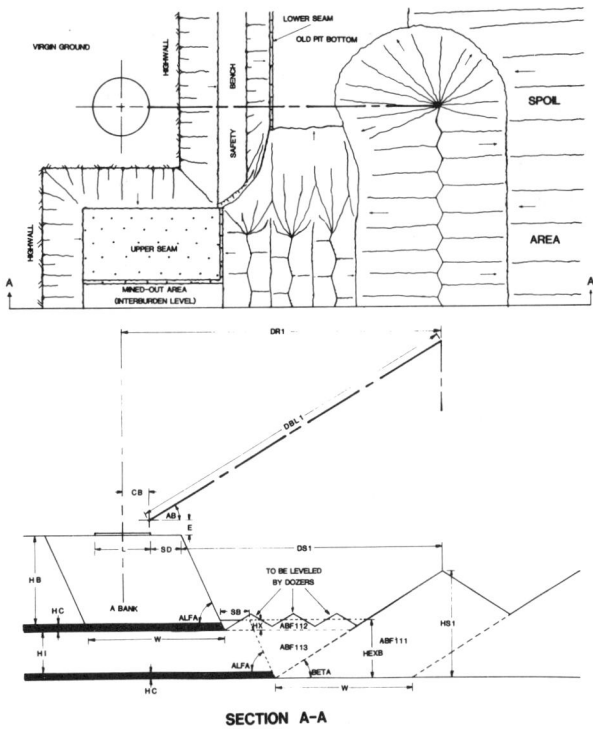

Figure 2.   Plan and cross section of a dragline
            pit, Mode 1, phase 1.

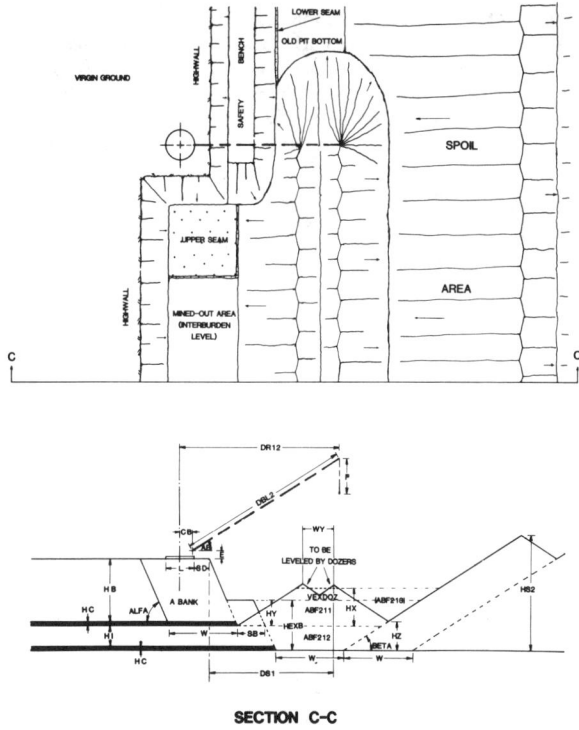

Figure 4.   Plan and cross-section of a dragline
            pit, Mode 2, phase 1.

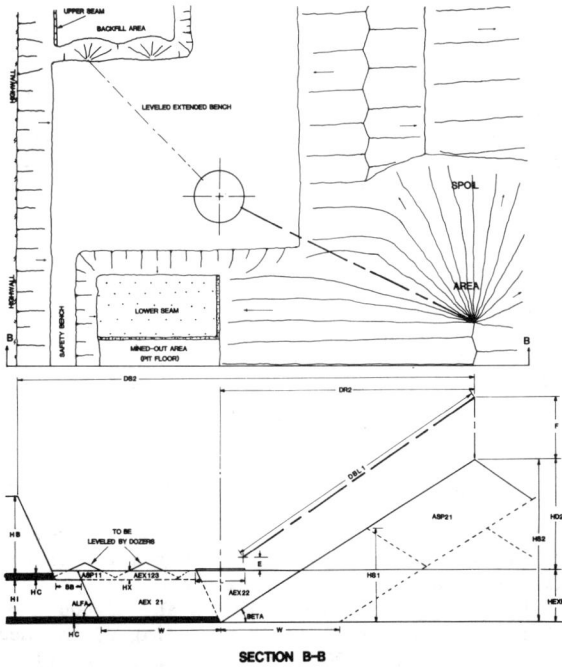

Figure 3.   Plan and cross-section of a dragline
            pit, Mode 1, phase 2.

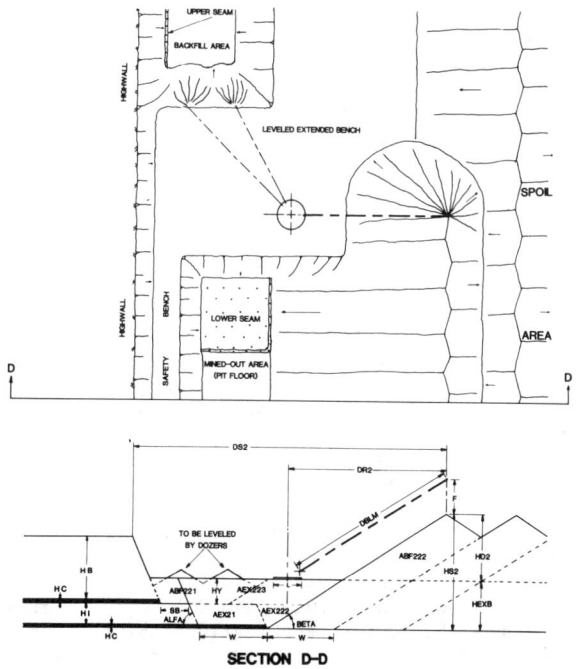

Figure 5.   Plan and cross-section of a dragline
            pit, Mode 2, phase 2.

because the vertical reach of the machine in phase 2 is sufficient to dump all excavated material onto the final spoil pile from the upper seam level. At overburden heights above 25 m, the dragline boom length reaches its maximum value, so the pad must be raised above the level of the upper seam to a height HX (m). To form this pad, the dragline dumps some of the material excavated in phase 2 into the backfill area ASP11, as shown in Fig. 3. This area is behind the dragline's position in the direction of mining. This process forms a movable ramp that progresses along with pit excavation. In Mode 1, the height of the spoil in phase 1 is greater than the level of the dragline pad in phase 2. The height of the dragline pad in phase 2 (HX) is calculated from two criteria: the maximum dragline boom length and the condition that the excavated area, AEX21 + AEX122 + AEX123, must equal the backfilled area, ASP11 + ASP21, as shown in Fig. 3.

Mode 2. As overburden thickness increases, the height of the dragline pad increases as well until finally all of the overburden excavated in phase 1 is used to build up the pad. At this point, the original spoil pile formed by the dragline in phase 1 has to be graded down by dozers to a new height, HY, which is the height of the dragline pad in phase 2. This is shown in Fig. 4. The height of the dragline pad XY (m) is calculated using the condition that the excavated area, AEX21 + AEX222 + + AEX223 must equal the backfill area ABP221 + + ABF222, as shown in Fig. 5.

The dozers are more important in Mode 2 than in Modes 0 or 1. Besides leveling the peaks of the final spoil pile and grading the phase 2 dragline pad, they have to reduce the height of the intermediate spoil pile in phase 1 from HX (its original height above the interburden) to HY. This operation involves a significant volume of material and requires additional machines.

Dragline Selection

The most important dragline parameters are the bucket volume BV (m^3), the dumping radius DR (m) and the boom length DBL (m).

The boom length is calculated from the horizontal reach required in both phases of the operation and from the vertical reach required in phase 2. This value must not exceed the assumed maximum boom length. For safety reasons, it is assumed that the minimum distance between the tub and the edge of the highwall in phase 1 is 9 m, and that the distance between the tub and the foot of the spoil pile in phase 2 is 15 m.

The longest existing dragline boom is 122 m (400 ft) long. It was installed on a Bucyrus-Erie 2570-W dragline. No other stripping draglines have booms more than 110 m (360 ft) long. Draglines with longer booms are technically feasible, but their mass and cost are considered prohibitive. In this model, two maximum boom lengths, 110 and 120 m, are considered. The results presented in Table 1 are for the 110-m boom.

Table 1. Selected Technical and Economic Data for Dragline Stripping (cut width 40 m, maximum dragline boom length 110 m).

Description	Unit 2)	Pit Depth m			
		31	41	46	56
General Data					
Excavation Rate	MBm^3/y	9.38	12.85	14.58	18.06
Stripping Ratio	Bm^3/t	6.25	8.56	9.72	12.04
Width of Safety Bench	m	4.0	9.0	12.3	20.8
Safe Highwall Angle	degrees	54.8	51.5	49.8	46.5
Spoil Pile Angle	degrees	29.2	27.4	26.7	25.8
Operation Mode 1)	-	0	1	2	2
Dragline Data					
Elev.of Dragline Pad	m	14.0	17.8	23.1	29.3
Prod.Rate, Phase 1	MBm^3/y	5.21	8.92	11.19	16.10
Prod.Rate, Phase 2	MBm^3/y	4.17	7.97	11.43	16.33
Dragl.Rehandl.Factor	%	0.0	31.4	55.1	79.6
Number of Draglines		1	1	1	2
Boom Length	m	100.6	110	110	110
Boom Angle	degrees	25.9	28.4	28.8	32.5
Dumping Radius	m	99.7	105.9	105.5	101.9
Bucket Volume	m3	41.5	73.1	96.0	69.6
Hourly Excav.Rate	Bm^3/h	1,554	2,798	3,749	2,688
Dragl.Delivered Mass	t	2,925	5,150	6,334	4,760
Dozer Data					
Number of Dozers		1	1	1	2
Rehandling Factor	%	7.8	5.3	13.3	17.4
Cost					
Dragline Capital Cost	M$	20.8	36.6	45.0	33.8
Dragl.Ownership Cost	$/h	317	558	686	516
Dragl.Operating Cost	$/h	556	969	1,258	907
Total Cost		894	1,546	2,028	3,003
Total Unit Cost	$/Bm^3	0.57	0.73	0.84	1.00

Note:
1) Operations modes: 0 - single side casting
                     1 - HS1 above the dragline pad
                     2 - HS1 below the dragline pad.

2) Bm^3 = bank m^3.

The size of a dragline is well described by the moment DM, formed by the load DSM (tons), suspended at the radius DR (m): DM = DR * DSM (t-m). This moment does not vary much for any particular dragline type. For a longer dumping radius, the suspended load must be reduced, and vice versa. The suspended load consists of the mass of the bucket and its rigging plus the payload mass. It can be expressed by the formula

$$DSM = \left(BMF + \frac{FF}{SFB} * BD\right) * BV \quad (t),$$

where
  BMF = bucket and rigging mass per unit of bucket volume, t/m^3;
  FF = bucket fill factor;
  SFB = material swell factor (1 + swell);
  BD = material density, t/ bank m^3; and
  BV = bucket volume, m^3.

BMF was assumed to be 1.20 t/m^3 in this model. It is valid for bucket volumes between 50 and 120 m^3.

The highest dragline moments are 39,350 t-m for the Bucyrus-Erie 3270-W and 36,740 t-m for the Marion 8950-8A. The next highest moments are 29,400 t-m for the Marion 8950-9A and 23,600 t-m for the Bucyrus-Erie 2570-W. Accordingly, the model uses a moment of 29,000 t-m for standard draglines with a 112 m^3 maximum bucket volume at a boom length of 110 m, and 39,000 t-m for very large draglines (136 m^3 at 120 m). In cases where the dragline moment would exceed these values, a two-dragline system is adopted. Two units were found to be necessary for a lignite production rate of 1.5 Mt/year at pit depths greater than 50 m.

Computer Program Description

Seven pit depths, from 31 to 61 m in 5-m increments, are analyzed. Plans and cross-sections of the pits are shown in Figs. 2-5.

Data File. The 58 variables are grouped in two NAMELIST files called "Data" and "Cost". The Data file contains the deposit, mine, and equipment parameters; the Cost file contains other variables necessary for calculating the equipment ownership and operating costs.

Dragline Choice. The program chooses the most appropriate dragline and calculates its cost. The basis for the cost calculation is the delivered mass of the dragline. The mass (tons/m of horizontal reach) is given by the following regression formula, derived from the analysis of 19 typical Marion and Bucyrus-Erie draglines:

$$UDRM = 9.136 * \sqrt{BV} - 29.50,$$

where BV is the bucket volume (m^3). This formula is valid for bucket volumes from 30 to 130 m^3 and dumping radii from 67 to 116 m. A regression correlation factor of R = 0.981 indicates that this formula is a good fit. The total capital cost of the dragline in millions of dollars is determined by the formula

$$AIDR = UDRM * DR * UCDR *10^{-3},$$

where DR is the dumping radius (m) and UCDR is the unit cost in dollars per kg. UCDR includes the FOB price and the delivery and erection costs and is equal to $7.10 per kg of delivered mass (average value in 1984 dollars).

Power Consumption. The power consumption per bank m^3 of excavated material is then calculated using the formula UCP = 0.17 + 0.0147 * DR (kWh/bank m^3). This cost relationship is shown graphically by Learmont (1984). For a boom length of 100 m, the unit power consumption is 1.64 kWh/bank m^3.

Maintenance and Repair Costs. The maintenance and repair cost calculations, also after a graph by Learmont (1984), are then calculated. An adjustment factor of 1.10 is included to account for cost escalation between 1981, when the data in the graph were derived, and 1984. These costs are calculated by the formula UCM = (32.0 + 1.3 DR)10^{-3} ($/Bm^3). The formula is valid for any dumping radius greater than 75 m. At a dumping radius of 100 m, the maintenance and repair cost is $0.16/bank m^3.

Direct Operating Personnel Cost. The direct operating personnel cost for the dragline is taken to be $63.70 per hour of equipment scheduled time. This assumes a crew of four, and fringe benefits equal to 30% of wages.

Dozers. The program calculates the number and cost of dozers needed. The total cost of a large dozer is assumed to be $120 per operating hour.

Ownership Costs. The capital cost of the dragline is increased by a factor of 1.06 to account for the estimated two years that it takes to deliver, erect, and commission a new dragline. A 12% annual interest rate and a 20-year dragline life are assumed.

Results

The main parameters for pit depths of 31, 41, 46, and 56 m are shown in Table 1. Given the assumed dragline size, operation at a pit depth of 61 m was not feasible. A graph of total unit cost vs. pit depth is shown in Fig. 6. The relationship between pit depth and the rehandling factor is shown in Fig. 7.

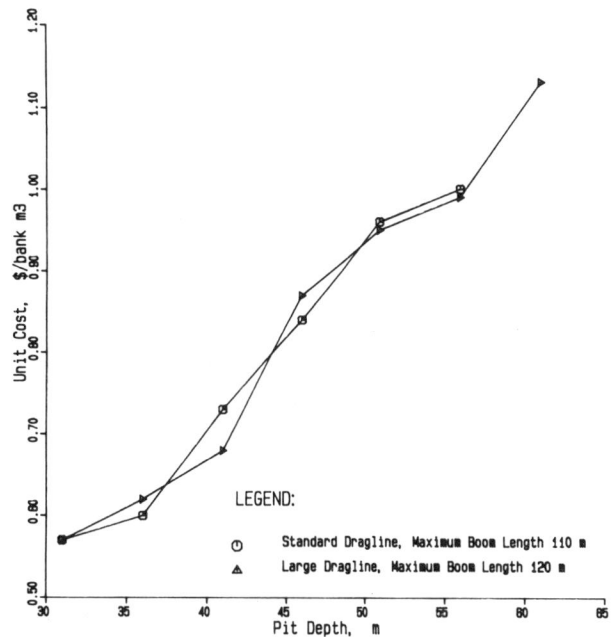

Figure 6.   Unit cost vs. pit depth for standard and very large draglines.

The simple side casting method (Mode 0) was found to be feasible at pit depths of up to 31 m with a 110-m boom and at depths of up to 36 m with a 120-m boom. As Fig. 7 shows, the use of a longer boom reduces material rehandling significantly at depths of up to 50 m, but the total cost is not affected. Obviously, the extra cost of the long boom is not justified by the drop in rehandling expense. As a rule, a very long boom is only justified at great pit depths. For example, a dragline with a 120-m boom can work at a pit depth of 61 m, a depth which is not feasible with a 110-m boom. Using the simple side casting method with a long-boom dragline is not necessarily more economical than using a dragline with a shorter boom and an extended bench in phase 2. At a pit depth of 31 m, the latter method costs $0.03/bank m^3 less than the former method.

LEGEND:

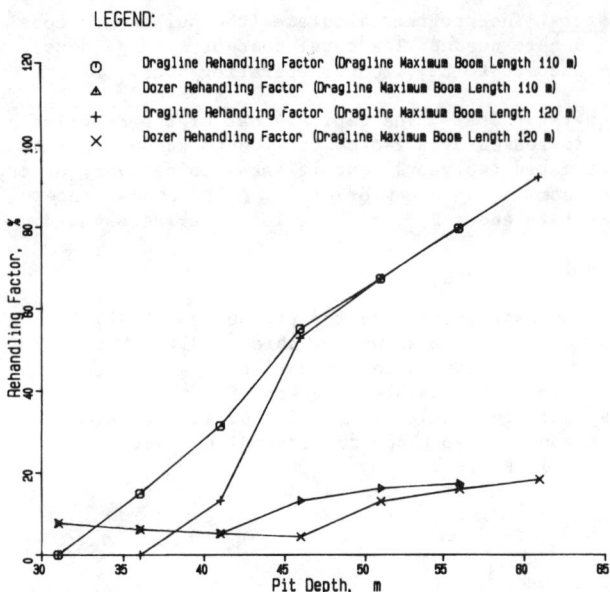

Figure 7. Rehandling factor vs. dragline pit depth

The cost of stripping is almost exactly
directly proportional to the pit depth. This
relationship is approximated by the formula

$$UC = 0.022 * HP - 0.20, \qquad (\$/\text{bank } m^3)$$

where HP, the total pit depth, ranges from 35 to
60 m. Of course, the results above refer to the
particular conditions specified in the input file.
Any of these values may be modified easily to
account for local conditions and available
dragline sizes.

SCRAPER-DOZER MODEL
General

A "V" scraper-dozer overburden stripping system
developed by the Caterpillar Tractor Co. (Morgan,
1979) was adapted for this study and modified for
two-seam stripping. A plan and cross-section of
this system at a pit depth of 46 m is shown in
Fig. 8. The dozers begin excavation by pushing
material over the highwall at the maximum possible
angle (taken here to be 30%). Then they build a
platform with a 15% grade across the pit, as
indicated by line DEALK in Fig. 8. The area
excavated by the dozers is defined by the contour
ABCDEA. The dimensions L, H1, and H2 depend on the
overburden and interburden thicknesses (HB and HI),
the cut width (W), and the highwall and spoil
angles (ALFA and BETA, respectively).

After the bulldozing operation, scrapers are
used to flatten the 15% grade zone and to form a
scraper-turning platform, which is indicated by the
line FALN. The scrapers haul excavated material
over this platform and place it along a dumping
ramp that also is maintained at a 15% grade. The
scrapers excavate the remaining overburden along
a ramp that has the same width (W) as the cut. The
lower portion of this ramp is used for hauling

lignite from the upper seam and is maintained at
a 12% grade. The length of the exposed upper
lignite seam and of the mined-out area should be
kept as low as possible to minimize the scraper
traveling distance. (In this model, this length
is assumed to be 60 m.) Thus, the mining of the
upper seam must be well synchronized with the
stripping operation. The total area excavated
by the scrapers is the sum of the new cut area
(line AEGH) and the area filled with material by
the dozers (line AHLA). Therefore, the rehandling
factor is defined here as the ratio of the area
AHLA to the area B'C'GHB'. Dimensions H3, H4, and H5
depend on the overburden thickness, the cut width,
and the highwall and spoil angles. The dimension L
is found from the condition that the areas excavated and filled by the dozers must be the same.

As the material is excavated, both the cut ramp
and the fill ramp advance along the pit, forming
the reversed "V" shape. This "reversed-V" system
has two main advantages over the conventional V

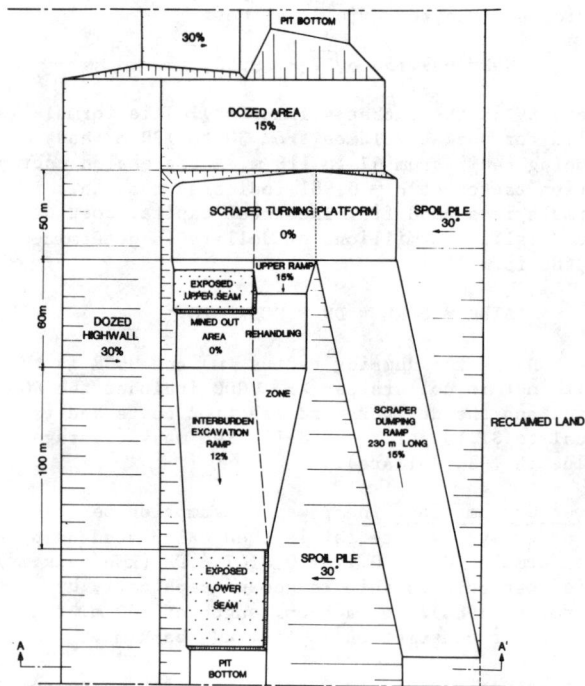

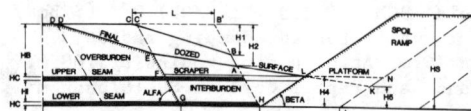

SECTION A-A'

Figure 8. Plan and cross-section of the scraper-
           dozer stripping system for the pit depth
           of 46 m.

system in which material is hauled across the bottom of the pit: shorter hauling distance and more efficient use of dozers. Inconveniences that should be noted, however, include the material rehandling involved and the need for close coordination among various stages of operation and pieces of machinery.

Program Description

Seven pit depths, ranging from 31 to 61 m in 5-m increments, are analyzed.

Dozer Calculations. The average dozing distance is taken to be the distance between the centroid of the excavated area (ABCDE) and the centroid of the backfill area (AHJKL). This distance increases from 63 to 95 m as the pit depth increases from 31 to 61 m at a constant cut width of 45 m. The dozer production rates are calculated using the Caterpillar (1982) data. The dozer production rate decreases from 348 to 240 bank m^3/hour and the number of dozers increases from 3 to 10 as the pit depth increases from 31 to 61 m.

Scraper calculations. First, the height of the scraper turning platform is set. It is equal to the elevation of the top of the upper seam (HI + 2HC) if H4 is less than HI + 2HC. If H4 is greater than HI + 2HC, then the platform height is set equal to H4. This minimizes the area AHL and thus minimizes the amount of material to be rehandled. Next, the scraper distances and cycle times are determined. The rolling resistance is assumed to be 4% in undisturbed areas and 6% in disturbed areas. The maximum speeds of the scrapers are found from the rimpull and gradeability curves provided by the equipment suppliers, and are then reduced by the speed factors to account for acceleration, braking, and turning. The average hauling distances and cycle times increase from 273 m and 5.0 minutes at a pit depth of 31 m to 365 m and 6.6 minutes at a pit depth of 61 m. The total number of scrapers increases from 7 at a pit depth of 31 m to 18 at a pit depth of 61 m. The scraper rehandling factor increases from 19% at a cut depth of 31 m to 23% at a cut depth of 61 m.

Other Equipment. The remaining equipment consists of the ripper/pushers, graders, and water trucks. It is assumed that, when tandem-motor scrapers are used under the overburden conditions found in Texas lignite mines, one ripper/pusher is needed for every two scrapers. One grader is required for every five scrapers, and one water truck for every seven scrapers.

Input and Output. 32 parameters are stored in the NAMELIST input file. Some selected data for the pit depths of 31, 41, 51, and 61 m are shown in Table 2.

Effects of Cut Width and Pit Depth

Fig. 9 shows how cut width and pit depth affect the unit cost of stripping. Cut width has little influence on the unit cost at pit depths between 31 and 61 m. Increasing the cut width causes a slight increase in the hauling distance, but at the same time it cuts down on rehandling, and the two effects tend to cancel each other out. In general, the optimum cut width is between 40 and 60 m for pit depths between 31 and 61 m. The pit depth has a strong influence on the unit stripping cost. As the pit depth increases from 31 to 61 m, the unit stripping cost increases from \$0.93 to \$1.19/bm^3. This cost increase is primarily due to the significantly greater dozing and scraper hauling distances in the deep pits. As the pit depth increases from 31 to 61 m, the dozing distance

Table 2.   Selected Technical and Economic Data for the Scraper/Dozer Stripping System (Cut Width 45 m)

Description	Unit 1)	Pit Depth (m) 31	41	51	61
General Data					
Overburden Thickness	m	15	25	35	45
Excavation Rate	MBm^3/y	9.38	12.85	16.32	19.79
Stripping Ratio	Bm^3/t	6.3	8.6	10.9	13.2
Rehandling Ratio	%	19	18	20	23
Dimensions: L	m	20.5	35.1	50.5	66.6
H1	m	7.4	12.7	18.3	24.2
H2	m	14.8	20.1	25.7	31.5
H4	m	11.7	15.1	18.3	21.4
Elevation of Turning Platform	m	16.0	16.0	18.3	21.4
Stripping with Dozers					
Average Dozing Distance	m	63	73	84	95
Hourly Production Rate	Bm^3/h	348	304	269	240
Annual Production Rate	MBm^3/y	1.40	1.23	1.08	0.97
Excavated Volume	MBm^3/y	3.87	5.71	7.65	9.67
Number of Dozers		3	5	7	10
Stripping with Scrapers					
Average Hauling Distance	m	273	261	334	365
Average Production Rate	Bm^3/h	276	253	230	210
Average Production Rate	MBm^3/y	1.11	1.02	0.93	0.85
Volume Excavated	MBm^3/y	7.34	9.41	11.97	14.60
Number of Scrapers		7	10	13	18
Additional Equipment					
Pushers and Rippers		4	5	7	9
Graders		2	2	3	4
Water Trucks		1	2	2	3
Cost					
Total Cost per Year	M\$/y	8.71	12.89	17.17	23.62
Total Average Unit Cost	\$/Bm^3	0.93	1.00	1.05	1.19

Note:
1) B - Bank, L - Loose

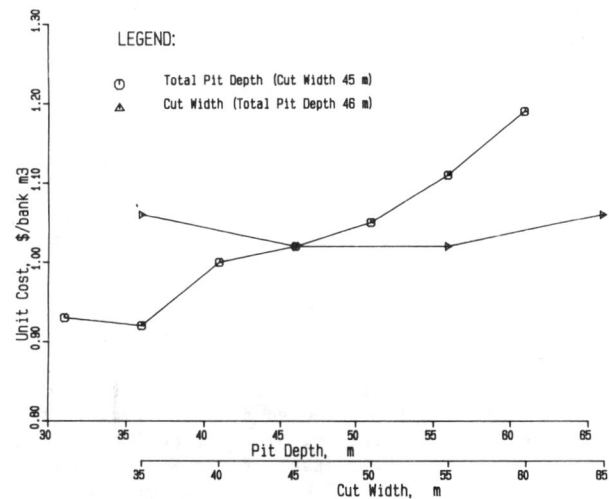

Figure 9.   Unit cost vs. cut width and pit depth for the scraper-dozer stripping system.

increases by 61%, and the scraper hauling distance
by 33%. As a result, more equipment is needed for
a given production rate, which in turn means higher
unit costs. This relationship can be roughly
approximated with the formula

$$UC = 0.0083 * HP + 0.65,$$

where

UC = total direct cost ($/bank m^3), and
HP = pit depth (m), between 33 and 57 m.

This formula is applicable only to the conditions and geotechnical and cost parameters used in
this study. However, it can be adapted easily to
any set of conditions.

SHOVEL-TRUCK MODEL

General

A typical open-pit stripping system using
shovels and trucks, as shown in Fig. 10, is used
in this model. A central lignite hauling ramps
divide the pit into two wings, only one of which is
shown in Fig. 10. Overburden and interburden
material is hauled around both sides of the pit to
the spoil benches via 8% ramps.

For this study, the height of the overburden
bench was set at 12 m. Overburden heights are 12,
24, 36, and 48 m; total pit depths are 28, 40, 52,
and 64 m; and 2, 3, 4, and 5 benches result.

The lignite production rate is held constant
as the overburden thickness is varied. As a
result, the excavation rate per bench is also
constant and equals 4.2 MBm^3/year.

Shovel and Truck Selection

First, the truck size is arbitrarily selected.
Then the shovel size is calculated, assuming five
shovel dippers per truckload. In this study the
truck payload is 108 t and the optimum shovel
dipper volume is 17.5 m^3. The selected shovel has
to be checked for its use-of-availability factor,
which should be between 0.75 and 0.90, a commonly
achieved range in mining operations. The reqired
shovel hourly production rate is calculated and, if
the use-of-availability falls outside the limits
given above, the program sends a warning that the
equipment size was not properly chosen. In that
case, the truck size must be increased or decreased
in order to return the use-of-availability to the
proper range. In the example used for this study,
the availability factor is 0.85, the use-of-availability factor is 0.78, and the utilization factor
is 0.66. The number of trucks that can be serviced
by one shovel is calculated with the formula
NTPS = CTT / CTSH, where CTT is the average truck
cycle time (min), and CTSH is the average truck
loading time (min). The maximum truck speed is
calculated from the rimpull and gradeability
curves and is limited to 40 km/hour for safety.
The speed is further reduced by speed factors that
account for acceleration, braking, and cornering.
The average hauling distance for a wing length of
1,000 m is 1.36 to 1.59 km, which gives a truck
cycle time of 14.5 to 16.1 min. The average number
of trucks per shovel (NTPS) is 4.8 to 5.4,
depending on the total pit depth. The total number

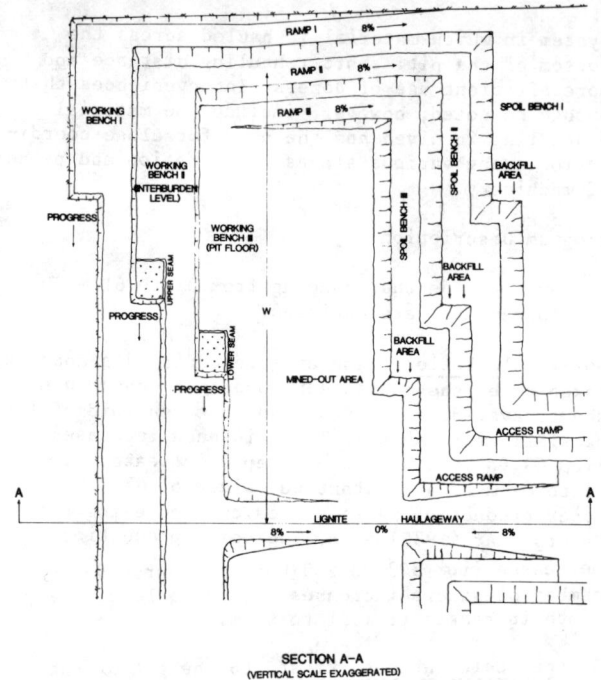

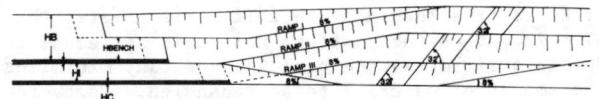

SECTION A-A
(VERTICAL SCALE EXAGGERATED)

Figure 10. Plan and cross-setion of the shovel-
truck pit

of trucks is calculated with the formula

$$NT = NS * NTPS * \frac{UFS}{UFT},$$

where NS is the total number of shovels. The
ratio UFS/UFT accounts for the difference between
the shovel utilization factor and the truck
utilization factor. In the example analyzed for
this study, the truck utilization factor UFT is
0.67.

Ramps and Roads

Ramps. The volume of ramp material required per
meter of pit progress is equal to the total spoil
pile height times the ramp width. This value,
multiplied by the pit progress per year, gives the
total volume of ramp material needed annually. The
cost of filling and compaction is estimated to be
$1.00/m^3.

Haulage Roads. The haulage roads consist of
the roads along the excavation benches and the
ramps and roads on the spoil pile. The total cost
of road maintenance equals the total road length
times the unit maintenance cost, which is taken
to be $50.00/km-day.

Remaining Equipment

The remaining equipment includes the dozers,
graders, and water trucks. One dozer per bench is
used for ramp construction, spoil dump leveling,
and other pit work. One grader and one water truck
per every two benches are used for road maintenance.

Program Input and Output

The 33 main variables are stored in a "NAMELIST" input file. A sample output file, Table 3, is shown.

Table 3. Sample Output of the Shovel/Truck Stripping System for the Pit Depth 52 m and Length 2000 m

GENERAL PIT PARAMETERS
****
OVERBURDEN HEIGHT	= 36.	m
TOTAL PIT DEPTH	= 52.	m
TOTAL NUMBER OF BENCHES	= 4	
HEIGHT OF EACH BENCH	= 12.	m
TOTAL VOLUME OF MATERIAL EXCAVATED	= 16.67	MBm^3/yr
ANNUAL PRODUCTION OF LIGNITE	= 1.50	Mt
STRIPPING RATIO	= 11.1	Bm^3/t
ANNUAL PIT PROGRESS	= 347.	m/yr
SCHEDULED EXCAVATION TIME	= 6375.	h/yr

SHOVEL PARAMETERS
****
TOTAL NUMBER OF SHOVELS	= 4	
NOMINAL DIPPER VOLUME	= 17.5	m^3
HOURLY EXCAVATION RATE	= 992.	Bm^3/h
ANNUAL EXCAVATION RATE	= 4.17	MBm^3/y
OPERATING TIME PER YEAR	= 4198.	h/yr
UTILIZATION FACTOR	= 0.66	

TRUCK PARAMETERS
****
AVERAGE NUMBER OF TRUCKS PER SHOVEL IN OPERATION	= 5.2	
TOTAL NUMBER OF TRUCKS	= 21	
TRUCK PAYLOAD	= 108.0	t
MEAN TRAVEL DISTANCE	= 1513.	m
MEAN CYCLE TIME	= 15.6	min
PRODUCTION RATE PER HOUR	= 202.	Bm^3/h
PRODUCTION RATE PER YEAR	= 0.87	MBm^3/h
OPERATING TIME PER YEAR	= 4303.	h/yr
UTILIZATION FACTOR	= 0.67	

RAMP CONSTRUCTION AND ROAD MAINTENANCE
****
VOLUME OF RAMP BACKFILL MATERIAL	= 0.62	Mm^3/yr
TOTAL LENGTH OF ALL ROADWAYS	= 10.1	km

AUXILIARY EQUIPMENT
****
NUMBER OF DOZERS	= 4
NUMBER OF MOTOR GRADERS	= 2
NUMBER OF WATER TRUCKS	= 2

COST DATA
****
UNIT COST OF BUILDING RAMPS	= 0.037	\$/Bm^3
UNIT ROAD MAINTENANCE COST	= 0.011	\$/Bm^3
UNIT EQUIPMENT COST	= 1.128	\$/Bm^3
TOTAL UNIT COST	= 1.18	\$/Bm^3

The influence of pit depth and length on direct unit cost is shown in Fig. 11. Pit depth has little influence on unit cost, since the bench heights remain constant, but increasing the pit length has a significant effect on the stripping cost because of the increase in truck travel time. As the pit length is increased from 1,000 to 3,000 m, the unit cost increases from \$1.05 to \$1.26 per bank m^3. Those values apply to the cost of hauling overburden and interburden, not to the cost of hauling lignite. But the cost of hauling lignite also increases with pit length.

Once the sizes of the trucks and shovels are selected, various design parameters of the pit can be adjusted. For example, the bench height and the

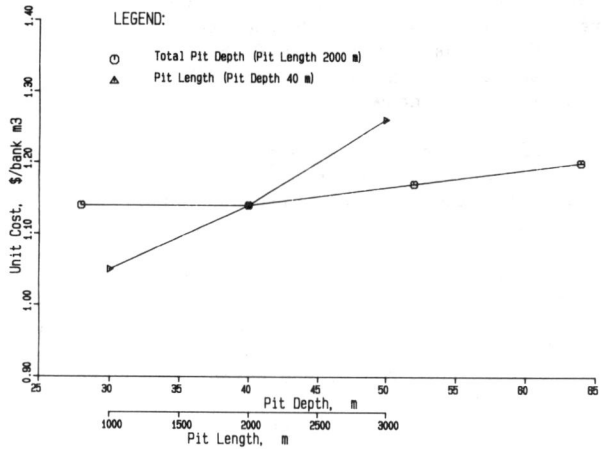

Figure 11. Unit cost vs. pit length and depth for the shovel-truck stripping system.

lignite production rate can be optimized.

CONTINUOUS EXCAVATION AND HAULING SYSTEMS

General

The continuous excavation and handling system model presented below was described in detail in the paper presented by Buddecke and Wolski at the 1985 SME-AIME meeting in New York. The model is described briefly here.

Due to the instability of the spoil piles, the bucket wheel excavator (BWE) is used only for overburden removal. The 12-m interburden is removed with a dragline. The highwall slopes are 55 degrees for the overburden and 60 degrees for the interburden.

Two types of continuous hauling systems are considered: the cross-pit conveyor (CPC) and the around-pit conveyor (APC).

BWE Selection

The required size of a BWE depends mostly on the required production rate and the bench height. Since the lignite production rate is held constant in this model, the BWE size here depends upon the BWE output efficiency and utilization factors, the overburden thickness, and the number of benches. The bench heights in this model are 15 to 30 m. These conditions require relatively large BWEs.

BWE and CPC Model

Bench Height. Although the overburden thickness in this model varies from 15 to 45 m, bench height is restricted to 30 m for safety reasons. This indicates a two-bench operation in the cases where overburden thickness is 30 m or greater. The cut width of 30 to 50 m is the same for both the dragline and the BWE.

Number of BWEs. One or two BWEs can be used in a two-bench operation. If only one BWE is used, the machine must ramp up and down between the two benches. Since it makes two cuts for each dragline cut, it would also interfere with the dragline twice as often. On the other hand, if two BWEs are used with one CPC, both BWEs must progress at the same pace. This diminishes the flexibility of the mining operation.

Cross-Pit Conveyor (CPC) Selection. In all of the BWE bench options described here, a single CPC is used to transfer the overburden material across the pit to the spoil pile. A two-support bridge receiving conveyor and a cantilevered swinging discharge conveyor are used. The length of the discharge boom conveyor, which depends on the pit depth, varies from 130 m for a pit depth of 31 m to 220 m for a pit depth of 61 m. As the boom length approaches 200 m, the weight (3,800 tons) and the capital cost ($32 million) of the CPC become very high. The length of the CPC can be reduced by decreasing the width of the bottom of the pit into which the dragline spoils material (WX). Reducing this width to one-half of the dragline pit width (W) results in a slight increase in dragline boom length, but allows a 15% decrease in CPC boom length. This solution is only feasible for a moderate dragline bench height and with the use of the simple side-casting dumping method. A cross-section of the BWE/CPC system with one BWE and two benches is shown in Fig. 12.

BWE and APC Model

The BWE/APC is composed of a BWE, a mobile transfer conveyor, a traveling hopper and cable reel car, a face conveyor, which discharges through another movable hopper onto a connecting conveyor, a shiftable spoil conveyor, a tripper car, and a mobile spreader.

In the BWE/APC system, one face conveyor may serve one, two, or more BWE benches. The overburden is transferred from one bench to another with mobile transfer conveyors (MTCs). A one-bench BWE/APC system for overburden thicknesses of 15, 20, 25, and 30 m was analyzed. Two or more benches could be used for overburden thicknesses greater than 30 m. In that case, either one or two BWEs could be used, just as in the BWE/CPC system.

The BWE/APC system described here could be used together with a BWE/CPC. The BWE/APC would operate on the upper bench, and the BWE/CPC would excavate the deeper overburden. The only difference between that case and the one analyzed here would be the length of the APC connecting conveyor at the end of the pit.

Modeling Results

Four FORTRAN-77 computer programs were written for the model. Three of them model the three options of the BWE/CPC system: one BWE-one bench (Option A); one BWE-two benches (Option B); and two BWEs-two benches (Option C). A fourth program deals with the BWE/APC system. A "NAMELIST"-type input file supplies 45 variables for these models and allows the user to change their values easily.

Total Unit Cost (TUC). In the single bench operation, the TUC decreases with the overburden height. In the case of the APC system this strong trend is justified by the fact that the conveyor is more economical at high production rates. In double-bench operations, the unit cost increases with increasing bench height and is 65 to 95% higher than the unit cost of the BWE/CPC system for a pit length of 3,000 m. Comparison between the double-bench options indicates that using two BWEs (Option C) is 10 to 12% more expensive than using one BWE (Option B).

Economic Influence of Pit Length. Pit lengths of 1000, 2000, and 3000 m were analyzed; the results are shown in Fig. 13. Pit length has very little economic influence on the CPC systems, but it has a great deal of influence on the APC systems. Even at a pit length of 1000 m, though, the APC cost is at least 12% higher than the CPC cost.

CONCLUSIONS

Stripping Methods

Fig. 14 shows the relationship between unit costs and pit depths for all stripping methods described above. For pit depths of up to 35 or 40 m, dragline stripping is the most economical technique. For depths between above and 60 m, stripping the overburden with a BWE/CPC and stripping the interburden with a dragline is the best choice.

Scraper-dozer and shovel-truck stripping systems were found to be the most expensive methods for all pit depths. Pit depth has a strong influence on

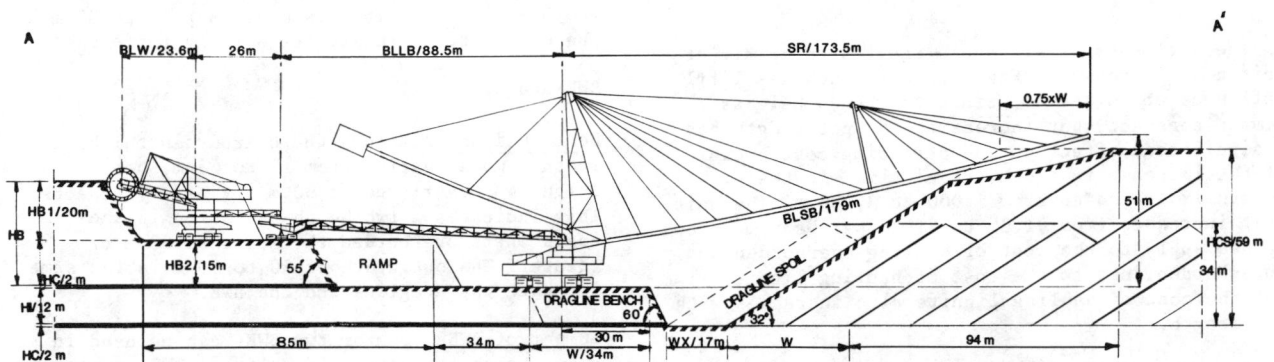

Figure 12. Cross-section for the BWE and CPC system.

the scraper-dozer system. At a pit depth of 31 m, the scraper-dozer system costs $0.93 per bank m^3 as compared with $1.14 for the shovel-truck system. But both systems cost the same, $1.19 per m^3, at the depth of 61 m. Nevertheless, the results indicate that the scraper-dozer stripping system is more economic than the shovel-truck system to the depth of 60 m. The cost of the latter was above $1.10/$bm^3$ for all pit depths analyzed, and it is the most expensive of all stripping methods.

Continuous overburden stripping methods can be successfully applied to lignite mines in Texas. They are the most economical systems for pit depths greater than 40 m. The use of draglines alone is not the best solution for pits deeper than 40 m. However, draglines are still the most economical systems for use in the lower parts of the deep pits (Kahle and Moseley, 1982).

Cross-pit conveyors provide the most economical means of hauling the material excavated by BWEs. The only limiting factors in their use are the pit depth and the angle of repose of the spoil. The maximum pit depth for CPC use is about 60 m. In the double-bench BWE/CPC operation, a single BWE seems to be a better solution than the application of two BWEs. Two BWEs feeding a single CPC form a rigid system which may cause many operational difficulties. Besides, this solution is 10 to 12% more expensive than a single BWE excavating two benches.

For work beyond the reach of a CPC system, an around-pit conveyor may be used. An APC is a complex system with a relatively low utilization factor. If APCs are used, both the pit and the spoil pile must be kept in a very straight line.

At very great pit depths, a combination of BWE/APCs, BWE/CPCs, and draglines may be used.

Computer Modeling

Computer models are powerful tools for the selection of stripping methods. With a computer, many options can be analyzed quickly and easily so that the best method can be selected. Once the proper stripping method is selected, the computer can then be used to perform sensitivity analyses to optimize the values of selected parameters. The preliminary selection of pit and equipment parameters can be performed in a few hours, resulting in a significant savings in engineering work.

LEGEND:

⊙	BWE/CPC - one BWE - Bench Height 15 m
△	BWE/CPC - one BWE - two Benches 30 and 15 m High
+	BWE/CPC - two BWEs - two Benches 30 and 15 m High
×	BWE/APC - one BWE - Bench Height 15 m
◇	BWE/APC - one BWE - Bench Height 30 m

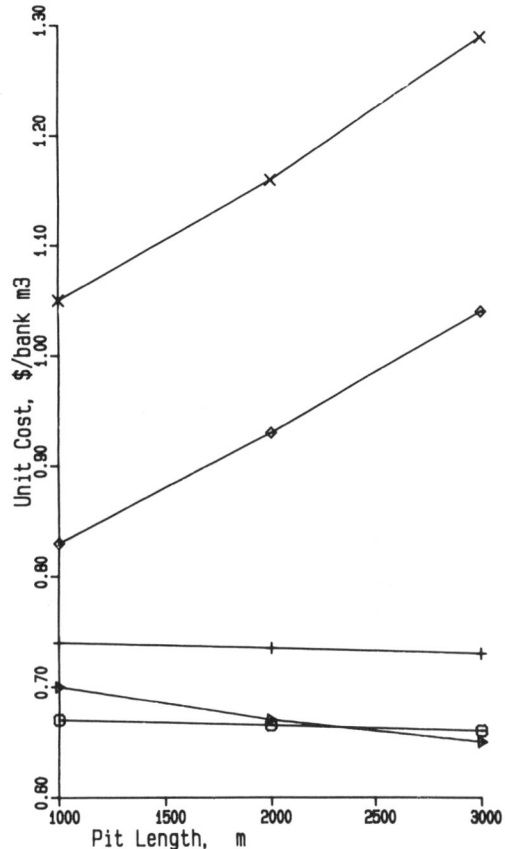

Figure 13. BWE/CPC and BWE/APC stripping costs for various pit lengths.

REFERENCES

Bhattacharyya, K.K. and S.Y.Chi, 1982, Geotechnical Aspects of Lignite Mining - Southeast U.S.A.", SME-AIME Annual Meeting, Feb.14 - 18

Buddecke, D.B. and Wolski, J., 1985,"Application of Bucket Wheel Excavator Using Cross-Pit or Around-Pit Conveyor Technology in Texas Lignite Mines", SME-AIME Annual Meeting, New York, February.

Caterpillar Performance Handbook, 1982, Caterpillar Tractor Company, Peoria, Illinois.

Kahle, M.B. and Moseley, C.A., 1982, "Development of Mining Methods in Gulf Cost Loignites". SME-AIME Annual Meeting, Dallas, Feb.14 - 18.

Learmont, T., 1984, The Walking Dragline, Part 2 - Cost Comparison of Stripping Methods", The Surface Miner, Bucyrus-Erie Company's Publication, Vol. 13.

Miercort, C.R., 1984, "Potential and Economics of Texas Lignite", American Mining Congres, International Coal Show, Chicago.

Morgan, W.C., 1979, "Overburden Stripping Systems - Production and Cost vs. Overburden Depth", Coal Conference, Louisville, Kentucky.

Prince, D.R.,1984 "The Application of Bucket Wheel Excavator and Cross-Pit Conveyor Technology to Overburden Stripping in Lignite Mines", M.Sc. Thesis, New Mexico Institute of Mining and Technology, Socorro, New Mexico.

Wolski, J., 1984, "Program for After-Tax Cost
    Analysis", FORTRAN-77 Programs for Engineering
    Economic Analysis, Engineering Press, San Jose,
    California, pp.5-1 to 5-21.

LEGEND:

⊙   SCRAPER/DOZER Stripping System

△   SHOVEL/TRUCK Stripping System (Pit Length 2000 m)

+   DRAGLINE Stripping System

×   BWE/CPC - one BWE - Interburden Stripped by a Dragline

◇   BWE/CPC - two BWEs - Interburden Stripped by a Dragline

♠   BWE/APC - CPC at the Lower Bench for Pit Depths 46 to 61 m

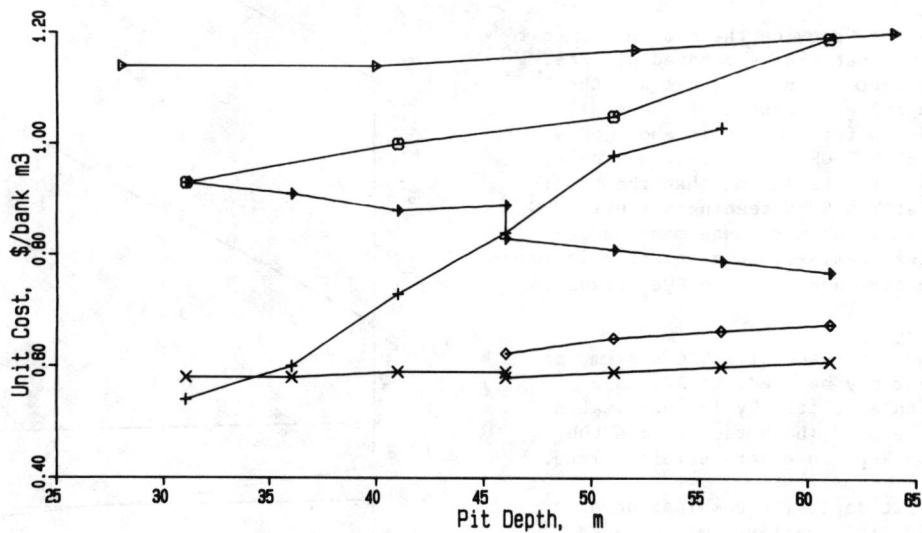

Figure 14.   Stripping costs for various excavation
             systems.

GROUND CONTROL

Chapter 39

COMPUTER APPLICATION TO ROOF ROCK QUALITY ANALYSIS

by

John L. Hill, III
Geologist

Eric R. Bauer
Mining Engineer
Bureau of Mines
U. S. Department of the Interior
Pittsburgh, Pennsylvania

ABSTRACT

The Bureau of Mines, in association with conducting basic research in ground control, is involved in the assessment of roof rock quality in coal mines using commercially available software packages. This paper illustrates the usefulness of a type of computer software package that dramatically increases the speed of performing calculations, constructing tables, and finally interpreting results through statistical analysis. The application has allowed for easy reduction of the vast amount of data produced during analysis of rock strength, as well as for Rock Quality Designation (RQD). Examples of both of these applications are presented in the paper along with a statistical analysis of the results.

INTRODUCTION

Hazardous roof conditions have long been a concern of the coal mining industry. Attempts have been made to systematically analyze mine roof conditions as a guide to roof support selection, but often the methods are tedious. These rock mass classifications often include detailed analysis of relative rock strengths which require testing of hundreds of individual specimens and eventually statistical analysis. As part of a continuing investigation by the Bureau of Mines on the causes of roof failure in coal mines, core drilling of roof rock was conducted in a coal mine of central Pennsylvania. Previous work at this mine by the authors (1984), demonstrated that roof failure was occurring adjacent to clastic dikes. Throughout the main entry system of the mine (fig. 1), the clastic dikes and associated roof failure increased in frequency near the present face position. The research also showed that maximum support in the form of cribbing or trusses in addition to the normal bolting pattern detered the formation of cutter failure if the supplemental support was installed shortly after mining. Although these measures controlled the roof failure condition, it was not determined if rock strength and/or fracture frequency in the roof were contributing to the failure mechanism.

This paper presents the results of an analysis of core taken from the roof of the main entry system. Ten holes were cored in the roof, roughly equally spaced between the shaft and face (fig. 1). They were drilled to a height of 3.05m (10 ft) above the coalbed-roof rock interface to ensure analysis of the rock through and above the roof bolt anchor horizon. The core was logged with respect to lithology, fractures, and fracture surfaces, and finally subjected to strength

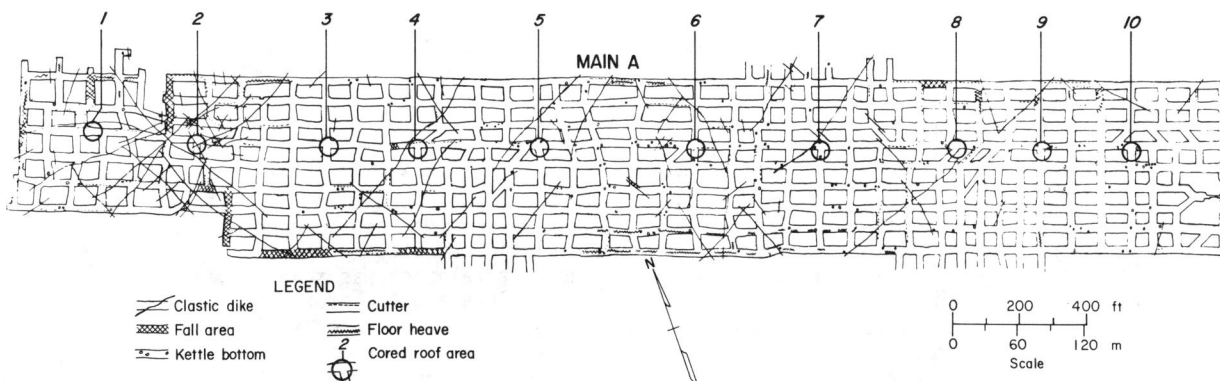

FIGURE 1. - Geologic and deformation map of Main 'A' showing core hole locations.

testing. Throughout this investigation and the investigation previously described, the computer played a major part in the process of data reduction and analysis. Details of this application of the computer to a ground control investigation are described herein, along with the results of strength testing and comparison of logged data to mapped geologic and deformational features.

The RS/1 software package as part of the VAX computer system designed by DEC was used for this investigation, reference to specific products does not imply endorsement by the Bureau of Mines. The package allows for easy manipulation of tables, programming, graphic display, and a limited amount of statistical analysis. Other systems by different manufacturers, offer similar features and would be suitable for the type of work discussed in this paper.

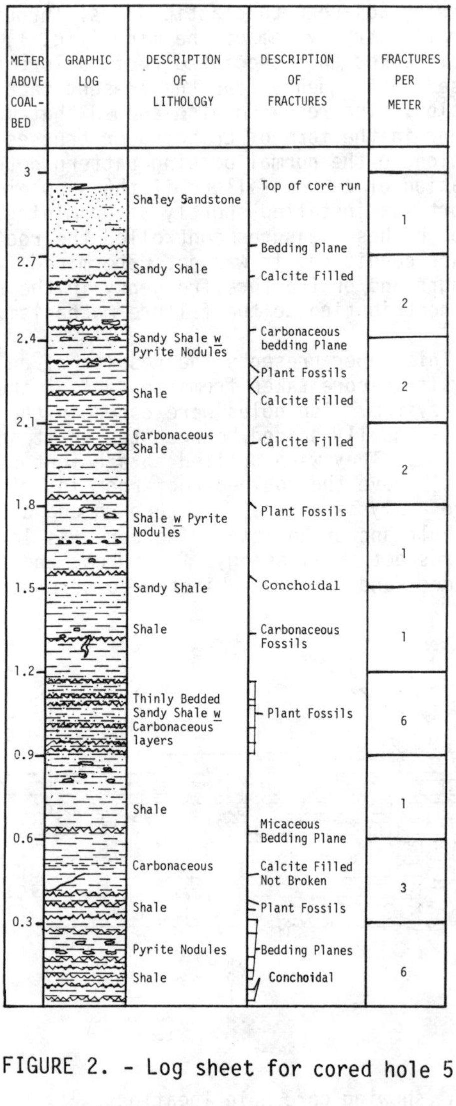

FIGURE 2. - Log sheet for cored hole 5.

QUALITATIVE CORE ANALYSIS

The first phase of the recovered-core evaluation was the process of descriptively logging the core with respect to lithology and the nature of fractures. This process was conducted as the core was removed from the barrels to avoid the erroneous logging of fractures induced by transportation. Each break in the core was investigated to determine if it was present before coring or if it was induced by coring. Subtle indicators such as weathering, infilling, and striations were used in this process and were logged along with fracture locations.

As an example of a logged core, figure 2 is the logged data for hole 5, the location of which is shown in figure 1. The first column of the log is the distance of the core above the coalbed. The second column is the lithology of the core shown graphically. The third column is a description of the lithology of the core, the fourth described the fracture surfaces, and the fifth lists the actual fractures per meter which do not appear to have been induced by coring. All of the fracture-per-meter data were stored in the computer as a table for further evaluation. To relate fracture frequency to roof stability, a program was developed to convert the fracture data to a unit devised by Derre, et al., 1969, referred to as Rock Quality Designation (RQD). This is a method used by civil engineers to describe rock stability, and Kalia, 1976, has described a correlation between RQD and roof support requirements for coal mines. Since the mine was using 1.8m (6 ft), point anchor resin, tension rebar type roof bolts, it was advantageous to look at the RQD data in four different ways: the overall RQD value for each hole, and then for each third of the hole to analyze below the anchor horizon of the bolt, in the anchor horizon, and above the anchor horizon. Figure 3 is the computer printout of these data.

	1 H1R	2 H2R	3 H3R	4 H4R	5 H5R	6 H6R	7 H7R	8 H8R	9 H9R	10 H10R
OVERALL	71	55	93	93	75	60	71	74	52	78
BOTTOM					45	45	36	86	22	82
MIDDLE	50	0	93	100	97	47	83	66	44	66
TOP	94	100	93	88	82	82	91	70	87	83

FIGURE 3. - RQD data for each cored hole. Bottom, middle, and top refer to each third of a hole.

Two distinct correlations were found between RQD and roof conditions. First, when the overall RQD values for each hole were compared with mapped geologic and deformational features around the hole, a relation was seen between holes with low RQD values and the proximity of these holes to a clastic dike or cutter failure (figs. 4 and 1). Secondly, a relation was seen between low RQD values in areas of each core run, and the presence of a carbonaceous sandy shale.

By combining these two correlations, the data would suggest that the anchor horizon should be at about the 1.8m (6 ft) height, or above, to ensure a solid anchor above the fractured zone. Kalia's, 1976, recommendations would indicate that maximum support should be employed when clastic dikes are present or where cutter failure has occurred since our data show low RQD values for these areas. Otherwise the roof is stable according to Kalia and should require only normal patterned bolting. The work conducted at the mine by the authors, as previously described, supports this conclusion.

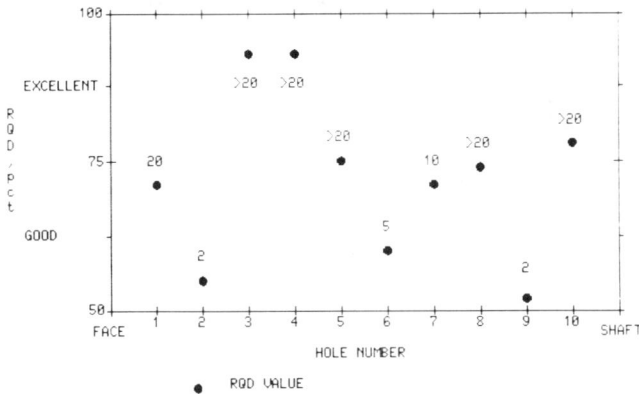

FIGURE 4. - RQD data accompanied by a number indicating the distance from the cored hole to a clastic dike or cutter failure in feet.

ROCK STRENGTH ANALYSIS

The next phase of the roof rock analysis consisted of rock strength testing of the logged core. Under normal core recovery conditions, uniaxial compressive strength tests would be conducted to obtain ultimate strength values. However, owing to the short lengths of core recovered (low RQD values), the necessary 2:1 ratio of length to diameter needed for conical failure in these tests was not obtained. Thus, the point load testing method as described by Broch and Franklin, 1972, was used with slight modifications as described by Peng, 1976. Bauer, 1984, also suggests less than a 1:1 ratio for length to diameter when testing highly anisotropic rocks and the axial tests (testing perpendicular to bedding) produce a better correlation to uniaxial compressive strength than does the diametral test (parallel to bedding) or a ratio of axial to diameteral.

For our purposes it seemed desirable to have a measure of the anisotropic nature of the rock in addition to the basic rock strength since the rock being tested is shale, which is highly anisotropic. The core specimens were thus subjected to axial and diametral tests. For each core specimen tested, the following information was recorded: rock type (given a numerical value), platen separation, point load failure, and the height

above the coal from which the core was taken. Once all the core data were tabulated in the computer, a short program was written which converted the data to strength indices and an anisotropic index based on Broch's formulas. By including the platen separation, we were able to normalize all of the results to a standard size using the size correction formula developed by Fritzhardinge, 1978. Equation (1) is used to calculate the diametral strength index. Since the platen separation is constant during diametral tests, the size factor is constant, and for 0.054m (2.1 in) diameter core the constant is approximately 1.0206 to convert the index to a standard size of 0.05m (1.97 in):

$$D(Is) = \frac{f \times A \times G}{(Ps)^2} \times 1.0206, \qquad (1)$$

where
f = failure pressure in kg/cm^2

A = platen area in cm^2

G = gravity in m/sec^2

Ps = platen separation in mm

$D(Is)$ = diametral strength index

For axial strength index calculations the size factor becomes more critical since the axial height of core specimens is not as easily controlled. Equation (2) illustrates the manner in which the size becomes a variable:

$$A(Is) = \frac{f \times A \times G}{(Ps)^2} \times \frac{Ps}{7.2} \qquad (2)$$

where:
f = failure pressure in kg/cm^2

A = platen area in cm^2

G = gravity in m/sec^2

Ps = platen separation in mm

$A(Is)$ = axial strength index

Figure 5 is a computer printout of the actual program. Although the program is relatively short and straightforward in content, for this project it conducted the operations for 593 individual point load tests. This would have been a very tedious task if performed by hand. Finally, by taking the ratio of the means of the axial and diametral indices, the anisotropic strength index is found. Figure 6 is the final table as calculated by the computer for cored hole 5. As figure 7 shows, rock strength varied very little, overall, for each hole, from one end of the main entry system to the other, and the rock is very strong by Broch and Franklin's standards, 1972.

Figure 8 is the statistical analysis of the axial strength indices (fig. 7) as calculated by the computer using the RS/1 system. The standard deviation (row 7 of fig. 8) indicates there is very little deviation between each of the cored holes. By taking the standard deviation, dividing

```
/* PROGRAM PNTLOAD */
PROCEDURE (PLHOLS);
ERASE;
TYPE ' THIS PROGRAM CALCULATES THE STRENGTH
INDEX VALUES FROM THE DATA IN PLHOLS. ';
TYPE ' ';
TYPE NOCR ' HIT <RETURN> TO BEGIN PROGRAM. ';
INPUT WAITING;
C=3; D=4; E=5; F=6; H=9; I=10; J=11; K=12; L=14; M=13;        /* INITIALIZING VARIABLES AND ARRAYS */
CNT=0; PS='PLHOLS';
SET AD[45,1] TO EMPTY;
SET AJ[45,1] TO EMPTY;
LOOPER:
CNT=CNT+1;
DO Z=1 TO LAST ROW OF PLHOLS;
    IF NOT EMPTY (PS[Z,C]) THEN DO;                           /* DIAMETRAL CALCULATIONS */
        AD[Z,1]=PS[Z,C]*141.414;                              /* FAILURE x PLATEN AREA x GRAVITY */
        AD[Z,1]=AD[Z,1]/2916;                                 /* DIVIDING BY PLATEN SEPARTION SQUARED */
        PS[Z,D]=RNDOFF((AD[Z,1]*1.0206207),-3);               /* STANDARD SIZE FACTOR */
    END;
    IF NOT EMPTY (PS[Z,H]) THEN DO;                           /* AXIAL CALCULATIONS */
        AJ[Z,1]=PS[Z,H]*141.414;                              /* FAILURE x PLATEN AREA x GRAVITY */
        AJ[Z,1]=AJ[Z,1]/(PS[Z,I]**2);                         /* DIVIDING BY PLATEN SEPARATION SQUARED */
        PS[Z,J]=RNDOFF((AJ[Z,1]*SQRT(PS[Z,I])/7.2),-3);       /* STANDARD SIZE FACTOR */
    END;
END;
PS[1,E]=RNDOFF((MEAN OF COL D OF PLHOLS),-3);                 /* DIAMETRAL STRENGTH INDEX */
PS[1,F]=RNDOFF((STDEV OF COL D OF PLHOLS),-3);
PS[1,K]=RNDOFF((MEAN OF COL J OF PLHOLS),-3);                 /* AXIAL STRENGTH INDEX */
PS[1,M]=RNDOFF((STDEV OF COL J OF PLHOLS),-3);
PS[1,L]=RNDOFF((PS[1,K]/PS[1,E]),-3);                         /* ANISOTROPIC STRENGTH INDEX */
C=C+14; D=D+14; E=E+14; F=F+14; H=H+14;
I=I+14; J=J+14; K=K+14; L=L+14; M=M+14;
IF CNT=10 THEN GO TO FINISH;
GO TO LOOPER;
FINISH:
DISPLAY PLHOLS;
END;
```

FIGURE 5. - Computer printout of program used to convert raw Point Load data to a strength index.

57 RTYPE HOLE 5	58 HEIGHT HOLE 5	59 DFAIL HOLE 5	60 D (Is) HOLE 5	61 M D(Is) HOLE 5	62 STDEV HOLE 5	63 RTYPE HOLE 5	64 HEIGHT HOLE 5	65 AXFAIL HOLE 5	66 PSmm HOLE 5	67 A (Is) HOLE 5	68 M A(Is) HOLE 5	69 STDEV HOLE 5	70 ANISO HOLE 5
5	0.50	5	0.247	1.146	0.522	5	0.50	85	47	5.181	5.484	1.248	4.785
5	0.75	10	0.495			5	0.75	85	50	4.722			
5	1.50	30	1.485			5	1.00	90	45	5.856			
5	1.70	20	0.990			5	1.25	110	57	5.020			
5	1.90	36	1.782			5	1.50	110	48	6.497			
5	2.10	25	1.237			5	1.70	105	48	6.201			
4	2.30	33	1.633			5	1.90	110	58	4.891			
5	2.50	34	1.683			5	2.10	130	51	7.010			
5	2.70	24	1.188			4	2.30	115	53	5.854			
6	2.90	47	2.326			5	2.50	95	56	4.452			
5	3.40	16	0.792			5	2.70	115	52	6.024			
6	3.50	23	1.138			6	2.90	95	58	4.224			
6	3.70	34	1.683			5	3.40	110	50	6.111			
6	3.80	19	0.940			6	3.50	100	48	5.906			
5	4.00	38	1.881			6	3.70	105	53	5.345			
5	4.20	11	0.544			6	3.80	100	54	4.950			
4	4.60	32	1.594			5	4.00	105	58	4.669			
5	4.80	27	1.336			5	4.20	115	50	6.389			
5	5.00	18	0.891			4	4.50	120	40	9.316			
5	5.30	18	0.891			5	4.60	110	59	4.767			
5	5.50	18	0.891			5	4.80	100	57	4.564			
5	5.70	23	1.138			5	5.00	80	49	4.581			
5	5.90	26	1.287			5	5.30	100	56	4.687			
5	6.10	29	1.435			5	5.50	75	45	4.880			
5	6.30	41	2.029			5	5.70	100	56	4.687			
5	6.50	42	2.079			5	5.90	100	45	4.506			
5	6.60	28	1.386			5	6.10	95	59	4.117			
8	7.00	18	0.891			5	6.30	110	43	7.662			
5	7.20	15	0.742			5	6.50	80	49	4.581			
5	7.40	17	0.841			5	6.60	80	56	3.749			
5	7.60	17	0.841			8	6.70	95	59	4.117			
5	7.80	12	0.594			6	7.00	100	49	5.726			
6	8.00	25	1.237			5	7.20	95	56	4.452			
6	8.20	10	0.495			5	7.40	95	52	4.976			
6	8.40	10	0.495			5	7.60	110	55	5.297			
5	8.60	20	0.990			5	7.80	95	58	4.224			
6	9.00	5	0.247			5	8.00	95	51	5.123			
2	9.20	6	0.297			6	8.20	105	50	5.833			
2	9.40	36	1.782			6	8.40	120	53	6.108			
2	9.60	23	1.138			5	8.60	115	52	6.024			
2	9.80	28	1.386			6	9.00	95	58	4.224			
						2	9.20	115	54	5.692			
						2	9.40	185	51	9.976			
						2	9.60	110	52	5.762			
						2	9.80	115	53	5.854			

FIGURE 6. - Computer printout of Point Load data and calculated strength indices as produced by the computer.

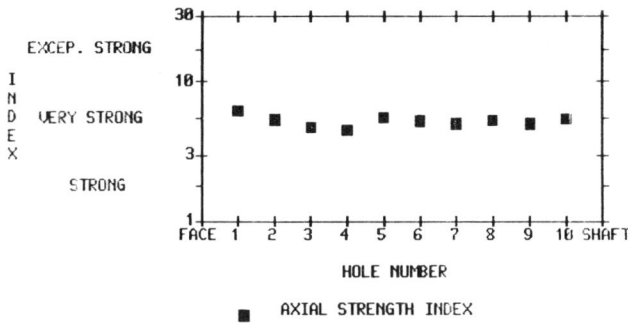

FIGURE 7. - Axial strength index values for each hole cored.

O Statistic	1 Value
1 Count (N)	10.000000
2 Sum	52.284000
3 Mean	5.228400
4 SEM (s.e. of mean)	0.144330
5 Median	5.253500
6 Variance	0.208311
7 StDev (sd)	0.456411
8 Maximum	6.191000
9 Minimum	4.541000
10 Range	1.650000
11 Skewness	0.675755
12 Kurtosis	1.440883

FIGURE 8. - Statistical analysis of axial strength index values shown in figure 7.

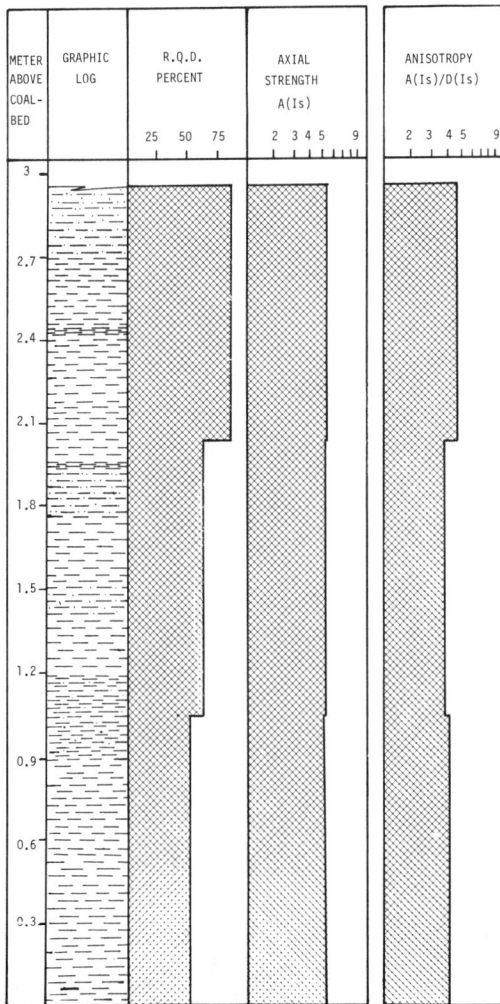

FIGURE 9. - Generalized stratigraphic column, RQD, axial strength, and anisotropy of the immediate 3.05m of roof rock for Main 'A'.

by the mean, and multiplying by 100, we find that there is only 8.7% variation, which is well below the limit of 30% suggested as a maximum allowable variation for a statistically significant result.

Figure 9 is a cumulative diagram of an assessment made of the immediate 3.05m (10 ft) of roof rock. The stratigraphic column, RQD, axial strength, and anisotropic value are all generalized to represent the entire main entry system from the shaft area to the face.

ATTRIBUTES OF THE COMPUTER

Obviously the application discussed in this paper is not the only application to roof rock quality analysis, nor are they very complicated. All of the calculations could have been performed on a simple hand held programmable calculator achieving the same end result. However, the advantages to using a computer for even simple applications becomes manifold with the ability to check the data as it is entered for erroneous numbers, and the ability to store the data for future analysis.

Data can be checked as it is entered, by running the input through a simple program which tests each value to determine if it falls within a particular range of values. If the input does not, the program will prompt the user to make a change or verify that the input is correct. Once the data is stored within the computer and the originally desired analysis conducted, a permanent record of the input is available for any further analysis which was not forseen as being applicable at the time of the initial analysis. Since the data is already in the computer there is no need to spend time re-entering the data and the data is now available in table form to be analyzed by any of the commercially available software packages.

CONCLUSIONS

The application of the computer to the calculation of rock strength, statistical analysis, and graphic display of data greatly reduced the time needed to produce results during the research effort. By simple programming and data manipulation, we were able to quickly show statistically that roof rock strength had not drastically changed between the shaft and face. By combining data, we were able to show that RQD decreases near clastic dikes although roof rock strength does not change.

This information supports our earlier findings and would suggest that clastic dikes should be supported immediately after mining with trusses or cribbing, to a distance of 3.05m (10 ft) inby and outby at the study mine.

REFERENCES

Bauer, R. A., 1984, "The Relationship of Uniaxial Compressive Strength to Point-Load and Moisture Content Indices of Highly Anisotropic Sediments of the Illinois Basin," Proc. 25th Symp. Rock Mechanics, Northwestern University, Evanston, IL, p. 398-405.

Broch, E. and J. A. Franklin, 1972, "The Point-Load Strength Test," Int. J. Rock Mech. and Miner. Sci. v. 9, p. 669-697.

Deere, D. U., et al, 1969, "Design of Tunnel Liners and Support Systems," PB-183 799, Clearing House for Federal Scientific and Technical Information, Springfield, VA, p. 138--144.

Fritzhardinge, C. F. R., 1978, "Note on Point-Load Strength Index Test," ISRM Paper S1021, p. 1

Hill, J. L. and E. R. Bauer, 1978, "An Investigation of the Causes of Cutter Roof Failure in a Central Pennsylvania Coal Mine: A Case Study," Proc. 25th Symp. Rock Mechanics, Northwestern University, Evanston, IL, p. 603-614.

Kalia, H. N., 1976, "Improving the Design of Coal Mines by Borehole Logging, Rock Mechanics, and Related Data," SME preprint 73F72, Transactions Vol. 260, p. 17-20.

Peng, S. S., 1976, "Stress Analysis of Cylindrical Rock Discs Subjected to Axial Double Point-Load," Int. J. Rock Mech. and Miner. Sci., v. 13, p. 97-101.

Chapter 40

COMPUTERIZED EVALUATION OF NEAR-SEAM INTERACTION

A. Grenoble
C. Haycocks
Wei Wu

Department of Mining & Minerals Engineering
Virginia Polytechnic Institute and State University
Blacksburg, Virginia 24061

ABSTRACT

The mining of seams in close proximity considerably amplifies interaction effects. Detailed analysis of the interaction phenomena has been carried out for mining under an existing operation. Finite element analysis was used to determine loadings on both upper and lower pillars and to establish firm criteria for lower seam pillar design and roof and floor stability. This work was supplemented by computer aided analysis of body-loaded photoelastic models. The photoelastic modeling was two-dimensional using a homogeneous material in a variety of layer thicknesses. The results of both types of modeling work were correlated with field data to verify the analysis. The design criteria established as a result of this work have been assembled on a microcomputer for engineering purposes. The resulting program is interactive in nature and can be implemented by an operator with minimal computer experience.

INTRODUCTION

The frequency of interaction problems between mines operating in a multi-seam environment continues to grow. The opening of new seams to underground mining frequently involves the threat of interaction between workings. This characteristic is particularly applicable to the Appalachian region where the majority of reserves exist in a multi-seam environment. Historically, the selection of mining sequence was based upon coal quality, seam ownership and accessibility rather than ground control considerations. The sequence of selecting seams for mining has and will continue to create major ground control problems as a result of interaction. The negative effects of interaction can include safety hazards, high mining costs, and even the total loss of a section of reserves.

The problems of interaction are accentuated as the interval between seams decreases. At distances of less than 33.5 m (110 ft) vertically, problems may be expected under some conditions in both the underlying and overlying seams (Stemple, 1956, Haycocks and Karmis, 1983). For an engineer in the process of designing a new operation, tools are needed to help predict when and where interaction might be expected to occur and how severe the problem is likely to be. This information can then be used to eliminate interaction problems altogether, minimize loading in the underlying seam, estimate the stability and support needs, or determine how much coal can ultimately be recovered in the affected areas.

Research into the mechanisms and effects of interaction due to multi-seam mining has provided many useful design criteria (Stemple, 1956, Peng and Chandra, 1980, Haycocks and Karmis, 1983). Unfortunately, these can be rather complicated both to understand and to use. As a method of conveying this data to the field engineer, much of the information has been combined into easy to use software packages suitable for use on a personal computer. The software is now available and provides a method of quickly developing a data base from which an engineer in the field can successfully design an operation in the vicinity of another working or mined out seam. An additional benefit of this approach is that it provides an input format for collecting data and allows the engineer to conduct a sensitivity analysis for various combinations of mining geometry and strata conditions.

FIELD STUDIES

Case studies that are available from the mining industry have greatly facilitated the understanding of multi-seam mining problems (Stemple, 1956; Hasler, 1951; Peng and Chandra, 1980). These case studies represent an immense variety of geologic conditions and mine layouts for both room and pillar and the longwall mining methods. Analysis for specific geologic settings has been sufficient in some instances to define both the trends and limits of interaction under a variety of ground control conditions. Initial analysis of field study data demonstrated that interaction problems could be evaluated by considering pillar load transfer, subsidence, arching and block failure into the lower excavation (Haycocks and Karmis, 1981). These four phenomena have formed a basis for further analysis and the field study results combined with subsequent model studies have been used to develop a computer model which incorporates design criteria for mining in multi-seam situations. An example of the application of field data may be seen in Figure 1 which establishes the innerburden limits for interaction problems due to pillar loading onto underlying seams. Subsequent analysis of this phenomena indicated that the degree of layering in the innerburden was much more important than the modulus (Ehgartner 1982). Research has also shown that relative extraction ratios are extremely important in determining whether or not pillar load transfer problems will be experienced in the lower seam (Figure 2). The significance of the ratio of overburden depth to innerburden thickness was recognized by Stemple (1956) and has been evaluated empirically as well as numerically. In addition to this type of application of field data, case studies have also been used to validate trends developed from finite element and photoelastic models and to correlate predicted values with phenomena observed underground.

MODEL ANALYSIS

To extend the information gained from case studies, analysis of various interaction phenomena has been carried out using both finite element and body loaded photoelastic models. To further enhance stress determination made using the finite element analysis, the Mohr-Coulomb failure criteria has been incorporated into the program to evaluate safety factors where shear failure might be anticipated (Barko, 1982).

Photoelastic models were used to evaluate stress phenomena in close proximity to the excavation since they could best incorporate the effects of layering and bed separation which is extremely difficult to accomplish using finite element methods. These models were used extensively to study pillar load transfer. Peng and Chandra (1980) predict that the load approaches the normal background value at a distance of approximately four times the pillar width below the floor. Ehgartner (1982) found that the distance stress was transferred depended on the nature of the rock below the pillar. In particular, he found that low modulus stratified materials tended to increase the distance through which stress is transferred while stiff isotropic materials had the opposite effect. Results of his models showed that the zero influence bulb could extend as deep as eight times the pillar width in a highly stratified material.

Finite element analysis was used to evaluate the influence of the thickness and location of sandstone layers in the innerburden and overburden on the size and shape of the pressure arch around the upper opening. This information was, in turn, used to locate the area in which the upper seam arch intersected the lower workings. Hudock (1983), modeling an isotropic overburden, found that the modulus of the material had little influence on arch dimensions or abutment pressures. The significance of a high modulus layer above an excavation, however, has been recognized and is docu-

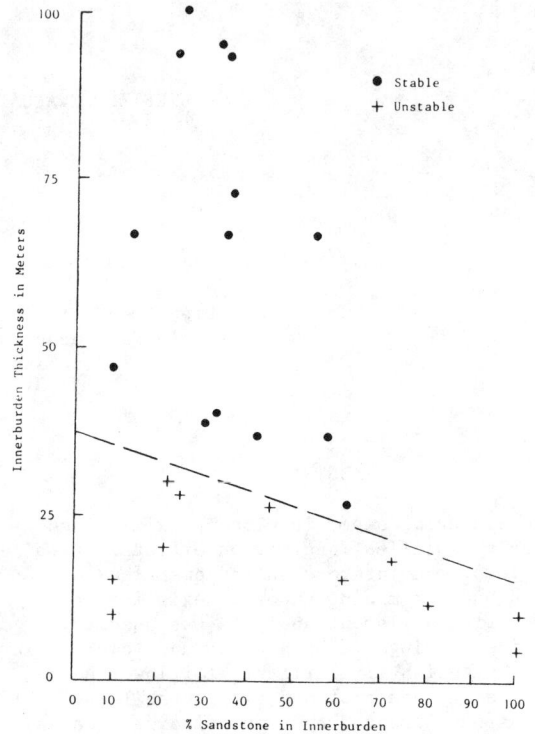

Fig. 1. Relationship between percent sandstone in the innerburden and required innerburden thickness to prevent interaction effects from upper seam pillar load transfer.

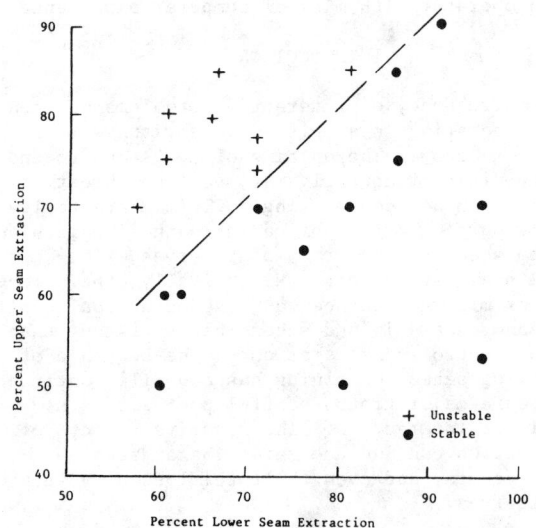

Fig. 2. Relationship between percent extraction in the upper and lower seams and lower seam stability (After Ehgartner, 1982)

mented in the literature (Randolph, 1915; Peng and Chandra, 1980).

The basis of this analytical work has been to establish relationships between geometry, material properties and innerburden thickness for selected points in the upper and lower openings where failure might be anticipated and where it has been observed in field studies. The results of these model studies are being incorporated with the field data to provide realistic design criteria and methods for evaluating a wide variety of mining geometries and geologic settings.

PROGRAM INPUT

The results to date have incorporated into a software package written in BASIC for an IBM PC. A typical input menu is illustrated in Figure 3. The program has been designed to be interactive and should greatly facilitate data input by leading the engineer through the process step by step. Variables such as roof and floor characterization factors are intended to be used in conjunction with the User's Guide for definition of these values. This package remains experimental and will be continued to be developed and refined as more information is gathered and analyzed. Field testing of all facets the program is helping to define what is available in terms of input data and also what can be used in terms of output. The K factor has been selected for defining coal strength because it does not involve specification of cube size. The incorporation of joint orientation facilitates a three-dimensional analysis.

PROGRAM DESIGN AND OUTPUT

A typical flow-chart for the program is shown in Figure 4. It should be noted that the initial inquiry is to whether interaction problems will or will not occur for the conditions as specified. Because of the extreme complexity of modeling ground behavior, the program is designed to be used as a tool rather than provide absolute answers and it is intended that the engineer will use it to evaluate different layouts based on existing or planned upper seam pillar configurations. Further enhancements to the software include sensitivity analyses which are carried out in terms of pillar and floor loading and roof stability for a variety of geometries.

Typical output is shown in Figure 5. In this situation it has been determined that interaction effects will occur and the program defines worst conditions for the geometry as it is layed out. It should be remembered that not all effects of interaction are negative. For example, location of an opening immediately below a mined out area will place it in a very low stress field and therefore probably enhance its stability.

It is anticipated that some of the most useful information will come from the graphical output. This includes contours of pillar load as well as anticipated roof and floor conditions for the

area under consideration. No attempt has been made to incorporate support factors or methods since the relationship between support and failure probabilities are at this time are unknown. Based on analysis of the output, the engineer has the choice of whether to work around the problems or to change his design to eliminate some of the anticipated difficulties. Additionally, joint orientation analysis will provide an opportunity to evaluate the relative merits of different opening orientations.

CONCLUSIONS

It is anticipated that the software package, as it currently exists, will provide a design tool for evaluation of lower seam conditions. The ability to evaluate entry orientation with respect to jointing introduces a three-dimensional factor into the analysis which is essential if the program is to realistically evaluate field phenomena. The interactive nature of the program should make it easy to use, while the package itself provides a good reproducible technique for evaluation of interaction phenomena. Finally, the input data format lays out clearly and specifically the type of information which is required.

REFERENCES

Barko, E., 1982, "Mechanics of Interseam Failure in Multiple Horizon Mining," M.S. Thesis Virginia Polytechnic Institute and State University, Blacksburg, Va.

Ehgartner, B. L., 1982, "Pillar Load Transfer Mechanisms in Multi-Seam Mining," M.S. Thesis, Virginia Polytechnic Institute and State University, Blacksburg, Va.

Hasler, H.H., 1951, "Simultaneous vs. Consecuative Working of Coal Beds," Trans. AIME, Mining Engineering, May, pp. 436-440.

Haycocks, C. and Karmis, M., 1983, Ground Control Mechanisms in Multi-Seam Mining, Office of Mineral Institutes, Bureau of Mines, U.S. Dept. of the Interior, 342 pp.

Haycocks, C. and Karmis, M., 1981, "Ground Control Mechanisms in Multi-Seam Mining," Research Performance Report, Office of Surface Mining, June.

Hudock, S.D., 1983, "The Effects of the Pressure Arch upon Multiple Seam Mining," M.S. Thesis, Virginia Polytechnic Institute and State University, Blacksburg, Va.

Peng, S.S. and Chandra, U., 1980, "Getting the Most from Multiple Seam Reserves," Coal Mining and Processing, November, pp. 78-84.

Randolph, B.S., 1915, "The Theory of the Arch in Mining," The Colliery Engineer, Vol 35, pp. 427-29.

Stemple, D.T., 1956, "A Study of the Problems
    Encountered in Multi-Seam Mining in the Eastern
    United States," M.S. Thesis, Virginia Polytech-
    nic Institute, Blacksburg, Va.

INPUT

Mining Site: Lee Co., Va.

Mine Under: 1. Remnant Pillar
              2. Solid/Gob Interface
              3. Pillars in Place
Enter Number: 1

UPPER SEAM DATA

Seam Name: Pocahontas No. 5
Depth of overburden (ft): 510
Density of overburden (lb/cu.ft.): 140
Mining Height (in.): 57
Residual Pillar Dimensions (length,width): 150,245
Coal K Factor: .27
Roof Character (10=excellent, 1=poor): 8
Floor Character (10=excellent, 1=poor): 6
Jointing Present? (y/n) y

Joint Set No.	Spacing(ft)	Orientation(deg)	Dip(deg)
1	26	170	85
2	40	80	70

INNERBURDEN DATA

Average Thickness (ft): 55
Total number of beds: 8
Sandstone present? (y/n)   y

Bed No.	Thickness	Distance Above Lower Seam (ft.)
1	10	30
2	4	39

Jointing Present? (y/n) y

Joint Set No.	Spacing (ft)	Orientation (deg)	Dip (deg)
1	40	80	70

LOWER SEAM DATA

Seam Name: Pocahontas No. 3
Seam Height (in): 48
Roof Rock Characterization (10=excellent, 1=poor) 4
Floor Rock Characterization (10=excellent, 1=poor) 9
Coal K Factor: .23
Jointing Present? (y/n) n

Fig. 3. Typical input menu.

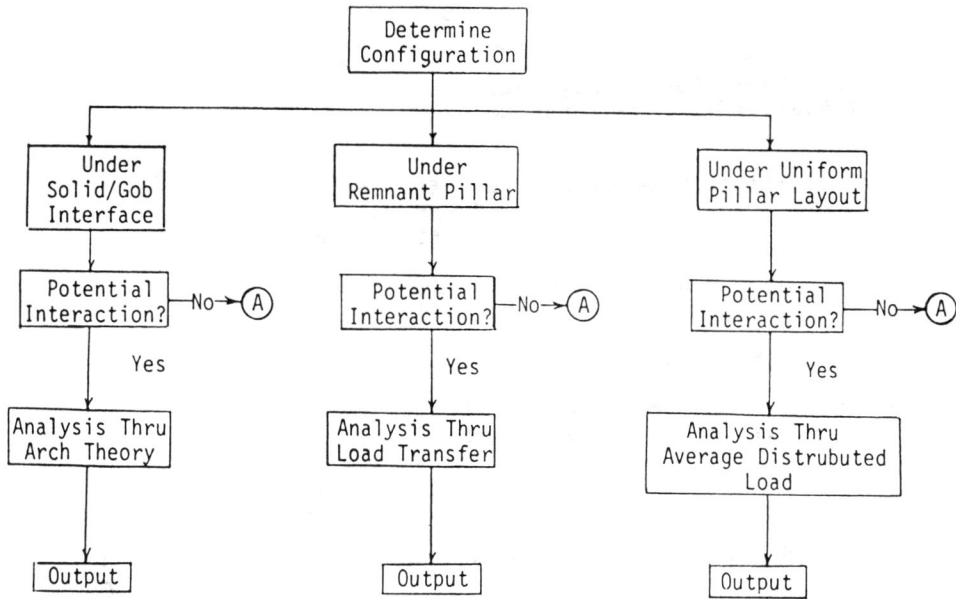

Fig. 4. Flow chart outlining basic program steps.

OUTPUT

```
Mining Site:  Lee County, Va.
Seam Analyzed:  Pocahontas No. 3


Optimum Pillar Dimensions:  50  ft. long
                            55  ft. wide
                            70  ft. centers

Optimum Room Width:  Entries    20  ft.
                     Crosscuts  15  ft.

Projected Extraction:  51%

Maximum Pillar Load:  4873  psi.
Average Pillar Load:  1763  psi.

Roof Index (10=excellent, 1=caved):  Worst    4
                                     Average  7

Floor Index (10=excellent, 1=serious heave):  Worst    5
                                              Average  8

Most Likely Failure Mechanism:  Roof Shear
Potential for Cracking to Upper Seam (0=none, 5=certain):  3
Probability for Success: Good


Graphics Options:  Enter Number

1. Plot of Upper Seam Geometry
2. Plot of Lower Seam Geometry
3. Plot of Superimposed Geometries
4. Contour of Pillar Load
5. Contour of Roof Index
6. Contour of Floor Index
7. Plot of Projected Failure Zone in Cross-Section
```

Fig. 5. Example output indicating anticipated lower seam conditions.

Chapter 41

STABILITY ANALYSIS OF ENTRIES IN A DEEP COAL MINE
USING FINITE ELEMENT METHOD

by
Duk-Won Park
and
Nadim F. Ash

Department of Mineral Engineering
The University of Alabama
University, Alabama

Abstract. Strain softening behavior of rock and coal layers around a coal mine opening was simulated using a quasi-elastic approach, which accounts for residual strength after failure. This simulation was achieved in such a way that material values were modified after each run according to stress distribution until it reached an equilibrium condition. MSC/NASTRAN FEM program was used for the simulation. The effect of yield pillar system was analyzed comparing with regular pillar system. The results agreed with the field measurement in terms of stress distribution and floor heave.

INTRODUCTION

In view of the increasing need for the U.S. coal industry to enhance productivity and improve coal recovery, longwall mining technique has been recently adopted for its large extracting capability of underground coal deposits. This method of extraction being used for deep underground mines requires the development of three or four entries running parallel to the longwall panel. This study treats only the four entry system.

The stability of developed entries in terms of floor and roof movements is greatly affected by the size of supporting pillars. The conventional system of pillars which consists of three equally sized pillars has created problems such as floor heave, and a high state of stress in the region adjacent to the longwall panel. A relatively recent development of a yield pillar system being considered has shown an improvement over the other in that is allows the pillar closer to the panel to yield, thus minimizing floor heave and allowing the high stresses to transfer to a relatively larger pillar away from the longwall panel.

The object of this paper is to make an analytical comparison using a computer simulation technique for the stress distribution in the pillars, and floor movements relating to both situations.

STATE OF STRESS IN COAL PILLARS

The study area is an underground coal mine in the Black Warrior Basin in Alabama where the depth of the coal seam is approximately 2000 feet. This mine uses two pillar systems at its longwall sections. The first one is the conventional system called the regular pillar system. When subjected to the high stresses, a yield zone is distributed around the edge of each pillar, followed by a peak stress cone. As the longwall face advances, the yield zone increases with the peak stresses transferring toward the inner core of the pillar. This being the case, high stresses will be maintained in the region near the longwall panel, with floor heave developing in the entries adjacent to the panel.

The second system is called the yield pillar system whereby the three pillars vary in size, in such a way that the stresses close to the panel will be transferred away toward the large pillar in the middle. This is achieved by yielding the first pillar adjacent to the panel.

Pillar stress monitoring was conducted although accurate stress measurements follow a rather laborious and time-consuming procedure. Vibrating wire gages had to be installed in the pillars at critical locations. The gages were placed at various depths varying from 10 to 30 feet, and were read at regular time intervals. The regular and yield pillars along with the gage locations in the study area are shown in Figure 1.

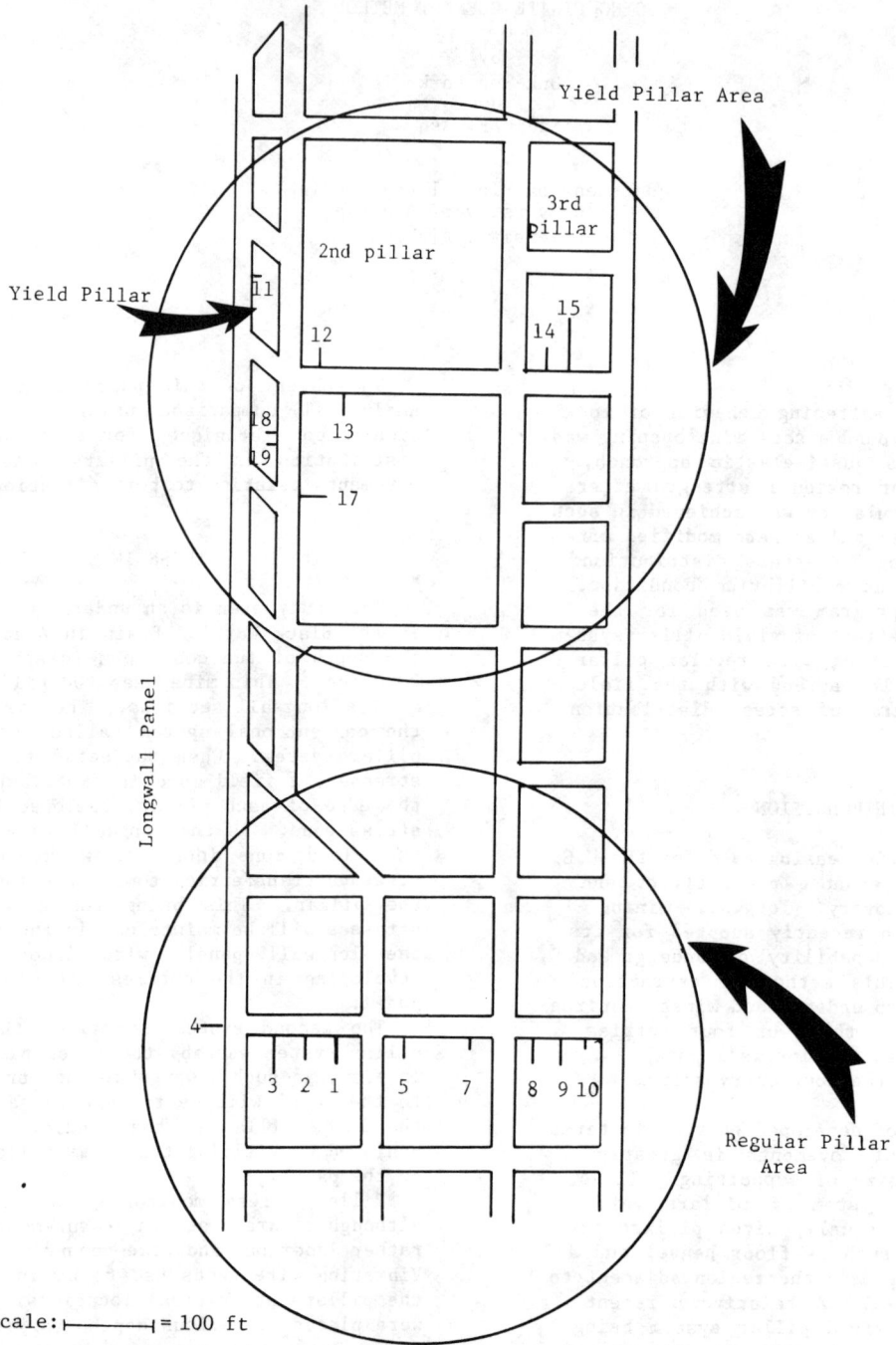

Figure 1. Regular Pillar and Yield Pillar Areas and Gage Locations

FINITE ELEMENT METHOD

In simulating both cases, a finite element program, MSC/NASTRAN, developed by NASA, was utilized. This package program is flexible enough to handle problems relating to static, dynamic, heat transfer and many others. Due to the large number of elements needed to form the mesh, two programs were developed; one for generating two dimensional grid points, the other to generate quadrilateral elements. Both programs are compatible with the MSC/NASTRAN input format. Figures 2 and 3 show the yield pillar and the regular pillar area models respectively.

In order to simulate the actual conditions existing in the area, extensive testing had to be conducted to determine the physical properties for each layer of rock and coal. Tables 1 through 4 show material properties necessary for the input. The regular pillar model consisted of 877 plate elements with 1,038 nodes, while the yield pillar model was formed by combining 984 elements and 1,158 nodes. Constraints for both models were placed along two edges of the models. Gravitational and horizontal loading have been imposed on the other two surfaces.

The major variable affecting the iterative procedure is the modulus of elasticity because it represents the relative stiffness of the layers in question. This procedure has been previously discussed by Kripakov (1), adopted from the theoretical concept developed by Kidybinsky and Babcock (2) and Pariseau and Eitani (3). Assuming a plain strain condition and a linearly elastic behavior, when an element exceeds the Mohr-Coulomb criteria for failure, the difference between the calculated and the admissible residual stress is redistributed to other elements. This redistribution can be made based on the stiffness of the system and its relation to the stiffness of the element. If the state of stress does not exceed the failure stress of the element, no modification in the properties of that element are required. If the state of stress exceeds the failure stress for the element in question, a redistribution of the stresses surrounding the element is made. This procedure is repeated until the system converges, that is, until no change in the state of stress in the element can be found.

Generally, failure can occur in one of three ways: tension, compression, or shear. In case where tension is governing, and this can occur in pillar ribs, floor and roof surfaces where those elements can no longer have a supporting capability. Thus almost all the stresses will be transferred to surrounding locations. In the case of shear and compression, the redistribution of stresses is achieved through a modification of the material property of the specific elements using the following relations:

$$\text{F.S. (shear)} = \frac{S_o + \sigma_n \tan \phi}{\tau_n} \tag{1}$$

or

$$\text{F.S. (compression)} = \frac{C_o + \sigma_{min} \tan \beta}{\sigma_{max}} \tag{2}$$

where S_o = shearing strength

C_o = uniaxial compressive strength

ϕ = angle of internal friction

$\beta = \dfrac{1 + \sin \phi}{1 - \sin \phi}$

$\sigma_{max}, \sigma_{min}$ = major and minor principal stresses

σ_n = normal stress

τ_n = shearing stress

Table 1

Physical Properties of Rock Samples
(Average Values)

Specimen	Direction	Strength (psi)	Modulus of Elasticity (psi x 10⁶)	Poisson's Ratio (ν)
roof	horizontal	8,880	3.55	0.38
	vertical	11,990	3.38	0.31
floor	horizontal	5,800	NA	NA
	vertical	NA	NA	NA
middleman	horizontal	5,053	4.60	0.43
	vertical	5,550	6.75	0.14

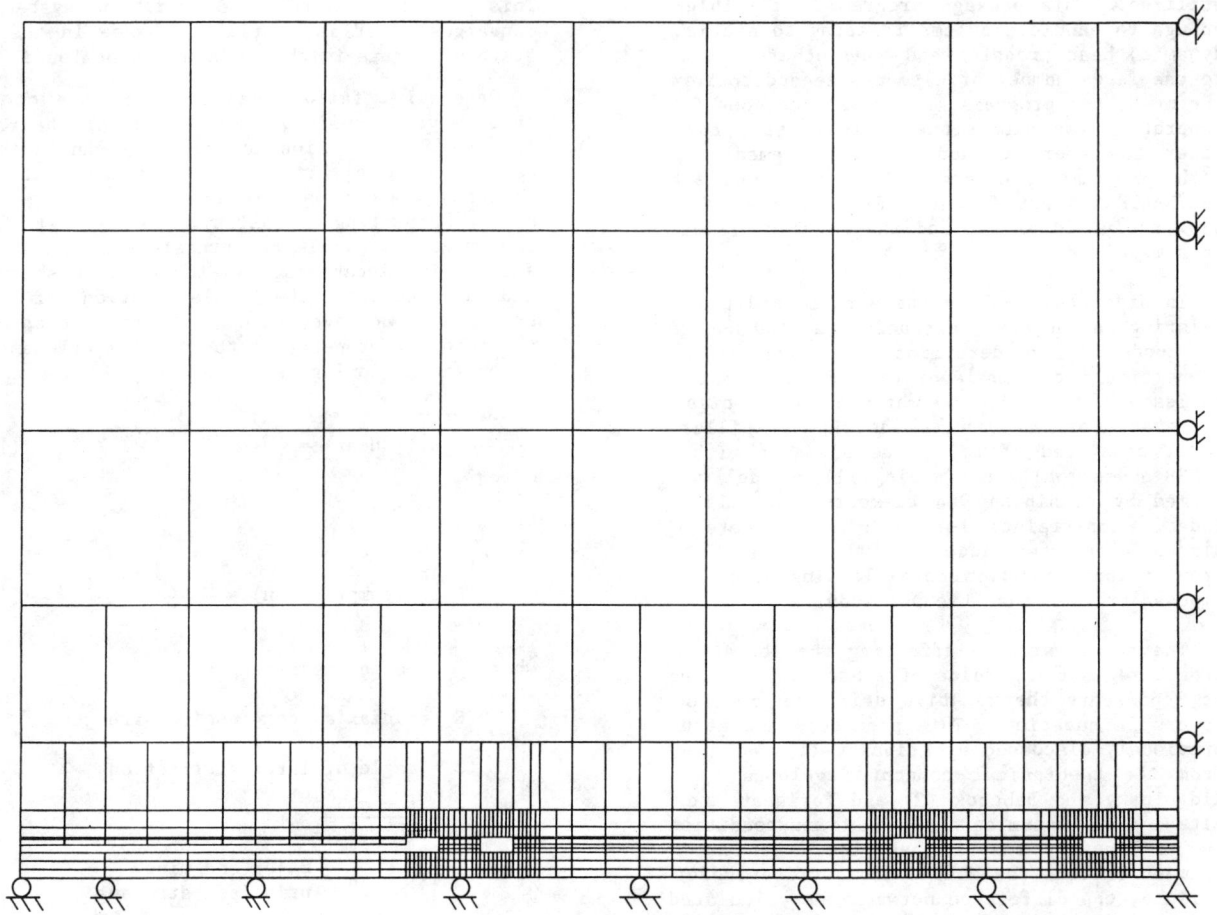

Figure 2. Yield Pillar Model

Figure 3. Regular Pillar Model

Table 2

Hardness Test Results (From 6 Specimens for Each
Direction, 10 Tests on Each Specimen)

Specimen	Direction	Average Shore Hardness Number
middleman	vertical	16.1
	horizontal	16.2
roof	vertical	24.2
	horizontal	27.6

Table 3

Average Tensile Strength (Brazilian)

Specimen	Direction	Tensile Strength (psi)
middleman	vertical	740
	horizontal	662
roof	vertical	1410
	horizontal	1254
floor	vertical	842
	horizontal	NA

Table 4

Triaxial Test Results on Rock Layers

Specimen Type	Horizontal Stress	Vertical Stress	Angle of Internal Friction (ϕ)	Cohesion (psi)
middleman vertical	500	8,570		
	1000	8,970		
	1500	8,530	40.8	1,250
middleman horizontal	500	10,290		
	1000	8,930		
	1500	10,800	56.8	650
floor horizontal	500	7,590		
	1000	9,450		
	1500	15,940	38.9	1,350
floor vertical	500	14,550		
	1000	21,800		
	1500	17,080	35.0	3,100
roof horizontal	500	15,420		
	1000	15,330	--	--

The redistribution of stresses is achieved through a modification of the moduli of elasticity of elements using equations (1) and (2). If the calculated factor of safety value is larger than 1, the same modulus of elasticity (E) is used. Otherwise a new modulus of elasticity (E_{new}) is calculated according to the following equation:

$$E_{new} = F.S. \times E_{original}$$

Modified moduli of elasticity (E_{new}) are used for the next iteration according to the above procedure. This process is repeated until the new modifications do not produce any considerable change in the stress values of the elements in question.

Regular Pillar System

Stress distribution for the initial run of the finite element program as seen in Figure 5 shows high peak stresses at the edges valued at approximately 7000 psi, and 4000 psi in the middle of the pillar. Due to the difficulty in getting readings in the mine from the vibrating wire gages, one can see from Figure 4 that only gages 5 and 7 were capable of registering the stresses in the middle pillar even after the longwall face passed the gage location.

Comparing the readings from the two gages and the simulated stress distribution for the three stages as seen in Figure 5 for the middle pillar, gage 5 shows a steady increase in stress while gage 7 shows a relative drop in both the analysis and actual gage reading. It is expected that the stress distribution generated by computer would give higher stress values than expected. The third iteration describes a distinct yield zone and a competent inner core area. A layer of rock nearly 3 feet of thickness tends to act as a reinforcement for the pillar as a whole thus slowing the development of the yield zone around the pillar.

Table 5

Comparison of Stresses for the Regular Pillar System

Stages	Computer Program Gage 5	Gage 7	Actual Readings Gage 5	Gage 7
1	3500	4000	–	–
2	4000	4000	2400*	2400*
3	4200	3800	3200**	2500**

* 350 feet away from face location
**face location at the gage

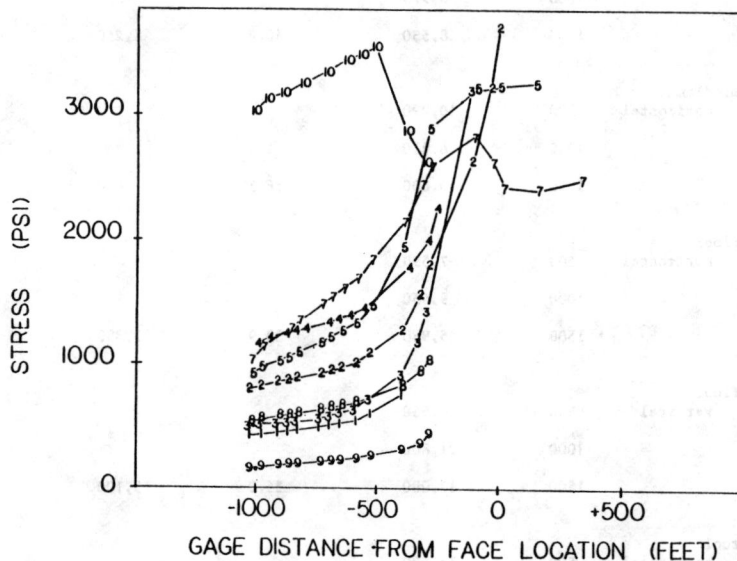

Figure 4. Stress Increase vs. Face Distance From Gages at Regular Pillar Area

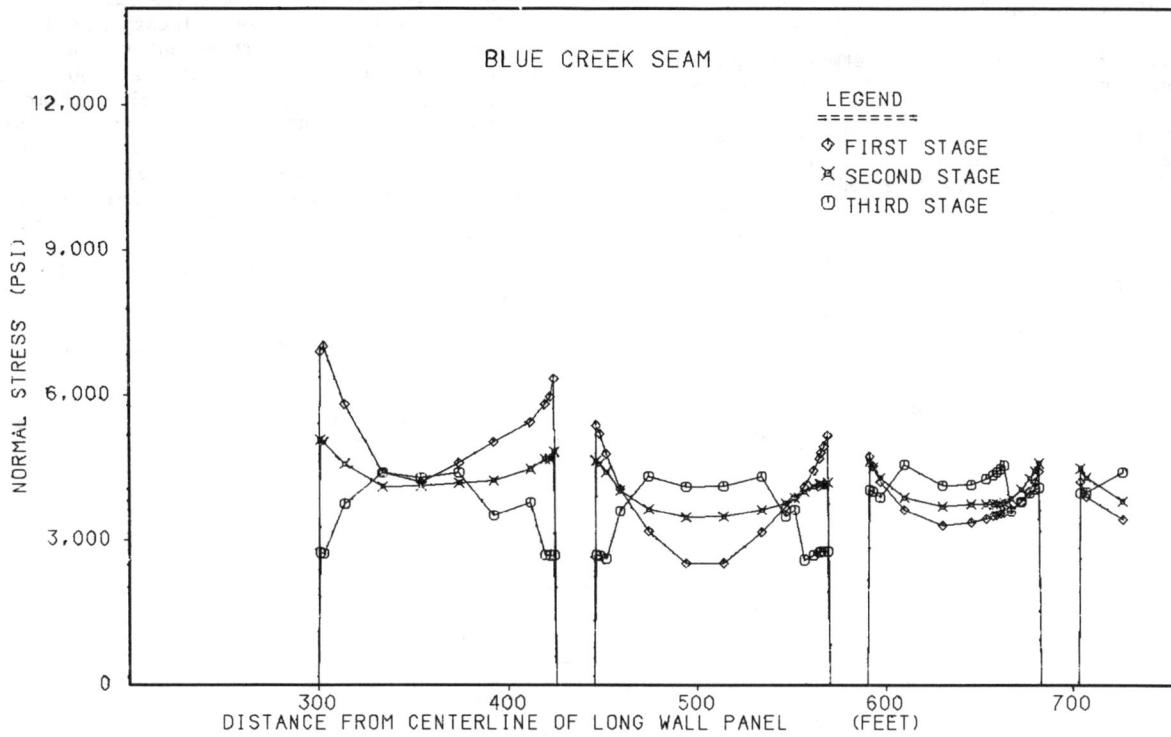

Figure 5. Stress Distribution for the Regular Pillar Area

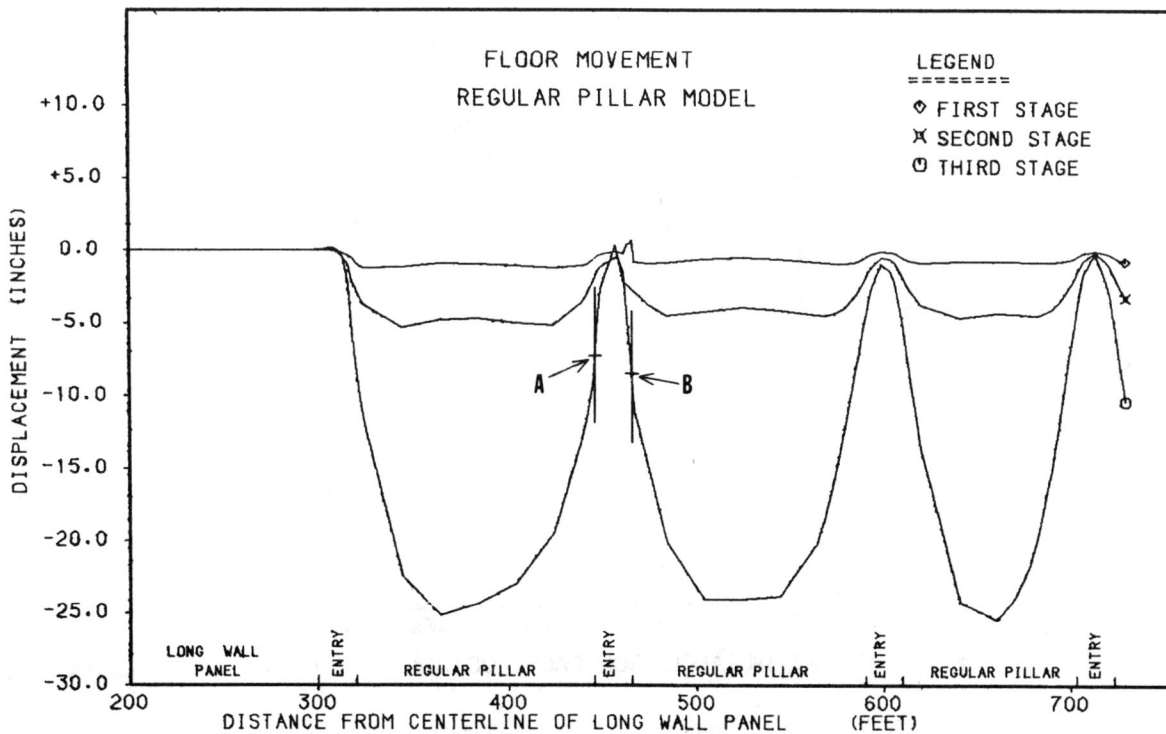

Figure 6. Floor Movement for the Regular Pillar Area

Yield Pillar System

A comparison can be made between pillar stresses from Figure 7, according to locations shown in Figure 1, with the stress distribution generated by the finite element program. Gage 18, which records stresses in the yield pillar, shows a steady fall in stress from 3000 psi to 2000 psi. On the other hand, the pillar stress distribution in Figure 8 shows an initial stage which is an instantaneous state of stress as high as 7000 psi, while stages 2 and 3 show a drop in stress from 3200 psi to 2800 psi, which are in close agreement with the measured values. A similar comparison can be made at locations in the large pillar and the third pillar represented by gages 17 and 14. Table 6 shows a comparison between the measured and simulated stresses at each location. The second and third stages represent the position of the face relative to the gage at 350 and 0 feet, respectively. In the yield pillar system it can be found that the large pillar yields on the side closer to the longwall panel, while an increase in stress on the other side is produced. That yielding occurs on the right side of the large pillar and through the third pillar. Thus time is a factor in the stress development through the large pillar.

The presence of a rock layer in the middle of the pillars played an important role in preventing large deteriorations in the pillar ribs, by acting as a reinforcement for the coal seam.

Table 6

Comparison of Stresses for the Yield Pillar System

Stages	Gage 18		Gage 17		Gage 14	
	simulated	actual	simulated	actual	simulated	actual
1	7000	–	4500	–	3000	–
2	3200	3000*	3000	3000*	4000	4000*
3	2800	2000**	4800	4500**	4200	5000**

* 350 feet away from face location
**face location at the gage

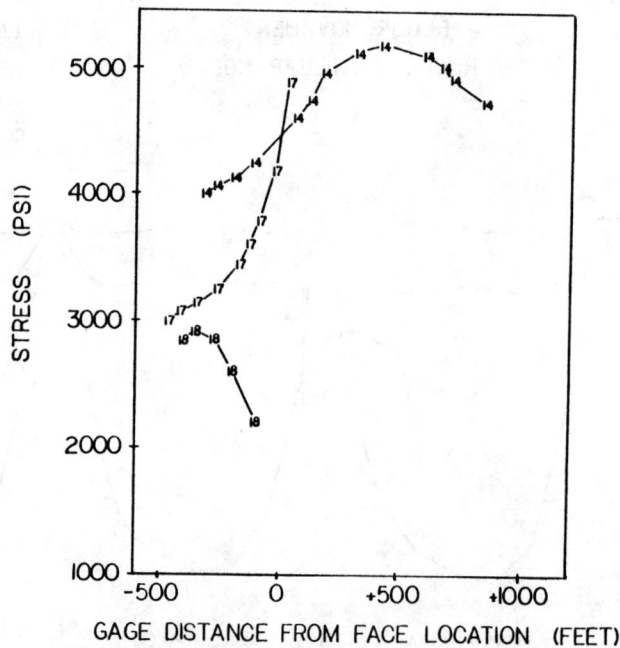

Figure 7. Stress Increase vs. Face Distance From Gages
at Yield Pillar Area

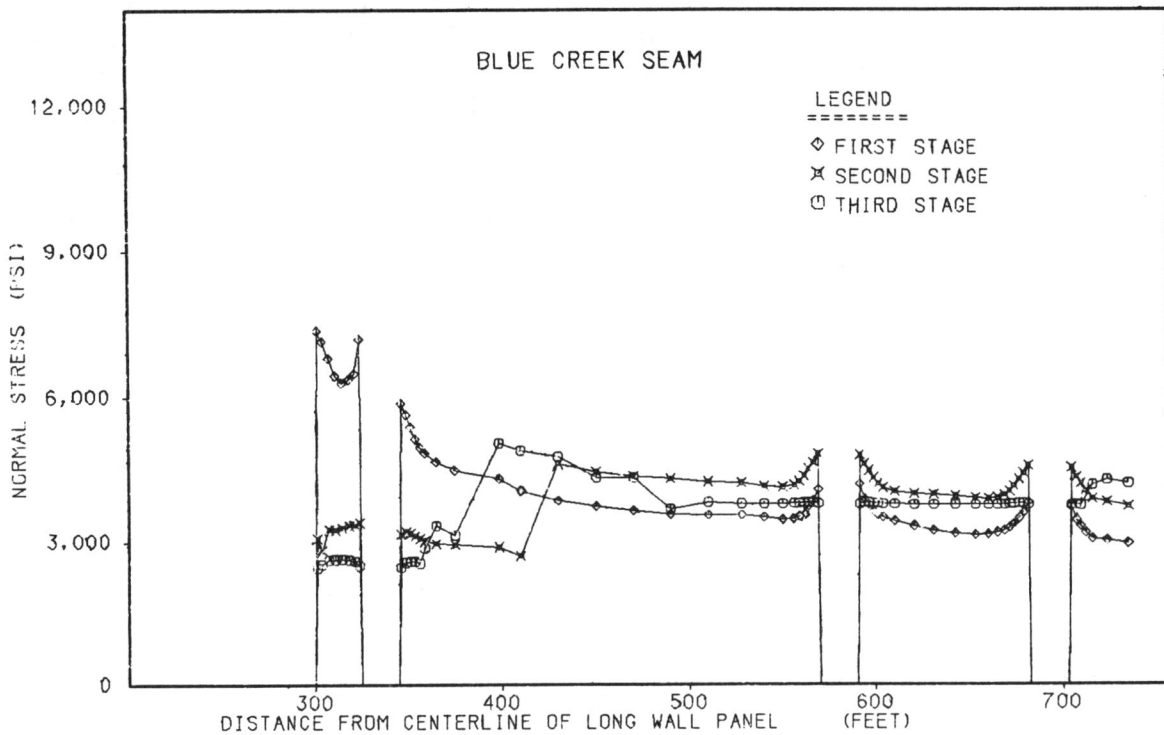

Figure 8. Stress Distribution for the Yield Pillar Area

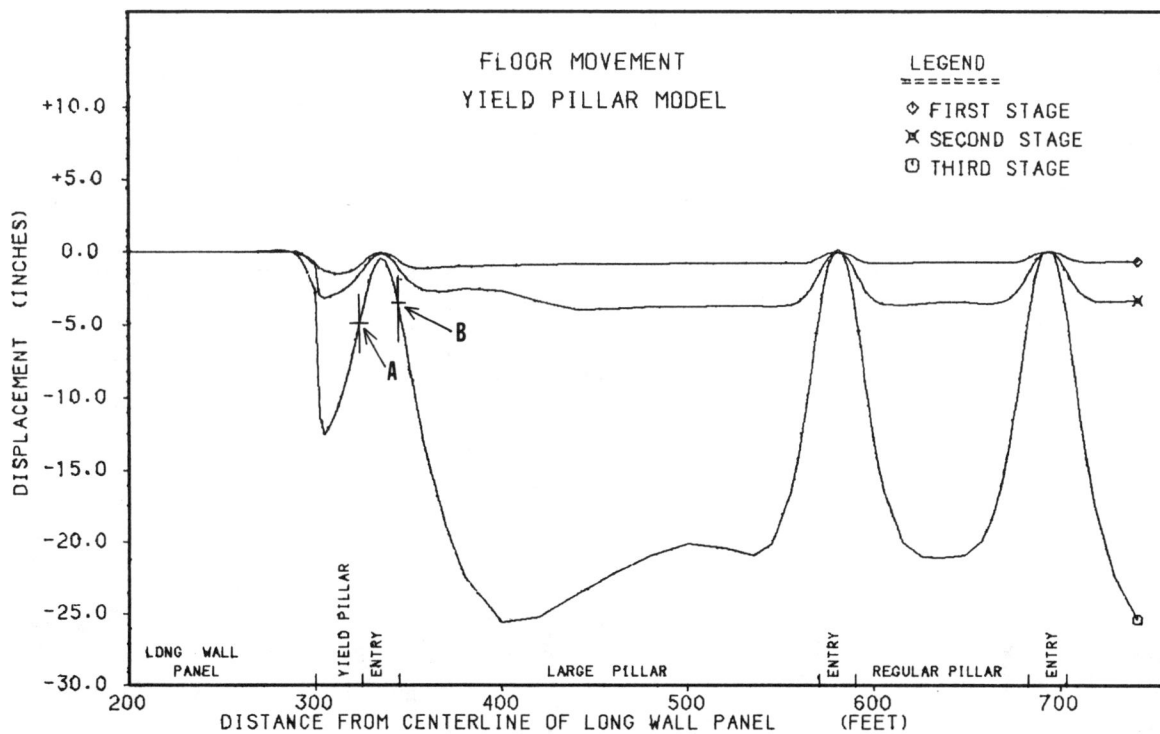

Figure 9. Floor Movement for the Yield Pillar Area

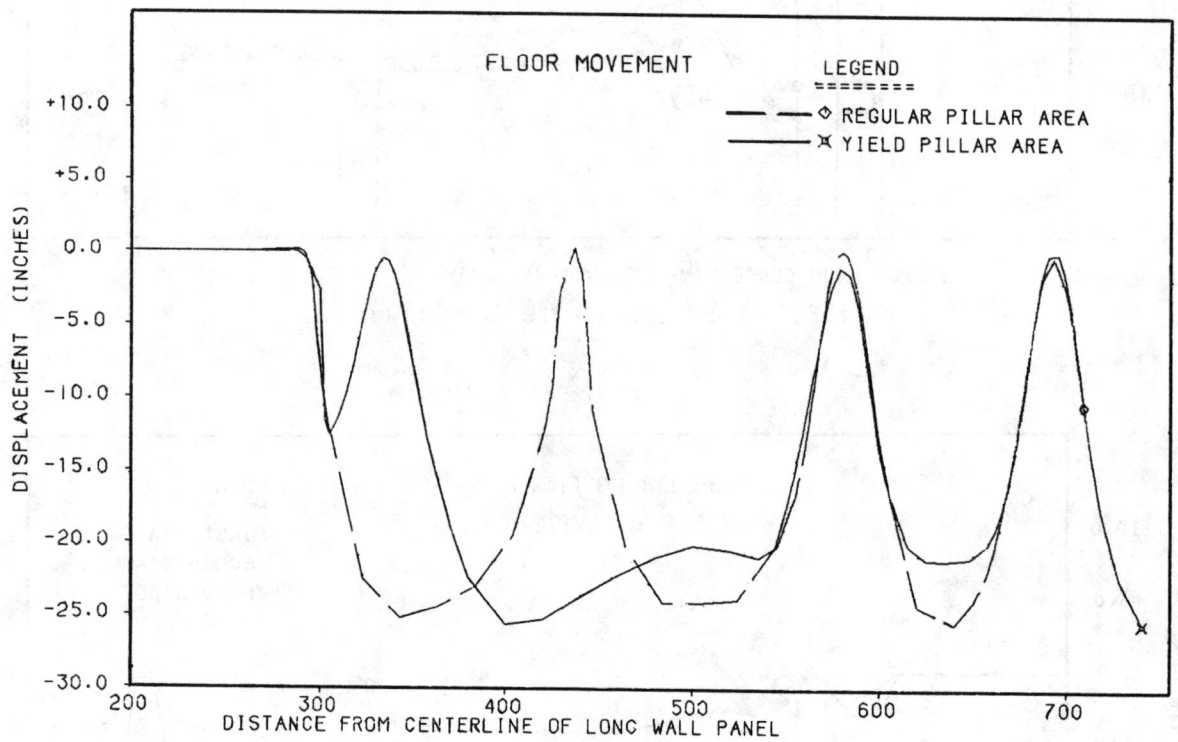

Figure 10. Final Stage Floor Movement Comparison for the
Regular and Yield Pillar Systems

Floor movement which is of great importance in the design of pillar systems can be very critical in a mine. Areas of high stresses have shown considerable floor heave which can be interpreted as a bearing capacity failure. Regular pillar systems have shown considerable floor heave in the first and second entries, which lead for the search to find an alternative system that can minimize this floor heave. In comparing Figures 6 and 9, the second entry in the yield pillar system showed considerably less floor movement than its corresponding entry in the regular pillar system. This can be mainly attributed to the yielding of the small pillar, thus shifting the high stresses to the middle of the second pillar. All through the three different iterations, the floor movement was much less in the yield pillar system than the regular pillar system. Floor movements in the regular pillar system plotted in Figure 6 show values of 4 and 5 inches at A and B, respectively, at the edges of the second entry while the yield pillar system whose values are plotted in Figure 9 show relatively larger floor movements. In the second entry at locations A and B, movements of 7 and 8 inches, respectively, can be seen. These values do not describe a floor heave accordingly because the computer simulation analysis is based on the linearly elastic concept while the actual failure occurring is plastic. On the other hand, the lower values in the yield pillar system predict the same trend if a plastic analysis were to be conducted.

CONCLUSION

The finite element technique presented in this paper showed very promising results that correspond very well with field measurement in stresses and predicts a less critical floor heave in the yield pillar system than the regular pillar system. The stress analysis, through the redistribution of stresses has actually simulated to a large extent the plastic behavior of the material, which resulted in values relatively close to actual field measurements. Large deformations in the field can be better simulated by plastically deforming the elements after each iteration. Nevertheless a relative comparison can be successfully made for the displacements in both cases. An extension of this work can be made by analyzing roof to floor convergence which is of importance to mining. The finite element method with its iterative method for progressive failure simulation has proven to be a valuable tool which can play an important role in the coal industry.

REFERENCES

1. Kripakov, N.P., and Melvin, M.T., 1983, "A Computer Procedure to Simulate Progressive Rock Failure Around Coal Mine Entries," Proceedings of the First Conference on the Use of Computers in the Coal Industry, West Virginia University, Morgantown, West Virginia.

2. Kidybinsky, A., and Babcock, C.O., 1973, "Stress Distribution and Rock Fracture Zones in the Roof of Longwall Face in a Coal Mine," Rock Mechanics Journal, Vol. 5/1.

3. Pariseau, W.G., and Eitani, I.M., 1981, "Comparisons Between Finite Element Calculations and Field Measurements of Room Closure and Pillar Stress During Retreat Mining," International Journal of Rock Mechanics, Mining Science and Geomechanics Abstracts, Vol. 18.

Chapter 42

DEVELOPMENT OF AN INTERACTIVE ROCK MECHANICS SOFTWARE PACKAGE FOR DATA PROCESSING AND ANALYSIS

G. Goodman
Department of Mining and Minerals Engineering
Virginia Polytechnic Institute and State University
Blacksburg, Virginia

W. Kane
Department of Civil Engineering
University of Alabama in Huntsville
Huntsville, Alabama

T. Triplett
U.S. Bureau of Mines
Twin Cities Research Center
Minneapolis, Minnesota

Abstract. The increasing power, availability, and use of micro-computers in the mining industry has provided a unique opportunity to develop software for practical application. This is particularly true in the field of rock mechanics where more emphasis is now being placed on the acquisition and analysis of geotechnical data.

To provide efficient analysis of rock test data, an interactive, BASIC language program has been developed. Utilizing a menu-driven format, this software package processes data obtained from the four laboratory tests considered essential by the International Society for Rock Mechanics (ISRM) for characterization of rock properties. These tests include uniaxial unconfined compressive, direct and indirect tensile, direct shear, and triaxial procedures. The user can then transfer the processed data to a statistical analysis package for determination of test variability.

INTRODUCTION

For many years, computer applications in the mining industry were virtually nonexistent. Only with the development of mine simulation programs in the early 1970s, did computers gain initial acceptance in this industry. Despite the overwhelming advantages of mine simulation, its use was highly dependent on access to a large mainframe computer network. The years following saw rapid advancements in computer technology and in the development of software applicable to the mining industry. Currently, personal or mini-computers are becoming increasingly popular as instruments for data acquisition, manipulation, and analysis. This is especially true in the field of rock mechanics where the emphasis is now placed on the accumulation and processing of large volumes of geotechnical data.

Several researchers have previously utilized computers for the manipulation and analysis of geotechnical data. Peyfess (1982) utilized a mini-computer to determine stresses and strains surrounding a non-circular opening. By specifying both trigonometric relationships to describe the underground opening and various geotechnical parameters, the magnitudes and orientations of deformations or opening perimeter were calculated. A second investigator, Cox (1982), developed interactive software for the graphic mini-computer. This program conducted an initial analysis of drill core data with a subsequent correlation of this information to the structural and lithologic characteristics of the rock. The user was then able to apply this information to several geostructural models of roof behavior, thus developing guidelines for coal mine design.

In order to provide an efficient and flexible means of manipulating and analyzing rock test data, an interactive, BASIC language program, ROCKTEST, was developed for use on the IBM Personal Computer. This software package processes rock test data obtained from the four laboratory tests considered essential by the International Society for Rock Mechanics (ISRM) for the characterizations of rock properties. These tests include uniaxial unconfined compressive (measuring either circumferential or diametrical strain), direct and indirect tensile, direct shear, and triaxial procedures. The user can then transfer the processed data to statistical analysis routines for determination of sample mean, range, standard deviation and variance.

PROGRAM DEVELOPMENT

The creation and manipulation of data files is critical to the performance of the ROCKTEST package. These random access files are maintained internally in the software and store all user-inputted information, as well as, the processed data. The first three blocks of each file are allocated to the test type, rock type, and specimen shape of the specific test conducted. The succeeding blocks contain all input and processed data gathered from tests having those characteristics. This mode of file construction was found to simplify both the program's logic and the subsequent application of the statistical analysis routine.

In the early stages of the research, the investigators discovered that establishing a file for every combination of test type and specimen shape would require the creation of six files, each with a different format. Using the BASIC language, this would lead to the coding of six separate file definition statements, six file format statements, and six buffer format statements. To avoid this repetition, it was decided to create one large file with sufficient storage to accommodate the test data of all combinations. In this manner, only one statement each was utilized for file definition, file format, and buffer format (Figure 1).

Realizing that files created in this manner will only be partially full for any combination of test type and specimen shape, an arbitrary value was allocated to the unused storage locations. In this manner, any gaps in the structure of the data file were eliminated, thus preserving its continuity.

The ROCKTEST program utilizes a multi-menu format that permits user control to be transferred to various routines by simply striking the appropriate numeric key. Audio-visual warnings are included to prevent an option selection out of the specified range. A modified flowchart is given in Figure 2. The first, or "main options," menu is initially displayed after starting ROCKTEST and provides the user with the options available within this package. These options include the following:

1) create files for rock test data,
2) process rock test data,
3) conduct statistical analyses of processed test data.

Selection of the first option transfers the user to the file construction routines of this program. A second menu is now displayed and prompts the user for his choice of test type (Figure 3). After making the appropriate selection, the user is further prompted for specimen shape and test characterization, as illustrated in Figure 4, for the example of unconfined compressive testing. Because such tests as triaxial and direct shear are limited to cylindrical and cubical specimens, respectively, the user is only given the one appropriate choice. With test type and specimen shape now determined, the user is prompted for an eight character designation of the file he is about to create. Any alpha-numeric combination is permitted as long as it conforms to the eight character maximum. Data input now begins with the user asked to supply both the rock type and the appropriate testing values. The data is echoed on the screen with the user able to interactively change any input values. A sample session for unconfined compressive testing, measuring

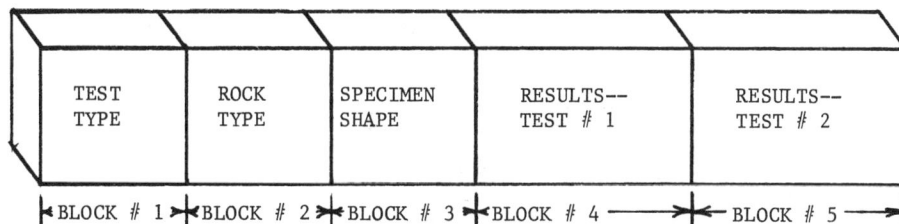

FIGURE 1: File format used in ROCKTEST

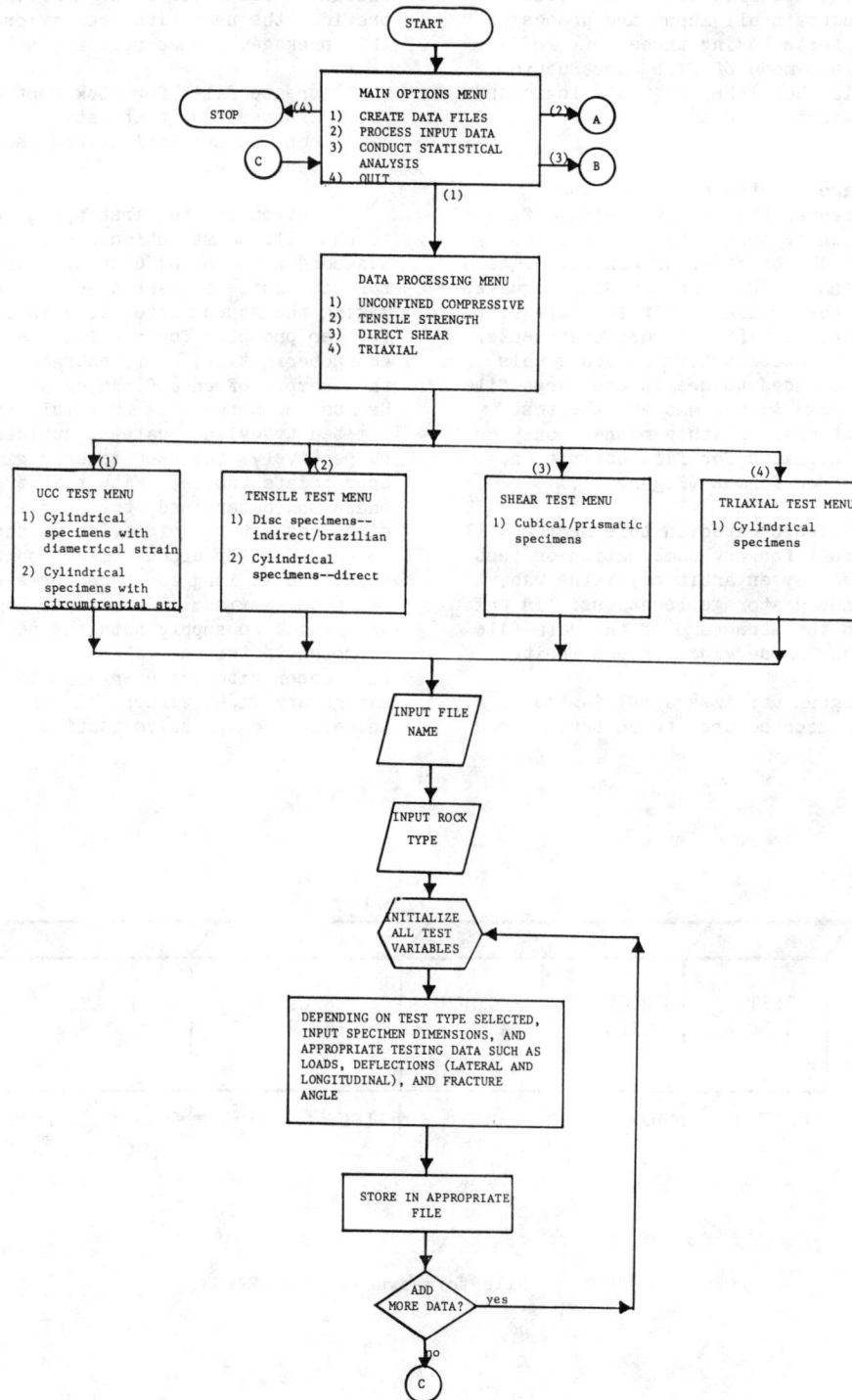

FIGURE 2: Flowchart of ROCKTEST program

```
                    ( A )
                      │
                      ▼
         ┌──────────────────────────┐
         │ ENTER RAW DATA FILE       │
         │ TO BE ANALYZED            │
         └──────────────────────────┘
                      │
                      ▼
           ┌──────────────────────┐
           │ ACCESS APPROPRIATE    │
           │ FILE                  │
           └──────────────────────┘
                      │
                      ▼
         ┌──────────────────────────┐
         │ DISPLAY TEST TYPE,        │
         │ ROCK TYPE, AND SPECIMEN   │
         │ SHAPE, AND NUMBER OF TESTS│
         │ PRESENT                   │
         └──────────────────────────┘
                      │
                      ▼
            ╱─────────────────╲
           ╱  INPUT FILE       ╲
          ╱   NAME FOR          ╲
          ╲   ANALYZED          ╱
           ╲  DATA             ╱
            ╲─────────────────╱
                      │
                      ▼
             ⬡  INITIALIZE
                ALL OUTPUT
                VARIABLES  ⬡
                      │
                      ▼
    ┌──────────────────────────────────┐
    │ CONDUCT ANALYSIS TO DETERMINE      │
    │ STRESSES (ULTIMATE, RESIDUAL,      │
    │ SHEAR), STRAINS (TRANSVERSE,       │
    │ LONGITUDINAL, VOLUMETRIC), AND     │
    │ ELASTIC CONSTANTS (POISSON'S       │
    │ RATIO, YOUNG'S MODULUS)            │
    └──────────────────────────────────┘
                      │
                      ▼
         ┌──────────────────────────┐
         │ STORE IN APPROPRIATE FILE │
         └──────────────────────────┘
                      │
                      ▼
                    ( C )
```

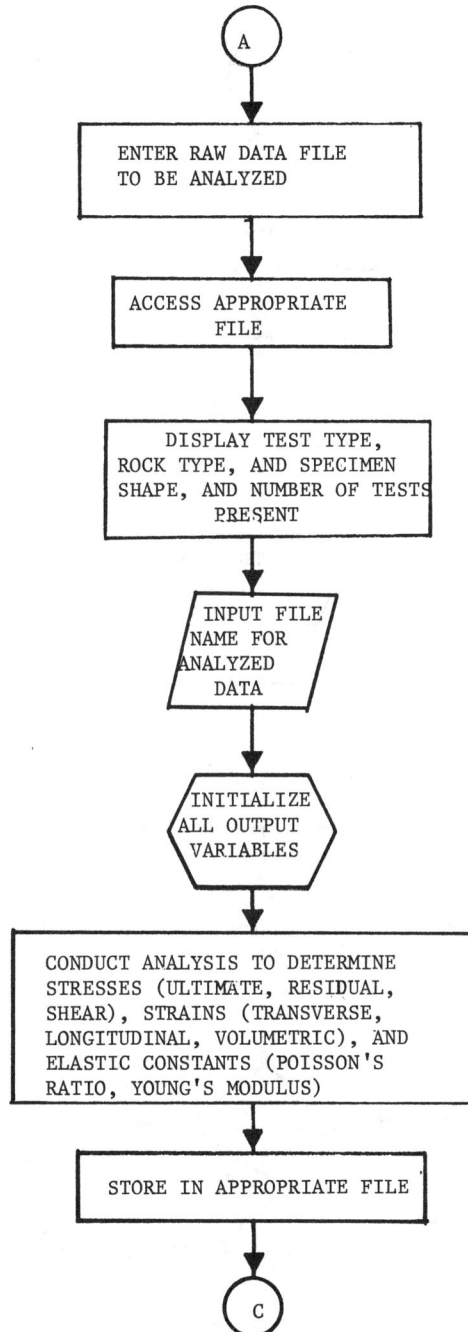

FIGURE 2: Continued

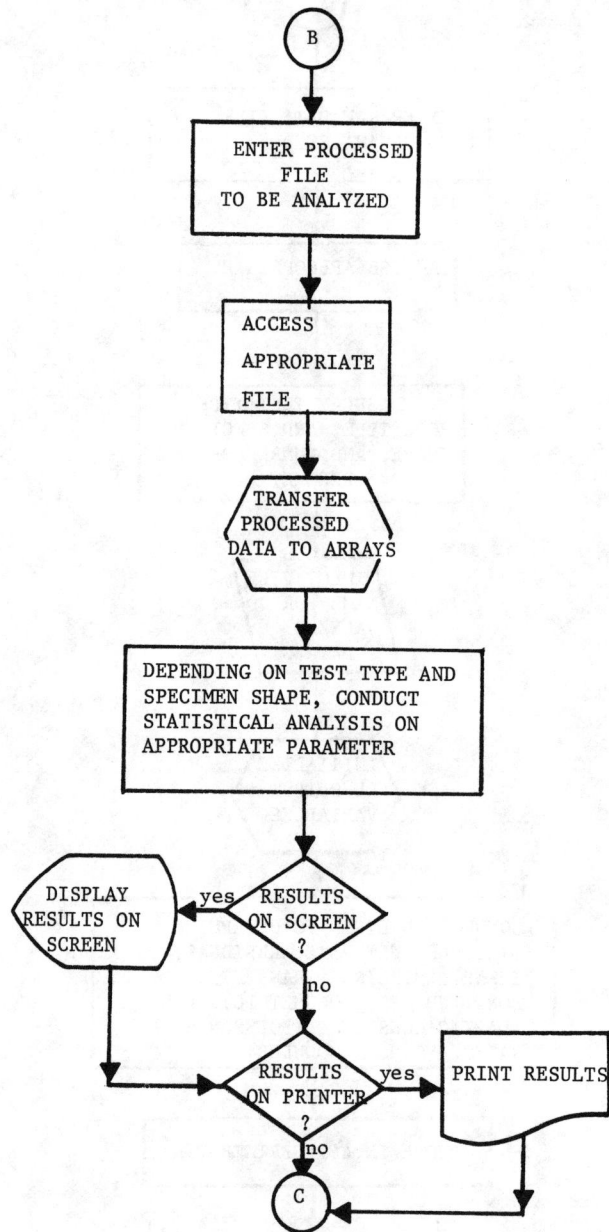

FIGURE 2:   Continued

RAW DATA PROCESSING MENU

1 = UNCONFINED COMPRESSIVE TESTING

2 = TENSILE STRENGTH TESTING

3 = DIRECT SHEAR TESTING

4 = TRIAXIAL TESTING

ENTER OPTION, PLEASE _

FIGURE 3: Test selection menu

TEST SPECIMEN CHARACTERIZATION FOR

UNCONFINED COMPRESSIVE TESTING

1=CYLINDRICAL SPECIMENS WITH CIRCUMFRENTIAL STRAIN

2=CYLINDRICAL SPECIMENS WITH DIAMETRICAL STRAIN

ENTER OPTION, PLEASE _

FIGURE 4: Specimen selection menu

diametrical deformation is shown in Figure 5. Additional series of testing data may be entered at the end of each input session.

Selection of the second, or data processing, option permits the user to analyze the testing data previously entered. ROCKTEST prompts for the eight character file name containing the rock testing data and subsequently displays test type, rock type, specimen shape, and the number of tests found in that file (Figure 6). Another eight character file name is required to specify the file containing the testing data read from the first file, analyzed according to ISRM standards, and then placed in the second file.

The third and final option encountered on the main menu conducts a statistical analysis of the processed data. As before, ROCKTEST prompts for the eight character file name of both the processed data and the statistically analyzed output. Statistical processing is then conducted to determine the mean, range, standard deviation, and variance of the data. The results are transferred to a line printer or to the display screen.

DATA ANALYSIS

This software utilizes those methods of calculation recommended by the Commission on Testing Methods of the International Society for Rock Mechanics (Mellor and Hawkes, 1971; Brown, 1981). These computations are standardized throughout the industry and are used to compute the various rock parameters in this program.

Uniaxial compressive testing is conducted to determine unconfined compressive strength, Young's modulus, and Poisson's ratio. Load data is used to compute the unconfined compressive strength by dividing the ultimate load by the original cross-sectional area of the specimen. Determination of Young's modulus and Poisson's ratio, on the other hand, requires knowledge of the stress-strain behavior of the specimen. In ROCKTEST, both axial, and either circumferential or diametrical, strains must be noted along with their corresponding loads. By measuring these values twice during the test, the slope of the axial stress-strain curve can be calculated as the average Young's modulus. Further, Poisson's ratio is determined by dividing the value of Young's modulus by the slope of the diametrical (or circumferential) stress-strain curve. This is equivalent to the quotient of diametrical or circumferential strain and axial strain. The calculations of Young's modulus and Poisson's ratio assume that the required values are located within the linear-elastic region of the stress-strain curve.

Tensile strength determination can be accomplished by either direct or indirect methods such as the Brazilian test. In the program, the direct tensile strength is found simply by dividing the load at failure by the original cross-sectional area of the specimen. Tensile strength from Brazilian test data is found from the following formula:

$$\sigma_t = \frac{2P}{\pi D t}$$

where,

σ_t = tensile strength (psi)
P = load at failure (lbs)
D = test specimen diameter (in)
t = specimen thickness (in)

Laboratory direct shear testing is conducted by applying a horizontal shearing force to a specimen that is confined by a normal load. Under a constant normal load, the shear force is applied in such a way as to maintain uniform shear displacement. Typical behavior shows shear stress increasing to a peak shear value before decreasing asymptotically to a residual shear value. These values of peak, residual, and normal load are inputted to ROCKTEST and the corresponding stresses outputted for plotting as illustrated in Figure 7. From this graph, values of apparent friction angle, ϕ_a, residual friction angle, ϕ_r, cohesion intercept, C', and apparent cohesion, C, can be determined (ASTM, 1981).

Program input for triaxial testing consists of specimen dimensions, confining stress, ultimate load, and failure angle. From this information, the axial stress is calculated, thus permitting a Mohr's circle diagram to be constructed (Figure 8). From this graph, the values of internal friction angle, ϕ, and apparent cohesion, C, can be determined.

Statistical processing within the ROCKTEST software determines such parameters as the sample mean, minimum and maximum values, range, standard deviation, and variance. Sample mean is calculated by dividing the sum of values by the number of tests conducted, n. A simple sorting routine is then performed on all values to determine both the maximum and minimum with the statistical range being the difference of the two. The standard deviation, s, is found from the following equation:

$$s = \left[\frac{\Sigma x^2 - (\Sigma x)^2 / n}{n - 1} \right]^{\frac{1}{2}}$$

Finally, the variance is calculated as the square of the standard deviation.

CONCLUSIONS

At the present, the need exists for a versatile software package to process and analyze accumulated geomechanical data. The ROCKTEST program described herein attempts to fill part of that need. With the ease of menu-driven, interactive operation, this package permits efficient and accurate data input and analysis.

```
INPUTTING DATA FOR UNCONFINED COMPRESSIVE TESTING

SPECIMEN DIAMETER (INCHES)                    2.125
SPECIMEN HEIGHT (INCHES)                      4.250
ULTIMATE SPECIMEN LOAD (LBS.)                 95000
LOWER LOAD FOR ELASTIC ANALYSIS (LBS.)        11500
UPPER LOAD FOR ELASTIC ANALYSIS (LBS.)        65000
LONG. STRAIN. AT LOWER LOAD (INCHES)          .007
LONG. STRAIN. AT UPPER LOAD (INCHES)          .039
DIAMETRICAL STRAIN AT LOWER LOAD (IN.)        .002
DIAMETRICAL STRAIN AT UPPER LOAD (IN.)        .012
```

FIGURE 5: Session with test data input

```
TEST CHARACTERSTICS FOR JOB TESTRUN
-------------------------------------




TEST TYPE:              UCC CYLD

ROCK TYPE:              SANDSTNE

SPECIMEN SHAPE:         CYLD

NUMBER OF TESTS ON FILE:     4




PRESS ANY KEY TO CONTINUE
```

FIGURE 6: Screen display of input file contents

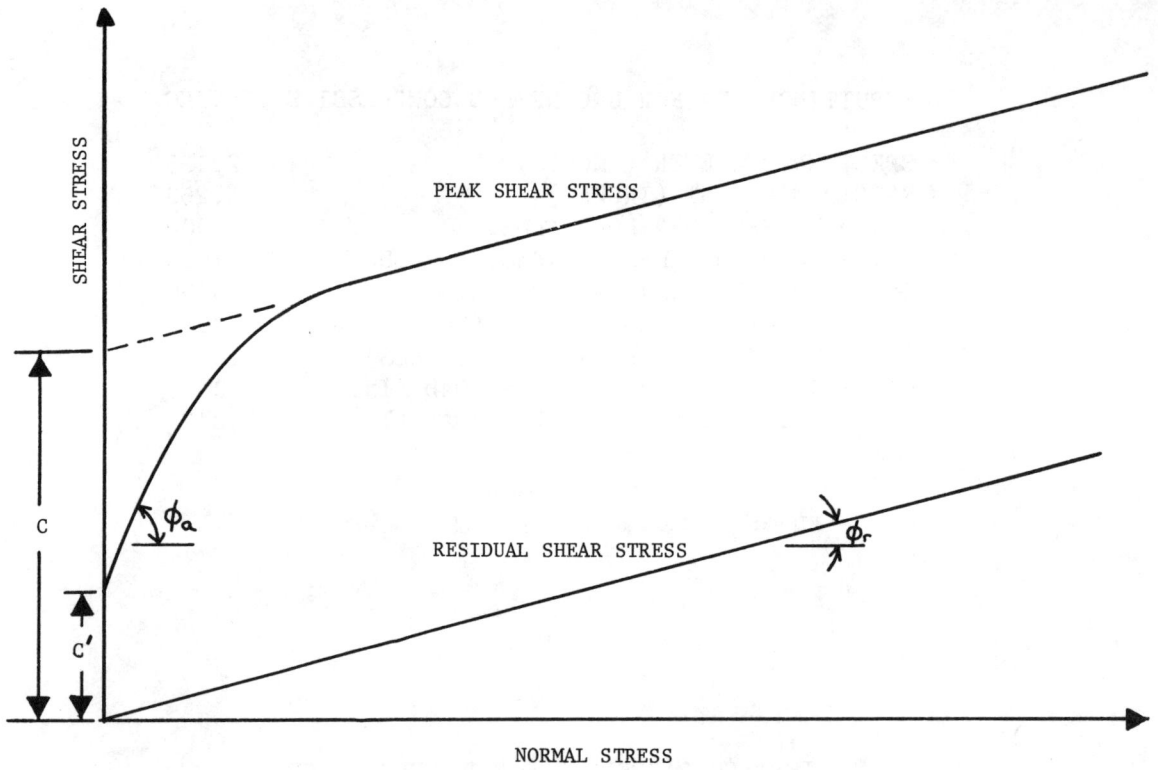

FIGURE 7: Graph of shear stress vs normal stress showing
           peak and residual shears

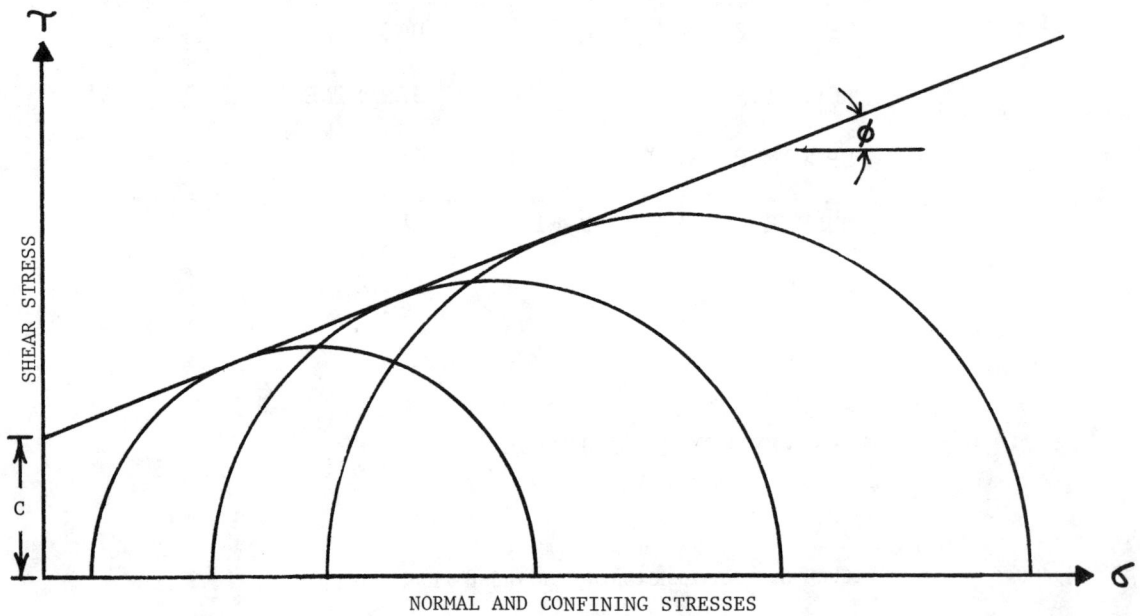

FIGURE 8: Mohr's envelope plot

REFERENCES

Brown, E. T., (ed.), 1981, <u>Rock Characterization Testing and Monitoring</u>, Pergamon Press, New York, 211 pp.

Cox, R. M., 1982, "The Use of a Graphics Mini-Computer System as a Rock Mechanics and Roof Support Design Tool," <u>Proceedings</u>, Second Conference on Ground Control, West Virginia Univ., July, pp. 159-170.

Mellor, M. and Hawkes, I., 1971, "Measurement of Tensile Strength by Diametrical Compression of Discs and Annuli," <u>Eng</u>. <u>Geol</u>., Vol.5, pp. 173-225.

Peyfuss, K. F., 1982, "A Simple Method for Calculating Displacements and Stresses Around Tunnel Openings--An Application for the Mini-Computer," <u>Rock Mechanics</u>, Vol. 15, pp. 199-207.

Chapter 43

ON THE SELECTION OF MATERIAL PARAMETERS FOR COMPUTER ANALYSIS OF SUBSIDENCE

H. J. Siriwardane

Department of Civil Engineering
West Virginia University
Morgantown, West Virginia 26506

ABSTRACT

Selection of appropriate material properties as input parameters to phenomenological models is an important aspect of subsidence prediction. This paper deals with some aspects of reducing material parameters in the finite element models for prediction of subsidence. In most of the finite element models, the material properties determined in the laboratory have been reduced by using certain reduction factors to match field measurements with predictions. This paper presents the variations in reduction factors required in a finite element model to obtain reasonable predictions for maximum subsidence at a number of long-wall coal mine panels.

INTRODUCTION

Subject of subsidence prediction has been a topic of interest to many researchers in the United States during the last decade. Subsidence is a major consequence of underground mining. It has been estimated that about 2 million acres of U.S. land has been affected by subsidence due to underground mining. The influence of subsidence causes damages to a number of civil engineering systems such as houses, highways, waterlines and surface structures. Prior planning of underground activities is necessary for minimizing the aforementioned damages. This in turn requires reliable predictive models for subsidence.

There have been two broad categories of methods for predicting surface subsidence, namely empirical and phenomenological models. Most of the empirical methods used in U.S. coal fields can be considered as extensions of methods developed for European coal fields. In general, empirical methods do not account for factors such as geologic profile, surface topography, and material properties, and hence their applicability is limited to areas where a prior knowledge on the influences of mining exist. The other category of predictive models falls under phenomenological methods which are based on principles of continuum mechanics. Depending on the complexity of the geometry and material models selected, either closed form or numerical solutions could be obtained for predicting subsidence. In recent years, these solutions for the most part have been obtained by using computer based analysis techniques such as finite element and boundary element methods. Selection of appropriate material properties as input parameters to these models is an important aspect since predictions may be highly sensitive to some of the properties. This paper deals with some aspects related to reduction of laboratory determined properties in order to match predictions with field observations.

METHOD OF ANALYSIS

From the viewpoint of an analyst, subsidence prediction can be thought to comprise two main factors. These are: (a) the shape of the subsidence profile or normalized subsidence profile, and (b) the magnitude of the maximum subsidence. As reported elsewhere (Siriwardane and Amanat, 1983, 1984a-c), the shape of the subsidence profile can be predicted reasonably well by using a nonlinear finite element procedure, but the prediction of maximum subsidence using these models cannot be considered satisfactory. Limited results obtained from finite element analyses of subsidence at a few longwall panels have been reported in the literature by several investigators. These results indicate that different investigators have used varying reduction factors for material properties in order to match the field observations with predictions from finite element models. The reductions in material properties have been justified as some of the laboratory-determined properties are higher than the true values in the field due to size effects. However, at the present time, no guidelines are available in determining the reduction factors in order to obtain realistic values of maximum subsidence from a computer based analysis technique. This paper presents a study involving computer analysis of a number of longwall mine panels. The commonly used material properties of such an analysis include elastic properties, cohesion, angle of internal friction, and coefficient of earth pressure at rest. Influence of these factors on the predicted subsidence profile is presented in detail elsewhere (Siriwardane and Amanat, 1984a).

RESULTS

A number of true case histories were analyzed in this study. Information pertinent to these mines was obtained from published literature. A complete list of references is not given herein to avoid undue length; however, these are cited elsewhere (Siriwardane and Amanat, 1983). The idealized and true geologic profiles and the extent of the finite element meshes used in the study are presented elsewhere (Siriwardane and Amanat, 1983, 1984c). The material properties for the analyses were selected on the basis of available information at the specific sites wherever possible. When such properties were not available, reasonable properties were estimated on the basis of available laboratory properties of similar materials as described elsewhere (Siriwardane and Amanat, 1983). In almost every case analyzed in this study, it was observed that the predictions based on linear elastic finite element analysis were not in good agreement with field observations. This was true for both the shape of the subsidence profile as well as the maximum subsidence values. A summary of elastic predictions of maximum subsidence is given in Table 1. A typical shape of the predicted profile is shown in Figure 1. As stated earlier, one of the ways to match field observations with predictions was to reduce elastic properties of all materials in the overburden by a certain reduction factor. The required reduction factors associated with a linear elastic analysis of each case are also shown in Table 1. The reduction factors appear to be highly case-dependent. Although the predicted maximum subsidence can be changed to match with the field observations when the reduction factors in Table 1 are used, the shapes of the subsidence profiles do not change significantly. That is, no significant improvement in the shape of the profile can be expected simply by reducing elastic properties in the overburden. With this realization, an improved procedure (Siriwardane and Amanat, 1983) for predicting the shape of the subsidence profile, that is, normalized subsidence profile was developed on the basis of nonlinear finite element analysis.

It was noticed in all cases, that the subsidence profiles obtained from a linear analysis and a nonlinear analysis, with the same material properties, were identical. It was also observed (Siriwardane and Amanat, 1983) that the material remained essentially elastic in the nonlinear analysis. This is not surprising since the properties determined in the laboratory are known to be much higher than the true values in the field. This has been commonly explained as "size effects" in laboratory test specimens. Thus, in order to obtain material properties under field conditions, it is necessary to reduce laboratory properties by some amount. In the past, different reduction factors have been used in predictive models (Brown, 1968; Dahl and Choi, 1973, 1981; Sutherland and Schuler, 1981). Information on the influence of reduced engineering properties on the subsidence profile can be extremely useful in deciding appropriate reduction factors for future analysis or in developing unified approaches for predicting maximum subsidence.

Prediction of the Shape of Subsidence Profile

As stated earlier, the initial nonlinear analysis with actual properties (properties based on laboratory data) indicated that the material response was essentially linear elastic for a typical mine. Therefore, in order to account for field conditions and to observe its influence, the cohesion, c, was reduced by using "Reduction Factors," R_f, of 1/10, 1/20, 1/100, 1/200. Additionally, a value of 1 psi for cohesion was also investigated (Siriwardane and Amanat, 1984a). The predicted subsidence profiles corresponding to reduction factors (R_f) of 1/100 and 1/200 were almost identical. Therefore, it appears that beyond a certain limit of reduced cohesion, the subsidence profile is not affected significantly. The predicted profiles corresponding to a R_f of 1/100 and to a cohesion of 1 psi (R_f = 1/4,000) appear to be identical right above the gob area. It can be concluded that the reduction in cohesion certainly improves the quality of predictions compared with field observations. This may be attributed to the fact that field parameters are smaller than their laboratory values. From the viewpoint of the analyst, however, such reductions can improve the predictions only to a certain limit as observed in Figure 1. Foregoing results were obtained by using 3 sequences in the nonlinear analysis with one iteration per step.

The study (Siriwardane and Amanat, 1984a) also revealed that the profile obtained on the basis of averaging predicted values at the mine roof and at the ground surface was in better agreement with field measurements. As presented in Refs (Siriwardane and Amanat, 1983, 1984a), the angle of internal friction within realistic limits, and the elastic modulus, E, do not appear to have a significant influence on shape of the profile, although those factors have significant influence on the magnitude of maximum subsidence. The study on the influence of elastic modulus, E, indicated that reduction in E value by a factor as low as 1/20 does not have a significant influence on the normalized subsidence profile. Some irregularities in the subsidence at roof level were noticed beyond a certain reduction factor for E values. This indicates that below a certain limit for the reduction factor, the analysis can predict some irregularities in the roof displacements; thus, any reduction in E by a factor below this limit must be performed with caution.

Prediction of Maximum Subsidence, S_{max}

As shown elsewhere (Siriwardane and Amanat, 1984a) the normalized average subsidence profile is in reasonable agreement with field observations. This needs to be supplemented by the maximum subsidence, S_{max}, in order to obtain complete information from an analysis. Although the reductions in c and E were seen to have an insignificant influence on normalized profiles, they have a significant influence on S_{max}. To obtain reasonable agreement between predicted and measured values of S_{max}, the elastic modulus of all layers can be reduced by a certain factor (denoted by α) or, more logically the elastic moduli of those elements in

Table 1:   Prediction of Maximum Subsidence Using Linear Elastic Analysis

Case	Mine	Predicted Maximum Subsidence at Ground Surface, Inches	Required Reduction Factor $(1/\beta)$
1	Shoemaker I	1.92	18.90
2	Shoemaker II	4.31	9.13
3	Old Ben No. 24	0.90	53.87
4	Ireland	4.28	9.39
5	Quarto 7A	3.08	10.60
6	Sigma (Face at 617 feet)	4.81	8.11
7	Sigma (Face at 480 feet)	2.60	1.06
8	Sufco	4.23	5.99
9	Monton	1.00	6.48
10	Quarto 7B	2.28	14.47
11	Blacksville No. 1 (Transverse)	2.45	9.80
12	Blacksville No. 1 (Longitudinal)	9.15	2.88
13	York	3.66	21.97

1 inch = 0.0254 meters

Table 2:   Predicted Maximum Subsidence for Ireland Mine

Vertical Displacements at Panel Center, Inches					
1	2		3	4	
	No Reduction to Elastic Modulus		Reduction Factor β	With Reduction to Elastic Modulus	
	Surface	Average of Roof and Surface		Surface	Average of Roof and Surface
Elastic	4.28	6.16	—	—	—
c/100	9.56	18.72	0.35	15.65	37.80
c = 1 psi	10.48	23.94	0.49	15.41	39.48

S_{max}(measured) = 40.2 inches
1 inch = 0.0254 meters

plastic/tensile zones could be reduced by a reduction factor (denoted by β). In several of the past studies (Brown, 1968; Dahl and Choi, 1973, 1981; Sutherland and Schuler, 1981), the moduli E for the entire overburden (domain) had been reduced by a certain factor in order to have reasonable agreement between predictions and measurements of S_{max}. In this study, preference is given to the idea of reducing elastic moduli of those elements falling in plastic/tensile zones. The required reduction factors, β, that leads to reasonable agreement with field data were found by a trial-and-error procedure, and are given in Table 2 for the case of Ireland mine. It appears that the value of E needs to be reduced by a factor of 0.35 corresponding to c/100 while a factor of 0.49 is suitable for constant c of 1 psi in order to have a reasonable agreement between computed maximum subsidence and the measured value which is equal to 40.2 inches (1.02 meters). The computed subsidence is taken as the average of the roof and the ground surface since the average profile (Siriwardane and Amanat, 1984a) was closer to field measurements than individual profiles. It can be concluded that for all cases studied, the normalized subsidence profile obtained by reducing cohesion c/100 or a constant c of 1 psi in the finite element analysis was in reasonable agreement with field observations, over the gob area. By reducing E values as shown in Table 3, similar agreement for true subsidence (not normalized) can be obtained over the gob area.

DISCUSSION AND CONCLUSION

As shown earlier (Siriwardane and Amanat, 1984a), the shape of the subsidence profile, that is the normalized subsidence profile, can be predicted reasonably well for several cases by using a nonlinear finite element analysis. However, the predictions of maximum subsidence in those cases could not be considered satisfactory. The shape of the subsidence profile itself can be very useful in locating the zones of influence and maximum damage. Besides, a knowledge of measured subsidence profile at one location together with the predicted shape of the subsidence profile could be used in calibrating the true subsidence profile. In most of the previous investigations, the elastic properties of the overburden materials were reduced by certain factors to match field observations with predictions from finite element models. The reductions in material properties have been justified as the laboratory determined parameters are higher than the values in the field due to size effects. No guidelines appear to be available for detering required reduction factors. This paper presented required reduction factors (Table 3) for the elastic moduli of those elements within plastic/tensile zones in order to obtain realistic predictions for maximum subsidence at a number of mine panels. These results do not show obvious trends in the required reduction factors. Reduction factors appear to be highly case-dependent.

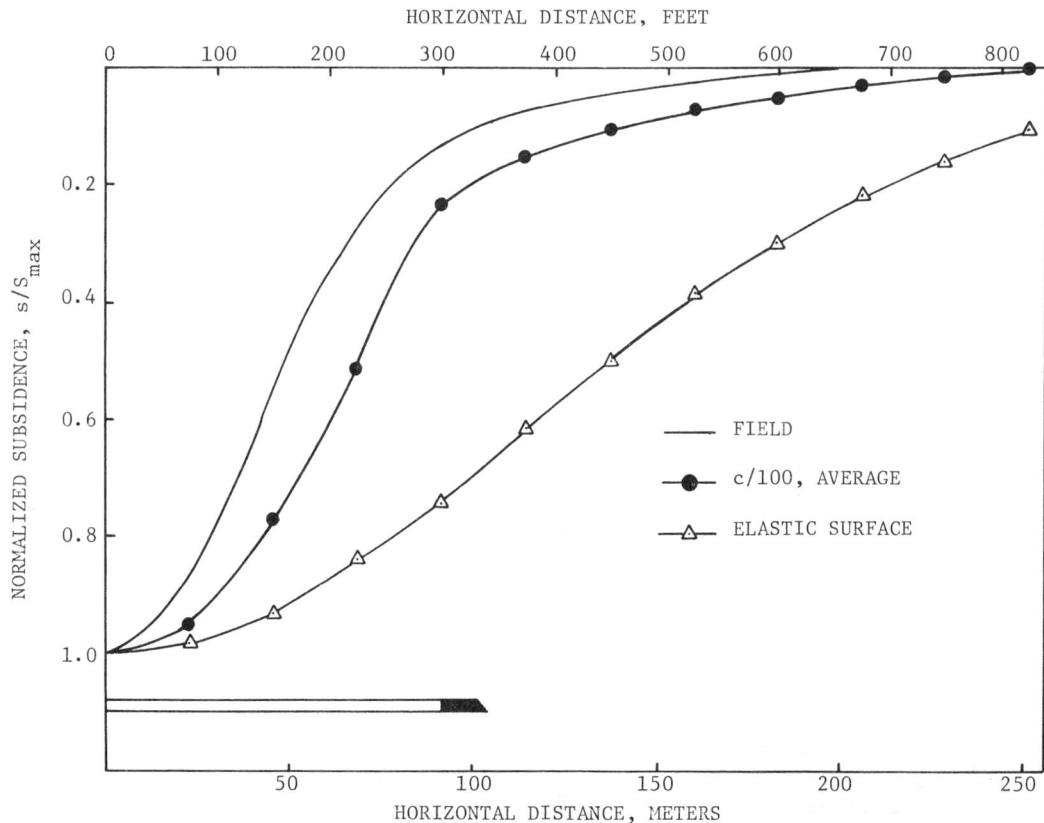

Figure 1. Comparison of Measured Surface Subsidence and Predicted Values Based on Linear and Nonlinear Analyses of Ireland Mine

Table 3. Summary of Reduction Factors for Material Properties at a Number of Mine Panels

	Predicted Displacements, Inches												Predicted Horiz. Strain				Computed Limit Angle (deg.)	
	No Reduction to Elastic Modulus						Reduction factors β		with Reduction Factors Shown in Columns 5 and 6				Maximum Horizontal Compr. and Tensile Strain x 10⁻³				Predicted	
	Displacement at Panel Center, inch								Displacement at Panel Center (inch)									
	Elastic		c/100		c = 1psi		c/100	c=1psi	c/100		c = 1psi		c/100		c = 1psi		γ	γ
Case	Surf.(in.)	Avg.(in.)	Surf.(in.)	Avg.(in.)	Surf.(in.)	Avg.(in.)			Surf.(in.)	Avg.(in.)	Surf.(in.)	Avg.(in.)	ε_{Cmax}	ε_{Tmax}	ε_{Cmax}	ε_{Tmax}	(c/100)	(c=1psi)
1	1.92	3.39	3.30	7.03	4.04	9.84	0.015	0.10	16.86	35.80	14.35	38.30	3.100	2.440	1.93	1.230	34.88	34.98
2	4.31	6.00	9.95	18.84	10.94	25.67	0.350	0.45	16.64	37.78	15.96	40.56	2.270	1.220	2.52	1.640	31.13	31.13
3	0.90	1.45	2.03	4.62	2.18	5.52	---	0.02	---	---	20.54	51.48	---	---	3.81	6.360	---	41.67
4	4.28	6.16	9.56	18.72	10.48	23.94	0.350	0.49	15.65	37.80	15.41	39.48	2.120	1.100	2.35	1.470	32.33	32.33
5	3.08	5.02	6.10	12.18	7.33	17.14	0.125	0.36	15.54	31.76	12.72	34.80	2.190	1.830	1.82	1.360	32.33	32.33
6	4.81	5.28	12.72	15.71	17.65	23.81	0.300	0.50	31.00	38.16	27.64	40.32	5.170	2.000	4.71	2.020	26.90	26.90
7	2.60	3.07	5.71	7.08	8.60	13.32	---	---	9.30	24.43	---	---	---	---	---	---	---	---
8	4.23	7.20	9.30	24.43	10.54	34.64	1.000	0.50	16.36	58.37	14.13	56.00	0.709	0.326	0.90	0.570	---	---
*8b	---	---	---	---	---	---	0.300	0.50	1.46	6.72	---	---	1.690	0.770	1.44	0.826	---	---
9	1.00	3.01	1.46	6.72	1.59	8.06	1.000	0.30	18.18	34.88	15.90	35.49	0.054	0.069	---	---	---	---
10	2.28	3.20	6.00	9.48	8.00	14.40	0.080	0.50	9.49	25.32	6.84	25.32	2.100	2.600	1.64	3.400	33.78	33.78
11	2.45	3.53	5.20	11.29	4.94	14.36	0.300	---	---	---	---	---	2.180	0.680	2.39	0.900	35.80	35.80
12	9.15	10.35	23.34	29.36	18.40	30.60	---	---	---	---	---	---	---	---	---	---	---	---
13	3.66	5.28	7.21	11.40	10.97	19.94	0.150	0.15	---	---	39.24	86.76	---	---	3.67	8.290	---	19.80 to 23.7
1	2		3		4		5	6	7		8		9	10	11	12	13	14

*Based on 60% of Seam Thickness for Maximum Subsidence Calculation

A study involving a statistical analysis is being carried out to investigate whether there are correlations with other geometric parameters.

ACKNOWLEDGEMENTS

The work reported herein was supported in part by the Energy Research Center of West Virginia University under the project no.: ST-83-CM6 on subsidence prediction. This support is gratefully acknowledged. J. Amanat contributed towards computations involved in this study.

REFERENCES

Brown, R. E. (1968), "A Multi-Layered Finite Element Model for Predicting Mine Subsidence," Ph.D. Dissertation, Carnegie-Mellon University, Pennsylvania.

Dahl, H. D. and D. S. Choi (1981), "Measurement and Prediction of Mine Subsidence Over Room and Pillar Workings in Three Dimensions," Proceedings, Workshop on Surface Subsidence Due to Underground Mining, West Virginia University, West Virginia.

Dahl, H. D. and D. S. Choi (1972), "Some Cases of Mine Subsidence and Its Mathematical Modelling," Proceedings, Fifteenth Symposium on Rock Mechanics, Custer State Park, South Dakota.

Siriwardane, H. J. and J. Amanat (1983), "Finite Element Analysis and Prediction of Subsidence Caused by Underground Mining," Report CE/GEO 83-2, Department of Civil Engineering, West Virginia University, Morgantown, West Virginia.

Siriwardane, H. J. and J. Amanat (1984), "Analysis of Subsidence Caused by Underground Mining," International Journal of Mining Engineering, (In Press).

Siriwardane, H. J. and J. Amanat (1984), "Prediction of Subsidence in Hilly Ground Terrain Using Finite Element Method," Proceedings, Second International Conference on Stability in Underground Mining, AIME, Lexington, Kentucky.

Siriwardane, H. J. and J. Amanat (1984), "Analysis and Prediction of Subsidence in Horizontal Ground Terrain," Submitted to the International Journal for Numerical and Analytical Methods in Geomechanics.

Sutherland, H. J. and K. W. Schuler (1981), "A Review of Subsidence Predicting Research Conducted at Sandia National Laboratories," Proceedings, Workshop on Surface Subsidence due to Underground Mining, West Virginia University, West Virginia.

Chapter 44

A COMPUTER AIDED AND PROBABILISTIC APPROACH TO COAL MINE ROOF STABILITY

W. F. Kane
Civil Engineering
University of Alabama in Huntsville
Huntsville, Alabama

M. Karmis
Department of Mining and Minerals Engineering
Virginia Polytechnic Institute and State University
Blacksburg, Virginia

ABSTRACT

Roof falls and roof instability still remain one the most serious problems in underground coal mining. In addition to safety considerations, the economic losses resulting from lost time, clearing costs and damaged equipment due to roof failures are considerable.

During this research an extensive underground mapping of several mines was conducted and a large data base was compiled for the southern Appalachian coal fields. This effort resulted in the evaluation of a number of geomechanical variables indicative of unstable roof conditions. Potential indicators of roof instability were analyzed using chi-square and linear regression analysis, and the most significant variables were determined. These factors were weighted and incorporated into a Roof Rating Index (RRI) which is based on the sum of the weighted ratings. The RRI for each location within a mine was compared for different roof types and the probability of a roof fall was determined.

A BASIC computer code was written for the IBM microcomputer in which the user inputs information drawn from the exploration program, or from underground mapping. Input data for each station includes location, entry and cross-cut trends, roof type, roof fracture types and orientations, fracture densities, and the presence or absence of slickensides, fossils, and faulted coal. The Roof Rating Index is then computed for that location and the probability of a failure is determined. The engineer then has the opportunity to decide if the planned roof support will be sufficient, or if additional support will be required.

INTRODUCTION

The impact of underground coal mine roof failures on the industry is enormous, in terms of injuries, lost time and equipment. Because of the extreme complexity of the underground environment, attempts at understanding the mechanism of roof collapse and at predicting such failures have met with limited success.

From a purely descriptive standpoint, most of the dangerous situations which can be encountered in underground coal mining have been delineated. These include the presence of faults (Iannachinone, et al., 1981; Shepherd and Burston, 1977 among others) and joints (Moebs, 1974, 1977, 1981; Moebs and Ellenberger, 1982). Slickensides associated with these fractures have been implicated in roof falls by Milici et al. (1982) and Moebs and Ellenberger (1982). The spacing of these fractures, or fracture density, can exert a control on roof stability (HRB-Singer, Inc., 1979).

Lithologic changes have also been shown to cause roof falls. For example, paleochannel boundaries, where a sandstone channel interfaces a shale or siltstone zone, are one such region where roof falls are common (Milici et al., 1982). Other situations in which changes of lithology are involved in roof failures are where crevasse splays, cyclic repetition of thin beds of sandstone, siltstone, and shale occur (Moebs and Ellenberger, 1982), or where rider coals, thin seams of coal above the mined seam, are present (Burgraff, 1981).

Other factors have also been suggested in the literature, as shown in Table 1.

Although most hazardous situations have been described, little attempt has been made to quantify these variables in such a manner as to be useful to the mine operator in planning his mine and carrying out day-to-day mine operations. The introduction of the microcomputer into mining operations has led to the possibility of processing geomechanical data quantitatively, in order to determine roof support needs in advance of mining.

Statistics have been used in the past with regard to roof instability. However, these studies have often been limited to determining such factors as the day of the week most likely to experience a roof fall (Smith, 1984). While interesting, such information provides little insight into the causes of roof failure, and are of little help to the mine engineer during layout of his mine.

TABLE 1. Factors Cited in Roof Instability within the Appalachian Basin.

PARAMETER	REFERENCE
Clay Veins	Krausse, et al., in press; Moebs, 1974; 1977, 1981; Moebs and Ellenberger, 1982; Peng, 1978
Climate	Boyer, 1964; Chenevert, 1969; Colbeck and Wild, 1965; Haynes, 1975; Parker, 1966; Stateham and Radcliffe, 1978
Crevasse Splays	Moebs, 1974, 1977, 1981; Moebs and Ellenberger, 1982
Expandable Clays	Wier, 1969
Faults	Iannacchione, et al.,1981; Nelson, 1981; Shepherd and Burston, 1977; Shepherd and Fisher, 1978
Fracture Density	HRB - Singer, Inc., 1979
Joints	Moebs, 1974, 1977, 1981; Moebs and Ellenberger, 1982
Kettlebottoms	Ferm, et al., 1978; Milici, et al., 1982
Lineaments	Dimick and Barnum, 1978; Jansky and McCabe, 1979; Oberlick, 1978; Rinkenberger, 1978
Lithology	Ferm, et al., 1978; HRB - Singer, Inc., 1979
Location Within the Mine	Peng, 1980
Overburden	Moebs, 1974; Overbey, et al., 1973
Paleochannels	Moebs, 1974, 1977, 1981; Moebs and Ellenberger, 1982
Percent Extraction	HRB - Singer, Inc., 1979
Pinch-outs	Moebs, 1974, 1977, 1981; Moebs and Ellenberger, 1982
Rider Coals	Burggraf, 1981; HRB - Singer, Inc., 1979
Slickensides	Milici, et al., 1982; Moebs, 1974, 1977, 1981; Moebs and Ellenberger, 1982
Tectonic Stresses	Blevins, 1982; HRB -Singer, 1979; Parsons and Dahl, 1971; Price, 1966
Time	Cox, 1974
Water or Gas Pressure	Headlee, 1944

One exception to this trend was by Ferm et al. (1978). They analyzed data from several mines in the Pocahontas No. 3 seam in Virginia and West Virginia and attempted to draw statistical correlations between geological parameters and good mine roof, a fair roof, etc.

RESEARCH IN THE APPALACHIAN COALFIELDS

Due to the complex nature of coal measure formation geological variability, and structural characteristics, it was felt that the best means to quantifiably approach the problem of roof instability was to use statistical methods. In this regard, an underground mapping program was carried out in southwestern Virginia (Figure 1).

Four mines were chosen and mapped in a systematic manner. At specific stations in each mine the variables shown in Table 2 were recorded. Roof type was categorized according to a classification proposed by Karmis and Kane (1984), as shown in Figure 2, and fall types were classified by geometry as suggested by Patrick and Aughenbaugh (1979). Measurements were recorded in a field notebook and transferred to data sheets for input into computer files. The emphasis here was on constructing as large a data base of geomechanical data as possible. This included areas where roof conditions were stable as well as those which were unstable. The data base eventually consisted of 260 actual roof falls and 286 instances of stable roof. With respect to roof type, 98 stations were in Interface Roof, 144 in Sandstone Roof, and 304 in Shale Roof.

Statistical procedures were carried out by means of the Statistics Analysis System package (SAS Institute, Inc., 1982). For this purpose, it was necessary to divide the parameters into two groups: classification variables and numerical variables. Classification variables have discrete levels which are treated as names rather than as numerical values on an ordered scale. Classification variables in this study included, among others, the presence or absence of slickensides and the type of roof fractures; numerical variables included such parameters as fall size and fracture dip angle.

SAS provides two, somewhat different, procedures for handling classification variables, FREQ and FUNCAT. FREQ produces frequency tables and performs chi-square testing, a measure of the degree of independence existing between two variables. For example, chi-square can determine if roof failure in a Shale Roof is dependent on the occurrence of fossils within that roof. FUNCAT can also test for independence, but to do so is unwieldy and difficult (SAS, 1982). The value of FUNCAT lies in its ability to analyze several variables and check for the mixed effects of those variables since some may be inter-related. It was felt that independence with roof failure was more important to determine in this study since it was already known that, within the list of variables, some were closely related such as slickensides and roof fracture type. In an actual study underground, it may be possible to determine only

the presence of slickensides and not fracture type. Thus, it was decided not to reject an easily measured variable because it was interrelated to another. For this reason, a more "correct" model using FUNCAT was discarded in favor of a more useful model. A summary table of chi-square measures of independence for different variables and roof types with respect to roof falls is shown in Table 3. A chi-square probability of less than 0.05 indicates strong dependence between the variable and roof failure.

Numerical variables can be handled in a number of ways. For this research, regression analysis was used to process these variables. Regression analysis attempts to fit a function to a set of values, i.e. whether an equation can be written to describe how fall size increases or decreases with a corresponding increase or decrease in dip angle. The regression procedure employed herein was RSQUARE. The latter fits all possible combinations of a list of variables in order to determine which parameters have the greatest effect on the dependent variable. A value of 1 signifies a strong relationship and a value of 0 demonstrates no relationship whatsoever. In this research, fall size was tested with regard to fracture strike and dip, coal thickness, fracture density, entry trend and deviation of fracture from entry trend. The RSQUARE procedure was used to determine correlations between the above-mentioned variables (Table 4).

DISCUSSION

Once the data was statistically analyzed, it was possible to weight the different variables according to their contribution to roof failure. Each variable was assigned a value on a scale of 0 to 10, where 0 was associated with a very strong chi-square correlation with failure and 10 was associated with total variable independence. Values from the RSQUARE phase of analysis were used to adjust the weighting factors further.

The result of this process is the Roof Rating Index (RRI) shown in Table 5. The index is used by assigning values to each variable present at a certain mine location. A minimum of six variables must have values. The index can be calculated by summing the individual values.

Once derived, the index was tested by means of the chi-square test in order to determine if a correlation existed between the Index and roof failure. As shown in Table 6, a very strong correlation existed between Sandstone and Shale Roofs and the RRI, while a weaker one was found between Interface Roof and the Index. With the relationship between RRI and roof instability demonstrated, a probability of failure could be calculated for each roof type and RRI (Table 6).

A computer code in BASIC was written to compute the Roof Rating Index given the variables used in this research. A simplified flowchart is shown in Figure 3. Data is read from data statements; missing data is accounted for. If the number of

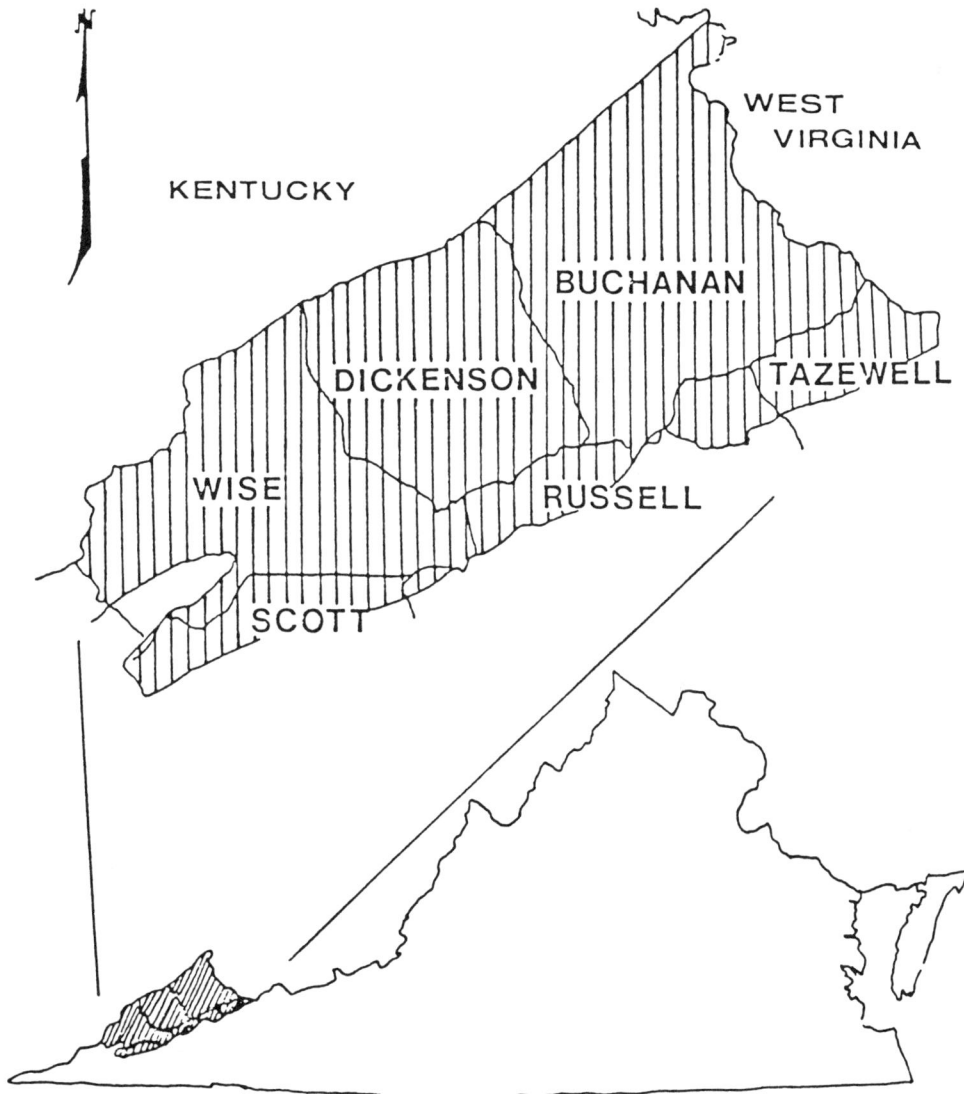

Figure 1. Location map of study area in southwestern Virginia.

TABLE **2.** Variables Mapped in Study Area.

GEOMECHANICAL VARIABLES	MINING VARIABLES

GEOMECHANICAL VARIABLES

ROOF
1. Type
2. Presence of Slickensides
3. Presence of Fossils
4. Strike and Dip of Bedding

FALL
1. Dimensions
2. Type
3. Location
4. Time (Before or After Bolting)
5. Trend

FRACTURES
1. Type
2. Strike and Dip
3. Density

COAL
1. Height
2. Presence of Faults
   a. strike and dip
   b. type
3. Cleat Strike and Dip

PRESENCE OF WATER

MINING VARIABLES

BOLT (SUPPORT) TYPE

ENTRY TREND AND WIDTH

CROSSCUT TREND AND WIDTH

TABLE 3. Chi-Square Measures of Independence for Study Variables and Roof Failure.

ROOF TYPE	LOCATION	FAULTED COAL	FOSSILS	SLICKENSIDES	ROOF FRACTURE TYPE	FRACTURE DEVIATION FROM ENTRY TREND	FRACTURE DIP	FRACTURE DENSITY
INTERFACE	0.4087	0.9025	1.0000	0.7189	0.0688	0.3402	0.4648	0.4396
SANDSTONE	0.0017	0.4565	0.0251	0.1430	0.2077	0.9556	0.0847	0.0001
SHALE	0.0001	0.0165	0.4347	0.0001	0.0096	0.0050	0.0003	0.0001

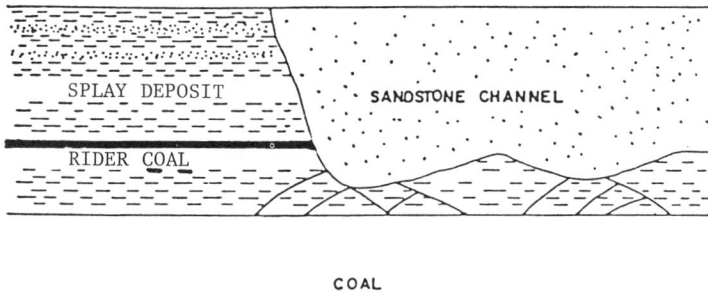

Interface Roof: Characterized by heterogenous strata in immediate and main roof and dependent on these differences in lithology for its degree of stability. Examples include channel boundaries where sandstone cuts down into the immediate roof or where slickensided fractures in shale intersect a channel bottom or sides. Interbedded sandstones, siltstones and shales such as splay deposits. Rider coals are also included but not smaller features such as carbonareous bedding planes.

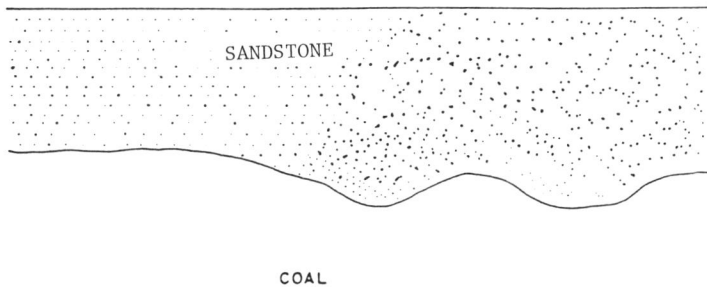

Sandstone Roof: Evenly bedded, smooth sandstone or a sandstone with an irregular base due to channel bottoms, load casts, and other features extending downward. No shales or siltstones are present or else the top would be classified as Interface Roof.

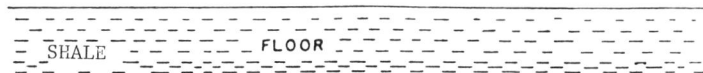

Shale Roof: Shales or siltstones which may be planar-bedded or faulted, slickensided, or deformed. Can also include coarsening upward sequences.

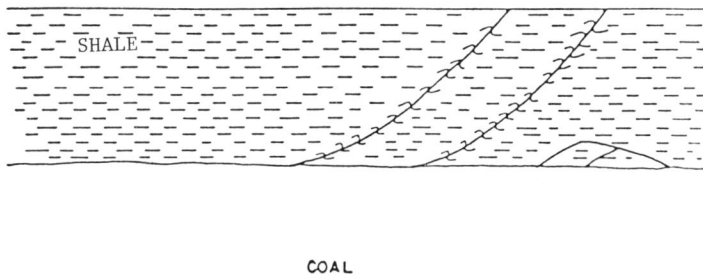

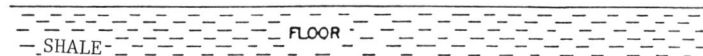

Figure 2. Roof classification (Karmis and Kane, 1984).

TABLE 4. Values of r-squared for Numeric Variables with Relation to Roof Fall Size.

Stepwise regression model for Interface Roof

NUMBER IN MODEL	R-SQUARE	VARIABLES IN MODEL
1	0.23441762	FRACDENS
1	0.29064297	DEVIATE
1	0.32318540	ANGLE
1	0.33085298	COALHT
2	0.32544520	DEVIATE ANGLE
2	0.32840403	FRACDENS ANGLE
2	0.33130317	FRACDENS COALHT
2	0.33640272	FRACDENS DEVIATE
2	0.35477719	COALHT DEVIATE
2	0.35955703	COALHT ANGLE
3	0.33807901	FRACDENS DEVIATE ANGLE
3	0.35673707	FRACDENS COALHT DEVIATE
3	0.36051607	FRACDENS COALHT ANGLE
3	0.36084991	COALHT DEVIATE ANGLE
4	0.36095987	FRACDENS COALHT DEVIATE ANGLE

Stepwise regression model for Sandstone Roof

NUMBER IN MODEL	R-SQUARE	VARIABLES IN MODEL
1	0.05262285	DEVIATE
1	0.15213809	COALHT
1	0.26037015	ANGLE
1	0.40096106	FRACDENS
2	0.15247066	COALHT DEVIATE
2	0.28156089	DEVIATE ANGLE
2	0.39171048	COALHT ANGLE
2	0.40100395	FRACDENS COALHT
2	0.40437115	FRACDENS ANGLE
2	0.40587023	FRACDENS DEVIATE
3	0.40806012	FRACDENS COALHT DEVIATE
3	0.41895789	FRACDENS DEVIATE ANGLE
3	0.41989535	COALHT DEVIATE ANGLE
3	0.43477256	FRACDENS COALHT ANGLE
4	0.45525587	FRACDENS COALHT DEVIATE ANGLE

Stepwise regression model for Shale Roof

NUMBER IN MODEL	R-SQUARE	VARIABLES IN MODEL
1	0.00339805	DEVIATE
1	0.08143211	COALHT
1	0.12962373	ANGLE
1	0.74240115	FRACDENS
2	0.08303584	COALHT DEVIATE
2	0.13261842	DEVIATE ANGLE
2	0.15817274	COALHT ANGLE
2	0.74274933	FRACDENS DEVIATE
2	0.76806741	FRACDENS COALHT
2	0.77016011	FRACDENS ANGLE
3	0.15905262	COALHT DEVIATE ANGLE
3	0.76943005	FRACDENS COALHT DEVIATE
3	0.77021717	FRACDENS COALHT ANGLE
3	0.77129353	FRACDENS DEVIATE ANGLE
4	0.77145440	FRACDENS COALHT DEVIATE ANGLE

ANGLE   = Fracture Dip
COALHT   = Coal Thickness
DEVIATE = Fracture Deviation From Entry Trend
FRACDENS = Fracture Density

TABLE 5. Weighted Values Assigned to Specific Variables in Order to Determine Roof Rating Index (RRI).

ROOF TYPE	A LOCATION			B FOSSILS		C SLICKENSIDES		D FAULTED COAL	
	4-Way Intersection	3-Way Intersection	Entry or Crosscut	Present	Absent	Present	Absent	Yes	No
INTERFACE	2	6	8	1	6	5	8	3	10
SANDSTONE	4	6	10	3	8	7	9	7	9
SHALE	3	4	8	6	10	3	10	2	10

ROOF TYPE	E FIRST FRACTURE TYPE			F SECOND FRACTURE TYPE			G THIRD FRACTURE TYPE		
	Joint	Tear Fault	Dip- or Oblique-Slip Fault	Joint	Tear Fault	Dip- or Oblique-Slip Fault	Joint	Tear Fault	Dip- or Oblique-Slip Fault
INTERFACE	2	8	8	2	8	8	2	8	8
SANDSTONE	1	8	8	1	8	8	1	8	8
SHALE	8	7	4	8	7	4	8	7	4

ROOF TYPE	H MINIMUM DEVIATION OF ROOF FRACTURES FROM ENTRY TREND									I MAXIMUM DIP ANGLE OF ROOF FRACTURES								
	0-9	10-19	20-29	30-39	40-49	50-59	60-69	70-79	80-90	0-9	10-19	20-29	30-39	40-49	50-59	60-69	70-79	80-90
INTERFACE	8	6	8	6	1	3	6	6	8	8	8	8	8	8	8	9	9	8
SANDSTONE	6	8	6	6	6	6	8	8	8	3	8	8	6	6	3	3	2	9
SHALE	6	6	6	3	4	4	4	4	6	4	4	6	3	6	4	4	3	4

ROOF TYPE	J FRACTURE DENSITY (Number of Fractures/3 Meters)					
	0-2	3-4	5-6	7-8	9-10	>10
INTERFACE	7	5	6	5	5	4
SANDSTONE	10	8	8	7	6	1
SHALE	10	7	6	5	2	0

ROOF RATING INDEX (RRI) = A + B + C + D + E + F + G + H + I + J

TABLE 6.   Roof Rating Index (RRI) and the Probability of Failure for each Roof Type.

INTERFACE ROOF
(Chi-square Probability = 0.1071)

ROOF INDEX	PROBABILITY OF FAILURE
0 - 19	(No Data Available)
20 - 29	58%
30 - 39	44%
40 - 49	83%
> 50	48%

SANDSTONE ROOF
(Chi-square Probability = 0.0001)

ROOF INDEX	PROBABILITY OF FAILURE
0 - 19	71%
20 - 29	69%
30 - 39	44%
40 - 49	33%
> 50	15%

SHALE ROOF
(Chi-square Probability = 0.0001)

ROOF INDEX	PROBABILITY OF FAILURE
0 - 19	77%
20 - 29	79%
30 - 39	77%
40 - 49	72%
> 50	28%

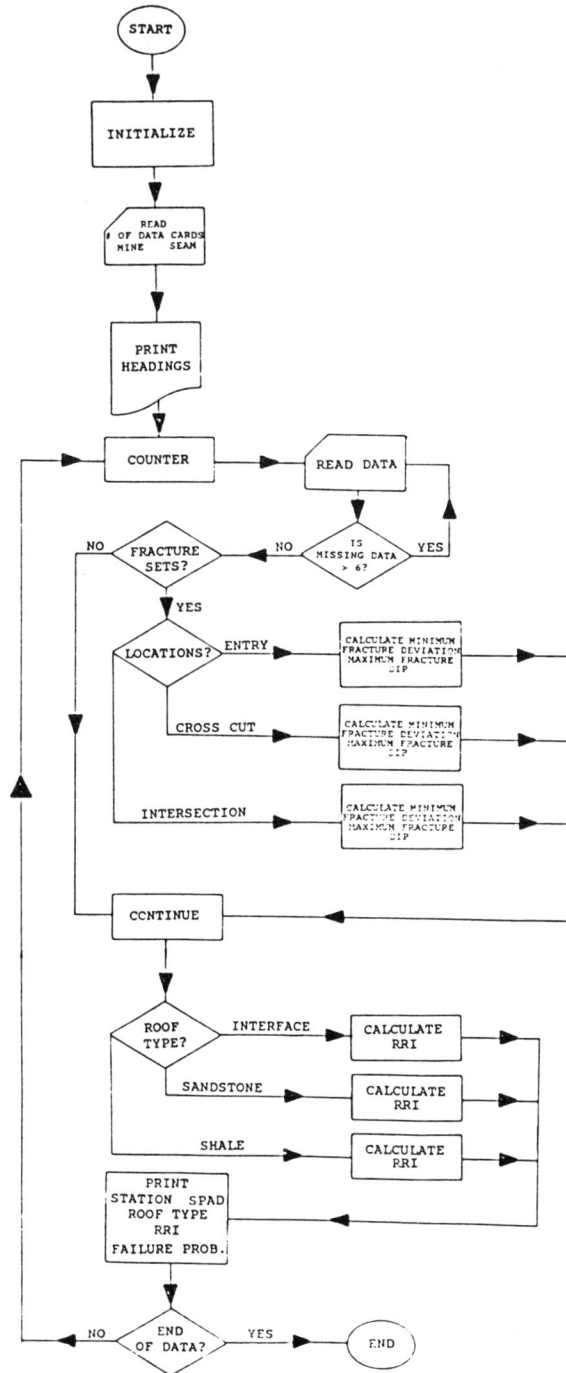

Figure 3. Flow chart of roof evaluation program.

missing variables is greater than six, a message is printed stating that there is not enough data to calculate a reasonable RRI, and the next data statement is read. The program then branches to different subroutines depending on station location within the mine. The minimum fracture deviation from entry or cross-cut trend is calculated along with the maximum fracture dip. Contol is then returned to the main program, where branching to subroutines takes place once again, in order to calculate the RRI and failure probability for each roof type. The main program then outputs the data and either returns to read another data statement or ends if all data have been read. Table 7 shows a sample output from the program.

Figure 4 shows a section of a mine map for which the failure probabilities derived from the computer program have been contoured and roof falls in that section of the mine plotted. The trend of more numerous and larger roof falls with increasing probability of failure is clear from this diagram. The larger falls are all grouped relatively close together within the 60 per cent failure probability contour; smaller falls are scattered randomly over a much larger area and tend to be found well below the 60 percent contour line.

CONCLUSION

By using the computer to process large amounts of data statistically, a Roof Rating Index can be determined for any region, seam, or mine. This Index allows the mine operator an opportunity to assess the chances of a roof fall in a particular section of his mine. The variables input can be readily determined, for the most part, from drill core and surface investigation, as well as from underground mapping, so that the Index can be employed in the design phases of mine layout. Once calculated, the RRI can be plotted on a map, the probability of failure determined, and appropriate support or avoidance measures taken.

The ready availability of statistical software packages for microcomputers, and the use of a program such as developed here, make this approach ideal for many ground control programs. A data base can be compiled from underground mapping programs. This data base could be company-wide or mine-wide depending on the judgement of mine personnel and local conditions. The failure probabilities could then be back-calculated as was done in this research.

These probabilities would then be used to predict, in advance of mining, the chance of a roof fall using only drill core and surface mapping data. The data base can continually be updated and refined, and variables can be added or deleted depending on experience. For example, in some areas coal cleat direction may be important, or the effect of linears can be incorporated. It should also be noted that, because the data from which the Roof Rating Index was derived comes primarily from the Virginia coal fields, other areas may experience different conditions and that failure probabilities may change.

It is felt that this approach, using the microcomputer and information that can quickly and easily be determined by mine personnel, has great potential in refining mine layout and planning, and in lessening the hazards of unfortunate roof failure in underground coal mines.

ACKNOWLEDGEMENTS

This paper is based upon work supported by the Generic Mineral Technology Center, Mine Systems Design and Ground Control, Bureau of Mines, U.S. Department of the Interior. Any opinions, findings, conclusions, or recommendations expressed in this publication are those of the authors and do not necessarily reflect the views of the sponsor.

The authors would especially like to thank Dr. R. C. Milici, Commissioner, Virginia Division of Mineral Resources, and N. N. Moebs, Pittsburgh Research Center, Bureau of Mines, for their valuable suggestions and insight. The cooperation of Clinchfield Coal Co., Paramont Mining Corp., and their respective staffs is gratefully acknowledged. Finally, appreciation is due to M. J. Kane who drafted the figures and T. L. Triplett and G. V. R. Goodman for their valuable discussions.

REFERENCES

Blevins, C. T., 1982, "Coping with high lateral stresses in an underground Illinois coal mine," Proc. 2nd Conf. on Ground Control in Mining, S. S. Peng and J. H. Delley, eds., West Va. Univ., July 19-21, Morgantown, pp. 137-141.

Boyer, R. F., 1964, "Coal mine disasters: frequency by month," Science, vol. 144, pp. 1447-1448.

Burggraf, C., 1981, "Roof control at Republic's Kitt mine," Min. Cong. Jour., Dec., pp. 15-17.

Chenevert, M. E., 1969, "Shale hydration mechanics," Proc. 4th Symp. on Drilling and Rock Mech., Austin, Tex., SPE 2401, Jan. 14-15, pp. 201-212.

Colback, P. S. B. and Wild, B. L., 1965, "The influence of moisture content of the compressive strength of rocks," Proc. 3rd Canadian Symp. on Rock Mech., Toronto, Canada, Jan. 15-16, pp. 65-84.

Cox, R. M., 1974, "The correlation of mine roof failure with the time elapse before support installation," Final Report USBM Contract No. H111413, Univ. of Alabama, NTIS PB-262, 478 pp.

Dimick, D. and Barnum, B., 1978, "Remote sensing at Southern Utah Fuel," Proc. 9th Ann. Inst. on Coal Mining Health, Safety and Research, Virginia Polytechnic Inst. and State Univ., Blacksburg, pp. 191-208.

TABLE 7. Sample Output from Computer Code.

MINE: SEAM: JAWBONE

STATION	SPAD	ROOFTYPE	RRI	FAILURE PROBABILITY (%)
M7	597A	Not have enough data to calculate Roof Rating Index		
M7A	658	SANDSTONE	16	71
M7B	658A	SANDSTONE	30	44
M7C	658A	SANDSTONE	27	69
M7D	799	SANDSTONE	27	69
M7E	799	SANDSTONE	28	69
M7F	799A	SANDSTONE	24	69
M7G	841	SANDSTONE	27	69
M7H	841A	SANDSTONE	27	69
M7I	812B	SANDSTONE	18	71
M7J	841C	SANDSTONE	32	44
M8	812C	SANDSTONE	28	69
M8A	878	SANDSTONE	28	69
M8B	957	SANDSTONE	24	69
M8C	957	SANDSTONE	28	69
M8D	1182	SANDSTONE	28	69
M8E	1241	SANDSTONE	25	69
M8E	1241	SANDSTONE	29	69
M8F	1321	SANDSTONE	19	71
M9	1504	SANDSTONE	19	71
M9A	1596	SANDSTONE	19	71
M9B	1606	SANDSTONE	19	71
M9C	1627	SANDSTONE	47	33
M9D	1648	SANDSTONE	73	15
M9E	1678	SANDSTONE	73	15
M9F	1697	INTERFACE	81	15
M9G	1697A	SHALE	57	28

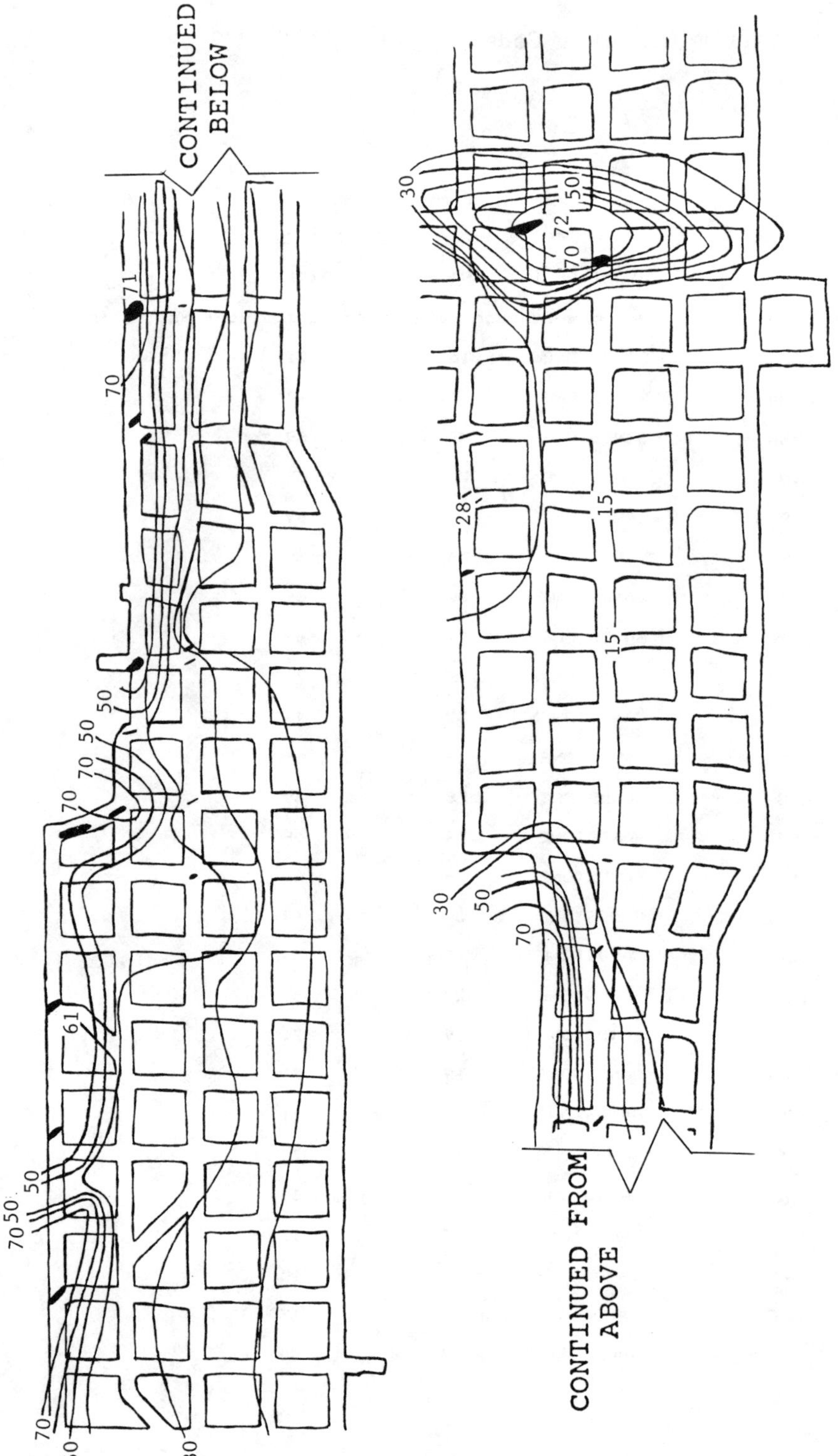

Figure 4. Mine map showing roof falls (shaded areas) and contours of failure probability (%).

Ferm, J. C., Melton, R. A., Cummins, G. D., Mather, D., McKenna, L., Miur, C., and Norris, G. E., 1978, "A study of roof falls in underground mines in the Pocahontas #3 Seam, Southern West Virginia," USBM Minerals Health and Safety Technology, Contract Number H0230028, 81 pp.

Haynes, C. D., 1975, "Effects of temperature and humidity variations on coal mine roof stability," Proc. Symp. on Underground Mining, Louisville, Ky., Oct. 21-23, publ. by National Coal Assoc., Wash., D.C., vol. 2, pp. 120-126.

Headlee, A. J. W., 1944, "Fracture zones in mine strata," Min. Cong. Jour., vol. 30, pp. 57-60.

Iannacchione, A. T., Ulery, J. P., Hyman, D. M., and Chase, F. E., 1981, "Geologic factors in predicting coal mine roof-rock stability in the Upper Kittanning Coalbed, Somerset County, Pa.," USBM RI 8575, 41 pp.

Jansky, J. J. and McCabe, K. W., 1979, "Evaluation of remote sensing techniques for planning mining operations," SME-AIME Mini-Symposium Series No. 79-07, Underground Coal Mine Design and Planning, New Orleans, La., Feb., pp. 39-45.

Karmis, M. and Kane, W., 1984, "An analysis of the geomechanical factors influencing coal mine roof stability in Appalachia," Proc. 2nd Int. Conf. Stability in Underground Mining, Lexington, KY, pp. 311-328.

Krause, H. F., and Damberger, H. H., in press, "Clay-dike faults and associated structures in coal-bearing strata—deformation during diagenesis," Compte Rendu, Ninth Int. Cong. of Carboniferous Geology and Stratigraphy.

Milici, R. C., Gathright, T. M., II, Miller, B. W., and Gwin, M. R., 1982, "Geologic factors related to coal mine roof falls in Wise County, Virginia," Report prepared for Appalachian Regional Commission, Contract No. CO-7232-80-I-302-0206, 103 pp.

Moebs, N. N., 1973, "Geologic guidelines in coal mine design," Ground Control Aspects of Coal Mine Design, Proc. Bureau of Mines Technology Transfer Seminar: Lexington, Ky., March 6, USBM IC 8630, pp. 63-69.

_____. 1977, "Roof rock structure and related roof support problems in the Pittsburgh coalbed of southwestern Pennsylvania," USBM RI 8230, 30 pp.

_____. 1981, "The geologic character of some coal wants at the Westland Mine in southwestern Pennsylvania," USBM RI 8555, 25 pp.

Moebs, N. N. and Ellenberger, J. E., 1980, "Hazardous roof structures in Appalachian coal mines," Unpublished Report, U.S. Bureau of Mines, 7 pp.

_____. 1982, "Geologic structures in coal mine Roof," USBM RI 8620, 16 pp.

Nelson, W. J., 1981, "Faults and their effect on coal mining in Illinois," Illinois State Geol. Surv., Circular 523, 40 pp.

Oberlick, G. J., 1978, "Remote sensing at Marrowbone Development Company," Proc. 9th Ann. Inst. on Coal Mining Health, Safety, and Research, W. E. Foreman, ed., Virginia Polytech. Inst. and State Univ., Blacksburg, Aug., pp. 163-189.

Overbey, W. K., Jr., Komar, C. A., and Pasini J., III, 1973, "Predicting probable roof fall areas in advance of mining by geological analysis," USBM Health and Safety Research Program, Technical Progress Report--70, May, 17 pp.

Parker, J., 1966, "How moisture affects mine openings," Eng. and Min. Jour., vol. 167, no. 11, pp. 95-97.

Parsons, R. C. and Dahl, H. D., 1971, "A study of the causes of roof instability in the Pittsburgh coal seam," Proc. 7th Canadian Symp. Rock Mech., Mar. 25-27, Alberta, Canada, 14 pp.

Patrick, W. C., and Aughenbaugh, N. B., 1979, "Classification of roof falls in coal mines," Min. Eng., March, pp. 279-283.

Peng, S. S., 1978, Coal Mine Ground Control, John Wiley & Sons, New York, 450 pp.

_____. 1980, "Roof falls in underground coal mines," Technical Report No. TR 80-4, West Virginia Univ., 42 pp.

Rinkenberger, R. K., 1978, "Operational application of remote sensing technology for predicting mine ground hazard areas," Proc. 9th Ann. Inst. on Coal Mining Health, Safety, and Research, Virginia Polytechnic Inst. and State Univ., Blacksburg, pp. 225-243.

Shepherd, J. and Burston, R. J., 1977, "Maps of faulting and roof failure at Aberdare East Colliery, Cessnock, N.S.W," Investigation Report 122, CSIRO Minerals Research Laboratories, 6 pp.

Shepherd, J. and Fisher, N. I., 1978, "Faults and their effect on coal mine roof failure and mining rate: a case study in a New South Wales colliery," Min. Eng., Sept., pp. 1325-1334.

Smith, A. D., 1984, "Relationships of assumed condition of mine roof and the occurrence of roof falls in eastern Kentucky coal fields," Proc. 2nd Int. Conf. Stability in Underground Mining, Lexington, Ky., pp. 329-345.

Stateham, R. M. and Radcliffe, D. E., 1978, "Humidity: a cyclic effect in coal mine roof stability," USBM RI 8291, 19 pp.

Statistical Analysis System, 1982, <u>SAS User's
    Guide: Statistics</u>, SAS Institute Inc., Cary, NC,
    584 pp.

Wier, C. E., 1969, "Factors affecting coal mine
    roof rock in Sullivan County, Indiana," Proc.
    Indiana Acad. Sci. for 1969, Indianapolis, vol.
    79, pp. 263-269.

MINE VENTILATION

Chapter 45

USE OF MICROCOMPUTERS IN MINE VENTILATION NETWORK DESIGN

Jozsef S. Gozon
and
Slobodan S. Lukovic

The Ohio State University
Mining Engineering Division
Columbus, Ohio 43210

Abstract. Although the use of computers in the
mining industry is not a new idea, it is only in
the last several years that microcomputers started
to appear in the offices of small mine companies
and classrooms of mining engineering colleges.
Improvements in microcomputer hardware, operating
systems and languages, accompanied by sharp price
reductions, have made an excellent base for the
development of software for a broad range of mining engineering applications. It is very likely
that mining schools around the country will soon
start requiring the possession of a microcomputer
from their newly enrolled students. It is obvious
that new concepts of teaching mining engineering
disciplines with the aid of microcomputers will
also have to be established.

This paper is dealing with the use and limitations of microcomputers in every-day work of mining engineers. In particular, attention is given
to the mine ventilation network analyses and design.

The final task of improving the safety, health
and working conditions in underground mine is also
one of concerns of this presentation. The interactive program developed by the authors enables
the user to create, modify, display and calculate
mine ventilation networks using the popular Commo-
dore 64 Computer.

The computation takes into account natural
ventilation and fan characteristics as a part of
convenient input block. High resolution color
graphics make the whole procedure easily under-
standable and manageable. Sound effects are also
available if desired.

The presentation, along with demonstration,
serves to illustrate capabilities and potential
power of microcomputers as teaching tools, using
a longwall installation with ventilation network
as an example.

INTRODUCTION

The mining industry began using computers in
the late 1950's. The rigid standards set by computer centers, the programmers without experience
in mining, budget limitations, delays in performance, and other shortcomings disillusioned the
user-pioneers and disappointed the mine managers.
Almost thirty years later came the microcomputers
with their new promises. The fresh graduates in
mining engineering enjoyed the patronage coming
from the engineering departments and became inita-
tors in microcomputer applications. The mine management also changed the way of thinking. Due to
unexpected recession in the coal market, the mining
companies were forced to make dramatic changes in
their approach with regard to planning operations
and evaluating the possible alternatives. One of
the topics requiring new concepts in engineering
design and operation, was the mine ventilation
system with its unsteadiness and high cost.

Only in the last 25 years have mechanical devices been used to analyze mine ventilation networks. Simple electrical networks were replaced by
compressed-air analog models which are more flexible and more expensive, however less accurate.
Complex mine ventilation networks have been studied
through the use of analog computers for less than
two decades, and then digital computers with their
data storage capabilities entered the field. Less
than 15 years ago, ventilation systems, in particular face ventilation with methane liberation and
dust generation, were analyzed by using airflow
patterns and actual readings from operation. All
of these steps in the evolution of analytytical
methods led to the development of sophisticated
computer hardware and complicated mathematical simulation methods. The result was a giant computer
with specially skilled operators who only performed
a routine program run, without dealing with the
engineering aspects of the mine ventilation network
design or analysis.

In the coming years, with extending availability of microcomputers, requirements will focus on
the development of user-friendly software to cover
all phases of mining [1]. Under difficult market
conditions, those coal mine companies which are
ready to fully utilize the new technological wave,
are likely to make some profit out of it. The
first step in this progress is the education of
mining engineers, who accept the challenge. The
mining college and coal industry partnership could
create the required environment for productive cooperation in this new field. The authors are

trying to encourage such an action by submitting this paper and offering demonstration of their program.

PROGRAM FEATURES

The design of an underground mining operation requires the integration and synchronization of transportation, ventilation, ground control, mine drainage, and mining methods to form a system which provides the highest possible degree of safety for miners. This concept has been used in teaching core courses, such as Mine Environmental Control Systems, at the Mining Engineering Division at The Ohio State University. Fortunately, the procedure in design of ventilation systems for longwall operations in underground coal mines is well-discussed in the adapted course textbook [2]. Assuming an ideal situation when a student is familiar with the basic laws and requirements related to the mine ventilation network design, the user of this program can complete the whole procedure within a short period of time compared to the manual method.

After having completed developing the general air distribution plan, including arrangement of main airways, flow direction, fan location, etc., as well as coordinating the mining and ventilation systems, the program user is ready to inputting data. A convenient input block allows for an easy data introduction through screen display. The ventilation system can be displayed in any phase of computation. As the resistance calculation progresses, the schematic of the ventilation system is getting less complicated; finally, the equivalent scheme will be reached and displayed if desired.

A block for the determination of the pressure drop over all splits, including the location and size of regulators and/or blowers required, is one of the major program parts. It ends with the calculation of the mine velocity head at the discharge.

Finding the total mine head for fan selection is the result of the use of a separate program block. It can be combined with another block providing efficient tools in handling the effects of the natural ventilation. In the first step, the atmospheric pressure changes between the two shafts collars will be neglected [3]. Thus, a simplified model will be used for working point determination. For detailed analysis, the consideration of elevation difference, changes in atmospheric pressure and moisture content also are included in the program.

By introducing the necessary data for fan characteristic curves, it becomes possible to determine the working point coordinates, including fan efficiency and speed and the pitch of the fan blades. The screen display supported by high resolution color graphics is one of the emphasized features throughout this program. Furthermore, series and parallel fan configurations can easily be handled by the program in the same manner. Also, provisions are made in the program for handling speed adjustment and air density alterations.

Computer assistance in cost estimation is also available.

APPARATUS

The determination of air quantity required for the entire coal mine is based on the air quantities assigned to each one of the working faces. The minimum air quantity in a working place is determined by human needs, legal requirements, and work-efficiency standards. In gassy coal mines, the legal requirements with regards to the air quality control are the most important aspects in air quantity calculations. The modern mining methods, such as longwall mining, demand a flexible mine ventilation system which can be adjusted in accordance with the development of the production system. As an example, illustrating the wide range of possible air quantity required, a coal mine with three longwall faces and a couple of developmental workings is shown in Figure 1.

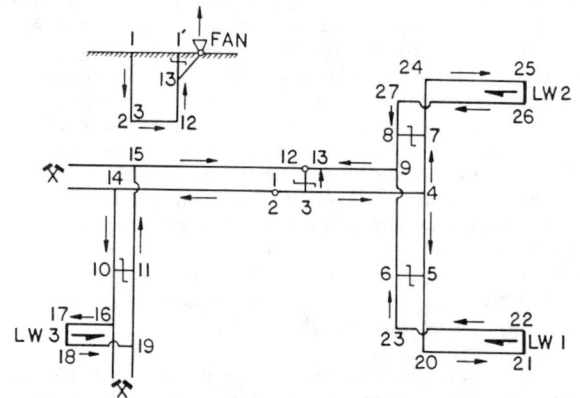

Figure 1. Schematic diagram of the mine-ventilation system for three longwall faces.

The methane dilution is the essential parameter in the ventilation air flow determination. Methane liberation from the newly exposed working faces is the major contributor to the problem. The secondary methane liberation due to the exposure to the ventilating air during the transportation has little effect on the air quantity passing through each section. Considering the operational requirements, comfort and work efficiency, as well as air velocity standards, the number and sizes of entries will be determined and entered in the program. The production scale is shown on the schematic of the ventilation system. As it can be seen on the sample display (Fig. 2), the effect of the developmental working faces on the equivalent resistances of the mine has been excluded from the investigation.

Theoretically, almost an infinite number of mine ventilators is available for the operating range of main fan during the life of the mine. Practically, however, the problem is in the selection of a most satisfactory ventilator which will be capable of meeting the safety, noise-level, and size requirements, while having the lowest overall cost per year. Before reaching the financial consideration phase of fan selection, the designer performs numerous calculations and simulations whose result is a preliminary selection of suitable commercial fans. By using the fan-rating tables, special selection charts, and fan characteristic curves, the designer faces the efficiency contours of the fans considered.

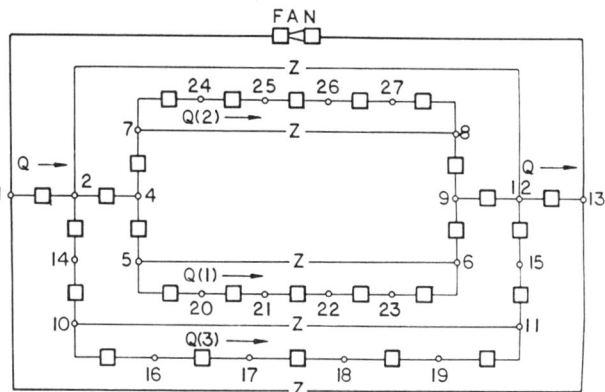

Figure 2. Sample screen display for the mine-ventilation network.

Centrifugal fans are capable of moving large volumes of air and gas mixture over the range of moderate to very high pressures encountered in some mining applications. Located on the surface, or just below ground level, these ventilators are particularly suitable for deep coal mines. Typical flow rate vs. static pressure characteristic curves of a radial-flow main ventilators are shown in Fig. 3. The aerodynamic characteristic of centrifugal fans facilitates parallel operation. When connected to a common fan drift, this feature permits a phased expansion of the ventilation capacity as the coal mine develops.

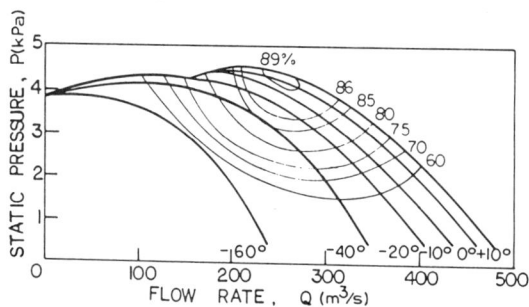

Figure 3. Radial-flow fan performance curve.

The investigation of parallel configuration of radial-flow mine fans is one of the tasks considered during the program development. Relatively simple mathematical apparatus was used to approximate the head-discharge curves with regard to the pitch angle range and increment of the fan blades with constant speed. The location of the optimum efficiency contours is controlled by second-degree polynoms. The program could handle any type of radial-flow fan which has a digitazable performance curve.

The axial-flow mine fans are designed for applications where large volumes of air and gas mixture need to be extracted at moderate to high pressures. As a coal mine develops, the flow rate of

the main fan can be gradually increased to meet the maximum air flow requirements. Also, during periods, when the coal mine is not working at its desired capacity, the ventilation air flow rate can be reduced accordingly by adjusting the pitch of the fan blades, either manually or automatically. In this way, the main ventilator delivers the optimum required output at all times and substantial savings in power consumptions are achieved by the coal mine company.

A typical performance curve of an axial-flow ventilator is shown in Fig. 4. The optimum efficiency contours are approximated by ellipses in the program. The user is capable of adjusting the mine resistance by altering the ventilation network circuit, so the optimum efficiency contours fall on the system resistance curve. The dotted line indicates the border of the instable region of the performance diagram.

The lack of availability of information on mathematical relation by which the effect of a certain change in pitch of the blades can be calculated, challenged the authors in finding an approximation offering considerable accuracy for optimum efficiency contours. The results obtained with the apparatus in the program, so far, are in good correlation with test data. Field test showed that air flow rate varies almost proportionally, and the head directly, to a fractional exponent with a limited change in pitch of the blades.

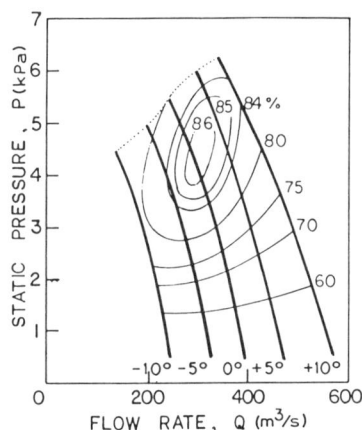

Figure 4. Typical characteristics of an axial-flow main fan (Courtesy of Aerex Ltd.)

CONCLUSIONS

The application of microcomputers in the everyday work of the mining engineers plays an increasing role especially when rapid changes in operational conditions of coal mines are concerned. This paper illustrated the possibilities of software use in microcomputer-assisted mine ventilation network design and analysis. The mining colleges are being challenged to consider necessary changes and improvements in their curricula and put emphasis on providing opportunities for their students in gaining experience, among the others, in ventilation network design problems particularly in gassy coal mines.

ACKNOWLEDGMENT

The authors are grateful to mine ventilator manufacturers for their cooperation in providing the required information for completion of this paper. We thank the staff of the Mining Engineering Division at The Ohio State University for their support, comments and many stimulating discussions on the subject. Our thanks and appreciation to Mrs. Patricia Coulson for her contribution in editing and typing.

REFERENCES

1. Gigliotti, S.J. - Haney, R.A.: "Computer Applications in the Analysis of Face Ventilation Systems."

2. Hartman, H.L. et al.: "Mine Ventilation and Air Conditioning. Second Edition." John Wiley & Sons. New York. 1982.

3. Hall, C.J.: "Mine Ventilation Engineering" SME-AIME. New York. 1981.

Chapter 46

INTERACTIVE GRAPHIC DISPLAY OF VENTILATION NETWORKS

N. d'ALBRAND, J.B. GUNTHER, J.P. JOSIEN

Mine Ventilation group - CERCHAR - FRANCE

The ventilation of underground mine workings is necessary to ensure the safety of personnel and equipment. Such safety is obtained by appropriate dilution of the various pollutants that are contained in the atmosphere of the mine ; the quality of the atmosphere is then controlled periodically.

Among the various pollutants that can be found in the atmosphere of a mine one may mention explosive, or intoxicating or asphyxiating gases eg firedamp, CO, NO_x, CO_2, N_2..., dust and heat. Some of these pollutants are inherent to the deposit, other are created by the method of mining.

The quantity of pollutants must of course be limited as much as possible, eg by methane extraction, drainage, control of diesel engines, but this is never sufficient and it is always necessary to provide forced ventilation, in order to supply fresh air to the mine. Mine regulations stipulate for the different pollutants maximum contents that must not be exceeded.

Once the regime of methane release in a deposit is known, that the method of mining is selected and the circuits followed by diesel vehicles in the mine are established, it is fairly easy to determine the quantity of fresh air that must be supplied. However, planning of the distribution of air among the different parts of the mine is practically impossible without using a computer.

The Centre d'Etudes et Recherches de Charbonnages de France (CERCHAR), during the past 30 years, has developed and applied two different methods for determining ventilation networks in underground mines, using :

- an analog calculator, or
- a computer software programme.

Each method has its advantages and disadvantages (1). Because of the increased power of the microcomputer, we have developed a method which combines the advantages of the two methods and excludes their disadvantages. Our aim was also to obtain an inexpensive system which could easily applied in the mine without necessitating the contribution of solftware specialists.

The decision was made to develop two software programmes that respond to two different needs of the mine operator, i.e routine management of the ventilation network of a small mine and the development of new projects. New projects can be either at short term, i.e aiming at improving the present situation or at longer term, i.e involving the opening up of new districts or the sinking of a new shaft, etc...

In order to facilate the possibility of conversions in the future, both software programmes were developed in the Fortran language.

The first of these programmes, PC VENT, determines the distribution of air in the mine ; the second, VENDIS, determines the distribution of air but also displays the ventilation network on a graphical display terminal ; this is most advantageous if one wishes to study a ventilation network conveniently.

PV VENT can be used with an inexpensive microcomputer such as an IBM-PC. VENDIS requires a more expensive equipment.

DESCRIPTION OF THE PC VENT PROGRAMME

This programme was developed on a microcomputer Micro-PDP/11 equipped with a rapid calculating card.

It can be used to calculate a ventilation network comprising 1000 nodes, 1000 branches and up to 9 different fan characteristic curves. These dimensions were adopted arbitrarily ; they could be increased if necessary.

The programme was then adapted to the IBM-PC which, equipped with the 8087 coprocessor, achieves equivalent performance.

This adaptation was made in association with the division "mines and infrastructures" of the Mining School of Paris, within the framework of a policy for the development of mine applications using microcomputers.

The programme is interactive and self documented. These means that in the course of execution of the programme, the user is prompted to supply the necessary data and receives indications on way in which such data are to be given. At the start, the user selects a system of units, international, US or French units, then a menu is displayed and it is possible to select different options.

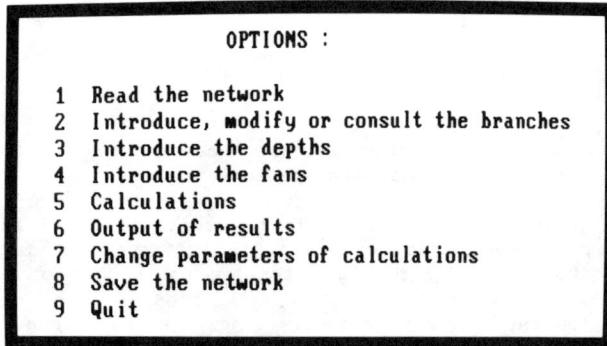

```
┌─────────────────────────────────────────────────────┐
│                                                     │
│               OPTIONS :                             │
│                                                     │
│   1  Read the network                               │
│   2  Introduce, modify or consult the branches      │
│   3  Introduce the depths                           │
│   4  Introduce the fans                             │
│   5  Calculations                                   │
│   6  Output of results                              │
│   7  Change parameters of calculations              │
│   8  Save the network                               │
│   9  Quit                                           │
│                                                     │
└─────────────────────────────────────────────────────┘
```

Enter your choice : 1_

FIGURE 1. MAIN MENU OF PC VENT

Among these options, one may mention :

- read (on a disk file) a previously created network,
- create, suppress or modify branches,
- create fans,
- calculate network,
- save (on a disk file) a created or calculated network,
- display results.

If one selects the option create, the programme requests the user to enter on the keyboard, the numbers of the end nodes of the branch, its aeraulic resistance, the average temperature of air, and possibly the maximum air flow which cannot be exceeded ; in the latter case, the programme adjusts the value of resistance (by introducing a restriction to the flow) in order to limit the air flow.

The calculations are performed with the method of Hardy Cross, after the meshing is established according to the algorithm of Sollin. The network is in fact necessary only for a first calculation, or when the structure of the network is to be modified (suppression or creation of a branch). A modification of aeraulic resistance does not change the meshing.

The calculation time depends upon the structure of the network. For example, the time required for a first calculation of the network of

the Gardanne Mine (in the Centre Midi Coalfields) which comprise of 182 branches, 110 nodes and 3 fans (see Figure 4) is approximately 90 s. After a modification of the resistance in one branch, the calculation times make it possible to examine a great number of alternatives.

In this programme the natural ventilation draught is also taken into consideration.

The saving of a network is obtained by requesting the save option in the menu. The programme then requests the name under which the network is to be saved and stores a file on the disk having this name and containing all the information that concerns this network i.e branches with their resistance and temperature, depth of the nodes, flows, characteristics of the fans.

For subsequent work on the same network, this file name has to be entered in order to read the network.

DESCRIPTION OF THE VENDIS PROGRAMME
(ventilation network display)

Equipment

This programme was developed on a microcomputer PDP/11 connected with a graphical display terminal AFIGRAF 6080 (CSEE).

This terminal has a 21" refresh screen ;it has a useful surface of 32 x 32 cm, a resolution of 1 024 x 1 024 and an image computation of 4 096 x 4 096.

A digitalizing tablet of the same resolution is connected with the graphical display terminal and is used to designate points on the screen by means of a mouse.

FIGURE 2.
OVERALL VIEW OF THE GRAPHICAL DISPLAY TERMINAL

VENDIS enables us :

- to enter a network using the ventilation plan of the mine,
- to calculate the ventilation network,
- to display the network with the results of calculations,
- to save the network with the results of calculation for subsequent uses;
- to modify the network.

All these operations are made by selecting options in the menus displayed on the alphanumerical terminal ; these menus guide the user and indicate him what he must do.

Creation of a network

Before calculating a network, a file containing all the geometrical and aeraulic data of the ventilation network of the mine must be created. By using a digitizing tablet on which a mine plan is placed, two points of the plan are designated and the underground coordinates of these two points are entered on the keyboard. Scaling is then done automatically by VENDIS.

A branch can be linear or consist of several linear sections ; therefore it is necessary once the starting node is designated, to indicate the number of sections and also, before designating the end node, to designate the pseudo nodes that is the ends of each section constituting the branch.

When a node is being created and designated, the programme requests the number that is to be attributed to this node. If no reply is given, then the programme will automatically allocate a number by selecting one in a table of available numbers.

Then, it is necessary, if the branch has a resistance, to enter on the keyboard, the values of aeraulic resistance, average temperature of air, and possibly the maximum flowrate that should not be exceeded. If the branch comprises a fan, its number must be indicated.

The characteristic curves of the fans are also entered using a digitizing tablet. A reproduction of the characteristic curve of the fan only has to be placed on the table ; then one designates the origin of the coordinates, the extreme points of the axes by indicating the numerical values of flow and total pressure which correspond to these extremes, and then 10 points of the curve between the extremes. The characteristic curve of the fan then appears on the graphical terminal screen and it is possible to modify some of points of the curve if required.

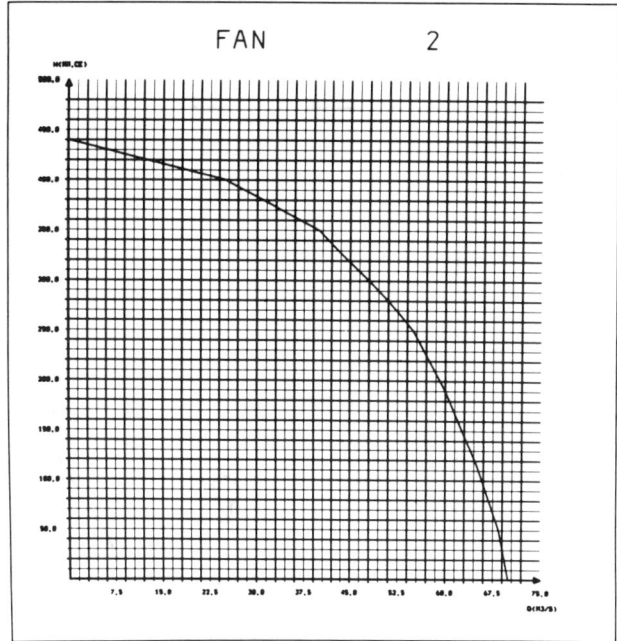

FIGURE 3. DISPLAY OF A FAN CHARACTERISTIC CURVE

After having entered into the system all the branches of the mine plan, the file must be saved by assigning to it a name with a length of 6 alphabetical or numerical characters.

Calculation

The calculation sub-routine of the programme uses the same algorithms as the PC VENT programme. The calculation times are the same. The size of the networks can be of 1 000 nodes and 1 000 branches.

Display option

By using this option, the network is displayed on the terminal screen, as it was entered i.e an image of the mine plan appears. Because the coordinates of each node are known, the programme calculates the coordinates of the intersections of the different branches, projected on a plane which is transverse to the direction of viewing, in order to interrupt the plotting of those parts of the branches which are hidden by other ones.

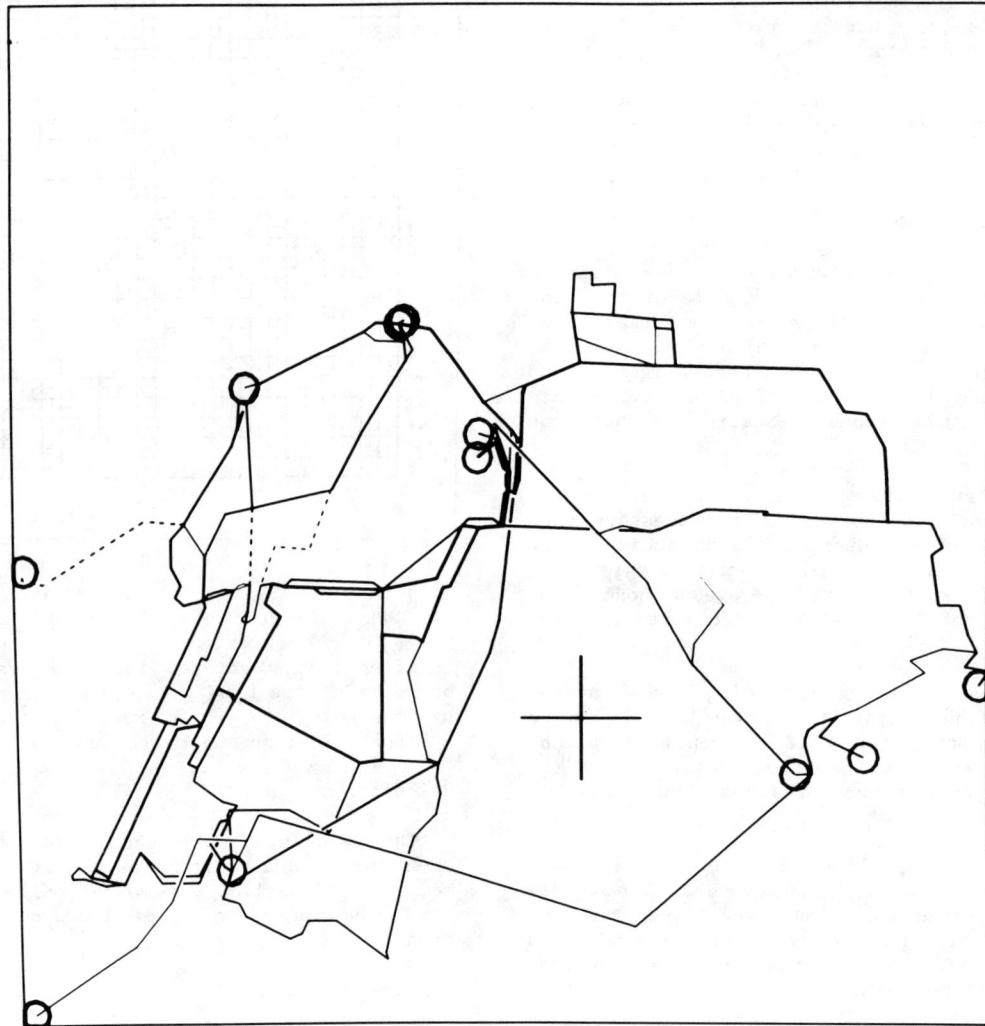

FIGURE 4. OVERALL VIEW OF THE VENTILATION NETWORK OF GARDANNE COLLIERY

The branches where the air flow is reversed, as compared with a previous situation, start blinking.

One option of the programme makes it possible to display nodes with their numbers and another one to display the branches with indication of the calculated flows.

When the number of branches of a network becomes too large, the use of the two above options can be difficult. Then it is possible to adopt the following procedure in order to display the results of the calculations :

designate successively each branch with the mouse. The flowrate is then displayed next to the position that has been designated and the corresponding branch starts blinking. The whole set of data concerning this branch is at the same time displayed on the alphanumerical terminal,

– zoom the zone of interest by defining a window with the mouse or by using the keyboard ; then use the option for the display of nodes and flow rates as mentioned above.

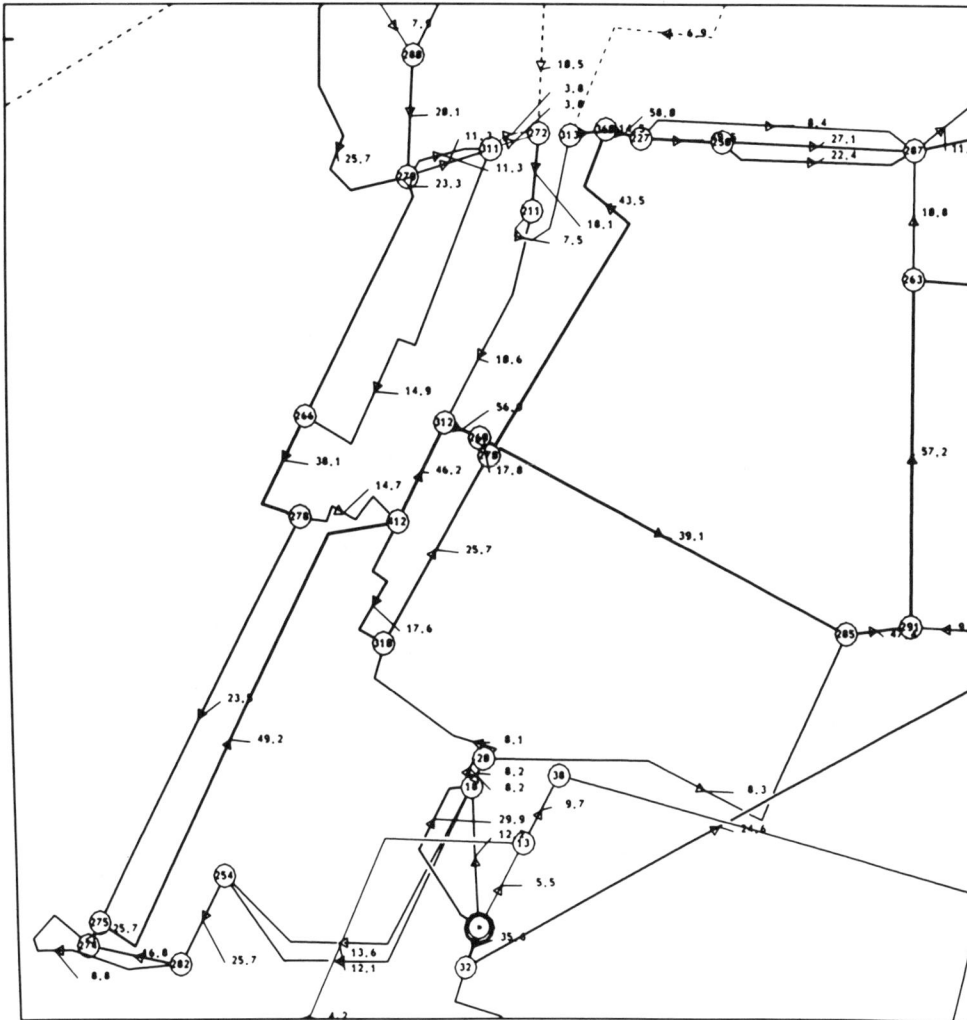

FIGURE 5. PARTIAL VIEW OF THE VENTILATION NETWORK OF GARDANNE COLLIERY
WITH THE RESULTS OF A CALCULATION

- list the complete set of results on a printer.

The designation of branches is most advantageous when one is concerned with fire hazards.
Indeed, when designating a particular point on a
branch, among all the data that are given by the
programme, there is the atmospheric pressure at
this point. Therefore by designating successively
two points, between two different ventilation circuits, it is possible to immediately know the
difference of pressure between them. The risk
that spontaneous heating may develop between the
two points can be assessed as a function of the
pressure difference, which can provoke a leakage
of air (eg through fissured strata, caved zones,
etc...).

```
  Characteristics of the pointed out branch
*******************************************************
*              * Number * X(m) * Y(m) * Z(m) * P(mmCE) *
*******************************************************
* Starting node *  140 * 9202 * 946 * -30 * 10370.8 *
* Ending node   *   38 * 3985 * 562 * -18 * 10390.6 *
* Designated point*       * 8404 * 271 * -27 * 10374.3 *
*******************************************************

  Name             :
  Length           :    5866 (meters)
  Resistance       :   55.35 (murgues)
  Nominal resistance :  52.00 (murgues)
  Temperature      :      19 (deg.C)
  Flow rate        :   -9.70 (m3/s)
  Head loss        :   -5.21 (mmCE)

 1 - Modify
 2 - Suppress
 3 - Create a new node for definite point
 4 - Prepare output for plotter BENSON
 5 - End of session

 Enter your choice : (Carriage return for continue)
```

FIGURE 6. CHARACTERISTICS OF A BRANCH

Change of the direction of viewing

By selecting a triaxial system of reference O
x y z, the mine plan represents a projection of
the mine on the horizontal plan x o y that is
transverse to a unit vector 0, 0, 1. With this
direction of viewing, vertical mine workings such
as main shafts or blind shafts or nearly vertical
workings such as faces in steep seams cannot be
shown on a horizontal plan. To show them it is
necessary to select a horizontal direction of
viewing i.e with a unit vector 0, 1, 0 or 1, 0, 0
in order to be able to visualize the mine in
elevation.

Any intermediate direction of viewing is also
possible.

AN EXAMPLE OF APPLICATION

The Gardanne colliery in the south of France
mines a coal seam of 2 to 3 m thickness by
longwall faces. From the mineable reserves it is
expected that mining operations will extend up to

year 2010 at a rate of production of 1.6 million
tons per annum. The distance of the faces from
the main shaft is increasing continously ;
because of increasing difficulties to ensure
proper ventilation of the mine, it was decided to
sink two new shafts Y and Z and to reverse first
the Gérard shaft which is presently air intake
and later shaft Z.

Shaft Y will be used both for air intake and
as a service shaft for men and supplies. Shaft Z
could be equipped if necessary for winding of
coal.

In close collaboration with the mine operator,
we applied our software calculation programme
with a view to determining :

- the optimum diameter of the shaft to be sunk,
- the operating conditions of the new main fans
  to be installed,
- the operation to be performed underground for
  reversing the ventilation.

In this particular application, we considered
a number of underground airways in addition to
the two new shafts, according to the mine
operators indications. The aeraulic resistance of
these airways was estimated by comparison with
the resistance of existing airways of identical
cross-section which were equipped in similar
manner.

The aeraulic resistance of the shafts was calculated for several diameters and different
characteristic curves of fans were proposed.

Successive calculations were made for different alternative solutions considered ; the
results enabled the mine operator to select the
optimum solution for the shafts i.e diameter
10 m, depth 1150 m for shaft Y ; diameter 6.5 m,
depth 860 m for shaft Z, as well as the type of
main fan to be installed in Gérard shaft and
Z shaft.

The selection of these parameters was most
important in view of the capital costs involved
for the implementation of this project of more
than 600.10^6 FF. It was therefore crucial to be
in a position to select optimum characteristics
of the new installations to be built as well as
of the new fans whose unit power would attain
1 500 KW.

Once the final solution was selected, we then
defined, step by step, the different operations
to be carried out underground i.e opening of
doors, installations of air-locks, etc... in
order to reverse the Gérard shaft from untake to
return.

The display, in the form of blinking sections,
of those branches in which the air flow was
reversing and the knowledge of air pressure at
different points of the network, enabled us to
calculate the methane contents which were to be
expected at the time of reversal of the air flow.

CONCLUSION

PC VENT and VENDIS are two software programmes for the calculation of ventilation networks ; they operate interactively and with micro-computers. These programmes have been written so as to be used without requiring any assistance from computer software specialists.

A mine operator in possession of one of these software programmes can control the ventilation system of a mine on a day to day basis. For example, it is possible, in only a few seconds, to evaluate how the connection of a gallery is going to affect the ventilation system or to determine the new operating conditions to be given to a fan. VENDIS further permits display of a ventilation network as well as all the districts that are planned in any underground mine. If a mine operator envisages to modify a network, the branches where there will be a reserval of the ventilation will start blinking.

The time required for calculations and display are so short that a great number of alternate solutions can be rapidly investigated.

A version of the PC VENT programme, which enables us to take into account the effect of a fire on the ventilation system of a mine, is about to be released. This programme called PC FIRE will enable us not only to know the modified distribution of air provoked by a fire, but also, in association with the VENDIS programme, to display the circuit followed by the smokes and gases on the graphical display terminal.

REFERENCE

(1) d'ALBRAND N. - FROGER C. - JOSIEN J.P.
    Practical use of micro-computer for ventilation calculations. Third international Mine Ventilation Congress. Harrogate - England (1984)

Chapter 47

A COMPUTER-IMPLEMENTED COALBED METHANE MODEL

by

David M. Hyman
Geologist

Peter A. Stahl
Physicist
Bureau of Mines
US Department of Interior
Pittsburgh, Pennsylvania

ABSTRACT

A computer-implemented methane model for coal mines has been developed and undergone preliminary field testing. The model was built under contract by the US Steel Technical Center and considers two-phase (gas and water) porous medium flow in two horizontal dimensions with an optional vertical equilibrium effect. Methane is allowed to sorb and desorb from coal as a function of pressure and time. Once methane has desorbed from unfractured coal blocks, the gas flows simultaneously with water through the fractures (cleat) of the coalbed down a pressure gradient into the pressure sink of the mine opening or a drainage well. The changes in permeability and porosity of the coalbed due to mining-induced stresses are considered, as are the effects of gas slippage, flow regime, and fluid compressibility. This paper presents computer code exercises of the one-dimensional version of the methane model, as well as a history-matching exercise with the results of prior Bureau in-mine research. The computer code exercises illustrate the results of running the four default base-case coalbeds, as well as simulating methane emissions during mining. The history-matching exercises show good agreement with observed coalbed gas pressure gradients. Obtaining site-specific data on coalbed permeability, reservoir pressure, and coal gas contents allows the user to make realistic simulations with a minimum amount of assumed input specifications.

INTRODUCTION

The Bureau of Mines has historically been involved in health and safety related research to mitigate explosive atmospheres in underground mines, particularly coal mines. Methane and other hydrocarbon gases, generally collectively termed methane or firedamp, are inherent in coalbeds. Methane gas is introduced into the mine atmosphere in varying amounts and rates as a complex function of coalbed geology and mining operations. Normally, the methane gas emitted into the mine workings is diluted to well below its lower explosive limit by air circulated through the mine workings by the mine's ventilation system. Methane can also be partially drained in advance of mining through boreholes drilled into the coalbed. This drained gas can be piped out of the working sections and be exhausted, or recovered as an unconventional energy source.

Essential components to effective mine engineering practices to mitigate methane hazards are the quantitative prediction of the amount of the methane contained in the coalbed, its desorption mechanism, and its migration through the coalbed and into the mine workings. A variety of models have been developed, at least as numerous as the countries with underground coal mines. European methane prediction models consider advancing longwall mining, where considerable proportions of methane are emitted from adjacent strata with the worked seam contributing gas mainly during coal cutting and haulage operations.

In the United States, the predominance of room-and-pillar and retreating longwall mining methods, as well as the nature of the coalbed properties, has given rise to a somewhat different modeling orientation in which methane emissions are considered to be primarily from the worked seam. The modeling of conventional petroleum and gas reservoirs is a relatively advanced art in the oil and gas industry, and coalbed methane models have drawn upon this.

Major US computer models have been based on an idealization of the coal seam, which is similar to a gas or oil sandstone reservoir. The fundamental difference between the characterization of petroleum and coalbed reservoirs is the nature of gas retention and release in the rock matrix. In coal, methane is sorbed onto the internal surfaces of the coal micropore structures, as well as compressed in the free pore and fracture space of the coal. Considerably more methane can be sorbed on coal surfaces than compressed into its pore space. A conventional petroleum reservoir model does not consider the sorption-desorption process. This process is reversible and is a function of coalbed reservoir pressure and coal rank. As the fluid pressure in a coalbed is reduced due to mining or drainage, methane is released from the internal surface area of the micropores and migrates into the pore free space. From there the gas diffuses into the coal cleat system and other fractures. The gas in these fractures then undergoes Darcy flow down the pressure gradient, moderated by relative permeability effects into the mine openings or drainage well.

The coalbed methane model presented in this paper was initiated under a Department of Energy contract that was later transferred to the Bureau of Mines (Contract J0333952). The research was conducted by the US Steel Corporation Technical Research Center, Monroeville, PA, with Dr. F. C. Schwerer as the Principal Investigator. The one-dimensional version of the model developed in Phase 1 of the contract is the subject of this paper. A complete treatment of the underlying physics and mathematics is presented by Boyer et al. (1982).

The Bureau had sponsored the development of two prior coalbed methane models through contract research efforts. The first of these was developed by Intercomp, Inc., and is described by Price et al. (1974). This two-dimensional model considered two phases (gas and water) undergoing Darcy laminar flow. A basic limitation of this model was that the gas desorption process was considered to be instantaneous and that essentially an infinite amount of gas could be desorbed from the coal. Furthermore, considerable tabular-form input data specifications for the coalbed physical properties and functional parameters were required from the user.

The second model was developed at Pennsylvania State University as a part of a much larger mine system simulation package and is reported by Ramani and Owili-eger (1976). This model considers two-dimensional single-phase (gas) Darcy flow in the coalbed and therefore ignores the relative permeability effect in moderating the flow of gas. The spacings of the computational grid are uniform in each of the two spatial dimensions, which limits resolution in areas of rapid pressure changes, such as occur near a mine face.

The new model developed by the US Steel Technical Center incorporates several important improvements over previous efforts. The model contains enough default parameter specifications so that any one of currently four base-case coalbeds can be modeled without further user input. This allows the model to be run with a minimum amount of input specifications. The physical parameters and relationships that describe the gas and water flow phenomena that are minimally required for model operation also have defaults contained in the source code. The source code is written in a structured dialect of FORTRAN called USSTRAN and is converted to ANSI-77 FORTRAN by a precompiler program. The program has been run on a CDC Cyber 750 computer. The source code contains extensive comment documentation.

Physically, this model considers two fluid phases flowing in two coplanar directions with a vertical equilibrium option which can be invoked to approximately describe flow in a third direction. The flow through the fractures is described by a generalized formulation of Darcy's law which includes laminar as well as turbulent flow of both the gas and water phases. In addition, a gas slippage, or Klinkenberg, effect is included for the gas phase.

The absolute permeability of the coalbed is altered by a general relative permeability relationship which is a function of the relative abundance or degree of saturation of each of the two fluid phases. The user can optionally specify alternative values for the absolute permeability of the coalbed and for the relative permeability-saturation relationship in the model. The model is relatively sensitive to the absolute permeability, and ideally the user should have some field determinations for site-specific coalbed permeability values.

Another important model component is the gas desorption source term, which is represented by a functional description of the pressure and time dependence of the gas sorption-desorption from within the unfractured elements of the coal. The time dependence of this process is probably determined by diffusional flow.

```
        ┌─────────┐                                    Ⓐ
        │  Start  │                                     │
        └─────────┘                                     ▼
             │                              ┌──────────────────────┐
             ▼                              │  Increment time step │
  ┌──────────────────────┐                 └──────────────────────┘
  │ Allocate COMMON blocks│                            │
  │  Set I/O default values│                           ▼
  └──────────────────────┘                 ╱────────────────────────╲
             │                              │ Output:                 │
             ▼                              │  -Fluid pressures and   │
  ╱────────────────────────╲               │   relative saturations  │
  │ Input:                  │              │  - Mine conditions      │
  │  -Type of simulation    │              ╲────────────────────────╱
  │  - Coal seam            │                           │
  ╲────────────────────────╱                            ▼
             │                              ┌──────────────────────────┐
             ▼                              │      Call EPISODE         │
  ┌──────────────────────┐                 │                           │
  │ Set default values for│                │ Solve differential equations│
  │ simulation type and coal│              │ for two phase fracture and│
  │ seam if not specified │                │     diffusion flows       │
  └──────────────────────┘                 │                           │
             │                              │ Update permeability and   │
             ▼                              │ porosity as function of   │
  ┌──────────────────────┐                 │  pressure and space in    │
  │     Call METHPAR      │                │      coal seam            │
  │ Input:                │                └──────────────────────────┘
  │  -Mining parameters   │                            │
  │  -Fluid properties    │                            ▼
  │  -Equation constants  │                 ╱────────────────────────╲
  │                       │                 │ Output:                 │
  │ Set default values if │                 │  -Current model conditions│
  │ not specified in input│                 │  -Fluid pressure profiles│
  │ stream                │                 ╲────────────────────────╱
  │                       │                            │
  │ Normalize parameters  │                            ▼
  │                       │                 ╱────────────────────────╲
  │ Set boundary conditions│                │ Read input stream for   │
  └──────────────────────┘                 │  simulation changes     │
             │                              ╲────────────────────────╱
             ▼                                          │
  ┌──────────────────────┐                             ▼
  │ Initialize time and model│                ◇─────────────◇
  │     variables         │                   │  Finished ? │
  └──────────────────────┘                    ◇─────────────◇
             │                           No         │
             ▼                    ┌───────────┘      │
            Ⓐ                     ▼                  ▼
                        ┌──────────────────┐  ╱────────────────────────╲
                        │ Update drainage well│ │ Output summary data for│
                        │   conditions     │  │ completed simulation    │
                        └──────────────────┘  ╲────────────────────────╱
                                                          │
                                                          ▼
                                                     ┌─────────┐
                                                     │  Stop   │
                                                     └─────────┘
```

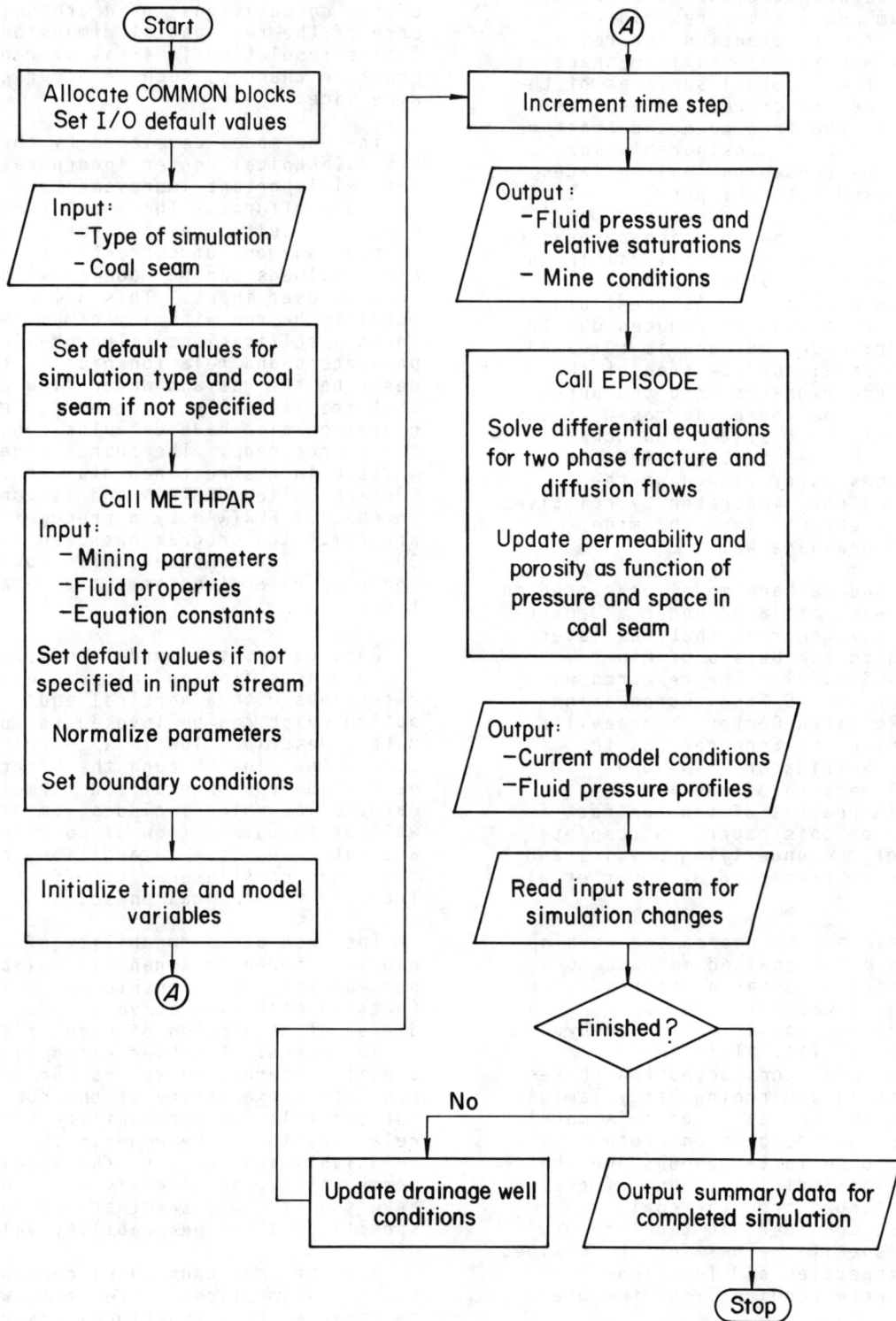

FIGURE 1. - Generalized Flow Chart of the Model

The model uses the relatively simple but physically realistic Langmuir relationship for monolayer sorption on the internal surfaces of the micropores. This relationship is used to describe the amount of gas that would be sorbed or desorbed at equilibrium as a result of a gas pressure change. This is combined with an equation which describes the rate at which the sorption-desorption process occurs. Combining the two equations results in a kinetic representation of the diffusion process which is computationally more expedient than the conventional diffusion equations used in prior models. The kinetic equation describes the rate that an equilibrium to a gas pressure change is restored and how much gas is involved in the sorption-desorption process. Other important features of the model include the option to have nonuniform computational grid spacings in the plane of the coalbed, the effects of horizontal and vertical gas drainage wells, and the simulation of an advancing mine face or a mine pillar.

THE ORGANIZATION OF THE MODEL

The model has two major parts: the initialization part and the time simulation part. Figure 1 is a generalized version of the main driver routine with calls to two major subroutines. A complete set of flowcharts as well as a source code listing of the one-dimensional version of the present model is presented by Boyer et al. (1983). The first task of the computer program is to dimension the common blocks and set the input-output (I/O) default parameters. The input stream is read for the type of simulation (mine face or coal pillar) to be performed and the coalbed base case desired. Default parameters provided are for a mine face simulation in the Blue Creek (lower bench of the Mary Lee) coalbed and are selected if not specified otherwise by the user. The other three base case coalbeds currently available in the code are the Mary Lee (upper bench), the Pocahontas No. 3, and the Pittsburgh.

The first major subroutine call is to one called METHPAR for the specification and calculation of the fluid and coalbed physical properties. The input data-job stream is searched for alternative or user-specified values for mining and other simulation parameters. If values are not specified in the input stream for physical constants for the gas and water fluid phases, as well as constants for the model equations, then default values are assigned. Boundary conditions for the differential equations are set, and the computational grid spacings and time step intervals are determined. Values

for variables such as pressure, distance, and time are normalized to an average value of each of the variables. When control returns to the main program, the time step and other required values are calculated.

The simulation portion of the model begins with an initial time increment. The current fluid pressures and saturations, as well as the mine conditions, are output. Various components of the model are updated. The second major subroutine call is to EPISODE, which solves the set of coupled partial differential equations for two-phase fracture and diffusional flows with sorption and desorption of methane. A novel feature is the description of the functional dependency of porosity and permeability on pressure. A complete treatment of this is presented by Smelser et al. (1984). When control is returned to the main program, the current model conditions are output for the time step just evaluated. The gas and water profiles output at this stage were used to produce the graphs in figures 2-8. The input job stream is next searched to determine whether simulation operating conditions such as mine advance rate or drainage well (if any are present) are to be changed for the next time period. If the end of the simulation has not yet been reached, then the operating conditions are updated and the program loops with a time incrementation for the next time step. If the end of the simulation has been reached, summary statistics of mine ventilation and drainage well conditions are output and the program execution terminates.

EXERCISES OF THE COMPUTER CODE

To illustrate the four coalbed base cases, graphs representing coalbed gas pressures versus distance from a mine face and in the direction of the entry are presented as figures 2-5. Two curves representing two different points in time are used to illustrate the coalbed reservoir pressure drop due to the presence of a mine opening. These base cases required from 9.4 to 26.0 CPU seconds to run on a CDC Cyber 750 computer. The depth and permeability default values are shown on the figures. It can be seen that lower absolute coalbed permeabilities support higher gas pressure gradients over time than the higher permeabilities do. The permeability of the coalbed represents its ability to transport fluids through the fracture and cleat system of the coalbed and down the pressure gradient established by the presence of the mine entry acting as a pressure sink. The original coalbed

USE OF COMPUTERS IN COAL INDUSTRY II

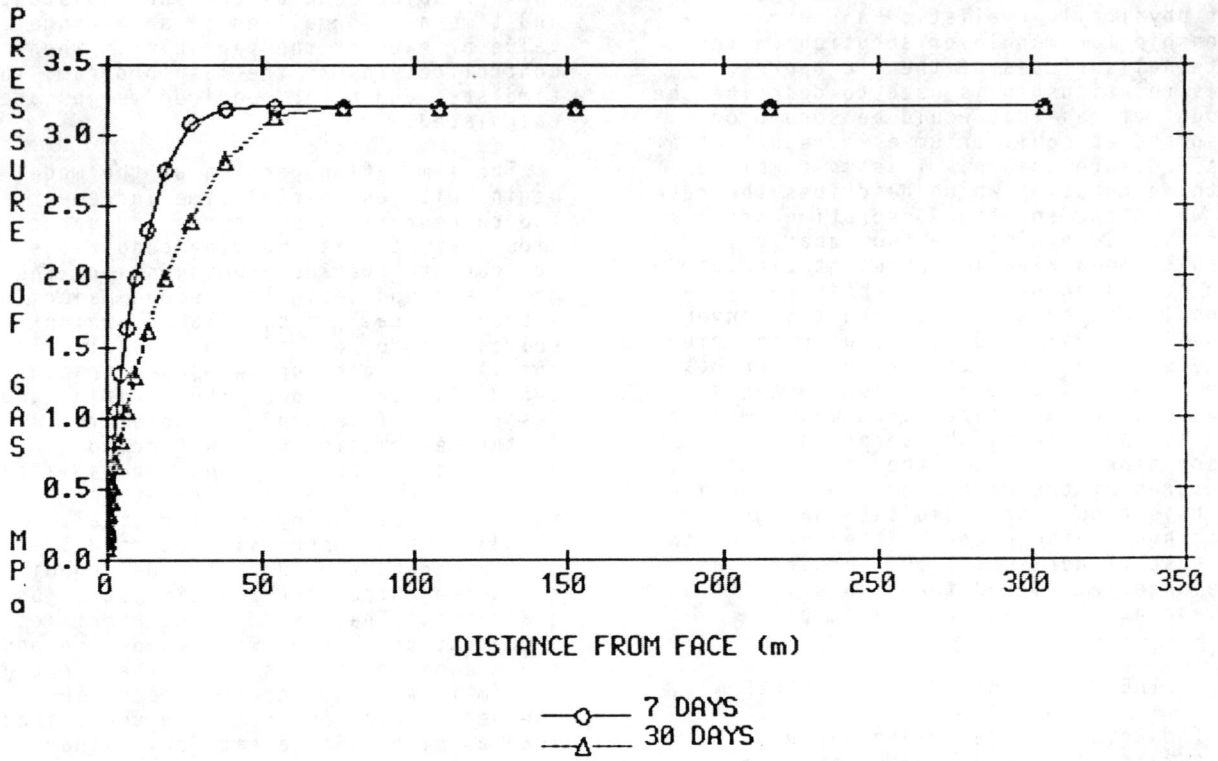

FIGURE 2. - Base Case for Blue Creek Seam (depth=366 m, perm.=3 md)

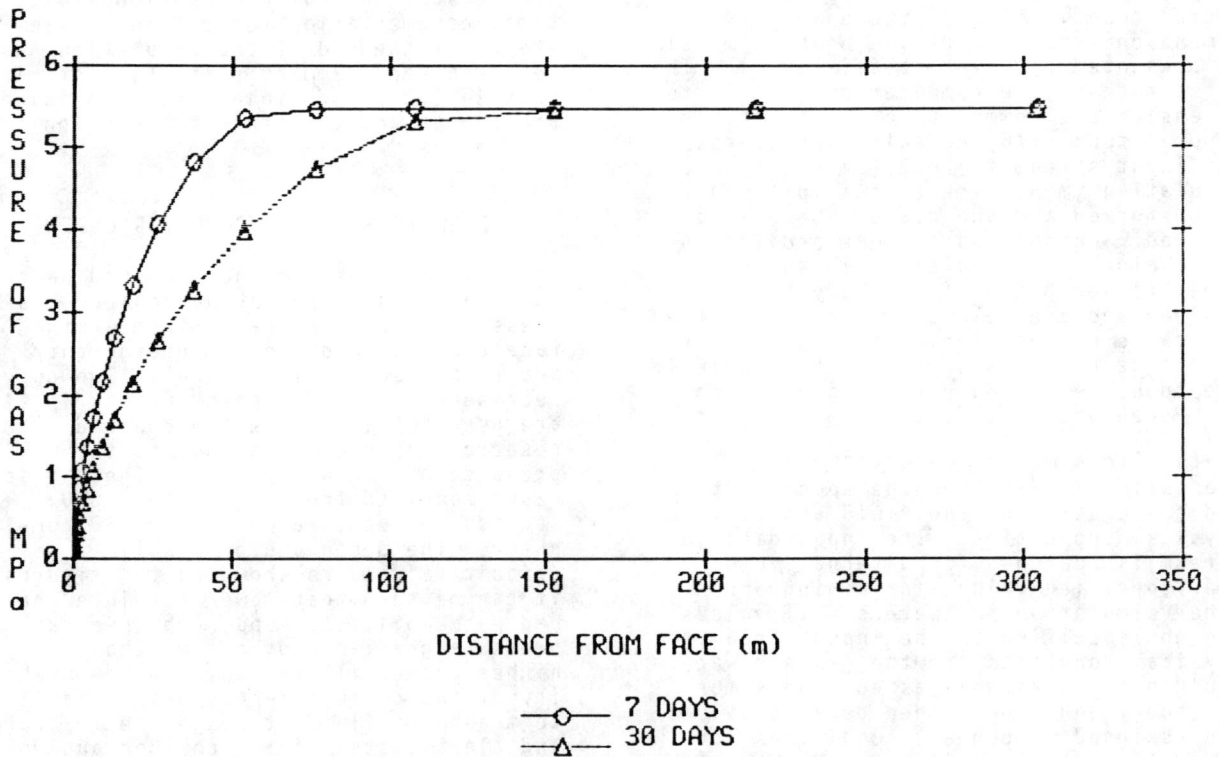

FIGURE 3. - Base Case for Pocahontas Seam (depth=549 m, perm.=5 md)

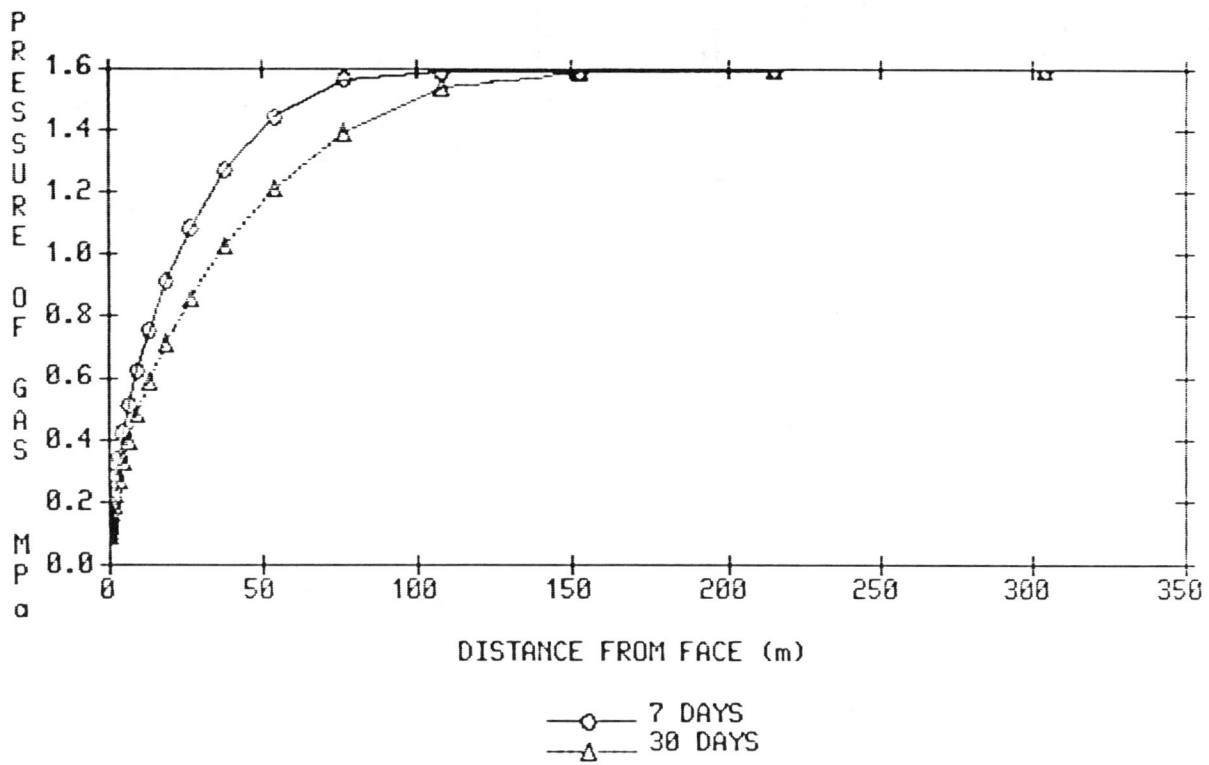

FIGURE 4. - Base Case for Pittsburgh Seam (depth=152 m, perm.=20 md)

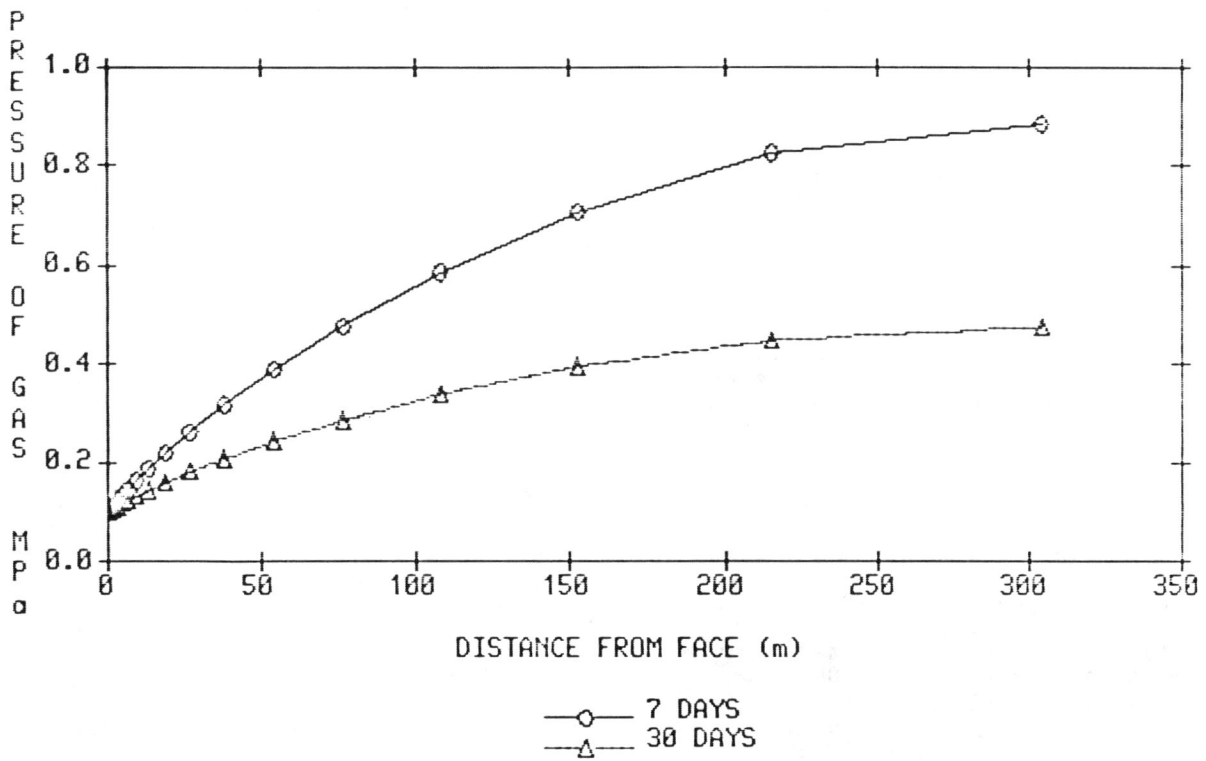

FIGURE 5. - Base Case for Mary Lee Seam (depth=335 m, perm.=75 md)

reservoir pressure is considered to be equal to the hydrostatic head for the depth of the coalbed, but it is more appropriate to use site-specific data if available. The depth or equivalent depth of the coalbed would be the first physical parameter that the user should specify, rather than using the default value of the computer code. The coalbed permeability could be the next parameter for which the user would provide a site-specific value. The permeability or hydraulic conductivity of the coalbed can be calculated from surface borehole pumping and/or injection test procedures. This value can alternatively be determined from the same type of hydraulic conductivity testing in an in-mine horizontal borehole drilled into the coalbed. Other parameters, for which the user initially might want site-specific or at least nominal values, describe the sorption capacity of the coal as a function of pressure. The two parameters involved can be obtained either from laboratory determinations (adsorption isotherms) or from published literature.

Essential to the complete development of a computer model of real world phenomena is the comparison of the model results with real world observations. The relative sensitivities of the model to the input parameters and model specifications should be determined. A discussion of the sensitivity of the new model is presented by Boyer et al. (1982). An example history-matching is presented in figure 6. Two in-mine horizontal boreholes were drilled into the Pocahontas No. 3 coalbed, and gas emissions from the boreholes were measured. One borehole was drilled into a face that had been exposed 15 days after mining, and the other was drilled into a face that had been exposed 180 days. This research is reported by Kissell and Edwards (1975). The example in figure 6 shows the observed gas pressure gradient for each of the two boreholes. Also shown on the figure are model simulations for corresponding times using the default parameters provided by the model code for the Pocahontas No. 3 coalbed base case. The calculated gas pressure gradients slightly underestimate the observed values for the 15-day period. The observed maximum is about 1.03 MPa (150 psi) less than the calculated maximum of about 5.52 MPa (800 psi). The default depth used is 549 m (1800 ft). The depth of the in-mine experiment was about 610 m (2000 ft). Comparison of the calculated and observed values at the 180-day time point is not particularly good. Permeabilities were calculated from the fluid flow data of the boreholes. When these calculated in situ permeabilities are substituted for

the default permeability, figure 7 shows that the 180-day curve matching is greatly improved, while the 15-day curve shows a model overestimate. The default permeability for this coalbed is 5 md. The borehole in the face that had been exposed 15 days had a calculated permeability of 0.57 md, while the other borehole exposed for 180 days had a calculated permeability of 280 md. These in situ permeability values were used for their respective boreholes in the simulation, while all other parameters retained their default values. For this example, using a permeability value within an order of magnitude of an in situ determination yielded relatively good history matches for gas pressure gradients. These examples required from 14.4 to 36.0 CPU seconds to run. Obviously, much more comprehensive history-matching and model exercises are required to validate the model output.

Model exercises and comparisons with the results of in-mine emission studies for three different coal mines representing three of the base cases are to be reported in the Final Report of the contract. This report will be available through the Bureau of Mines and the NTIS.

The examples presented thus far have been in terms of gas pressure gradients in the coal seam in advance of the working face. Of more practical interest to the mining industry is methane emission at a working face, so that the ventilation demand for the face can be anticipated and designed for. This "bottom line" output is presented graphically in figure 8, where the volume of daily methane emission is shown as a function of mining activity. For the first 30 days, the working face is inactive. Then mining begins at a constant rate of about 517 tonnes per day (509 tons per day) for 40 days. Mining in this continuous mining section then ceases, and emissions are modeled for 50 more days. The coalbed used for this example is the Pittsburgh, and default coal parameter values were used. The user input for this example consisted of specifying the time periods for mining activity (30-70 days), the mining rate of 30.5 m per day (100 ft per day), and the mine entry width of 6.1 m (20 ft). This simulation took 57.7 CPU seconds to run. Output options provide summary statistics on ventilation requirements as well as information on the state of various physical parameters such as fluid saturation, gas pressure, and sorbed gas content as functions of space and time.

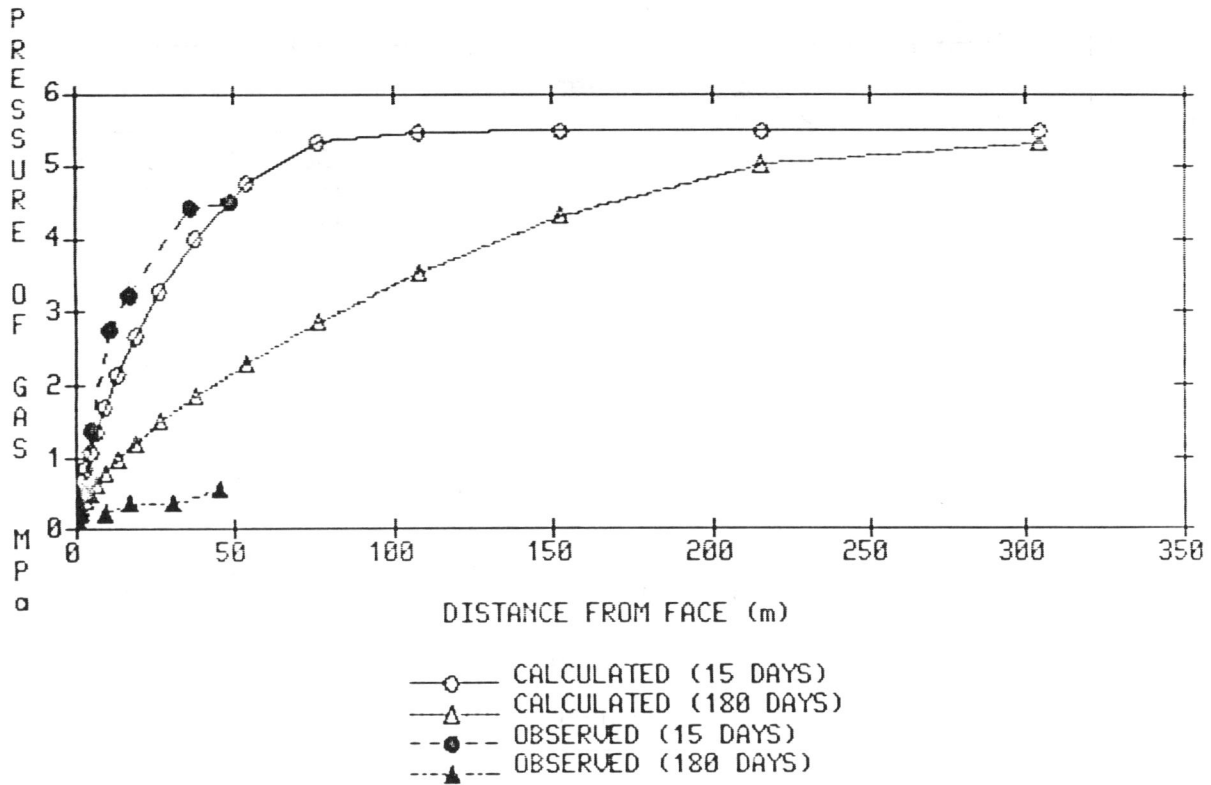

FIGURE 6. - History-Matching Exercise Using Default Permeability Value

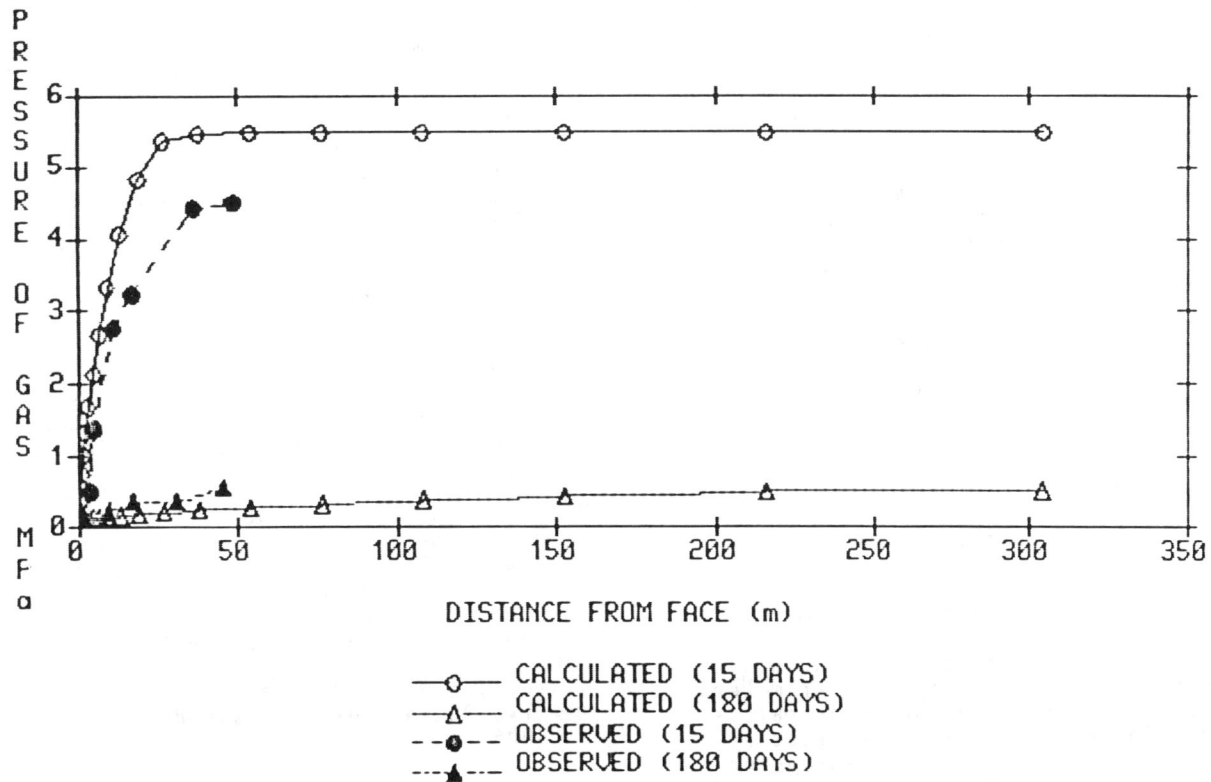

FIGURE 7. - History-Matching Exercise Using Measured Permeability Values

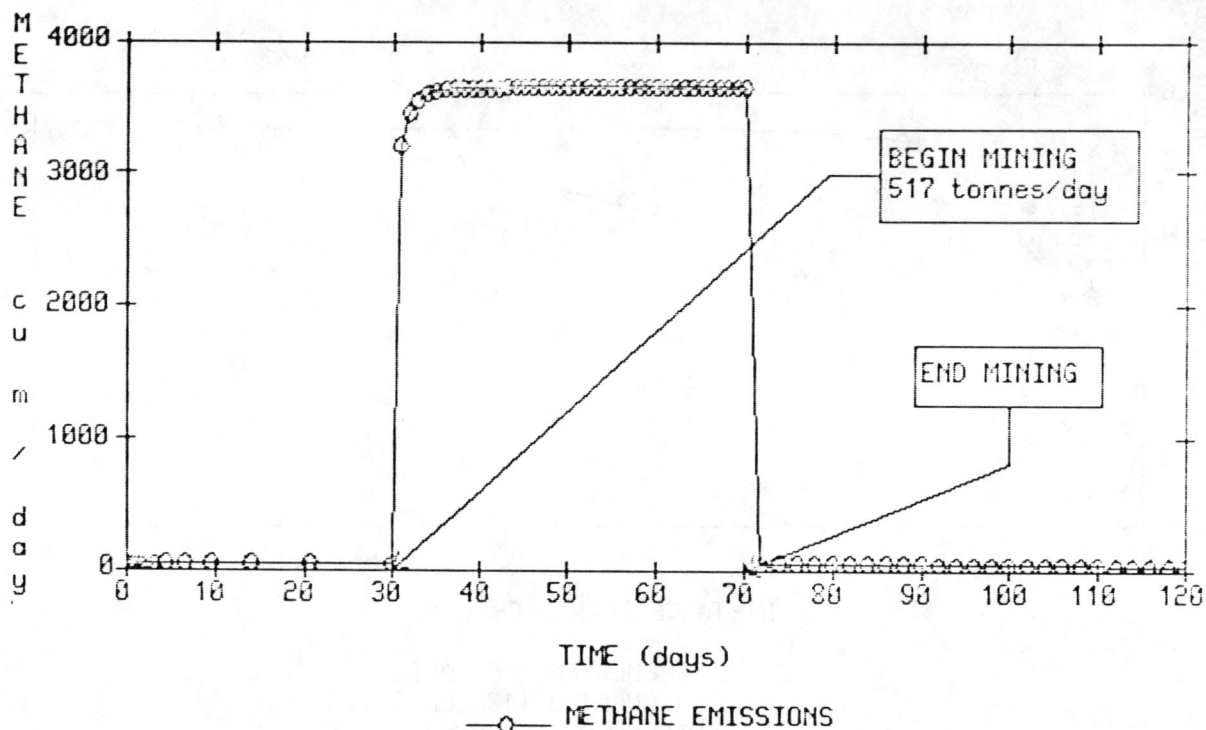

FIGURE 8. - Mining Simulation for Pittsburgh Seam

SUMMARY AND CONCLUSIONS

A computer-implemented coalbed methane model is available through Government-sponsored contract research performed by the US Steel Technical Center. The model simulates simultaneous gas and water flow through a coalbed in two spatial dimensions with an optional vertical equilibrium effect to simulate flow in a third dimension. From a user point of view, a very positive aspect of the code is that it contains default values for each input parameter. Default values are specified for four base case coalbeds:

Pittsburgh

Pocahontas No. 3

Blue Creek (lower bench of the Mary Lee)

Mary Lee (upper bench of the Mary Lee)

The model is most sensitive to absolute permeability, coal desorption parameters, and reservoir pressure, which is calculated from depth. Preliminary history-matching exercises show good agreement between model output and in-mine observations. The computer code is written in a structured FORTRAN dialect and is extensively documented internally through liberal comments. To date, the code has been successfully run on a CDC Cyber 750. The code required about 1 minute to execute a basic mining simulation. While further validation and field testing are required, the use and evaluation of this methane model by the mining industry are encouraged.

REFERENCES

Boyer, C. M., et al., 1982, "Methane Modeling--Predicting the Inflow of Methane Gas Into Coal Mines, Phase 1 One-Dimensional Computer Models and Technical Literature Review," Report No. DOE/PC/30123-T4.

Boyer, C. M., et al, 1983, "Methane Modeling--Predicting the Inflow of Methane Gas Into Coal Mines, Quarterly Technical Progress Report for the Period October 1, 1982-December 31, 1982." (Available through NTIS)

Kissell, F N., and Edwards, J. C., 1975, "Two Phase Flow in Coalbeds," USBM RI 8006.

Price, H. S., et al., 1973, "A Computer
    Model Study of Methane Migration in
    Coalbeds", Presented at CIM Ann.
    Western Meeting., Saskatoon, Sask.,
    October 1972, The Canadian Mining and
    Metallurgical Bulletin, September
    1973, pp. 103-112.

Ramani, R. V., and Owili-eger, A. S. C.,
    1976, Methane Generator, Vol 3 of "A
    Master Environmental Control and Mine
    System Design Simulation for Under-
    ground Coal Mines," USBM OFR
    84(3)-76.

Smelser, R. E., Richmond, O., Schwerer,
    F. C., 1984, "Interaction of Compac-
    tion Near Mine Openings and Drainage
    of Pore Fluids From Coal Seams,"
    International Journal of Rock
    Mechanics, Mineral Science and
    Geomechanics Abstracts, v. 21, pp.
    13-20.

Chapter 48

The Newton Method of Calculating the Ventilation Network

by

Wang Yizhang
Gui Zhou Engineering Institute
Guiyang, Guizhou
The People's Republic of China

Li Shuhe
Gui Zhou Engineering Institute
Guiyang, Guizhou
The People's Republic of China

Introduction

Many scholars have studied the methods of calculating ventilation networks for some time, and have already produced many methods. Among them, the Hardy Cross method was a principal method. Its use has been widespread thus far. In this paper, we apply a graph to get a mathematical model of this problem, and obtain the solution by the Newton Method.

Problem Formulation

Any ventilation system may constitute a graph, $G=(V.E)$ V is a set of nodes, $|V| = m$ (m is the number of nodes). E is a set of branches, $|E| = n$ (n is the number of branches). Then the number of independent circuits is $b=n-m+1$. Suppose $Q_y=(q_1, q_2,\ldots\ldots q_b)$ is air quantity vector of independent circuit. According to Kirchhoff's law, we get a system of equations:

$$\sum_{\substack{j=1 \\ j\neq i}}^{b} r_{ij}(q_i-q_j)^2 + r_{ii}q_i^2 = p_i \qquad (1)$$

$$i=1, 2,\ldots\ldots b$$

Here, r_{ij} is the resistance of common branch between the ith circuit and the jth circuit. r_{ii} is the sum of resistance of branches contained only in the ith circuit. p_i is the sum of fan pressure and natural air pressure contained in ith circuit. q_i is the air quantity in the ith circuit. The subscripts refer to the circuit number.

Let $f_i(q_1, q_2,\ldots\ldots q_b)= \sum_{\substack{j=1 \\ j\neq i}}^{b} r_{ij}(q_i-q_j)^2 +$

$$r_{ii}q_i^2 - p_i \qquad (2)$$

$$i=1, 2,\ldots\ldots b$$

Equation (1) may be written as follows

$$f_1(q_1, q_2,\ldots\ldots q_b)=0$$

$$f_2(q_1, q_2,\ldots\ldots q_b)=0$$

$$\ldots\ldots\ldots\ldots\ldots \qquad (3)$$

$$f_b(q_1, q_2,\ldots\ldots q_b)=0$$

Hardy Cross solved the system of equation (3) with iterative computation method:

Let $Q_y^k =(q_1^k, q_2^k,\ldots\ldots q_b^k)$ be kth iterative solution. According to Taylor's expansion and neglecting higher order part, we obtain

$$f_i(q_1^{k+1}, q_2^{k+1},\ldots\ldots q_b^{k+1})=f_i(q_1^k, q_2^k \ldots\ldots$$

$$q_b^k)+ \frac{\partial f_i}{\partial q_1}\Big|_{Q_y=Q_y^k} \Delta q_1^k + \frac{\partial f_i}{\partial q_2}\Big|_{Q_y=Q_y^k} \Delta q_2^k +$$

$$\ldots\ldots + \frac{\partial f_i}{\partial q_b}\Big|_{Q_y=Q_y^k} \Delta q_b^k=0 \qquad (4)$$

$$i=1, 2,\ldots\ldots b$$

When $\dfrac{\partial f_i}{\partial q_i} \cdot \Delta q_i^k > \sum_{\substack{j=1 \\ j\neq i}}^{b} \dfrac{\partial f_i}{\partial q_j} \cdot \Delta q_j^k \qquad (5)$

$$i=1, 2,\ldots\ldots b$$

Simplifying, Equation (4) becomes

$$\frac{\partial f_i}{\partial q_i}\cdot \Delta q_i^k = -f_i(q_1^k, q_2^k\ldots\ldots q_b^k) \qquad (6)$$

$$i=1, 2\ldots\ldots b$$

By substituting function (3) for f_i of Equation (6), and considering the correction of the direction of air flow we obtain:

$$\triangle q_i^k = -\frac{\sum\limits_{\substack{j=1\\j\neq i}}^{b} r_{ij}(q_i^k-q_j^k)\left| q_i^k-q_j^k\right| +r_{ii}q_i^k\left| q_i^k\right| -p_i}{2\left(\sum\limits_{\substack{j=1\\j\neq i}}^{b} r_{ij}\left| q_i^k-q_j^k\right| +r_{ii}\left| q_i^k\right| \right)-\left.\dfrac{dp_i}{dq_i}\right|_{q_i=q_i^k}}$$

$$i=1, 2\ldots\ldots b \qquad (7)$$

(q_i-q_j) is air quantity of common branch between circuit i and circuit j. So we can write out results:

$$\triangle q_i^k = -\frac{\sum\limits_{s\epsilon c_i} r_s q_s^k\left| q_s^k\right| -p_i}{2\sum\limits_{s\epsilon c_i} r_s\left| q_s^k\right| -\left.\dfrac{dp_i}{dq_i}\right|_{q_i=q_i^k}} \qquad (8)$$

$$i=1, 2\ldots\ldots b$$

C_i is a set of branches belonging to circuit i. Then we can get (k+1)th iterative solution:

$$q_i^{k+1}=q_i^k+\triangle q_i^k \qquad i=1, 2\ldots\ldots b$$

$$\text{Until} \quad \max_{1\leq l\leq b}\left\{\left| q_i^k\right| \right\} < \varepsilon$$

From the above-mentioned analysis, we find conditions 5) very important to Hardy Cross's method, but they are untenable.

The Newton Method

Suppose C_f is an independent circuit matrix

$$C_f=(c_{ij})_{b\times n}$$

R_d is a resistance diagonal matrix

$$R_d=\begin{bmatrix} r_1 & & & & \\ & r_2 & & 0 & \\ & & \ddots & & \\ & 0 & & \ddots & \\ & & & & r_n \end{bmatrix}$$

We can write out resistance matrix about independent circuit:

$$R=(r_{ij})_{b\times n}=C_f \cdot R_d$$

$$\text{Let} \quad Q_y=(q_1, q_2\ldots\ldots q_b)$$

q_i is the air quantity of independent circuit c_i. So we can say Q_y is the air quantity vector of independent circuit. In fact, q_i can be expressed by air quantity (including direction) of chord in the independent circuit c_i. According to Kirchhoff's law, we write out the system of equations of independent cut:

$$Q_N=Q_y \cdot C_f \qquad (9)$$

Q_N is air quantity vector of set E

$$\text{Let} \quad P=P_F+P_H+P_Z$$

$P_F=(p_{f1}, p_{f2}\ldots\ldots p_{fn})$ is the fan pressure vector of set E, if there is a fan j in branch j, its characteristic of air pressure is $p_{fj}=a_jq_j^2+b_jq_j+c_j$, if there is no fan in branch j, then $p_{fj}=0$. $P_Z=(p_{z1}, p_{z2}\ldots\ldots p_{zn})$ is the natural air pressure vector of set E. $P_H=(p_{h1}, p_{h2}\ldots\ldots p_{hn})$ is the fire air pressure vector of set E. $P_{zj}=0$, when there is no natural air pressure in branch j. P_{hj} is similar to p_{zj}. So we can say $P=(p_1, p_2\ldots\ldots p_n)$ is the pressure vector of set E. Applying Kirchhoff's law, we have

$$\sum_{j=1}^{n} r_{ij}q_j\left| q_j\right| = \sum_{j=1}^{n} c_{ij}p_j \qquad (10)$$

$$i=1, 2\ldots\ldots b$$

Applying Equation (9), the system of equation (10) is nonlinear, about (n-m+1) independent variables $(q_1, q_2\ldots\ldots q_b)$ and (n-m+1) independent equations. So the problem changes into solving the system of equation (10).

$$\text{Let} \quad f_i(q_1, q_2\ldots\ldots q_b)= \sum_{j=1}^{n} r_{ij}\left| q_j\right| q_j - \sum_{j=1}^{n} c_{ij}p_j$$

$$i=1, 2\ldots\ldots b$$

Applying Newton Method we have

$$\begin{bmatrix} \dfrac{\partial f_1}{\partial q_1} & \dfrac{\partial f_1}{\partial q_2} & \cdots & \dfrac{\partial f_1}{\partial q_b} \\[2mm] \dfrac{\partial f_2}{\partial q_1} & \dfrac{\partial f_2}{\partial q_2} & \cdots & \dfrac{\partial f_2}{\partial q_b} \\[2mm] \cdots\cdots\cdots\cdots \\[2mm] \dfrac{\partial f_b}{\partial q_1} & \dfrac{\partial f_b}{\partial q_2} & \cdots & \dfrac{\partial f_b}{\partial q_b} \end{bmatrix} \cdot \begin{bmatrix} \Delta q_1^k \\[2mm] \Delta q_2^k \\ \vdots \\ \Delta q_b^k \end{bmatrix} = - \begin{bmatrix} f_1 \\[2mm] f_2 \\ \vdots \\ f_b \end{bmatrix} \quad (11)$$

Suppose $Q_y^k=(q_1^k, q_2^k \cdots q_b^k)$ is Kth iteration solution of equation (10). Then $Q_N^k = Q_y^k \cdot C_f$.

The system of equation (11) is a linear equation about vector $\Delta Q_y^k=(\Delta q_1^k, \Delta q_2^k \cdots \Delta q_b^k)$. Here:

$$f_i \Big|_{Q_N^k} = \sum_{j=1}^{n} r_{ij} q_j^k |q_j^k| - \sum_{j=1}^{n} c_{ij} p_j \quad (12)$$

$$i=1, 2, 3, \ldots\ldots b$$

When $i \neq j$
$$\frac{\partial f_i}{\partial q_j}\Big|_{Q_N^k} = \sum_{s=1}^{n} c_{js} r_{is} |q_s^k| \quad (13)$$

When $i=j$
$$\frac{\partial f_i}{\partial q_i}\Big|_{Q_N^k} = \sum_{s=1}^{n} |r_{is}| \cdot |q_s^k| -$$

$$c_{ii}(2A_i q_i + B_i) \quad (14)$$

$$j, i = 1, 2 \ldots\ldots b$$

Note $\dfrac{dp_{fi}}{dq_i} = 0$ when the air pressure of fan i is constant. Solving (11) we get

$$\triangle Q_y^k=(\triangle q_1^k, \triangle q_2^k \cdots \triangle q_b^k)$$

So (k+1)th iterative solution is

$$Q_y^{k+1}=Q_y^k + \triangle Q_y^k$$

When $\max_i \left\{ |\triangle q_i^k| \right\} < \varepsilon$ calculation converges, Q_y^{k+1} is the solution, else calculation will be iterative again.

Algorithm of Newton Method

We present an algorithm according to the above mathematical model. Algorithm N is called the algorithm of Newton Method.

Algorithm N

N1: [establish matrix of network] establish matrixes C_f, R;

N2: [give initial value] Let k=1, give initial value, $Q_y^1=(q_1^1, q_2^1 \cdots q_b^1)$;

N3: [calculate Q_N^k] $Q_N^k = Q_y^k \cdot C_f$;

N4: [calculate coefficient] calculate from 13) or 14);

N5: [calculate free term] calculate from 12);

N6: [solve equations 11)] find solutions $\triangle Q_y^k=(\triangle q_1^k, \triangle q_2^k \cdots \triangle q_b^k)$;

N7: [calculate Q_y^{k+1}] (k+1)th iterative solution $Q_y^{k+1}=Q_y^k + \triangle Q_y^k$;

N8: [distinguish] if $\max_i \left\{ |\triangle q_i^k| \right\} < \varepsilon$ go to N10;

N9: [next iteration] Let k=k+1 go to N3;

N10: [find solutions] $Q_N = Q_y^{k+1} \cdot C_f$;

N11: [stop] finish calculating.

A Numerical Example

Our computational experience is discussed in this section with an example problem for the network of Fig. 1 and data in Tables 1 and 2. The problem consists of 19 branches, 11 nodes and 2 fans. We have b=n-m+1=19-11+1=9 independent circuits and independent air quantities q_1, $q_2 \cdots q_9$.

We solved the problem, choosing different spanning trees and using Newton Method and Hardy Cross's method respectively. Results are shown in Tables 3, 4, and 5.

Conclusion

In this paper our theoretical analysis and sample calculations proved that Hardy Cross's method of calculating the mine ventilation network is not accurate enough, and its convergence rate and accuracy is affected greatly by the choosing of the "spanning tree." But the Newton Method does not have such problems. The example indicated that the mathematical model which we present is correct. In this paper, we present an algorithm about the Newton Method of calculating the mine ventilation network, that has been realized by computer programming.

Table.1 Data for the Example problem

Branch i	Resistance $r_i (kg \cdot s^2/m^8)$	Branch i	Resistance $r_i (kg \cdot s^2/m^8)$
1	0.168012409	11	0.0502
2	0.287305363	12	0.0103
3	0.355303599	13	0.0108
4	0.00825	14	0.1716
5	0.0058	15	0.0526
6	0.0232	16	0.0383
7	0.0528	17	0.0081
8	0.00338	18	0.025
9	0.2588	19	0.0395
10	0.00589		

Table.2 Data of Fans for the Example problem

Branch i	TYPe	Fan Ratio of speed (r.p.m)	angle of vane
4	70B2-21-NO18	1000	30
5	70B2-21-NO24	750	25

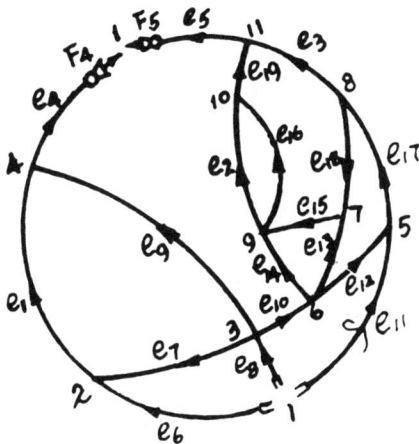

Figure .1

Table.3 Calculating Results of Newton Method for Example problem
Maximal Spanning Tree

$$T_2 = (e_9, e_1, e_6, e_2, e_{11}, e_{12}, e_{14}, e_{19}, e_3, e_{18})$$

Branch i	Quantity $q_i (m^3/s)$	Branch i	Quantity $q_i (m^3/s)$
1	31.200075	11	25.925013
2	13.600114	12	3.3507847
3	24.199966	13	26.704892
4	56.355296	14	18.827643
5	74.808333	15	31.780724
6	29.186874	16	37.008253
7	2.0132010	17	29.275798
8	76.051741	18	5.0758316
9	25.155220	19	50.608367
10	48.883320		

Table.4 Calculating Results of Newton Method for Example problem
Minimal Spanning Tree

$$T_1 = (e_8, e_{10}, e_{12}, e_{13}, e_6, e_{18}, e_{16}, e_{19}, e_{15}, e_1)$$

Branch i	Quantity $q_i (m^3/s)$	Branch i	Quantity $q_i (m^3/s)$
1	31.200075	11	25.925013
2	13.600114	12	3.3507847
3	24.199966	13	26.704892
4	56.355296	14	18.827643
5	74.808333	15	31.780724
6	29.186874	16	37.008253
7	2.0132010	17	29.275798
8	76.051741	18	5.0758316
9	25.155220	19	50.608367
10	48.883320		

Table 5. Rate of Convergence for the Example problem

precision	No. of Iterations
	minimal spanning tree T_1
	Newton method Hardy Cross method
10^{-5}	5 10
	maximal spanning tree T_2
	Newton method Hardy Cross method
10^{-5}	6 248

References

1. Cress, H., 1936, "Analysis of Flow in Network of Conduits or Conductors," Bull. 286, Engineering Experiment Station, University of Illinois.
2. Scott, D.R. and Hinsley, F.B., 1951, "Ventilation Network Theory," Colliery Engineering, Vol. 28, 67, 159, 229, 497.
3. Harary, F., Graph Theory, 1969.
4、平松良雄，日本矿业会志 69, 15(1953)
5、《煤矿通风与安全》编写组．煤矿通风与安全．煤炭工业出版社， 1979
6、清华大学、北京大学《计算方法》编写组．计算方法，科学出版社， 1975

Chapter 49

Mine Ventilation Network Theory

by

Li Shuhe

Guizhow Institute of Technology
Guiyang, Guizhow
The People's Republic of China

Wang Yizhang
Guizhow Institute of Technology

Guiyang, Guizhow

The People's Republic of China

Any mine ventilation system may constitute a graph $G=(V,E)$. V is a set of nodes ($|V|=m$), E is a set of branches ($|E|=n$). It's called a mine ventilation network $N=(G,f)$, when each branch is given power by parameters of ventilation, such as air quantity q_j or pressure loss h_j or resistance r_j etc. In $N=(G,f)$ f is called power function. The subscripts refer to the branch number. In addition, the branch whose resistance between inlet node and outlet node is zero, is provided to "close" the ventilation network, so we can think of a mine ventilation network $N=(G,f)$ as a strongly connected directed graph.

According to the graph theory, $m-1$ nodes must be independent of each other among m nodes, and $b=n-m+1$ circuits must be independent of each other in graph G. We can choose a positive direction for each circuit arbitrarily. Arranging all branches and nodes in order, we can express a mine ventilation network with matrixes, the jth column of the matrixes refers to the branch e_j, the ith row of the matrixes refers to the node v_i or circuit c_i.

1. Basic incidence matrix (reference node is v_k)

$$B_k=(b_{ij})_{(m-1)xn}$$

where $b_{ij}=1$ when $e_j=(v_i,v_e)$ v_i is the starting point of branch e_j

$b_{ij}=-1$ when $e_j=(v_e,v_i)$ v_i is terminal point of branch e_j

$b_{ij}=0$ when e_j and v_i are not incident.

2. Independent circuit matrix
$$C_f=(c_{ij})_{(n-m+1)xn}$$

among $c_{ij}=1$ when $e_j \in c_i$ and their directions are the same

$c_{ij}=-1$ when $c_j \subset c_i$ and their directions are opposite

$c_{ij}=0$ when $e_j \bar{\in} c_i$.

We choose a spanning tree of the network according to fixed demand. The spanning tree contains $m-1$ branches, called tree branches; the other $b=n-m+1$ branches are called chords. We know there is only a chord for any one of the independent circuits in the graph G, and the chords between different circuits are different. So we can define the direction and air quantity of an independent circuit with the air quantity and the direction of the chord which is contained in the circuit; the air quantity of the chord is called circuit air quantity.

Let us arrange an order about all branches like this: branches 1 to b are contained in the chord set, branches $b+1$ to n are contained in the tree branch set. Then B_k and C_f become form of block matrix:

$$B_k=(B_{11}, B_{12});$$

$$C_f=(C_{11}, C_{12})=(I, C_{12});$$

B_{11} is $(m-1)x(n-m+1)$;

B_{12} is $(m-1)x(m-1)$;

C_{11} is a unit matrix of $n-m+1$;

C_{12} is $(n-m+1)x(m-1)$.

We can calculate C_f from B_k

$$C_f=(I, -B_{11}^{-1}(B_{12}^{-1})^T)$$

The parameters of ventilation are expressed in vectors:

Pressure losses vector $H=(h_1,h_2,\ldots h_n)$

$$h_j=r_j \cdot q_j^2$$

Air quantity vector $Q_N=(q_1,q_2,\ldots q_n)$

Resistance vector $R=(r_1,r_2,\ldots r_n)$

Air quantity vector about independent circuit $Q_y=(q_{y1}, q_{y2}\ldots q_{y(n-m+1)})$

fan pressure vector $P_F=(p_{f1},p_{f2}\ldots p_{fn})$

p_{fj} is fan pressure when there is a fan in the branch e_j, if not $p_{fj}=0$.

Natural air pressure vector $P_Z=(p_{z1},p_{z2}\ldots p_{zn})$ p_{zj} is natural air pressure when there is natural air pressure in the branch e_j, if not $p_{zj}=0$.

fire air pressure vector $P_H=(p_{h1},p_{h2}\ldots p_{hn})$

p_{hj} is fire air pressure in the branch e_j when there is a fire in the mine, if not, when there isn't any fire air pressure in the branch e_j, $p_{hj}=0$.

Let $P=P_F+P_H+P_Z$

We get two matrix equations:
   Matrix equation of independent circuit:
       $C_f(H^T-P^T)=0$
   Matrix equation of independent cut:
       $Q_N=Q_y \cdot C_f$

The above two equations are fundamental rules of air flow in the ventilation network. They provide scientific basis for analyzing ventilation state, controlling air flow, analyzing and adjusting working state of fans, and selecting fans.

According to the two equations we come to two conclusions:

A. In a mine ventilation network, air quantity of each tree branch can be shown linearly by air quantity of chords.

B. Necessary adjusting appliance quantity for efficiently controlling air quantity assignment is b (b=n-m+1); these appliances must be located on chords respectively.

According to the above-mentioned conclusion

we can use computers to perform all the calculations necessary for a mine ventilation network.

Calculation About Mine Fan

For the characteristics of mine fan when rotational speed or angle of vane is definite, we can write out the expressions:
Characteristic of air pressure

$$p_f=Aq_f^2+Bq_f+c$$

Characteristic of efficiency

$$\eta_f= a_0+a_1q_f+a_2q_f^2 +a_3q_f^3$$

Characteristic of axis power

$$N_f= \frac{p_f}{102. \, \eta_f} \, q_f$$

In the expressions, p_f, η_f, N_f are working pressure, efficiency, and axis power when the working air quantity of the mine fan is q_f.

Coefficients A, B, C and a_0, a_1, a_2, a_3 are calculated by the method of least squares.

When the rotational speed of fan changes from ω_1 to ω_2 ratio of speed is $t= \omega_2/\omega_1$, then every characteristic coefficient would change respectively.

$A_2=A_1,$ $B_2=t \cdot B_1,$ $C_2=t^2 \cdot C_1$

$a_{02}=a_{01},$ $a_{12}=a_{11}/t,$ $a_{22}=a_{21}/t^2, a_{32}=a_{31}/t^3.$

We can do the following calculations:

1. If we know that total resistance of the subnetwork of the fth fan working area is R_c, and rotational speed of the fan is ω_1, coefficients are A_1, B_1, C_1, a_{01}, a_{11}, a_{21}, a_{31} then we have an equation:

$$(A_1-R_c) \cdot q_f^2+B_1 \cdot q_f+C_1=0$$

This quadratic equation has two solutions q_{fa} and q_{fb}. We then obtain working air quantity of the fan as follows:

$$q_f=\max (q_{fa}, q_{fb})$$
and we can calculate other working parameters in the network.

2. If we don't know the total resistance, then we have to calculate the working air quantity of the fan by the mathematical method of ventilation network.

But we should say: In calculating independent circuit equations, a term of fan air pressure is expressed by an equation characteristic of air pressure:

$$P_{fj}=A_j q_{fj}^2+B_j q_{fj}+C_j$$

This quadratic equation has two solutions q_{f1} and q_{f2}. In fact, we would express the fan air pressure characteristic as follows for calculating air quantity convergence to the correct direction:

$$P_f=\begin{cases} Aq_f^2+Bq_f+C & \text{when } q_f \geqq q_{f1} \\ \\ P_{f1} & \text{when } q_f < q_{f1} \end{cases}$$

P_{f1} and q_{f1} are bounds of pressure and air quantity when the fan comes into unstable running area or working efficiency less than 0.6.

3. When the working air quantity of fan is not satisfied with productive needs, the rotational speed of the fan should be calculated and adjusted according to ratio of air quantity with proportion law, and considering network controlling, we find an optimum scheme.

Ventilation Network State Analysis

Having had two matrix equations about independent circuit and independent cut, we can write out the following system of equations:

$$\sum_{j=1}^{n} c_{ij} r_j |q_j| q_j = \sum_{j=1}^{n} c_{ij} p_j$$

$$i=1,2,\ldots n-m+1$$

$$q_j = \sum_{i=1}^{b} q_{yi} c_{ij} \qquad j=1,2\ldots\ldots n$$

Solving the system of equations with the Newton Method, we can obtain some results about ventilation state; for example, direction of air flow for each branch, air quantity for each branch, and working state for each fan.

Control of the Air Flow in the Mine Ventilation Network

If we choose an optimum spanning tree for controlling air flow in the mine ventilation network, then the optimum combination distribution of adjusting appliance is determined. A

further step is to choose an adjusting method and decide adjusting quantity through calculation. It is obvious that some branches need fixed air quantity in the ventilation network, but more branches do not need it. We define a subnetwork of natural ventilation as the subnetwork that contains all these circuits where every branch does not need fixed air quantity. According to the condition of fixed air quantity, solving the subnetwork with Newton Method we get air quantity in each branch of the subnetwork, and ventilation pressure losses are out of balance in each circuit of the subnetwork. Through calculating the subnetwork of natural ventilation, we can find pressure loss in each branch of network when the air quantity is fixed. We can get a set H_y whose element is pressure loss of chords and a set H_T whose element is pressure loss of tree branches, then we have pressure loss vector:

$$H=(H_y, H_T)$$

In independent circuit equations, we write out $P_c=C_f \cdot P^T$, then

$$P_{ci}=\sum_{j=1}^{n} c_{ij} p_j$$

$$i=1,2,\ldots.(n-m+1)$$

P_{ci} expresses total ventilation energy in the independent circuit c_i. So we have

$$C_f \cdot H^T=P_c$$

$$(I, C_{12})\begin{pmatrix} H_y^T \\ \\ H_T^T \end{pmatrix}=P_c$$

Because $h_j=r_j|q_j|q_j$ h_j is an element of vector H, q_j is the necessary air quantity of branch j, and the equation is out of balance. An increment must be added, Δh_{yi} (i=1,2.....(n-m+1)) to h_{yi} to place the equation in balance.

Let vector $\Delta H_y=(\Delta h_{y1}, \Delta h_{y2}\cdots \Delta h_{y(n-m+1)})$

ΔH_y is called an increment vector about pressure loss of chords, then we can write out the independent circuit equation as:

$$(I, C_{12})\begin{pmatrix} H_y^T + & H_y^T \\ & H_T^T \end{pmatrix} = P_c$$

Solving the equation, find and discuss the solution

$$\Delta H_y^T = P_c - (H_y^T + C_{12}H_T^T)$$

We can get results:

a. When $\Delta h_{yi} > 0$, the chord e_i needs adjusting appliance for increasing pressure loss.

b. When $\Delta h_{yi} < 0$, the chord e_i needs adjusting appliance for decreasing pressure loss.

c. When $\Delta h_{yi} = 0$, the chord e_i does not need any adjusting appliance.

Though a fan is a power in ventilation network, it is also an adjusting appliance, because it is adjusted immediately according to the ventilation state. So we would choose a branch which contains a fan as chord. In addition, because the increment of pressure loss is zero in an independent circuit contained by a subnetwork of natural ventilation, the chords do not need an adjusting appliance contained by above independent circuit. So if there are f fans in a ventilation network, and there is one branch contained by the subnetwork of natural ven-

tilation, then we have an optimal number of necessary adjusting appliances for controlling the air flow in the mine ventilation network:

$$S = n - m + 1 - f - 1$$

The results of this paper are also applicable in case of mine fire, roof falling. The algorithm has been solved through computer programming.

References

1. Cress, H., 1936, "Analysis of Flow in Network of Conduits or Conductors," Bull. 286, Engineering Experiment Station, University of Illinois.
2. Scott, D.R. and Hinsley, F.B., 1951, "Ventilation Network Theory," Colliery Engineering, Vol. 28, 67, 159, 229, 497.
3. Harary, F., Graph Theory, 1969.
4、平松良雄，日本矿业会志 69, 15(1953)
5、《煤矿通风与安全》编写组。煤矿通风与安全。煤炭工业出版社， 1979
6、北京大学、吉林大学、南京大学合编，计算方法 人民教育出版社。1961
7、李慰萱 图论 湖南科学技术出版社。1980

SPECIAL APPLICATIONS/SUPPORT FUNCTIONS

Chapter 50

ENVIRONMENTAL LABORATORY DATA MANAGEMENT SYSTEM

Jane Koch

Peabody Coal Company

ABSTRACT

The Environmental Lab Management System is an easily accessible data base system which performs two major functions. One function is to handle the laboratory's needs for data storage, sample batching and tracking, calculations, and automatic or requested reports on the data. The second function is to handle the field users' needs with respect to data retrieval methods and access of quality results, which are interfaced with existing, related systems and software.

The Environmental System is available company wide through terminals at division offices, labs, mines, and prep plants. It can currently handle data for the qualitative analysis of water, coal, soil, overburden, and vegetation as separate systems, which are integrated with Peabody's Drill Hole System.

* Computer Network

* Software Environment

* Integration with Related Systems

* Lab Management

* Sample Point Processing

* Field User Access

* End User Involvement

INTRODUCTION

Two years ago Peabody Coal Company implemented a computer system to meet the information processing needs of hydrologists and lab technicians involved with the analysis of water samples. This system was a success in fulfilling the requirements of lab and field users. The system has recently been expanded to include coal quality analysis, overburden analysis, soils analysis, and vegetation analysis (for trace elements).

FUNDAMENTAL FACTORS

Six fundamental factors which have contributed to the success of the environmental systems at Peabody Coal Company are:

COMPUTER NETWORK

Peabody currently utilizes two large IBM compatible mainframe computers in the St. Louis Data Center. Both computers are linked to engineering, environmental, geology, design, financial, and other users via an extensive telecommunication network.

The communications network uses IBM's System Network Architecture (SNA) to support micro and mini computers, graphics display terminals, digitizers, engineering plotters, low speed laser printers and approximately seven hundred (700) 3270 terminals. The network services over eighty (80) business, operating and engineering locations throughout the company.

The computer network allows for the interchange of information, data and laboratory analyses between Central Laboratory facilities at Freeburg, Illinois, local laboratory facilities in Kentucky, Ohio and West Virginia, and engineering and environmental users at mine locations.

SOFTWARE ENVIRONMENT

The software environment at Peabody Coal Company as it pertains to the Lab Management System consists of the following:

    o Operating System

    o Application Development Language

    o Data Base Language

    o End User Programming Languages

OPERATING SYSTEM

The operating system used at Peabody for end user computing is VM/CMS. This operating system provides an ideal climate for an end user to use a mainframe computer as though it were his own personal computer. The range of end user functions, from file maintenance thru program initiation, may be presented in a full screen mode that is menu driven and eliminates the need for voluminous end user documentation.

APPLICATION DEVELOPMENT LANGUAGE

The Environmental System has been developed in a programming language called PL/I. PL/I has combined the best of FORTRAN and COBOL into a single language that includes an interactive debug facility which has increased programmer productivity by an order of magnitude. The system consists of over 500,000 executable lines of code.

DATA BASE LANGUAGE

The data base language used for the environmental systems is Cullinet's Integrated Data Base Management System (IDMS). IDMS provides an integrated data dictionary that manages data resident on the data base as well as programming variables. The major strength of IDMS is the ability to describe a network of data relationships as they exist in the real world. Other significant features are the ability to protect data from hardware failures, complete journalizing of all transactions against the data base, roll forward and roll back of transactions.

END USER PROGRAMMING LANGUAGES

Realizing that it is impossible to meet 100% of the end user needs with application programs the environmental systems provide an interface to the SAS Institute's "Statistical Analysis System" (SAS). SAS provides a complete set of statistical functions from simple statistics (mean, standard deviation) through multiple regression and analysis of variance models. SAS also provides report writing capabilities and a full range of graphic displays. The SAS interface has been designed to provide a menu of predefined SAS routines that are accessed by a menu selection or thru extracted data written with a corresponding generic SAS program which then allows the user to write his own SAS applications.

INTEGRATION WITH RELATED SYSTEMS

Integration of the Environmental System with existing related systems (concerning software, data storage, and data access) is of primary importance in the design. The factors involved are:

    o Variety of Functional Users

    o Data Security

    o Elimination of Redundant Data

    o Integration with the Drill Hole System

VARIETY OF FUNCTIONAL USERS

A wide variety of functional users must have access to the data. These functional users include hydrologists, geologists, mining engineers, laboratory personnel and operations personnel. The system had to be flexible enough to meet the needs of this range of users but restrictive enough to preserve the data integrity.

DATA SECURITY

Data security was implemented at two levels. First the user must be explicitly authorized to execute any of the approximately 500 menu functions that comprise the system. For instance the capability to add new mines or change calculation equations is restricted to a very small number of users. Some users are authorized to view the data but cannot make any changes. Users are also authorized to view the data by data type (e.g. field data vs. lab data). There are separate authorization codes for Water Quality, Vegetation, Soils, Overburden or Coal users. Users are also restricted by company, division or laboratory.

ELIMINATION OF REDUNDANT DATA

Elimination of redundant data is of paramount importance in maintaining corporate data. Any drill hole in the system has exactly one set of coordinates. These coordinates apply for engineering, geology, and laboratory personnel as well as for the systems - Drill Hole, Mine Planning, Overburden Analysis, Soil Core Analysis and Coal Quality Analysis. If the identifier for the hole changes (e.g. transferred to another mine number) then the network structure of the data base allows for all data relationships associated with the hole to be preserved.

INTEGRATION WITH THE DRILL HOLE SYSTEM

Integration with the Drill Hole System was also necessary to preserve data integrity and to take advantage of existing programs for graphics representation and volumetric calculations. Integration of the Overburden and Soils Components of the system with the Drill Hole System provides the following capabilities:

- Ad Hoc Tabular Reports

- Contour Maps

- Isopac Maps

- Drill Hole Cross Section Maps

- Volume Calculations

LAB MANAGEMENT

The Lab Management system contains the following functions:

o Sample Creation

o Sample Batching

o Calculations from Observed Values

o Daily Lab Reports

o Committing to End Users

o Cost Accumulation

o Accounting Records

SAMPLE CREATION

After a sample has been collected and sent to the laboratory, it may be created within the Environmental System in one of three ways.

The collector, engineer, hydrologist, or any field user may create the sample by entering the location and observable data which he knows. He does this from his terminal at the mine or division office.

When the lab receives the sample and associated paper work, the sample is logged in by specifying some key information such as location, depth, or time. Subsequently, a lab number is generated. If the sample has not been entered by a field person, the lab person may enter any data which he has on the paper work sent in with the sample.

The third method of sample creation is that of batch input from a file created directly by an outside laboratory or by a field user with either outside lab results or other data which is now being input to the new system.

SAMPLE BATCHING

Sample Batching allows the lab to receive samples from several locations and group similar samples together as a "batch" in order to do the analysis of the same parameters (e.g. Iron, pH, Sulfur, etc.) for 10-20 samples simultaneously. This allows the lab to keep their analysis procedure efficient and cost effective. Even though batch samples are associated, they may still be accessed as individual samples, and the results of the analyses are still stored as unique samples.

CALCULATIONS FROM OBSERVED VALUES

Calculations from observed values from laboratory equipment are often required in order to determine the actual quantity present in the sample for the parameter being analyzed (e.g. dilution facors, concentrations, temperature, volumes of titrants present, etc.).

Using standard notation, the lab user may explicitly define standard and alternate nontrivial algebraic equations for each parameter in each system. The equation may consist of variables, constants, functions and results of other analyzed parameters from the same sample. The equation is stored on the data base and may be updated or modified at any time by users with the proper authorization.

DAILY LAB REPORTS

On a daily basis, up to four lab reports may be produced automatically for each Environmental System, at each laboratory.

LAB LOG REPORT

The Lab Log report provides the lab with a list of all samples which have been logged in for analysis on the previous day. Associated information such as locational or date and time information about the sample, date received, charge numbers, cost for analysis and status of the sample are printed in the report.

LAB LIMIT EXCURSION

The Lab Limit Excursion report provides the lab with the lab numbers and excursion values of any samples for which the analyzed value exceeds a laboratory determined upper limit. The primary purpose for the laboratory limits is to catch any results that appear incorrect and should be re-calculated, re-analyzed, or at least investigated.

LAB PRE-COMMIT

The Lab Pre-Commit report provides the lab with the complete results and all data associated with the sample which has been analyzed but still resides within the control of the lab. The purpose is to give the lab an opportunity to scrutinize and compare results before sending the data to the end user in the field.

LAB COMMIT

The Lab Commit report contains the lab numbers and dates during which the sample was within the control of the lab. It tells the lab which samples were completed and committed during the previous day and are now accessible to the field user.

COMMITTING TO END USERS

After the analysis is complete, and the results have been checked on the daily reports, the results are (selectively) made available to the end user in the field (Most of the results are made available to the end user; however, results prepared for lab purposes only, might be misleading if made available to field users.)

COST ACCUMULATION

Cost accumulation is done for the completed analysis at commit time. The lab user with authorization determines the cost for analyzing each parameter within his system. At commit time, the cost for each parameter analyzed is accumulated for a total cost for the sample.

ACCOUNTING RECORDS

Accounting records are maintained by the system for each parameter and each mine during the lab processing. The costs are associated with charge numbers which refer to various contractual and permitting agencies, and are accumulated in a monthly report by division, mine, and charge number for each lab.

SAMPLE POINT PROCESSING

Sample point processing integrates the flow of information between the laboratory and the field user. The Sample Point Processor provides the following functions:

   o Sample Point Definition & Maintenance

   o Required Parameter List

   o Value Excursion Limits

SAMPLE POINT DEFINITION & MAINTENANCE

Sample point definition & maintenance allows the field user to define individual sample points (e.g. streams, wells, lakes for Water Quality). All data that pertains to the description of that sample point (e.g. coordinates, section, township, storage coefficient, remarks, etc.) are maintained by the user.

REQUIRED PARAMETER LIST

The required parameter list is determined maintained by the field user and determines which of the many parameters shall be analyzed for at the lab. For each mine, and within each system, there are various types of samples which may be analyzed at the lab. The analysis list may vary depending on the type of sample (e.g. a pond sample may have a different list of required parameters from a well or a river sample). Whether the sample is taken before, during, or after the mine's active period, may also determine the required parameter list. The field user may specify which parameters are to be analyzed normally, monthly, quarterly, semi-annually, or annually. The flexibility of these required parameter lists, allows the field user and the lab to know what is to be analyzed for every sample with a minimum of effort.

VALUE EXCURSION LIMITS

Value excursion limits are defined, within each individual mine and sample type, for each parameter to be analyzed. The excursion limits may vary depending on the mine or type of sample they will be analyzed for. The upper limits are used for comparison with each parameter analyzed, and will be reported on a daily basis to the field user as a value excursion in order that corrective measures may be taken and the excursion can be reported to various contractual or permitting agencies.

FIELD USER ACCESS

The field user has access to the results of the analysis when the lab has completed the analysis and committed the results for their viewing. In addition to menu selections of specific reports, inquiries, and SAS functions available to view the data, there are three very nice features at his disposal:

    o Daily Field Reports

    o SAS Graphics

    o Piper Trilinear Diagrams

DAILY FIELD REPORTS

On a daily basis, reports may be automatically generated and sent to the various mine and division sites to keep the user informed as to the complete results of samples which have been committed by the lab, or samples which have been deleted. Value excursion reports may also be produced daily depending on the existance of an excursion. The final value of each analyzed parameter is compared with standards set by regulatory agencies. Whenever a standard limit is exceeded, a value excursion report will be produced and sent to the field users so that corrective measures may be taken and the excursion can be reported immediately.

SAS GRAPHICS

SAS graphics allows the user to graphically represent scatter plots, trend lines, and correlation of multiple values with respect to time. The user may extract the data for statistical analysis by choosing mine number, mine type, major minor types, sample point numbers over a date range. The user may also define both axes in terms of parameters on the data base. Standard graphs may be generated by the user through established menu options. Specialized graphs may be produced by the technician for non-standard applications utilizing their own graphics procedures.

PIPER TRILINEAR DIAGRAMS

The Piper Trilinear diagram allows the user to graphically display the relationships between ions in water which were determined by analysis. The data may be selected by mine, sample point, date and time. The user may also choose which of the cations and anions, on the system, he wishes to compare.

One triangle displays the relationships between the major cations (calcium, magnesium, and sodium + potassium). The percentages of each are derived from the appropriate side of the triangle.

The other triangle displays the relationships between the major anions (bicarbonate, chloride, and sulfate). The percentages of each are derived from the appropriate side of the triangle.

The location of points in the diamond shape region are calculated by determining the intersection of projecting rays from the points in the individual triangles.

END USER INVOLVEMENT

The definition of success for any system is directly proportional to the amount of user involvement when the system enters a production status.

Peabody's Central Laboratory realized an annual savings of over $150,000 utilizing only the Water Quality component of the system. The system has become an integral part of the lab's function and day to day operation.

Field technicians have also integrated the system into the routine of their day to day operations. Using the system, one western Peabody mine has been able to produce approximately 85 percent of the content of quarterly and annual Water Quality reports required by their state regulatory agency. There are similar success stories at other mines. Beyond the direct successes of the field and laboratory users of the system, Peabody anticipates further benefits as the data is now stored in a form that can be easily accessed or reported on in a standard and repeatable manner making the permit reporting process acceptable to state agencies minimizing the burden on those who produce the reports.

Chapter 51

COMPUTER-AIDED PUMP SELECTION FOR MINE DEWATERING

by
Jozsef S. Gozon
Robert R. Britton
Mining Engineering Division
The Ohio State University

and

Theodore R. Myers
CompuServe
Columbus, Ohio

Abstract. The objective of any mining course should be to create a solid background in the use of fundamental applications drawing upon enough theory to afford understanding of methods of solution to mine design problems. Therefore, the purpose of this paper is to address the advantage of using micro-computers in teaching engineering design, application, and equipment selection for mine dewatering systems.

Since wet mines offer the operator a continuous problem, the manner in which this dilemma is handled will impact mining costs. At The Ohio State University, a computer program has been developed which aids the student in solving mine dewatering problems. Pipe sizing, pump selection, medium characteristics, sump dimensioning and optimum location determination of the components are interrelated. To maximize utilization of the system, the flow parameters are generated as part of the output. Also included are the pressure and flow rate conversion, optimum flow determination, pipe fittings, pump characteristics, and the appropriate mathematics.

The methodology incorporates an exemplified demonstration with an instructional paper. This presents the student with a systematic approach to solving complex dewatering problems.

Once the basic problem is solved, various parameters may be re-entered into the program for evaluation and optimization of the system and its cost. It must be remembered, however, the main thrust here is to use a micro-computer to increase teaching and learning efficiency.

PROBLEM ADUMBRATION

Mining engineers are continually challenged by dewatering problems as they effect operations, system design or equipment selection. Unfortunately, operators have either ignored or have not recognized the need for overall mining efficiency by not optimally controlling mine water. Such parameters as anticipated inflows and unexpected large inrushes from water-bearing strata adversely effect productivity, as well as health and safety of the miners. In any case, the total cost to mine a ton of coal is affected, and the absence of a properly sized and reliable dewatering system can cause extended property damage, even flooding of part or all of a mine.

Although this is the extreme, a certain level of knowledge may be acquired through available literary sources [1,2] and outside experts. Technical writings in the area of computer-aided analysis for mine drainage systems [3] has been scarce. Even though there is a major breakthrough on the horizon, pencil-plotted graphs disseminated by manufacturers are the primary mechanisms for determining pump characteristics and pipe resistances. Unless their use can be expanded and dealt with within the complexity of the system, design optimization will be hampered.

The computer and its software afford the new generation of mining engineer a means for system optimization never before available. Therefore, programs are being developed to enhance the Ohio State students' skills in solving mine dewatering design problems.

Some accompanying topics that may serve as a guide for the students analysis is given in the sections that follow. The mentioned equations and discussion aid in explaining some of the items encountered in the program. First suction and discharge pipes, pumping stations and sump design will be discussed. Second, the steps in approaching a solution will be reviewed. A third topic, economic aspects, is present in this program, but it will not be included here except to say that a model has been developed for the F.O.B. prices of the centrifugal pumps and their electric motors. The cost function has the following elements:

- ownership cost and
- operating cost

SUCTION AND DISCHARGE PIPES

The centrifugal pump is very popular in coal mines because of its ability to handle a large quantity of highly contaminated water with suspended particles. The principal features of the centrifugal pump include:

- simple construction
- small floor space requirement,
- easy maintenance,
- low capital cost,
- smooth non-pulsating discharge,
- possible automatic control,
- easy to make corrosion resistant, and
- relatively high suction lift.

There is an inherent weakness in centrifugal pump installations. The unit will cavitate, especially where high suction lift is encountered. Early mine pump stations were designed for high velocities in the suction pipes. Velocities of 3 m/s were common. Also, unnecessary fittings were installed in the suction pipe producing unwanted head losses. New pump station designs call for straight suction lines, tight connections, and air pocket elimination. If a foot valve with strainer is used on the suction side, a convenient arrangement for pump priming is a small-diameter by-pass pipe and valve which is installed between the column pipe and the suction pipe and above the foot valve. Water is drawn from the high-pressure column to fill the suction pipe and the pump.

The discharge pipe of a centrifugal pump is connected to a horizontal main collector pipe of larger diameter. Each pump discharge is provided with a check valve and gate valve. The check valve is necessary to prevent backflow through the pump in the case of power failure. A gate valve is interposed between the header and the column pipe. Two or more shaft columns are necessary to provide a connection to the collector pipe in such a manner that any pump can be shut down for repairs. Shaft columns should have expansion joints placed in the pipe line. Thus, the upper section is carried upon bearers above the expansion unit. The lower section is supported by a tee fitting at the bottom.

Standard and extra-strong steel pipe possessing various coatings and linings can be used for most cases to increase corrosion resistance. Pipe sections are joined by welding, and at certain intervals flanged sections are installed for easy maintenance. Velocities in high-pressure columns should be kept as low as possible. For ordinary pumping, pipe velocities should be less than 2.5 m/s. An increase in velocity would result in a proportional rise in delivery, but this would require more power to overcome non-linearly growing head losses due to friction. All pipe lines must be designed with the fewest possible changes in both direction and cross section.

PUMPING STATIONS

Present practice favors a rectangular pump room in which centrifugal pumps are arranged in a single or double row with short branches to the main line collector. The width of the pump room is designed to provide access to the units and sufficient area for dismantling. The number of units and the area necessary for the electrical control panel determine the length of the station. For automatic operation, the space requirement in the main room can be reduced, but provision should be made for a central control panel. The energy supply requires the installation of an underground sub-station next to the main pumping station, and the ventilation of these openings is of primary concern in coal mines. The maximum power demand determines the heat factor which is inversely proportional to motor efficiency. Large pump stations are best ventilated by an overcast or ventilator tube connecting them to the return air entry. Under certain conditions, fan ventilation may become necessary.

Economic details of pump station design include the desired life of the station, the water flow ranges, the suction lift, the discharge head, the installation costs, and the efficiency of the electric motors. Labor costs are negligible compared to energy costs.

SUMP DESIGN SUMMARY

A sump is classically defined as an excavation in the coal or rock below the gangway or in the bottom of a shaft to collect mine water. It is a catch basin that receives the gravity drainage, pumped water from auxiliary sumps, and supplies the pump inlets. The sump also serves as a settling reservoir to remove solid particles and sludge from the water before it enters the pumps. To keep pump maintenance to a minimum, clean water is necessary. For this reason the design of a sump should provide a sufficient settling capacity with baffles to still and break up currents as much as possible.

Caved gobs were frequently used as a suitable collecting reservoir for mine water drainage. However, the water residing in the gob often caused acid contamination when water contacted the pyritic impurities of waste material. The stringent environmental regulations with regard to acid mine drainage resulted in a cost increase of mine dewatering and lessened the popularity of using the gob as a water collecting area or sump.

New methods of handling mine drainage include the application of plastic piping which replaces open water channels minimizing water contamination by preventing contact with acidic materials present in the mine. Subsequently, the water is pumped out of the sump as quickly as possible. Recent mining legislation has demanded the establishment of treatment ponds on the mine surface so that acid mine water may be neutralized and the suspended solids removed before the mine water can be discharged into streams.

For small water flow rates in relatively dry coal mines, the sump is usually placed below the level of the pump station and in close proximity to it. The suction pipe extends to the pump station having few fittings, such as, foot valve, basket strainer and elbow. Since the suction lift is limited to about 5 meters in order to avoid cavitation in the first pumping stage, the sump depth and its connecting ditch are limited to 4.5 meters. Two other arrangements are advantageous where heavy flows are encountered. In one, the pumps are placed below the level of the bottom of the sump and receive gravity flow through their inlets. A pipe and valve control the flow to each pump. In the other arrangement, the main sump is placed below the pump station and a smaller sump is located above to afford relatively constant pressure flow to each pump.

Sumps are designed to have sufficient capacity to allow a considerable time interval before the flooding of a pump station can take place. Where flooding is anticipated, development openings at pump stations are provided with pressure doors and dams designed with valved outlets. With these arrangements, sudden water inrushes can be controlled and/or shut off. Main sump capacity must be properly sized especially in a wet mine. Early practice in sump design considered at least an 8-hour flow as the determining parameter for storage capacity. In many cases, a 24-hour inflow was taken as a basic requirement. Large sump capacity is advantageous in that pumping can be scheduled at times of off-peak power loads of the mine plant. Under such conditions, pumping time may vary from 8 to 16 hours and the cummulative pumping capacity can be arranged accordingly.

If the bottom of an over-drilled vertical shaft is used as a main sump, the pump system can be of several designs.

Submersible pumps can be hung in the shaft on support beams. There should be a minimum of two pumps, each one capable of handling the water inflow. These pumps usually are connected to a common pump discharge line with check valves so that either pump may be kept in operation.

Positive displacement pumps can be installed on the mining level with a suitable size sump for water storage. It is used in conjunction with a small pump at the shaft bottom which discharges water to the main sump to keep the shaft itself dry. These pumps should be equipped with automatic starting and stopping switches enabling reliable operation to maintain the water depth in the main sump between pre-determined levels. It should be mentioned, that these two systems are not often used in modern coal mining practice.

Where drainage in small wet coal mines is almost continuous with two or more pumps in operation while others are being repaired, the large-scale mines manage a cyclic operation of its main sump pumps. In this situation, a preventive maintenance program for all pump units is a governing factor in the operational sequence. To facilitate removal of deposited solid material and sludge, sump capacity is divided so that one part can be decanted and cleaned, while the other part is in operation, and vice versa.

In the past, the presence of an auxiliary power source was included in the main pump station design criteria. Nowadays, two independent energy distribution systems are available for all mines. Although the power distribution systems are reliable in operation, malfunctions and total power outage of limited duration can be anticipated. The length of such energy supply failures is usually less than one hour. The modern sump design criteria should include allowances for such failures by providing additional sump capacity.

PROGRAM DEVELOPMENT

The first block of the program is designed to eliminate misunderstanding in pressure units and flow rates. Although SI is now the standard system, pressures and discharge rates are still expressed in Imperial units. This is no problem since the program will accept any commonly used units and immediately display the desired measures on the screen. That is, conversions are made internally. The geometric parameters are handled in the same way.

In order to check the completion of related reading assignments, a quiz-like segment can be included. By answering the questions, the student accesses the working sector of the main program. Should a student fail to enter the correct answer, a tutoring session becomes available.

The essential segment of the mine dewatering program contains the following features:

- possibility of entering design parameters through screen data input,
- rapid calculation of flow parameters based on number and type of pipe fittings required,
- accurate computation of pump efficiency and corresponding motor capacity,
- graphical display of pump layout including fittings,
- graphical display of working point zone with an opportunity to analyze changes in speed and/or number of stages,
- choose by analysis, subsystems in series or parallel, and
- assistance in main sump engineering design.

METHODOLOGY

The characteristics that define how a centrifugal pump performs are summarized by relationships between

- head and discharge,
- brake horsepower and discharge, and
- efficiency and discharge.

The point of peak efficiency corresponds to the pump design parameters. Losses occur both from hydraulic and mechanical sources; so that, for a centrifugal pump, only a fraction of the input power produces useful fluid energy. In many cases, the cavitation effect is another limiting factor. If a pump is required to raise water from a sump to the surface, it must generate sufficient head to overcome:

- the height difference between the two water levels and
- the frictional and shock losses in the associated pipe system.

The governing equation is of the form:

$$H = DH + R \times Q^2 \qquad 1)$$

where DH - is the static lift,
    Q - is the flow rate,
    H - is the head, and
    R - is the system resistance.

In coal mine drainage practice, turbulent flow is usually encountered where the friction factor is a function of the Reynolds number and the relative roughness. For commercial pipes, Moody's diagram or his approximate formula for friction factor is widely used. Its accuracy of 5% is totally acceptable for this program.

It is required that the system characteristic and the pump characteristic at any given speed must intersect near the pump design point in order to be operating with reasonable efficiency. The characteristics of a centrifugal pump can be represented by a quadrant of an ellipse [4]:

$$\left[\frac{H}{H(0)}\right]^2 + \left[\frac{Q}{Q(0)}\right]^2 = 1 \qquad 2)$$

where H(0) - is maximum theoretical flow
           head present, and
    Q(0) - is maximum theoretical flow
           rate of a pump at zero head.

A common method for varying head and/or flow rate of centrifugal pumps is by adjusting the impeller speed. Within reasonable limits, it is possible to derive new characteristic curves for a centrifugal pump from a given head versus discharge rate relationship by applying the following pump laws:

- flow rate varies in direct proportion to the speed,
- head varies in proportion to the square of the speed,
- horsepower varies in proportion to the cube of the speed.

An equation for the affin parabolic curve can be included in the program so that parameters for the ellipses will be available for speed adjustment if desired. This principle is illustrated in Fig. 1. Here, the location of the maximum heads with varying speed, is shown for a centrifugal pump.

It is very important that the head and discharge for a centrifugal pump be established by matching the pump to the given pipe columns. Also, the efficiency of the whole system, when at the desired operating point, must be at a maximum or at least reasonably high, and the input power requirements for the pump must be met. The variation of efficiency with discharge for a centrifugal pump can be approximated by:

$$EF = EF(max) - C(ef) \times [Q - Q(ef)]^2 \qquad 3)$$

where EF(max) - is the peak efficiency,
    Q(ef) - is the discharge rate for peak
           efficiency, and
    C(ef) - is a constant determined by regression.

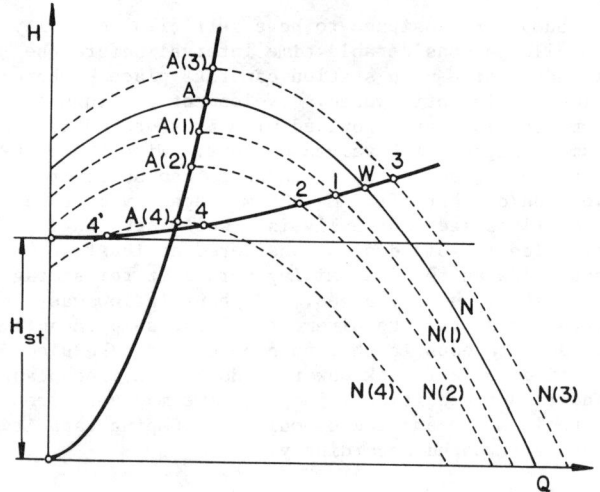

Figure 1. Speed adjustment of a centrifugal pump.

For pump selection purposes, it is necessary to construct the contours of equal efficiency while allowing for reasonable speed adjustment. The maximum efficiency values are subjected to the affinity law. This provides a new set of ellipses to be introduced into the program. Fig. 2, depicts several of these iso-efficiency curves for a typical centrifugal pump.

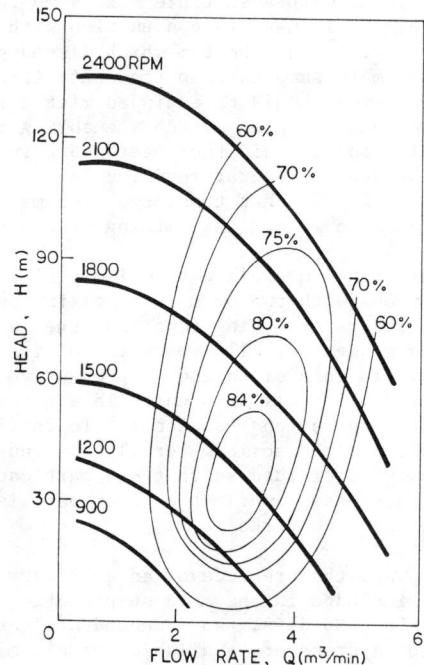

Figure 2. Iso-efficiency curves of a centrifugal pump.

In order to avoid cavitation of a centrifugal pump, the net positive suction head should be kept above the minimum. The following well-known equation includes all the important factors for controlling cavitation:

$$HS(np) = \frac{P(at) - P(vp)}{9.81 \times RO(w)} - H(s) \qquad 4)$$

where P(at) - is the atmospheric pressure,
       P(vp) - is the vapor pressure,
       RO(w) - is the density of water, and
       H(s) - is the suction head.

The suction head required to lift the water from the sump level up to the level of the centrifugal pump inlet is comprised of a static component and dynamic head loss terms. It should be less than that which the pump can generate. The cavitation coefficient suggested be Thoma [4] and the vapor head can be entered in the program to calculate the net positive suction head of the centrifugal pump being considered.

Finally, the horsepower curves can also be approximated by second-degree polynomials if the linear relationship is not acceptable for orientational calculations.

SUMMARY

The traditional methods of pump selection for mine drainage require a strenuous calculation and graphical solution of equations. The discussed program offers an adequate and quicker method for handling this apparatus with the aid of a microcomputer.

The approximations in this teaching program can be replaced by equations of higher accuracy with minimal increase in running time. The high-resolution graphics applied in this program is a useful tool for gaining insight and understanding of the pump selection procedure needed for design of coal mine pump stations.

REFERENCES

1. Given, I.A., ed. "SME Mining Engineering Handbook", AIME-SME. New York, N.Y. 1973.

2. Stefanko, R., "Coal Mining Technology - Theory and Practice", AIME-SME. New York, N.Y. 1983.

3. Bise, C.J. and Van Scyoc, R.L., "Computer-Aided Analysis of Mine Drainage Systems", Mining Engineering. September 1984. pp. 1346 - 1349.

4. Smith, P.D., "BASIC Hydraulics", Butterworth Scientific. London, 1984.

5. Snow, R.E. and Cosler, D.J., "Computer Simulation of Ground Water Inflow to Underground Mines". Proceedings of the First Conference on Use of Computers in the Coal Industry, August 1-3, 1983. Morgantown, W.Va., pp. 587-593.

ACKNOWLEDGMENT

Mrs. Patricia J. Coulson's help in typing and editing this paper is greatly appreciated.

Chapter 52

TOOTH MODIFICATION ON BUCKET WHEEL EXCAVATORS

by

Farid Sadeghipour
The Ohio State University
Lima, Ohio
and
Jozsef S. Gozon
The Ohio State University
Mining Engineering Division
Columbus, Ohio

Abstract. Today, bucket wheel excavators of all sizes and designs are used throughout the world for overburden removal and raw material digging.

Hard strata conditions make the removal of overburden dependent upon the performance of bucket wheels. Efficient performance contributes to low cost. Bucket wheels perform dependent on the cutting edge of the bucket tooth. The design of a bucket teeth directly effects cutting rate and service time. In this paper, various shapes of bucket teeth were designed, built, and evaluated to define a bucket tooth with a higher penetration force, cutting rate and longevity. Variables included wedge type, blade and pyramidal models.

Using the finite element method combined with computer-aided design, the stress concentration factor was estimated for three different designs of bucket teeth. The effect of selected bucket tooth shapes on a block of simulated overburden was studied in the laboratory, and the preferred tooth was chosen.

INTRODUCTION

Increased surface mining efficiency can be obtained by a continuous mining technique utilizing the bucket wheel excavator (BWE) as the primary means of earth moving.

The bucket wheel excavator was first built and used to excavate only soft and easy-digging materials. However, the increasing demand for energy materials will require reconsideration of the BWE's application to hard and consolidated overburden [1,2]. This, in turn, would reduce the operational cost of materials such as coal. That was the subject of the research work described herein. Bucket wheel excavators can be used for harder materials by completing some or all of the following items:

(a) Development of a satisfactory cutting geometry which will reduce the specific energy consumption.
(b) Using the finite element method [3,4,5] combined with computer aided design to estimate the stress concentration and displacement for different tooth shapes.

(c) Measure the cutting force of the bucket tooth.

INVESTIGATION OF VARIOUS TOOTH SHAPES

Here, the purpose is to estimate the stress concentration factor of a tooth using a finite element method.

Three different shapes of bucket teeth were studied. Teeth were modelled mathematically and loaded to define their physical behavior and determine their break strength. Three different tooth shapes were modelled and subjected to a set stress load in three directions normal to each other.

Three geometric files for the models were made, named regular, blade and pyramid model. Six critical points on each tooth were tested to determine their stress and displacement as shown in Figures 1 through 3. The models were subjected to certain loads in the X, Y, and Z directions. These forces were applied to the left corner edge of the tooth for regular and blade models.

1. After the models were created, the nodes were checked by the computer for coordinate modification.
2. Some nodes were restricted by the user. These nodes did not have any displacement or rotation.
3. The calculations of all the normal, principal and shear stresses for all the nodes were performed by computer.

Many FEM program packages have been developed. The input to all of these programs includes:
- physical properties,
- material properties,
- nodal coordinates,
- element table,
- constraints,
- loading,
- and a few other properties.

Data read from specified files FEM2 and analyzed the program by assembling stiffness matrix, reducing the stiffness matrix, solving for displacement, and calculating element stresses.

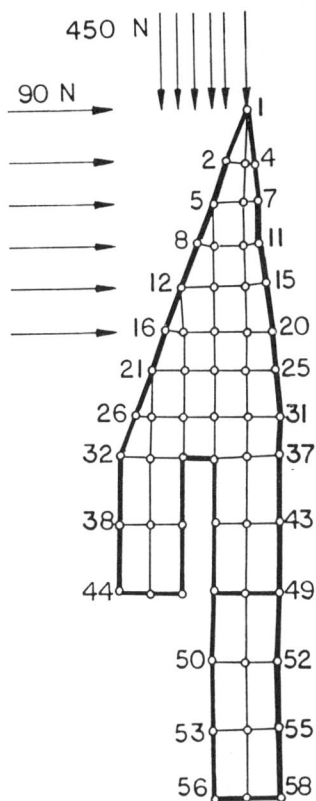

FIGURE 1. Load diagram for the wedge tooth model.

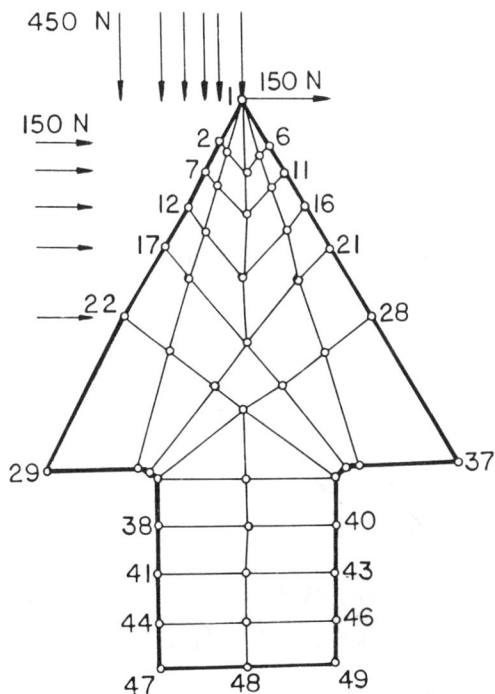

FIGURE 2. Load diagram for the blade tooth model.

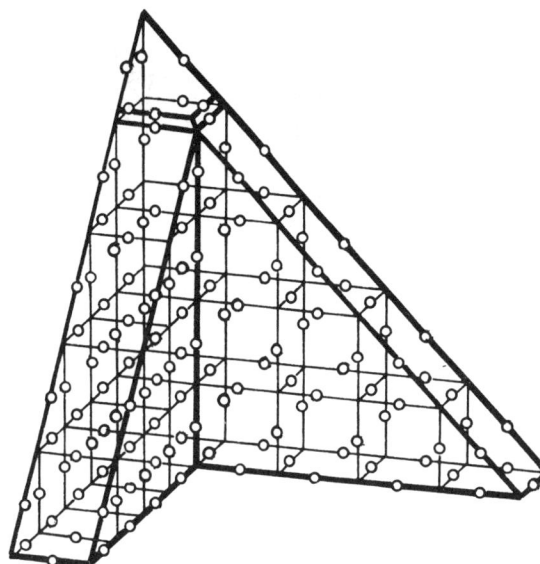

FIGURE 3. Three dimensional view of the pyramid tooth model.

FORCE MEASUREMENT AND STRESS

CALCULATION FOR THE MODEL TEETH

Results Obtained From the Laboratory Experiments

In the laboratory experiment, a strain gauge was placed 25 mm from the tip of each model tooth. For each individual cut, the strain was measured. Since the force value could be calculated in this experiment, the model teeth were generated through the finite element method aided by the computer.

Overburden removal using a bucket wheel excavator was simulated in the laboratory by means of an apparatus incorporating a small cutting tooth (Fig. 4). The cutting tooth was driven through a sample block at a constant speed of 6 rpm which resulted in different cutting depths for each tooth.

The cutting force was measured by means of a strain gauge attached to the model. The strain gauge output was amplified in a signal conditioner and amplifier, which possesses a built in power supply for the gauges. The signal was then recorded by an oscilloscope.

Through the use of an oscilloscope, which monitored strain of each cutting tooth, data were obtained for each model tooth.

The stress applied to each tooth was obtained through the results from the oscilloscope. The electric motor used in the laboratory could generate a maximum load of 1.53 kN. Calculations of the torque characteristics of the electric motor were completed considering an overall efficiency of 0.85 a cutter speed of 6 rpm, and a rated horsepower of 3.1 kW.

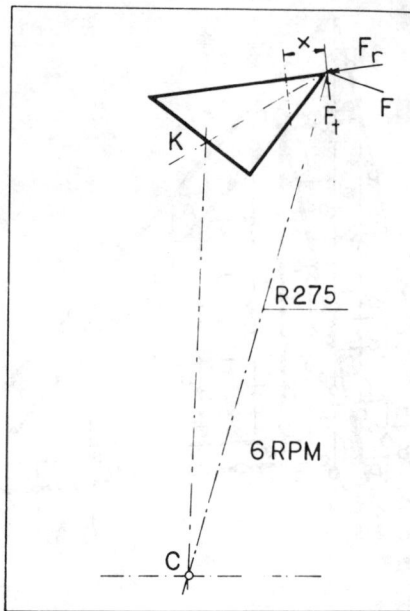

FIGURE 4.   Schematic of the penetration force
            components.

Results Obtained From the Finite Element Method

The measured stress distribution and the deflection for the three different shapes of bucket
teeth were compared by using the finite element
method [5].   Six critical points on each tooth
were tested to determine stresses and displacements.

In the final phase of analysis the magnitude of
loads used in finite element method ranged from
44 to 178 N for tangential forces and 13 to 53 N
for lateral forces.   Later, the loads were increased to 1.33 kN.   The results obtained from
this investigation lead to the conclusion that the
pyramidal shaped model had higher stress at the
critical node points than the other two shapes.
The deflection was higher for the blade shaped
model than for the other shapes.

Deflection appeared to be shape dependent.   The
regular model, which had a smooth cutting edge,
showed less deflection on the tip than did the
blade shape or the pyramid shaped model.   The
pyramid model, with its sharp cutting edge developed a bent-tip shape after deformation.

Different loads ranging between 0.45 to 1.3 kN
were applied to the tip of each tooth.   Three
specific nodes (nos. 3, 6 and 7) at distances
ranging to 33 mm from the tip were chosen, and
their strains were calculated on the computer.
The laboratory test results obtained by use of
strain gauges and monitored by a memory type
oscilloscope were considered as random time functions and analyzed accordingly.   The strains versus forces for each of these nodes was then plotted.   Through the use of these plots and the

strain estimated through experimentation, the
loads applied to each tooth were found.   As seen
in Figures 5 through 7, the range of forces is
different for the pyramid and regular models due
to their different construction.

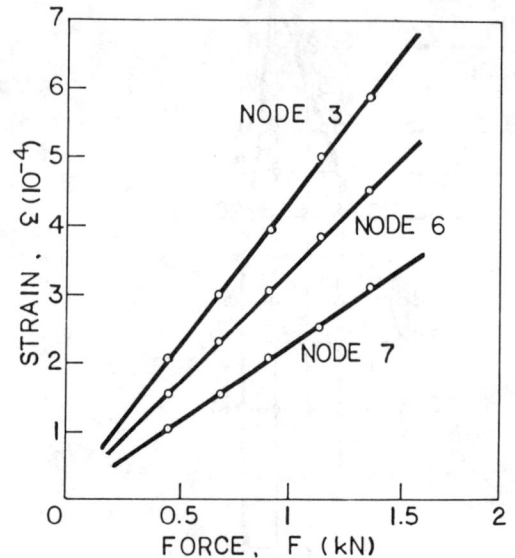

FIGURE 5.   Strain versus force for specified nodes
            in pyramidal tooth.

COMPARISON BETWEEN THE RESULTS OF TWO METHODS

The data obtained from the laboratory experiments show that:

1.   The model laboratory-size cutting used in this
     research program was suitable for penetration
     force measurements.

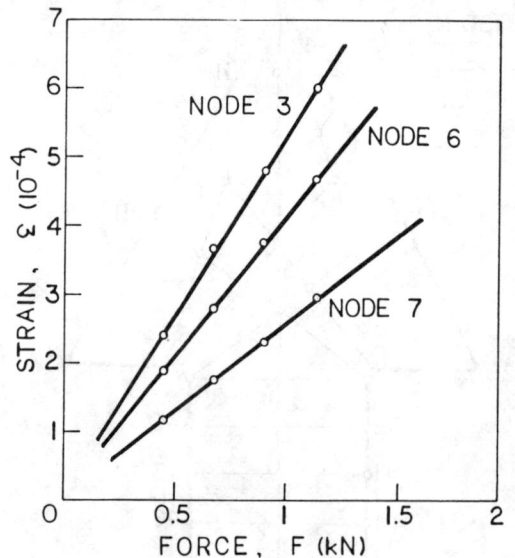

FIGURE 6.   Strain versus force for specified nodes
            in regular tooth.

2. Three different cutting bits were used with varying depth of cut to yield results comparable to those which were reported by numerous researchers.
3. The combination of the signal conditioner and amplifier with dual-beam memory-type oscilloscope permitted:

   a) easy data acquisition and monitoring.
   b) magnification and reduction of the amplitudes of the strain gauge output.
   c) modification of the time scale of an experiment's record for detailed analysis.
   d) storage of six records for comparison of the results from photographs.

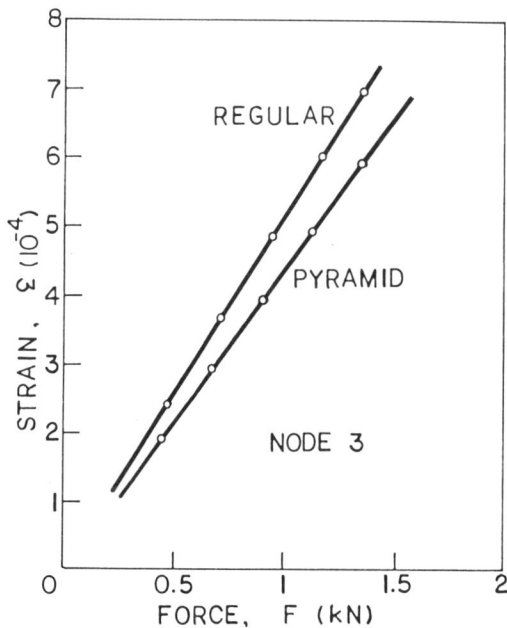

FIGURE 7.   Strain versus force for regular (wedge) and pyramid teeth.

4. The pyramidal shaped tooth had a higher penetration force to cutting length ratio as it was cutting the material. Therefore, the pyramidal shaped teeth conserved energy in the cutting process and reduced load fluctuation on the wheel drive system.
5. Measurement of the material dug by different teeth in the laboratory showed that the pyramidal shape has a lower production rate than the other shapes due to the regular model's larger cutting area.
6. Cutting resistances calculated for particular blocks are in the range of data obtained by large-scale in-situ measurements and by smaller scale measurements in laboratory observations.
7. The measurement method does not require expensive instrumentation, and it can be a source for modern data acquisition systems.
8. Various types of cutting elements, such as, teeth on bucket wheel excavators, draglines and shovels, picks on continuous miners, and a wide variety of drill bits, can be considered for quick and accurate investigation by use of the laboratory equipment and instrumentation developed in this work along with the application of finite element method of computer modelling.

RECOMMENDATIONS FOR FUTURE RESEARCH

In order to increase the flexibility and the applicability of the method of cutting resistance measurements, the following comments and suggestions deserve consideration.

1. Blocks should be made of various blends of materials including very soft material such as gypsum.
2. Prior to in-situ measurement planning, natural blocks obtained and conserved should be introduced into laboratory investigation.
3. In order to obtain reliable data, a number of strain gauges should be used to insure measurement accuracy.
4. The fractured material should be weighed and seived to measure size distribution and calculate newly created specific surfaces as indicators of energy consumption.
5. A device for block advancement monitoring should be designed and attached to the measuring devices.
6. The effect of other cutting speeds and depths should be examined in a more detailed investigation.
7. Vibrating the cutting tooth causes a reduction in the magnitude of the cutting force. Work should be undertaken to qualify this effect.
8. A simplified erosion model should be described as a function of the cutting length, the strength and the abrasiveness of the material investigated, the hardness and the geometry of the cutting element, and the cutting speed.
9. Further work should define energy consumed and expended in laboratory cutting simulations.

CONCLUSIONS

The overall results of this research may be summarized as follows: Two methods were applied to measure the stress and forces applied on the bucket tooth. The result from the finite element method showed that the pyramid model at a specified node has a higher penetration force than the other models. The blade shaped model witnessed more deflection than the other two shapes.

The stresses for the three models were calculated by using a computer assisted finite element program with forces ranging between 44 to 1330 N. However, in the laboratory experiment the forces were similar since the force range was between 0.18 to 1.33 kN.

The calculation of stresses from the laboratory experiment produced similar results to these formulated by finite element method. These comparisons showed that a higher penetration force was produced by the pyramid shaped model and more deflection was obtained by the blade shaped model.

ACKNOWLEDGEMENTS

The authors wish to thank the staff of the
Mining Engineering Division at The Ohio State
University for their support. Also, special
appreciation to Mrs. Patsy Coulson for her
patience and efforts in typing and proofing this
paper.

REFERENCES

1. Gozon, J., "Possibilities and Limitations in
    the Application of Bucket Wheel Excavators"
    Proceedings of the Second MMIMT Symposium on
    Mining Techniques and Mining Equipment selec-
    tion, Vol. 5, Socorro, New Mexico, April 21-
    23, 1983, p. 24.

2. Kasturi, T.S., "Development of Bucket Wheel
    Teeth in the Neyvell Lignite Mines to Meet the
    Hard Overburden Condition,"
    Neyveli, India, 1978.

3. Becker, E.B., Carey, G.F., and Oden, J.T.:
    "An Introduction to F.E.M."
    Vol. 1, 1980, p. 115-126.

4. Graphics System User Manual, SUPERTAB, 1980,
    F.A.E., 1-B, Jan. 1981.

5. Foster, T.C. and Shah, J., "Interactive Finite
    Element Program"
    FEM-OSU Package, The Ohio State University,
    Columbus, Ohio, June, 1982.

6. Keast, D., and Bolt, B., "Measurements in
    Mechanical Dynamics"
    WAVEFORMS AND WAVEFORM GENERATORS.
    Newman, Inc., 1967, Vol. 1, pp. 5-25.

Chapter 53

A PROTOTYPE DATA AND COMMAND SYSTEM FOR ROBOTIZATION

Irving J. Oppenheim and Daniel R. Rehak

Department of Civil Engineering
Carnegie-Mellon University
Pittsburgh, PA   15213

ABSTRACT

The prospect exists for robotization of certain mining tasks such as roof bolting. However, such robotization will require a central data and control system for management of the robots within the workspace. A prototype system has been developed and demonstrated on a personal workstation. The data structure must contain a geometric model of the mine space itself, together with the geometry of the robot and other equipment within the workspace. The control system must recognize interference between objects, the act of moving an object along a path, and the act of modifying the environment by adding or removing material. The computer must also process the commands by which the robot is controlled, whether by a human operator or by another computer program. The software package was developed by students at Carnegie-Mellon University. It represents a first phase demonstration of all the software elements needed to create such a data and command system. It is successfully demonstrated, but is limited by the storage mode of its data structure.

INTRODUCTION

There is widespread interest in bringing robotic capabilities to underground mining. Increased safety is one immediate motivation, while improved productivity is the important long-range objective. We have further observed that robotization generally brings a positive synergy to a problem. Once robotic tools are brought to a problem they suggest new approaches and new abilities which tend to improve the fundamental process itself. This paper describes a first contribution to robotics for mining, the development of a prototype software system for robot control.

The modern robot manipulator was created by the coupling of low cost microprocessors to mechanical actuators. Our interest now is in "intelligent" robot systems which can sense, interpret , and act . Developing such a system for mining or construction entails problems which are generally beyond those encountered in manufacturing applications. Factory robots are fixed base, permitting reference of all motions to a fixed frame. The geometry within which the factory robot works is essentially designed around the robot. This design having been accomplished, the factory robot is free to perform its moves without danger of hitting anything. In fact, the typical factory robot is oblivious of its environment, and contains no model of it within its software intelligence.

In contrast, mining robots will have mobile bases and must move within a workspace (the mine) which contains numerous other solid objects. The robot must have a description of this three-dimensional geometry if it is to move and execute its tasks without interference. This need, for a model of the workspace with the robot inside of it, is the central problem addressed in this paper. While various procedures for "solids modelling" exist in modern computer methodology, the program for the robot system must also contain certain primitive functions which are not part of these common data structures. (Those needed primitive functions include interference checking, distance measurement, and so on.) Therefore, a new data structure was envisioned, designed from the outset to serve as the central database for robot operation. A new data structure was selected since conventional modelling schemes are not effective when dealing with the dynamic range of information present in a mine model. This same program must also contain the command set by which the robot is operated ensuring that one program (one database) controls all actions. In this example, the program can be run by a human operator. Equivalently, the program can be run by a higher level program, one which links numerous work stations together.

Another motivation for robotics in mining comes from the fact that mine equipment is already highly sophisticated and capable. In other words, the existing hardware is quite impressive. It may be possible to develop robot systems largely through the creation of sensors and software, utilizing existing equipment with relatively moderate-cost alteration.

The hardware example chosen in this study is a (hypothetical) roof bolter. A prototype data and command system was developed to direct its operation in a mine space. The problems addressed include modelling the solid geometry of mine and equipment, checking for interference before commencing moves, checking distances to nearest solids along any vector, controlling bolter elevation, adjusting for irregular geometries, and maintaining an indelible record of all actions.

PROBLEM STATEMENT

A program was written to control the actions of a hypothetical roof bolter. The bolter was represented by an object shown in Figure 1. It has a base, 2m x 2m x 1m in dimension. It has a manipulator modelled as a cylinder (0.15 m diameter) which extends normal to the base, over an extension range of 0 to 5m. "Bolting" is represented as positioning the robot in plan, extending the manipulator to contact the roof, and then executing a bolt emplacement.

The bolter has three degrees of freedom: two of base movement, one of manipulator extension. The bolter should be imagined to have treads, permitting it to make a general x-y move through a sequence of rotations and translations. The control of such a mobile base is a major research problem in itself. While further discussion is beyond the scope of this present paper, we do note that actual laboratory and field studies are currently underway at our institution.

The mine is represented by a general collection of connected "rooms." While each room may tend to be prismatic, the program was developed to permit considerable dimensional irregularity on any surface. For instance, floors may have sloping sections and walls or roofs may be irregular in section, containing non-planar facets, and so on. Regular geometry can, of course, be modelled easily using this more general capability.

The program was developed as a graduate project undertaken by three students. It was written in PASCAL on a PERQ Systems personal engineering workstation. One objective was to make the program user-friendly, envisioning it as a console from which an operator could control a mine area. As such, the program operates with screen graphics input-output and the robot commands are presented by menu to the user. Nonetheless, the program was equally designed to be driven by a higher-level program, in which case the command set forms the interface between the two programs. This paper addresses the user mode of operation, as it is more demonstrative.

DISCUSSION OF SCOPE

The problem has been idealized to a great degree, taking as given many capabilities which do not yet exist. A number of these capabilities are themselves the subject of research, and a short discussion is appropriate at this time. A first research area is in the sensing of dimensional data. In its present form, the program assumes that dimensional data are available as input; for example, points and elevations in an existing mine could be taken manually by a survey crew and reduced to input data for the program. However, work is also underway on the robotic mapping of general three-dimensional spaces. This research takes some sensor type (vision, sonic or laser ranging, etc.) which is mounted on a mobile robot base, and sent to map a space. The sensors are managed by software which at the same time controls the robot base doing the mapping. The software, which employs procedures developed in artificial intelligence research, then assembles a map of the space. Such studies are underway at Carnegie-Mellon University, including some preliminary tests on actual mapping of mine spaces.

Another research area, already cited, is in providing for software control of mobile equipment bases. Modern microprocessor boards are available which permit controlling track movement as if it were under digital control. At present,

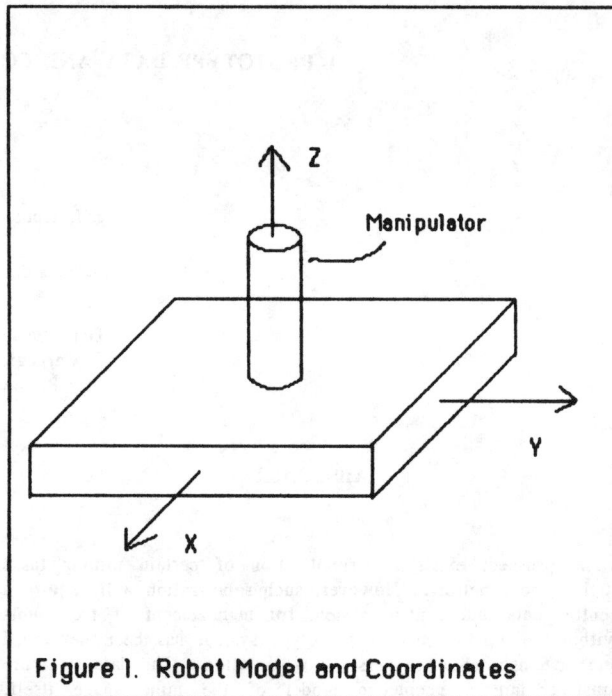

Figure 1. Robot Model and Coordinates

it is still a research and development exercise to assemble the processors and encoders, but the steps to prototype implementation should not present any insurmountable difficulties.

However, dead reckoning positioning (that is, "step counting") is limited in its accuracy, especially with skid-steer equipment. A corollary research need is in position control enhanced by sensors. Those sensors include extrapolative devices such as directional gyroscopes, or interpolative devices such as monuments from which bearings and distances can be measured autonomously. It is likely that dead-reckoning control will not be offered as an end-result in itself, but will be packaged with some sort of advanced positioning control.

A somewhat related problem is in local real time sensing of the immediate environment around the robot. While the central data and command program will determine the amount of extension needed to reach the roof, such manipulator extension would not be performed blindly. Instead, there would be local sensors to check the manipulator approach. In the same fashion, any robot will likely have "bump sensors" on all corners to curtail movement if an unexpected contact occurs. Integration of all such sensing channels is a major task.

Another area of research activity is in software at the strategic level. One example is for management of multiple pieces of equipment (miner, bolter, haulers) for production efficiency. Such programs would run at a higher level than the system which was developed, and would issue commands to it.

In brief summary, there are several areas of related research. The program described herein took as given numerous capabilities which are in fact still in the laboratory or research stage.

MODELLING ISSUES

General Solids Modelling

A variety of procedures are available to create computer models of three-dimensional objects, almost invariably linked to their graphic display. A summary paper (Requicha, 1980) reviews the major approaches and issues. A very powerful procedure in general is Constructive Solid Geometry (CSG), in which rigid solids are represented as Boolean constructions. Such an approach permits a full and accurate representation of such actions as union of solids (positive or negative), intersections, and so on. This very advanced approach is useful for problems such as architectural modelling (Woodbury 1983; Weiler, 1979) and forming or machining. It is unclear whether or not it would be suited to the modelling of a robot workspace.

A simple but potentially effective method (for certain problems) is spatial volume occupancy or "oct tree" encoding. A space can be divided into many cubes, and a solid is represented in binary fashion by giving each cube a 1 (if it is "occupied" by the solid) or 0. This approach was not pursued, but is mentioned in way of overview.

Another powerful procedure, the one which was employed in this study, is called a "boundary representation." The surfaces are described as a set of vertices, connected so as to describe the surface and to form the object. This is essentially the concept behind finite element modelling of solids. It can be "set up" by the programmer to best represent the particular problem under investigation, and for this reason was chosen in this study. While it permitted a quick "customization" to suit the problem as outlined, it may not be the most ideal procedure in the long run; such a "top-down" study has not yet been performed.

Modelling Used in This Study

Figure 2 shows one domain segment (one room) modelled in a boundary element representation as used for this study. That representation is expedient if, as in this case, the surfaces can be pictured as a set of vertices on a planar grid, supplemented by elevations normal to the plane. This may be considered a 2½-D (as opposed to 3-D) representation. It permits local irregularities to be modelled to any accuracy while not necessitating extensive storage for a detailed model of the entire mine map (often several square miles). A rectangular or trapezoidal wall can be defined using only 4 corner points. In Figure 2 the roof and right-hand wall are represented on a finer grid because they contain out-of-plane features modelled as faceted surfaces, as shown.

A data structure was created using a small set of different domain segments (channel, tee, cross, boxend, elbow, object) which connect (at vertices) by simple rules. The basic mine is represented by a directed graph, with each node being a domain element and the links representing the faces of the elements. A general mine layout can be built out of this set. An individual domain segment can be represented only by corner points (if prismatic). Vertex data for each face is stored in an array. Thus, the data structure can model accurately while providing fast access to the data. Since extensive memory management was not implemented, the prototype program is limited in the size and detail which can be represented. In this first phase demonstration that limitation was accepted.

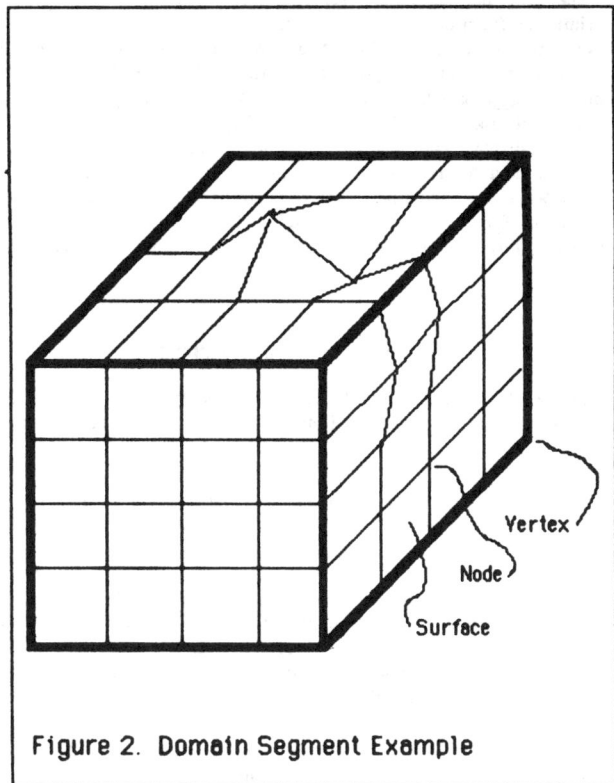

Figure 2. Domain Segment Example

Primitive Functions and Display Features

A number of primitive functions were built into the program and they are the central features needed for creating a robotic command system. The most important is interference checking, to determine if the robot bolter interferes with the solid environment or with other objects within the space. The interference check operates during robot moves to signal a collision as a robot moves along a path. A move which does not trip an interference flag is one which can indeed be accomplished. If interference is found, it shows the point in the intended path at which the interference occurred. For user convenience in assembling sample problems, the interference checking can be turned off; this permits the user to lift a robot and place it elsewhere in the space without having to route it through the actual passages.

Another function is distance measurement along any ray from the robot. For instance, this command could be issued normal to the robot base (locally upwards) to determine the distance at which the roof is encountered. A bolting command would then order the manipulator to extend that distance. Similarly, if bolting is to be performed at a preset spacing, distance measurement to the closest wall would be useful in determining whether another bolt placement is indicated.

Another primitive function is to place or remove additional objects into the space. Face removal can be modelled in this manner, as can the introduction of another equipment into the workspace.

The program has other useful features which are not actually primitive functions but are more related to I/O capabilities. For instance, it is possible to preview an operation. While the robot remains at its present location, the operator can also picture its presence at a second location to study whether its movement there would be practical or not. Similarly, the graphic display can change its view between plan, longitudinal section, and transverse section; all views are accompanied by windowing and zooming display capabilities. Sequences can be recorded for future replay, and if the environment has been altered (through removal of material) the updated domain can be stored.

PROGRAM DESCRIPTION

Figure 3 shows the PERQ screen in its default configuration. The screen is divided into four areas, which will be described in turn.

The upper left region contains a plan view of the overall mine on an x-y coordinate system as shown. A symbol (*) denotes the present location of the robot. The untypical mine shape in Figure 3a is modelled from eight domain segments, and was selected to demonstrate the capacity for modelling irregular volumes. It is possible to window and zoom on the plan view. A robot move is reflected as a translation of the cursor in the plan view. In ordering a move, it is possible to direct the robot by moving the cursor with the mouse.

The large lower region (Figure 3b) is a transverse section, through the robot, showing the tunnel cross-section and the robot. The faceted nature of the roof is seen in this section. The view in this main window can be changed from the transverse section (called simply section) to a longitudinal section (called an elevation). A sloping floor or roof would be evident in such an elevation view. Robot movement or manipulator extension is seen in this main window, which also serves as the window when a preview is requested. The plan view can be moved into this main window also, all under control of the operator.

The area on the right above the main window (Figure 3c) is the status window. It contains the global coordinates of the robot, the manipulator extension status, the tunnel dimensions, and the error messages, if any.

The area above that (Figure 3d) is the menu window. The main menu is shown in Figures 3 and 3d. Sub-menus appear if an appropriate command from the main menu is chosen. Menu choices can be made with a cursor and mouse (not shown). The MOVE choice brings up a sub-menu to direct the robot movement. The INTERFERER ON choice permits switching the interference checking function on or off. The EXT/RETRACT (ext end/retract manipulator) choice generates a request for a command and argument to govern manipulator movement. The CHANGE choice brings up a sub-menu to permit altering of the display, windowing and zooming.

A series of sample exercises were run, demonstrating all the program capabilities and the ease of use. While this constituted only a first phase software package, it did demonstrate the various elements needed for a data and command system for robotic applications.

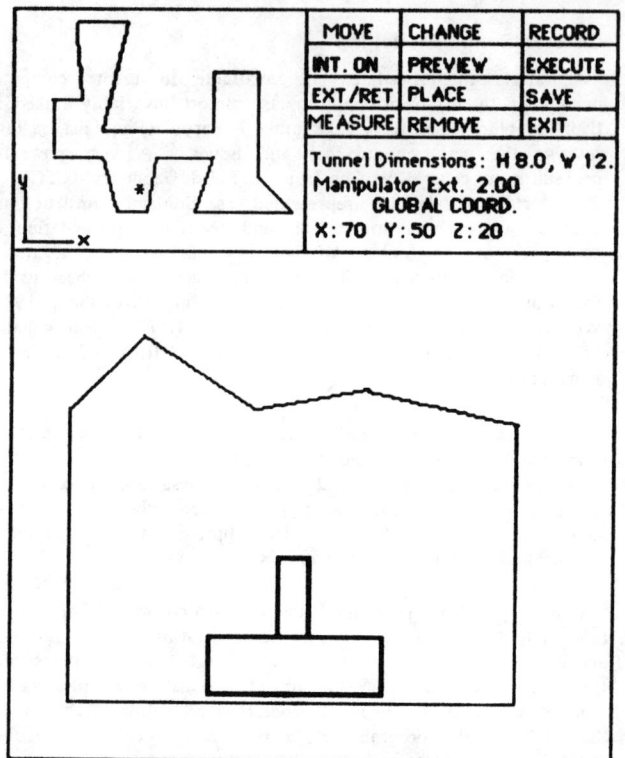

Figure 3. Typical Output Display

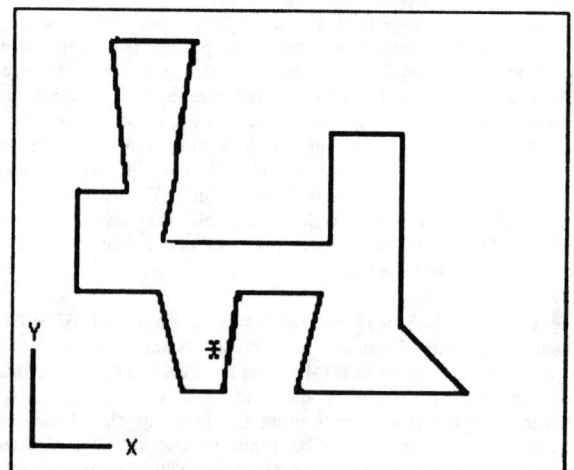

Figure 3a. Plan View of Mine

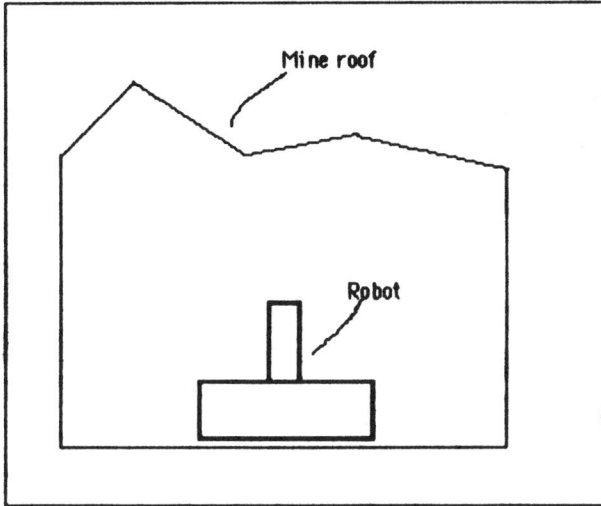

Figure 3b. Sectional View

Tunnel Dimensions: H 8.0, W 12.

Manipulator Ext. 2.00

GLOBAL COORD.

X: 70   Y: 50   Z: 20

Figure 3c. Status Window

MOVE	CHANGE	RECORD
INT. ON.	PREVIEW	EXECUTE
EXT/RET	PLACE	SAVE
MEASURE	REMOVE	EXIT

Figure 3d. Menu Window (Main Menu)

DISCUSSION: PRESENT AND FUTURE WORK

The program is not meant for use in its present form but serves as an outline for future versions. Work is presently underway on a basic data structure for the environment. It will be hierarchical in nature, presenting a coarse model for overall planning and for gross movements, and a fine model for detailed manipulator movement. This capability is not present in existing data structures, and will require a multi-year research effort.

As outlined earlier, many of the hardware, sensor, and control issues are currently under investigation. The research team at Carnegie-Mellon University, Civil Engineering and Construction Robotics Laboratory, is undertaking a series of test-bed experimental studies. The test-bed scale is established by the smallest test space (typical dimensions in tens of meters) which contains the same unresolved spatial issues as the real application environment.

ACKNOWLEDGEMENTS

The computer program was written under the direction of the authors by three students: Mark Borland, Scott Swetz and Leslie Veach. We are indebted to them for their efforts. Facilities and computers were supplied by the Civil Engineering and Construction Robotics Laboratory and several faculty colleagues shared their previous work with us. Finally, the project was enhanced greatly by the participation of Mr. Tom Fisher and others from the U.S. Bureau of Mines, Pittsburgh Research Center, serving as intended users for the student project effort.

REFERENCES

Requicha, A.A.G., 1980, "Representations for Rigid Solids: Theory, Methods and Systems," Computing Surveys, Vol. 12, No. 4, 1980, pp. 437-464.

Weiler, K. and Eastman, C.M., 1979, "Geometric Modelling Using the Euler Operators," Technical Report DA-78-03, Carnegie-Mellon University, February.

Woodbury, R.F. and Glass, G.J., 1983, "VEGA: A Geometric Modelling System," Proc. GRAPHICS INTERFACE '83, NCGA, Canada, pp. 103-109.

Chapter 54

THE USE OF COMPUTERS IN RE-EVALUATION OF THE
WEST VIRGINIA MINE ELECTRICIANS CERTIFICATION PROGRAM

John T. Grasso

Office of Research and Development
West Virginia University
Morgantown WV 26506

ABSTRACT

In recent years, West Virginia University has
assisted the WV Department of Mines and State mine
training board to improve the program for certify-
ing mine electricians in the State. A new certification program incorporating computer-based pools
of test questions and computer-produced tests was
implemented in September 1982, and the new program
was described at the First Conference on Use of
Computers in the Coal Industry, in August 1983.

A major feature of the new program is continual
monitoring, in order to ascertain whether training
goals and greater safety are being achieved. By
June 1984, results for over 650 persons had been
monitored and statistically analyzed. Results of
analysis, combined with industry feedback, suggest
that the new program has stimulated impressive
improvements in mine electrical training in West
Virginia.

Even greater initiatives in electrical training
and safety are now underway with the coal industry
making new contributions to the pool of test
questions, while ongoing computer-assisted analysis will ensure the validity of incremental change
to the item pool and tests, to safeguard the validity of the certification program. The discussion
concludes with potential application of the same
principles to other testing and training endeavors
within the state and nationwide.

STATEMENT OF THE PROBLEM

The WV Board of Miner Training, Education and
Certification is responsible for establishing the
criteria and standards for programs of education,
training and examination required of miners prior
to their certification in several mining occupa-
tions, including mine electrical work. The State
mining laws specify that "No electrical work shall
be performed in low-, medium-, or high-voltage
distribution circuits or equipment, except by a
qualified person or by a person trained to perform
electrical work and to maintain electrical equipment under the direct supervision of a qualified
person" (WV Code 22-2-40, 19).

In 1980 the Board determined that the current
Mine Electrician Certification Program was in need
of revision, and promulgated rules and regulations
governing new standards for mine electricians.
The Department of Mines then sponsored a project
engaging West Virginia University, and specifically the Mining Extension Service (College of
Mineral and Energy Resources), the Department of
Electrical Engineering (College of Engineering),
and Office of Research and Development (Extension
and Continuing Education). In this developmental
project, MES and EE addressed the content and
substance of the new written examinations, while
ORD addressed computer-based implementation and
carrying out a validation study to ensure compliance with equal employment opportunity requirements (e.g., as given by the Federal Register,
which recognizes the Standards for Educational
and Psychological Tests). Results of the development project and validation study are available
from the author (Grasso, 1982).

Components of the Certification Examination

Under the revised certification program, the
Board has subdivided mine electricians' classifications into three categories: apprentice, low
and medium voltage electrician, and low-, medium-
and high-voltage electrician. Aside from the new
apprenticeship requirements, the Board determined
a need for new and upgraded examinations, consisting of a written and a practical (hands-on)
portion for each electrician classification.

Written examination. For persons desiring certification as low and medium voltage electrician,
the new written examination contains six parts:
direct current theory and application, alternating
theory and application, electrical equipment and
circuits, permissibility of electrical equipment,
state and federal legal requirements and the
National Electrical Code. For persons desiring
certification as low-, medium-, and high-voltage
electrician, the new written examination consists
of the six parts named above, plus a seventh part
on the dangers of high voltage.

Practical examination. In addition to these
written-test components, practical tests in three

areas were implemented with assistance from Joy Manufacturing and Ensign Electric, which developed special electrical panels for testing purposes, consisting of AC and DC panel boards and an AC power center simulator panel.

Special cases. Since the Board's new standards and criteria included establishing new electrician classifications, as well as the ugraded examination, the testing program has provisions for persons already possessing electrical certification in previously-existing classifications (underground, preparation plant, surface) who wished to upgrade their certification within the new classifications, without submitting to the entirety of the new examination.

The Development Project

The new certification examination content was developed initially with substantial contributions from EE and MES at WVU, and greatly aided by welcome assistance provided by the coal industry, especially throughout the field test and validation study. Information about the content and substance is available from the author (Grasso, 1982 and 1983b; also see Klishis, 1983).

Beyond this, computer-based procedures to manage the pools of test questions, to generate the written tests, and to provide for analysis of the test results were described at the First Conference on Use of Computers in the Coal Industry (Grasso, 1983a). Essentially, the system permits:

Storage and retrieval of single questions-and-answers (and related material, e.g., reference or source information)

Computer production of study guides distributed by the Department of Mines to new potential candidates for certification

Computer production of actual written tests, including randomly-selected and randomly-sequenced test questions and individual answer keys

A computer-based system for analyzing results of tests administered to candidates

The system was used first during the validation study, but was also designed to serve the purpose of monitoring and re-evaluation of the certification program, through periodic re-analysis of results from tests that are administered to certification candidates over time.

The Second Year

Since 1981, the WV Department of Mines has been implementing a new computer-based system for managing information on companies' comprehensive mine safety programs (the Safety Information System, SIS), which was developed with assistance from WVU's College of Human Resources and Education, with computing services provided by MIS, Inc., of Buckhannon WV, on its VAX 11/750 running VAX/VMS,

POISE data base management software (The POISE Co. Inc., Roswell NM) and the Statistical Analysis System (SAS Institute, Inc., Cary NC).

During 1982-83 the WVU Office of Research and Development moved the mine electricians certification data files and computer programs from the computer system available at WVU to the SIS, to reside along with the wide variety of other information on safety, production, employment, injuries, etc.

During the same period, WVU/ORD collected results of over 650 test administrations that were given to persons desiring to become certified under the new program, in preparation for the first periodic re-evaluation of the certification program. In addition, this work included steps necessary to begin to incorporate the "hands-on" or practical portion of the certification program within the periodic re-evaluation.

DATA

During 1983-84, results for both written tests and hands-on practical tests for over 650 administrations have been monitored and statistically analyzed. Available data are discussed below.

Parts of written test

At maximum, there are seven parts to the written test:

    AC theory and application
    DC theory and application
    Electrical equipment and circuits
    High voltage
    Legal requirements
    The National Electrical Code
    Permissibility

Not all persons are examined on all seven parts. All seven portions apply only to persons who wish to obtain certification in low-, medium- and high-voltage. All portions except the part on high voltage apply to persons who wish to obtain certification in low- and medium-voltage.

In the vast majority of test administrations, candidates were examined on all seven parts (or, on all except high voltage). These two groups account for 75 percent of the cases. The next most sizable subgroup (18 percent) were examined on the National Electrical Code only, which applies to those previously certified as an underground electrician, who wished to expand their certification to include the most recently-added certification topic. A few of the remaining cases resulted from administrative inconsistency, primarily on questions of identifying the specific sections of the test to administer to given persons who already were certified under several varying previous certification categories. During 1983-84, administrative inconsistencies of this type were uncovered during the re-evaluation and new, uniform practices were implemented in order to avoid any further inconsistency.

At the beginning of the certification examination, each candidate is given one of the written test "forms," and is informed of the sub-parts to complete. In order to pass the written portion of the certification examination, a candidate must achieve a score of at least 80 percent correct on every subpart of the written test that is administered.

Parts of practical test

As originally implemented, there were three parts to the practical examination: an AC panel board, a DC panel board, and a power center simulator, each with an accompanying diagram. Once a specific problem has been selected for administration, the State electrical inspector administering the exam will "program" the problem into board by setting switches (located within enclosures on the backs of panels). Problems then consist of the candidate using meters and identifying the problem and providing a solution verbally and in writing. Soon to be implemented is a new high voltage panel board, to expand the examination.

Nearly all the data for practical examination administrations were analyzed, but about 50 sets of results could not merged with the written tests for the same persons; future analysis will incorporate all these results to a greater extent than possible thus far.

In order to pass the practical portion of the certification examination, the candidate must pass one practical problem on each board.

Different written test forms administered

During the period, twenty-four different computer-produced test forms were in use. Although not shown in this report, the different written test forms were used approximately equally across all Department of Mines Division Offices. (Analysis of variance was performed to ascertain whether the specific test form was related to pass/failure: the ANOVA F-ratio 0.90 and p-value 0.60 indicated there was no statistically significant relation between pass/failure and the specific test form thatlwas adminsitered).

Different practical problems administered

Persons taking the examination are required to work one practical problem, presumably determined at random, on each of three panel boards. On administration of the practical examination parts, analysis of available information revealed that there was somewhat greater inequality in frequencies of individual practical examination problems than there was for the individual written test forms. The primary reason was determined during review meetings with the State electrical inspectors who administer the exams: some discretion had been used in screening practical problems that were selected, when the problem applied only to certain equipment, and the specific candidate was not exposed to that equipment; in such cases, an analogous problem (with different equipment) was administered. Passing rates for the different practical problems varied, statistically non-significantly.

ANALYSIS AND FINDINGS

The statistical analysis of the results of test administrations was concerned with two dimensions of the certification program: for candidates, to identify factors related to passing and failure of the certification examination; and for the test components, to identify any portions or procedures that may require review.

Overall

Table 1 contains results for the written test alone. Approximately 71 percent of cases passed all parts of the written which they attempted. Sections of the written varied only slightly in rate-of-failure: at the extremes, only twelve percent failed Electrical Equipment and Circuits, while almost nineteen percent failed the National Electrical Code. (Note, however, that the National Electrical Code was not previously a part of the mine electricians certification program; while this part had the highest failure rate initially, more recent of these data indicate substantial improvement in passing rate.)

Table 1.

Passing Rates, Written Portion Only

		Percentage
TOTAL WRITTEN	Failed	28.7% (of all)
	Passed	71.3% (of all)
EACH SUB-PART:		Percent failing
AC theory & applic.		16.8% (those taking)
DC theory & applic.		15.6%
Elec. eqt. & circuits		12.0%
High voltage		15.1%
Legal req'ts.		13.6%
N.E.C.		18.8%
Permissibility		15.3%

Passing the practical examination was evidently more difficult: Approximately four cases out of every nine failed (43.2 percent, versus 28.7% for the written test, shown above). The failure rates on individual panel boards varied, from approximately twenty percent failing the AC and DC panel board problems, to almost thirty percent failing the Power Center Simulator panel board.

Table 2 contains results for the written and practical tests combined. Of all persons who passed the written, about 66 percent also passed the practical; while, of all persons who passed the practical, about 84 percent also passed the written. From all these data, between 45 and 50 percent of test-takers are passing all parts and becoming certified in mine electrical work.

Table 2.

Results for Written and Practical Portions
(Frequencies)

	Written Test		
Practical Test:	Failed	Passed	Total
Didn't complete all 3 boards	3	5	8
Failed all 3 parts	19	12	31
Failed 2 boards	38	46	84
Failed 1 board	55	94	149
Passed all 3 boards	56	301*	357
Results unknown**	18	12*	30
TOTAL	189	470	659

* A total of 301 (plus up to 12 more) passed both portions, out of 659: estimate of percentage passing both: range of 45.7 to 47.5 percent.
** Denotes odd assortment of cases in which the practical test results were available, but could not be associated one-to-one with the appropriate written examination results.

Correlation analysis (Tables not shown) were completed to investigate the relationship among passing likelihoods of the different parts of the certification examination. In this analysis, correlation coefficients were computed for (a) all seven parts of the written test (each measured as number correct, and as pass/fail dichotomous variables), (b) the overall written test result (pass/fail, dichotomous), and (c) each part, plus the overall practical test (pass/fail, dichotomous variables).

Between the overall written pass/fail and overall practical pass/fail measures, the (point-biserial) correlation is only .30, indicating only a mild tendency for test-takers passing one of these to also pass the other. Moreover, among all these pass/fail dichotomous measures, the correlations ranged from about .17 to about .35. Even the correlations between individual parts that might have been expected to show consistency were not impressively large: only .27 between the AC part of the written and the AC practical panel board, and only .29 between the DC part of the written and the DC practical panel board. All these correlations support the view that the different parts of the test, written and practical, are providing rather broad coverage on mine electrical knowledge and skills.

Analysis of factors related
  to candidates' passing

Analysis of the entire group of test results pointed to several variables found to be related to passing or failure.

Candidate type. As might be expected, passing rates varied between (a) candidates seeking first-time certification, and (b) those already certified in a previously-existing category who were

taking only selected sub-parts of the written test to expand their certification under the new categories (Tables not shown).

Times tested. Persons failing the certification examination may apply to be (re-)tested not more than two more times, without retraining requirements. Table 3 contains results by the numbers of times candidates were tested, which clearly show that repeaters are more successful than first-time candidates.

Table 3.

Passing Rates for WRITTEN, PRACTICAL, and BOTH,
by Number of Times Tested

	Number of times tested		
	1	2	3
WRITTEN portion	69%	76%	91%
PRACTICAL portion	52%	54%	77%
BOTH portions together	43%	46%	68%

Administrative differences. In order to investigate potential differences in test results from one Division Office to another, or among the different test forms being administered, a series of one-way analyses of variance were performed, with some factors serving as limitations on ability to interpret findings with complete confidence. E.g., it would have been desirable to control for characteristics of the persons taking the examination, e.g., to separate the persons with out-of-state prior certification (who were wishing to become certified in West Virginia); this was not possible. With use of all available data, findings indicate virtually no basis for concern at this time, over differences between Division Offices or among test forms used.

Trends over time. It is important to monitor results over time, to ascertain the presence of significant trends. Table 4 reveals findings of a series of analyses of variance on date-of-examination. Results indicate a clear trend toward increasing passing rate. For the entire examination, the rate of increase in passing is almost 18 percent per year. For the written section only, the rate is 17.2 percent. For the practical section only, the rate is 10.6 percent. All three of these rates are estimated while statistically controlling for the influence of re-testing.

Also contained in the Table are estimates of time trends for each part of the written and practical tests. For the written tests, every part shows an increasing and statistically significant trend; for the practical, every part shows a positive trend, but the trend is statistically significant only for the DC panel board part. For the AC panel and the power center simulator, the estimated trend is not statistically significantly different from zero trend.

Table 4

Time Trends for Results of
Written and Practical Tests

Test part	Estimate of time trend	SEE
Total test pass/fail *	+17.4 % per year	4.7
Total written test *	+17.2 % per yr.	4.2
Total practical test *	+10.6 % per yr.	4.7

Written portions:
   number of correct responses, each sub-part

AC theory & applic.	+1.1 per yr.	0.4
DC theory & applic.	+1.7 per yr.	0.5
Elect. eqt. & ckts.	+0.8 per yr.	0.4
High voltage	+1.0 per yr.	0.5
Legal req'ts.	+1.0 per yr.	0.4
N.E.C.	+1.6 per yr.	0.4
Permissibility	+0.8 per yr.	0.5

Practical portions,
   pass/fail, each panel board

Pass, AC panel	+2.2 % per yr.	3.5 (ns)
Pass, DC panel	+8.8 % per yr.	3.7
Pass, Power ctr. sim.	+1.9 % per yr.	4.1 (ns)

* Estimates for the total test, overall written
  test, and overall practical test shown above
  were computed while statistically controlling
  for the number of times attempting to pass the
  test. As for times tested, that estimate was
  an increase of about 8% likelihood of passing,
  per time tested, with a standard error of estimate of approximately (+/-) 3.5%, and alpha
  level between .01 and .05.

In our judgement, interpreting this set of findings requires taking account of a few factors.
First, the nonsignificance of the trend in the
practical examinations seems to be a parallel to
earlier findings (presented in an unpublished mid-
year progress report) on the written examination.
Trends for the written were earlier non-signific-
ant, but later estimates based on greater numbers
of cases proved to be significantly positive.
Findings on the significance of the practical
trends may change, based only on better data when
temporary difficulties (e.g., with the one
Division's results) are overcome.

Also, during the year, meetings with the State
electrical inspectors who administer the exams and
communicate with electrical training personnel at
local mines revealed reports of casual information
that local electrical training programs have been
improving since the new electrical certification
program was implemented. The findings of a positive time trend is consistent with these reports,
and suggests that desired improvement in safety
and mine electrical work are being accomplished.

Analysis related to
   the pool of written test questions

Comprehensive analysis also required attention to
results for individual written test questions,
regardless of the different test forms on which
they appeared, or their sequence of appearance.
Thus, the analysis system permits accumulation of
the following type of information for each item in
the master pool of written test questions:

Pool Number   0004

Number of times administered	305
Number correctly answered	218 (71%)
Number first incorrect alternative	54
second, etc.	...
Number of times left blank	1
Number of times multiple response	2

Correlation between this item and category subtotal	0.40

Further, a cumulative distribution of test results
indicated that very few items in the written test
pool were very likely to be answered incorrectly,
or correlated with subtotal sub-scores (i.e., same
subject-matter written test questions) perversely.
Results of this analysis provide a basis for
reviewing individual test questions, e.g., for
some of them, changes occurring in the practice of
mine electrical work, or in law or the National
Electrical Code, necessitate changes to the test
questions.

Other findings

During the period, feedback from State electrical
inspectors also included information about errors
in the testing materials and study guide for the
written tests. A new version of all these
materials is being planned for 1984-85, which will
incorporate the necessary corrections and improvements. Other feedback also indicated that some
types of analysis originally planned for the
results of the certification examinations would
not be possible to conduct, due to limitations in
the original data and record-keeping procedures.
Future planning efforts will include a review of
these limitations and should produce recommendations for improvement of the monitoring and
reporting system.

FUTURE WORK

Highlights of all these findings have been reported to the Director of Mines, and to the Board of
Miner Training, Education and Certification. In
November 1984, the first of series of meetings
with industry trainers was held, to report findings reported and implications discussed.

Outcomes of these and related discussions include
the following future steps:

1. Continuing monitoring and analysis of results of the new West Virginia mine electricians certification program, including correction of temporary data difficulties, including:

   review of types of analysis and reporting that have been completed to date, to ensure that appropriate and useful types of information are available to those with decision-making responsibilities.

   analysis of results for both written and practical test items.

2. Planned improvements to the written test:

   a. further expansion of the pool for all seven sub-parts of the written examination, primarily through new items donated by experts and trainers in the coal industry.

   b. implementation of continuous validation of new test questions: methods to include the new test questions within "live" test administrations, undistinguished from actual test questions and thereby unknown to candidates, with special scoring key provisions by which answers are used only for pre-test purposes until results provide a valid basis for adding the item to the actual certification pool.

3. Planned improvements to the practical exam:

   expansion of the practical examination, specifically to include a new part based on a high voltage panel board/simulator.

4. Continuing improvement to the computer-based test generating system, to begin to accomodate the electrical diagrams that currently are incorporated only manually, while the computer-generation of written tests is confined now only to the text of questions and answers.

5. Continuing research, in order to ensure the validity of the expanded and improved certification program.

Each of these steps is consistent with the design of the West Virginia mine electricians certification program, which envisioned a process of periodic evaluation in order to achieve the goals of improved safety and electrical work in the mines of this State.

ADDITIONAL REFERENCES

Grasso, J.T., 1982, "The Field Test of the West Virginia Mine Electricians Written Examination," final report to the WV Board of Miner Training, Education and Certification, and WV Department of Mines. Morgantown WV: WVU Office of Research and Development, April 30, 1982.

Grasso, J.T., 1983a, "The Use of Computers in the West Virginia Mine Electricians Certification Program." In Y.J. Wang and R.L. Sanford (eds.), Proceedings of the First Conference on Use of Computers in the Coal Industry. New York: Society of Mining Engineers of AIME, 1983.

Grasso, J.T., 1983b, "Implementation of the West Virginia Mine Electrician Certification Program," TRAM 10: Training Resources Applied to Mining, proceedings of a conference held at University Park PA, August 14-17, 1983. University Park PA: The Pennsylvania State University, 1983.

Klishis, M.J., 1983, "Design and Development of the West Virginia Mine Electrician Certification Program," TRAM 10: Training Resources Applied to Mining, proceedings of a conference held at University Park PA, August 14-17, 1983. University Park PA: The Pennsylvania State University, 1983.

RESERVE ANALYSIS

Chapter 55

DRILL HOLE DATA MANAGEMENT ON A LARGE SCALE COMPUTER

M. Mark Davydov

Peabody Coal Company

ABSTRACT

This paper describes the essential characteristics of an information system developed at Peabody Coal Company for entering, editing, reporting, extracting, and mapping drill hole data. The system is utilized by Peabody's mining engineers, geologists, drillers, and laboratory personnel to manage more than 160,000 drill holes.

The system runs on an IBM compatable mainframe that supports a network of over 700 terminals, any of which may access the data at any time. The software was written to provide a user-friendly interactive environment for editing and checking drill hole data. This software has been built around Cullinet's Integrated Data Base Management System (IDMS) which aids in assuring data integrity and protects the data from being lost due to system malfunctions.

INTRODUCTION

Peabody Coal Company developed an information system for entering, editing, reporting, extracting, and mapping drill hole data. The system, referred to as the Drill Hole Data Base system (DHS), is utilized by Peabody personnel to manage more than 160,000 drill holes.

The system handles different types of data elements pertaining to a drill hole which are broken down functionally into the following five categories:

HOLE — This data describes the location (coordinates, collar elevation) of the drill hole, the status of the drill hole (whether the hole is active or not, if the coordinates and elevation are estimated or actual).

HEADER — This data describes the additional location (township, range, etc) of the drill hole, general information about the drilling (e.g. date drilled, driller initials, drilling equipment, driller remarks), and verification status for certain parameters.

STRATA — The strata description data include : depth, thickness, rock type, description codes and seam correlation codes for each strata in the drillers log.

LOGS — This data include the additional descriptions or interpretations of the drillers log; the system allows the user to overlay part of the strata with his own or more detailed description (i.e. geophysical or geological interpretation).

QUALITY — The quality parameters consist of the analysis of a strata which include: ash, sulfur, BTU, moisture, PH, carbon, volatility and free swell indices; these values are carried on both a washed and a raw basis.

The data elements mentioned above comprise a data base which is structured within the Integrated Database Management System (IDMS) environment. IDMS centralizes the data and performs all input/output operations requested by the programs. The emphasis in the IDMS environment is both on maintaining data integrity and on making the data readily accessible. The programs can use existing data with minimum effort. IDMS supports a central storage facility called the integrated data dictionary/directory for maintaining source information on data descriptions and relationships. IDMS provides a full compliment of backup procedures and journaling that protects the data from being lost due to system malfunctions.

Five essentional characteristics that determine the success of the system at Peabody Coal Company are:

- User/machine interface

- Data protection

- Powerful data updating capabilities

- Data retrieval and reporting

- System integration

These characteristics are discussed below. Additionally, it should be highlighted that the development of DHS as a monolithic data base system for the Peabody Corporation has eliminated data redundancy (for example, there is exactly one set of coordinates for a drill hole and all the information related to that drill hole share those coordinates).

USER/MACHINE INTERFACE

The hallmark of an effective computerized system like DHS is its user interface. The focus of the user interface is on the following:

- Interactive access

- Flexibility

- Assisting the user

Interactive access

DHS is a completely interactive system. It is capable of sending as well as receiving information in an interactive mode using appropriate input/output terminals (teletypes or CRT terminals). Interactive access to the computer has a number of obvious advantages for the users. It allows them to receive immediate feedback. For example, an engineer or geologist can load the data into the system and then generate all required reports and maps to verify the data. The data can then be changed interactively and the process repeated if necessary. Fast turnaround makes the system more usable. The dialog with the user is organized in two fashions: line prompts (for example, "ENTER MINE NUMBER:", "ENTER TRANSACTION LINE:", and etc) and menu-prompts.

Flexibility

Flexibility means that there are a variety of ways for the user to perform a task. For example, there are two options for copying information from one set of holes to another, three options for updating the strata description data, and etc. This allows the user to select the method most appropriate to his needs. Ideally the user will begin to think of DHS as a tool rather than just an interactive system from which to make rote menu selection. Flexibility allows the system be adaptive to changing needs and situations, and to be capable of supporting users with different backgrounds.

Assisting the user

Assisting the user refers to the degree in which the system can help the user interact with the computer successfuly. Within DHS at every prompt the user may enter "HELP" and the system will respond with a message describing the computer prompt or provide detailed information about the parameter being entered.

DATA PROTECTION

A wide variety of functional users must have access to the data. These functional users include geologists, mining engineers, laboratory personnel and operations personnel. In this situation preserving the data integrity is of primary importance. The factors involved are:

- Data security

- Data locking

- Audit trails capabilities

Data security

To protect the data base against unauthorized disclosure, alteration, or destruction, a security subsystem has been developed. Data security was implemented based upon vertical and horizontal structure of the application. First the user must be explicitly authorized on a regional basis (e.g. corporate office, division, mine, etc). Within a particular mine he may access only selected data items (e.g. coordinates, seam codes, ash, etc). For each user the system maintains a record (called the user log record) specifying the regions the user is authorized to access and the operations the user is authorized to perform within those regions. To identify the users (that is, say who they are) DHS uses the operating system user logon number. Also the users need to authenticate their identification by supplying a password which is specific to DHS. Thus being logged on to a some account does not automatically gain access to DHS.

Data locking

The security enforcement scheme protects the the data base only against unauthorized access. But in a multiuser environment some sort of control mechanism is needed to protect the data base against unnecessary or accidental alteration by an user. In DHS this is handled by a data locking technique. To identify a lock the initials of the person are used who verified the particular data on the data base. The initials can be up to three characters long. Locks can be applied to larger or smaller units of data: to a specific field (seam code, coordinates, collar elevation) or to a transaction group (in DHS terms-that is, the set of all fields of a given type, for example, "hole", "header", "strata", and "quality"). Locked data cannot be changed easily.

Audit trails capabilities

An audit trails facility has also been developed to track certain types of activities such as transferring and deleting holes, changing seam codes and the location fields, unlocking the verified data. The audit trails subsystem provides specific information such as user ID, name and business address, the date and time when an action happened, and action description. For example, this information is used by laboratory personnel. It allows them to monitor transactions against holes with quality data, to keep track of the holes they use to make quality projections and update those projections when major transfers/changes take place.

POWERFUL DATA UPDATING CAPABILITIES

The data entry facility allows the entering of data into the data base, modification of the existing data, and deletion of the existing data from the data base. It was designed to meet the needs of the user under a wide variety of conditions. Data may be entered using the following options:

- On-line interactive input

- Batch input

- On-line input from a computer file

On-line interactive input

On-line inretactive input is an option when the user communicates with the programs by a transaction line using CRT terminals or teletypes. The user's data entries are grouped together; there may be up to 25 entries in a transaction line, and commas are used as delimiters to set up the boundaries of an entry. The format list of a transaction line is predefined (e.g., the position of a particular entry in the line and their sequence are permanent for a transaction). The on-line input offers free form input (absolutely no need to "zero fill", "left justify", etc) and interactive editing. As data are entered and errors are detected the system immediately responds with a error message and quizzes for a correct response.

Batch input

DHS also allows keypunched data to be entered in a traditional batch mode, usually overnight. A detailed error report is provided which contains input images, invalid fields, and error messages.

On-line input from a computer file

The main difference between this option and the on-line interactive input is that instead of processing one transaction line at a time the system is capable of processing a large number of the transaction lines from a external computer file. This file can be created by the user or by the system (for example, if there is a need to move quality data from one hole to another the user may request the file containing the required data to be produced in a format that can be used for on-line input).

DATA RETRIEVAL AND REPORTING

Obtaining information by request is an important characteristic of DHS. Instead of looking upon system output as periodic reports, DHS allows the user to obtain information immediately, when it is needed, rather than tomorrow or next week.

In the DHS environment the system output may be fully specified by the user, therefore a set of retrieval keywords has been defined to allow selection of certain drill holes, specification of data items to be output, and to sort the output in any order.

The main features of the reporting facility are:

- Selective retrieval

- Formal report generator

- Ad hoc queries

- Flexible sorting

Selective retrieval

The system provides selective retrieval of information using different search criteria. Drill holes to be considered in creating outputs may be selected by one or more of several schemes including specific hole numbers, rectangular coordinate boundaries, seam codes, particular attributes possessed by a hole, existence of quality data for the hole, or words found in any remark associated with the hole.

Formal report generator

DHS provides two basic master reports : the drill log content report and the drill log quality report.

The drill log content report includes drill log location information, drilling information, stratigraphic description, lithological description, quality data adjacent to interval, and driller remarks.

The drill log quality report includes a snapshot of the drillers log, raw analytical data, multiple washed analytical data, and remarks.

Ad hoc queries

DHS is capable of producing ad hoc reports with only those data items that are of immediate user concern. Two types of queries are supported: fully undefined (tabular reports, the system provides only those data items that are specified by the user) and partially predefined (the user is able to choose a data group to be printed, for instance, HEADER, STRATA, or QUALITY, etc).

The keywords used for selecting and sorting the data are used to generate columns for the inquiry tabular report. For example, to obtain for a group of holes their coordinates, collar elevation, and list of correlated seam codes the user request is: HOLE,EAST,NORTH,ELEV,LSEAM.

Using this facility, the user may retrieve information in such a way that meaningful patterns and correlations among the data can be discerned. Going one step further, the query capabilities of DHS help the user produce a data base upon which the project evaluation and preliminary mine planning can be confidently based.

The inquiry reports can be routed to any of the following (or combination of) destinations: user's terminal, high speed printer, and computer file.

Basic statistics for certain data items such as number of observations, maximum value, minimum value, standard deviation are provided at the end of the inquiry tabular reports. Calculated values are the major outcome of the query facility. For example, seam thickness is calculated as the distance from the top of the first strata containing the seam code to the bottom of the last strata that contains the seam code, or seam parting is calculated as the sum of the thicknesses of all non-coal strata within a coal seam, quality values are a weighted average (based on thickness) of all coal quality intervals that intersect the seam interval.

Flexible sorting

As part of the flexible reporting structure, the users are able to sort a report (master or inquiry) by any (or combination of) fields from the "hole" and "header" groups as well as by seam structure values (top, bottom, thickness, overburden, innerburden, hole depths, and etc), and by seam quality values (ash, sulfur, BTU, moisture, PH, etc).

SYSTEM INTEGRATION

DHS is integrated with other systems and subsystems so that they can function as one unified environment. Information is processed according to functional needs rather than along system lines. The integration is accomplished in two ways: directly or using extraction file techniques.

The following subsystems and systems are integrated with DHS:

- Coal quality subsystem

- Interactive graphics subsystem

- SEAMSYS system

- Lab Management system

Currently, the design process is underway to integrate a Traverse System that aids in establishing coordinates for drill holes, property corners or any other surveyed points.

Coal quality

Coal quality is a subsystem of DHS that was developed in conjunction with Peabody's analytical services department that allows the coal quality to be reported on either a raw, washed or a received basis. Default contamination parameters are stored for each mine, subarea, and seam. Authorized users can very the contamination parameters that allows them to play "what if" with the data. The following reports are provided: dry inseam core analysis report, at equilibrium moisture core analysis report, and raw or washed contaminated core analysis report.

Interactive graphics

The interactive graphics facility allows the user to select maps that give a graphical display of the data. The user may either preview the maps at a graphics tube or have the map plotted at a hard copy plotter at a later time. The following maps are available: drill hole location, drill hole cross section, contour/isopach map, linear contour/isopach map, and multi-seam drill hole location.

The Contour/Isopach mapping programs produce a map which graphically represents the variation of a drill hole or correlated strata related parameter over a particular area using smoothed curved isolines. The programs generate grid values by the "Weighted Inverse Square Distance" method of interpolation, and use these values to generate the contour lines.

The Linear Contour mapping program produces a map which graphically represents the variation of a drill hole or correlated strata related parameter over a particular area using straight line segments. The contours are generated by the traditional "triangulation" method.

The Drill Hole Location mapping program plots the location of drill holes in a defined area along with a single drill hole or correlated strata related parameter.

The Multi-Seam Drill Hole Location mapping program plots the location of drill holes in a defined area along with multiple drill hole and/or seam related parameters.

The Drill Log Cross Sections programs produce a graphic description of drillers' logs. The information that is selected from the logs to be displayed is user defined as is the representation of the log.

SEAMSYS

SEAMSYS is a very large seam modeling and evaluation system that is independent of DHS; however, an interface is provided that allows an easy transfer of clean and edited data to SEAMSYS.

Lab Management system

Peabody has developed a lab management system that tracks various stages of laboratory analysis. The lab management system handles all types of analysis such as ultimate analysis, composites, metallurgical analysis. As the analyses are completed they are "linked" to the corresponding drill hole and are made accessible to DHS.

CONCLUSION

Incorporating a large mainframe environment and general data base management system with in-house developed application software package the Drill Hole Data Base system provides a sophisticated set of high level data processing functions and is a significant tool for managing drill hole data.

Chapter 56

AN INTERACTIVE GEOSTATISTICS PACKAGE FOR MICROCOMPUTERS: GEOSTAT

Ahmet Unal
Christopher Haycocks
Arun Kumar

Department of Mining and Minerals Engineering
Virginia Polytechnic Institute and State University
Blacksburg, Virginia

ABSTRACT

A geostatistics package designated as GEOSTAT has been developed at the Mining and Minerals Engineering Department of Virginia Polytechnic Institute and State University. GEOSTAT can be used to obtain point or block estimates or simulated values of spatially correlated variables which may in turn be used for contour mapping, reserve estimation, mine planning and production scheduling. In present form, it can readily be used for spatial variability modeling and providing input to FRAPS (Friendly Room and Pillar Simulator) and GROCON (Ground Control Evaluation Program), which are two computer programs developed as part of the same suite. In many cases when used with coal, it eliminates the need for large, highly sophisticated and expensive computer packages that run only on mainframes.

The interactive environment and on-line help option keep the required computer skills at minimum. Much effort has been spent to implement a design philosophy which would appeal to a user with little geostatistical background. GEOSTAT was constructed to be used on a medium sized personal computer with color graphics capabilities, which enhances and facilitates the analysis.

Menu-driven downward flow from easy to use data base management options to semivariogram analysis and estimation procedures eliminates the lengthy batch processing environment of typical geostatistical packages. Modular structure provides flexibility and allows for upward or downward flow without losing information entered previously.

INTRODUCTION

Until recently, estimation of coal seam thickness and quality parameters for coal reserve estimation and production scheduling were done using polygonal, triangular, inverse distance, inverse distance squared, weighted moving average, and polynomial approximation methods. The weighting concepts inherent in these methods often proved most unsatisfactory since a particular method may serve to define one variable for a specific purpose while causing catastropic

results for some other variable or purpose in the same deposit. In addition, none of these methods provide means of quantifying the estimation error that is unavoidable with any estimation technique. Essentially, even a good estimator alone is not sufficient to make the best decision in every phase of a project. One also needs to know how good that estimator is and what might be the risks involved in using it.

The geostatistical approach which uses the concept of random functions provides the optimum estimation procedure and quantifies the error of estimation (Journel, and Huijbregts, 1978). Despite the obvious benefits, the first application of geostatistics to coal was published almost two decades after its theoretical foundation (Sabourin, 1975). The immediate application of these methods was to study blending possibilities of high sulfur coal in the presence of strict EPA regulations. The considerable success of the work greatly helped geostatistics gain recognition in the coal industry (Kim, Martino, and Chopra, 1981).

The interdisciplinary nature of geostatistics which combines probability, statistics, geology and mining kept many people from using it. In addition, the necessary use of computers to implement almost all the available techniques created a handicap for the non-computer oriented engineer. Today, there are hardly any (coal) reserve estimation software packages without a geostatistics option. Even though most of these high-cost software packages are sophisticated, they are equally complex, and run on mainframe computers only and demand expertise to install and use. Although some of the packages are partially interactive, all of them require batch processing for a thorough analysis. These are major disadvantages for the engineer with minimum computer background or who wants to use widely available microcomputers instead of the costly company mainframe or time-sharing system.

As part of an effort to develop a suite of modern programs for microcomputers designed to facilitate planning and improve productivity of room and pillar mining operations, GEOSTAT was created. This program fulfills the need to put geostatistics in the reach of the field engineer.

GEOSTATISTICS AND APPLICATIONS IN COAL

Geostatistics is an interdisciplinary science which provides a methodology for the probabilistic study of spatially distributed variables, often called regionalized variables. Seam and coal thickness, seam elevation, overburden thickness, quality parameters, sulfur, ash, reject, Btu are examples to regionalized variables that may be studied using geostatistics.

In every estimation method, the average value in a deposit or a part of it is associated with the observations inside or in the proximity of the deposit. How to quantify the variability (or equivalently correlation) in the domain of interest and how to weight the observations to get the best estimator and determine how reliable the estimator is are the major problems which are best answered by geostatistical methods.

Simple non-probabilistic weighting concepts used in polygonal, triangular, and other conventional estimation methods are not adequate. Neither statistics nor geology alone has provided satisfactory answers. Geological methods are not sufficiently quantitative for mine planning. Until recently, no statistician has applied statistics to the unique characteristics of geovariables and the methods that use them for mine valuation and planning. Geostatistics combines geology and statistics for use in mining.

Quantification of variability of a regionalized variable is handled by variogram analysis and modeling. Random behaviour and the extent of continuity of a variable can be defined by fitting a theoretical semivariogram model. Once a model is selected, the error of estimation for any weighting method can be defined. However, depending on the degree of the weighting equation used, it is possible to use the very well known optimization techniques to minimize the error of estimation. These optimization techniques which were adopted to specific mining problems in the realm of geostatistics have been given names with "kriging" at the end as a convention.

The following type of problems have been solved successfully by using geostatistical techniques:
1. Global estimation of the total tonnage of coal and average quality for feasibility studies (Pierce, F. W., et al., 1983),
2. Local estimation of the total tonnage of coal and average quality for use in short-term and long-term production scheduling (Lonergan, Kim, and Martino, 1983),
3. Simulation of in-seam coal characteristics to study the quality fluctuation for washing plant input or blending (Kim, Martino, and Chopra, 1981),
4. Cartography and generation of contour maps (Pierece, F. W., et al., 1983), and
5. Selection of optimal drill hole locations (Scheck and Chou, 1983).

DESIGN OBJECTIVES

Initially, GEOSTAT was planned to be a dedicated coal reserve estimation tool which would utilize the design philosophy of CONTUR and RESERV-COAL, but upgrade them to the latest microcomputer technology (Haycocks, C. et al., 1981). However, as the programming techniques were reviewed, a decision was made to interface GEOSTAT with two other packages, GROCON and FRAPS (Haycocks, C. et al.,1984). The idea of creating a much more diversified and flexible package to be used for reserve estimation as well as for any other phase(s) of a geostatistical analysis was adopted.

The hardware requirements were restricted to a microcomputer with a single diskette drive, preferably but not necessarily a hard disk, 64K of RAM, a dot matrix printer, RGB monitor, and a color graphics card so that it would be affordable even by the smallest mining companies, and readily available in most universities. The biggest challenge with regard to programming was to keep the program size and memory requirements at 64K while not sacrificing from the precision of the calculations. As a solution, it was decided to implement a modular structure and take advantage of the overlaying and chaining capabilities of the Advanced BASIC.

One of the main objectives was to offer a tool for any mining engineer, geologist or student wishing to use geostatistics but lacking sufficient background. Another important consideration was to create a highly ergonomic software package that would require almost no computer experience.

FEATURES AND APPLICATION FIELDS

GEOSTAT is comprised of eight main modules:
1. MAINMENU   5. STAT
2. HLP   6. VARIOGM
3. INPUT   7. GRID
4. PLOTS   8. RESERV

MAINMENU is the executive of the package. A specific task starts and ends here. All options of the package are summarized in the display, and the user is asked to make a choice (Figure 1). If an option is chosen before visiting the prerequisite option(s), a warning is issued. INPUT option is the prerequisite for all the rest of the modules except for instructions, listing of the files, and returning to the operating system (option 1, 8, and 9 in the main menu). Likewise, statistical analysis is the prerequisite of variogram analysis which is in turn the prerequisite for gridding and estimation. Thus, skipping the necessary steps in a geostatistical analysis is prohibited.

The user is provided the convenience of reaching the help option independent of what module is being executed. HLP finds the relevant page in the help file and displays it. When finished, it is returned to where it was left. All this is done by pressing a key. Help file is both the user's manual with error codes in the appendix, and a geostatistics tutorial at the same time. Once

```
                        MAIN MENU

        1   GEOSTAT Instructions
        2   Input of Drill Hole Data and Boundary Points
        3   Plot of Drill Holes and Boundary Lines
        4   Statistical Analysis
        5   Spatial Modeling-Variogram Analysis
        6   Griding
        7   Global and/or Block Estimation
        8   List Files on Default Disk
        9   Exit from GEOSTAT


Your Choice: 6


        Option 2  is the  prerequisite
        for any of the options 3 to 7.
_____
PRESS:  F1 (Help)  F2 (Proceed)  F3 (Previous Menu)  Shift+PrtSc (Print Screen)
Invalid entry! Enter the number of your choice from the menu between 1 and 9.
```

Figure 1. Main menu of GEOSTAT.

```
                    2.3. DRILL HOLE DATA
No value at a drill hole =) ? in the first column of that   Variable 1   field.
      Drill Hole Label      E-Coordinate      N/S-Coordinate       Variable 1
1     10-A                  7820              2180                 1822
2     10-B                  12280             1840                 1859
3     10-C                  15600             2840                 1876
4     10-D                  10240             3120                 ?
5     10-E                  13880             6120                 1873
6     10-F                  7540              8060                 1828
7     10-G                  9980              8280                 1832
8     A-73                  9060              11220                1752
9     B-73                  11540             11020                1759
10    C-73                  10340             10160                1786
11    D-73                  13600             9160                 1814
12    E-73                  13240             12820                1729
13    A-74                  13400             12220                1742
14    B-74                  15420             14880                1696
15    C-74                  19180             12920                1742
16    AA-75                 22000             14180                1678
17    AB-75                 19520             15320                1659
_____
F1 (Help)     F2 (Done;proceed)   F3 (Previous Menu)    Shift+PrtSc (Print Screen)
F7/F8 (Page up/down)    F9/F10 (Move left/right between fields)
  /  (Move up/down between fields)    /  (Move right/left within a field)
This value is not in the range you have specified ! Please correct it.
```

Figure 2. Built-in page editor of GEOSTAT and its functions.

into the help option, it is possible to move to any page.

The drill hole data file may be edited or re-trieved by the INPUT module. In its present form, up to 129 drill holes with up to 8 attributes for each drill hole can be handled. A built-in page editor which handles each attribute one at a time, creates an easy to use data base management system (Figure 2). Provisions have been made so that drill holes without values may be excluded from the analysis instead of taking them as zeros. All the inputted values may be checked for possible errors with regard to user-defined ranges.

PLOTS is a graphics module which displays drill holes and property boundary lines. Due to the limitations of the size of the RGB monitors and dot matrix printers, the user is provided numerous options to create useful plots. Among many others, the capability of zooming any section of the plot renders obtaining clear plots (Figure 3.a and Figure 3.b). On the other hand, it is also possible to fit more information on the screen by panning (Figure 3.c).

STAT implements basic statistical routines which would help in variogram analysis. The major sta-tistics calculated are mean, median, and variance of the samples. For test of normality, Kolmogorov-Simirnov 2-sided goodness of fit test is used (Stephens 1977). Box-plots are preferred over histograms, which are more useful in inspecting

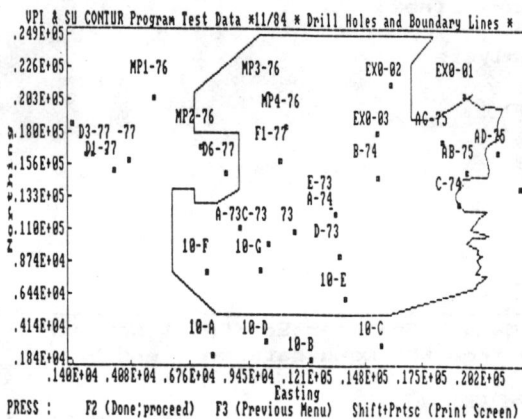

Figure 3.a. Display of drill holes (with codes)
             and property boundary lines.

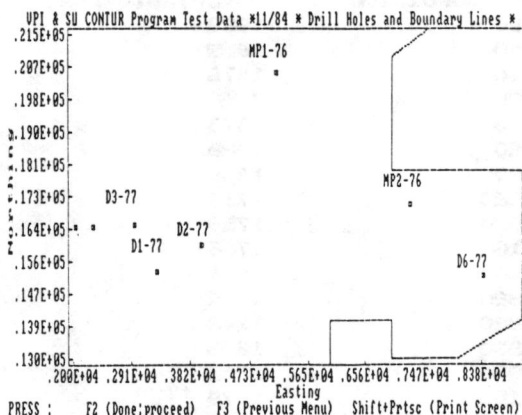

Figure 3.b. Zoomed display of NW quadrant.

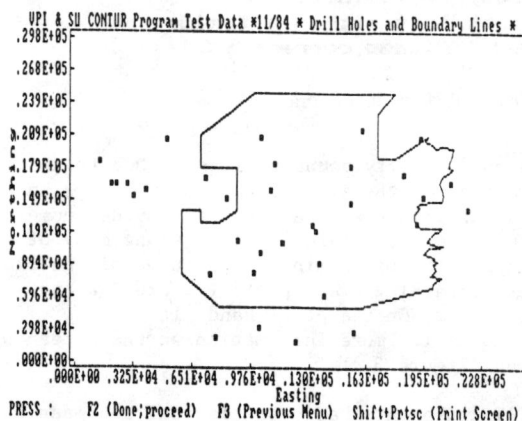

Figure 3.c. Panned display of drill holes.

the normality and log-normality of data.

Two-dimensional semivariogram modeling is carried out by the VARIOGM module using the color graphics capabilities and the interactiveness. Thus, fitting a theoretical variogram to the experimental variogram has been accelerated and enhanced. The user may display up to five theoretical models and the experimental model at once and compare them. Figure 4 shows consecutive display of three different theoretical semivariograms to select the best fitting one. Experimental semivariograms may be drawn for up to five directions. If anisotropic conditions are prevailing, a geometric anisotropy model may be fitted whereas logarithms of sample values are used to construct the semivariogram in case of proportional effect. If the user does not specify the path to be followed at the beginning, he is asked to use his judgement to answer the questions by referring to what is displayed on the screen at the end of each step of the default path. Finally, a cross-validation procedure is applied to test the model chosen.

The module GRID was incorporated to generate a data file containing regularly spaced grid of data generated from the drill hole data. Grid values are calculated by simple linear kriging. The output file is intended to be used by a contour or surface plot program. Unfortunately, no contour or surface plotting routines are incorporated in GEOSTAT at the present time. GRID uses the same point kriging routine which is used in VARIOGM for cross-validation.

Global reserve estimation and local block estimation is done by the module RESERV using simple linear block kriging. In the case of global reserve estimation, the total tonnage of coal and the estimated mean value of each attribute inside the property boundary lines are calculated. The confidence limits for the estimators are also reported according to alpha-risk chosen by the user. For local block estimation, the same procedure is repeated for each of the blocks that the deposit was divided into. In addition, the tonnage above or below a fixed value of an attribute is reported if the distribution of the average value of the attribute in the blocks is verified to be normal or log-normal.

GEOSTAT is menu-driven and interactive but not in the usual sense. The user is not expected to answer a question immediately after the question is asked. Instead, all the relevant questions whether they be yes/no type or else are collected in screen pages in the form of questionnaire sheets. Thus, repeated question-answer-error message sequence has been avoided, plus the user has the flexibility of going back and checking what information he has supplied up to that moment.

At the end of every menu or submenu, each entry is checked with regard to type and possible range. Consequently, the probability of program interruption due to an entry error is eliminated.

All the graphics and plots are designed for four colors in (extended) high-resolution mode. Hence, it has been possible to present more information in a display without sacrificing from clarity. A character set has been added to print special characters and labels in the vertical direction.

An uninterrupted analysis is very desirable in a lengthy reserve analysis in that one should be able to retrieve or edit the data file, process it, and get the estimated values when he loads a software package. Trying to figure out how and which program to load, and what is wrong when an error occurs is scary for the non-computer oriented engineer and presents an unnecessary loss of time. Batch processing has been avoided in GEOSTAT unless it is desired to use gridded data by a contouring or surface plotting software. In this case, the user has to leave the GEOSTAT environment and load the contouring software.

Any of the problems addressed to be solved by geostatistical techniques in the previous section can be solved by GEOSTAT. While it may be best to use it in full capacity for global reserve and block estimation, it may also be used as a data base management and graphics system for drill holes alone. The gridding option may be interfaced with a contour and surface plotting package. Modules up to and including variogram analysis may provide the input to a computer program which selects optimum drill hole locations.

FURTHER ENHANCEMENTS

GEOSTAT is not the most comprehensive geostatistics software available in that it is restricted to two-dimensional analysis and linear geostatistics. An important tool which is not incorporated yet is a coal seam simulator which may be used for: (1) educational purposes - may be used to study a hypothetical deposit with known characteristics and then make a retrospective evaluation; and (2) to study different production scheduling policies in the early stages of a project when the data is not ample enough to make an analytical study. A conditional simulator was planned to be incorporated in GEOSTAT, but has not yet been implemented.

Three-dimensional analysis might only be necessary if the height of estimation blocks has to be smaller than the thickness of the coal seam. This may only arise in the analysis of thick coal seams. Revision for three-dimensional analysis is being considered.

Commonly, the regionalized variables of coal are stationary, that is, they do not increase or decrease systemmatically in any direction over a region (Pierce et al., 1983). Thus, simple linear kriging is satisfactory for most occasions and there are ways to get around non-stationarity in case of a simple global drift (trend). Nevertheless, universal kriging or the superset of it, the intrinsic functions of order-k, might be the direct solution for the non-stationary case. The latter is usually desirable in cartography-oriented studies and has been deemed out of the

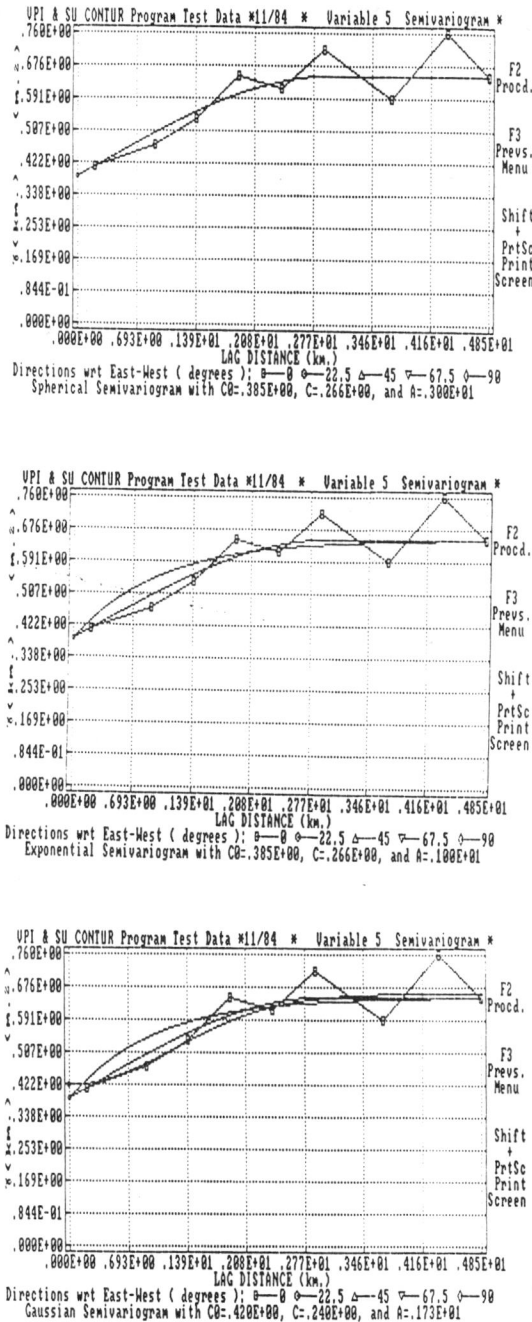

Figure 4.   Semivariogram modeling with GEOSTAT

scope of GEOSTAT. The addition of an universal kriging routine is amongst the items on the list.

Use of a light pen has been thought to be very beneficial, especially in defining geological structures and stationary zones in a geostatistical analysis. However, the resolution of an average RGB monitor today does not permit meaningful use of a light pen where high scale factors apply in plots. Recently, very high resolution RGBs have been marketed which would make efficient use of light pens possible. The potential advantage of using a light pen with super-high resolution RGBs is being investigated.

At the moment, the number of drill holes and the parameters that might be attributed to them are being increased without increasing memory requirements.

CONCLUDING REMARKS

A general purpose geostatistical software package for microcomputers has been developed which may readily be used to solve most estimation problems encountered in coal mining. The interactive environment, on-line help option, and color graphics capabilities made it an excellent tool for a non-computer oriented engineer with little or no geostatistics background. It may also serve as a tutorial in geostatistics for educational purposes.

REFERENCES

Haycocks, C., et al., 1981, "Design Optimization in Underground Coal Systems Vol. 1: The CONTUR Program", FE-1231-19, Feb., U.S. Department of Energy, pp. 1-57.

Haycocks, C., et al., 1981, "Design Optimization in Underground Coal Systems Vol. 2: The RESERV-COAL Program", FE-1231-20, Feb., U.S. Department of Energy, pp. 1-77.

Haycocks, C., Unal, A., Kumar, A., and Osei-Agyemang, S., 1983, "The Ground Control Systems Interface in the Room and Pillar Design", Proceedings, Annual Workshop-Generic Mineral Technology Center Mine Systems Design and Ground Control, pp. 189-200.

Haycocks, C., Kumar, A., Unal, A., and Osei-Agyemang, S., 1984, "Interactive Simulation for Room and Pillar Mines", 18th International Symposium-Application of Computers and Mathematics in the Mineral Industries, IMM, London, pp. 591-597.

Journel, A. G., and Huijbregts, Ch. J., 1978, Mining Geostatistics, Academic Press, London, pp. 7-25.

Kim, Y. C., Martino, F., and Chopra, I. K., 1981, "Application of Geostatistics in a Coal Deposit", Mining Engineering, Vol. 33, No. 10, Oct., pp. 1476-1481.

Lonergan, E. J., Kim, Y. C., and Martino, F., 1983, "Computerized Mine Planning as an Aid to Emmision Control", Proceedings, First Conference on Use of Computers in Coal Industry, SME/AIME, pp. 113-120.

Pierce, F. W., Grundy, W. D., Connor, C. W., and Marbeau, J. P., 1983, "Geostatistical Study of Three Coal Deposits in the Western United States", Proceedings, First Conference on use of Computers in Coal Industry, SME/AIME, pp. 359-373.

Sabourin, R., 1975, Chapter 1-2, "Geostatistical Evaluation of the Sulfur Content in Lingan Coal Mine, Cape Breton," 13th International Symposium on the Application of Computers and Mathematics for Decision Making in the Mineral Industries, Dorstewitz, G., et al., ed., Vol. 1, Gluckauf, Essen, pp. I-2 1-16.

Scheck, D. E., and Chau, D. R., 1983, "Geostatistical Anaylses of Coal Reserves", Proceedings, First Conference on Use of Computers in Coal Industry, SME/AIME, pp. 374-384.

Stephens, M. A., 1977, "EDF Statistics for Goodness of Fit and Some Comparisons", Technometrics, Vol. 19, pp. 205-210.

Chapter 57

AUTOMATED DRAFTING OF DRILL-HOLE DATA

by

James P. Reed
RockWare, Incorporated
Geologist/Programmer
Denver, Colorado

ABSTRACT

When used in conjunction with portable computers, the latest generation of log-plotting software is capable of drafting graphic strip-logs directly at the well-site, mine office, or base camp. These logs may also be transferred (via a phone modem) to a home or client office for subsequent re-plotting or transmission to other locations.

Data input has been simplified by incorporating a technique called "natural language data entry", which allows the field scientist to type in qualitative data in a conventional, "as-is" format without pre-coding. In addition, abbreviated lithologic descriptions are automatically translated into longhand.

The user is able to create new lithologic patterns and vary the plotting scales. This concept of "adaptive software" also provides the user with the ability to change the logic by which the software associates keywords with graphic patterns.

The costs associated with these new hardware and software systems are now low enough to promote widespread acceptance within the coal, petroleum, industrial materials, and metals industries at both the corporate and the individual level.

INTRODUCTION

Historically, geological applications software has been accessible only to professional data processing personnel within large corporations. Geologists and engineers would present their data to these individuals for entry and processing on large, mainframe computers. As a consequence, true computing power was controlled by a select group. Each new application required a substantial development and maintenance cost which virtually eliminated all but the most essential software (e.g. accounting, payroll).

Today, inexpensive yet powerful microcomputers are being purchased by individual geologists and engineers. Due to the correspondingly small software development overhead and the increased market size, a new generation of high PPB (power per buck) geological software has emerged. Unlike most mainframe software, microcomputer-based programs are designed for the novice user who has little experience with computers.

This paper will focus on a description of the relatively new field of microcomputer-based log-plotting and will illustrate how these systems have been used within the mining and petroleum industries. An emphasis will be placed upon defining the fundamental requirements that apply to log-plotting software. In addition, several new approaches concerning data preparation and user-adaptable software will be presented.

The following text is not a "wish list" of what should be. Instead, the described capabilities have already been implemented within currently available log-plotting software packages and are used by geologists and engineers throughout the world.

TERMINOLOGY

It should be noted from the onset that the terms "automated log-plotting" and "log-analysis" are not synonymous. In general, log-plotting software is limited to performing tasks that would otherwise be done by a draftsperson. In contrast, log-analysis software provides the user with information that is not otherwise apparent from a visual review of the raw data. A notable exception to these generalizations involves log-analysis software that is also capable of plotting both raw and processed data.

An example of using log-analysis software in conjunction with log-plotting software is well illustrated by a program in use by the United States Geological Survey. This software package examines down-hole geophysical data from test holes and determines the locations of coal seams. Once these seams have been delineated, the files are transferred to a log-plotting software package where they are subsequently plotted as graphic strip-logs.

The remainder of this paper will be devoted solely to automated log-plotting software. However, many of the concepts concerning data preparation and user adaptable software also apply to log-analysis software.

ON-SITE LOG-PLOTTING

An increasing number of geologists and engineers are now using "transportable" computers at drill-sites and mine-sites to plot the data as it is being collected. This approach eliminates the need to manually plot several copies of the same log (e.g. rough-draft, daily log, final draft). Considering that the bulk of drilling-related geologic analysis involves manual drafting, this new technology provides the field or mine-site scientist with more time for observation and analysis.

A typical on-site computer system includes a microcomputer about the size of a portable sewing machine and a dot-matrix printer smaller than an attache case. These two devices are placed inside a logging trailer or field vehicle where the site geologist or technician can enter the data on an "as-collected" basis. When required, a graphic strip-log of any portion of the hole can then be plotted on the dot-matrix printer at a rate of three to five minutes per page.

Recently, a second generation of portable computers has been introduced by several of the major computer vendors. These battery-powered machines are smaller and lighter than an attache case, yet they possess computing power equivalent to their desktop counterparts. Geologists within the coal and hard-rock industries who do not enjoy the luxury of logging trailers have been quick to embrace this new technology.

TRANSFERRING DATA

Most individuals within the mining and petroleum industries consider the quality of telecopied drill-hole logs to be unacceptable for final presentations. In some cases, the operators have chosen to employ a courier to transport the daily logs from the field to the home office. Other companies often have the well-site geologist verbally dictate the field notes over the phone.

As an alternative, data which has been stored in a computer file may be electronically trans-ferred instantaneously via a variety of tele-communications mediums (e.g. land-line phone networks, radio phones, direct satellite "up-links", cellular phones). After the data has been transferred from a well-site to a computer within an office, the log may be re-plotted or transferred to other locations.

The telecommunications capability is not with-out cost. At least one additional computer system must be purchased for plotting the log in the office. Both the sending and the receiving computer must also be equipped with electronic

devices called "modems" (modulator/demodulators), which allow the data to be transferred over conventional phone lines. However, when compared with alternate means of transferring data from site to office, the computer system pays for itself within a very short period of time.

Many well-site geological consulting firms within the Rocky Mountain region now use phone modems to update their clients on a daily basis. By plotting the logs within the office as they are being drilled in the field, the project managers are able to make informed decisions without waiting for couriers or mail delivery.

NATURAL LANGUAGE DATA ENTRY (NLDE)
- AN ALTERNATIVE TO PRE-CODING -

Until recently, automated log-plotting software required that all qualitative data (e.g. lithologic descriptions) be "pre-coded" prior to entry into the computer system. In order to pre-code the data, a user would compare the geologic field notes with a code index table. For example, if a geologist wanted to plot a "silicified meta dolostone", he/she would refer to the code index table. If a match was found between the observed rock type and a code number, then the user would enter the code number into the computer. If a match was not found, then the program would have to be modified by a software engineer in order to recognize the new rock type.

Many individuals within the industry accused this process of being laborious and counter-productive. Others found the pre-coding scheme to be effective provided that a computer programmer was assigned to the project on a full-time basis in order to continually update the program to recognize new rock types.

An alternative approach that has recently gained wide acceptance within the geosciences is a combination of two techniques called "keyword recognition" and "pattern association". Keyword recognition involves a process whereby lithologic descriptions are scanned for specific keywords. If a keyword is found within a description, then the appropriate pattern is plotted.

Table No. 1
Keyword/Pattern-Association Table

Keyword	Pattern	Pattern Description
LIMESTONE	1	brick pattern
SANDSTONE	2	offset dots
SHALE	3	horizontal lines

For example, in the simple keyword index table as depicted by Table 1 (above), numbers are associated with specific keywords. These numbers refer to patterns described within the right-hand column. Keyword/pattern-association tables define the logic by which log-plotting software systems convert lithologic descriptions into graphic charts.

As an example of how Table 1 might be used, consider the following lithologic description: "LIMESTONE: arenaceous, gray, very fine grained, fossiliferous." A typical log-plotting program would scan this description to determine if it contains any of the keywords defined within Table 1. A match would be found for the keyword "LIMESTONE", causing the program to plot pattern number 1 (brick symbol) within the designated interval.

This new approach eliminates the pre-coding requirement. Instead, the geologist or engineer simply enters the lithologic descriptions in a manner identical to conventional field notes. This technique is also referred to as "natural language data entry". It has been found that the natural language data entry technique reduces the amount of training that a geologist must undergo to use an automated log-plotting program. This is especially important in a multiple-user environment where new and/or inexperienced employees must be able to quickly enter the data without specialized training.

ABBREVIATED DESCRIPTION TRANSLATION (ADT)

Most geologists and engineers are accustomed to recording their data in an abbreviated format. It is therefore reasonable to assume that automated log-plotting software must be capable (as a user option) of converting abbreviations into longhand prior to scanning for keywords and plotting the graphic log. This translation process is accomplished by means of an "abbreviation equivalence table" similar to the keyword/pattern-association table described in the previous section.

Table No. 2
Abbreviation Equivalence Table

Abbreviation	Longhand Equivalent
arg.	argillaceous
f.	fine
fos.	fossiliferous
gr.	grained
gy.	gray
ls.	limestone
v.	very

In the sample abbreviation equivalence table depicted by Table 2 (above), geologic shorthand terms are associated with "longhand equivalents". The abbreviation-translation portion of the log-plotting software will scan each description for the abbreviations within Table 2 and replace them with the associated longhand word(s).

To illustrate the role of an equivalence table, consider the following abbreviated description: "Ls., arg., gy., v. f. gr., fos.". The log-plotting program would scan this description and replace the abbreviations with their longhand counterparts, resulting in the following description: "Limestone, argillaceous, gray, very fine grained, fossiliferous".

This abbreviation-translation capability reduces the amount of adjustment that a user must undergo in order to use the log-plotting software. For example, a geologist in Tulsa, Oklahoma wanted to plot strip-logs for thousands of abbreviated driller's logs within the state archives. By modifying the abbreviation index table within the log-plotting program, the project was completely automated once the drill-hole descriptions were entered, without modification, into the computer.

USER DEFINABLE PATTERNS

Despite many noble attempts to establish a a norm, there are no universally accepted standards for symbol usage within the geosciences. This problem is well illustrated by a comparison of petroleum, coal and mining logs. The petroleum geologist might require twenty different patterns for depicting various types of carbonates, while the coal geologist might want ten different coal patterns, and the metals geologist needs twelve different rhyolite patterns. It is therefore impossible for a programming team to anticipate every possible combination of lithologic, alteration, and mineralization patterns. What's more, the programmer would need to impose a standard upon the user to eliminate conflicting symbol usage.

Therefore, of particular importance in evaluating or designing log-plotting software is the capability for creating new patterns. Unless the user is able to add new patterns, the system is worthless. Additionally, the actual process of creating a new pattern must be relatively easy (i.e. by means of interactive graphic pattern editors).

Case in point: A mining company in Florida purchased a log-plotting software package to draft phosphate test holes. Although the system did not initially contain the desired phosphate patterns, the built-in "pattern editor" allowed the users to rapidly design new, project-specific symbols within one day.

VARIABLE SCALES AND PLOTTING PARAMETERS

The vertical scales at which drill-hole logs are plotted vary considerably. As a result, automated log-plotting software must be adjustable to suit any desired output scale.

For example, a geologist working in the Powder River Basin was having trouble correlating drill-core logs with electric logs because the scales did not match due to distortions produced by the photocopying of the electric log. By simply adjusting the scale within the plotting program to the same distorted scale as the geophysical plot, the geologist was able to quickly generate a new "working" log for direct correlation with the photocopied log.

Variable plotting scales also reduce the amount of redundant work within a mine operation. For example, a mine geologist in Colorado uses

GOLD ASSAY VALUES (PPM)
0.0 0.2 0.4 0.6 0.8 1.0   SILVER ASSAY VALUES (OZ/TON)

ALTERATION MINERALIZATION LITHOLOGY
DEPTH

LITHOLOGIC DESCRIPTIONS AND/OR REMARKS

(1130-1136) LIMESTONE: medium gray to dark gray, thinly bedded, no visible fossils, secondary CALCITE filling en echelon fractures.

1135

(1136-1140) CHERT: cryptocrystalline, bedded, alternating black, gray and white, 1/4 inch beds.

1140

(1140-1144) LIMESTONE: ARGILLACEOUS, buff, minor echinoids.

1145

(1144-1149) SANDSTONE: white orthoquartzite, very well rounded, well cemented, silica matrix, frosted, no visible alteration.

1150

(1149-1153) Quartz LATITE: crystal fragments in DEVITRIFIED glass matrix.

1155

(1153-1161) DOLOMITE: CALCAREOUS, gray to cream, finely crystalline, no visible mineralization or alteration.

1160

(1161-1166) DOLOMITE: light gray, fine grained, QUARTZ stringers, minor SKARN, very minor brecciation throughout interval.

1165

(1166-1169) SHALE: gray, abundant PYRITE, CHLORITIZED, odiferous.

ZONE A

1170

(1169-1173) LIMESTONE: ARENACEOUS, abundant disseminated PYRITE, minor FeOx along fractures has been OXIDIZED.

(1173-1178) LIMESTONE: DOLOMITIC, abundant ADULARIA within vugs, sand grains display PHYLLIC alteration.

1175

ZONE B

FAULT AT 1178'

1180

(1178-1183) QUARTZITE: sucrosic, white with gray banding, superface is fault contact. No alteration or mineralization.

1185

(1183-1192) RHYOLITE: ARGILLIZED, porphyritic (feldspar casts), abundant CHALCOPYRITE and chalcedonic quartz concentrated along fractures.

1190

FAULT AT 1191'

(1192-1200) TUFF: lithic, andesitic, sub-rounded to sub-angular, poorly sorted porphyritic andesite fragments, 6% qtz. phenos., 65% plag., highly sheared, PHYLLIC alteration, abundant GALENA.

1195

ZONE C

1200

(1200-1210) ANDESITE: dark-gray to black, shattered, OXIDIZED (biotite altered to hematite), minor hornblende, abundant stockwork QUARTZ veining.

ZONE D

1205

TOTAL DEPTH = 1210 FEET

1210

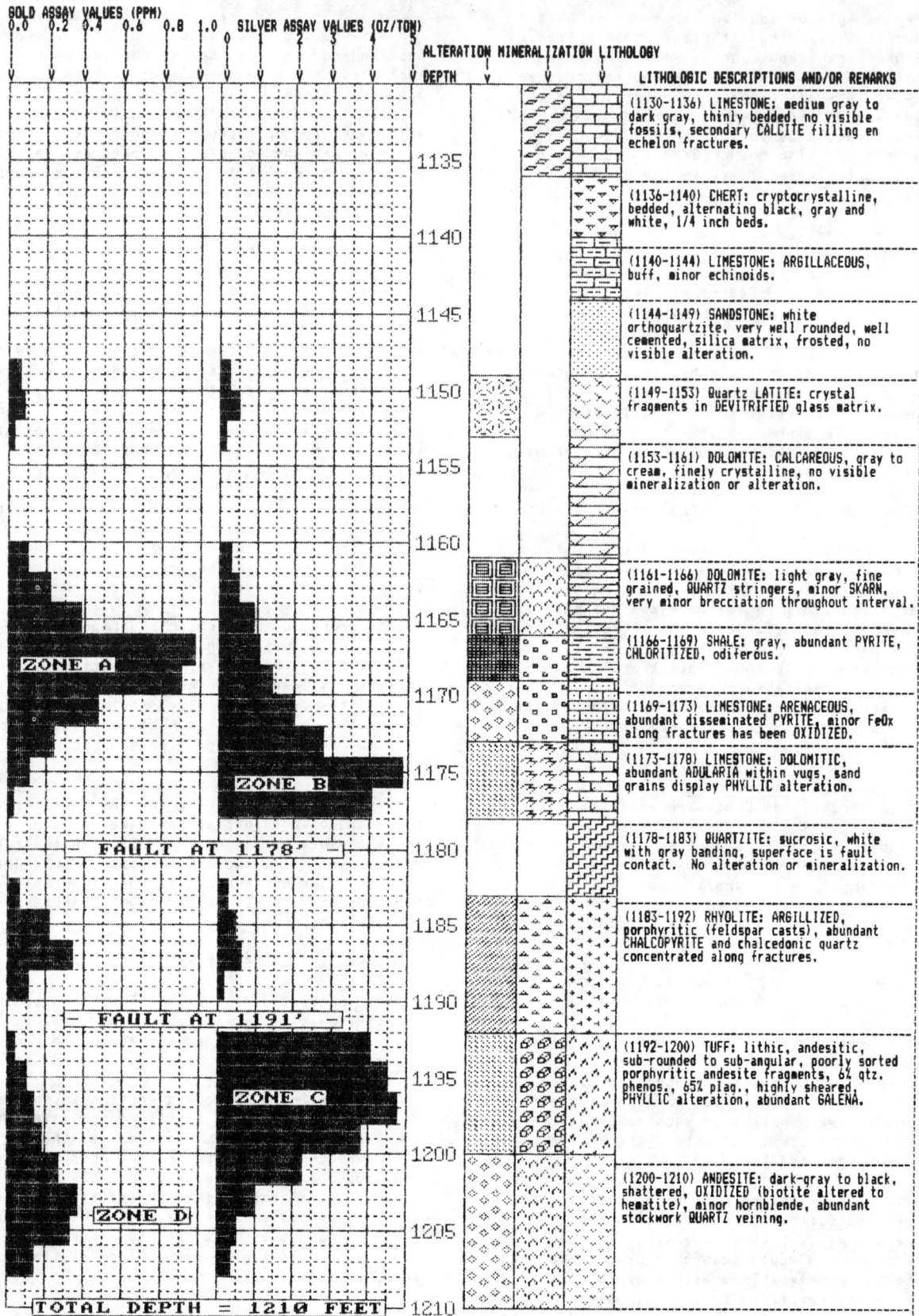

FIGURE 1. Example of computer-generated mining log.

the variable scale capability of the log-plotting software to produce two logs for every hole. A log plotted at 1:50 is used as a "working copy" for detailed correlation with adjacent test holes and cross-section generation, while a 1:600 log is printed for inclusion in monthly reports. By entering the data only once, and changing the plotting scales, the additional log is plotted with just a few minutes of additional effort.

Some users have discovered new applications for the variable plotting scale capability within automated log-plotting systems. A geologist logging a precious-metal hole in the Mojave desert plotted detailed portions of a drill-log at a scale of 1:1 for direct comparison with cored intervals of importance. Copies of these "sub-logs" were then pasted inside the core boxes for archival purposes.

In addition to variable vertical scales, it is also necessary for the user to be able to vary the horizontal scales at which quantitative data is being plotted. Many coal geologists use this capability to change the horizontal scale at which the BTU values are plotted depending on the overall range within a project area.

Another important consideration involves the general format of the drill-hole log. For example, many coal geologists plot a lithologic column on the left side of a log and display the quantitative data (e.g. BTU's, ash content, moisture content) to the right of the lithology column. Other geologists prefer the opposite convention. Figure 1 (metals exploration log) and figure 2 (petroleum exploration log) illustrate two possible formats which may be generated by a single log-plotting software package. This variable format capability is a fundamental requirement within any log-plotting program in order to prevent immediate obsolesence if a new format is required.

ADAPTIVE SYSTEMS

Adaptative log-plotting software provides the user with a capability to modify the logic by which the program associates patterns with keywords (i.e. the keyword/pattern association table). This "self-tailoring" feature may be used to configured the software to suit almost any type of specialized application (e.g. coal, petroleum, uranium, metals, industrial minerals) or personal and corporate preferences. In addition, the user is able to readily convert the system to recognize a foreign language.

For example, while drilling over 400 test-holes on a talc exploration project in the Trans-Pecos of West Texas, the site geologists encountered a number of rock types (e.g. phyllite, meta-dolostone, and numerous talc types) that had not been included within the initial keyword set. Because the log-plotting software which was used for this project included a keyword/pattern-association editor, modifying the system to recognize the new rock types took less than two minutes per modification.

The same adaptability must also extend to the abbreviation equivalence table and the pattern definitions for the same reasons as those outlined for the keyword/pattern-association table. This provides the user with the ability to modify the very logic which is used by the log-plotting software.

OUTPUT DEVICES (PRINTERS VERSUS PLOTTERS)

The decision as to what type of output device is to be used for plotting a strip-log usually involves the choice between a dot-matrix printer and a pen-plotter. When comparing these two alternatives, there are four major factors that must be considered: cost, portability, speed and the quality of graphic output:

COST: In general, a plotter that is capable of plotting continuous geologic logs will cost an order of magnitude more than a dot-matrix printer.

PORTABILITY: Pen-plotters are not readily transportable, being rather bulky and delicate, whereas dot-matrix printers are smaller and better suited for transportation.

SPEED: Pen-plotters require more time to generate graphic products than do dot-matrix printers. Pen-plotters must either move the pen or the paper from one line endpoint to another, while dot-matrix printers transfer an image to the printer as a series of horizontal strips. The resultant difference in the time required to print a detailed drill-hole log is considerable. Logs that require hours on a printer may require days on a plotter.

QUALITY: The quality of graphic output from a pen-plotter is superior to that of a dot-matrix printer. This difference is due to the fact that pen-plotters are capable of drawing relatively straight diagonal lines while dot-matrix printers construct diagonal lines by offsetting a series of small dots (hence the "stair-step" appearance). In relation to log-plotting, many users consider the quality of dot-matrix graphic products (see Figures 1 and 2) to be acceptable in light of the other advantages of cost, portability and speed.

COMPATIBILITY WITH OTHER SOFTWARE

Log-plotting software must be capable of reading data from an ASCII (American Standard Code for Information Interchange) text file. Some users store the data for more than one drill hole by using "off-the-shelf" data base programs. Most of these programs allow the data for an individual drill-hole to be copied into an ASCII text file. This ASCII compatibility minimizes the problems associated with managing multiple drill-hole files.

For example, a user in Nevada who had been measuring geologic sections at outcrops and entering the data into spreadsheet data files

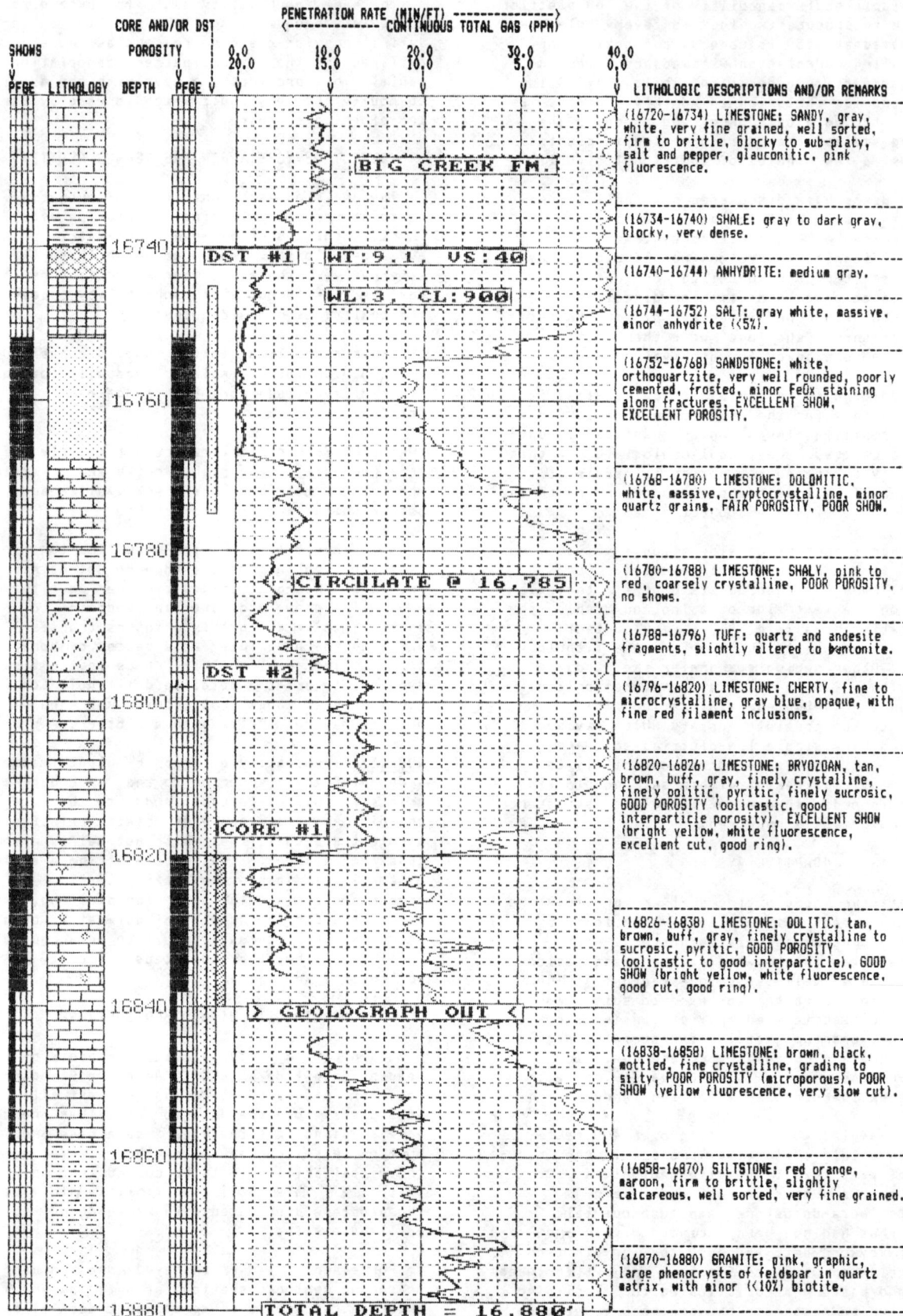

FIGURE 2.   Example of computer-generated petroleum log.

used the "print file" command from within the spreadsheet program to create ASCII files for each section. This data was then read by a log-plotting program and automatically plotted in a strip-log format.

The capability to read ASCII text files also provides a means by which microcomputer logging programs can access data within minicomputer data bases. For example, many corporations and governmental offices are using minicomputers to store large drill-hole data bases. In order to plot a selected log, the data is transferred to a microcomputer (via a phone modem) where it is plotted by inexpensive microcomputer software.

The ability of the log-plotting software to read ASCII files also provides the user with a means of upgrading to new log-plotting software without re-entering the data. This capability is especially important when large amounts of data are to be archived.

DIRECT ANALOG DATA ACQUISITION

Many software/hardware developers have created or are currently working on systems that directly interface to analog measurement devices at the drill-site (e.g. gas chromatographs, geolographs, down-hole probes). These systems minimize the amount of data that must be manually keyed into the computer.

A major technological difficulty involved in direct analog data acquisition involves the lack of compatibility among analog collection devices. A system designed to interface with a certain type of chromatograph will not necessarily operate with another brand of chromatograph. As a solution, some vendors are marketing packaged systems which include the computer hardware and software as well as the analog data collection devices.

USER REQUIREMENTS

Many geologists and engineers are not familiar with the use of a typewriter keyboard. Tasks such as selecting a character that corresponds to a menu option are therefore difficult and frustrating. The recent advances in "touch-screens" and pointing devices (e.g. "mice") may provide many first-time users with a means of utilizing selected software until they become accustomed to using the keyboard. Unfortunately, these techniques do not help the log-plotting software user who needs to enter textual lithologic descriptions into a computer.

Any individual who needs to use microcomputer software must first become familar with the "disk operating system" (DOS). The DOS is a program that is used for a variety of funda-mental computer operations (e.g. file copying, diskette preparation, file renaming, diskette copying). Unless the user is competent in the use of the disk operating system, almost all programs (with the exception of games) will be inaccessible.

On a philosophical note, the most serious impediment concerning the use of log-plotting software involves computer-phobia. Because graphics applications such as log-plotting are relatively advanced, it is recommended that the first-time user become accustomed to micro-computers by initially experimenting with general purpose software (e.g. word-processors, spreadsheets, data-base management). By slowly introducing oneself to this new technology, the well-site adjustment period will be minimized.

COSTS AND PORTABILITY

A typical computer system for generating geologic logs directly within the field would involve the hardware and software listed within Table 3 (below).

Table No. 3
Hardware/Software Costs

Item	Costs
Portable computer	2,300
Dot matrix printer	500
Power filter	120
Word-processing software ...	60
Log-plotting software	1,000
Totals:	$3,980(US$)

The configuration described by Table 3 assumes that 120VAC (volts-alternating current) is available at the drill site. In the event that 120VAC is not available, either a battery operated computer or the addition of a generator would be required.

These costs also assume that the log-plotting software is to be licensed from a software vendor. If the software is to be developed "in-house", a time allotment of 2,000 to 3,000 hours for development and "de-bugging" will be required. For example, RockWare's "LOGGER" program was conceived three years ago. It was modified on a case-by-case basis for two years before the decision to mass-market was made. The last year has been devoted solely to an intensive re-write of the entire package in order to meet the requirements outlined in this paper. In short, if you plan on writing a log-plotting package, be sure to allot a generous amount of development and de-bugging time.

SUMMARY AND CONCLUSIONS

The new generation of microcomputers and user-adaptable software has provided the petroleum and mining industries with a cost-effective method for automatically plotting drill-hole logs in a fraction of the time normally required for manual drafting. These systems are now small enough to allow geologists to plot logs directly at the drill-site. Data from the drill-site may be instantaneously transferred to a home or client office via conventional phone lines for subsequent re-plotting.

Chapter 58

STAND-ALONE GEOLOGIC WORKSTATIONS: APPLICATIONS TO THE COAL INDUSTRY

Geoffrey W. Mathews

TERRASCIENCES Inc.
7555 W. 10th Avenue
Lakewood, Colorado 80215

Abstract. Personal computer-based geologic workstations soon will impact analysis and modelling of coal occurrences. Workstations minimize manual activities and maximize time spent on interpretation.

A personal computer, stand-alone workstation with an integrated software system focuses on geologists' two principal concerns:

o   Ease of Use: The user should not have to be versed in "computerese." All stages of the system should be easy to use: from data entry and editing, through file handling to data analysis and presentation.

o   Technically Complete: The system must perform all functions normally carried out manually and include routines that are not commonly undertaken. The system should be expandable.

With his own workstation, the coal geologist can more efficiently:

1.   Interactively analyze geophysical borehole logs, including estimates of coal quality. These data are retained in the database for mapping;

2.   Interactively correlate boreholes from logs, cores, and/or chemical data; stratigraphic picks are automatically entered into the database;

3.   Access and subset large databases resident on corporate mainframes or commercial data vendors; and

4.   Draw and overlay isopleth maps such as:

    a)   Structural contours
    b)   Isopach
    c)   Stripping ratios
    d)   Coal quality isograds.

The geologic workstation requires little computer experience. Its menu-driven format allows users to concentrate on analysis and interpretation by providing advanced file handling capabilities and interactive graphics solutions.

INTRODUCTION

Stand-alone microcomputer-based geologic workstations are rapidly gaining acceptance and will soon become important adjuncts to coal exploration and production. These workstations will free the coal geologist from manual chores such as shifting and correlating logs, measuring seam thicknesses, posting points, calculating and contouring isopach data, making structural contour maps, calculating volumes and tonnages and a myriad of other time consuming mechanical tasks. A stand-alone geologic workstation will increase a geologist's efficiency in all of these routine functions; it will allow the geologist to devote much needed time to the more interesting and important aspects of his job. Instead of being inundated with tasks of data accumulation and display, he will have time to spend on interpretation, prediction and planning. By letting the workstation do the mundane, the geologist will be free to devote time to geology.

In addition to freeing the geologist from manual chores, microcomputer geologic workstations will enable him to expand his interpretation base and interpretive capabilities. Geologic workstations will let the geologist perform interactively with the computer those tasks that previously had been found to be either impossible or too time consuming to do manually. The microcomputer workstation will give the geologist the capability to experiment with different ways of looking at data; it will allow him to play "what if" games; it will allow him to make geologic models, test hypotheses, evaluate prospects and properties. In short, the geologic workstation will assist the geologist in escaping tedium and allow him to move into the more challenging arena of data interpretation and analysis. With the geologic workstation, and with adequate data, the only real limits to a geologist's capabilities are the limits of his imagination.

The purpose of this paper is to summarize some of the capabilities of microcomputer-based geologic workstations and how they can be applied to the coal industry. Included in this discussion is my perception of the design, configuration and attributes of such a workstation. I have included a few examples of the application of such a workstation. All of the figures in this report were produced on a geologic workstation; they were printed with a dot matrix printer. As with any practicing geologist, the limits of applicability discussed herein are determined by the limits of my own imagination. With a geologic workstation and the proper data, a geologist's mind can soar.

THE GEOLOGIC WORKSTATION

A stand-alone geologic workstation must be multifaceted in order to become a permanent, useful tool to the coal geologist. The workstation must be capable of performing, in minimal time and under the geologist's guidance, all those manual tasks that previously have occupied so much of the geologist's time. The workstation should do for the geologist chores such as computing data (e.g., isopach, ratios, regression equations), database management, and contouring all types of maps; it must display logs and allow the geologist to interactively slip one relative to others; it should display user defined stratigraphic correlations and cross sections. It must be technically complete. The geologic workstation must let the geologist interact with it; the geologist must guide the system.

In order to let the geologist expand his analytical capabilities, the workstation must be capable of performing tasks that are not commonly undertaken by coal geologists. This might include computerized volume or tonnage calculations, trend surface analysis (including mapping residuals) and three dimensional, perspective block diagrams of contour maps. The workstation must provide the geologist with adequate graphic output on the monitor that can be readily reproduced as working hardcopies. It must be capable of producing high quality finished-product maps and diagrams for presentation.

It is essential that the workstation be easy to use. Geologists should not have to be computer hardware/software specialists; they should not have to become versed in "computerese." The workstation must allow easy data entry and editing; it must provide versatile file handling capabilities and interactive data analysis and data presentation. The geologic workstation must be an extension of the user, not an adversary.

Configuration

Several components are essential to a microcomputer geologic workstation. First is the microcomputer itself. Although numerous micros are commercially available, there is a limited number that meet the suggested minimal requirements for a workstation. The microcomputer must have a fairly large memory and sufficient storage capabilities for both data and programs. We have found that 512 Kbytes of memory are sufficient and a 10 megabyte hard disk is adequate for most operations. We have installed on a few workstations an additional storage device that contains two 10-megabyte removable hard disks. This greatly enhances storage, back-up capabilities and data transfer.

A high resolution color graphics monitor should be part of the workstation. Although color monitors are more expensive than monochrome monitors, the added dimension of color is more than worth it. A digitizer is important for entering non-digital data into the system (base map data, XYZ data, log data, and so forth).

A printer and a plotter are both essential. The printer is used to make listings of data files and directories. It lets the geologist make page-size working copies of maps, cross-sections and diagrams. With these he can preview, study and modify maps and diagrams without having to make large copies on the plotter. Only the final versions need to be routed to the plotter. A small, color ink jet printer is helpful for previewing and working with intermediate products, although it is not essential. A modem will be needed if it is necessary to communicate with other computers. If much of the data is stored on magnetic tapes, a tape drive will be necessary.

The physical configuration of the hardware is a matter of personal preference. We have found the most comfortable arrangement is to have the monitors about eye level and to have the digitizer near the keyboard. Most digitizing software requires user input from both the digitizing tablet and the keyboard. The user can operate both input devices simultaneously when the digitizer is near the keyboard. It is not necessary to have the printers and plotters within immediate reach; they should be readily accessible however. A switch box may be needed so the user can drive more than one peripheral device off a single port.

APPLICATIONS

The capabilities of a geologic workstation are a function of the software. Workstation software must be coherent; it must allow easy transitions from one task to another, such as from correlations to mapping. All operations must work on the same database. Above all, the software must allow the geologist to easily work with, analyze, edit, interpret and display relevant data.

The two principal uses of a geologic workstation are in the study and interpretation of borehole logs and in mapping. Borehole logs are used in stratigraphic correlation, isopach determinations, constructing cross-sections and, as discussed below, can be used for estimating coal quality without having to resort to extensive, expensive, time consuming laboratory analysis of borehole cores. Mapping with a stand-alone geologic workstation puts the geologist in control; he can interactively make all types of maps in minimal time.

Logs

Geophysical borehole logs exhibit characteristic patterns in coal. Logs typically run in coal exploration and development programs are natural gamma, gamma-gamma density, and resistivity (short normal). These commonly are used to get fairly accurate depth-to-coal measurements, apparent coal thickness, and to indicate the presence of intraseam rock. The top of a coal seam is characterized by a sharp leftward deflection (decrease) of the density and gamma logs and an equally sharp deflection to the right (increase) on the resistivity logs, as shown in Figure 1. These sharp deflections define the top of the coal. Opposite deflections in the logs at the base of the coal fairly precisely defines the top of subcoal rock unit(s). The apparent thickness of the coal is easily obtained. If the dip of the coal bed is known, either from nearby outcrops or from calculations made using nearby boreholes or structural contours, the true stratigrapic

thickness of the coal is easily calculated. These data are stored in the workstation database and are used in making isopachous maps and coal volume estimates. A microcomputer geologic workstation can perform all of these chores. Stratigraphic tops are stored in XYZ data files when the correlations are made. Apparent thickness values are calculated from the top data, and, given the angle of dip, true thicknesses are calculated and mapped.

These activities are time consuming when done manually. A significant amount of time is invested in manually picking stratigraphic tops from paper strip logs, computing apparent thicknesses, correcting isopach data for dipping beds, posting structural contour and isopach data, and contouring the data. It is much more efficient, much more productive to let the geologic workstation do the calculating, posting, plotting, contouring and organizing the database (X, Y, multiple Z values). The geologist is then free to do what geologists do best: interpret the data; generate and apply predictive models.

The workstation lets a geologist display side by side borehole logs from several different locations; he can shift the logs and make stratigraphic picks and correlations. Top data of selected horizons are stored automatically in the project database to be used for mapping. Figure 2 is an example of borehole correlations as shown on a microcomputer monitor. Figure 2a shows the correlation for two boreholes using gamma, density and resistivity logs whereas Figure 2b is a four borehole correlation diagram

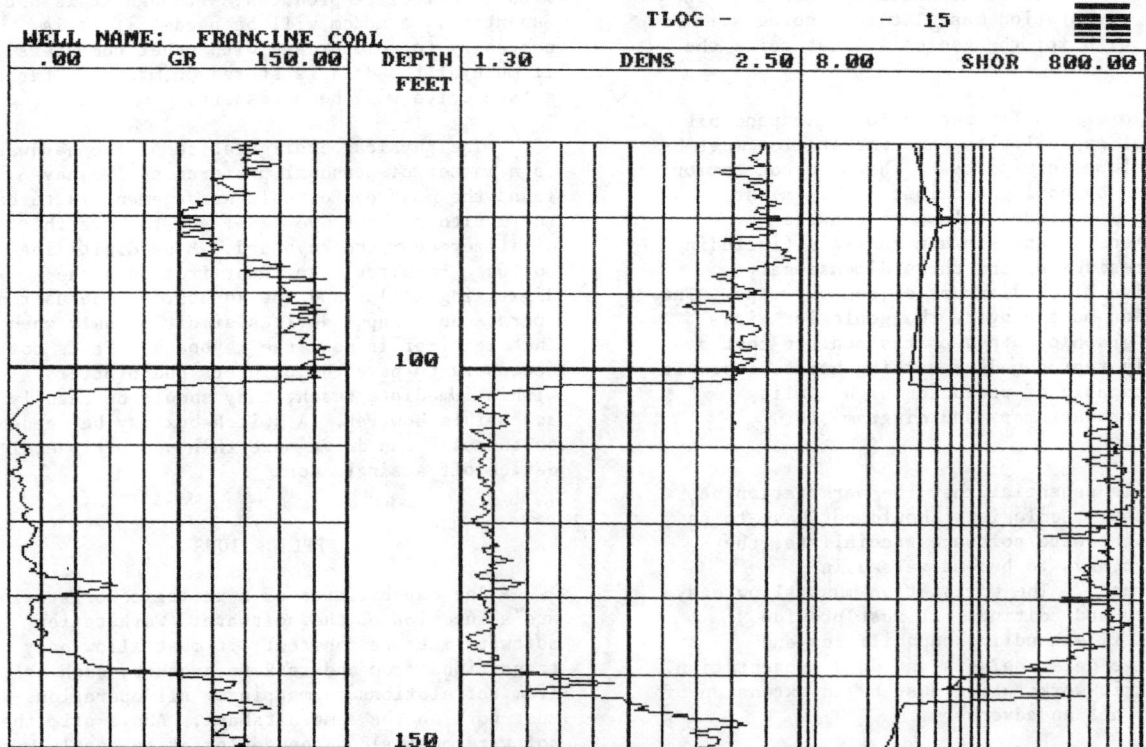

FIGURE 1. Example Logplots of Coal

made using a single (density) log. Because
stratigraphic picks are stored automatically in
an X-Y-Z data file, the geologist does not have
to post and plot the data manually; he can
generate structural contour maps using any of the
top data in minimal time. He can readily map the
top of a coal seam and the surface upon which the
coal rests (structural contour on the subcoal
rock unit(s)). Using the computer, he can
subtract the top and bottom surfaces and draw an
isopach map. He can use the microcomputer to
make thickness corrections that may be necessary
because of the dip of the beds or the azimuth and
plunge of directional holes, or both. True
isopachous maps are made quickly and easily.
Figure 3 is an example of a structural contour
map. This map was created on a microcomputer
geologic workstation and printed with a dot
matrix printer.

Manual correlation of multiple coal seams
among several scattered boreholes is an arduous
task that requires handling several sets of paper
logs, shifting the logs and tentatively penciling
in suspected correlations, only to have to change
them when more information becomes available.
The geologic workstation will let the geologist
make tentative correlations, interactively change
selected picks, draw cross-sections and
structural contour maps on any specified horizon,
all within a short time. Because the workstation
is so fast at data display and data manipulation,
the geologist can try several variations; he can
experiment; he can work with and analyze the
data. The rapidity of graphical display of logs,
cross sections and maps greatly expands the

geologist's capabilities in thoroughly analyzing
the data.

Crossplots commonly are used to infer
composition, porosity, and hydrocarbon effect.
These plots can be very useful in the coal
industry. Figure 4, for example, is a gamma-
density crossplot of the data from the borehole
shown in Figure 1. As might be expected, the
crossplot points occur in two clusters. The
tight cluster in the low density, low gamma
portion of the plot is the coal, whereas the
loose cluster in the higher density, higher gamma
portion of the crossplot is shale and siltstone.
The scatter in the shale-siltstone cluster
reflects the variable nature of nonmarine fine-
grained sediments. A feature of this borehole
that is not readily apparent on the log plots of
Figure 1 is the broad band of points that extend
from the coal point cluster to that of the shale-
siltstone. This scatter of points represents a
progression from ash-rich, argillaceous coal
through carbonaceous shale to the nonmarine shale
of the shale-siltstone cluster. The geologic
workstation makes simple the task of determining
where in the borehole these transition
lithologies occur. In this example, most of the
transition lithologies occur immediately below
the coal and represent a progressive change in
the depositional environment from nonmarine shale
into the euxinic environment of coal formation.
The change away from the coal-forming
environment, however, was much more rapid,
perhaps caused by a rapid influx of argillaceous
material, inundating the coal-forming
environment.

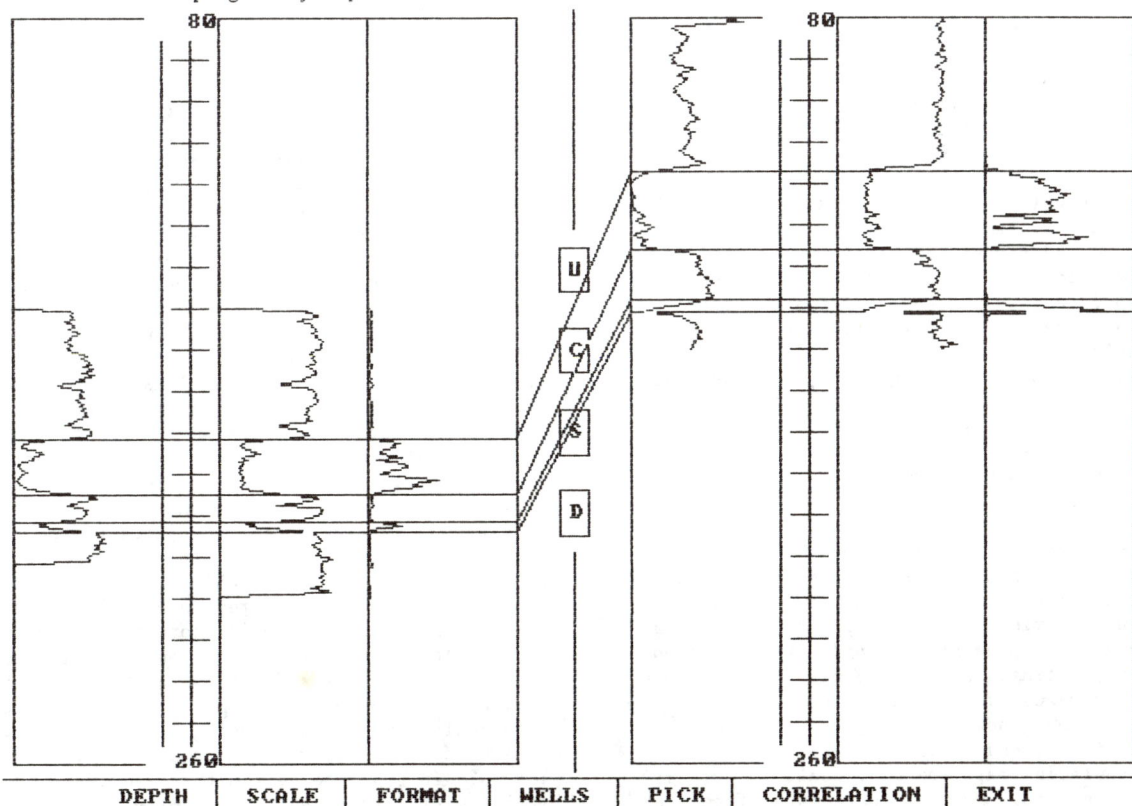

FIGURE 2a. Example Two-Well, One-Log Correlation

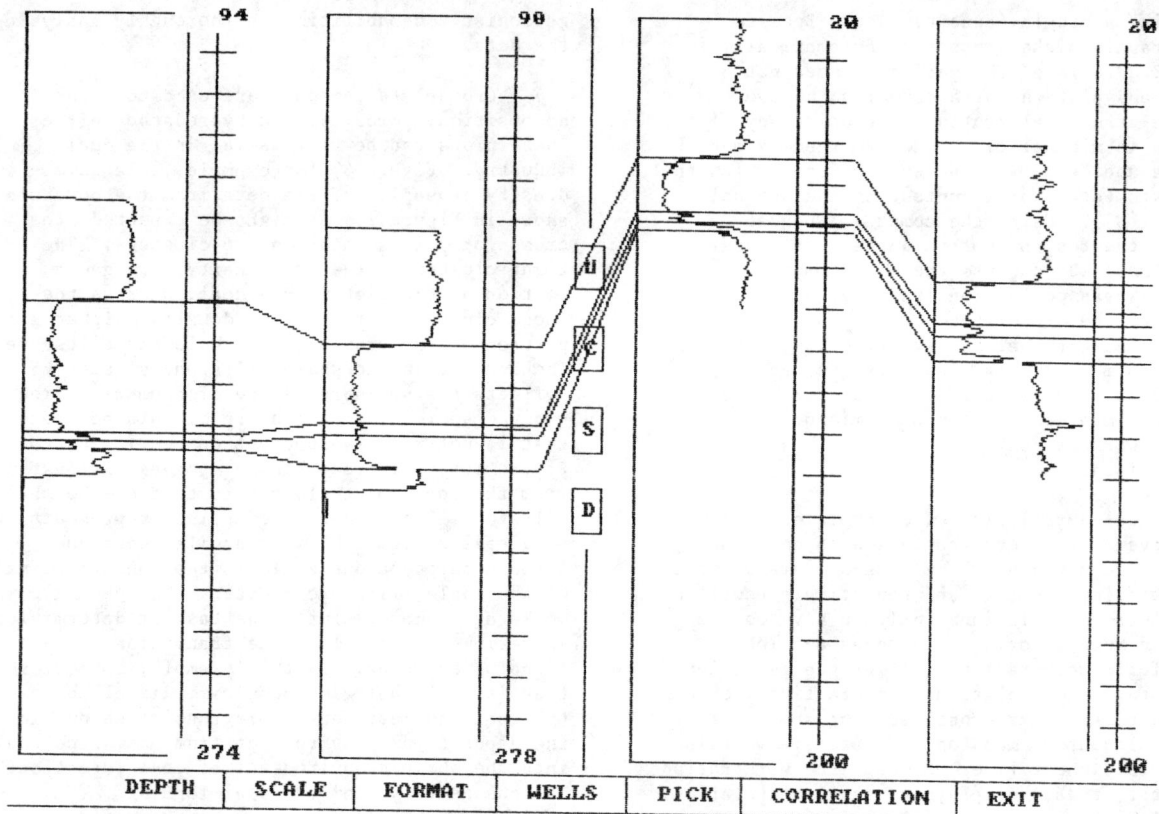

FIGURE 2b. Example Four-Well, One-Log Correlation

Maps

One of the most useful features
of a geologic workstation is its ability to make
reliable working maps in minimal time. It is
this feature that allows geologists to do all
ordinary tasks and still have time for
experimentation.

Structural contour and isopachous maps,
discussed above, are essential parts of
exploration and production. However, each time
more data become available, the maps must be
updated and redrawn. Manual updates are costly,
and, as a consequence, they often are not made
immediately. Because of these constraints, the
geologist seldom gets sufficient time to use his
imagination to create and generate other,
potentially useful maps. Maps of selected
attributes can be used to infer and demonstrate
much more than can be shown on simple structural
contour maps.

Structural contour and isopach maps are
important tools in describing the geometry and
orientation of a stratigraphic unit. However,
they do not provide sufficient information for a
thorough areal representation and interpretation
of a unit or stratigraphic sequence. Other types
of maps are needed for the more thorough
analysis. Among these may be trend surfaces,
residuals from trend surfaces, various ratio
maps, lithofacies maps and coal quality
distribution maps. Many of these maps often are
not made because of time constraints.

For example, it is sometimes difficult to
make and interpret a coal isopach map. Although
simple in concept, isopach data determinations
are not always straight forward. By definition,
true stratigraphic thickness is the thickness of
a unit measured normal to the stratigraphic
contacts. When correlating a coal seam
laterally, lenses of sand or shale sometime
appear within the coal seam. If a shale lens
appears in the center of a seam being traced
laterally, thereby splitting the seam into two
subseams, the problem of correlation arises. Is
the original coal seam to be correlated with the
upper, lower or both of the two subseams? Is the
thickness of the shale included in or excluded
from the coal isopach? Do we map pure coal
isopach, coal plus intra-seam lenses, or should
it be converted to a coal:rock ratio or a
coal:coal + rock ratio? If we convert to a
ratio, should the map values be expressed in
terms of thickness or tonnage? These are
decisions the geologist must make while working
with the data. The decision depends, of course,
on the purpose to which the map is to be put.
Immediate decisions are usually valid,
particularly when restricted to a limited area.
However, decisions or definitions made in one
portion in a coal seam are not necessarily the
proper decisions or definitions elsewhere in the
same seam. The geologist may choose to ignore a
thin shale lens in calculating isopachs, but what
does he do with several thin shale intracoal
lenses? In order to help with these decisions,
it is helpful to be able to display the
distribution of shale seams within the coal.

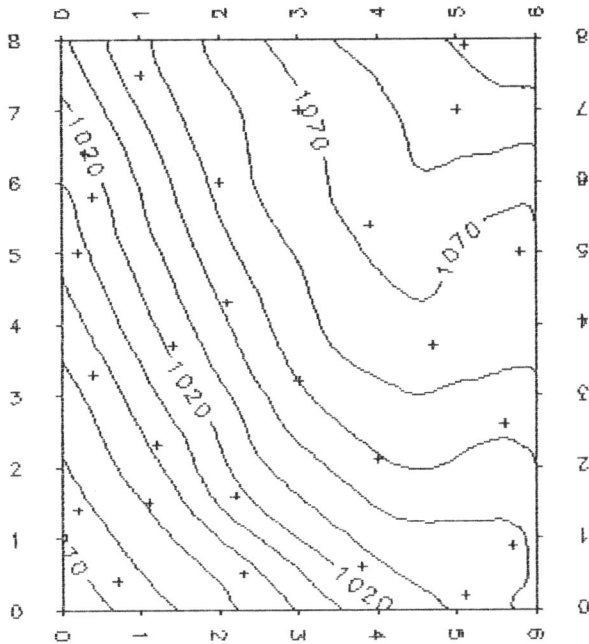

STRUCTURAL CONTOUR: COAL

FIGURE 3. Example Structural Contour Map

The ability to rapidly make different types of maps will assist the geologist in more thoroughly describing and depicting the different vagaries of coal seams, such as those discussed above. A ratio map of coal:shale expressed in terms of thicknesses of the two lithologies, may bring out the presence and "contribution" of shale to the "coal" seam. A similar ratio map expressed in terms of the weight of the two units may give a different interpretation, particularly if floatation is to be used in processing the coal. Figure 5 is a series of maps that depict the same data, mapped differently. Figure 5a is a coal plus interbedded shale isopach map. Figure 5b is a coal only isopach whereas 5c is an isopach map of only the intracoal shale. Figure 5d is a intracoal shale:coal ratio map expressed in terms of thickness. We could use the workstation to calculate volumes of the coal and intraseam shale. By weighting the different volume estimates by their respective densities, we could then calculate and map an intraseam shale:coal ratio map expressed in terms of tonnages. Data calculations and all four maps of Figure 5 were made on a geologic workstation and plotted on a dot matrix printer in less than an hour.

Obviously, different mine plans will be developed depending upon whether 48 inches of intracoal shale is distributed in four roughly evenly spaced twelve inch thick lenses or whether all 48 inches of shale are in one layer near the base of the coal.

An entropy map, such as those described by Krumbein and Sloss (1963), could be made to show the degree of mixing of coal and shale. Doveton and others (1984) have suggested an innovative means of studying vertical variation through maps. This method of three dimensional trend mapping would provide an excellent way to depict and analyze coal data distributions areally.

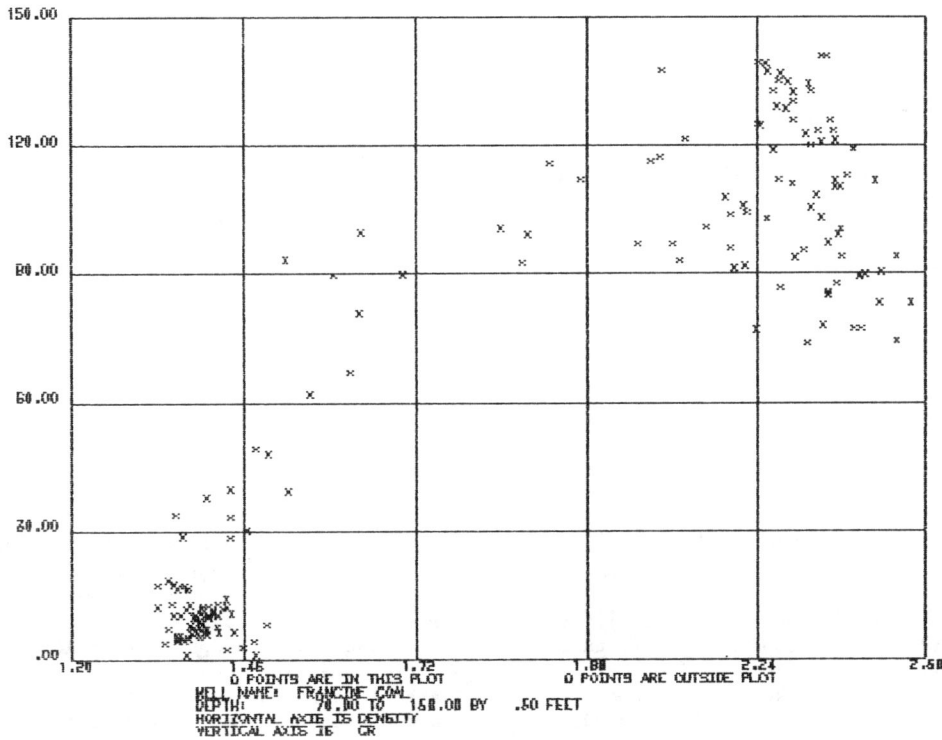

FIGURE 4. Gamma-Density Crossplot

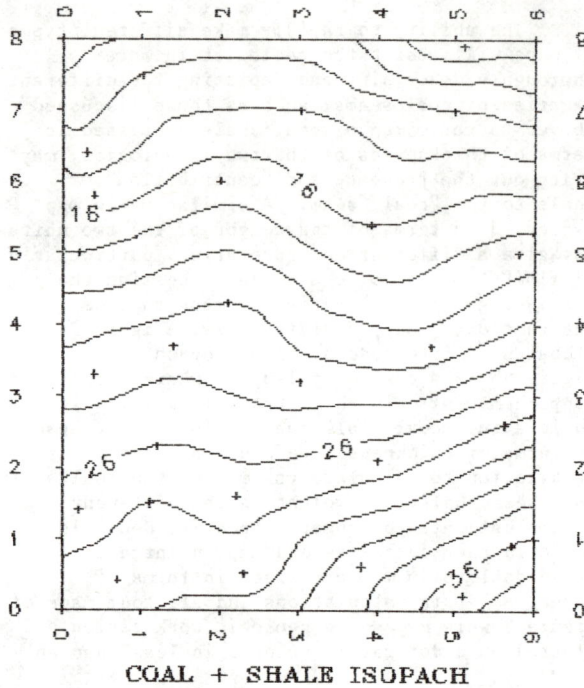

COAL + SHALE ISOPACH

FIGURE 5a. Example Isopach Map:
Coal + Intraseam Shale

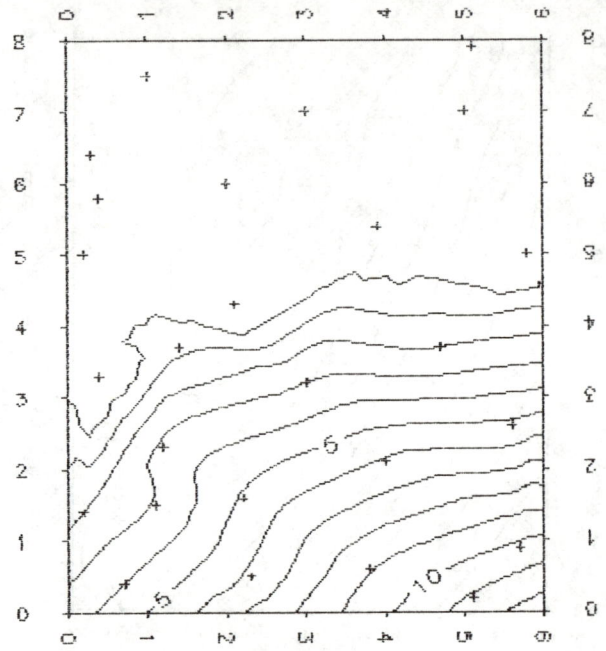

INTRASEAM COAL ISOPACH

FIGURE 5c. Example Isopach Map:
Intraseam Shale Only

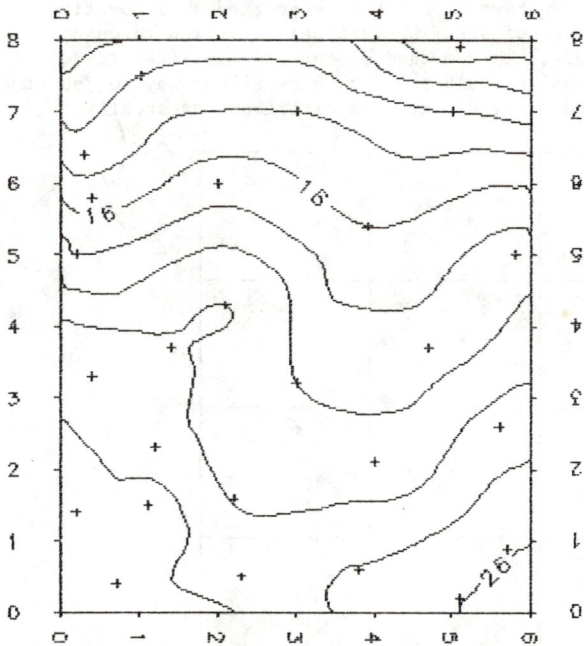

COAL ONLY ISOPACH

FIGURE 5b. Example Isopach Map: Coal Only

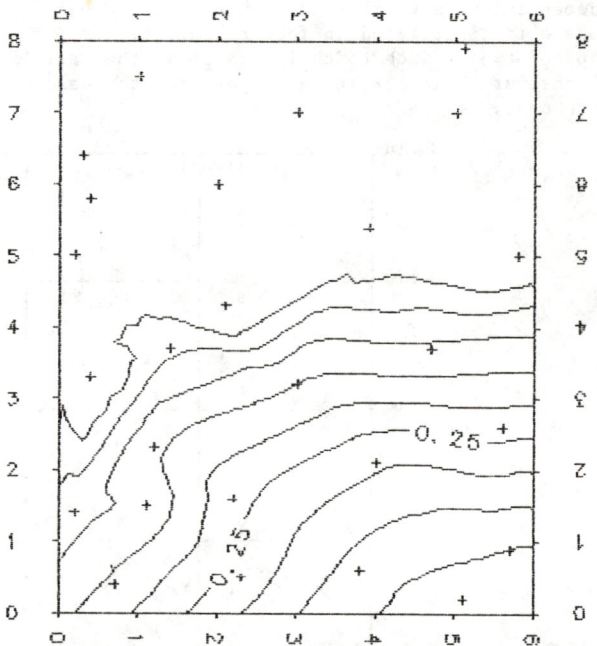

INTRASEAM SHALE:COAL RATIO: THICKNESS

FIGURE 5d. Example Thickness Ratio Map:
Intraseam Shale:Coal

Superimposition of maps can be useful. For example, by overlaying an isopach map onto an isograd map gives important information pertaining to quality - thickness relations. Alternatively, coal:rock ratio map (expressed in whatever terms are appropriate) may be overlayed on a clean coal to total thickness ratio map. If the task is to identify mineable tonnage within

areas that have, say, a greater than 4:1 coal:rock ratio and a clean coal:total thickness ratio of greater than 8:1, one identifies the area that satisfies both criteria, measures the acceptable area, and, using the coal thickness, estimates the coal volume within the acceptable limits. This is very time consuming when done manually. With a microcomputer-based geologic workstation, the task is performed easily and in minimal time. The workstation can be used to define acceptable data ranges through minimum/maximum windowing procedures, thereby eliminating the time consuming process of overlaying, tracing, redrawing, and so forth. With the workstation, the geologist will be able to make reserve calculations based on several different sets of parameters. He will be able to experiment with different coal:rock ratios; different quality isograds; different isopach and entropy cutoffs. These all can be done in less time than it would take to make the same calculations manually on a single set of defining parameters.

Coal Quality

Until recently, most coal quality evaluations have been undertaken by quantitatively analyzing borehole cores for moisture, ash, sulfur, and BTU. Recently, investigations have been undertaken into the use of borehole logs to quantitatively estimate coal quality (e.g., Norris and Thomas, 1980; Weltz, 1976). Reasons for the slow acceptance of coal quality estimates from borehole logs include concerns regarding calibration, accuracy and precision. Calibration of a particular type of borehole log from one area may differ significantly from the calibration of the same log in a different area. The need for readily available, interactive computer power may be another reason for the slow acceptance of quantification of borehole logs for coal quality estimations. With microcomputer geologic workstations, data calibration and analysis are nearly immediate. The advantages of being able to estimate fairly precisely coal quality from borehole logs are obvious. It is much less expensive than having to core all the holes. Also, it is much more timely; the geologist need not wait for laboratory analysis. One important advantage is that coal quality within user-specified zones would be expressed in terms of confidence limits rather than as point estimates as in the case of laboratory analyses.

Borehole logs traditionally have been used in coal seam identification and delineation, thickness measurements and correlation, as discussed above. In order to be used to estimate coal quality, borehole logs must be calibrated with quality data obtained from the laboratory. This is done on a relatively few, selected boreholes scattered throughout the area of interest.

Laboratory analyses are reported for zones within a coal seam, zones in which the coal samples have been blended and homogenized. This has the effect of eliminating within-zone variance of coal quality; the reported value can be thought of as an average value for that zone. In order to compare zone point values (zone values) to log values for that zone, the average log values for the zone must be used. The variance in log responses with a given zone are readily calculated with a geologic workstation,

but without a comparable measure of the laboratory quality data, the value of the within-zone log variance is not known. The relationship between the average quality over a coal zone, as measured in the laboratory, and the variance of log values for that zone might be worth investigating. One might expect quality to be somehow inversely related to log variance. If this suspected relationship can be shown to be systematic and can be properly quantified within a given area, log variance might be used as an indicator of coal quality in that area. The workstation will allow the geologist to readily test and investigate such relationships and their applicability.

Simple regression and multiple regression techniques can be used to generate predictive lines with quality as a function of a single log (e.g., Norris and Thomas, 1980), as planes with quality as a function of two logs, or as hyperplanes with quality as a function of three or more logs. After calibrating the logs to quality values using borehole core data, the geologist can estimate coal quality values at specified depth intervals or over specified zones. Confidence lines, surfaces and hypersurfaces can be calculated at any user-specified level. These are used to bracket coal quality estimates in boreholes that have not been cored. Given the map overlaying capabilities of the microcomputer workstation, the user can overlay quality variance maps onto ratio maps or maps of the lower confidence bounds with stripping ratio maps to obtain conservative volume/quality estimates or, conversely, upper confidence limit bounds to obtain optimistic volume/quality maps. Estimates of both mean and variance, as well as confidence limits can be used in the versatile mapping routines of the geologic workstation.

CONCLUSION

Standard procedures of manually working with coal data soon are going to be outmoded. Microcomputer-based geologic workstations will improve the thoroughness of data analysis, and will increase geologists' analytical probing. They will provide freedom for innovation. The workstation does not, can not perform tasks that the geologist cannot do manually himself. The workstation however, can perform these tasks in a small fraction of the time needed to do them manually. Because it takes so much time to contour data, geologists are disinclined to make extensive revisions, modifications or play what-if games with the data. The geologist normally has standard overlays with which he works. Because of the time needed to make these standard maps and overlays, he generally does not experiment with other variations and combinations.

Geologic workstations will help coal geologists. Geologic workstations will give them the capabilities to undertake their normal tasks with alacrity. Plus, they will provide geologists with the freedom to be imaginative, to

be innovative. Geologic workstations will put
the fun back into exploration and development.

REFERENCES

Doveton, J. H., Ke-An, Z, and Davis, J. C., 1984,
    Three-Dimensional Trend Mapping Using Gamma
    Ray Logs: Simpson Group, Smith-Central
    Kansas, A.A.P.G. Bulletin, Vol. 68, No. 6,
    June, pp. 690-703.

Krumbein, W. C., and Sloss, L. L., 1963,
    Stratigraphy and Sedimentation, 2nd edition,
    W. H. Freeman, San Francisco, 660 p.

Norris, J. O., and Thomas, R., 1980, An In Situ
    Coal Quality Prediction Technique, SPE-AIME,
    55th Annual Fall Technical Conference and
    Exhibition, Dallas.

Weltz, L. S., 1976, Log Evaluation of
    Sub-Bituminous Coals in Magallanes-Chili,
    Trans. SPWLA Seventeenth Annual Logging
    Symposium, Vol. 76-K, pp. 1-13.

Chapter 59

Ore Reserve Calculation: A Partial Block Reserve Area Algorithm

Ralph W. Barbaro

The Pennsylvania State University

Abstract. This paper presents an algorithm to
efficiently calculate the exact area of each block
that lies inside the reserve area. This algorithm
eliminates the approximation error associated with
the centroid and four-corner methods, and thus,
provides more accurate ore reserve estimates.
Because of its efficiency, the algorithm can
compute the exact areas in a fraction of the time
taken by less efficient approximation methods. The
paper first discusses the centroid and four-corner
methods and their associated approximation errors.
The algorithm is then fully described followed by a
step-by-step description of the algorithm.
Finally, comments about the efficiency,
capabilities, and applications of the algorithm are
presented. The appendix contains a listing of the
program PARBRA (PARtial Block Reserve Area), which
is a Fortran version of the algorithm.

INTRODUCTION

Ore reserve calculation is a common task in the
mining industry. It is performed (1) before
purchasing or selling a reserve, (2) to assess the
amount of reserves currently owned, and (3) in mine
planning to determine the amount of reserves that
will be mined with alternative mine plans.

Increasingly, reserve calculations are
performed by computers because they can calculate
the ore reserves in a fraction of the time that it
would take using manual computations, while at the
same time providing more accurate results. In
addition to calculating the ore reserves, computer
systems can draw useful contour, cross-section, and
3-D maps of the reserve.

The standard procedure for a computer
calculation of the amount of reserves in an area is
to (1) overlay the area with a grid that divides
the reserve into blocks, (2) estimate the average
reserve thickness for each of the blocks from known
sample data, (3) calculate the amount of ore in
each block, and (4) sum the appropriate proportion
of ore in each block within the reserve area to
obtain the total ore reserves.

The simplest way to determine the proportion of
ore in each block is to use an approximation
method. Various approximation methods exist but
all use a similar approach. A specified number of
points inside the block are evaluated to determine
if they lie within the reserve area. The
proportion of the block that is inside the reserve
area is assumed to have the same ratio as the
number of points inside the reserve area to the
number of evaluated points. As a result,
approximation methods introduce additional error
into the reserve calculation.

During the past few years, much attention has
been placed on improving accuracy in Step 2.
However, this paper concentrates on improving
accuracy in Step 4 and presents an algorithm that
completely eliminates the block area approximation
error. The end result is a significant improvement
in the overall accuracy of reserve estimates.
Hence, this algorithm allows geologic block models
(from a geostatistical analysis for example) to be
used in situations where the block area
approximation error in Step 4 can be large (i.e.,
in mine planning and other ore reserve calculations
involving relatively small areas).

APPROXIMATION METHODS AND THEIR ASSOCIATED ERROR

In ore reserve calculations, approximate
block areas are used instead of exact areas because
of their computational simplicity. It requires
fewer steps to determine the approximate proportion
of a block within the reserve area than the exact
partial block reserve area.

The two most common approximation methods are
the centroid and the four-corner method. The
centroid method uses one point (the center of the
block) to determine the status of the block. This
method assumes the entire block is inside the
reserve area if the block center point is inside.

The four-corner method uses four points (the
corners of the block) to determine the status of
the block. This method assumes that the fraction
of the block inside the reserves is proportional to
the number of corners inside the reserve area. For
example, if three of four corners are inside, then
the reserve area of the block is assumed to be
three-fourths.

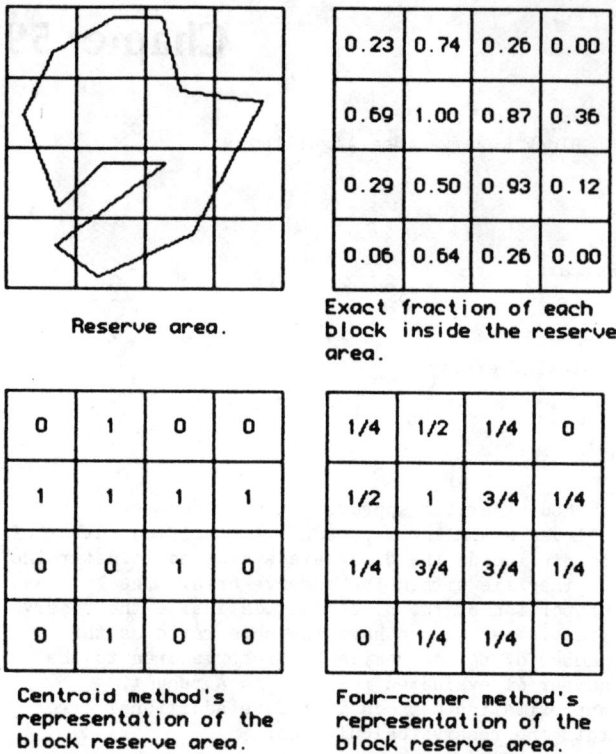

Reserve area.

0.23	0.74	0.26	0.00
0.69	1.00	0.87	0.36
0.29	0.50	0.93	0.12
0.06	0.64	0.26	0.00

Exact fraction of each block inside the reserve area.

0	1	0	0
1	1	1	1
0	0	1	0
0	1	0	0

Centroid method's representation of the block reserve area.

1/4	1/2	1/4	0
1/2	1	3/4	1/4
1/4	3/4	3/4	1/4
0	1/4	1/4	0

Four-corner method's representation of the block reserve area.

Figure 1.   Block reserve area representation with the centroid and four-corner methods versus the exact block reserve area.

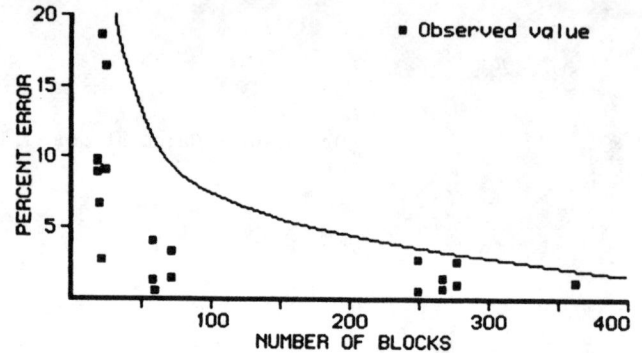

Figure 2.   Range of approximation errors for the centroid and four-corner methods versus the number of blocks covering the reserve area (Barbaro, 1982).

Figure 1 shows the reserve area representation of the centroid and four-corners methods on a block-by-block basis for a small sample reserve. While it appears that the four-corner method has four times the accuracy of the centroid method, in reality it has essentially the same error. This is because the error is directly related to the number of points evaluated. Since the corner points are shared by the four surrounding blocks, the apparent increase in accuracy is not realized. For example, suppose a grid of 100 rows and 100 columns is used. The centroid method would evaluate 10,000 points whereas the four-corner method would evaluate 10,201 points. Thus, the four-corner method is only a slight improvement.

With both methods, the approximation error for an individual block is either an overestimation or an underestimation. For the centroid method, overestimation occurs when the center of the block is inside the reserve area but the entire block is not; whereas underestimation occurs when the center of the block is outside the reserve area but part of the block is inside.

Although the errors for individual blocks can be large, the underestimation errors tend to offset the overestimation errors. As the total number of blocks covering the reserve area increases, these two types of error become closer, causing the combined approximation error of the total reserves

to decrease. However, even with 20 blocks, errors as large as 18 percent can occur (Barbaro, 1982). To be fairly confident that the overall error is less than one percent, approximately 200 blocks or more are needed; but even then, errors of two to three percent are possible although unlikely. Figure 2 shows the approximate upper limit for errors that can occur with respect to the number of blocks overlaying randomly-shaped reserve areas.

Even larger errors than those shown in Figure 2 are likely if the reserve area has a regularly shaped outline, where the long straight lines tend to cause the same types of errors to occur repeatedly. Thus either the over or underestimation error can dominate and cause the overall reserve estimation error to be larger than with a random shape.

WHY PARTIAL BLOCK RESERVE AREAS?

The key question that should be asked of any alternative method is, "Why should this method be used instead of the existing methods?"

In this case, there are two answers to this question. The first is improved accuracy. As Figure 2 depicts, the reserve calculation error caused by block area approximation for large reserves (over 200 blocks overlaying the reserve area) is about one percent. If the reserve estimation error is also one percent, half of the overall error is eliminated when the exact area of each block is used instead. However, for reserve areas covered by fewer than 200 blocks, the improvement in accuracy can be much greater. The second reason for adopting this new method is decreased computation time. This can be very important if several different reserve areas have to be evaluated or if a microcomputer is used.

The partial block reserve algorithm and the centroid method were evaluated on the same reserve to compare their computation times. A reserve area with 248 boundary outline points and a 20-by-25 block grid was used with an IBM 3081 mainframe performing the calculations. The centroid method took 20.72 cpu seconds whereas the partial block reserve area algorithm took only 0.62 cpu seconds--over 33 times faster.

This improvement becomes even larger with more boundary points and/or blocks. If these methods were run on an IBM-PC with Microsoft Fortran, the computation times would be approximately 830 times slower (Knoble, 1984), or 4 hours and 47 minutes for the centroid method versus only 8 minutes and 34 seconds for the partial block reserve area method. Hence, both "turn around" time and computation costs are greatly reduced.

With these two advantages, the partial block reserve area method allows geologic block models to be used for mine planning. Frequently, reserve estimation for a reserve area is done with a hand-drawn contour map. Each contour interval inside the reserve area is planimetered and multiplied by the average value of that contour interval. Not only is this a time consuming task, but it is subject to error both in preparation of the contour map and the planimetering of the contour intervals.

The solution is to use geostatistics or some other reserve estimation technique to provide both unbiased and accurate estimates for the reserve on a block-by-block basis. Then by digitizing (or entering the coordinates of the corners of the reserve area), the partial block reserve area method will provide the exact area of each block inside the reserve area. This information can then be used to calculate the reserves for that block. And finally, summing up the reserves for all of the blocks will provide a more accurate estimate of the reserve.

Not only is this procedure more accurate, but it eliminates the manual planimetering and hand calculations. Because evaluation of a reserve area is virtually instantaneous, several different mining plans can be evaluated quickly. The extra time spent planimetering and performing the calculations can then be used for other tasks.

Another related advantage of the partial block reserve area method is that it allows larger blocks to be used in the reserve block model. A smaller block size is frequently adopted in an effort to minimize the approximation error of the centroid or four-corner methods. Since the partial block reserve area method eliminates the block area approximation error, larger blocks can be used during the reserve estimation step, thus reducing computation and storage requirements for the reserve block model.

PARTIAL BLOCK RESERVE AREA CALCULATION

Two phases are needed to calculate the exact area of each block that lies inside the reserve area. The first phase is to calculate the reserve area of all blocks that are partially within the reserve (these are the blocks through which the reserve boundary line passes). The second phase is to determine whether each of the remaining blocks is either inside or outside the reserve area.

Phase 1: Partial Block Area Calculation

The goal is to calculate the exact reserve area of each block that overlays the reserve boundary outline. The basic procedure is to start at any point on the reserve boundary outline, traverse clockwise around the outline, and calculate the partial reserve area of each block as the boundary line enters and exits.

Four basic operations are required:

1. Calculate the point of intersection between the reserve boundary outline and the edge of the block.

2. Obtain the corner points of a polygon that describes the partial reserve area of the block.

3. Calculate the area of the polygon.

4. Determine the reserve area of the complex blocks that have several different partial areas; i.e., blocks intersected more than once by the reserve boundary.

The method used for each of these operations is discussed separately.

Calculating the Point of Intersection. The point of intersection between the boundary outline and the edge of the block can be found by different methods; the method used in this algorithm was selected because of its speed.

The basic procedure (refer to Figure 3) is to calculate the slope of the line from the point inside the block (a,b) to the point outside (c,d), and the quadrant it lies in. Given the quadrant that contains the line, only two sides of the block can possibly be intersected. Figure 3 illustrates this; since the line is in the third quadrant (lower left), only the bottom or the left side could be intersected.

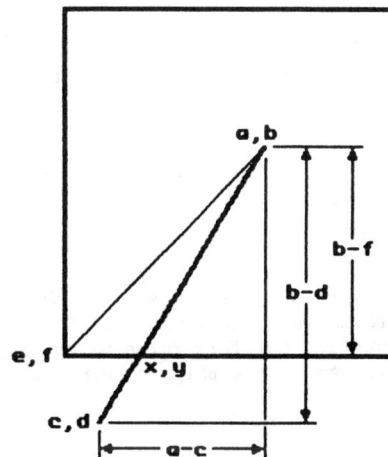

Figure 3

The next step is to calculate the slope from the point inside the block (a,b) to the corner point of the block in the same quadrant (e,f), and determine the side of the block that the line intersects. In this case, because the slope to the corner point is less than the slope to the outside point, the bottom of the block must be the intersected side.

Finally, knowing the intersected side and the slope to the second point, the point of intersection (x,y) can be calculated. In this example, y=f since the bottom of the block is a horizontal line, and x=a-(b-f)/s, where the slope s=(b-d)/(a-c).

Two sets of equations are needed to calculate the point of intersection, depending on the side of the block that is intersected:

Top and Bottom Sides:

$$x = a + (f-b)/s$$
$$y = f$$

Left and Right Sides:

$$x = e$$
$$y = b + (e-b)*s$$

where point (a,b) is the point inside the block, point (c,d) is the point outside the block, point (e,f) is the corner point of the block in the quadrant containing the line, and s is the slope of the line from point (a,b) to point (c,d), i.e., s=(d-b)/c-a).

Obtaining the Partial Reserve Area Polygon. The procedure (refer to Figure 4) to determine the corner points of a polygon that describes the partial reserve area of the block as shown in Figure 4a consists of four steps:

1. Calculate the point where the boundary line first enters the block (Point 1),

2. Moving clockwise, add all boundary points inside the block (Points 2 thru 6),

3. Calculate the exit point of the boundary line (Point 7), and

4. Moving clockwise, add all corners of the block between the exit point and the entrance point (Points 8 and 9).

The polygon in Figure 4a has nine corner points, as shown in Figure 4b: one entrance point, five interior boundary points, one exit point, and two block corner points.

Calculating the Area of the Polygon. The method used to calculate the polygon area is the area-by-coordinates method. This section derives the basic equation for this value using only the coordinate values of the polygon corners.

The area enclosed by an outline of any curved shape can be calculated with the following line integral when the perimeter is traversed in the clockwise direction (Sokolnikoff and Redheffer, 1966),

$$A = \frac{1}{2} \int_C (y \, dx - x \, dy), \tag{1}$$

where A is the area of the polygon, x and y are the coordinate values of a point on the perimeter of the shape, and the region C is the perimeter of the shape in the clockwise direction.

By discretizing, x_i can be substituted for x, y_i for y, $x_{i+1} - x_i$ for dx, $y_{i+1} - y_i$ for dy, and n distinct segments for the continuous perimeter C. Discretizing Equation 1 produces,

$$A = \frac{1}{2} \sum_{i=1}^{n} (y_i(x_{i+1} - x_i) - x_i(y_{i+1} - y_i)), \tag{2}$$

Expanding Equation 2 yields,

$$A = \frac{1}{2} \sum_{i=1}^{n} (y_i x_{i+1} - x_i y_i - x_i y_{i+1} + x_i y_i), \tag{3}$$

The second and fourth terms cancel leaving,

$$A = \frac{1}{2} \sum_{i=1}^{n} (y_i x_{i+1} - x_i y_{i+1}), \tag{4}$$

By shifting terms, Equation 4 can be rewritten as,

$$A = \frac{1}{2} \sum_{i=1}^{n} x_i(y_{i-1} - y_{i+1}), \tag{5}$$

Equation 5 uses y_0 and y_{n+1} which are not in the range of points from 1 to n. Because the polygon is a continuous loop, set $y_0 = y_n$ and $y_{n+1} = y_1$.

Equation 5 can be used to compute the area of any polygon provided the points are specified in the clockwise direction. If the points are counterclockwise, the area will be correct but it will have a negative value. [Note that Equation 5 is the general form of the method used to calculate area by coordinates that is used in surveying texts (Brinker and Wolf, 1984)].

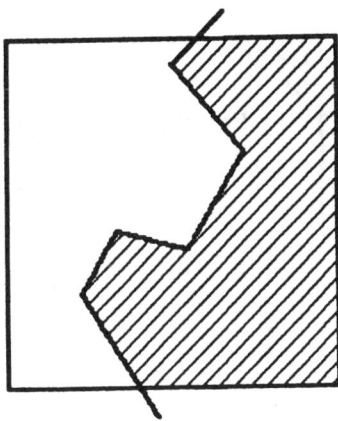

Figure 4a

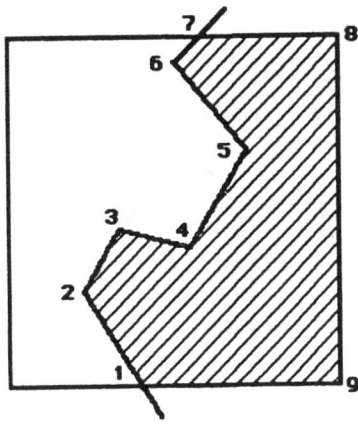

Figure 4b

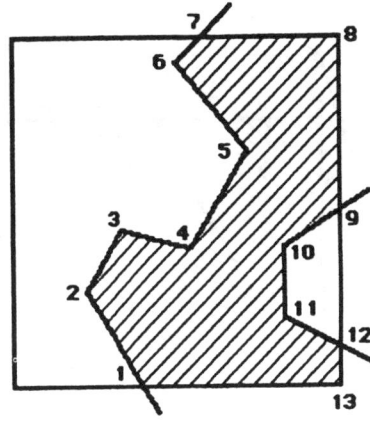

Figure 4c

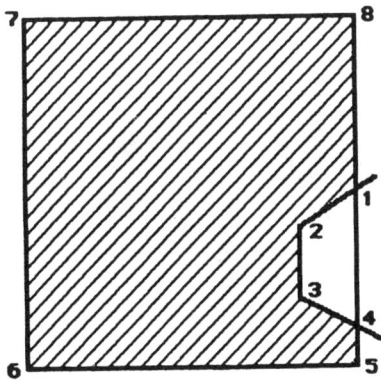

Figure 4d

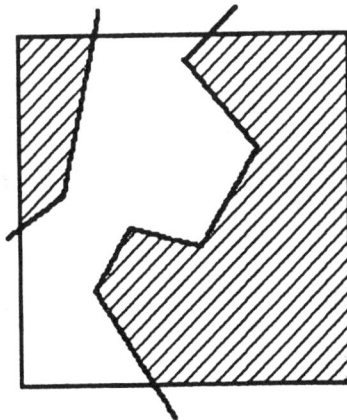

Figure 4e

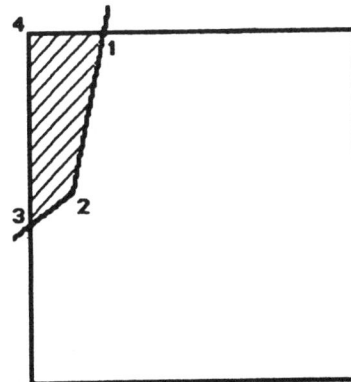

Figure 4f

Several Partial Areas Within the Same Block.
A complicating situation occurs whenever the
boundary outline exits and then reenters the block
(Figure 4c). The problem arises because with this
procedure the reserve outline is traversed blindly,
so there is no way to know whether or not the
boundary line reenters the block. The simplist
solution is to calculate the first area as shown in
Figure 4b and use a different approach if the
outline reenters the block. By visual inspection,
the partial block reserve area can be defined as
the polygon shown in Figure 4c.

The first time the boundary line enters the
block, the partial area of the block is computed
normally. Each time the boundary line reenters the
block, a secondary area for the polygon is defined
(Figure 4d) using the normal procedure. But visual
inspection shows that the non-crosshatched area in
Figure 4d should be subtracted from the
crosshatched area in Figure 4b. If the reserve
outline was as shown in Figure 4e, the secondary
area (crosshatched in Figure 4f) should be added to
the crosshatched area in Figure 4a to obtain the
partial area for that block.

The general rule on whether to add the secondary
area or to subtract the difference between the
total area of the block and the secondary area
depends on where the boundary line:

1. first enters the block,

2. last exits the block, and

3. reenters the block.

Given these three locations, only two cases can
occur:

Case 1. If the boundary line reenters the block
clockwise between the first entrance
point and the last exit point, then add
the secondary area.

Case 2. If the boundary line reenters the block
clockwise between the last exit point and
the first entrance point, then subtract
the complement of the secondary area
(difference between the total area of the
block and the secondary area).

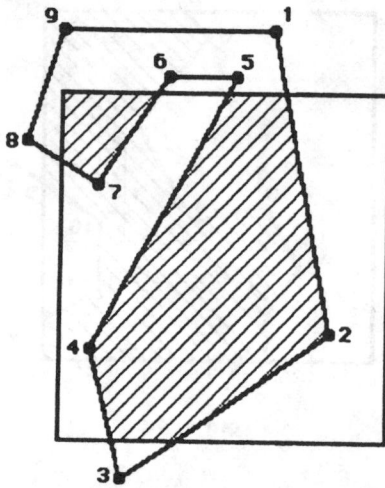

Figure 5a

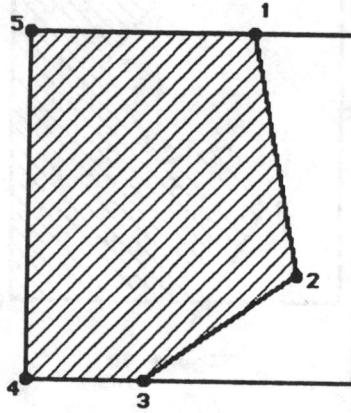

Figure 5b

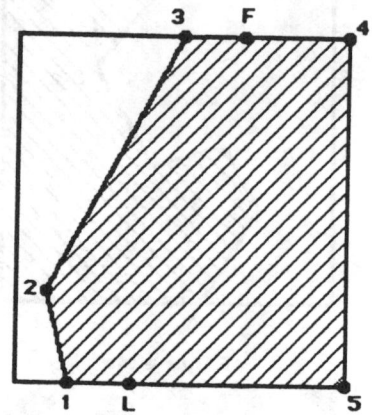

Figure 5c

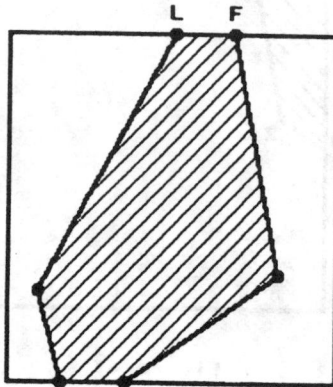

Figure 5d

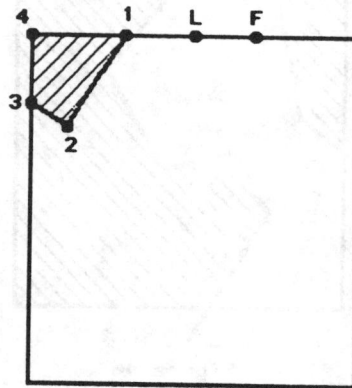

Figure 5e

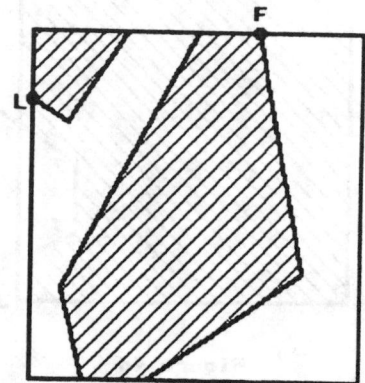

Figure 5f

It is important to remember that this rule applies only when the outline is traversed in the clockwise direction.

As an example, consider the reserve area shown in Figure 5a in which the reserve boundary consists of nine corner points. The procedure is to start at Point 1 and progress clockwise; hence, the polygon shown in Figure 5b is the first partial block reserve area. Note that in this figure, the first entrance point is Point 1 and the last exit point is Point 3.

The next intersection produces the polygon shown in Figure 5c. Because the new entrance point (Point 1) lies clockwise between the last exit point (Point L) and the first entrance point (Point F), subtract the noncrosshatched area of Figure 5c from the crosshatched area in Figure 5b to arrive at the new partial block reserve area shown in Figure 5d.

The last intersection with the block is shown in Figure 5e. Since the new entrance point (Point 1) is clockwise between the first entrance point (Point F) and the last exit point (Point L), add the crosshatched area to obtain the new partial reserve area shown in Figure 5f. Because the reserve outline does not intersect the block again, the partial block reserve area shown in Figure 5f represents the area of the block inside the reserve area. This area compares favorably with the original reserve outline shown in Figure 5a.

Note that the first enter point did not change. The point where the reserve outline first enters the block is always the first entrance point. However, the last exit point changes every time the reserve outline reenters the block. The last exit point is the exit point of the most recent partial reserve area of the block.

Phase 2: Evaluation of the Remaining Blocks

After the boundary line has been traversed and the partial reserve areas of all blocks that intersect the reserve outline have been calculated, the remaining blocks must be either completely inside or completely outside the boundary area. This status must be determined.

Evaluating if a Block is Inside the Reserve. For this evaluation, the center of the block is commonly used as the evaluation point, although any other point within the block can be used. There are two methods available to determine if a point is inside or outside a polygon: the interior angle method and the linear intersection method. The interior angle method is commonly used in the centroid and four-corner methods, even though it is computationally slower and has no advantage over the linear intersection method. Because of its faster speed, the linear intersection method is used in this algorithm, but the interior angle method is discussed because of its popularity.

INTERIOR ANGLE METHOD: In this method, the sum of all possible interior angles is found (see Figure 6), using the point of interest as the vertex and two adjacent boundary points to complete the angle. The clockwise angles are assigned a positive sign; and the counterclockwise angles are assigned a negative sign. If the sum of the interior angles is 360 degrees, the point is inside the polygon. If the sum is zero, the point is outside the polygon. When the boundary line has n points, this method requires n interior angles to be calculated for each block to be evaluated.

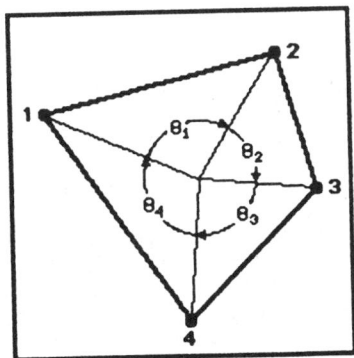

Figure 6. Interior angle method.

LINEAR INTERSECTION METHOD: In this method, the number of intersections between the reserve boundary outline and a line extending from the point is counted. Figure 7 demonstrates this method with an extremely erratic outline. If there is an odd number of intersections (Figure 7a), the point is inside the reserve area; whereas an even number of intersections (Figure 7b) means the point is outside. Since the direction of the line extending from the point can be chosen, it is convenient to select a vertical or horizontal line to simplify computation.

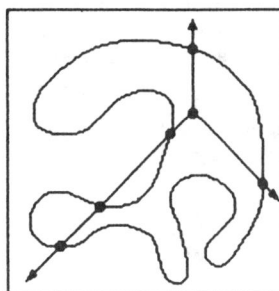

Figure 7a. Odd number of interserctions, point in.

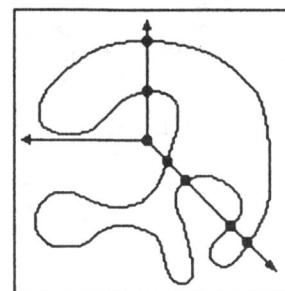

Figure 7b. Even number of intersections, point out.

Figure 7. Linear intersection method to determine if a point is inside or outside the reserve area.

Even though the linear intersection method is faster, it is still a time-consuming process because each side of the reserve outline must be evaluated for an intersection with the line extending from the point. The computation time is significant because this evaluation must be performed independently for each block.

Similarity of Neighboring Blocks. The application of the following concept reduces the number of blocks that need to be evaluated with the linear intersection method. The basic idea is that blocks completely inside the reserve area are always adjacent to other blocks partially or completely inside. Similarly, blocks completely outside are always adjacent to other blocks partially or completely outside.

The first application of this concept is to identify a large number of blocks outside a reserve area that lies entirely inside the block grid. Once all blocks that intersect the boundary line have been assigned a partial reserve area, any row or column can be traversed inward from the outside edge of the grid, and each block encountered that has not already been assigned an area must be outside.

Using the example boundary outline in Figure 8, all unmarked blocks were assigned a partial reserve area by the partial area calculation procedure in Phase 1. The right crosshatched blocks were identified as outside the reserve area, using the above-described procedure on a row-by-row basis. This procedure can greatly reduce the number of blocks outside the reserve area that must be evaluated. In this example, the number was reduced from 31 to 9, or a 71 percent reduction.

The second application is to extend the information obtained from the linear intersection method for a single block to the neighboring blocks. When one of the remaining blocks (left crosshatched blocks in Figure 8) has been evaluated with the linear intersection method, the result can be extended to all similar neighboring blocks. The procedure can be extended in every direction until a block that intersects the reserve boundary is encountered.

The second application further reduces the number of blocks that must be evaluated with the linear intersection method. This search procedure also reduces the computation time significantly.

When this concept is integrated with the linear intersection method, the efficiency of the algorithm is greatly improved. Because the algorithm searches from left to right and from bottom to top, only those blocks to the upper right are considered. When a block is determined to be inside (outside), all remaining blocks directly to the right must be inside (outside) and all remaining blocks directly above those blocks must be inside (outside). Referring to the example in Figure 8, only three blocks actually need to be determined with the linear intersection method, i.e., the lower left block of each of the three groups of unmarked blocks. Thus by applying the similarity of neighboring blocks concept, the number of blocks that are evaluated with the linear intersection method is reduced from 31 to 3, or a 90 percent reduction. In general, the percentage of blocks that are evaluated with the linear intersection method decreases when the total number of blocks in the grid increases.

PARTIAL BLOCK ALGORITHM

This algorithm describes the step-by-step procedure to calculate the exact area of each block that lies inside the reserve area. Given a description of the grid layout and the reserve outline, the algorithm utilizes the two-phase procedure discussed in this paper to calculate the reserve area for each block.

The most important design criteria for the algorithm was generality and efficiency. The algorithm needed the capability to handle any shape or size of reserve area, and in addition, to allow the grid layout to be any size or location.

At the same time, the computation time and memory requirements were minimized. In those instances where the computation time and memory requirements conflicted, the computation time was given priority. Every attempt was made to make the algorithm as fast as possible.

The step-by-step algorithm as presented in this section is intended to be generic, that is, not specific to any particular computer language. However, a FORTRAN program of the algorithm is presented in the Appendix to provide potential users with a complete version that has been extensively validated.

The algorithm requires seven different categories of information (arrays) to be stored. Two arrays are for the x- and y-coordinates of the reserve corner points, and two are temporary arrays for the x- and y-coordinates of the polygon that describes the current partial block reserve area. The last three arrays are required to store the following information each block: (1) reserve area, (2) location of the first entrance point, and (3) location of the last exit point.

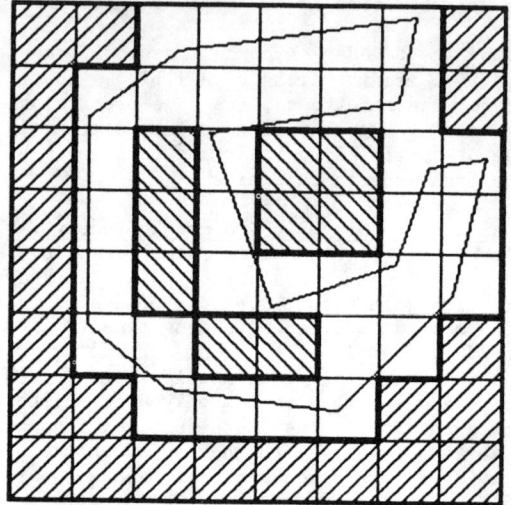

Figure 8.   Indentifying similar blocks.

Algorithm Procedure

1. Input data:
   a. Input the minimum and maximum coordinate values of the overall block grid and the block size.
   b. Input the coordinates of the reserve boundary corners. Count the number of points in the reserves boundary as the coordinates are input and find the minimum and maximum coordinate values.
   c. Check if all reserve boundary points are inside the grid area. If not, stop and enlarge the grid area.

2. Initialize values:
   a. Calculate the number of blocks in the x- and y-directions.
   b. Calculate the area of a block.
   c. Initialize the following arrays to -1: (1) area, (2) first entrance point, and (3) last exit point.
   d. If the first and last points in the reserve boundary coordinate array do not have identical coordinates, add the first point coordinates to the end of the array.
   e. Calculate the area of the reserve. If the sign is negative, the order of the boundary points must be counterclockwise, and therefore, reverse the order of the points so that they are clockwise.
   f. Calculate the row and column of the block that contains the first reserve boundary point. This block is the starting block.
   g. Modify the reserve boundary outline so that the first point lies on the edge of the starting block. Start at the first point of the input reserve boundary, and sequentially check each counterclockwise point to find the first one that lies outside the block. Once the last point inside and the first point outside is known, calculate the coordinates of the point of intersection. Use this point

as the new starting and ending point for the reserve boundary outline. This point is also the entrance point for the starting block.

3. Calculate the area of the blocks on the outline: Traverse the reserve boundary outline and calculate the partial block area of each block as they are encountered. Use a temporary array to save the points that define the partial reserve area for the block. (Refer to Figure 5b).
   a. Set the first point in the temporary arrays equal to the coordinate values of the entrance point (Point 1).
   b. Add the coordinate values of all boundary points clockwise from the entrance point that are inside the block (Point 2) to the temporary arrays.
   c. Calculate the coordinate values of the exit point (Point 3) and add to the temporary arrays.
   d. Complete the partial block polygon by adding all corners of the block clockwise from the exit point to the entrance point (Points 4 and 5) to the temporary arrays.
   e. Calculate the partial block area with the area-by-coordinates method using the coordinate values stored in the temporary arrays.
   f. If this is the first time that the reserve outline intersects this block, set the reserve area of the block equal to the partial block area and save the clockwise perimeter distance from the upper left corner of the block to the entrance point in the first entrance point array.
   g. If the block was previously intersected by the reserve outline, then add the partial area if the current entrance point is clockwise between the first entrance point and the previous exit point. Otherwise, subtract the complement of the partial area. Save the clockwise perimeter distance from the upper left corner of the block to the current exit point in the last exit point array.
   h. Calculate the row and column number of the next block that the boundary line enters and set the exit point of the last block equal to the entrance point of the next block.
   i. Repeat Steps 3a through 3h until the exit point of the last block is equal to the last point in the modified reserve boundary outline.

4. Identify blocks outside the reserve area:
   a. For each row in the block grid, traverse from the left side of the grid towards the right and assign all blocks with an area equal to -1 an area of zero (completely outside the reserve boundary) until a block with an area not equal to -1 is reached.
   b. Repeat the procedure in Step 4a except traverse from the right side of the grid towards the left.

5. Determine the area of the remaining blocks:
   a. Scan all blocks row-by-row from left-to-right.
   b. If the block has an area equal to -1, determine if its center is inside or outside the reserve area with the linear intersection method and assign the corresponding area to the block.
   c. Assign the same area to all blocks that can be reached by moving to the right and up provided that no block with an area not equal to -1 area is encountered.
   d. During the scan, sum the area of each blocks to obtain the total area of the reserves.

6. Print results:
   a. Print the total reserve area of all of the blocks.
   b. Print a table of the partial reserve area for each block that is not completely outside the reserve area.

SUMMARY

This paper presents an algorithm to calculate the exact area of each block inside the reserve area. The exact block areas can be used in place of the approximate block areas provided by common heuristic methods to provide more accurate ore reserve calculations. The accuracy of heuristic methods decreases dramatically as the number of blocks that cover the reserve area decreases below 200. Thus this algorithm will allow ore reserve block models to be used for mine planning and other applications that require ore reserve calculations of relatively small areas.

The algorithm was designed to be as efficient as possible, and thus the algorithm can determine the exact reserve area of each block with computation times many times faster than that of a typical approximation method. The computation time was minimized by incorporating the linear intersection method and the "similarity of neighboring blocks" concept along with efficient procedures for calculating the point of intersection between a line and a block, and for determining the polygon that describes the partial reserve area for blocks intersected by the reserve boundary outline.

The partial block reserve area algorithm is superior to the centroid and four-corners methods because it not only provides the exact area of each block, but does it in a fraction of the time required by the approximation methods. Hence the resulting ore reserve calculations are not only more accurate but the computation times are faster and the computing costs are lower.

ACKNOWLEDGEMENTS

The author wishes to thank the Department of Mineral Engineering at The Pennsylvania State University for their cooperation with this work. Also a special thanks to Judy Kiusalaas for her help in editing and proofreading this paper.

REFERENCES

Barbaro, R. W., 1982, "An Evaluation of the
    Centroid and Four-Corner Methods for Area and
    Ore Reserve Approximation," Unpublished report
    submitted for Mining 590, Mining Section,
    Department of Mineral Engineering, The
    Pennsylvania State University, University Park,
    PA.

Brinker, R.C., and Wolf, P.R., 1984, Elementary
    Surveying, Seventh Edition, Harper & Row,
    Publishers, New York, NY, pp. 249,250.

Knoble, H. D., 1984, "Computing Awareness in Time,
    Space, and Precision," Microcomputer
    Information Exchange Conference, March 10,
    1984, The Pennsylvania State University,
    University Park, PA.

Sokolnikoff, I. S., and Redheffer, R. M., 1966,
    Mathematics of Physics and Modern Engineering,
    Second Edition, McGraw-Hill Book Company, New
    York, NY, pp. 372.

APPENDIX

```
C***********************************************************************
C
C   PARBRA.V1.1  -  SEPTEMBER 24, 1984  -  COPYWRITE 1984
C
C   WRITTEN BY:   RALPH W. BARBARO
C                 THE PENNSYLVANIA STATE UNIVERSITY
C                 104 MINERAL SCIENCE BUILDING
C                 UNIVERSITY PARK, PA   16802
C
C      THIS PROGRAM CALCULATES THE PARTIAL RESERVE AREA OF EACH BLOCK
C   GIVEN THE RESERVE OUTLINE POINTS AND THE ORIENTATION OF THE BLOCKS.
C   THE PROGRAM IS CURRENTLY DIMENSIONED TO HANDLE A 100-BY-100 BLOCK
C   GRID AND UP TO 1000 RESERVE OUTLINE POINTS.  THESE DIMENSIONS CAN BE
C   INCREASED IF NECESSARY.
C
C      NO PART OF THIS PROGRAM MAY BE REPRODUCED, STORED IN A RETRIEVAL
C   SYSTEM, OR TRANSMITTED IN ANY FORM OR BY ANY MEANS, ELECTRONIC,
C   MECHANICAL, PHOTOCOPYING, RECORDING OR OTHERWISE WITHOUT THE PRIOR
C   WRITTEN PERMISSION OF THE AUTHOR WITH THE EXCEPTION THAT THE PROGRAM
C   MAY BE ENTERED, STORED, AND EXECUTED IN A COMPUTER SYSTEM FOR
C   NONCOMMERCIAL USE.
C
C***********************************************************************
C
C                             DISCLAIMER
C
C      THE AUTHOR HAS TAKEN DUE CARE IN PREPARING THIS PROGRAM, INCLUDING
C   RESEARCH, DEVELOPMENT, AND TESTING TO ASCERTAIN ITS EFFECTIVENESS.
C   THE AUTHOR MAKES NO EXPRESSED OR IMPLIED WARRANTY OF ANY KIND WITH
C   REGARD TO THIS PROGRAM.  IN NO EVENT SHALL THE AUTHOR BE LIABLE FOR
C   INCEDENTAL OR CONSEQUENTIAL DAMAGES IN CONNECTION WITH OR ARISING
C   OUT OF THE PERFORMANCE OR USE OF THIS PROGRAM.
C
C***********************************************************************
C
C   I N P U T    D O C U M E N T A T I O N
C
C***********************************************************************
C   CARD TYPE 1 - TITLE FOR THE COMPUTER RUN
C
C      TITLE = TITLE FOR THE RESERVE AREA (MAXIMUM LENGTH IS
C              72 ALPHANUMERIC CHARACTERS)
C
C      FORMAT(9A8)
C
C*********
C   CARD TYPE 2 - INPUT INFORMATION FOR GRID LAYOUT
C
C      XMIN = MINIMUM X-COORDINATE FOR THE GRID
C      XMAX = MAXIMUM X-COORDINATE FOR THE GRID
C      YMIN = MINIMUM Y-COORDINATE FOR THE GRID
C      YMAX = MAXIMUM Y-COORDINATE FOR THE GRID
C      XWID = DISTANCE BETWEEN GRIDS IN THE X-DIRECTION
C      YWID = DISTANCE BETWEEN GRIDS IN THE Y-DIRECTION
C
C      ***IMPORTANT NOTE*** ==> ALL BOUNDARY OUTLINE POINTS MUST BE
C         INSIDE THE MINIMUM AND MAXIMUM GRID VALUES.
C
C      FORMAT(6F10.0)
C
```

```
C*********
C   CARD TYPE 3 - FORMAT SPECIFICATION FOR THE RESERVE OUTLINE
C
C      IFMT = FORMAT OF THE RESERVE BOUNDARY DATA
C
C      FORMAT(9A8)
C
C*********
C   CARD TYPE 4 - RESERVE BOUNDARY OUTLINE COORDINATES
C
C      POINTX(I) = X-COORDINATE VALUES FOR THE ITH BOUNDARY POINT
C      POINTY(I) = Y-COORDINATE VALUES FOR THE ITH BOUNDARY POINT
C
C      REPEAT CARD TYPE 4 FOR EACH BOUNDARY POINT.
C
C      NOTE:  NO SPECIAL CARD IS NECESSARY TO DESIGNATE THE END OF DATA.
C
C      FORMAT(IFMT) <==SPECIFIED IN CARD TYPE 4
C
C**********************************************************************
C   DEFINITION OF ARRAYS:
C
C      AREA(I,J)   = ACTUAL AREA OF BLOCK I,J THAT IS INSIDE THE OUTLINE
C      ARRAYX(M)   = SET OF X-COORDINATES DEFINING THE PARTIAL AREA OF
C                    OF THE BLOCK WHERE M IS THE POINT NUMBER.
C      ARRAYY(M)   = SET OF Y-COORDINATES DEFINING THE PARTIAL AREA OF
C                    OF THE BLOCK WHERE M IS THE POINT NUMBER.
C      DENTER(I,J) = DISTANCE AROUND THE PERIMETER OF THE BLOCK
C                    FROM THE UPPER LEFT CORNER TO THE POINT THE THAT
C                    BOUNDARY LINE FIRST ENTERED THE BLOCK I,J
C      DEXIT(I,J)  = DISTANCE AROUND THE PERIMETER OF THE BLOCK
C                    FROM THE UPPER LEFT CORNER TO THE POINT THE THAT
C                    BOUNDARY LINE LAST LEFT THE BLOCK I,J
C      POINTX(I)   = X-COORDINATE VALUES FOR THE ITH BOUNDARY POINT
C      POINTY(I)   = Y-COORDINATE VALUES FOR THE ITH BOUNDARY POINT
C**********************************************************************
C   DEFINITION OF MAJOR VARIABLES:
C
C      BLARE   = AREA OF AN ENTIRE BLOCK
C      BX1     = X-COORDINATE OF THE LEFT SIDE OF THE CURRENT BLOCK
C      BX2     = X-COORDINATE OF THE RIGHT SIDE OF THE CURRENT BLOCK
C      BY1     = Y-COORDINATE OF THE BOTTOM SIDE OF THE CURRENT BLOCK
C      BY2     = Y-COORDINATE OF THE TOP SIDE OF THE CURRENT BLOCK
C      IENTSD  = SIDE OF THE BLOCK THAT THE PROPERTY OUTLINE ENTERED THE
C                CURRENT BLOCK: 1=NORTH, 2=EAST, 3=SOUTH, 4=WEST
C      IEXSD   = SIDE OF THE BLOCK THAT THE PROPERTY OUTLINE EXITED THE
C                CURRENT BLOCK
C      IX      = COLUMN INDEX FOR THE CURRENT BLOCK, LEFT COLUMN VALUE IS 1
C      IY      = ROW INDEX FOR THE CURRENT BLOCK, BOTTOM ROW VALUE IS 1
C      NARRAY  = NUMBER OF POINTS IN THE TEMPORARY BLOCK AREA RRAY
C      NUMPTS  = NUMBER OF POINTS IN THE BOUNDARY OUTLINE
C      NXBLOC  = NUMBER OF BLOCKS IN THE X-DIRECTION (COLUMNS)
C      NYBLOC  = NUMBER OF BLOCKS IN THE Y-DIRECTION (ROWS)
C      XCMIN   = MINIMUM X-COORDINATE FOR THE RESERVE OUTLINE
C      XCMAX   = MAXIMIM X-COORDINATE FOR THE RESERVE OUTLINE
C      YCMIN   = MINIMUM Y-COORDINATE FOR THE RESERVE OUTLINE
C      YCMAX   = MAXIMUM Y-COORDINATE FOR THE RESERVE OUTLINE
C      XMIN    = MINIMUM X-COORDINATE FOR THE GRID
C      XMAX    = MAXIMUM X-COORDINATE FOR THE GRID
C      YMIN    = MINIMUM Y-COORDINATE FOR THE GRID
C      YMAX    = MAXIMUM Y-COORDINATE FOR THE GRID
C      XWID    = DISTANCE BETWEEN GRIDS IN THE X-DIRECTION
C      YWID    = DISTANCE BETWEEN GRIDS IN THE Y-DIRECTION
C**********************************************************************
```

```
      IMPLICIT REAL*8 (A-H,P-Z)
      REAL*4 DENTER,DEXIT
      REAL*8 TITLE(9),IFMT(9)
      COMMON /BLOCK1/ POINTX(1000),POINTY(1000),ARRAYX(100),ARRAYY(100)
      COMMON /BLOCK2/ AREA(100,100),DENTER(100,100),DEXIT(100,100)
      COMMON /BLOCK3/ XMIN,XWID,XMAX,YMIN,YWID,YMAX,NYBLOC,NXBLOC,BLARE
C***INPUT DATA
      READ(5,10) TITLE
      WRITE(6,15) TITLE
      READ(5,20) XMIN,XMAX,YMIN,YMAX,XWID,YWID
      NXBLOC=(XMAX-XMIN)/XWID
      NYBLOC=(YMAX-YMIN)/YWID
      BLARE=XWID*YWID
      WRITE(6,25) XMIN,XMAX,YMIN,YMAX,XWID,YWID,NXBLOC,NYBLOC,BLARE
      READ(5,10) IFMT
      XCMIN=1.0D50
      XCMAX=-1.0D50
      YCMIN=1.0D50
      YCMAX=-1.0D50
C**INPUT THE COORDINATES OF A BOUNDARY POINT
      I=0
100   I=I+1
      READ(5,IFMT,END=110) POINTX(I),POINTY(I)
      IF(POINTX(I).LT.XCMIN) XCMIN=POINTX(I)
      IF(POINTX(I).GT.XCMAX) XCMAX=POINTX(I)
      IF(POINTY(I).LT.YCMIN) YCMIN=POINTY(I)
      IF(POINTY(I).GT.YCMAX) YCMAX=POINTY(I)
      GOTO 100
110   NUMPTS=I-1
C**PRINT THE MINIMUM AND MAXIMUM COORDINATES VALUES FOR THE OUTLINE
C**  AND CHECK THAT THEY ARE INSIDE THE INPUTTED GRID AREA
      WRITE(6,30) XCMIN,XCMAX,YCMIN,YCMAX
      IF(XCMIN.GT.XMIN.AND.XCMAX.LT.XMAX.AND.YCMIN.GT.YMIN.AND.
     *   YCMAX.LT.YMAX) GOTO 115
      WRITE(6,35)
      STOP
C***INITIALIZE OR CALCULATE THE VALUES FOR SEVERAL VARIABLES
115   CALL INIT(NUMPTS,AREA1)
C***PRINT THE POINTS IN THE CLOCKWISE ORDER AND THE AREA OF THE PROPERTY
      WRITE(6,40)
      DO 120 I=1,NUMPTS
         WRITE(6,45) I,POINTX(I),POINTY(I)
120   CONTINUE
      WRITE(6,50) AREA1
C***SET UP THE PARAMETERS FOR THE FIRST BLOCK
      NBACK=1
      X1=POINTX(NUMPTS)
      Y1=POINTY(NUMPTS)
      J=NUMPTS-1
      X2=POINTX(J)
      Y2=POINTY(J)
      IX=(X1-XMIN)/XWID+0.9999999999D0
      IY=(Y1-YMIN)/YWID+0.9999999999D0
      BX1=XMIN+DFLOAT(IX-1)*XWID
      BX2=BX1+XWID
      BY1=YMIN+DFLOAT(IY-1)*YWID
      BY2=BY1+YWID
C***CHECK IF NEXT COUNTERCLOCKWISE POINT IS OUTSIDE OF THE FIRST BLOCK
130   IF(X2.LT.BX1.OR.X2.GT.BX2.OR.Y2.LT.BY1.OR.Y2.GT.BY2)  GOTO 140
      NBACK=NBACK+1
      X1=X2
      Y1=Y2
      IF(NBACK.EQ.NUMPTS) AREA(IX,IY)=AREA1
      IF(NBACK.EQ.NUMPTS) GOTO 250
      J=NUMPTS-NBACK
      X2=POINTX(J)
      Y2=POINTY(J)
      GOTO 130
```

```
C***FIND THE POINT WHERE THE LINE EXITS THE BLOCK, AND STORE IT AS THE
C***   FIRST POINT IN THE BLOCK ARRAY
140       CALL PCALC(X1,Y1,X2,Y2,BX1,BX2,BY1,BY2,X,Y,ISIDE)
          IENTSD=ISIDE
          ARRAYX(1)=X
          ARRAYY(1)=Y
          IF(IENTSD.EQ.1) DISST=X-BX1
          IF(IENTSD.EQ.2) DISST=XWID+(BY2-Y)
          IF(IENTSD.EQ.3) DISST=XWID+YWID+(BX1-X)
          IF(IENTSD.EQ.4) DISST=XWID+XWID+YWID+Y-BY1
          DENTER(IX,IY)=DISST
C***ADD THE REMAINING POINTS INSIDE FIRST BLOCK TO THE BLOCK ARRAY
          IF(NBACK.EQ.1) GOTO 160
          J=NUMPTS-NBACK
          DO 150 I=2,NBACK
             J=J+1
             ARRAYX(I)=POINTX(J)
             ARRAYY(I)=POINTY(J)
150       CONTINUE
160       NUMPTS=NUMPTS-NBACK+1
          POINTX(NUMPTS)=X
          POINTY(NUMPTS)=Y
          NARRAY=NBACK+1
          ARRAYX(NARRAY)=POINTX(1)
          ARRAYY(NARRAY)=POINTY(1)
C***BEGIN MOVING CLOCKWISE FROM THE FIRST POINT
          NLAST=0
          X1=POINTX(1)
          Y1=POINTY(1)
          DO 240 I=2,NUMPTS
             X2=POINTX(I)
             Y2=POINTY(I)
C*** CHECK IF THE NEXT POINT IS OUTSIDE THE BLOCK
170          IF(X2.LT.BX1.OR.X2.GT.BX2.OR.Y2.LT.BY1.OR.Y2.GT.BY2)  GOTO 180
             IF(I.EQ.NUMPTS) NLAST=1
             IF(I.EQ.NUMPTS)  GOTO 180
             NARRAY=NARRAY+1
             ARRAYX(NARRAY)=X2
             ARRAYY(NARRAY)=Y2
             X1=X2
             Y1=Y2
             GOTO 240
C***FIND THE POINT WHERE THE LINE EXITS THE BLOCK, AND STORE IT AS THE
C***   LAST POINT IN THE BLOCK ARRAY
180          CALL PCALC(X1,Y1,X2,Y2,BX1,BX2,BY1,BY2,X,Y,ISIDE)
             NARRAY=NARRAY+1
             IEXSD=ISIDE
             ARRAYX(NARRAY)=X
             ARRAYY(NARRAY)=Y
             IEN=IENTSD
             IEX=IEXSD
             IF(IEXSD.EQ.1) DISEND=X-BX1
             IF(IEXSD.EQ.2) DISEND=XWID+(BY2-Y)
             IF(IEXSD.EQ.3) DISEND=XWID+YWID+(BX2-X)
             IF(IEXSD.EQ.4) DISEND=XWID+XWID+YWID+Y-BY1
C***ADD ALL OF THE CORNER POINTS OF THE BLOCK TO THE BLOCK ARRAY THAT
C***   ARE CLOCKWISE FROM THE EXITING POINT TO THE ENTERING POINT
             CALL COMBL(NARRAY,BX1,BX2,BY1,BY2,IEN,IEX,IX,IY,DISST,DISEND)
C***CALCULATE THE AREA IN THE BLOCK
             ARRAYX(NARRAY+1)=ARRAYX(1)
             ARRAYY(NARRAY+1)=ARRAYY(1)
             CALL AREAC(ARRAYX,ARRAYY,NARRAY+1,AREA1)
C*** IF THE BLOCK HAS A NONZERO AREA, THEN THE OUTLINE HAS PREVIOUSLY
C***   ENTERED THIS BLOCK
             IF(AREA(IX,IY).LT.0.0D0) AREA(IX,IY)=0.0D0
             IF(AREA(IX,IY).EQ.0.0D0)  GOTO 190
```

```
C***CHECK WHETHER TO ADD OR DELETE FROM THE CURRENT AREA
         IF(DISST.LT.DEXIT(IX,IY).AND.DISST.GT.DENTER(IX,IY)) GOTO 190
         IF(DENTER(IX,IY).GT.DEXIT(IX,IY).AND.(DISST.GT.DENTER(IX,IY).
     *      OR.DISST.LT.DEXIT(IX,IY))) GOTO 190
         AREA1=-(BLARE-AREA1)
190      AREA(IX,IY)=AREA(IX,IY)+AREA1
         DEXIT(IX,IY)=DISEND
C***QUIT THIS LOOP IF THE LAST POINT HAS BEEN REACHED
         IF(NLAST.EQ.1) GOTO 240
C***SET UP THE PARAMETERS FOR THE NEXT BLOCK
         X1=X
         Y1=Y
         NARRAY=1
         ARRAYX(1)=X
         ARRAYY(1)=Y
         GO TO (200,210,220,230),IEXSD
C***MOVE UP TO NEXT BLOCK
200      IENTSD=3
         IY=IY+1
         BY1=BY2
         BY2=BY1+YWID
         DISST=XWID+YWID+(BX2-X)
         IF(DENTER(IX,IY).LT.0.0) DENTER(IX,IY)=DISST
         GOTO 170
C***MOVE RIGHT TO NEXT BLOCK
210      IENTSD=4
         IX=IX+1
         BX1=BX2
         BX2=BX1+XWID
         DISST=XWID+XWID+YWID+(Y-BY1)
         IF(DENTER(IX,IY).LT.0.0) DENTER(IX,IY)=DISST
         GOTO 170
C***MOVE DOWN TO NEXT BLOCK
220      IENTSD=1
         IY=IY-1
         BY2=BY1
         BY1=BY2-YWID
         DISST=X-BX1
         IF(DENTER(IX,IY).LT.0.0) DENTER(IX,IY)=DISST
         GOTO 170
C***MOVE LEFT TO NEXT BLOCK
230      IENTSD=2
         IX=IX-1
         BX2=BX1
         BX1=BX2-XWID
         DISST=XWID+(BY2-Y)
         IF(DENTER(IX,IY).LT.0.0) DENTER(IX,IY)=DISST
         GOTO 170
240   CONTINUE
C***DETERMINE IF THE REMAINDER OF THE BLOCKS ARE INSIDE OR OUTSIDE
C***   AND FIND THE TOTAL AREA OF ALL OF THE BLOCKS
250   CALL FINISH(NUMPTS,TOTARE)
C***PRINT OUT THE TOTAL AREA OF ALL OF THE INDIVIDUAL BLOCKS
      WRITE(6,55) TOTARE
C***PRINT SUMMARY OF THE RESERVE AREA OF EACH BLOCK
      WRITE(6,60)
      DO 260 J=1,NYBLOC
         Y=DFLOAT(J)*YWID+YMIN-YWID/2.0D0
         DO 260 I=1,NXBLOC
            X=DFLOAT(I)*XWID+XMIN-XWID/2.0D0
            FRACT=AREA(I,J)/BLARE
            IF(AREA(I,J).GT.0.0D0) WRITE(6,65) X,Y,FRACT,AREA(I,J)
260   CONTINUE
      STOP
```

```
C*****************************************************************
10      FORMAT(9A8)
15      FORMAT('1',65('*')/T2,'*',T66,'*'/
    * T2,'*',T17,'PARTIAL BLOCK RESERVE AREA PROGRAM',T66,'*',/
    * T2,'*',T66,'*'/
    * T2,'*',T18,'VERSION 1.1   SEPTEMBER 21, 1984',T66,'*'/
    * T2,'*',T66,'*'/T2,65('*'),
    * ///1X,'TITLE: ',9A8///)
20      FORMAT(6F10.0)
25      FORMAT(' INPUT DATA FOR BLOCK GRID SETUP:'/
    * '    MINIMUM X-COORDINATE = ',T35,F12.2/
    * '    MAXIMUM X-COORDINATE = ',T35,F12.2/
    * '    MINIMUM Y-COORDINATE = ',T35,F12.2/
    * '    MAXIMUM Y-COORDINATE = ',T35,F12.2/
    * '    BLOCK SIZE (X-DIRECTION) = ',T35, F12.2/
    * '    BLOCK SIZE (Y-DIRECTION) = ',T35,F12.2/
    * '    NUMBER OF BLOCKS IN X-DIRECTION =',T43,I4/
    * '    NUMBER OF BLOCKS IN Y-DIRECTION =',T43,I4/
    * '    AREA OF A BLOCK =',T35,F12.0//)
30      FORMAT(' MINIMUM AND MAXIMUM RESERVE OUTLINE COORDINATES:'/
    * '    MINIMUM X-COORDINATE =',T35,F12.2/
    * '    MAXIMUM X-COORDINATE =',T35,F12.2/
    * '    MINIMUM Y-COORDINATE =',T35,F12.2/
    * '    MAXIMUM Y-COORDINATE =',T35,F12.2//)
35      FORMAT(///,' ***FATAL ERROR*** ALL RESERVE OUTLINE POINTS ARE',
    * ' NOT INSIDE THE SPECIFIED GRID AREA')
40      FORMAT(' LIST OF BOUNDARY COORDINATES:'/5X,'NUMBER',8X,
    * 'X-COORD        Y-COORD'/)
45      FORMAT(6X,I5,2F15.2)
50      FORMAT('0TOTAL AREA OF ORIGINAL POLYGON:',T35,F15.2)
55      FORMAT(' TOTAL AREA OF ALL PARTIAL BLOCKS:',T35,F15.2/)
60      FORMAT('1S U M M A R Y   O F   B L O C K   R E S E R V E S'
    * ///'    X-COORD  Y-COORD    PERCENT RESERVE          AREA')
65      FORMAT(1X,2F10.0,6X,F8.6,3X,F15.2)
        END


        SUBROUTINE INIT(NUMPTS,AREA1)
C*****************************************************************
C       THIS SUBROUTINE DOES THE FOLLOWING:
C  1.   CALCULATES THE VALUES OF VARIABLES BASED ON THE INPUT DATA,
C  2.   INITIALIZES THE ARRAYS AREA, DENTER, AND DEXIT TO -1,
C  3.   COMPLETES THE PROPERTY OUTLINE IF NECESSARY,
C  4.   CALCULATES THE AREA OF THE PROPERTY OUTLINE, AND
C  5.   REVERSES THE ORDER OF THE POINTS TO CLOCKWISE IF THE AREA
C       IS NEGATIVE.
C*****************************************************************
        IMPLICIT REAL*8 (A-H,P-Z)
        REAL*4 DENTER,DEXIT
        COMMON /BLOCK1/ POINTX(1000),POINTY(1000),ARRAYX(100),ARRAYY(100)
        COMMON /BLOCK2/ AREA(100,100),DENTER(100,100),DEXIT(100,100)
        COMMON /BLOCK3/ XMIN,XWID,XMAX,YMIN,YWID,YMAX,NYBLOC,NXBLOC,BLARE
C***INITIALIZE THE VARIABLES AREA AND BLAREA TO -1.0
        DO 100 J=1,NYBLOC
            DO 100 I=1,NXBLOC
            AREA(I,J)=-1.0D0
            DENTER(I,J)=-1.0
100     CONTINUE
C***CLOSE THE BOUNDARY IF NECESSARY
        IF(POINTX(1).EQ.POINTX(NUMPTS).AND.POINTY(1).EQ.POINTY(NUMPTS))
    * GOTO 110
        NUMPTS=NUMPTS+1
        POINTX(NUMPTS)=POINTX(1)
        POINTY(NUMPTS)=POINTY(1)
C***CALCULATE THE AREA INSIDE THE ENTIRE POLYGON
110     CALL AREAC(POINTX,POINTY,NUMPTS,AREA1)
```

```
C***REVERSE THE ORDER OF THE POINTS TO CLOCKWISE IF THE AREA IS
C***    NEGATIVE
        IF(AREA1.GE.0.0D0) RETURN
        AREA1=-AREA1
        N1=NUMPTS/2
        DO 120 I=1,N1
            J=NUMPTS-I+1
            TEMP=POINTX(I)
            POINTX(I)=POINTX(J)
            POINTX(J)=TEMP
            TEMP=POINTY(I)
            POINTY(I)=POINTY(J)
            POINTY(J)=TEMP
120     CONTINUE
        RETURN
        END


        SUBROUTINE PCALC(X1,Y1,X2,Y2,BX1,BX2,BY1,BY2,X,Y,ISIDE)
C*********************************************************************
C     THIS SUBROUTINE CALCULATES THE COORDINATES OF THE POINT
C   (X,Y) WHERE THE BOUNDARY OUTLINE EXITS THE BLOCK AND THE SIDE
C   THAT IT EXITS FROM (ISIDE: 1=TOP, 2=RIGHT, 3=BOTTOM, AND
C   4= LEFT SIDE).  THIS SUBROUTINE ASSUMES THAT THE FIRST POINT IS
C   INSIDE AND THAT THE SECOND POINT IS OUTSIDE THE BLOCK.
C
C   THE SUBROUTINE REQUIRES THE FOLLOWING INFORMATION:
C       X1  = X-COORDINATE OF THE POINT INSIDE THE BLOCK
C       Y1  = Y-COORDINATE OF THE POINT INSIDE THE BLOCK
C       X2  = X-COORDINATE OF THE POINT OUTSIDE THE BLOCK
C       Y2  = Y-COORDINATE OF THE POINT OUTSIDE THE BLOCK
C       BX1 = X-COORDINATE OF THE LEFT SIDE OF THE BLOCK
C       BX2 = X-COORDINATE OF THE RIGHT SIDE OF THE BLOCK
C       BY1 = Y-COORDINATE OF THE BOTTOM SIDE OF THE BLOCK
C       BY2 = Y-COORDINATE OF THE TOP SIDE OF THE BLOCK
C*********************************************************************
C        Y
C        A
C        |
C   Y2 | - - - - - - - - - - - - - - - - 0
C        |                ISIDE=1            *
C   BY2 | - - - ----------------------- ---------    *    |
C        |      |                       |        *
C        |      |                       |  *
C        |      |                       ?  |
C        |      |                    *  |
C        |      |                 *     |
C        |    4 |              *        |2
C        |      |           *           |
C        |      |        *              |
C        |      |     *                 |
C   Y1 | - - - | - - - 0                |       |
C        |      |                       |
C   BY1 | - - - - - - - - - - - - -     |       |
C        |            ISIDE=3           |
C        |      |          |            |       |
C        |      |          |            |       |
C        |      |          |            |       |
C        ----------------------------------------------->
C             BX1        X1            BX2      X2     X
C                                       X
C*********************************************************************
        IMPLICIT REAL*8 (A-H,P-Z)
```

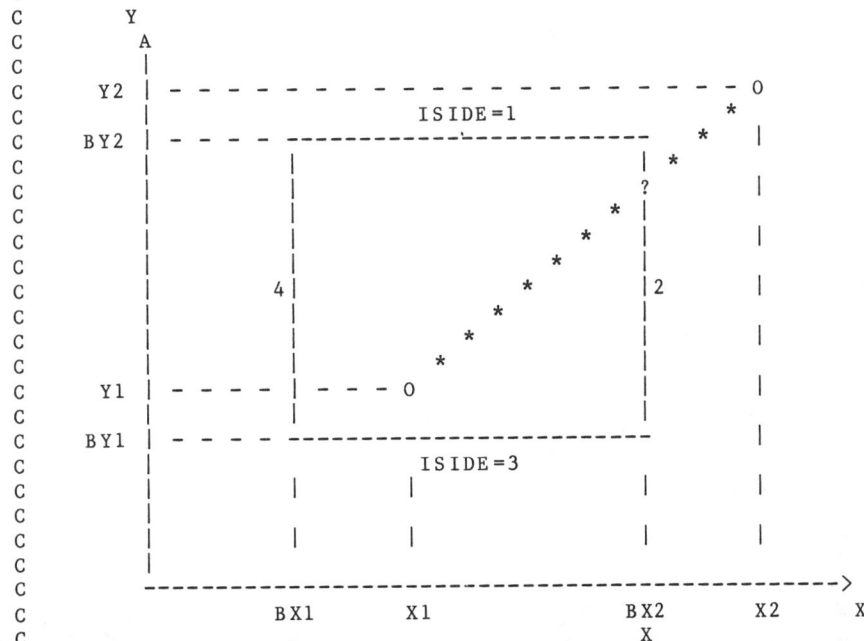

```
C***CALCULATE THE SLOPE
      DX=X2-X1
      DY=Y2-Y1
      IF(DX.EQ.0.0D0) GOTO 500
      SLOPE=DY/DX
      SLOPE=DABS(SLOPE)
      IF(DX.GE.0.0D0.AND.DY.GE.0.0D0)   GOTO 100
      IF(DX.LT.0.0D0.AND.DY.GE.0.0D0)   GOTO 200
      IF(DX.LT.0.0D0.AND.DY.LT.0.0D0)   GOTO 300
      IF(DX.GE.0.0D0.AND.DY.LT.0.0D0)   GOTO 400
C***FIRST QUADRANT
100   IF(X1.EQ.BX2) GOTO 110
      SLOPEC=(BY2-Y1)/(BX2-X1)
      IF(SLOPE.GE.SLOPEC)  GOTO 120
110   ISIDE=2
      X=BX2
      Y=Y1+(BX2-X1)*SLOPE
      RETURN
120   ISIDE=1
      X=X1+(BY2-Y1)/SLOPE
      Y=BY2
      RETURN
C***SECOND QUADRANT
200   IF(X1.EQ.BX1)  GOTO 210
      SLOPEC=(BY2-Y1)/(X1-BX1)
      IF(SLOPE.GE.SLOPEC)  GOTO 220
210   ISIDE=4
      X=BX1
      Y=Y1+(X1-BX1)*SLOPE
      RETURN
220   ISIDE=1
      X=X1-(BY2-Y1)/SLOPE
      Y=BY2
      RETURN
C***THIRD QUADRANT
300   IF(X1.EQ.BX1)  GOTO 310
      SLOPEC=(Y1-BY1)/(X1-BX1)
      IF(SLOPE.GE.SLOPEC)  GOTO 320
310   ISIDE=4
      X=BX1
      Y=Y1-(X1-BX1)*SLOPE
      RETURN
320   ISIDE=3
      X=X1-(Y1-BY1)/SLOPE
      Y=BY1
      RETURN
C***FOURTH QUADRANT
400   IF(X1.EQ.BX2)  GOTO 410
      SLOPEC=(Y1-BY1)/(BX2-X1)
      IF(SLOPE.GE.SLOPEC) GOTO 420
410   ISIDE=2
      X=BX2
      Y=Y1-(BX2-X1)*SLOPE
      RETURN
420   ISIDE=3
      X=X1+(Y1-BY1)/SLOPE
      Y=BY1
      RETURN
C***VERTICAL LINE
500   IF(DY.GT.0.0D0)  GOTO 510
      ISIDE=3
      X=X1
      Y=BY1
      RETURN
510   ISIDE=1
      X=X1
      Y=BY2
      RETURN
      END
```

```
      SUBROUTINE COMBL(NARRAY,BX1,BX2,BY1,BY2,IEN,IEX,IX,IY,DISST,
     * DISEND)
C*****************************************************************
C     THIS SUBROUTINE ADDS THE CORNERS THAT ARE CLOCKWISE BETWEEN
C  THE EXITING POINT AND THE ENTERING POINT.  THESE POINTS COMPLETE
C  THE OUTLINE OF THE POLYGON DESCRIBING THE PARTIAL AREA OF THE
C  BLOCK.
C*****************************************************************
      IMPLICIT REAL*8 (A-H,P-Z)
      COMMON /BLOCK1/ POINTX(1000),POINTY(1000),ARRAYX(100),ARRAYY(100)
      NCORN=IEN-IEX
      IF(NCORN.EQ.0.AND.DISST.GE.DISEND) RETURN
      IF(NCORN.EQ.0.AND.DISST.LT.DISEND) NCORN=4
      IF(NCORN.LT.0)  NCORN=NCORN+4
      DO 140 I=1,NCORN
        J=NARRAY+I
        GOTO (100,110,120,130),IEX
100     ARRAYX(J)=BX2
        ARRAYY(J)=BY2
        IEX=2
        GOTO 140
110     ARRAYX(J)=BX2
        ARRAYY(J)=BY1
        IEX=3
        GOTO 140
120     ARRAYX(J)=BX1
        ARRAYY(J)=BY1
        IEX=4
        GOTO 140
130     ARRAYX(J)=BX1
        ARRAYY(J)=BY2
        IEX=1
140     CONTINUE
      NARRAY=NARRAY+NCORN
      RETURN
      END


      SUBROUTINE AREAC(PX,PY,N,AREA)
C*****************************************************************
C     THIS SUBROUTINE CALCULATES THE AREA INSIDE THE POLYGON WHICH IS
C  STORED IN ARRAYS PX AND PY, (X- AND Y-COORDINATES).  N IS THE NUMBER
C  OF POINTS, AND AREA IS THE CALCULATED AREA OF THE POLYGON.
C*****************************************************************
      REAL*8 PX(N),PY(N),AREA
      AREA=0.5D0*PX(1)*(PY(N-1)-PY(2))
      J=N-1
      DO 10 I=2,J
        AREA=AREA+0.5D0*PX(I)*(PY(I-1)-PY(I+1))
10      CONTINUE
      RETURN
      END


      SUBROUTINE FINISH(NUMPTS,TOTARE)
C*****************************************************************
C     THIS SUBROUTINE DETERMINES IF THE BLOCKS THAT DO NOT INTERSECT
C  THE PROPERTY OUTLINE ARE INSIDE OR OUTSIDE THE PROPERTY.  THE FIRST
C  PART USES THE INFORMATION THAT IF MOVING FROM THE OUTSIDE OF THE GRID
C  AREA TO THE INSIDE, ALL OF THE BLOCKS ENCOUNTERED BEFORE A BLOCK
C  THAT DOES INTERSECT THE PROPERTY OUTLINE ARE OUTSIDE THE PROPERTY
C  OUTLINE.  THE SECOND PART CHECKS ALL OF THE REMAINING BLOCKS BY
C  CALCULATING THE NUMBER OF TIMES THAT THE PROPERTY OUTLINE CROSSES
C  DIRECTLY ABOVE (NORTH) OF THE BLOCK.  IF THIS NUMBER IS ODD, THE
C  BLOCK IS COMPLETELY INSIDE, IF THE NUMBER IS EVEN THE BLOCK IS
C  COMPLETELY OUTSIDE THE BLOCK.  THE SECOND PART ALSO TOTALS THE AREA
C  OF ALL OF THE BLOCKS.
C*****************************************************************
```

```
            IMPLICIT REAL*8 (A-H,P-Z)
            REAL*4 DENTER,DEXIT
            COMMON /BLOCK1/ POINTX(1000),POINTY(1000),ARRAYX(100),ARRAYY(100)
            COMMON /BLOCK2/ AREA(100,100),DENTER(100,100),DEXIT(100,100)
            COMMON /BLOCK3/ XMIN,XWID,XMAX,YMIN,YWID,YMAX,NYBLOC,NXBLOC,BLARE
C***SET THE AREA OF THOSE BLOCKS COMPLETELY OUTSIDE THE OUTLINE TO ZERO
C***FROM THE LEFT
            DO 130 J=1,NYBLOC
               DO 100 I=1,NXBLOC
                  IF(AREA(I,J).GE.0.0D0) GOTO 110
                  AREA(I,J)=0.0D0
100            CONTINUE
C***FROM THE RIGHT
110            DO 120 K=1,NXBLOC
               I=NXBLOC-K+1
               IF(AREA(I,J).GE.0.0D0) GOTO 130
               AREA(I,J)=0.0D0
120            CONTINUE
130         CONTINUE
C***TOTAL THE AREA OF ALL OF THE BLOCKS.  IF A BLOCK STILL HAS THE
C*** VALUE -1, DETERMINE IF IT IS INSIDE OR OUTSIDE THE POLYGON.
            POINTX(NUMPTS)=POINTX(1)
            POINTY(NUMPTS)=POINTY(1)
            TOTARE=0.0D0
            DO 190 J=1,NYBLOC
               Y=DFLOAT(J)*YWID+YMIN-YWID/2.0D0
               DO 190 I=1,NXBLOC
                  IF(AREA(I,J).GE.0.0D0)  GOTO 180
                  X=DFLOAT(I)*XWID+XMIN-XWID/2.0D0
                  NA=0
                  X2=POINTX(1)
                  Y2=POINTY(1)
                  DO 150 K=2,NUMPTS
                     X1=X2
                     Y1=Y2
                     X2=POINTX(K)
                     Y2=POINTY(K)
                     XXMIN=X1
                     XXMAX=X2
                     IF(XXMIN.LT.XXMAX) GOTO 140
                     TEMP=XXMIN
                     XXMIN=XXMAX
                     XXMAX=TEMP
140                  IF(XXMIN.GT.X.OR.XXMAX.LT.X) GOTO 150
                     SLOPE=(Y2-Y1)/(X2-X1)
                     YY=(X-X1)*SLOPE+Y1
                     IF(YY.GT.Y) NA=NA+1
150               CONTINUE
                  NB=NA/2*2
                  AREA1=0.0D0
                  IF(NA.NE.NB)  AREA1=BLARE
                  DO 170 K=I,NXBLOC
                     IF(AREA(K,J).GE.0.0D0) GOTO 180
                     DO 160 L=J,NYBLOC
                        IF(AREA(K,L).GE.0.0D0) GOTO 170
                        AREA(K,L)=AREA1
160                  CONTINUE
170               CONTINUE
180               TOTARE=TOTARE+AREA(I,J)
190         CONTINUE
            RETURN
            END
```

AUTHOR INDEX